ADVANCED BIOFUEL TECHNOLOGIES

ADVANCED BIOFUEL TECHNOLOGIES
Present Status, Challenges and Future Prospects

Edited by

DEEPAK TULI
Former Chair, Bioenergy-Biosciences, DBT-IOC Bioenergy Centre, Faridabad, India

SANGITA KASTURE
Scientist F, Department of Biotechnology, Ministry of Science and Technology, New Delhi, India

ARINDAM KUILA
Department of Bioscience & Biotechnology, Banasthali Vidyapith, Vanasthali, India

ELSEVIER

Elsevier
Radarweg 29, PO Box 211, 1000 AE Amsterdam, Netherlands
The Boulevard, Langford Lane, Kidlington, Oxford OX5 1GB, United Kingdom
50 Hampshire Street, 5th Floor, Cambridge, MA 02139, United States

Notices

Knowledge and best practice in this field are constantly changing. As new research and experience broaden our
understanding, changes in research methods, professional practices, or medical treatment may become
necessary.

Practitioners and researchers must always rely on their own experience and knowledge in evaluating and using
any information, methods, compounds, or experiments described herein. In using such information or methods
they should be mindful of their own safety and the safety of others, including parties for whom they have a
professional responsibility.

To the fullest extent of the law, neither the Publisher nor the authors, contributors, or editors, assume any liability
for any injury and/or damage to persons or property as a matter of products liability, negligence or otherwise, or
from any use or operation of any methods, products, instructions, or ideas contained in the material herein.

Library of Congress Cataloging-in-Publication Data
A catalog record for this book is available from the Library of Congress

British Library Cataloguing-in-Publication Data
A catalogue record for this book is available from the British Library

ISBN: 978-0-323-88427-3

For information on all Elsevier publications
visit our website at https://www.elsevier.com/books-and-journals

Publisher: Joseph Hayton
Acquisitions Editor: Susan Dennis
Editorial Project Manager: Lindsay Lawrence
Production Project Manager: Sujatha Thirugnana Sambandam
Cover Designer: Mark Rogers

Typeset by STRAIVE, India

Contents

16. Biorefinery approach for production of some high-value chemicals

Andrea Komesu, Johnatt Oliveira, Débora Kono Taketa Moreira,
Ali Hassan Khalid, João Moreira Neto, and
Luiza Helena da Silva Martins

17. Biofuels and bioproducts from seaweeds

Karuna Nagula, Himanshu Sati, Nitin Trivedi, and C.R.K. Reddy

18. Anaerobic gas fermentation: A carbon-refining process for the production of sustainable fuels, chemicals, and food

Melvin Moore, Vicki Z. Liu, Chih-Kai Yang, Zachary Cowden, and
Sean D. Simpson

19. Cyanobacteria as a renewable resource for biofuel production

Deepti Sahasrabuddhe, Annesha Sengupta, Shinjinee Sengupta,
Vivek Mishra, and Pramod P. Wangikar

20. Current technical advancement in biogas production and Indian status

S.D. Sawale and A.A. Kulkarni

21. Regulations and specifications for biofuels

T. Nagamani and Sangita Kasture

Contributors

Deepika Awasthi Lawrence Berkeley National Laboratory, Berkeley, CA, United States

Sougata Bardhan School of Natural Resources, University of Missouri, Columbia, MO, United States

Thallada Bhaskar Material Resource Efficiency Division (MRED), CSIR-Indian Institute of Petroleum (IIP), Dehradun; Academy of Scientific and Innovative Research (AcSIR), Ghaziabad, Uttar Pradesh, India

Bijoy Biswas Material Resource Efficiency Division (MRED), CSIR-Indian Institute of Petroleum (IIP), Dehradun; Academy of Scientific and Innovative Research (AcSIR), Ghaziabad, Uttar Pradesh, India

Ipsita Chakravarty Department of Chemical Engineering, VNIT Nagpur, Nagpur, India

Zachary Cowden LanzaTech Inc., Skokie, IL, United States

Luiza Helena da Silva Martins Institute of Animal Health and Production, Federal Rural University of the Amazon (UFRA), Belem, PA, Brazil

Piyali Das The Energy and Resources Institute (TERI), New Delhi, India

S. Dasappa Center for Sustainable Technologies, Interdisciplinary Centre for Energy Research, Indian Institute of Science, Bangalore, India

Diptarka Dasgupta Material Resource Efficiency Division (MRED), CSIR-Indian Institute of Petroleum (IIP), Dehradun; Academy of Scientific and Innovative Research (AcSIR), Ghaziabad, Uttar Pradesh, India

Aviraj Datta International Crops Research Institute for the Semiarid Tropics, Patancheru, India

Saleem A. Farooqui CSIR-Indian Institute of Petroleum, Dehradun, India

Dipali Gahane Department of Chemical Engineering, VNIT Nagpur, Nagpur, India

Debashish Ghosh Material Resource Efficiency Division (MRED), CSIR-Indian Institute of Petroleum (IIP), Dehradun; Academy of Scientific and Innovative Research (AcSIR), Ghaziabad, Uttar Pradesh, India

Soumyajit Sen Gupta Department of Chemical Engineering, Indian Institute of Technology (Indian School of Mines) Dhanbad, Dhanbad, India

Dheeban Chakravarthi Kannan The Energy and Resources Institute (TERI), Navi Mumbai, India

Sangita Kasture Department of Biotechnology, Ministry of Science and Technology, New Delhi, India

Ali Hassan Khalid National University of Science and Technology (NUST), Islamabad, Pakistan

Andrea Komesu Department of Marine Sciences, Federal University of São Paulo (UNIFESP), Santos, SP, Brazil

Bhavya B. Krishna Material Resource Efficiency Division (MRED), CSIR-Indian Institute of Petroleum (IIP), Dehradun; Academy of Scientific and Innovative Research (AcSIR), Ghaziabad, Uttar Pradesh, India

Arindam Kuila Department of Bioscience & Biotechnology, Banasthali Vidyapith, Banasthali, Rajasthan, India

A.A. Kulkarni Praj Matrix—R & D Center (Division of Praj Industries Limited), Pune, Maharashtra, India

M. Kiran Kumar   Microbial Processes and Technology, CSIR-National Institute for Interdisciplinary Science and Technology, Trivandrum, India

Vicki Z. Liu   LanzaTech Inc., Skokie, IL, United States

Chaitanya Sampat Magar   The Energy and Resources Institute (TERI), Navi Mumbai, India

Sachin Mandavgane   Department of Chemical Engineering, VNIT Nagpur, Nagpur, India

Anu Jose Mattam   Bioprocess Division, Hindustan Petroleum Corporation Limited, HP Green R&D Centre, KIADB Industrial Area, Tarabahalli, Devanagundi, Hoskote, Bengaluru, India

Vivek Mishra   Department of Chemical Engineering, Indian Institute of Technology Bombay, Mumbai, India

S. Venkata Mohan   Bioengineering and Environmental Sciences Lab, Department of Energy and Environmental Engineering, CSIR-Indian Institute of Chemical Technology (CSIR-IICT), Hyderabad; Academy of Scientific and Innovative Research (AcSIR), Ghaziabad, India

Melvin Moore   LanzaTech Inc., Skokie, IL, United States

Débora Kono Taketa Moreira   Department of Food Technology, Federal Institute of Brasilia (IFB), Campus Gama, Brasília, DF, Brazil

Triya Mukherjee   Bioengineering and Environmental Sciences Lab, Department of Energy and Environmental Engineering, CSIR-Indian Institute of Chemical Technology (CSIR-IICT), Hyderabad; Academy of Scientific and Innovative Research (AcSIR), Ghaziabad, India

T. Nagamani   Petroleum, Coal and Related Products Department (PCD), Bureau of Indian Standards, New Delhi, India

Karuna Nagula   DBT—ICT Centre for Energy Biosciences, Institute of Chemical Technology, Mumbai, Maharashtra; Parul Institute of Pharmacy & Research, Parul University, Vadodara, Gujarat, India

Abhilek K. Nautiyal   Material Resource Efficiency Division (MRED), CSIR-Indian Institute of Petroleum (IIP), Dehradun; Academy of Scientific and Innovative Research (AcSIR), Ghaziabad, Uttar Pradesh, India

João Moreira Neto   Department of Engineering, Federal University of Lavras, Lavras, MG, Brazil

Johnatt Oliveira   Institute of Health Sciences, Faculty of Nutrition, Federal University of Pará (UFPA), Belem, PA, Brazil

Shailja Pant   Department of Bioscience & Biotechnology, Banasthali Vidyapith, Banasthali, Rajasthan, India

Anjan Ray   CSIR-Indian Institute of Petroleum, Dehradun, India

B. Suresh Reddy   Centre for Economic and Social Studies, Hyderabad, India

C.R.K. Reddy   DBT—ICT Centre for Energy Biosciences, Institute of Chemical Technology, Mumbai, Maharashtra; Indian Centre for Climate and Societal Impacts Research, Shri Vivekanand Research and Training Institute (VRTI), Kutch, Gujarat, India

Ritika   Department of Bioscience & Biotechnology, Banasthali Vidyapith, Banasthali, Rajasthan, India

Deepti Sahasrabuddhe   Department of Chemical Engineering; DBT-Pan IIT Center for Bioenergy; Wadhwani Research Center for Bioengineering, Indian Institute of Technology Bombay, Mumbai, India

Jitendra Kumar Saini   Department of Microbiology, Central University of Haryana, Mahendergarh, Haryana, India

Himanshu Sati   DBT—ICT Centre for Energy Biosciences, Institute of Chemical Technology, Mumbai, Maharashtra, India

S.D. Sawale   Praj Matrix—R & D Center (Division of Praj Industries Limited), Pune, Maharashtra, India

Annesha Sengupta   Department of Chemical Engineering, Indian Institute of Technology Bombay, Mumbai, India

Shinjinee Sengupta   DBT-Pan IIT Center for Bioenergy, Indian Institute of Technology Bombay, Mumbai, India

K.T. Shanmugam   Department of Microbiology and Cell Science, University of Florida, Gainesville, FL, United States

Tripti Sharma   Material Resource Efficiency Division (MRED), CSIR-Indian Institute of Petroleum (IIP), Dehradun; Academy of Scientific and Innovative Research (AcSIR), Ghaziabad, Uttar Pradesh, India

Yogendra Shastri   Department of Chemical Engineering, Indian Institute of Technology Bombay, Mumbai, India

Anand M. Shivapuji   Center for Sustainable Technologies, Indian Institute of Science, Bangalore, India

Sean D. Simpson   LanzaTech Inc., Skokie, IL, United States

Anil K. Sinha   CSIR-Indian Institute of Petroleum, Dehradun, India

Rajeev K. Sukumaran   Microbial Processes and Technology, CSIR-National Institute for Interdisciplinary Science and Technology, Trivandrum, India

Chiranjeevi Thulluri   Bioprocess Division, Hindustan Petroleum Corporation Limited, HP Green R&D Centre, KIADB Industrial Area, Tarabahalli, Devanagundi, Hoskote, Bengaluru, India

Nitin Trivedi   DBT—ICT Centre for Energy Biosciences, Institute of Chemical Technology, Mumbai, Maharashtra, India

Deepak Tuli   Bioenergy-Biosciences, DBT-IOC Bioenergy Centre, Faridabad, India

A.V. Umakanth   ICAR - Indian Institute of Millets Research, Hyderabad, India

Harshad Ravindra Velankar   Bioprocess Division, Hindustan Petroleum Corporation Limited, HP Green R&D Centre, KIADB Industrial Area, Tarabahalli, Devanagundi, Hoskote, Bengaluru, India

Pramod P. Wangikar   Department of Chemical Engineering; DBT-Pan IIT Center for Bioenergy; Wadhwani Research Center for Bioengineering, Indian Institute of Technology Bombay, Mumbai, India

Chih-Kai Yang   LanzaTech Inc., Skokie, IL, United States

About the editors

Dr. Deepak Tuli is currently working as a lead consultant in Mission Innovation. Previously, he was Chair of Energy Biosciences of Department of Biotechnology. Prior to this, he was Executive Director of Indian Oil Corporation Limited, R&D centre, Faridabad. For 5 years, he headed the DBT-IOC Centre for Advanced Bioenergy Research, Faridabad, India. He completed his PhD in synthetic organic chemistry in 1977 and pursued his postdoctoral research at the University of Liverpool during 1978–81. He was an SERC senior research fellow at Robert Robinson labs, UK, during 1986–88. He headed alternative energy R&D in IOC comprising areas of biofuels and bioenergy & solar energy for 12 years and has 14 US patents, 2 European patents, 24 Indian patents, and 78 research publications. His major areas of research interests are bioenergy, LCA, technology scale-up, and additives for fuels. He has 34 years of research experience both in lab and as a team leader. He is Adjunct Professor at two Australian universities.

Dr. Sangita Kasture is Director (Scientist F) of the Department of Biotechnology, Ministry of Science and Technology, Government of India. She is associated with the Department of Biotechnology (DBT), Ministry of Science and Technology, Government of India, since 2010. She has been promoting innovations in bioenergy, environment biotechnology, and secondary agriculture through various R&D schemes of Government of India. She is the country focal point for multilateral programs like Mission Innovation, India, Biofuture Platform and bilateral R&D collaborations with Canada, Denmark, and the Netherlands. As a member of Indian delegation, she visited several countries including the United States, the United Kingdom, China, Sweden, Denmark, Canada, Brazil, Europe, Chile, etc., to represent the country strongly at various international platforms/meetings. She has industrial experience from two leading Indian biotech companies and postdoctorate research experience from Lund University (Sweden) and Polish Academy of Sciences (Poland). She holds a doctorate in chemical engineering from UDCT Mumbai and has done research in areas such as enzyme catalysis, green chemistry projects, liquid–liquid extraction, and postharvest technologies.

Dr. Arindam Kuila is currently working as Assistant Professor at the Department of Bioscience & Biotechnology, Banasthali Vidyapith (Deemed University), Rajasthan. Previously, he worked as a research associate at Hindustan Petroleum Green R&D Centre, Bangalore. He did his PhD in the area of lignocellulosic biofuel production, Agricultural & Food Engineering Department, Indian Institute of Technology Kharagpur, in 2013. His major areas of research are enzyme technology, bioprocess development for biofuel production, and biopolymer production from lignocellulosic biomass. He has one Indo-Brazil collaborative project funded by DBT, New Delhi. He has guided 2 PhD students; currently, 4 PhD students are working under his guidance. He has published 7 edited books, 8 book chapters, 28 papers in peer-reviewed journals, and filed 5 patents. He is currently acting as Guest Editor of the *Journal of Food Process Engineering* (Wiley, Impact Factor: 2.356) and *Biomass Conversion and Biorefinery Journal* (Springer, Impact Factor: 4.987).

Preface

Biofuels, mostly ethanol and biodiesel, have been blended for several years, in fossil transport fuels and their production has seen a robust growth. However, bioethanol is produced predominantly from sugarcane and corn, and biodiesel from plant oils and animal fat. Production of some of these raw materials is competing for the limited arable land against food and feed production. It is not feasible to exponentially increase biofuel production using the current technologies and feedstocks. Therefore, it is critical to investigate advanced biofuel production technologies to overcome feedstock constraints. Second-generation bioethanol is not commercialized on a large scale as it is expensive to produce due to technological challenges in pretreatment, enzymatic hydrolysis, and fermentation processes. Other advanced biofuels like drop-in fuels, bio-oil, and green diesel are being promoted based on appropriate feedstock. Innovations in technology are happening at a very fast pace to reduce the production cost of biofuels in a sustainable manner.

This book discusses the status of advanced biofuel technologies. This book comprises 22 chapters all of which are designed to cover both biochemical and thermochemical routes for advanced biofuel production. There are chapters dedicated to covering important topics such as feedstocks and their characterization, production technologies, and commercial status of advanced biofuel production in India and across the globe. Chapter 1 provides an overview of advanced biofuel conversion technologies. Chapter 2 describes the current global scenario of advanced

biofuels. One chapter exclusively discusses the advances in feedstock pretreatment technology, which is the heart of 2G ethanol process. Xylose fermentation requires designed microbes and there is a dedicated chapter to discuss the latest advances. There are two chapters that cover advanced biofuels and biochemical production through microalgae and macroalgae.

Regarding thermochemical routes to advanced biofuels, a chapter on biomass gasification and another on bio-oil production by pyrolysis are included. Some specific products (such as aviation jet fuels) and futuristic biofuels (such as methanol, butanol, and FT fuel) are adequately covered in the book. Lignin is a valuable by-product and its valorization is discussed in a separate chapter. Chapters dealing with biohydrogen production technologies and an overview of biorefinery approach for production of industrially important chemicals will give readers an idea about advances in these areas. Recently developed gas fermentation technology for advanced biofuel production has also been included and so are chapters on cyanobacteria-mediated biofuel production and current technological advancement in biogas production. Two important chapters cover an overview of regulation and specification of advanced biofuel production and life cycle and techno-economic assessment of advanced biofuel production.

We believe that this book will be useful for researchers and graduate students by assisting them in understanding the current status and future prospects of advanced biofuel production. The chapters have been contributed by eminent scientists and researchers who are both actively engaged for several years and have extensively published in this area. We extend our gratitude to the authors for accepting our invite to contribute to the book and also for their excellent efforts.

We thank Susan Dennis and Lindsay Lawrence of Elsevier Publishing Inc. for extending their full cooperation and help in the timely publication of the book. Our special thanks go to authors and publishing team for their efforts during the challenging times of the Covid-19 pandemic.

Dr. Sangita Kasture is thankful to Dr. Renu Swarup, Secretary, Department of Biotechnology, Ministry of Science and Technology, Government of India for her constant encouragement and support. Dr. Arindam Kuila is thankful to Department of Biotechnology (DBT), New Delhi for providing mission innovation grant. Dr. Deepak Tuli is thankful to Department of Biotechnology (DBT), New Delhi for award of Chair in Bio-energy Bio-Sciences.

Dr. Deepak Tuli
Dr. Sangita Kasture
Dr. Arindam Kuila

1

An overview of some futurist advanced biofuels and their conversion technologies

Deepak Tuli[a] and Sangita Kasture[b]

[a]Bioenergy-Biosciences, DBT-IOC Bioenergy Centre, Faridabad, India [b]Department of Biotechnology, Ministry of Science and Technology, New Delhi, India

1 Introduction

Biofuels, for transport applications, are classified as first-generation (Conventional) or advanced biofuels depending upon feedstock used for production, technology maturity, GHG reduction capability, and their product type. Generally, biofuels produced from feedstock that could be used as food are conventional and those from nonfood feedstock are advanced biofuels having higher GHG emission reduction potential as compared to conventional biofuels. Technologies for some of the advanced biofuels are still evolving while conventional biofuels are produced from mature commercial technologies. Biofuels can be either used as blends with petroleum fuels or as such if these are of fungible type. Conventional biofuels are produced from sugar/starch (ethanol) and from lipids of seeds (biodiesel) while advanced biofuels are from lignocellulosic residues (forest and agricultural), MSW, or algal materials. Advanced biofuels are produced by adopting cellulose fermentation, gasification followed by Fischer-Tropsch for jet fuel/diesel, hydro-processing of pyrolysis oil, and from syn. Gas. The biofuel production options from biomass by biochemical and thermochemical routes are shown in Fig. 1.

Conventional biofuels, based on food-based feed, would not be able to meet the growing demand of biofuels as per the national mandates of countries and advanced liquid biofuels would be essentially required to be produced at much larger scales. However, in the present scenario, the volumes of advanced biofuels are still small as these are much more expensive to

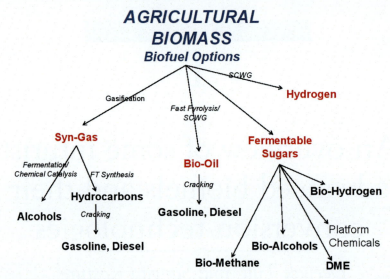

FIG. 1   Biofuel options from biomass.

produce as compared to conventional biofuels. Technology and policy interventions are necessary so as to bring down the cost and create sufficient market demand in the initial phase.

Advanced liquid biofuels have been the focus of government-supported research, development, and demonstration programs because they greatly expand the range of sustainable feedstocks. In particular, they have gained significant importance for climate change concerns. The transport policies promote the use of advanced liquid biofuels over conventional ones, mainly driven by their sustainability and better GHG reduction potential; 50%–70% relative to fossil fuels, compared to 30%–50% for conventional biofuels.

2 Biofuels for transport applications

According to the IEA report of global energy outlook-2018, the transportation sector, which consumed 32% of the produced oil in 2017 (International Energy Agency (IEA), 2018), was responsible for 24% of the global CO_2 emissions. Diesel, gasoline, and aviation kerosene constitute most of the fossil fractions used in the transport sector, and these, especially diesel and gasoline, were first to be targeted for part/full replacement by biofuels, which have lower GHG footprints. Although transport today accounts for less than a quarter of the energy-related global CO_2 emissions, the sector's emissions are growing by more than any sector of the global economy and according to IEA estimates transport sector CO_2 emissions from fossil fuels will increase by over one-third by 2030.

IEA technology status report published in 2020 indicates that road travel accounts for 76% of transport emissions. Most of this comes from passenger vehicles—cars and buses—which contribute 45.1%. The other 29.4% comes from trucks carrying freight. Road transport is 16% of global GHG emissions (IEA, 2020). Aviation—while it often gets the most attention in discussions on action against climate change—accounts for around 2.5% of total global emissions. The report predicts that with electrification—and hydrogen—technologies some of these subsectors could decarbonize within few decades although emissions from cars and buses are not completely eliminated until 2070 and thus would still make transport the largest contributor to energy-related emissions in 2070. As the IEA states, "Reducing CO_2 emissions in the transport sector over the next half-century will be a formidable task." Databases covering advanced biofuel (and biochemical) production plants are compiled and managed by IEA tasks (Cui and Kær, 2019; Hobson and Márquez, 2018), and (Dieterich et al., 2021; International Energy Agency Task33, n.d.; International Energy Agency Task34, n.d.; International Energy Agency Task39, n.d.). These have been used as a basis for the data presented in this chapter.

India's 2018 National Policy on Biofuels sets ambitious biofuel blending targets and aims to source biofuels only from sustainable feed-stocks so as not to threaten food security. The policy aims to achieve ethanol blending in gasoline to 20% by 2030, and blend 5% biodiesel in diesel (National Biofuel Policy, 2018) In the past, India did not meet its biofuel blending mandates. For example, the target set in 2009 of 20% ethanol blending by 2017 was missed almost completely and only 1.9% ethanol was blended in gasoline (U.S. Department of Agriculture, 2018) and biodiesel blending in diesel was less than 0.1% (Planning Commission, Government of India, 2003). The new biofuel policy provides for use of extended feedstock like agricultural residues and used cooking oil, excess sugarcane, and spoilt grains for biofuel production. Financial assistance for the production of advanced biofuels is also built into the policy.

Some of the advanced biofuels which are under various stages of development and which may become commercial in near future are discussed in this chapter. It is aimed to give the readers a fair idea about their production technologies, current status, and future projections.

3 Methanol

Methanol is single carbon alcohol with energy density approximately 50% lower than gasoline due to its 50% of oxygen content. Methanol can be blended with gasoline for use in road transport or converted to methyl *tert*-butyl ether (MTBE) for blending with gasoline. Possible applications are in rail and shipping and in fuel cells for generating electric power. Methanol may be converted to dimethyl ether (DME) for use as a diesel replacement in automobiles.

Methanol is used as a gasoline component or in the form of ether (MTBE) in gasoline and Fatty Acid Ester (FAME) in diesel. High concentration methanol blends, such as M85, are used in special Flexible Fuel Vehicles (FFVs). In some markets, the focus is on gasoline/ethanol/methanol blends (GEM) fuel blends. In many regions, for example in Europe and North America, blending of methanol is limited up to a few percentages in gasoline. Automakers are designing high-efficiency engines to compensate for low energy density of methanol (Araya et al., 2020).

Global methanol production has steadily increased in the past few decades from around 5 million metric tons in 1975 to around 110 million metric tons in 2018 (Dalena et al., 2018). Methanol is already a key component of several chemical products and its demand is increasing at an average annual growth rate of around 7% (Alvarado, 2016; Market Watch, 2019).

In Europe, max. 3 volume-% methanol is allowed to be blended in gasoline under the Fuel Quality Directive (2009/30/EC) and CEN standard (EN 228). In the U.S., ASTM D 4814-10a limits methanol up to 0.3 vol-% or up to 2.75 vol-% with an equal volume of butanol, or higher molecular weight alcohol. ASTM D7920–21 Standard Test Method for Determination of Fuel Methanol (M99) and Methanol Fuel Blends (M10 to M99) by Gas Chromatography is used to certify the methanol quality.

India also is considering methanol as an alternative fuel source that offers environmentally friendly characteristics associated with its use. Hence, in India, **IS 17075** specification for anhydrous methanol used as a blending component in methanol-based fuels, used in engines designed for running on such fuels for captive power generation, industrial, marine, locomotive, automotive, and off-road applications is published. This standard is applicable only for fuel applications and not for other industrial applications.

Another Indian Standard for 15% blending of this anhydrous methanol in motor gasoline **IS 17076** has also been published. This fuel is an admixture of anhydrous methanol meeting IS 17075 at 15% by volume in 85% gasoline, for use in vehicles equipped with spark-ignition engines, stationary and industrial engines specially designed for using such fuel. The engine control unit adjusts engine fueling relating to the oxygen content and reduced energy content of M15 fuel to maintain the proper air/fuel ratio under the various engine operating loads and conditions. The vaporization characteristics of M15 fuel tend to be different from normal petrol and therefore the engine fueling strategies under engine cold-start and warm-up conditions need to be addressed for this fuel. If M15 fuel has to be used in a conventional vehicle, fuel system materials and components may be affected over time and may lead to concerns like material compatibility, drivability, performance, and emissions. Since this fuel has to be used in spark-ignition engine vehicles with relevant modification, the fuel properties of M15 should be very similar to motor gasoline.

Dimethyl ether (DME) is made by dehydration of methanol and used as heating fuel, industrial fuel, and to replace diesel fuel or gas oil. DME can also be a substitute for diesel in slow RPM diesel engines and therefore DME is considered as a potential fuel. **IS 16704: 2018/ ISO 16861** Petroleum products—Fuels (class F)—Specifications of dimethyl ether (DME) specify the characteristics of DME used as fuel of which the main component is the dimethyl ether synthesized from methanol. The fuel can also be used in specifically designed compressed ignition engines, in place of diesel. Due to no emissions of NO_X, SO_X, and carbon monoxide, it is promoted as an advanced biofuel.

In China, the largest producer of methanol, a national standard for M85 fuel containing up to 85% of methanol is approved to be used as motor fuel. In addition, standards are in place in several provinces in China that govern the use of methanol in various blends with gasoline ranging from 5% to 100%, while a national standard for M15 fuel is also approved. Methanol accounts for 7%–8% of China's transportation fuel pool. China is leading the world with the largest production of Methanol & DME. China with 47 Million Tons (MT) of production in 2015 accounted for 55% of the global methanol production (85MT). China also produced 3.8 Million Tons of DME. It is to be noted that China produces 70% of its methanol from coal as it has the third largest coal reserves in the World, although, other countries (US, South

America, Iran) are largely producing methanol from natural gas due to its abundant availability in those countries at low prices.

China, which is rich in coal but heavily relies on imported oil for its transportation sector, has recently decided to use methanol nationwide as a transportation fuel to tackle both energy security. A rapid increase of methanol production through coal gasification has been achieved by China, though the production process is not considered environmentally beneficial. The use of methanol in the energy sector is rapidly growing due to China's policy of its use in the transport sector, consuming about 40% of the total ethanol production (Methanol Institute, n.d.; Zhao, 2019).

When considering methanol use as a gasoline component, corrosion inhibitors, co-solvents, and alcohol compatible materials in vehicles are needed to resist phase separation, to maintain stability and safety. The Table 1 indicates the approval status of methanol as the blend component in gasoline and the need of co-solvents. It may be noted that there is a limit on how much oxygen can be present in the final fuel blend.

3.1 Emissions reductions by use of methanol

The combustion of pure methanol gives no nitrogen oxide (NOx) emissions, no sulfur oxide (SOx) emissions, and very low particulate matter (PM) and carbon dioxide (CO_2) emissions as compared to gasoline or diesel. A mixture of methanol and petroleum fuel will result in lower emissions than regular petroleum fuel. Emissions from methanol-powered vehicles are quite less from automobiles, however, the LCA emissions for methanol produced from coal is quite high i.e. 190 g of CO_2e/MJ of fuel in comparison with gasoline/diesel run vehicles for which the number is in the range of 95–100. If carbon capture technologies are used, the LCA emissions of coal derived can be reduced to 85 g CO_2 e/MJ. None of the

TABLE 1 Methanol blends for automotive applications.

Market region		Introduction year	Maximum volume % methanol	Minimum volume % co-solvent	Maximum Wt % oxygen	Corrosion additives
Europe	EC Directive	1985	3.0	≥Methanol	3.7%	
U.S.A	Sub Sim*	1979	2.75	≥Methanol	2.0%	
U.S.A	Fuel waiver	1981	4.75	≥Methanol	3.5%	Required
U.S.A	Fuel waiver	1986	5.0	2.5	3.7%	Required
China, Shanxi	M15 Standard	2007	15.0	For water tolerance	~7.9%	Required

methanol plants in China are presently having carbon dioxide capture technologies and therefore methanol use in China is not lowering overall GHG emissions.

Even when methanol is produced from natural gas, carbon dioxide emissions occur during the reforming process. Methanol production can be GHG compliant if renewable hydrogen from water electrolysis via renewable electricity, from solar or wind is used for its production (Kiaee et al., 2013; Guinot et al., 2015; Mohammadi and Mehrpooya, 2018; Sherif et al., 2005).

3.2 India status—Methanol economy

NITI Aayog, the planning arm of the Government of India has been spearheading the production and use of methanol in India. According to discussion paper (Saraswat and Bansal, 2021) presently there are 5 main producers of methanol in India and there is a very large unused capacity in methanol production. Despite having unutilized capacity in India, it imports 85% of methanol requirements as production costs in India are higher due to the high cost of imported natural gas. Indian imports of methanol are from Iran & Saudi Arabia where cheap gas is available. China produces its methanol almost exclusively from coal for which they have very large availability. India also is blessed with large coal deposits but unfortunately most of the Indian coal is high ash and therefore needs modified coal gasification technology. The technology for methanol production by gasification of high ash Indian coal is under development and is currently at a pilot scale.

Under the Indian methanol economy program five methanol plants based on high ash coal, 5 DME plants and 1 natural gas-based methanol production plant with a total production of 20 MMT/annum in joint venture, are planned to be set up.

As a part of R & D program, work is in progress to set up coal to methanol plants in the country using the indigenous technology which is currently being demonstrated at a pilot scale of 1 to 40 TPD scale. India is also exploring the possibility of oxy gasification of abundantly available agricultural residue, to syngas and then its catalytic conversion to methanol.

3.3 Methanol as shipping fuel

Global demand for marine fuels is large. It has been estimated that international shipping consumes around 300 million tons of HFO annually (Buhaug et al., 2009). These figures highlight the potential market for low sulfur fuels such as methanol.

Methanol is readily available worldwide and every year over 70 million tons are produced globally. Currently, the main feedstock in methanol production is natural gas, however, methanol could be 100% renewable, as it can be produced from a variety of renewable feed-stocks or as an electro-fuel. This makes it an ideal pathway fuel to a sustainable future in which shipping is powered by 100% renewable fuels. From an environmental point of view, methanol performs well. Methanol readily dissolves in water and is biodegraded rapidly, as most micro-organisms have the ability to oxidize methanol. In practice, this means that the environmental effects of a large spill would be much lower than from an equivalent oil spill.

The renewable methanol sources like agricultural residues provide a pathway to compliance with the IMO's 2050 carbon emissions targets, without further ship-owner investment.

Methanol is a cost-effective alternative marine fuel in terms of the fuel itself, the dual fuel engine and the shoreside storage and bunkering infrastructure. The cost to convert vessels to run on methanol is significantly less than other alternative fuel conversions with no need for expensive exhaust gas after-treatment and, as a liquid fuel, only minor modifications are needed for existing storage and bunkering infrastructure to handle methanol.

Methanol is a clean-burning fuel that produces fewer smog-causing emissions than conventional fuels—such as SOx, NOx and particulate matter. It can help ships meet environmental fuel regulations and improve air quality and related human health issues. Methanol marine fuel complies with the most stringent regulations in emission control areas and would comply with the most stringent emissions regulations currently being considered. Methanol can reduce emissions of carbon monoxide, hydrocarbons, and nitrogen oxides when compared to gasoline and methanol is biodegradable. However, formaldehyde emissions tend to increase, especially at cold starts. Barriers to its use include concerns about human toxicity and corrosive effects on conventional engines.

Because of the demand, renewable methanol fuel substitution for heavy fuel oils in shipping could have great potential to reduce greenhouse gas emissions. Supply infrastructure is often in place for methanol, as it is already available and shipped through many ports around the world. Although methanol is less energy dense than traditional fuels, requiring more storage space on board, it's a liquid fuel, so it can be stored in ballast and "slop" tanks. This is in contrast to liquefied natural gas (LNG) which demands a large cryogenic tank, absorbing substantial on-board space.

Consider methanol as a good alternative for fuelling ships, an international cooperation under mission innovation has been launched in 2021 under the leadership of Denmark and India is participating in this mission.

Well-to-wheel CO_2 emission of various possible fuels which could be used for shipping is given below. Long-term total life cycle emissions in the case of methanol were lowest, as shown in Table 2.

However there is a possibility of emission of formaldehyde from methanol-based systems is which is an intermediate species in the reaction pathway of methanol oxidation and can constitute health concern (Verhelst et al., 2019).

Renewable Methanol Report covers the possible shipping fuelling alternatives for decarbonization and methanol is considered as the one which can be used in near terms.

TABLE 2 Life cycle CO_2 emissions from different fuelling options.

Type	Current status	Green scenario
Diesel	132 g/km	100 g/km
Gasoline	176 g/km	123 g/km
Hybrid	142 g/km	80 g/km
Battery electric	98 g/km	2 g/km
Hydrogen	178 g/km	3 g/km
Methanol	**83 g/km**	**2 g/km**

From: JENSEN, Mads Friis in Danish Department of Energy, Alternative Drivetrains 2014.

TABLE 3 Properties of methanol vs other shipping fuels.

Properties	Methanol	Methane	LNG	Diesel fuel
Molecular formula	CH_3OH	CH_4	C_nH_m; 90%–99% CH_4	$C_nH_{1.8n}$; C_8-C_{20}
Carbon contents (wt%)	37.49	74.84	≈75	86.88
Density at 16 °C (kg/m³)	794.6	422.5	431 to 464	833 to 881
Boiling point at 101.3 kPa(°C)	64.5	−161.5	−160 (−161)	163 to 399
Net heating value (MJ/kg)	20	50	49	42.5
Net heating value (GJ/m³)	16		22	35
Autoignition temperature (°C)	464	537	580	257
Flashpoint (°C)	11		−136	52 to 96
Cetane rating	5		0	>40
Flammability limits (vol% in air)	6.72–36.5	1.4–7.6	4.2–16.0	1.0–5.0
Water solubility	Complete	No		No
Sulfur content (%)	0	0	<0.06	Varies, <0.5 or <0.1

www.methanol.org/energy/; www.methanol.org/renewable-methanol/

Table 3 compares the possible low carbon fuels like methanol, methane and LNG with existing heavy diesel vehicles. Complete water solubility of methanol and its higher flash point than LNG is advantageous but then it has nearly half of the calorific value of diesel and therefore would need double the amount for the same distance.

3.4 Methanol production methods

Synthesis gas, a mixture of carbon monoxide, carbon dioxide and hydrogen, is first produced in a reformer. This is carried out by passing a mixture of the hydrocarbon feedstock, mostly natural gas or naptha, and steam through a heated tubular reformer. The ratio of hydrogen and carbon in the syngas may need to be adjusted by purging excess hydrogen or adding carbon dioxide. Developments here include the use of autothermal reforming, either alone or in combination with a primary reformer, in which oxygen is mixed with the steam.

The reactions of interest are:

$$2H_2 + CO \rightarrow CH_3OH$$

$$CO_2 + 3H_2 \rightarrow CH_3OH + H_2O$$

$$CO + H_2O \rightarrow CO_2 + H_2$$

The syngas is cooled and then compressed before being fed to the methanol converter. All three reactions are highly exothermic. The conventional commercial gas-phase process carries out the conversion in fixed-bed reactors at high pressure. Depending on the catalyst supplier.

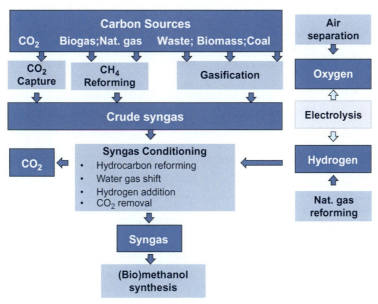

FIG. 2 Overview of major methanol production processes from various carbon sources. *From www.etsap.org, www. irena.org*

The methanol synthesis takes place in the presence of copper-based catalysts at 250–260°C. The crude methanol is recovered and purified by distillation.

Major technology providers include: Toyo Engineering Corporation; Lurgi Chemie GmbH and Foster Wheeler. General scheme of products which are possible are given in Fig. 2.

The market price of methanol was highly variable in the past ranging from 150 € per ton to 525 € per ton in the last years with prices of around 260 € per ton at the beginning of 2020. Regional benchmark methanol production costs were identified (Boulamanti and Moya, 2016) to vary between 51 € per ton in Saudi Arabia and 408 € per ton in Europe in 2013 mainly dependent on the regional feedstock costs. Middle East oil and gas producing countries are able to produce methanol at a lower cost as the main input, natural gas, is priced very low for their own plants.

3.5 Catalysts for methanol production

Currently, all commercially applied low-pressure catalysts are based on CuO and ZnO in most cases on a carrier of alumina with variable stabilizing additives and promoters like Zr, Cr, Mg, and rare earth metals. A detailed discussion of alternative catalyst formulations and promoters is given in several reviews (Wang et al., 2011; Ali et al., 2015; Jadhav et al., 2014; Bertau et al., 2014; Hansen and Nielsen, 2008).

The main suppliers of conventional methanol catalysts are Johnson Matthey; Clariant; Haldor Topsøe and Mitsubishi Gas Chemicals (Bertau et al., 2014). In industrial applications, the common catalyst life time is 4–6 years with up to 8 years being reported (Hirotani et al.,

1998; Grimm, 1991; Matthey, 2021). Life time is limited by catalyst deactivation caused by poisoning and thermal sintering. Major poisons for copper catalysts are sulfur compounds and chlorides.

3.6 Methanol from CO_2

Recent advances in heterogeneous catalysis of CO_2-based feed gases to methanol have been discussed in detail by several authors (Dang et al., 2019; Guil-López et al., 2019).

Methanol can be one of these carbon-neutral electro-fuels when produced from hydrogen via electrolysis and CO_2 from the atmosphere, biomass or from the exhaust of industrial processes (Goeppert et al., 2012; Brown and Le Feuvre, 2017; Cui and Kær, 2019; Hobson and Márquez, 2018). While both hydrogen and methanol can be CO_2 neutral if derived from renewable sources, methanol is easier to handle and a more direct substitute for oil in the chemical synthesis industry (Alberico and Nielsen, 2015). This method called e-fuel is the ultimate environmentally safe process and several groups are working on this and success would need availability of renewable green hydrogen at about one-third of the present cost.

Unlike in the traditional methods, where "black" methanol is produced from natural gas, coal, and oil; renewable methanol, also known as "green" methanol is produced from renewable H_2 and CO_2 sources with net-zero CO_2 emission, such as hydrogen from renewable electricity (Gahleitner, 2013) and CO_2 from biogas or the atmosphere (Bui et al., 2018). When methanol is produced from waste and by-products, which are only renewable to a degree, it is referred to as "gray" methanol (Serenergy A/S, 2021).

$$CO_2 + 3H_2 \rightleftharpoons CH_3OH + H_2O \; ; \Delta H = -49.4\,\text{kJ}\,\text{mol}^{-1} \tag{1}$$

While CO_2 hydrogenation incorporates the formation of water, CO is converted to CO_2 via the reverse water gas shift reaction (RWGS) by consumption of water as shown by Eq. (1):

The overall reaction is a result of reactions (2) and (3)

$$CO_2 + H_2 \rightleftharpoons CO + H_2O \quad \Delta H = 41.2\,\text{kJ}\,\text{mol}^{-1} \tag{2}$$

The CO conversion can be expressed as:

$$CO + 2H_2 \rightleftharpoons CH_3OH \quad \Delta H = -90.6\,\text{kJ}\,\text{mol}^{-1} \tag{3}$$

Both synthesis reactions are exothermal and involve a decrease in volume. Thus, methanol formation is favored by low temperatures and elevated pressures. CO hydrogenation is significantly more exothermic than CO_2 hydrogenation resulting in higher cooling demand. The maximum conversion is determined by the chemical equilibrium (Dieterich et al., 2021).

3.7 Commercial scale methanol production from CO_2/H_2

Mitsui Chemicals constructed a 100 t per a pilot plant in 2009 based on the established fixed bed synthesis loop scheme for direct carbon dioxide reforming to methanol. (Mitsui Chemicals Inc, 2009; Goto et al., 2016). It converts by-product hydrogen and CO_2 separated from the exhaust gas both originating from naphtha cracking. For future applications, hydrogen production from photo-catalysis and biomass are investigated. (Mitsui Chemicals Inc,

2010; Mitsui Chemicals Inc, 2015). Low-cost supply of renewable hydrogen is currently identified to be the major hurdle for commercialization.

4 Butanol

Four carbon alcohol has an energy density similar to gasoline. It can be blended with gasoline or diesel for use in road transport. US fuel standard allows up to 16% butanol in gasoline (ASDM D 4814) while EU fuel standard allows up to 15% in gasoline (EN 228). Presently we do not have a standard for butanol added fuels.

Butanol is a better blend component than ethanol and biodiesel due to its high energy density. Like ethanol, butanol is a biomass-based renewable fuel that can be produced by the alcoholic fermentation of the sugars derived from biomass feedstocks (Hansen et al., 2009; Fortman et al., 2008; Ezejia et al., 2005). Butanol contains more oxygen content compared with biodiesel, leading to further reduction of soot. NOx emissions can also be reduced due to its higher heat of evaporation, which results in a lower combustion temperature (Rakopoulos et al., 2010).

The advantages of butanol compared to ethanol and biodiesel are: (1) Easier distribution as butanol tolerates water contamination better than ethanol (2) is less corrosive than ethanol therefore it is more suitable for distribution through existing pipelines, whereas ethanol is not suitable for pipeline transportation. In blends with diesel or gasoline, butanol is less likely to separate from the base fuel than ethanol if the fuel is contaminated with water. This facilitates the storage and distribution of blended fuels. It has a higher viscosity, equal to that of diesel fuel, so it will not cause potential wear problems in diesel engines.

Since butanol is a four carbon alcohol, doubling the carbon of ethanol and containing 25% more energy and thus higher calorific value as compared to ethanol. Therefore, the fuel consumption will reduce and a better mileage can be obtained. It can be blended in diesel fuel without co-solvent and does not cause vapor lock problems.

Butanol is generally produced by fermentation of sugars, via ABE (acetone-butanol-ethanol) pathway by using microorganism, bacteria (specifically Clostridium). However, there is a serious problem that is the inhibition of fermentation organisms by the fermentation products which means that after butanol reaches a particular concentration the fermentation stops. The highest concentration achieved, on an industrial scale, for butanol by fermentation of sugar is $20\,g/L$. This means that product separation must take place from dilute fermentation broth and this consumes a large amount of energy in the product separation stage (Chen et al., 2013; Kujawska et al., 2015; Raganati et al., 2012; Quereshi et al., 2013). Acetone and ethanol are produced as co-products limiting the yield of butanol and increasing the complexity of product separation (e.g., liquid-liquid extraction), which can be very energy-intensive. Technology development needs include improved fermentation organisms which can give increased product selectivity and also improved low energy separation processes.

Butanol fermentation has thus far been optimized for high sugar content feedstocks, which were more expensive. However, some technology developers, such as Celtic Renewables, are also looking at the conversion of low sugar content substrates (and therefore lower-cost feedstocks) to butanol. This means combining lignocellulosic feedstock pretreatment and hydrolysis with the fermentation and purification processes in the existing cellulosic ethanol plants. There are several biotechnology companies, such as Butyl Fuel, Cathay Industrial Biotech, Cobalt Biofuels, Green Biologics, Metabolic Explorer, Tetravitae Bioscience, and others around the world dedicated to providing strains and process solutions for ABE fermentation for industrial customers and extensively reviewed in the literature (Kharkwal et al., 2009; Chao Jina et al., 2011).

Improvements in butanol production need to happen for making the biotechnology process economical. Three major factors determine the economics of the biotechnological butanol production and these are substrate costs, low product yields versus solvent toxicity, and costs for downstream processing. Metabolic engineering of *Clostridium acetobutylicum*, is main step for improving the fermentation process. Systems-level metabolic engineering of strains based on the recent availability of complete genome sequences and new metabolic engineering tools for clostridia are being adopted.

Butanol recovery by distillation, from a low titer broth, is another energy-intensive process. High product recovery cost is a problem in biological butanol production. Several other processes including pervaporation, adsorption, liquid–liquid extraction, gas stripping, and reverse osmosis have been developed to improve recovery performance and reduce costs (Ezeji et al., 2004). The industry is trying to scale up the fermentation process with improved bacteria strains and yeast. Green Biologics is testing, validating and optimizing its process based on Clostridium at a pilot scale. It is also a method for the in situ removal of butanol and separation using pervaporation (membrane separation).

5 Ammonia as a marine fuel

In 2018, the International Maritime Organization (IMO) agreed to reduce greenhouse gas (GHG) emissions from international shipping by at least 50% by 2050 and phase them out by 2100. Green ammonia has been identified as one of the most promising low-emission fuels, with the IEA predicting that its use for shipping will reach 130 m tons by 2070, twice as much as was used worldwide for fertilizer production in 2019.

Ammonia has been proposed as a potential marine fuel as it would not give emissions of CO_2 from the ship since it is a carbon-free molecule (Kirstein et al., 2018; Klüssmann et al., 2019; Maritime Knowledge Centre, TNO and TU, 2017). Ammonia can be converted to power using established internal combustion engine technology (Kroch, 1945; Gray et al., 1966; Pearsall and Garabedian, 1967; Liu et al., 2003).

However, ammonia is less energy-dense than oil, meaning ships will consume up to five times as much fuel by volume. Ammonia production would have to rise by 440 million tons—more than three times current production—requiring a huge amount of renewable energy, according to the International Chamber of Shipping. By developing decarbonization pathways for international shipping that combine different technologies including ammonia as fuel and reduction of CO_2 emissions (Zero-Emission Vessels 2030, 2017).

Japan is planning to introduce ammonia as fuel into the shipping industry and make it commercially available in the late 2020s as part of its efforts to go carbon-neutral by 2050. Being already mainly used for fertilizer applications, it has an international trading infrastructure in place, however, there is still a need for some technical issues to be resolved, including safe pressurization, liquefaction and combustion before it can become widely used as a fuel (Hansson et al., 2020). Ammonia can be stored in higher temperatures in a liquid form under atmospheric pressure. What is more, it is less flammable than bunker fuels.

Ammonia from natural gas, for hydrogen production, with CCS is another option with potentially low climate impact and to support a demand for ammonia as fuel in shipping, the production of renewable ammonia needs to increase substantially. As ammonia is currently mainly used for fertilizer production, this might lead to competition with potentially higher ammonia prices and higher food prices at least in the short term. However, compared to biofuels, large-scale production of renewable-based electricity is needed for renewable ammonia production.

Although more energy-dense than hydrogen, ammonia still occupies significantly more space than diesel for the same amount of propulsion. New vessel designs might be able to accommodate this, however major infrastructure changes will be required in existing ships.

Major ship engine builder, Wärtsilä is working on four-stroke engine designs, hoping to reach the stage of field tests as soon as 2022 (Schönborn, 2020). Finnish marine-to-energy giant is developing ammonia storage and supply systems to install ammonia fuel cells on a supply vessel, Viking Energy by 2023.

A disadvantage of ammonia over existing liquid bunker fuels is that it requires pressurization to 0.43 MPa at standard temperature (Valera-Medina et al., 2018). A further and much more severe disadvantage is its toxicity to humans, and to aquatic life (Merck, 2018). Human short-term exposure limits to ammonia vapor are typically regulated to be as low as 25–50 ppm and therefore very elaborate safety measures will be required adding to the shipping costs.

NH3 is still one of the most attractive fuels due to the following facts:

(i) It has a high octane rate of 110–130 and thus is a good fuel for internal combustion engines (Zamfirescu and Dincer, 2009).
(ii) The distribution infrastructure already exists for ammonia to deliver it in amounts larger than 100 million tons yearly or more.
(iii) It can be thermally cracked into hydrogen and nitrogen using low energy, to produce hydrogen for fuel-cells.

It is still safer than various fuels and hydrogen due to: In case of leakages/spills, it escapes into the atmosphere and dissipates rapidly because its density is lighter than that of air. It can be detected by nose at a low ppm of 5 and is nonflammable.

6 The Fischer–Tropsch process for liquid hydrocarbons

The Fischer–Tropsch process is a combination of chemical reactions that converts a mixture of carbon monoxide and hydrogen (syngas) into liquid hydrocarbons. Fischer–Tropsch synthesis is a well-known process that has been developed for almost a century. Big plants with coal and natural gas as a feedstock exist and can be economic, depending on the framework conditions. Smaller plants with biomass-based feedstock have been built in the last decade.

Generally, the Fischer–Tropsch process is operated in the temperature range of 150–300°C. Higher temperatures lead to faster reactions and higher conversion rates but also tend to favor methane production and leads to increase in input energy. For this reason, the temperature is usually maintained at the low to middle part of the range. Increasing the pressure leads to higher conversion rates and also favors formation of long-chained alkanes, both of which are desirable. Typical pressures ranges from one to 10 of atmospheres. Even higher pressures would be favorable, but the economics may not justify the additional costs of high-pressure equipment. The Fischer–Tropsch (FT) process is a heterogeneously catalyzed to convert syngas to liquid hydrocarbons. The products are a variety of simple hydrocarbon chains of different lengths, depending on the process conditions. They can be applied as a low-sulfur diesel substitute in transport applications. FT from coal was developed in Germany during world war and very large amount of transport fuel was made by this route.

Generally, syngas with a H2/CO ratio of 2–2.2 is processed on a Fe- or Co-catalyst at temperatures of at least 160–200°C at normal pressure. Higher pressures and temperatures are advantageous.

Reactions (4)–(6) take place during FT-synthesis with n as the resulting carbon chain length (typically ranging between 10 and 20). Reaction (4) is the desired one for the production of alkanes. Water, which is always a by-product, can be converted via water gas shift reaction with CO to CO_2 and H_2 by using Fe-catalysts.

$$nCO + (2n+1)H_2 \rightarrow CnH2n + 2 + nH_2O \qquad (4)$$

$$nCO + (2n)H_2 \rightarrow CnH2n + nH_2O \qquad (5)$$

$$nCO + (2n)H_2 \rightarrow CnH2n + 1 + H_2O \qquad (6)$$

Low-Temperature-Fischer–Tropsch (LT-FT) reactors contain liquid hydrocarbons and can hence be operated below condensation temperature. The main products in these reactors are waxes, whereas High-Temperature-Fischer–Tropsch (HT-FT) produces mainly alkenes and gasoline (Dry, 2002). The wax type saturated hydrocarbons need secondary treatment like cracking to obtain liquid hydrocarbons in fuel range of diesel and gasoline. This cracking can be done by existing cracking units in a typical petroleum refinery. Liquid products (in the range of 10–23 carbon atoms per chain) can be used to substitute gasoline or diesel and have advantages compared to conventional fuels, due to their low aromatics content and the absence of sulfur. This results in lower emission values regarding particles, SO_2 and aromatics.

6.1 FT on biomass

Production of renewable fuels, such as gasoline, diesel and jet fuel, using the Biomass to Liquid via Fischer–Tropsch Synthesis (BTL-FT) process has been gaining increasing attention during recent years. Renewable fuels from BTL-FT are usually much cleaner and environmentally friendly, and they contain little or even no sulfur and other compounds.

For gasoline, ethanol is a good blending component and it can be obtained from a variety of feedstocks. However, for diesel biodiesel, FAME has been promoted as a blend component. Due to poor oxidation properties and gum formations FAME %age in diesel is being limited to about 7%–10%. Otherwise also the feedstock for FAME, lipids, are limited. In view of this production of clean diesel and gasoline, fully fungible, by FT synthesis is being seen as a much-preferred route.

In the BTL-FT process, biomass, such as woodchips and straw stalk, is firstly converted into biomass-derived syngas (bio-syngas) by gasification. The biomass has to be converted into a smaller size for better conversion. Then, a cleaning process is applied to remove impurities from the bio-syngas to produce clean bio-syngas which meet the Fischer–Tropsch synthesis requirements. Cleaned bio-syngas is then conducted into a Fischer–Tropsch catalytic reactor to produce green gasoline, diesel, and other clean biofuels. (Demirbas, 2006; Tijmensen et al., 2002). To meet the environmental emission conditions, an elaborate gas cleaning process is generally used and it adds to the overall capital cost of the process.

Bio-syngas resulting from biomass gasification contain CO, H_2, CO_2, CH_4, and N_2 in various proportions (Klass, 1998; Beenackers and Swaaij, 1984). The average bio-syngas from a downdraft gasifier with air as the oxidant contains 22% CO, 18% H_2, 12% CO_2, CH_4, with N_2 and other gases makes the balance (Wei et al., 2009). Before FT reaction some purification is required but total removal of CO_2 is seldom needed.

Pretreatment before gasification is necessary and generally includes screening, size reduction, and drying (Faaij et al., 1998). Smaller biomass particle size will provide more surface area and porous structures per unit biomass, which will facilitate heat transfer and biomass conversion during the gasification process. However, in most gasifiers, feed particle sizes are most often in the range of 20–80 mm (McKendry, 2002). Reduction of size adds to the operating cost of the process and some biomass like agricultural residues needs more energy for grinding. Drying is the most important process in the pretreatment and this is generally carried out by passing recycled hot gases for biomass drying. Drier biomass can improve the efficiency of gasification, but also reduces the hydrogen content in the gas product, which is unfavorable in the subsequent Fischer–Tropsch synthesis. Drying can reduce the moisture content of the biomass feedstock to about 10% (van Ree et al., 1995).

6.1.1 Product selectivity and distribution in FT synthesis

In general, the product distribution of hydrocarbons formed during the Fischer–Tropsch process follows an Anderson–Schulz–Flory distribution (Flory, 1936) which can be expressed as:

$$W_n/n = (1-\alpha)^2\alpha^{n-1}$$

where W_n is the weight fraction of hydrocarbons containing n carbon atoms, and α is the chain growth probability or the probability that a molecule will continue reacting to form a longer chain. In general, α is largely determined by the catalyst and the specific process conditions.

Increasing α increases the formation of long-chained hydrocarbons and formation of waxes. Methane will be the largest single product if α is less than 0.5; however, by increasing α close to one, the total amount of methane formed can be minimized. The very long-chained hydrocarbons are waxes, which are solid at room temperature. Therefore, for the production of liquid transportation fuels it may be necessary to crack some of the Fischer–Tropsch products. In order to avoid this, some researchers have proposed, with limited success, using zeolites or other catalyst substrates with fixed sized pores that can restrict the formation of hydrocarbons longer than some characteristic size say C15·In this cracking gasoline and diesel range hydrocarbons are obtained.

6.1.2 Catalysts

The catalyst is the heart of Fischer–Tropsch synthesis. Higher activity with desired product selection and longer lifetime with less catalyst decay should be the priority of the FT catalyst design in future research. Coke formation at higher operating temperatures can lead to catalyst coating and deactivation. More attention should be paid to increase the carbon utilization in the bio-syngas conversion, to reduce greenhouse emissions and to promote the overall rate of carbon conversion into liquid fuels. In commercial processes, Fe- and Co-catalysts are used. Porous metal oxides with large specific surfaces, for example zeolite and aluminum oxide, are used as catalyst carriers and are reviewed (Khodakov et al., 2007).

A variety of synthesis-gas compositions can be used. For cobalt-based catalysts the optimal H_2:CO ratio is around 1.8–2.1. Iron-based catalysts can tolerate lower ratios. This reactivity can be important for synthesis gas derived from coal or biomass, which tend to have relatively low H_2:CO ratios (<1). New catalysts which can be effective at lower temperature and pressure and are robust for action of carbon coating are being researched.

In general, four different types of reactor systems have been used on commercial scale FT plants and these are (1) Fixed bed (trickle bed; 2) slurry phase (3) fluidized bed, and (4) circulating fluidized bed (CFB).

There are a large number of commercial plants based on Sasol and Shell technologies and in most of these plants syngas is generated by reforming of natural gas. Some of the commercial plants are Sasol. (2500 bpd); Shell (12,500 bpd); Qatar Petroleum on Shell technology (140,000 bpd): Qatar Petroleum on Sasol technology (34,000 bpd). The low-temperature FT facility at Ras Laffan, Qatar, is the largest FT plant in the world using cobalt catalyst at 230°C, converting natural gas to petroleum liquids at a rate of 140,000 barrels per day employing Shell FT technology. Most of the FT plants are based on abundant and cheaply priced natural gas in gulf countries and some in the US based on shale gas. The FT plants based on biomass residues are still in precommercial stage as these are much more capital intensive than gas-based plants due to very elaborate gas cleaning requirements.

In the case of currently installed FT plants, 60%–70% of the costs are spent for syngas production (coal gasification, methane reforming), 22% for the FT synthesis itself and around 12% for product upgrade and refining (Choi et al., 1996; Gregor, 1991; Choi et al., 1997). FT can be an economic alternative if some requirements are fulfilled. First of all the carbon source must be large enough and cheap, for example, stranded or associated natural gas, which is mostly flared now, or to be economically friendly based on carbon capture or biomass gasification.

7 Conclusions

Apart from conventional biofuels like ethanol, biodiesel several other low carbon fuels are being promoted especially for hard to abate transport sectors like shipping and aviation. Liquid biofuels offer advantage like easy storage, handling and dispensing. Existing automobile infrastructure can easily be adopted for liquid biofuels. Methanol as a fuel component of gasoline and diesel has attracted a lot of attention. China has started producing and using methanol at massive scales, though it may not lead to life cycle GHG reduction as the feedstock is coal. Methanol production from coal or even natural gas would need to be coupled with carbon dioxide capture and sequestration (CCS) to make it an environmentally sound proposition. Ammonia produced from green hydrogen may be a fuel for shipping as it does not give CO_2 on combustion. However major changes are required in engines, storage, and for safe use. Butanol is a better and energy-dense fuel that can be a much better blend component than ethanol. It has better water tolerance and can be transported by pipelines. Its production from biomass sugars is of interest and several genetic microbial manipulations are being done to increase its titer. FT liquids offer an advantage as these can be produced from biomass and give completely fungible clean diesel and gasoline.

In near future, the basket of low carbon transport fuels will get bigger and some of the fuel components discussed in this chapter will see large-scale commercialization.

References

Alberico, E., Nielsen, M., 2015. Towards a methanol economy based on homogeneous catalysis: methanol to H_2 and CO_2 to methanol. Chem. Commun. 51, 6714–6725.

Ali, K.A., Abdullah, A.Z., Mohamed, A.R., 2015. Recent development in catalytic technologies for methanol synthesis from renewable sources. Renew. Sust. Energ. Rev. 44, 508–518.

Alvarado, M., 2016. The Changing Face of the Global Methanol Industry. Technical report, IHS, London, UK.

Araya, S.S., Liso, V., Cui, X., Li, N., Zhu, J., Sahlin, S.L., Jensen, S.H., Nielsen, M.P., Kær, S.K., 2020. A review of the methanol economy: the fuel cell route. Energies 13, 596.

Beenackers, A.A.C.M., Swaaij, W.P.M., 1984. Thermochemical Processing of Biomass. Butterworth, London, UK.

Bertau, M., Offermanns, H., Plass, L., Schmidt, F., Wernicke, H.-J., 2014. Methanol. The Basic Chemical and Energy Feedstock of the Future: Asinger's Vision Today. Springer, Heidelberg.

Boulamanti, A., Moya, J.A., 2016. Production costs of the chemical industry in the EU and other countries: Ammonia, methanol and light olefins. Renew. Sust. Energ. Rev. 68, 1205–1212.

Brown, A., Le Feuvre, P., 2017. Technology Roadmap: Delivering Sustainable Bioenergy. International Energy Agency (IEA), Paris, France, p. 94.

Buhaug, Ø., Corbett, J.J., Endresen, Ø., Eyring, V., Faber, J., Hanayama, S., Lee, D.S., Lee, D., Lindstad, H., Markowska, A.Z., Mjelde, A., Nelissen, D., Nilsen, J., Pålsson, C., Winebrake, J.J., Wu, W., Yoshida, K., 2009. Second IMO GHG Study 2009. International Maritime Organization, London.

Bui, M., Adjiman, C.S., Bardow, A., Anthony, E.J., Boston, A., Brown, S., Fennell, P.S., Fuss, S., Galindo, A., Hackett, L.A., et al., 2018. Carbon capture and storage (CCS): the way forward. Energy Environ. Sci. 11, 1062–1176.

Chao Jina, B., Yaoc, M., Liuc, H., Chia-Fon, F., Leed, E., Jib, J., 2011. Progress in the production and application of n-butanol as a biofuel. Renew. Sustain. Energ. Rev. 15, 4080–4106.

Chen, Y., et al., 2013. Production of Butanol from Glucose and Xylose with Immobilized Cells of *Clostridium Acetobutylicum*. In: Biotechnology and Bioprocess Engineering. vol. 18. The Korean Society for Biotechnology and Bioengineering, Korea, pp. 234–241. no 1.

Choi, G.N., Kramer, S.J., Tam, S.T., Fox, J.M., 1996. Design/Economics of a Natural Gas Based Fischer–Tropsch Plant. Houston,.

Choi, G.N., Kramer, S.J., Tam, S.S., Fox, J.M., Carr, N.L., Wilson, G.R., 1997. Coal Liquefaction & Solid Fuels Contractors Review Conference.

Cui, X., Kær, S.K., 2019. Thermodynamic analyses of a moderate-temperature carbon dioxide hydrogenation to methanol via reverse water gas shift process with in situ water removal. Ind. Eng. Chem. Res. 58 (24), 10559–10569.

Dalena, F., Senatore, A., Basile, M., Knani, S., Basile, A., Iulianelli, A., 2018. Advances in methanol production and utilization, with particular emphasis toward hydrogen generation via membrane reactor technology. Membranes 8.

Dang, S., Yang, H., Gao, P., Wang, H., Li, X., Wei, W., Sun, Y., 2019. A review of research progress on heterogeneous catalysts for methanol synthesis from carbon dioxide hydrogenation. Catal. Today 330, 61–75.

Demirbas, A., 2006. Global biodiesel strategies. Energy Educ. Sci. Technol. 17, 27–63.

Dieterich, V., Buttler, A., Hanel, A., Spliethoffab, H., Fendt, S., 2021. Power-to-Liquid *Via* Synthesis of Methanol, DME or Fischer–Tropsch-Fuels: A Review.

Dry, M.E., 2002. The Fischer–Tropsch process. Catal. Today 71, 227–241.

Ezeji, T.C., Qureshi, N., Blaschek, H.P., 2004. Acetone butanol ethanol (ABE) production from concentrated substrate: reduction in substrate inhibition by fed-batch technique and product inhibition by gas stripping. Appl. Microbiol. Biotechnol. 63, 653–658.

Ezejia, T.C., Qureshib, N., Blascheka, H.P., 2005. Continuous butanol fermentation and feed starch retrogradation: butanol fermentation sustainability using *Clostridium beijerinckii* BA101. J. Biotechnol. 115, 179–187.

Faaij, A., van Ree, R., Meuleman, B., 1998. Long Term Perspectives of Biomass Integrated Gasification with Combined Cycle Technology: Costs and Efficiency and a Comparison with Combustion; EWAB Report 9840. The Netherlands Agency for Energy and the Environmen (NOVEM), Utrecht, The Netherlands.

Flory, P.J., 1936. Molecular size distribution in linear condensation polymers. J. Am. Chem. Soc. 58, 1877–1885.

Fortman, J.L., Chhabra, S., Mukhopadhyay, A., Chou, H., Lee, T.S., Steen, E., et al., 2008. Bio- fuel alternatives to ethanol: pumping the microbial well. Trends Biotechnol. 26, 375–381.

Gahleitner, G., 2013. Hydrogen from renewable electricity: an international review of power-to-gas pilot plants for stationary applications. Int. J. Hydrog. Energy 38, 2039–2061.

Goeppert, A., Czaun, M., Surya Prakash, G.K., Olah, G.A., 2012. Air as the renewable carbon source of the future: an overview of CO_2 capture from the atmosphere. Energy Environ. Sci. 5, 7833–7853.

Goto, Y., Takahashi, N., Yoshinaga, M., Murakami, M., 2016. US Pat. 9314774,.

Gray, J., Dimitroff, E., Meckel, N., et al., 1966. Ammonia fuel-engine compatibility and combustion. In: SAE Technical Paper Series No. 660156.

Gregor, J.H., 1991. Fischer-Tropsch products as liquid fuels or chemicals. Catal. Lett. 7, 317–331.

Grimm, P., 1991. Six years successful operation of Linde isothermal reactor. Rep. Sci. Technol. 49, 57–59.

Guil-López, R., Mota, N., Llorente, J., Millán, E., Pawelec, B., Fierro, J.L.G., Navarro, R.M., 2019. Methanol synthesis from CO2: a review of the latest developments in heterogeneous catalysis. Materials 12, 3902.

Guinot, B., Montignac, F., Champel, B., Vannucci, D., 2015. Profitability of an electrolysis based hydrogen production plant providing grid balancing services. Int. J. Hydrog. Energy 40, 8778.

Hansen, J.B., Nielsen, P.E.H., 2008. Methanol synthesis. In: Handbook of Heterogeneous Catalysis, pp. 13–2949.

Hansen, A.C., Kyritsis, D.C., Lee, C.F., 2009. Characteristics of biofuels and renewable fuel standards. In: Vertes, A.A., Blaschek, H.P., Yukawa, H., Qureshi, N. (Eds.), Biomass to Biofuels—Strategies for Global Industries. John Wiley, New York.

Hansson, T., Brynolf, S., Fridell, E., Leher, M., 2020. The potential role of ammonia as marine fuel—based on energy systems modeling and multi-criteria decision analysis. Sustainability 12, 3265.

Hirotani, K., Nakamura, H., Shoji, K., 1998. Optimum catalytic reactor design for methanol synthesis with TEC MRF-Z reactor. Catal. Surv. Jpn. 2, 99–106.

Hobson, C., Márquez, C., 2018. Renewable Methanol Report. Technical report, ATA Markets Intelligence S.L. on behalf of the Methanol Institute, Madrid, Spain.

IEA, 2020. Energy Technology Perspectives 2020.

International Energy Agency (IEA), 2018. World Energy Outlook 2018. https://www.iea.org/weo/weo2018/.

International Energy Agency Task33, Database—Gasification of Biomass and Waste. http://task33.ieabioenergy.com/content/.

International Energy Agency Task34, Pyrolysis Demoplant Database. http://task34.ieabioenergy.com/publications/pyrolysis-demoplant-database/.

International Energy Agency Task39, Database on Facilities for the Production of Advanced Liquid and Gaseous Biofuels for Transport. https://demoplants.bioenergy2020.eu/.

Jadhav, S.G., Vaidya, P.D., Bhanage, B.M., Joshi, J.B., 2014. Catalytic carbon dioxide hydrogenation to methanol. Chem. Eng. Res. Des. 92, 2557–2567.

Kharkwal, S., Karimi, I.A., Chang, M.W., Lee, D.Y., 2009. Strain improvement and process development for biobutanol production. Biotechnology 3, 202–210.

Khodakov, A.Y., Chu, W., Fongarland, P., 2007. Advances in the development of novel cobalt Fischer—Tropsch Catalysts for synthesis of long-chain hydrocarbons and clean fuels. Chem. Rev. 107 (5), 1692–1744.

Kiaee, M., Cruden, A., Infield, D., Chladek, P., 2013. Improvement of power system frequency stability using alkaline electrolysis plants. Proc. Inst. Mech. Eng. A. J. Power Energy 227, 115.

Kirstein, L., Halim, R., Merk, O., 2018. Decarbonising maritime transport—pathways to zero-carbon shipping by 2035. In: OECD International Transport Forum: Paris, France.

Klass, D.L., 1998. Biomass for Renewable Energy, Fuels and Chemicals. Academic Press, San Diego, CA, USA.

Klüssmann, J.N., Ekknud, L.R., Ivarsson, A., Schramm, J., 2019. The Potential for Ammonia as a Transportation Fuel—A Literature Review. The Technical University of Denmark (DTU), Lyngby, Denmark.

Kroch, E., 1945. Ammonia—a fuel for motor buses. J. Inst. Pet. 31, 214–223.

Kujawska, A., et al., 2015. ABE fermentation products recovery methods—a review. In: Renewable and Sustainable Energy Reviews. vol. 48. Elsevier, pp. 648–661.

Liu, R., Ting, D.S.-K., Checkel, M.D., 2003. Ammonia as a fuel for SI engine. In: SAE Technical Paper Series No. 2003-01-3095.

Maritime Knowledge Centre, TNO and TU, 2017. Framework CO2 Reduction in Shipping. Maritime Knowledge Centre, TNO and TU, Delft, The Netherlands.

Market Watch, 2019. Methanol market. In: Analysis and Technological Innovation by Leading Key Players 2026.

Matthey, J., 2021. Catalysts. Katalco. http://www.jmcatalysts.cn/en/pdf/Methanol-top-level.pdf.

McKendry, P., 2002. Energy production from biomass (part 3): gasification technologies. Bioresour. Technol. 83, 55–63.

Merck, 2018. Ammonium hydroxide. In: Safety Data Sheet.

Methanol Institute, Available online https://www.methanol.org.

Mitsui Chemicals Inc, 2009. CSR Report. http://www.mitsuichem.com/csr/report/index.htm. Accessed 28 June 2016.

Mitsui Chemicals Inc, 2010. CSR Report. http://www.mitsuichem.com/csr/report/index.htm. Accessed 28 June 2016.

Mitsui Chemicals Inc, 2015. CSR Report. http://www.mitsuichem.com/csr/report/index.htm.

Mohammadi, A., Mehrpooya, M., 2018. Acomprehensive review on coupling different types of electrolyzer to renewable energy sources. Energy 158, 632–655.

National Biofuel Policy, 2018. https://MoPNG.gov.in.

Pearsall, T., Garabedian, C., 1967. Combustion of anhydrous ammonia in diesel engines. In: SAE Technical Paper Series No. 670947.

Planning Commission, Government of India, 2003. Report of the Committee on Development of Bio-Fuel. http://planningcommission.nic.in/reports/.

Quereshi, N., et al., 2013. An economic evaluation of biological conversion of wheat straw to butanol: a biofuel. In: Energy Conversion and Management. vol. 65. Elsevier, pp. 456–462. n 1. [online].

Raganati, F., et al., 2012. Butanol production from lignocellulosic-based hexoses and pentoses by fermentation of *Clostridium Acetobutylicum*. In: Chemical Engineering Transactions. vol. 27. The Italian Association of Chemical Engineering (AIDIC), pp. 91–96. no 1. [online].

Rakopoulos, D.C., Rakopoulos, C.D., Giakoumis, E.G., Dimaratos, A.M., Kyritsis, D.C., 2010. Effects of butanol–diesel fuel blends on the performance and emis- sions of a high-speed DI diesel engine. Energy Convers. Manag. 51, 1989–1997.

Saraswat, V.K., Bansal, R., 2021. India's Leapfrog to Methanol Economy. Niti Aayog, Government of India.

Schönborn, A., 2020. Aqueous solution of ammonia as marine fuel. J. Eng. Marit. Environ., 1–10.

Serenergy A/S. Methanol Production. Available online 2021. http://www.serenergy.com.

Sherif, S., Barbir, F., Veziroglu, T., 2005. Wind energy and the hydrogen economy—review of the technology. Sol. Energy 78, 647–660.

Tijmensen, M.J.A., Faaij, A.P.C., Hamelinck, C.N., van Hardeveld, M.R.M., 2002. Exploration of the possibilities for production of Fischer Tropsch liquids and power via biomass gasification. Biomass Bioenergy 23, 129–152.

U.S. Department of Agriculture, 2018. India Biofuels Annual 2018. https://gain.fas.usda.gov/.

Valera-Medina, A., Xiao, H., Owen-Jones, M., et al., 2018. Ammonia for power. Prog. Energy Combust. Sci. 69, 63–102. 23.

van Ree, R., Oudhuis, A., Faaij, A., Curvers, A., 1995. Modelling of a Biomass Integrated Gasifier/Combined Cycle (BIG/CC) System with the Flowsheet Simulation Programme ASPEN+; ECN Report ECN-C–95-041. Energy Research Centre of the Netherlands (ECN), Petten, The Netherlands.

Verhelst, S., Turner, J.W., Sileghem, L., Vancoillie, J., 2019. Methanol as a fuel for internal combustion engines. Prog. Energy Combust. Sci. 70, 43–88.

Wang, W., Wang, S., Ma, X., Gong, J., 2011. Recent advances in catalytic hydrogenation of carbon dioxide. Chem. Soc. Rev. 40, 3703–3727.

Wei, L., Thomasson, J.A., Bricka, R.M., Sui, R., Wooten, J.R., Columbus, E.P., 2009. Syngas quality evaluation for biomass gasification with a downdraft gasifier. Trans. ASABE 52, 21–37.

Zamfirescu, C., Dincer, I., 2009. Ammonia as a green fuel and hydrogen source for vehicular applications. Fuel Process. Technol. 90, 729–737.

Zero-Emission Vessels 2030, 2017. How Do We Get There. Lloyd's Register Group Limited, London, UK. Halim, R.; Kirstein, L.; Merk, O.; Martinez, L. Decarbonization Pathways for International Maritime Transport: A Model-Based Policy Impact Assessment. Sustainability 2018, 10, 2243.

Zhao, K., 2019. A Brief Review of China's Methanol Vehicle Pilot and Policy. Technical report, Methanol Institute, Alexandria, VA, USA.

Advanced biofuels: Perspectives and possibilities

Deepika Awasthi[a] *and K.T. Shanmugam*[b]

[a]Lawrence Berkeley National Laboratory, Berkeley, CA, United States [b]Department of Microbiology and Cell Science, University of Florida, Gainesville, FL, United States

1 Introduction

The economic engine that drives human welfare is dependent on fossil fuels. Among the various fossil fuel sources, coal, crude oil, and natural gas are the dominant sources of energy contributing to about 80% of the total world energy consumption in 2019 (IEA, 2020). The transportation industry that requires energy-dense liquid fuels consumes >70% of the crude oil extracted from the ground. Proven world reserves of crude oil (~1.65 trillion barrels in 2016) are expected to be about 50-times the current annual rate of consumption of this resource (about 35 billion barrels in 2016) assuming no new reserves are identified. It should be noted that this estimated oil reserve represents the total economically recoverable and not technically attainable resource. As the economically recoverable oil reserves dwindle, the cost of liquid transportation fuels is expected to rise due to an increase from the current average worldwide breakeven price of oil at about US $47 in 2018. Besides the putative negative impact of this potential increase in energy cost on the world economy, the use of fossil fuels contributes to adverse climate change due to the accumulation of atmospheric CO_2, the product of energy extraction from fossil fuels. It is imperative that renewable liquid fuels with no net CO_2 emission are developed to replace petroleum at a cost that is competitive with petroleum-derived transportation fuels.

Today's fossil fuels are transformed biomass produced millions of years ago and this temporal separation of production and consumption is the cause of current environmental pollution leading to significant global climate change. By short-circuiting this process, an environmentally sustainable process can be developed for the production of liquid fuels with no net CO_2 evolution (Fig. 1). Among the various potential fuel molecules, ethanol is an attractive alternative to petroleum-derived liquid fuels owing to the long history of the fermentation

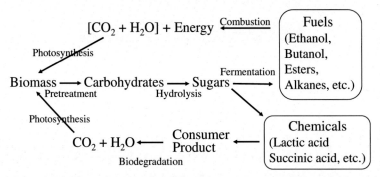

FIG. 1 A generalized scheme for conversion of lignocellulosic biomass to fuels and chemicals. Pretreatment of biomass by acid and heat generates free sugars, such as xylose and arabinose, while making cellulose accessible to enzyme hydrolysis. Other pretreatment processes using base or ionic liquids releases mostly carbohydrates. Only few examples of fuels and chemicals are listed. Combustion of the fuel molecules releases energy for transportation and the CO_2 released is reassembled into sugars, carbohydrates and biomass using the solar energy captured through photosynthesis. Sugars are also fermented to chemicals that are converted to products such as plastics. Upon end of life, these are converted to CO_2, for recapture by photosynthesis.

industry producing ethanol from grains and fruit juices (McGovern et al., 2004). The main polysaccharide in cereal grains is starch, an α-1,4 or α-1,6-linked homopolymer, that yields glucose upon enzyme hydrolysis. This glucose and the disaccharide sucrose from sugarcane and sugar beets are fermented effectively by *Saccharomyces cerevisiae* to ethanol accounting for more than 95% of the current world production of about 29 billion gallons (110 billion liters) in 2019 (https://afdc.energy.gov/data/10331). Volumetric productivity of ethanol as high as 5 g/(L.h) has been reported for yeast in simple batch fermentations indicating the effectiveness of this microbial biocatalyst in ethanol production (Rajoka et al., 2005).

The need for feeding an increasing world population negates the use of grains and sugars for the production of ethanol and other fuel molecules leaving nonedible lignocellulosic biomass as the preferred feedstock (Zilberman et al., 2013). As presented in Fig. 1, the carbohydrates in the recalcitrant biomass need to be hydrolyzed to sugars before fermentation by various native and/or engineered microorganisms to the desired biofuel or other chemicals that are currently derived from petroleum. This chapter will focus on critical factors that need to be addressed as an economically viable cellulosic biomass-based biofuel and biochemical industry is developed. It should be noted that Governments use mandates and subsidies to support biofuel industries. However, for long-term survival, these biorefineries need to develop technologies that can support effective competition in the marketplace without Governmental intervention. Specific discussions on pretreatment of biomass, enzyme hydrolysis of carbohydrates to sugars and fermentation are covered in other chapters in this book and the readers are referred to these chapters (Chapters 8, 9, and 10).

2 Cellulosic biomass as a feedstock for microbial fermentation to fuels and chemicals

The production cost of any biofuel, produced from grains and sugars or from cellulosic biomass, needs to match that of gasoline (petrol) for economic viability, now and in the near future, due to the dominance of petroleum-derived liquid fuels in the world economy. The cost of production of gasoline varies with the price of crude oil and in October 2020 it averaged at about US$1.66/gal (about $0.44/L) in California, USA that uses a special blend (at a higher cost of production), at a world crude oil price of about $40.00 per barrel. Cost of distribution, taxes, and other margins account for the rest of the retail price. It should be noted that in August 2021, the cost of gasoline production in California, USA was $2.96 per gal (about $0.61/L) due to the higher price of crude oil (West Texas Intermediate) at $67.73 per barrel (https://ww2.energy.ca.gov/almanac/transportation_data/gasoline/margins/index_cms.php). In comparison, the cost of ethanol production from corn in 2019 in a commercial plant in Iowa, USA was calculated at about $1.70/gal ($0.45/L) (Irwin, 2020) and due to the low energy content of ethanol compared to gasoline (∼0.7), this price increases to $2.40/gal ($0.63/L) on an equal energy basis (GGE, gallon gasoline equivalent). In August 2021, production cost of ethanol increased to $2.31 per gal (about $0/61/L) (https://www.extension.iastate.edu/agdm/energy/html/d1-10.html) but was still lower than the cost of gasoline production at that time. Apparently, market price of corn ($3.85 in 2019 to $6.33 per bushel (56 pounds) in August 2021; https://www.macrotrends.net/2532/corn-prices-historical-chart-data) is a major contributing factor to the production cost of ethanol. However, based on energy content (GGE), the production cost of ethanol in August 2021 ($3.30/gal; $0.87/L) was still higher than that of gasoline. Co-products generated during the ethanol production process, such as DDGS (distiller's dried grains with solubles) and corn oil, help improve the economics of the corn ethanol biorefinery. In addition, government subsidies ($0.45 per gallon of ethanol blended with gasoline in the USA) make ethanol a viable fuel.

In contrast to grains, lignocellulosic biomass is a complex mixture of several polymers with a composition that varies by plant species. Cellulose, a β-1,4-linked glucose homopolymer, is the dominant polysaccharide accounting for about 30%–50% of the total mass. Hemicellulose, a heteropolymer of hexoses, pentoses, and their derivatives, is another carbohydrate in this biomass (19%–25% of total weight) (Williams et al., 2016). These two carbohydrates, lignin, and other minor components form the structural part of plants providing rigidity and resistance to multiple attacks. Since cellulose and hemicellulose in the raw biomass are not readily accessible to appropriate enzymes to produce fermentable sugars, various physicochemical treatments of biomass are employed to improve enzyme access (Blanch et al., 2011; Kim et al., 2011; Tao et al., 2011).

Due to the complexity of sugar extraction from such biomass, the cost of production of cellulosic ethanol is expected to be higher than that of corn ethanol (Tao et al., 2011). Various analyses project a production cost of ethanol using dilute acid and steam pretreatment process between US$1.97 and $4.16 per gallon ($0.50–1.10/L) in a commercial scale biorefinery (Cheng et al., 2019; Gubiczaa et al., 2016; Johnson, 2016; van Rijn et al., 2018). For cellulosic ethanol to compete with gasoline, current biomass pretreatment and enzyme hydrolysis processes need significant improvements to lower the production cost of sugars (Brown et al., 2020).

Among the various pretreatment methods used for fermentative production of biofuels, dilute acid at elevated temperatures can be universally applied to plant materials, with appropriate modifications to suit plant density, to partially hydrolyze hemicellulose to hexoses, pentoses, and oligosaccharides (Kim et al., 2011). The cellulose in the posttreatment solids fraction is also readily accessible to cellulases. With sugarcane bagasse, this process (pretreatment and enzyme hydrolysis) can yield as high as 85% of the total sugars in biomass for fermentation (Geddes et al., 2013; Nieves et al., 2011). This process also led to a theoretical ethanol yield of 86% with wheat straw as feedstock (Saha et al., 2015). However, a disadvantage of the dilute acid process is the co-generation of microbial growth inhibitors that hamper both enzyme hydrolysis and fermentation (Franden et al., 2013; Geddes et al., 2015; Kumar et al., 2013; Martin et al., 2018; Miller et al., 2009). In an alternate process, lignin can be selectively solubilized by alkali and the solids containing cellulose and hemicellulose are separated for hydrolysis by a mixture of enzymes that hydrolyze both cellulose and hemicellulose (Jin et al., 2010). Depending on the source of biomass, ammonia treatment is also reported to generate microbial growth inhibitors (Ong et al., 2016). Although a multistep pretreatment process that generates cellulose without contaminating solids and inhibitors can improve the efficiency of enzyme hydrolysis and allow enzyme recycling to lower the cost of enzymes in the overall process, the increased fixed and operating costs of such a pretreatment process in the production cost of ethanol needs to be addressed. A promising biomass pretreatment process is based on the use of low-cost protic ionic liquids for dissolving lignin away with or without also hydrolyzing hemicellulose (ionoSolv) from biomass leaving cellulose for effective enzyme hydrolysis (Brandt-Talbot et al., 2017; Sun et al., 2017). The advantages of this ionoSolv process are, (1) lower pretreatment reactor cost due to the less corrosive characteristics of the ionic liquid compared to other pretreatments discussed above, (2) lower enzyme loading for cellulose hydrolysis with potential enzyme recycling, and (3) minimal toxicity to enzymes and microorganisms based on the chemistry of the protic ionic liquid used in the process. However, recycling of ionic liquid and water is critical in this process. The estimated minimum selling price of ethanol with the low-cost protic ionic liquid pretreatment process at US$3.5/gal (Sun et al., 2017) indicates that this process is yet to meet the challenge of economically competitive biofuel from biomass. It would suffice to state, the pretreatment process including enzyme hydrolysis developed for generating sugars from lignocellulosic biomass for fermentation to fuels needs to be simple, minimal, and efficient (both fixed and operating costs) (Geddes et al., 2011; Gubiczaa et al., 2016).

The cost of enzymes is the second or third highest variable cost component of a cellulosic ethanol biorefinery after feedstock and chemicals and this can be 20%–25% of the total production cost of ethanol (Cheng et al., 2019; Gubiczaa et al., 2016; van Rijn et al., 2018). This is more than 10-times the cost of enzymes used in a corn to ethanol process (McAloon et al., 2000). It is unrealistic to accept that the cost of cellulases can be lowered to that of starch hydrolyzing enzymes per unit of ethanol produced, due to the differences in the complexity of the substrates and the variations in the specific activities of the two enzyme systems (Lynd et al., 2002). However, cellulases are another target for cost reduction in cellulosic ethanol production (Klein-Marcuschamer et al., 2012).

To further lower the cost of ethanol, all the released hexoses and pentoses need to be rapidly fermented to completion by a new class of microbial biocatalysts since *S. cerevisiae* and *Zymomonas mobilis* lack the ability to ferment pentoses to ethanol (Scalcinati et al., 2012; Yang et al., 2016). Although diverting part of the sugar stream, especially the pentoses, for the production of high-value co-products is attractive from an economic standpoint, it should be noted that the worldwide demand for liquid fuels far outstrips the demand for chemical feedstocks. In 2019, an average of 41% of petroleum was consumed in the USA as gasoline while less than 2% of petroleum was used for the production of petrochemical feedstocks for the manufacture of various chemicals, synthetic rubber and plastics (https://www.eia.gov/

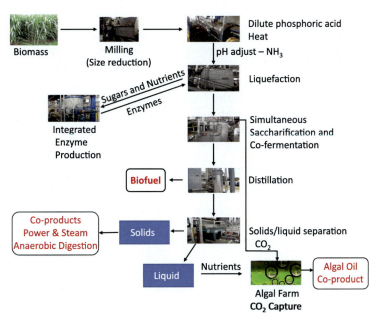

FIG. 2 An idealized biofinery.

dnav/pet/pet_cons_psup_dc_nus_mbblpd_a.htm). This shows the importance of R&D for maximizing the yield of liquid fuel from biomass while attempting to lower production cost.

Based on the discussion above, idealized integrated biomass to ethanol process can be visualized (Fig. 2) (Geddes et al., 2011; Gubiczaa et al., 2016). In this process, biomass after size reduction is treated with dilute phosphoric acid at high temperature and the slurry is liquified with enzymes to lower viscosity to decrease the capital cost associated with pumping and mixing. The liquified slurry at the highest solids loading is simultaneously saccharified and all the released sugars are fermented to ethanol (SScF) in a single vessel, preferably in a continuous mode. After removing ethanol by distillation or other means, liquid fraction of the stillage is separated and used as a nutrient-rich material to support an algal farm that captures the CO_2 released during fermentation. Algal oil is a secondary product of the biorefinery for conversion to biodiesel. Glycerol generated during this process can be fermented to ethanol or other desired chemicals by appropriate microbial biocatalysts. The solids from the stillage, rich in lignin, is a valuable feedstock for the production of other products (Abdelaziz et al., 2016; Beckham et al., 2016) as well as a source of power and steam, either directly or after anaerobic digestion (Khan and Ahring, 2019). Enzymes needed for the hydrolysis of carbohydrates in plant biomass are produced and consumed at site in this integrated biorefinery. This design achieves the following.

1. Simplified and integrated process lowers the capital cost since this is the highest contributor to cellulosic ethanol production cost (Brown et al., 2020).
2. Use of a less corrosive phosphoric acid (or low-cost protic ionic liquids) further lowers capital cost by eliminating the need for expensive alloys associated with the use of

corrosive acids like sulfuric acid. After pretreatment and pH adjustment with NH_3, the phosphate and ammonium salts are used as nutrients for various microorganisms, thus eliminating the cost of removal and disposal of the spent acid (Gubiczaa et al., 2016).

3. On-site production of enzymes lowers capital cost and eliminates the cost of concentration and transportation of enzymes (Johnson, 2016).
4. Liquefaction followed by SScF of all the sugars in one vessel at a high rate by an effective microorganism minimizes capital cost and also retention time (Geddes et al., 2010).
5. Additional co-products derived from algal oil and lignin improve the economics of the biorefinery while also lowering waste treatment cost, including CO_2 mitigation.

What are the challenges in achieving this idealized biorefinery for cost-competitive biofuel production?

1. Maximize conversion of biomass to sugars without generating growth-inhibitory side-products.
   (a) Maximize biomass pretreatment and enzyme hydrolysis to a hexose and pentose yield of $\geq 85\%$.
   (b) Lower the amount of enzyme needed for hydrolysis of carbohydrates by increasing catalytic efficiency of fungal enzymes and/or production of enzymes with high specific activity by the ethanol-producing microbial biocatalyst (consolidated bioprocessing) (Davison et al., 2020; Lopes et al., 2021; Lynd et al., 2005).
   (c) Lower the cost of fungal enzymes by increasing the rate and titer of enzyme production by the fungi beyond current levels (Agrawal et al., 2017; Fonseca et al., 2020; Ogunyewo et al., 2020).
2. Fermentation of sugars to ethanol.
   (a) Co-fermentation of both hexoses and pentoses at the same rate and product yield, either by appropriately engineered single or multiple microorganisms (Demeke et al., 2013; Xie et al., 2020).
   (b) Microbial biocatalysts are not inhibited by side-products generated during the pretreatment process (Martin et al., 2018).
   (c) Highest yield of ethanol (≥ 0.46 g ethanol/g sugar fermented) both on a total sugar and dry biomass basis.
3. Algal process for oil production
   (a) Cost-effective algal photobioreactors for continuous operation using solar energy for illumination (Anto et al., 2020). Although open ponds are inexpensive to construct and operate compared to photobioreactors, this culture system has the advantage of high algal productivity due to its ability to maintain axenic cultures in a controlled environment (Hannon et al., 2010). The higher energy demand of photobioreactors, in comparison to open ponds, is a challenge that can be met by appropriate integrated solar power stations.
   (b) Engineered algal strains for higher than current photosynthetic and CO_2 fixation efficiency (Gimpel et al., 2015).
   (c) Engineered algal strains convert a major fraction of photosynthate into neutral lipids without compromising growth and biomass production (Gimpel et al., 2015).
   Several strategies toward achieving these objectives in algal oil production are discussed in another chapter (Chapter 4) and are not detailed here.

4. Products from lignin

(a) Extraction and purification of lignin for the production of chemical feedstocks and/or end products of significant commercial value from lignin (Wang et al., 2019b).

(b) Chemical and microbial catalysts for cost-effective production of high-value products from lignin (Abdelaziz et al., 2016; Wang et al., 2019b).

Overcoming a part of these of challenges, especially at the level of sugar production and fermentation, can lower the production cost from the current estimates and close the gap to that of gasoline from petroleum (Shanmugam and Ingram, 2021; van Rijn et al., 2018).

In addition to ethanol, several other chemicals (fuels and chemical feedstocks) can be produced from sugars using microbial biocatalysts, although the rate, titer, and yield of these chemicals are not high enough for industrial deployment. As the microbial biocatalysts for production of these chemicals are being developed, it should be noted that the cost of production is expected to be the lowest in an anaerobic process compared to aerobic or O_2-limitation conditions. Besides the increase in capital cost associated with maintaining the needed O_2 concentration in large industrial-scale fermenters, microenvironments with differing O_2 levels within the large vessels have the potential to introduce unexpected side-products with an additional cost of product purification.

Ethanol and lactic acid are currently produced by anaerobic fermentation and butanol is another fuel molecule that was previously produced by fermentation. Improvements in microbial biocatalysts that can significantly lower the production cost of these chemicals are discussed in the following sections.

3 Cellulosic ethanol

Although several cellulosic ethanol biorefineries are being designed and constructed worldwide, only 5 plants are reported to be operational in 2019 and contribute less than 1% of the total ethanol produced (IEA, 2020; Padella et al., 2019). These biorefineries use the steam explosion of biomass as pretreatment with/without dilute acid followed by enzyme hydrolysis to generate fermentable sugars. Depending on the plant part and species, hemicellulose accounts for about 25%–44% of the total carbohydrates in lignocellulosic biomass (Robak and Balcerek, 2020) and the dilute acid and heat pretreatment hydrolyzes a fraction of the hemicellulose with pentoses as dominant monomeric sugars reaching a yield as high as 85%. Although glucose from cellulose can be readily fermented by *S. cerevisiae* or *Z. mobilis*, these microorganisms lack the metabolic potential to ferment pentoses in hemicellulose (Scalcinati et al., 2012; Yang et al., 2016). It should be noted that *S. cerevisiae* has the genetic capability of converting xylose to glyceraldehyde-3-phosphate, an intermediate in glycolysis, but fails to grow using xylose as C-source due to very low level of expression of these genes, especially xylitol dehydrogenase (Moyses et al., 2016; Toivari et al., 2004). Availability of a microbial biocatalyst that can ferment pentoses released from biomass to ethanol at a rate that is comparable to that of glucose fermentation by *S. cerevisiae*, an industrially preferred microbial biocatalyst, is a major challenge in the cellulosic ethanol biorefinery.

Two alternative approaches are used to construct microbial biocatalysts for fermentation of both hexoses and pentoses in the pretreated biomass slurries: endowing *S. cerevisiae* with pentose fermentation potential by introducing an active pathway and metabolic engineering of a

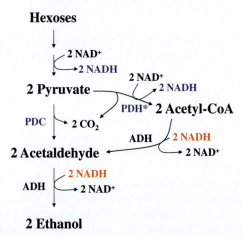

FIG. 3 Two alternative pathways for conversion of pyruvate to ethanol. PDC, Pyruvate decarboxylase; ADH, Alcohol dehydrogenase; PDH*, engineered pyruvate dehydrogenase.

microorganism with a native ability to ferment pentoses, such as *Escherichia coli*, for homo-ethanol production. A pioneering example of the second approach was introducing genes encoding pyruvate decarboxylase (*pdc*) and alcohol dehydrogenase (*adhB*) from *Z. mobilis* into *E. coli* (Ohta et al., 1991) (Fig. 3). In this example, the engineered *E. coli* converted close to 90% of hexoses and pentoses to ethanol while ethanol accounted for less than 10% of the sugar fermented in batch cultures of the wild-type *E. coli*. Further metabolic engineering removed the co-products resulting in ethanol as the only product. Ethanol production by an alternate pathway resulting from engineering native genes was also demonstrated in various pentose-fermenting bacteria (Jilani et al., 2017; Kim et al., 2007; Su et al., 2010; Zhou et al., 2008) (Fig. 3). However, the average productivities of ethanol that is ≤2.0g ethanol/(L.h) by these engineered microbial biocatalysts including yeast is significantly lower than the rate of eth-anol production from glucose by yeast of about 4.0g/(L.h) (Alterthum and Ingram, 1989; Demeke et al., 2013; Gonzalez et al., 2002; Jilani et al., 2017; Ohta et al., 1991; Wang et al., 2019c; Xie et al., 2020). In addition, low ethanol tolerance of many of these bacterial biocatsalysts restricted ethanol titer to about 6% and limited the use of this group of engineered microbial biocatalysts by ethanol biorefineries.

In pentose fermenting microorganisms, sugar is transported into the cell by a specific transporter or by nonspecific sugar transporters. Pentose in the cytoplasm is phosphorylated, transformed in the pentose phosphate pathway (PPP) and the resulting metabolic intermedi-ates enter the glycolytic pathway at the fructose-6-phosphate and glyceraldehyde-3-phosphate level for further conversion to the desired product. Since glucose and pentose fermentation pathways are common from fructose-6-phosphate to ethanol, the rate-limiting steps in pentose fermentation are in transport and conversion of pentoses to fructose-6-phosphate and glyceraldehyde-3-phosphate. Various strategies in the construction of a xylose-fermenting *S. cerevisiae* and improving the rate of fermentation are discussed in other chapters and are not discussed here. Toward this objective, a co-culture of two *S. cerevisiae* strains, each fermenting only one sugar (glucose or pentose), has the advantage of rapidly and simultaneously fermenting all the sugars released from biomass without the negative effect of glucose on pentose fermentation (Eiteman et al., 2008; Hanly et al., 2012; Wang et al., 2019c). It is sufficient to say, that pentoses can account for as much as 25%–30% of biomass by weight and these readily available sugars need to be fermented to the primary product by the biorefinery to lower production cost of ethanol.

4 Butanol

Butanol, a 4-carbon alcohol, is more reduced than ethanol and as a result has a higher energy content than ethanol (ΔHc of -2.68 vs -1.41 MJ/mole, respectively) (Domalski, 1972). In addition to reaching about 94% of the energy content of gasoline, it is also less volatile, less hygroscopic, and less corrosive than ethanol and is considered a drop-in liquid fuel with gasoline. Several anaerobic bacteria do ferment sugars to butanol and this fermentation process was commercially exploited during the early- to mid-20th Century (Jones and Woods, 1986). However, current industrial production of butanol is based on petrochemical feedstocks since the minimum selling price of butanol produced by fermentation of biomass-derived sugars ($0.85/L; $3.22/gal) is higher than that of gasoline as seen with cellulosic ethanol as a fuel (Qureshi et al., 2020).

In contrast to ethanol, native butanol-producing bacteria, such as Clostridia, do not produce butanol as the sole fermentation product due to the make-up of the metabolic pathway that is not redox balanced (Fig. 4). Redox imbalance of the native pathway can be overcome by metabolic engineering of the organism leading to butanol as the sole organic product (Fig. 5) (Abdelaal et al., 2019; Atsumi et al., 2008; Kim et al., 2007; Shen et al., 2011; Tucci and Martin, 2007). In Clostridia, fermentative production of butanol from acetyl-CoA requires 5 NADHs while glycolysis and pyruvate-ferredoxin oxidoreductase (PFOR) combined only produce 2

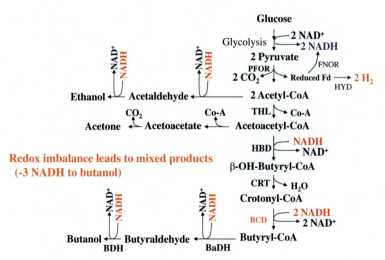

FIG. 4 Native butanol pathway is redox imbalanced. A combination of products are produced to maintain redox balance during growth and fermentation, as seen with *Clostridium acetobutylicum*. *BaDH*, butyraldehyde dehydrogenase; *BCD*, butyryl-CoA dehydrogenase complex; *BDH*, butanol dehydrogenase; *CRT*, crotonase; *FNOR*, ferredoxin-NADH oxidoreductase; *HBD*, hydroxybutyrate dehydrogenase; *HYD*, hydrogenase; *PFOR*, pyruvate-ferredoxin oxidoreductase; *THL*, thiolase.

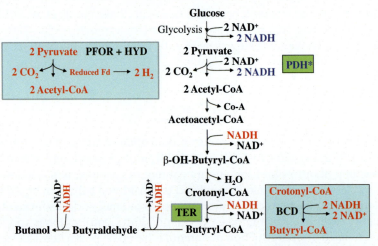

FIG. 5 An engineered fermentation pathway for production of butanol as the sole liquid fermentation product. In this pathway the native enzymes, pyruvate-ferredoxin oxidoreductase and butyryl-CoA dehydrogenase complex in the butanol pathway (reactions enclosed in a box) are replaced by an engineered pyruvate dehydrogenase (PDH*) and transenoyl-CoA reductase (TER), respectively. See Fig. 4 for other enzymes in the pathway.

NADHs during the conversion of glucose to 2 acetyl-CoAs (Fig. 4). Additional reductant available during the oxidative decarboxylation of pyruvate to acetyl-CoA is lost as H_2 with either reduced ferredoxin (Clostridia) or formate (enteric bacteria, such as *E. coli*) as intermediates. To capture this reductant, a formate dehydrogenase that generates NADH has been used in engineered *E. coli* (Fig. 6) (Shen et al., 2011). An alternate strategy was to capture the reductant directly as NADH during pyruvate decarboxylation using an engineered pyruvate dehydrogenase (PDH*), either by mutational alteration of the enzyme or by promoter exchange in *E. coli* (Fig. 5) (Abdelaal et al., 2019; Kim et al., 2008; Zhou et al., 2008).

The enzyme complex butyryl-CoA dehydrogenase (BCD) in native Clostridia utilizes 2 NADHs to reduce crotonyl-CoA to butyryl-CoA, a 2-electron/proton step. The reductant in the second NADH is released and lost as H_2 (Li et al., 2008). Replacing this enzyme with a trans-enoyl-CoA reductase (TER) from other microorganisms that utilize one NADH for this reduction eliminates this loss of reductant (Atsumi et al., 2008; Tucci and Martin, 2007). The engineered butanol fermentation pathway incorporates these two changes: PDH* and TER (Fig. 5) (Abdelaal et al., 2019).

Another challenge in butanol production, in contrast to ethanol production, is the low tolerance of microbial biocatalysts to this solvent that limits butanol titer to ≤20 g/L during batch fermentations irrespective of the microbial biocatalyst (Qureshi et al., 2020; Wilbanks and Trinh, 2017). Several attempts to increase butanol tolerance of microbial biocatalysts have resulted in small increments in tolerance and final butanol titer. Process modifications enabling continuous product removal, such as gas stripping, vacuum fermentation, distillation in situ, etc. have the potential to minimize product toxicity and increase butanol titer but at an increase in capital and operating costs (Qureshi et al., 2020).

An alternative to process modifications to lower butanol toxicity is to alter the final product to butyrate (Fig. 6). Butyrate is comparatively less toxic than butanol to the microbial biocatalyst and the butyrate titer is reported to reach about 45 g/L (Jang et al., 2014; Luo et al., 2018;

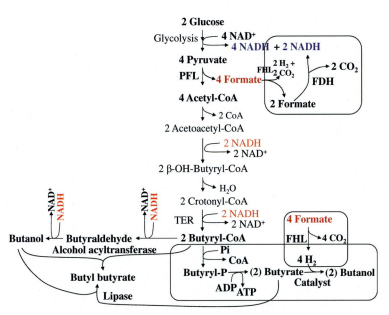

FIG. 6 Alternate pathways and products for production of butanol and butyl esters. The formate produced at the pyruvate formate lyase (PFL) reaction can serve as H_2 donor for chemical reduction of butyrate to butanol. NADH derived from formate is the needed additional reductant (NADH) for redox balanced production of butanol. One butanol and one butyrate can be combined to produce butyl butyrate in vivo using either enzyme or outside the cell using lipase. FDH, formate dehydrogenase; FHL, formate hydrogen lyase.

Wang et al., 2019a; Wang et al., 2015). Potential to increase this titer to over $100 \, g/L$ can be realized by engineering butyrate-tolerant bacteria that grow in the presence of $70–80 \, g/L$ butyrate. Butyrate can be chemically reduced to butanol to very high concentration using the H_2 generated during the production of butyrate (Fig. 6) (Lee et al., 2014). This chemical reduction process overcomes the toxicity associated with butanol production by microbial biocatalysts while also eliminating ethanol as a co-product of butanol since the enzymes that convert butyryl-CoA to butanol also catalyzes the reduction of acetyl-CoA to ethanol.

Under appropriate conditions, butyrate can be combined with butanol as produced or separately to generate butyl butyrate, an eight-carbon ester as a drop-in biofuel for diesel (Noh et al., 2018; Sjoblom et al., 2017; Xin et al., 2016). Although butyl-butyrate production by microbial biocatalysts has been demonstrated, toxicity of this ester to the microorganism precludes the production of this ester at high concentration by fermentation (Wilbanks and Trinh, 2017).

5 Thermotolerant microbial biocatalyst for production of fuels

One of the major cost components of ethanol or other product production using biomass as feedstock is the fungal enzymes with a pH and temperature optima for the activity of about 5 and 50–55°C, respectively (Patel et al., 2005). In an idealized biorefinery, the microbial biocatalysts ferment the sugars as released by enzymes to product (SScF) to minimize the inhibitory effect of sugars on enzymes and to lower the capital cost (Lynd et al., 2002; Pemberton and Crawford, 1980). In cellulosic ethanol production process, the optimum temperature for yeast growth and fermentation activity is 30–35°C (Salvado et al., 2011). Because of the differences in temperature optimum for the yeast and cellulase, SScF is conducted at an intermediate temperature that lowers the specific activity of the enzymes and increases enzyme cost. A microbial biocatalyst that optimally grows and ferments sugars to ethanol at 50–55°C (*Bacillus coagulans, B. stearothermophilus,* etc.) can significantly lower enzyme loading and cost without compromising activity (Ou et al., 2009; San Martin et al., 1992; Su et al., 2010). Several attempts to evolve *S. cerevisiae* for growth at about 50°C are yet to produce a derivative appropriate for SScF. Yeast strains that can grow at 40°C have been described; however, at this temperature the fungal cellulase activity is only about 60% of the activity at the optimum temperature indicating the need for higher cellulase loading and enzyme cost compared to SScF at 50°C (Ogunmolu et al., 2017; Pandey et al., 2019; Patel et al., 2005). Metabolic engineering of thermotolerant microorganisms for production of ethanol at a rate and yield observed with *S. cerevisiae* is critically needed for an idealized biorefinery.

In addition to lowering enzyme cost in SScF at 50–55°C, cooling cost of fermentation tanks is also expected to be lower compared to fermentations at 37°C. Higher operating temperature can also lower energy cost associated with continuous product removal as suggested for butanol to minimize product toxicity and maintain high productivity. Additional advantage of fermentation at 50–55°C is lower risk of contamination of fermentation vessels by mesophilic microorganisms (Firmino et al., 2020) that can lower product yield.

6 Lactic acid

Besides ethanol, lactic acid is a fermentation product that is easy to produce at high titer, yield and productivity. Like ethanol, lactic acid fermentation by microorganisms also has a long history (Prajapati and Nair, 2017). Traditionally, lactic acid is used in food industry as a preservative and flavoring agent, and additionally has applications in cosmetics and pharmaceutical industries. Production of poly-lactic acid (PLA), a biodegradable and sustainable polymer, is a recent use of optically pure lactic acid (Auras et al., 2010). The market size of lactic acid in 2019 was about US$1.3 billion and PLA accounts for about 45% of this total. Demand for PLA is expected to increase by an annual rate of about 10% during the next 5 years as a bio-based alternative to petroleum-derived plastics (Ahuja and Mamtani, 2020).

Although lactic acid can be made from petroleum, fermentation is the preferred production method due to the enantiomeric purity of the product needed by the plastics industry. The calculated production cost of lactic acid from corn grain is US$ 844–1251 per ton depending on the microbial biocatalyst used and associated fermentation/product purification steps (Manandhar and Shah, 2020). The higher cost of PLA at about $0.85/lb. compared to PET at about $0.65/lb. in 2018 makes PLA less attractive (https://packaging360.in/insights/polylactic-acid—a-sustainable-bioplastics-packaging-option). This disadvantage in cost can be overcome by the sustainability and biodegradability of PLA. However, to

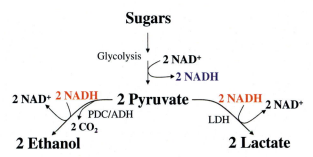

FIG. 7 Comparison of ethanol and lactate fermentation pathways. Pyruvate produced by glycolysis of sugars is either converted to ethanol or lactate as per the microbial biocatalyst to maintain redox balance during growth and fermentation. *ADH*, alcohol dehydrogenase; *LDH*, lactate dehydrogenase; *PDC*, pyruvate decarboxylase.

realize the shift from petroleum-based plastics to bio-based plastics, the production cost of lactic acid and other feedstocks needs to match that of petroleum-derived starting materials. Considering lactic acid and PLA as prime examples, the challenge is in minimizing product toxicity and simplifying product purification.

Various challenges in lactic acid production are listed below and some are already addressed successfully.

1. Fermentation of sugars to lactic acid is a simple process like that of ethanol (Fig. 7) but at a higher product yield of 1 g lactic acid per g sugar fermented (ethanol, 0.51 g/g). Considering the yield and ease of production of lactic acid, a significant part of production cost is associated with purification of lactic acid. Although ethanol can be readily removed by distillation from a postfermentation beer containing various solids and liquids, purification of lactic acid from such a complex mixture is difficult and expensive. Mineral salts medium with sugars as feedstock is preferred by the lactic acid industries to simplify purification.

2. Growth inhibition of lactic acid bacteria, such as *Lactobacillus*, by lactate lowers product titer (Goncalves et al., 1997) and requires a base for neutralization that needs to be removed and disposed. Extractive fermentation increases product titer, but the associated cost of this process is prohibitive at an industrial scale (Othman et al., 2017).

3. High titer of lactic acid is achieved by neutralization with $Ca(OH)_2$. The cost of disposal of gypsum generated during purification with the use of lime is an addition to the production cost. The use of yeast as a microbial biocatalyst that can tolerate a high concentration of lactic acid is an alternative to lactic acid bacteria (Manandhar and Shah, 2020).

4. Postfermentation purification of lactic acid is based on separation of lactic acid as calcium lactate, recovery of lactic acid from the calcium salt, esterification of the free acid form with methanol followed by distillation and hydrolysis (Filachione and Fisher, 1946; Manandhar and Shah, 2020). A microbial biocatalyst that can ferment sugars to lactic acid at a pH that is lower than the pK_a of lactic acid (3.86) that is also tolerant to a higher concentration of the product is preferred to lower production cost. Lactic acid-producing yeast strains appear to meet these requirements yielding a cost advantage over lactic acid bacteria as microbial biocatalysts (https://www.foodingredientsfirst.com/news/cargill-awarded-for-innovation-in-lactic-acid-production.html).

5. A thermotolerant microbial biocatalyst that grows and ferments sugars in mineral salts medium at a pH that is less than 4.0 at high rate and yield is an ideal microorganism for industrial production of optically pure lactic acid that in addition will also lower cooling cost of fermentation vessels as discussed above.

Although the fermentation conditions, yield and purification steps in the production of other organic acids may vary from that of lactic acid, product tolerance, thermotolerance, and mineral salts medium are process steps that are challenges shared with lactic acid production to achieve cost-effective industrial production of other bulk chemicals as chemical feedstocks.

7 Conclusion

As the world strives to wean away from fossil fuels as the dominant source of energy by envisioning alternate sustainable biofuels and biochemicals, numerous challenges stand in the way of achieving cost parity with petroleum-derived fuels and chemicals. An idealized biorefinery that has the potential to yield fuel ethanol that is cost-competitive with gasoline is presented and discussed. Minimal challenges associated with achieving this idealized lignocellulosic biomass biorefinery that uses nonfood carbohydrates as feedstock are identified and presented. Overcoming these obstacles to achieve a green economy is vital for the continued growth of the world economy without further deterioration of the environment.

Acknowledgment

This work by D.A. was supported by the Office of Science, Office of Biological and Environmental Research, of the U.S. Department of Energy under Contract No. DE-AC02-05CH11231 between Lawrence Berkeley National Laboratory and the U. S. Department of Energy.

References

Abdelaal, A.S., Jawed, K., Yazdani, S.S., 2019. CRISPR/Cas9-mediated engineering of *Escherichia coli* for n-butanol production from xylose in defined medium. J. Ind. Microbiol. Biotechnol. 46, 965–975.

Abdelaziz, O.Y., Brink, D.P., Prothmann, J., Ravi, K., Sun, M.Z., Garcia-Hidalgo, J., Sandahl, M., Hulteberg, C.P., Turner, C., Liden, G., Gorwa-Grauslund, M.F., 2016. Biological valorization of low molecular weight lignin. Biotechnol. Adv. 34, 1318–1346.

Agrawal, R., Satlewal, A., Sharma, B., Mathur, A., Gupta, R., Tuli, D., Adsul, M., 2017. Induction of cellulases by disaccharides or their derivatives in *Penicillium janthinellum* EMS-UV-8 mutant. Biofuels 8, 615–622.

Ahuja, K., Mamtani, K., 2020. Lactic Acid Market Size by Application (Industrial, Food & Beverage, Pharmaceuticals, Personal Care), Polylactic Acid (PLA) Market Size by Application (Packaging, Agriculture, Transport, Electronics, Textiles), Industry Analysis Report, Regional Outlook, Application Potential, Price Trends, Competitive Market Share & Forecast, 2020–2026. https://www.gminsights.com/industry-analysis/lactic-acid-and-polylactic-acid-market.

Alterthum, F., Ingram, L.O., 1989. Efficient ethanol production from glucose, lactose, and xylose by recombinant *Escherichia coli*. Appl. Environ. Microbiol. 55, 1943–1948.

Anto, S., Mukherjee, S.S., Muthappa, R., Mathimani, T., Deviram, G., Kumar, S.S., Verma, T.N., Pugazhendhi, A., 2020. Algae as green energy reserve: technological outlook on biofuel production. Chemosphere 242, 125079.

Atsumi, S., Cann, A.F., Connor, M.R., Shen, C.R., Smith, K.M., Brynildsen, M.P., Chou, K.J., Hanai, T., Liao, J.C., 2008. Metabolic engineering of *Escherichia coli* for 1-butanol production. Metab. Eng. 10, 305–311.

Auras, R., Lim, L., Selke, S.E.M., Tsuji, H., 2010. Poly(Lactic Acid): Synthesis, Structures, Properties, Processing and Applications. Wiley, New York.

Beckham, G.T., Johnson, C.W., Karp, E.M., Salvachua, D., Vardon, D.R., 2016. Opportunities and challenges in biological lignin valorization. Curr. Opin. Biotechnol. 42, 40–53.

Blanch, H.W., Simmons, B.A., Klein-Marcuschamer, D., 2011. Biomass deconstruction to sugars. Biotechnol. J. 6, 1086–1102.

Brandt-Talbot, A., Gschwend, F.J.V., Fennell, P.S., Lammens, T.M., Tan, B., Weale, J., Hallett, J.P., 2017. An economically viable ionic liquid for the fractionation of lignocellulosic biomass. Green Chem. 19, 3078–3102.

Brown, A., Waldheim, L., Landälv, I., Saddler, J., Ebadian, M., McMillan, J.D., Bonomi, A., Klein, B., 2020. Advanced biofuels—potential for cost reduction. In: IEA Bioenergy: Task 41:2020.01.

Cheng, M.H., Wang, Z.Q., Dien, B.S., Slininger, P.J.W., Singh, V., 2019. Economic analysis of cellulosic ethanol production from sugarcane bagasse using a sequential deacetylation, hot water and disk-refining pretreatment. Processes 7, 642.

Davison, S.A., den Haan, R., van Zyl, W.H., 2020. Exploiting strain diversity and rational engineering strategies to enhance recombinant cellulase secretion by *Saccharomyces cerevisiae*. Appl. Microbiol. Biotechnol. 104, 5163–5184.

Demeke, M.M., Dietz, H., Li, Y., Foulquie-Moreno, M.R., Mutturi, S., Deprez, S., Den Abt, T., Bonini, B.M., Liden, G., Dumortier, F., Verplaetse, A., Boles, E., Thevelein, J.M., 2013. Development of a D-xylose fermenting and inhibitor tolerant industrial *Saccharomyces cerevisiae* strain with high performance in lignocellulose hydrolysates using metabolic and evolutionary engineering. Biotechnol. Biofuels 6, 89.

Domalski, E.S., 1972. Selected values of heats of combustion and heats of formation of organic compounds containing the elements C, H, N, O, P and S. J. Phys. Chem. Ref. Data 1, 221–277.

Eiteman, M.A., Lee, S.A., Altman, E., 2008. A co-fermentation strategy to consume sugar mixtures effectively. J. Biol. Eng. 2, 3.

Filachione, E.M., Fisher, C.H., 1946. Purification of lactic acid - production of methyl lactate from aqueous solutions of crude acid. Ind. Eng. Chem. 38, 228–232.

Firmino, F.C., Porcellato, D., Cox, M., Suen, G., Broadbent, J.R., Steele, J.L., 2020. Characterization of microbial communities in ethanol biorefineries. J. Ind. Microbiol. Biotechnol. 47, 183–195.

Fonseca, L.M., Parreiras, L.S., Murakami, M.T., 2020. Rational engineering of the *Trichoderma reesei* RUT-C30 strain into an industrially relevant platform for cellulase production. Biotechnol. Biofuels 13, 93.

Franden, M.A., Pilath, H.M., Mohagheghi, A., Pienkos, P.T., Zhang, M., 2013. Inhibition of growth of *Zymomonas mobilis* by model compounds found in lignocellulosic hydrolysates. Biotechnol. Biofuels 6, 99.

Geddes, C.C., Peterson, J.J., Mullinnix, M.T., Svoronos, S.A., Shanmugam, K.T., Ingram, L.O., 2010. Optimizing cellulase usage for improved mixing and rheological properties of acid-pretreated sugarcane bagasse. Bioresour. Technol. 101, 9128–9136.

Geddes, C.C., Mullinnix, M.T., Nieves, I.U., Peterson, J.J., Hoffman, R.W., York, S.W., Yomano, L.P., Miller, E.N., Shanmugam, K.T., Ingram, L.O., 2011. Simplified process for ethanol production from sugarcane bagasse using hydrolysate-resistant *Escherichia coli* strain MM160. Bioresour. Technol. 102, 2702–2711.

Geddes, C.C., Mullinnix, M.T., Nieves, I.U., Hoffman, R.W., Sagues, W.J., York, S.W., Shanmugam, K.T., Erickson, J.E., Vermerris, W.E., Ingram, L.O., 2013. Seed train development for the fermentation of bagasse from sweet sorghum and sugarcane using a simplified fermentation process. Bioresour. Technol. 128, 716–724.

Geddes, R., Shanmugam, K.T., Ingram, L.O., 2015. Combining treatments to improve the fermentation of sugarcane bagasse hydrolysates by ethanologenic *Escherichia coli* LY180. Bioresour. Technol. 189, 15–22.

Gimpel, J.A., Henriquez, V., Mayfield, S.P., 2015. In metabolic engineering of eukaryotic microalgae: potential and challenges come with great diversity. Front. Microbiol. 6, 1376.

Goncalves, L.M.D., Ramos, A., Almeida, J.S., Xavier, A.M.R.B., Carrondo, M.J.T., 1997. Elucidation of the mechanism of lactic acid growth inhibition and production in batch cultures of *Lactobacillus rhamnosus*. Appl. Microbiol. Biotechnol. 48, 346–350.

Gonzalez, R., Tao, H., Shanmugam, K.T., York, S.W., Ingram, L.O., 2002. Global gene expression differences associated with changes in glycolytic flux and growth rate in *Escherichia coli* during fermentation of glucose and xylose. Biotechnol. Prog. 18, 6–20.

Gubiczaa, K., Nieves, I.U., Sagues, W.J., Bartaa, Z., Shanmugam, K.T., Ingram, L.O., 2016. Techno-economic analysis of ethanol production from sugarcane bagasse using a liquefaction plus simultaneous saccharification and co-fermentation process. Bioresour. Technol. 208, 42–48.

Hanly, T.J., Urello, M., Henson, M.A., 2012. Dynamic flux balance modeling of *S. cerevisiae* and *E. coli* co-cultures for efficient consumption of glucose/xylose mixtures. Appl. Microbiol. Biotechnol. 93, 2529–2541.

Hannon, M., Gimpel, J., Tran, M., Rasala, B., Mayfield, S., 2010. Biofuels from algae: challenges and potential. Biofuels 1, 763–784.

IEA, 2020. World Energy Outlook 2020. https://www.iea.org/reports/world-energy-outlook-2020.

Irwin, S., 2020. 2019 ethanol production profits: Just how bad was it? In: Farmdoc Daily. Vol. 10. University of Illinois, p. 16. https://farmdocdaily.illinois.edu/2020/01/2019-ethanol-production-profits-just-how-bad-was-it.html.

Jang, Y.S., Im, J.A., Choi, S.Y., Lee, J.I., Lee, S.Y., 2014. Metabolic engineering of *clostridium acetobutylicum* for butyric acid production with high butyric acid selectivity. Metab. Eng. 23, 165–174.

Jilani, S.B., Venigalla, S.S.K., Mattam, A.J., Dev, C., Yazdani, S.S., 2017. Improvement in ethanol productivity of engineered *E. coli* strain SSY13 in defined medium via adaptive evolution. J. Ind. Microbiol. Biotechnol. 44, 1375–1384.

Jin, M., Lau, M.W., Balan, V., Dale, B.E., 2010. Two-step SSCF to convert AFEX-treated switchgrass to ethanol using commercial enzymes and *Saccharomyces cerevisiae* 424A(LNH-ST). Bioresour. Technol. 101, 8171–8178.

Johnson, E., 2016. Integrated enzyme production lowers the cost of cellulosic ethanol. Biofuels Bioprod. Biorefin. 10, 164–174.

Jones, D.T., Woods, D.R., 1986. Acetone-butanol fermentation revisited. Microbiol. Rev. 50, 484–524.

Khan, M.U., Ahring, B.K., 2019. Lignin degradation under anaerobic digestion: influence of lignin modifications -a review. Biomass Bioenergy 128, 105325.

Kim, Y., Ingram, L.O., Shanmugam, K.T., 2007. Construction of an *Escherichia coli* K-12 mutant for homoethanologenic fermentation of glucose or xylose without foreign genes. Appl. Environ. Microbiol. 73, 1766–1771.

Kim, Y., Ingram, L.O., Shanmugam, K.T., 2008. Dihydrolipoamide dehydrogenase mutation alters the NADH sensitivity of pyruvate dehydrogenase complex of *Escherichia coli* K-12. J. Bacteriol. 190, 3851–3858.

Kim, Y., Mosier, N.S., Ladisch, M.R., Pallapolu, V.R., Lee, Y.Y., Garlock, R., Balan, V., Dale, B.E., Donohoe, B.S., Vinzant, T.B., Elander, R.T., Falls, M., Sierra, R., Holtzapple, M.T., Shi, J., Ebrik, M.A., Redmond, T., Yang, B., Wyman, C.E., Warner, R.E., 2011. Comparative study on enzymatic digestibility of switchgrass varieties and harvests processed by leading pretreatment technologies. Bioresour. Technol. 102, 11089–11096.

Klein-Marcuschamer, D., Oleskowicz-Popiel, P., Simmons, B.A., Blanch, H.W., 2012. The challenge of enzyme cost in the production of lignocellulosic biofuels. Biotechnol. Bioeng. 109, 1083–1087.

Kumar, R., Hu, F., Sannigrahi, P., Jung, S., Ragauskas, A.J., Wyman, C.E., 2013. Carbohydrate derived-pseudo-lignin can retard cellulose biological conversion. Biotechnol. Bioeng. 110, 737–753.

Lee, J.M., Upare, P.P., Chang, J.S., Hwang, Y.K., Lee, J.H., Hwang, D.W., Hong, D.Y., Lee, S.H., Jeong, M.G., Kim, Y.D., Kwon, Y.U., 2014. Direct hydrogenation of biomass-derived butyric acid to n-butanol over a ruthenium-tin bimetallic catalyst. ChemSusChem 7, 2998–3001.

Li, F., Hinderberger, J., Seedorf, H., Zhang, J., Buckel, W., Thauer, R.K., 2008. Coupled ferredoxin and crotonyl coenzyme A (CoA) reduction with NADH catalyzed by the butyryl-CoA dehydrogenase/Etf complex from *clostridium kluyveri*. J. Bacteriol. 190, 843–850.

Lopes, A.M.M., Martins, M., Goldbeck, R., 2021. Heterologous expression of lignocellulose-modifying enzymes in microorganisms: current status. Mol. Biotechnol. 63, 184–199.

Luo, H., Yang, R., Zhao, Y., Wang, Z., Liu, Z., Huang, M., Zeng, Q., 2018. Recent advances and strategies in process and strain engineering for the production of butyric acid by microbial fermentation. Bioresour. Technol. 253, 343–354.

Lynd, L.R., Weimer, P.J., van Zyl, W.H., Pretorius, I.S., 2002. Microbial cellulose utilization: fundamentals and biotechnology. Microbiol. Mol. Biol. Rev. 66, 506–577.

Lynd, L.R., van Zyl, W.H., McBride, J.E., Laser, M., 2005. Consolidated bioprocessing of cellulosic biomass: an update. Curr. Opin. Biotechnol. 16, 577–583.

Manandhar, A., Shah, A., 2020. Techno-economic analysis of bio-based lactic acid production utilizing corn grain as feedstock. Processes 8, 199.

Martin, C., Wu, G., Wang, Z., Stagge, S., Jonsson, L.J., 2018. Formation of microbial inhibitors in steam-explosion pretreatment of softwood impregnated with sulfuric acid and sulfur dioxide. Bioresour. Technol. 262, 242–250.

McAloon, A., Taylor, F., Yee, W., Ibsen, K., Wooley, R., 2000. Determining the Cost of Producing Ethanol From Corn Starch and Lignocellulosic Feedstocks. National Renewable Energy Lab. (NREL), Golden, CO, United States. NREL/TP-580-28893.

McGovern, P.E., Zhang, J.H., Tang, J.G., Zhang, Z.Q., Hall, G.R., Moreau, R.A., Nunez, A., Butrym, E.D., Richards, M.-P., Wang, C.S., Cheng, G.S., Zhao, Z.J., Wang, C.S., 2004. Fermented beverages of pre- and proto-historic China. Proc. Natl. Acad. Sci. U. S. A. 101, 17593–17598.

Miller, E.N., Jarboe, L.R., Turner, P.C., Pharkya, P., Yomano, L.P., York, S.W., Nunn, D., Shanmugam, K.T., Ingram, L.-O., 2009. Furfural inhibits growth by limiting sulfur assimilation in ethanologenic *Escherichia coli* strain LY180. Appl. Environ. Microbiol. 75, 6132–6141.

Moyses, D.N., Reis, V.C., de Almeida, J.R., de Moraes, L.M., Torres, F.A., 2016. Xylose fermentation by *Saccharomyces cerevisiae*: challenges and prospects. Int. J. Mol. Sci. 17, 207.

Nieves, I.U., Geddes, C.C., Mullinnix, M.T., Hoffman, R.W., Tong, Z., Castro, E., Shanmugam, K.T., Ingram, L.O., 2011. Injection of air into the headspace improves fermentation of phosphoric acid pretreated sugarcane bagasse by *Escherichia coli* MM170. Bioresour. Technol. 102, 6959–6965.

Noh, H.J., Woo, J.E., Lee, S.Y., Jang, Y.S., 2018. Metabolic engineering of *Clostridium acetobutylicum* for the production of butyl butyrate. Appl. Microbiol. Biotechnol. 102, 8319–8327.

Ogunmolu, F.E., Jagadeesha, N.B.K., Kumar, R., Kumar, P., Gupta, D., Yazdani, S.S., 2017. Comparative insights into the saccharification potentials of a relatively unexplored but robust *Penicillium funiculosum* glycoside hydrolase 7 cellobiohydrolase. Biotechnol. Biofuels 10, 71.

Ogunyewo, O.A., Randhawa, A., Joshi, M., Jain, K.K., Wadekar, P., Odaneth, A.A., Lali, A.M., Yazdani, S.S., 2020. Engineered *Penicillium funiculosum* produces potent lignocellulolytic enzymes for saccharification of various pretreated biomasses. Process Biochem. 92, 49–60.

Ohta, K., Beall, D.S., Mejia, J.P., Shanmugam, K.T., Ingram, L.O., 1991. Genetic improvement of *Escherichia coli* for ethanol production: chromosomal integration of *Zymomonas mobilis* genes encoding pyruvate decarboxylase and alcohol dehydrogenase II. Appl. Environ. Microbiol. 57, 893–900.

Ong, R.G., Higbee, A., Bottoms, S., Dickinson, Q., Xie, D., Smith, S.A., Serate, J., Pohlmann, E., Jones, A.D., Coon, J.J., Sato, T.K., Sanford, G.R., Eilert, D., Oates, L.G., Piotrowski, J.S., Bates, D.M., Cavalier, D., Zhang, Y., 2016. Inhibition of microbial biofuel production in drought-stressed switchgrass hydrolysate. Biotechnol. Biofuels 9, 237.

Othman, M., Ariff, A.B., Rios-Solis, L., Halim, M., 2017. Extractive fermentation of lactic acid in lactic acid bacteria cultivation: a review. Front. Microbiol. 8, 2285.

Ou, M.S., Mohammed, N., Ingram, L.O., Shanmugam, K.T., 2009. Thermophilic *Bacillus coagulans* requires less cellulases for simultaneous saccharification and fermentation of cellulose to products than mesophilic microbial biocatalysts. Appl. Biochem. Biotechnol. 155, 379–385.

Padella, M., OConnell, A., Prussi, M., 2019. What is still limiting the deployment of cellulosic ethanol? Analysis of the current status of the sector. Appl. Sci. 9, 4523.

Pandey, A.K., Kumar, M., Kumari, S., Kumari, P., Yusuf, F., Jakeer, S., Naz, S., Chandna, P., Bhatnagar, I., Gaur, N.A., 2019. Evaluation of divergent yeast genera for fermentation-associated stresses and identification of a robust sugarcane distillery waste isolate *Saccharomyces cerevisiae* NGY10 for lignocellulosic ethanol production in SHF and SSF. Biotechnol. Biofuels 12, 40.

Patel, M.A., Ou, M., Ingram, L.O., Shanmugam, K.T., 2005. Simultaneous saccharification and co-fermentation of crystalline cellulose and sugar cane bagasse hemicellulose hydrolysate to lactate by a thermotolerant acidophilic *Bacillus* sp. Biotechnol. Prog. 21, 1453–1460.

Pemberton, M.S., Crawford, S.D., 1980. Method for Ethanol Fermentation. United States Patent 4,224,410.

Prajapati, J.B., Nair, B.M., 2017. The history of fermented foods. In: Farnworth, E.R. (Ed.), Handbook of Fermented Functional Foods, second ed. CRC Press, Boca Raton, Florida, pp. 1–24.

Qureshi, N., Lin, X., Liu, S., Saha, B.C., Mariano, A.P., Polaina, J., Ezeji, T.C., Friedl, A., Maddox, I.S., Klasson, K.T., Dien, B.S., Singh, V., 2020. Global view of biofuel butanol and economics of its production by fermentation from sweet sorghum bagasse, food waste, and yellow top presscake: application of novel technologies. Fermentation 6, 58.

Rajoka, M.I., Ferhan, M., Khalid, A.M., 2005. Kinetics and thermodynamics of ethanol production by a thermotolerant mutant of *Saccharomyces cerevisiae* in a microprocessor-controlled bioreactor. Lett. Appl. Microbiol. 40, 316–321.

Robak, K., Balcerek, M., 2020. Current state-of-the-art in ethanol production from lignocellulosic feedstocks. Microbiol. Res. 240, 126534.

Saha, B.C., Nichols, N.N., Qureshi, N., Kennedy, G.J., Iten, L.B., Cotta, M.A., 2015. Pilot scale conversion of wheat straw to ethanol via simultaneous saccharification and fermentation. Bioresour. Technol. 175, 17–22.

Salvado, Z., Arroyo-Lopez, F.N., Guillamon, J.M., Salazar, G., Querol, A., Barrio, E., 2011. Temperature adaptation markedly determines evolution within the genus *Saccharomyces*. Appl. Environ. Microbiol. 77, 2292–2302.

San Martin, R., Bushell, D., Leak, D.J., Hartley, B.S., 1992. Development of a synthetic medium for continuous anaerobic growth and ethanol production with a lactate dehydrogenase mutant of *Bacillus stearothermophilus*. J. Gen. Microbiol. 138, 987–996.

Scalcinati, G., Otero, J.M., Van Vleet, J.R., Jeffries, T.W., Olsson, L., Nielsen, J., 2012. Evolutionary engineering of *Saccharomyces cerevisiae* for efficient aerobic xylose consumption. FEMS Yeast Res. 12, 582–597.

Shanmugam, K.T., Ingram, L.O., 2021. Principles and practice of designing microbial biocatalysts for fuel and chemical production. J. Ind. Microbiol. Biotechnol. https://doi.org/10.1093/jimb/kuab016. In press.

Shen, C.R., Lan, E.I., Dekishima, Y., Baez, A., Cho, K.M., Liao, J.C., 2011. Driving forces enable high-titer anaerobic 1-butanol synthesis in *Escherichia coli*. Appl. Environ. Microbiol. 77, 2905–2915.

Sjoblom, M., Risberg, P., Filippova, A., Ohrman, O.G.W., Rova, U., Christakopoulas, P., 2017. In situ biocatalytic synthesis of butyl butyrate in diesel and engine evaluations. ChemCatChem 9, 4529–4537.

Su, Y., Rhee, M.S., Ingram, L.O., Shanmugam, K.T., 2010. Physiological and fermentation properties of *Bacillus coagulans* and a mutant lacking fermentative lactate dehydrogenase activity. J. Ind. Microbiol. Biotechnol. 38, 441–450.

Sun, J., Konda, N.V.S.N.M., Parthasarathi, R., Dutta, T., Valiev, M., Xu, F., Simmons, B.A., Singh, S., 2017. One-pot integrated biofuel production using low-cost biocompatible protic ionic liquids. Green Chem. 19, 3152–3163.

Tao, L., Aden, A., Elander, R.T., Pallapolu, V.R., Lee, Y.Y., Garlock, R.J., Balan, V., Dale, B.E., Kim, Y., Mosier, N.S., Ladisch, M.R., Falls, M., Holtzapple, M.T., Sierra, R., Shi, J., Ebrik, M.A., Redmond, T., Yang, B., Wyman, C.E., Hames, B., Thomas, S., Warner, R.E., 2011. Process and technoeconomic analysis of leading pretreatment technologies for lignocellulosic ethanol production using switchgrass. Bioresour. Technol. 102, 11105–11114.

Toivari, M.H., Salusjarvi, L., Ruohonen, L., Penttila, M., 2004. Endogenous xylose pathway in *Saccharomyces cerevisiae*. Appl. Environ. Microbiol. 70, 3681–3686.

Tucci, S., Martin, W., 2007. A novel prokaryotic trans-2-enoyl-CoA reductase from the spirochete *Treponema denticola*. FEBS Lett. 581, 1561–1566.

van Rijn, R., Nieves, I.U., Shanmugam, K.T., Ingram, L.O., Vermerris, W., 2018. Techno-economic evaluation of cellulosic ethanol production based on pilot biorefinery data: a case study of sweet sorghum bagasse processed via L+SScF. Bioenergy Res. 11, 414–425.

Wang, L., Ou, M.S., Nieves, I., Erickson, J.E., Vermerris, W., Ingram, L.O., Shanmugam, K.T., 2015. Fermentation of sweet sorghum derived sugars to butyric acid at high titer and productivity by a moderate thermophile *Clostridium thermobutyricum* at 50°C. Bioresour. Technol. 198, 533–539.

Wang, L., Chauliac, D., Moritz, B.E., Zhang, G., Ingram, L.O., Shanmugam, K.T., 2019a. Metabolic engineering of *Escherichia coli* for the production of butyric acid at high titer and productivity. Biotechnol. Biofuels 12, 62.

Wang, H.L., Pu, Y.Q., Ragauskas, A., Yang, B., 2019b. From lignin to valuable products-strategies, challenges, and prospects. Bioresour. Technol. 271, 449–461.

Wang, L., York, S.W., Ingram, L.O., Shanmugam, K.T., 2019c. Simultaneous fermentation of biomass-derived sugars to ethanol by a co-culture of an engineered *Escherichia coli* and *Saccharomyces cerevisiae*. Bioresour. Technol. 273, 269–276.

Wilbanks, B., Trinh, C.T., 2017. Comprehensive characterization of toxicity of fermentative metabolites on microbial growth. Biotechnol. Biofuels 10, 262.

Williams, C.L., Westover, T.L., Emerson, R.M., Tumuluru, J.S., Li, C.L., 2016. Sources of biomass feedstock variability and the potential impact on biofuels production. Bioenergy Res. 9, 1–14.

Xie, C.Y., Yang, B.X., Wu, Y.J., Xia, Z.Y., Gou, M., Sun, Z.Y., Tang, Y.Q., 2020. Construction of industrial xylose-fermenting *Saccharomyces cerevisiae* strains through combined approaches. Process Biochem. 96, 80–89.

Xin, F., Basu, A., Yang, K.L., He, J., 2016. Strategies for production of butanol and butyl-butyrate through lipase-catalyzed esterification. Bioresour. Technol. 202, 214–219.

Yang, S., Fei, G., Zhang, Y., Contreras, L.M., Utturkar, S.M., Brown, S.D., Himmel, M.E., Zhang, M., 2016. *Zymomonas mobilis* as a model system for production of biofuels and biochemicals. Microb. Biotechnol. 9, 699–717.

Zhou, S., Iverson, A.G., Grayburn, W.S., 2008. Engineering a native homoethanol pathway in *Escherichia coli* B for ethanol production. Biotechnol. Lett. 30, 335–342.

Zilberman, D., Hochman, G., Rajagopal, D., Sexton, S., Timilsina, G., 2013. The impact of biofuels on commodity food prices: assessment of findings. Am. J. Agric. Econ. 95, 275–281.

Biomass feedstocks for advanced biofuels: Sustainability and supply chain management

A.V. Umakanth[a], Aviraj Datta[b], B. Suresh Reddy[c], and Sougata Bardhan[d]

[a]ICAR - Indian Institute of Millets Research, Hyderabad, India [b]International Crops Research Institute for the Semiarid Tropics, Patancheru, India [c]Centre for Economic and Social Studies, Hyderabad, India [d]School of Natural Resources, University of Missouri, Columbia, MO, United States

1 Introduction

The use of agro-waste as fuel wood is an age-old practice. Presently, biomass supplies about 14% of the world's energy demand. Developing countries account for 75% of this energy from biomass where it is primarily used for domestic cooking and heating purposes (Parikka, 2004). In a few countries such as Brazil, the largest producer and export of sugar in the world, where the sugar industry assues a significant part of its economy, the use of bioethanol for transportation and for electricity generation has seen the wide-scale application. Developed countries use 25% of global biomass energy mainly toward domestic heating needs and for power generation purposes. Today, bioethanol is seen as an important way of reducing our dependency on imported fossil fuels in India. The sustainable and environmentally friendly way of utilizing this vast quantity of biomass offers a win-win situation covering both environmental problems as well as quenching ever increasing per capita energy demand. Bioethanol production from these cellulosic agro-waste residues would certainly bring down the cost pf production and availability of bioethanol without competing with food crops for land and/or water resources. Availability of low-cost bioethanol is prerequisite for enhancing the blending ratio for petrol as a national policy. Till 2014 the blending of bioethanol in petrol in India was less than 1%, however, today the sector witnessed a sharp rise in recent years and presently the ratio is between 8.5% and 10.0% in India. Interestingly, the Govt. of India has preponed the target of achieving 20% bioethanol blending in petrol from 2030 to 2025 recently, indicating the strong tail wind the sector is experiencing. According to various assessment reports, surplus crop residue availability in India is about 50–60 million per annum which is theoretically equivalent to 10–15 billion liters of 2-G ethanol. The quantity is sufficient to achieve the 20% cent national ethanol blending target of India. As total petrol consumption in India was estimated to reach 381 billion liters by 2020 (before the pandemic), 20% blending by bioethanol would to lead to foreign exchange savings in a range between $8 billion and 10 billion per annum.

1.1 Biomass energy

Biomass typically consists of cellulose, hemicellulose, and lignin. Cellulose is the most abundantly available organic polymer on earth though its content may vary from 90% in cotton to about 30% in wood. Being a polysaccharide consisting of a linear chain of several hundred to many thousands of β linked D-glucose units, cellulose is susceptible to enzymatic degradation. Hemi-cellulose, typically constitutes about 15%–40% of the biomass, being more amenable to hydrolysis its content plays important role in determining the biorefining potential of biomass. Lignin which constitutes about 15%–35% of biomass. This randomly cross-linked aromatic polymer of phenylpropane units joined by different linkages (ex. Ether or covalent), resists biochemical conversion (Kaushik and Biswas, 2007). Depolymerization of lignocellulosic material to smaller molecules is critical for biorefining process which converts these smaller molecules into biofuels. Hydrolysis processes aim to liberate sugars from biomass containing predominately cellulose or hemicellulose whereas thermal processes such as pyrolysis and gasification, is more common for biomass containing predominately lignin. Bio-power or heat can be generated by the release of energy stored in biomass. Fluidized bed combustion (FCC) is most efficient biomass combustion process as it generates high temperature, allows a good air-fuel mixing ratio and long residence time. Bed material agglomeration remains a critical technical bottleneck for more wide-scale applications of FCC. Renewable electricity can be generated through combustion or gasification of biomass (dry) and also through controlled anaerobic digestion of biogas. Cofiring of biomass and fossil fuels (usually coal) is a low-cost means of reducing greenhouse gas emissions, improving cost-effectiveness, and reducing air pollutants in existing power plants. Pulverized fuel combustor (PFC) is the preferred technology due to its easy adaptability to minimum requirement for equipment modification. Bed agglomeration remains a critical challenge for co-combustion of biomass, as bed de-fluidization often leads to unplanned shutdown (Shimizu et al., 2006). Thermal energy can also be generated through the gasification process, where auxiliary fuel is converted to a gaseous product, termed as producer gas, where major components are carbon oxides, hydrogen and methane, as other hydrocarbon species. Tars in the producer gas pose a significant technical challenge for the gasification process. Generally, higher gasification temperature leads to lower tar production hence, optimized use of auxiliary fuel is desired (Holfbauer and Knoef, 2005). Pyrolysis is a process in which biomass gets heated at high temperature in absence of oxygen to generate solid char, vapors, and

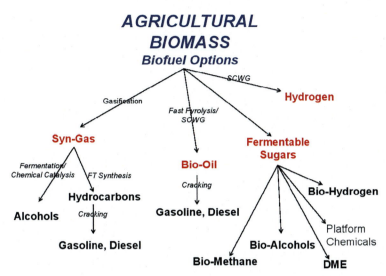

FIG. 1 Biofuel options from biomass.

noncondensable gases, separation, and condensation of gaseous compounds leads to the production of bio-oil. Direct application of this bio-oil in engines or turbines as fuel becomes challenging due to its acidity, high oxygen and moisture content and viscous nature. The usability of bio-oil can be augmented through hydrotreating and hydrocracking which reduces its density and viscosity by hydrotreating or hydrodeoxygenation (Pang, 2016).

Biomass, like agricultural residues, can be converted to various advanced biofuels and biochemicals by adopting thermo-chemical, catalytic, and biochemical platforms. Various possible products from biomass are shown in Fig. 1.

1.2 Biofuels

"Biofuel" is short for "biomass fuel," a term used for liquid fuels produced from biomass (Table 1), such as ethanol, bio-oil, and biodiesel that help to alleviate demand for petroleum products and improve the greenhouse gas emissions profile of the transportation sector. To promote biofuels as an alternative energy source, the Govt. of India in December 2009 announced a comprehensive National Policy on Biofuels which was revised during 2018 calling for blending at least 20% of biofuels with fossil fuels by 2030. In India, against the requirement of 3.3 billion liters of ethanol which is the prime source of biofuel for 10% blending in the country, ethanol supply contracts have been signed for 2.37 billion liters during 2018–19. The government has allowed sugar mills to manufacture ethanol directly from sugarcane juice or an intermediate product called B molasses by amending Sugarcane Control Order, 1966. The production of ethanol directly from sugarcane juice or B-molasses will address the issue of sugar overproduction and stabilize sugar prices. In addition, sweet sorghum has huge untapped potential for ethanol production in India. Large-scale mill crushing tests have successfully demonstrated that existing sugar mills can be used effectively, for sweet sorghum juice extraction. For biodiesel production, the cultivation of *Jatropha curcas* on

TABLE 1   Key feedstocks for biofuel production in different countries.

Country	Major feedstocks	
	Bioethanol	Biodiesel
United States	Corn	Soybean oil/diverse other oils
European Union	Corn/Wheat/Suagrbeet	Rapeseed oil/waste oils
Brazil	Sugarcane/Sweet sorghum	Soybean oil
China	Corn/Sweet sorghum	Waste oils
India	Sugarcane molasses	Palm oil
Canada	Corn	Waste oils
Indonesia	Molasses	Palm oil
Argentina	Corn/Sugarcane	Soybean oil
Thailand	Molasses/Cassava	Pam oil

Source: Modified from OECD/FAO (2019), OECD-FAO Agricultural Outlook. OECD Agriculture Statistics (Database). https://doi.org/10.1787/agr-outl-data-en.

wastelands is promoted by the government. However, substantial research on developing improved cultivars of Jatropha and Pongamia as well as management practices need to be developed before biodiesel from nonedible plants becomes economically viable. The National Biofuel Policy, 2018 envisages 40% reduction in carbon emissions by 2030. The new biofuel policy shifts the focus from first-generation (1G) biofuels which are made from molasses and vegetable oils to 2G biofuels reality. The Oil Marketing Companies (OMCs) are setting up 12 advanced 2G biofuel refineries in several states. The Indian Oil Corporation is currently operating three 2G biofuel plants and plans to increase its capacity from 100 tons to 1200 tons per day in the next 2 years. Third-generation biofuel from algae also has potential and research is on to grow algae using wastewater through decentralized constructed wetland as a business model in rural areas. In addition, the government has top priority for harnessing wind and solar energy for bioenergy and India is already second after China in renewables production with 208.7 Mtoe in 2016. India has already become the world's single largest renewable energy auctions market and the second biggest attracter of clean energy investments. Biofuels remain the principal source of clean and renewable transportation fuels till renewably produced electricity is used to run a significant number of electric vehicles. The international biofuel sectors are strongly influenced by national policies with three major goals: farmer support reduced greenhouse gas emissions, and/or reduced energy independency.

1.3 Bioproducts

It is well established by researchers that compound which can be synthesized from fossil fuels can also be obtained from biomass. Such renewable source would reduce the environmental footprints of products such as antifreeze, plastic materials, glue, artificial sweeteners, and toothpaste. Other bioproducts formed during biomass heating in presence of oxygen such as biosynthesis gas, which is an important precursor for the production of photo films, textile and synthetic fibers. Compounds such as phenol an important precursor for wood-sticks, plastic molds, insulating foam, etc. can be extracted from bio-oil produced by pyrolysis. "Bioproduct" is short for "biomass products" and can be used to describe a chemical, material, or other (nonenergy) product such as composites, plastics, and adhesives fertilizers, lubricants, industrial chemicals, etc. Bioproducts are widely used in our day-to-day lives today such as various cosmetics such as skin cream, nail polish remover (acetone), shampoo, hair conditioner (palmitic acid), mascara, etc. Renewed demand for biobased cosmetics has resulted in market size of $3 billion in 2019. Microalgal oil producers have the potential to generate 120 barrels (1 barrel equivalent to 159L) of oil per acre which can be used as a renewable fuel. Valuable bioproducts such as omega-3 fatty acids can be sourced from algal cultivation on a commercial scale. Bio-based surfactants and solvents can be used to produce detergents and other cleansers.

2 Biomass feedstocks

Biomass feedstocks for energy production can result from plants grown directly for energy or from plant parts, residues, processing wastes, and materials from animal and human activities. This makes biomass, a flexible and widespread resource that can be adapted locally to meet local needs and objectives. Every region has its own locally generated biomass feedstocks from agriculture, forest, and urban sources and most feedstocks can be made into liquid fuels, heat, electric power, and/or biobased products. In general, the classification of feedstocks may be based on categories of plants or residues and by the energy products they produce. Major energy crops available to fulfill the feedstock demand are Switchgrass, Miscanthus, high biomass or energy sorghum, as well as crop residues, such as rice straw, wheat straw, corn stover, corn cobs, etc. The second-generation (2G) of biofuels can be generated by using the nonfood parts of plants such cell walls, composed of structural polysaccharides, such as cellulose and hemicelluloses. This is considered to be advantageous over the first generation of biofuels as it has a higher energy production potential, lower cost, sustainable CO_2 balance, lack of competition with the food production and availability of a wide range of plant biomass sources at affordable costs to a biorefinery. In recent years, much emphasis is given to the production of ethanol from agricultural wastes/residues which contain cellulose (most abundant on earth) and hemicelluloses, the carbohydrates that can be converted to ethanol by fermentation. Cellulose has earlier been taken into account for chemical/biological saccharification and subsequent biological conversion of the monomeric sugars to ethanol. Advanced technologies based on cellulosic feedstocks are often seen as relevant technologies for the future as they are supposed to cause less competition with food products and emit safer levels of greenhouse gas emissions.

2.1 Sugar crops

These include sugarcane, energy cane, sugarbeet, sweet sorghum, high biomass sorghum, etc.

2.1.1 Sugarcane

Sugarcane (*Saccharum* spp.) is the main feedstock for bioenergy production, especially in the tropical and subtropical regions of the world (Long et al., 2015). In addition, to its adaptability for use in sugar and ethanol production, sugarcane crop residues, such as straw and bagasse, have increasingly been used in electricity cogeneration during burning of residues in boilers and in the second-generation of ethanol production (Dias et al., 2011; Sordi and Manechini, 2013). Ethanol produced from sugarcane, is a renewable fuel derived from sugarcane that grows typically in tropical and subtropical climates. The harvested stalks are roughly 70% moisture and the dry matter is composed basically by sucrose and lignocellulose. Approximately, one-third of the total energy in the above-ground biomass of today's sugarcane cultivars, is captured as the sugars (mostly sucrose) fraction present in the stalk while another third is present in the fibrous sugarcane bagasse and the last third is the trash left in the field after harvesting. Both last fractions are essentially lignocellulosic materials. Compared to other types of ethanol available today, using sugarcane ethanol to power cars and trucks yields greater reductions in greenhouse gases. In 2010, the EPA designated Brazilian sugarcane ethanol as an advanced biofuel due to its 61% reduction of total life cycle greenhouse gas emissions, including direct indirect land use change emissions.

Countries like Brazil has replaced more than half of its fuel needs with sugarcane ethanol,—making ethanol the standard and gasoline as the alternative fuel.

2.1.2 Energy cane

Energy cane (Fig. 2) is an interspecific hybrid arising from backcrossing two species, *S. spontaneum* (high fiber content) and *S. officinarum* (high sugar content), thereby producing a plant with higher fiber and lower sugar content when compared to sugarcane (Matsuoka et al., 2014). Among all the dedicated bioenergy crops so far analyzed, existing sugarcane cultivars are outstanding in terms of annual, renewable productivity per unit area, in terms of either wet or dry matter and energy cane has the potential to produce two to three times more than this (Burner and Legendre, 2000). As productivity is the main driver for the sustainability of any energy biomass source (economic, environmental, social, etc.), energy cane has the potential to effectively contribute to the world demand of bioenergy. From energy cane, ethanol is not the only combustible liquid that can be produced; jet fuel, biobutanol, biodiesel, biogas, methanol, syngas, and others are all forms of fuel that could potentially be obtained from energy cane (Tao and Aden, 2009). The introduction of new advanced low-carbon technologies with the addition of sugars converted from cellulosic materials and the development of high-biomass sugarcane (energy cane) has opened a new agroindustrial path. In energy cane, the carbon partition is more oriented toward fiber production instead of soluble sugars accumulation, resulting in a biomass index greater than $300\,ton\,ha^{-1}$ Greater growth seen in energy cane might be attributed to a vigorous rate of nocturnal growth with angulation in relation to the time of 11.89°, which is lower in sugarcane, with angulation of 5.47° (de Abreu et al., 2020). The perspective to improve the potential yield of bioethanol to almost 25,000 L per hectare is real (from 6900 today). In Brazil, some energy cane genotypes, such as VX12–1022 and VX12–0646 which have potential for greater dry biomass production than sugarcane in both the plant cane and first ratoon crop cycles have been identified (Boschiero et al., 2019). Considering a projected global consumption of gasoline of 1.7 trillion litters in 2025, energy cane-based bioethanol would be able to replace 10% of total gasoline

Energy cane   Sugarcane
(SP791011)

FIG. 2   Energy cane (Diniz et al., 2019).

consumed in the world using less than 10 million hectares of land. Furthermore, the world would quickly experiment with expressive carbon dioxide CO_2 emissions reduction in the transport sector, responsible for one-quarter of the total CO_2 emissions (Center for strategic studies and management, 2017).

The joint Louisiana State University (LSU) and the Houma-USDA program succeeded in producing some energy cane cultivars like US79–1002 which recorded fiber percentage as high as 28% with exceptionally high productivity: five harvests from a single planting averaged $211 \, t \, ha^{-1}$ per harvest, with continual yield increase from plant cane to the fourth ratoon (total biomass, wet basis) against $58 \, t \, ha^{-1}$ for a conventional sugarcane cultivar (Giamalva et al., 1985). A steady linear increase in productivity from $182 \, t \, ha^{-1}$ in plant cane to $247 \, t \, ha^{-1}$ in the fifth ratoon of energy cane US79–1002 was observed by Bischoff et al., 2008. Averaged across the three crops and two locations, energy cane had significantly higher biomass yield, lower nonstructural carbohydrate (reducing sugars and sucrose) concentrations, and higher concentrations of cellulose, hemicelluloses, and lignin than sugarcane. Although there were no differences between sugarcane and energy cane in total carbohydrate concentration (839 to 842 g/kg DW), energy cane had 80% higher cellulose, 63% higher hemicelluloses, and 76%

higher lignin; 69%, 64%, and 56% lower sucrose, glucose, and fructose concentrations, respectively, than sugarcane, when averaged across the three crops and two locations (Zhao et al., 2020).

2.1.3 Biomass sorghum

Sorghum is a short duration crop of about 3–4 months and produces higher biomass yield with less inputs. Energy sorghum, including biomass and sweet type varieties, has recently gained favor as bioethanol feedstock among numerous candidate crops (Rooney et al., 2007; Xie, 2012). Biomass sorghum does not produce grain until very late in the growing season. Instead, the plant puts all its energy in growing tall and can reach 4–5 m at the end of the growing season (Fig. 3). This sorghum type usually has more number of leaves, fibrous roots, greater potential for vegetative growth, and is suitable for mechanization (Venuto and

FIG. 3   Biomass sorghum in India.

Kindiger, 2008). This form has long been used as forage, but it has recently attracted attention as a potential source of domestic, environmentally sustainable, renewable and affordable biofuel. Besides producing second-generation ethanol, biomass sorghum also releases energy during biomass combustion (da Silva et al., 2018). It is a good substitute to corn and sugarcane with additional benefit of less water consumption. It is an annual grass having higher dry matter yield like perennial crops but in less duration, thus facilitating cheaper crop rotation. The convertibility of high biomass lines of sorghum to bioethanol is of special interest as the use of sorghum biomass for biofuel production will not lead to food price increase. In some sorghum genotypes, proportion of cellulose can vary between 27% and 52%, while the range of hemicellulose content is 17%–23% and lignin content is 6.2%–8.1%. Along with the biomass yield, low lignin, high cellulose, and hemicellulose contents are also the desirable selection attributes for energy sorghum genotypes (Mahmood and Honermeier, 2012). The natural attributes like abiotic stress tolerance, diverse genetic base, viable seed industry, and sound breeding system make sorghum a perfect candidate for establishing an efficient and low-cost biofuel industry. The convertibility of high biomass lines of sorghum to bioethanol is of special interest as the use of sorghum biomass for biofuel production will not lead to food price increase. Dry biomass production of several potential bioenergy sorghum crops can be impressive: $18–32\,Mg\,ha^{-1}$ for sweet sorghum, $16–24\,Mg\,ha^{-1}$ for forage sorghum, and $32\,Mg\,ha^{-1}$ for photoperiod-sensitive sorghum (Rooney et al., 2007). The potential exists to further develop sorghum as a bioenergy crop because it possesses an array of traits such as brown midrib, sweet stalks, staygreen, and high biomass that can be combined via plant breeding and genetic manipulation to maximize the conversion of biomass to ethanol (Vermerris et al., 2007). Total dry biomass yields in the energy sorghum hybrids EJ7281 and ES5200 were observed to be fluctuating between 22.2 and $37.5\,t\,ha^{-1}$ (Bartzialis et al., 2020). In Northern China, the most productive sorghum biomass hybrid GN-4, exhibited biomass and theoretical ethanol yields $>42.1\,t\,ha^{-1}$ and $14{,}913\,L\,ha^{-1}$, respectively (Tang et al., 2018). Bioethanol yields were estimated to be in the range of 223–506 L/ton in the sorghum straw dry matter in a study on Bioethanol Production from Biomass of Selected Sorghum Varieties Cultivated as Main and Second Crop is given in Table 2 (Batog et al., 2020).

In India, under a US-India Joint Clean Energy Research and Development Center project on Development of Sustainable Advanced Lignocellulosic Biofuel Systems, several high biomass sorghum genotypes with a dry biomass of $>25\,t\,ha^{-1}$ were developed at ICAR-Indian

TABLE 2 Bioethanol yield from sorghum straw ($L\,Mg^{-1}$ of straw DM).

	Year I	Year II	Year I	Year II	Year I	Year II
Variety	Crop					
	Main	Second	Main	Second	Main	Second
Rona 1	430	403	456	425	506	474
Santos	243	223	266	240	258	235
Sucrosorgo 506	413	365	484	428	451	397

Source: Batog, J., Frankowski, J., Wawro, A., Łacka, A., 2020. Bioethanol production from biomass of selected sorghum varieties cultivated as main and second crop. Energies 13, 6291. https://doi.org/10.3390/en13236291.

TABLE 3   Structural carbohydrate content in high biomass sorghum variety ICSV 25333.

Treatment	Cellulose % (w/w)	Xylose	Arabinose	Total Sugars	Lignin	Ash	Total of components
Raw biomass (triplicate)	46.29	27.26	8.22	81.77	14.56	2.42	98.75
	45.19	28.01	7.99	81.19	14.77	2.88	98.84
	46.01	28.11	8.09	82.21	13.90	1.75	97.86
Acid treatment	57.39	17.99	2.21	77.59	8.01	3.17	98.77
Alkali treatment	74.26	20.58	–	94.84	2.15	0.75	97.74

Source: ICRISAT 2016.

Institute of Millets Research (ICAR-IIMR) and International Crops Research Institute for the Semiarid Tropics (ICRISAT). Further, a high biomass sorghum entry ICSV 25333, promising for biomass yields in multilocation trials across India was investigated for its structural carbohydrate content (Table 3) and ethanol production potential which was 288 L/ton when C5 and C6 sugars were fermented together.

2.1.4 Brown midrib sorghum

The major impediment of converting biomass to biofuels is high pretreatment costs for removal of lignin besides high cost of enzymes used for saccharification. An advantageous feature of sorghum, which has been exploited worldwide for bioenergy, is the presence of brown midrib (*bmr*) mutations that can reduce lignin content. Lowered lignin has been shown to increase the conversion efficiency of biomass into ethanol. In an 11-year long-term Biomass and Potential Ethanol Yields study of Annual and Perennial Biofuel Crops, Roozeboom et al., (2018) reported $15.1 \, tha^{-1}$ dry biomass yields and $4.7 \, m^3/ha$ estimated total ethanol yields for BMR sorghum. Rivera-Burgos et al. (2019) reported a theoretical ethanol yield of 383 L/ton of dry biomass from brown midrib sorghum for control variety Atlas bmr and 403 L/ton for "brown-sweet" double mutant RIL group. In India, ICAR-IIMR has been in the forefront in development of feedstocks for lignocellulosic biofuel development. During 2019, CSV 43 BMR, which is India's first public sector bred brown midrib-low lignin sorghum variety with $16 \, tha^{-1}$ of dry biomass has been released from this institute for commercial cultivation. This line offers promise as a lignocellulosic biofuel feedstock for second-generation biofuel production because of the higher yield of fermentable sugars during pretreatment and enzymatic saccharification owing to its low lignin content.

2.2 Energy crops

Dedicated energy crops include herbaceous plant species like miscanthus (*Miscanthus* spec.), switchgrass (*Panicum virgatum*), Johnson grass (*Sorghum halepense*) and other fast-growing woody plant species like willow (*Salix* spec.), poplar (*Populus* spec.), eucalyptus (*Eucalyptus* spec.). A steady supply of uniform and consistent-quality biomass feedstock is necessary for large-scale viability of cellulosic ethanol production. Feedstocks for lignocellulosic biofuels can be divided into two main categories: dedicated energy crops and residues.

The potential of dedicated energy crops to increase farm profits and/or decrease the variability of profits will largely dictate the extent to which farmers will plant dedicated energy crops.

2.2.1 Switchgrass

Switchgrass is currently at the center of considerable attention and research. Switchgrass (*Panicum virgatum* L.) is a perennial plant native to North America that is well adapted to marginally productive croplands. Switchgrass has excellent potential as a bioenergy feedstock for cellulosic ethanol production, for heat and electricity production through direct combustion, gasification, and pyrolysis. It has consistently high yields relative to other species in varied environments and it requires minimal agricultural inputs. It is relatively easy to establish from seed, and a seed industry already exists (McLaughlin and Kszos, 2005; Sanderson et al., 2007). Switchgrass grows 3–10 ft tall, typically as a bunchgrass, but the short rhizomes can form a sod over time. In addition to potential bioenergy production, switchgrass finds it utility in soil and water conservation, carbon sequestration, and wildlife habitat. In the first year after seeding, it is common for fields to produce 75%–100% of potential yield, producing 8–13 Mg ha^{-1} on a dry matter (DM) basis (Mitchell et al., 2010). Switchgrass yields in Saunders County Nebraska ranged from 11.2 to 16.8 DM Mg ha^{-1}, with potential ethanol yields of 3740–5620 L ha^{-1} (Mitchell et al., 2012). In an 11-year long-term Biomass and Potential Ethanol Yields study of Annual and Perennial Biofuel Crops, Roozeboom et al. (2018) reported 11.3 t ha^{-1} dry biomass yields and 3.8 m^3/ha estimated total ethanol yields for Switchgrass. Average greenhouse gas (GHG) emissions from switchgrass-based ethanol were 94% lower than estimated GHG emissions for gasoline (Schmer et al., 2008).

2.2.2 Miscanthus

Giant miscanthus (*Miscanthus x giganteus Greef* et Deu.) is a perennial, warm-season Asian grass with the C4 photosynthetic pathway. It is a cold-tolerant and capable of high biomass yields at cool temperatures. Further, it tolerates marginal lands and some flooding. It has been extensively studied in the European Union and is now used commercially there for bedding, heat, and electricity generation. Its production currently occurs in Europe apart from the USA. Recently, Japan and China have taken renewed interest in this native species and started multiple research and commercialization projects. In the United States, it is also a leading feedstock for cellulosic ethanol. It is more amenable to thermochemical conversion to biofuel than biochemical conversion, with good potential for the heat and power as well as animal bedding industries. Miscanthus Giganteus is distinguished from other biomass crops by its high yields, particularly at cool temperatures, which can be more than double those typical of switchgrass. Harvestable yields for the standard M × g range from 10 to 30 Mg DM ha^{-1} depending on location and interannual weather variations during the growing season (Kalinina et al., 2017). Heaton et al., 2008 reported *M. × giganteus* peak dry biomass yields of 60.8 Mg ha^{-1} in a single site-year in central Illinois, USA, and a 3-year average of 38.2 Mg ha^{-1} over three locations in the state. The ranges for hemicellulose (295–303 g/kg), cellulose (446–458 g/kg) and lignin (70–80 g/kg) were reported by Battaglia et al., 2019. Fermentation of hydrolyzed Miscanthus using *Saccharomyces cerevisiae* resulted in an ethanol concentration of 59.20 g/L at 20% pretreated biomass loading (Han et al., 2011). In response to biomass production, total ethanol production was greater for miscanthus than for switchgrass—5594 vs 3699 L ha^{-1} (Scagline-Mellor et al., 2018).

2.2.3 Hybrid poplar

Hybrid poplars (*Populus* spp.) are among the fastest-growing trees in North America and are well suited for a variety of applications such as biofuels production, pulp and paper and other biobased products, such as chemicals and adhesives. Poplars are popular trees for landscape and agriculture use worldwide. They are known as "the trees of the people" (Gordon, 2001) and are considered one of the most important families of woody plants for human use. Poplars are more desirable for biofuels than many other woody crops because of their fast growth, their ability to produce a significant amount of biomass in a short period of time, and their high cellulose and low lignin contents. Yields of first-generation hybrid poplar planted on croplands in the Lake States of the USA have been estimated to be in the range of 7.9–11.8 dry Mg ha^{-1} year^{-1}. Poplar species and hybrids have cellulose contents ranging from ~42% to 49%, hemicellulose from 16% to 23%, and total lignin contents from 21% to 29%. The cellulose content of poplar is higher than that of switchgrass and corn stover and comparable to other hardwood feedstock such as eucalyptus, making it a desirable feedstock for the production of ethanol (Sannigrahi et al., 2010).

2.2.4 Bamboo

Bamboo is distributed in the tropics and subtropics and is the most widely utilized flowering perennials of the Poaceae family, with nearly 1500 species under 87 genera (Ohrnberger, 1999). The strong and flexible woody stem of bamboo is also used as a construction material and is frequently called "timber of the poor." In recent years, modern technology has expanded the use of bamboo beyond the traditional uses and currently, it can be utilized in many ways; in fact, it has more than 1500 applications (Lobovikov et al., 2007). Bamboo stands are dense and productive, with an average above ground net biomass production in the order of 10–20 t ha^{-1}/year (Scurlock, 2000). Due to their high growth rate which has been reported to be the highest on the planet, reaching 120 cm in 24 h, there is a fast turnover of harvest and regrowth from the same stand without damage to the plant (Tripathi and Khawlhring, 2010). The entire plant, of which includes the stem, branch and its rhizome, can be used to produce biofuel in the form of charcoal and briquette. Due to its fuel characteristics, high productivity, and short rotation, bamboo is now being explored as a potential feedstock to generate electricity through power plants and biofuels to substitute fossil fuels (Singh et al., 2017). Compared to other feedstocks, bamboo biomass has a relatively high cellulose and low lignin content which makes it suitable for bioethanol production. The chemical composition of bamboos have been reported to contain approximately 40%–48% cellulose, 24%–28% hemicellulose and 20%–26% lignin (as a percentage of dry matter), suggesting that with the appropriate technology there is an abundant pool of cell wall sugars available for bioethanol production (Yamashita et al., 2010). Sadiku et al., 2016 reported that the chemical composition of *Bamboo vulgaris* was in the range of 4%–7% for extractives, 61%–78% for cellulose, and 39%–46% for the lignin. *Bambusa emiensis* and *Phyllostachyus pubescens* are the two bamboo species that are potentially suitable to be used as a fuel in biomass fired combustion (Engler et al., 2012). As with other bioenergy crops, energy can be recovered from bamboo biomass in three main ways: thermal, thermochemical, and biochemical conversion

(Sharma et al., 2018). Direct combustion in power plants is the cheapest and most reliable route to producing power from biomass in standalone applications (IEA Bioenergy, 2009). Solid fuels (charcoals), liquid fuels, and gas (syngas) can be produced from bamboo biomass through pyrolysis. The liquid fuels or pyrolysis fuels can be processed in a biorefinery to produce biofuels. Biomass can be transformed into biogas or biofuels through biochemical conversion (Sharma et al., 2018). India is the second largest producer of bamboo in the world with an annual production of about 32 million tons. About 5.4 million tons of bamboo residues are generated in the country every year by the bamboo processing industries of which about 3.3 million tons remains as surplus. Dilute alkali pretreatment of the biomass resulted in efficient removal of lignin, effectively increasing the concentration of cellulose to 63.1% from 46.7% (Table 4) (Kuttiraja et al., 2013). Enzymatic saccharification and direct fermentation of the enzymatic hydrolysate of pretreated bamboo biomass has the potential to generate 143 L of ethanol per dry ton of bamboo process waste (Kuttiraja et al., 2013). In India, the eight states that lie at the foot of the Himalayas together make up about two-thirds of India's total bamboo production. Apart from keeping up with the country's surging demand for fuel, the Indian government is also trying to fulfill its pledge to meet a 10% reduction in the nation's energy imports by 2022. As a result, the biofuels industry is set to explode into a $15 billion market by 2020 with government backing. Indian oil companies are investing in biofuel refineries to boost ethanol production from nonmolasses sources. A $200 million joint venture between Numaligarh Refinery Ltd. and Finnish technology firm Chempolis Oy will crush bamboo, the longest of the grass family, to produce 60 million liters of ethanol every year in the tea producing state of Assam. This refinery is planning to use 5 lakh MT bamboo as raw material annually to produce 49,000 MT ethanol per annum as the main product. The major bioproducts from this plant will include acetic acid and furfural besides the production of biodegradable plastic out of furfural in collaboration with IIT Guwahati. The plant is scheduled to be commissioned by December 2021. That's enough to meet mandatory requirements for blending with gasoline in the entire northeastern region. This Bio Refinery has selected the National Small Industries Corporation (NSIC) to facilitate the supply of bamboo from farmers to different chipping centers around the Northeast states of Assam, Arunachal Pradesh, Nagaland and Meghalaya. NSIC will be responsible for developing the entrepreneurs that will be at the heart of the supply chain. More than 6000 direct and indirect jobs are expected to be created by 2021 with that increasing to more than 15,000 by 2026.

TABLE 4 Biochemical composition of native and pretreated bamboo biomass.

Parameter	Native biomass	Alkali pretreated biomass
Cellulose (%)	46.68 ± 0.03	63.11
Hemicellulose (%)	16.43 ± 0.29	14.19
Lignin (%)	17.66 ± 0.39	5.25
Water and ethanol extractives and others (%)	19.17 ± 1.17	16.75

Source: Kuttiraja, M., Sindhu, R., Varghese, P.E., Sandhya, S.V., Binod, P., Vani, S., Ashok Pandey., Rajeev, K.S., 2013. Bioethanol production from bamboo (Dendrocalamus sp.) process waste. Biomass Bioenergy 59, 142–150.

2.3 Agricultural residues

Agricultural crop residues are the carbon-based materials that are generated as a byproduct during the harvesting and processing of crops. The residues produced during harvest are field-based or primary residues while the residues produced during processing are secondary residues or agro-industrial residues. The most common residues include rice straw, wheat straw, barley straw, corn stover, corn cobs, cotton stalks, etc. Because of their immediate availability, agricultural residues are expected to play a key role in the development of the cellulosic ethanol or advanced biofuels industry. In general, the residues are utilized in several ways, as a source of fodder, for preventing soil erosion, as a fertilizer, etc. However, almost half of these resources are burnt on the farm itself before the planting of the next season crop. It is estimated that roughly one ton of residue is produced for every ton of grain harvested (Virmond et al., 2013). Cereal straw may represent an ideal resource for biofuel production, as it is a co-product of food production, and thus, its production does not compete with food generation (Townsend et al., 2017). India has enormous potential in the production of biofuels from crop residues whose use varies by region and depends on various factors viz., nutritive value, calorific values, lignin content, density etc. While a lot of cereals and pulses have fodder value, the woody nature of rice straw, rice husk, corn stover, corn cobs, cotton stalks etc. makes them a natural choice to be used as feedstock in the production of biofuels. According to a recent study "Availability of Indian Biomass Resources for Exploitation" jointly by Technology Information Forecasting and Assessment Council (TIFAC) and CSIR – National Institute for Interdisciplinary Science and Technology (NIIST), Sugarcane tops is the most surplus residue in India which is usually burnt in the fields itself. Other crops like cotton, chili, pulses and oilseeds generate surplus as they do not have any other use apart from being used as fuel. The agri. residues are usually burnt in the fields or used to meet household energy needs of the farmers. A potential of 61.1 MMT of fuel crop residue and 241.7 MMT of fodder crop residue are being consumed at farmer level and this can be freed on provision of alternatives to the farmers. The estimated total amount of residues used as fodder was 360 Mt in 2010–11 (Purohit and Fischer, 2014). This accounts for approximately 53% of total residue (Purohit and Dhar, 2015). Agricultural residues available for energy applications were estimated at 150 Mt in 2010–11 (Purohit and Fischer, 2014). The Biomass Atlas of India-BRAI 2015 estimates that an additional 104 Mt of biomass is available in India in forest and wastelands that can be converted into biofuels. Under the assumption that 20% of agricultural residue is lost in collection, transportation and storage and that ethanol yields of 214 L/ton dry matter for cellulosic-ethanol, 130 Mt of residue could be used to produce approximately 28 BL of ethanol annually.

2.3.1 Rice straw

Rice straw is one of the most potential lignocellulose sources for producing bioethanol because of its surplus availability around the globe. It is a major food crop around world with enormous biomass residues, and it is also a silica-rich C3 crop grown in wetlands. The world's rice area touched 162 million hectares with a record production of 755 million tons (FAOSTAT, 2019) which is distributed in Africa, Asia, Europe, and America. This generates approximately 1132 million tons of rice straw considering the fact that approximately 1.5 tons of rice straw is generated per ton of rice (Satlewal et al., 2018). About 50% of rice straw is burnt in the field while the remaining is utilized as fodder or used in the wood composite industry or left, as such, to decompose in landfills. Kim and Dale (2004) reported that 667.59 million tons rice straw were produced in Asia, and Binod et al. (2010) calculated that this could theoretically be converted into 281.72 billion liters of ethanol. In India, agricultural residues, including wheat and rice straw, are featured as feedstocks for producing biofuels in the National Policy on Biofuels 2018. Rice straw is a particularly attractive biofuel feedstock in northern states including Punjab, Haryana and Western Uttar Pradesh. For farmers, rice straw presence on a harvested field makes it difficult to sow the next crop, wheat. In India, 23% of rice straw residue produced is surplus and is either left in the field as uncollected or to a large extent open-field burnt to quickly get rid of the residues. Due to very small window (15–20 days) between rice harvest and wheat sowing in Punjab and unavailability/costly labor, high costs of renting machinery to mechanically harvest rice straw, farmers resort to burning of the straw. Inefficient burning of rice straw releases large amounts of harmful gases including carbon monoxide, polycyclic aromatic hydrocarbons, volatile organic compounds and nitrous oxides, along with suspended particulate matter. The solution to these problems is developing high-volume, value-added conversion technologies to harness the energy potential of surplus rice straw and also providing remunerative prices to the straw so that farmers can collect the straw and pay a higher wage to attract the needed labor during the short rice harvest time window. By doing so, tremendous waste of lignocellulosic biomass resource can be arrested.

Structural constituents

Rice straw predominantly contains cellulose 32%–47%, hemicelluloses 19%–27%, lignin 5%–24% and ashes 18.8% (Belal, 2013). Rice straw mainly composes of hexoses (i.e., glucose, galactose, mannose), hemicelluloses (i.e., xylose, arabinose), lignin (both acid soluble and insoluble), ash, silica and extractives (Table 5) (Satlewal et al., 2018). Extractives (nonstructural components) are mainly composed of proteins (about 3%) and pectins (about 2.8%) along

TABLE 5   Chemical composition of rice straw.

Component	Quantity (weight %)
Glucose	34.0–43.7
Xylose	19.0–22.0
Arabinose	2.0–3.6
Mannose	1.8–2.0
Galactose	0.4
Acid soluble lignin	2.2–6.0
Acid insoluble lignin	13.0–22.7
Ash and silica	7.8–20.3

Source: Satlewal, A., Agrawal, R., Bhagia, S., Das, P., Ragauskas, A.J., 2018. Rice straw as a feedstock for biofuels: availability, recalcitrance, and chemical properties. Biofuels Bioprod. Biorefin. 12 (1), 83–107.

with minor amounts of free sugars, chlorophyll, fats, oils and waxes. The presence of silica in general had a positive correlation with the amount of cellulose, hemicellulose and lignin in the cell walls of rice plants and increases the biomass formation of rice. The ash content has a proportion of up to 20% of the total biomass in rice straw (Zhang et al., 2015). In one recent study (Narra et al., 2015), ethanol produced from rice and wheat straw has been compared under same pretreatment and simultaneous saccharification and fermentation conditions and It showed that relatively high ethanol concentration was produced with rice straw (55.49 g/L) in comparison of wheat straw (38.19 g/L). These studies suggested that rice straw produced high bioethanol yield in comparison of wheat straw. For rice straw, pretreating at severities of between 3.65 and 4.25 would give a glucose yield of between 37.5 and 40% (w/DW, dry weight of the substrate) close to the theoretical yield of 44.1% w/DW, and an insignificant yield of total inhibitors. At a pretreatment severity of 3.65, twice as much ethanol was produced from rice straw (14.22% dry weight of substrate) compared with the yield from rice husk (7.55% dry weight of substrate) (Wu et al., 2018).

2.3.2 Wheat straw

Wheat straw is also a potential feedstock for ethanol production. European Union, China, India, USA, and Canada are the leading wheat cultivating countries in the world. Considering a ratio of 1.3 residue and crop, about 850 million metric tons wheat straw produced annually worldwide can be considered as a huge agricultural reside. As per the report from Otero et al. (2007) surplus wheat straw is able to produce 120 billion liters bioethanol annually, which can replace 93 billion liters gasoline. For wheat straw, about 400 million tons may be globally available for biofuel production (Tishler et al., 2015). In some studies, cellulose contents of wheat straw were found to reach almost 50% (Brandenburg et al., 2018). Wheat straw is also composed of cellulose, hemicellulose, and lignin in the range of 33%–40%, 20%–25%, and 15%–20% (w/w), respectively. Ash content in the wheat straw is almost three to four times lower than rice straw, which makes this substrate most suitable for the bioethanol production compared to rice straw (Prasad et al., 2007).

2.3.3 Corn stover

Corn stover, i.e., leaves, stalks and bare cobs from maize plants, is the most abundant straw generated in the USA (Panoutsou et al., 2017). Stover yields are generally closely related to grain yields. Since the ratio of grain to total plant biomass (harvest index) is usually near 0.50, the mass ratio of grain to stover in corn is close to 1:1 (Graham et al., 2007) and thus potential stover ethanol yield on a land area basis would be expected to have a direct correlation with grain yield. The upper part of the corn plant is generally less lignified and more digestible than the lower portion of the plant. As such, it is a more desirable fraction for cellulosic ethanol. In a comparative study of ethanol production using dilute acid (DA), ionic liquid (il) and AFEX™ pretreated corn stover, the ethanol yields calculated for DA, IL and AFEX pretreated residual solids were 14, 21.2 and 20.5 kg of ethanol per 100 kg of corn stover, respectively (Uppugundla et al., 2014). Corn cobs are currently being used for heat in some parts of Europe, while in the United States, this feedstock is rapidly being developed as a feedstock for cellulosic ethanol, co-firing, and gasification projects.

2.3.4 Cotton stalks

Cotton stalks are the residues left in the field following harvest which are usually buried or burnt to prevent pest build up. It is a potential raw material for conversion to ethanol because it is rich in cellulose (32%–46%) and hemicellulose (20%–28%) (Wang et al., 2016). Bioethanol from cotton stalk was produced utilizing two-stage dilute acid hydrolysis and fermentation of detoxified hydrolysate (Keshav et al., 2016). Highest ethanol concentration of $22.93 \pm 1.74 \, g/L$ with $0.36 \, g/g$ ethanol yield was achieved after 48h of incubation (Shahzad et al., 2019). To effectively utilize cotton stalk as a feedstock for ethanol production, alkaline pretreatment would be more effective (Silverstein et al., 2007).

2.3.5 Sugarcane bagasse

Using sugarcane bagasse as a feedstock for second-generation biofuels would lead to doubling the current output of biofuel production. To maximize the conversion efficiency of sugarcane biomass to biofuels, it is imperative to have sugarcane genotypes with improved total biomass: more cellulose and less lignin, resulting in less enzymatic recalcitrance and better saccharification yield. Dual purpose energy canes with more than 20% fiber and 15% brix can be used for both energy and alcohol production. Considerable technical progress has been made in the production of 2G ethanol and scaling up to commercial scales is underway but no industrial plant has operated yet at full capacity. Energy balance and overall costs need to be improved. Integration of second-generation (2G) with 1G ethanol production provides an option for fully renewable production of energy without the use of fossil fuels for thermal processes and electricity in the conversion process (Center for strategic studies and management, 2017).

3 Biomass availability

Biomass availability depends on several factors such as location, availability of land vis a vis competing land uses, competing uses of agricultural residues, market demand, sustainability requirements and policy interventions. These factors have made few states leading in terms of biomass power projects viz. Maharashtra, Uttar Pradesh and Karnataka, each having more than 1 GW of grid interacted biomass power. Other states with favorable policy and opportunities in Biomass are Punjab and Bihar.

4 Biomass supply chain structure and characteristics

The biomass supply chain incorporates several components of bioenergy production, which in turn consist of several activities. To obtain the critical mass of biomass residues needed for sufficient energy production, multiple suppliers are often involved in the biomass residues supply chain. This supply chain, also known as the Biomass supply chain, is composed of four main components, including (i) Biomass harvesting/collection (from single or several locations) and Preprocessing/pretreatment; (ii) storage (in one or more intermediate locations), (iii) transport (using a single or multiple levels) and (iv) final conversion in the biorefineries,

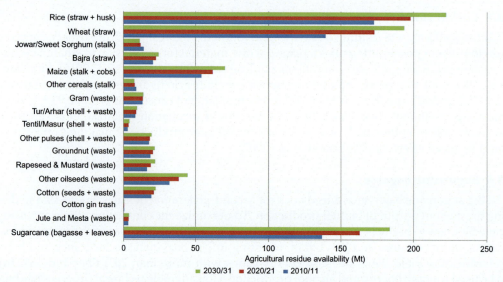

FIG. 4   Example of supply chain to produce bioethanol from switchgrass (Zhang et al., 2013).

as shown in Fig. 4 (Zhang et al., 2013). The biomass-to-energy supply chain can be classified into three parts: upstream, midstream and downstream (An et al., 2011), which is similar to the division made by Sandersson (1999) who identifies upstream supply, conversion and downstream provision. Upstream, is viewed as the part that supplies biomass to energy production. Midstream refers to energy conversion in power plants and downstream refers to energy distribution to consumers of energy. The single largest limiting factor for the production of bioenergy is the unavailability of biomass. The structure of the global market for biomass and the associated supply chains is evolving quite dynamically. Traditionally, biomass has been used for energy (mainly thermal) production in areas close to its production sites. However, an emerging practice for energy producers is to procure biomass from several suppliers in order to develop the critical mass necessary for the justification of an energy production facility.

4.1 Feedstock supply and logistics

This is one of the key components of a supply chain which provides the biorefineries with diverse feedstocks which are infrastructure-compatible and stable. For biomass feedstocks, emphasis would be on development of supply chain for economical and time-sensitive collection, pretreatment, storage, and transport.

4.1.1 Seasonal availability

Agricultural biomass types are usually characterized by seasonal availability, thus becoming a critical challenge in the operation of biofuel supply chain and dictating the need of storing large amounts of biomass for lengthy time periods increasing the operational costs of the biorefinery. In India, rainy season rice and maize are harvested during Sep-Oct, while wheat which is grown during the winter season is harvested during April where the residues would be available. The corn stover in the U.S. Corn Belt is mainly harvested from September through November. The analysis of a database of switchgrass biomass productivity studies showed that single harvest systems were the most practical and economically feasible for bioenergy systems and it should be harvested once in the fall, after killing frost, for biofuel production (Wullschleger et al., 2010). Wood residues are less seasonal unlike crop residues. Meticulous planning of the harvesting and scheduling of the biomass is needed for ensuring an uninterrupted supply to the industry. Perishability of the biomass products increases the complexity of biomass supply chains affecting the transportation and length of storage time. The greatest operational challenge is to manage the biomass storage to ensure an uninterrupted supply to the biorefineries. Biomass Supply Chains need to be robust with inbuilt flexibility to adapt to unforeseen market volatility, as the demand of the produced energy depends on the price of competitive fuel substitutes.

5 Exploring nuisances of sustainable biomass supply

The establishment of sustainable bioenergy supply chains would have to be remuneratively attractive to a wide range of stakeholders in the long term. Equitable and sensitive distribution of economical gains in different strata of the value chain would increase its social acceptance among communities. Lastly, the supply chain must be sensitive to specific ecosystem services pertaining to the area. Centre for Economic and Social Studies (CESS), Hyderabad, India conducted an empirical analysis of High Biomass Varieties (HBV) promoted by ICRISAT and ICAR-IIMR in the farmers' fields at different locations of Indore and Gwalior region of Madhya Pradesh, India with the assistance of scientists of Rajmata Vijayaraje Scindia Krishi Vishwa Vidyalaya (RVSKVV), Gwalior, India to elucidate these aspects. These HBV varieties were meant for use as feed stocks for biofuel production. Surveys undertaken by CESS tried to address the suitability of HBVs of sorghum and pearl millet feedstocks with regard to crop economics, socioeconomic dynamics, potential upscaling, issues with regard to use of wasteland, and finally the carbon neutrality. Farmers of Gwalior, Khargone, Dewas and Morena districts, who had taken up the varieties for high biomass production developed by ICRISAT and ICAR-IIMR in their lands (part of MLT) were surveyed during 2014–15 *kharif*. Focused group discussions (FGD) have also been conducted by CESS team in Gwalior and Indore region with MLT farmers during the years 2015 and 2016. Major finding of the 2014 *kharif* trials was that the average income collectively from both grain and fodder yield was relatively lower for the new variety than compared to the ones being cultivated in the previous year. The HBV sorghum grain yield (2015–16 *kharif*) in Nagzari was less due to less rains and some of it was eaten by birds and the fodder yield too was less. The reason for high HBV sorghum yield (*kharif* 2015–16) in Nagdha are fertile soils and one supplemental irrigation in September month (in the event of no rains). As grain yield was high, there was a reduction in fodder yield. The reason for less HBV sorghum yields in Palnagar is due to excess rains and failure of seed to germinate and the farmers had to go for second sowing which led to delay in sowing period and eventual low yields. In Nahardonki HBV crop height was very good but no grains were harvested due to multiple cuttings for fodder purpose. In Bijoli (Gwalior region) during 2015–16 *kharif*, there was very less rain and it was almost like a drought and hence low yields in HBV sorghum. However Hybrids and Traditional sorghum varieties did reasonably well (Table 6). In Palnagar and Bijoli, HBV sorghum yielded a fodder quantity of around 3000 kgs. In Nahardonki village of Morena District (Gwalior region),

TABLE 6 Sorghum and pearl millet crops and their year-wise grain and fodder yeilds.

Region	Year	Village	Crop	Variety	Avg. grain yeild in Q/acre	Avg.grain value in Rs./Q	Dry fodder yield in kgs/acre	Value of fodder in Rs./kg
Indore	2015–16	Nagzari	Sorghum	Existing varieties (Hybrids)	10–12	1300–1500	1600–2000	2
	2015–16		Pearl millet	Existing varieties	4	1300–1400	700–800	2–2.5
	2014–15		Sorghum	HBV	4.5 to 5	Consumed	400	–
	2015–16		Sorghum	HBV	1	Consumed	350	–
	2015–16	Nagdha	Sorghum	Existing varieties (hybrids)	14	1200–1300	1000–1250	2
			Sorghum	HBV	14	Consumed	1000	2
	2015–16	Palnagar	Sorghum	HBV	1–1.4	Consumed	2800	2
Gwalior	2015–16	Nahardonki	Pearl millet	HBV	Nil	–	6000	Own use
			Pearl millet	Existing varieties (mostly hybrids)	12	1200	2000	5
	2014–15		Pearl millet	Existing varieties	12	900–1000	1000	3
	2013–14		Pearl millet	Existing varieties	12	1100	1000	2
	2015–16	Bijoli	Sorghum	Hybrids	12	1000–1200	1600–2000	1
				Peeli sorghum	8	2000–2500	3200	1.5–2
				Desi safed sorghum	8	4500	3200	1.5–2
				HBV	4	Consumed	3200–4000	1.8–2
			Pearl millet	Existing varieties	8–10	–	1600–2000	1
Baseline Survey	2013–14	Average of all villages	Sorghum	Traditional sorghum	12.06	–	950	1.5–2
				Hybrid sorghum	11.41	–	890	1–1.5
		Average of all villages	Pearl millet	Traditional pearl millet	10.50	–	1000	1.5–2.0
				Hybrid	22		925	1–1.25

Source: FGD with sampled farmers of Indore and Gwalior region during 2014–15 and 2015–16 and Baseline survey of 2013–14.

despite zero grain yield in HBV Pearl millet crop, the fodder yield was highest with 6000 kgs/acre. The value of sorghum dry fodder changed from village to village. However, it generally ranged between 1 and 2 rupees/kg. In the case of pearl millet crop fodder, there was wide range during 2015–16 *kharif* as it varied between Rs. 1 per kg in Bijoli to Rs. 5/kg in Nahardonki of Gwalior region. The cost of fodder has implications for biofuel production as it is this material that is used a raw material. The lower the fodder price the more economical will be the biofuel production from these crops. From last 2 years, there is huge increase in market price of *safed* sorghum (traditional variety of the region) due to its utility for some industrial purpose. Hence, farmers are increasing the area under this crop in Gwalior region and there is a growing demand for the seed of this crop. In the case of both pearl millet and sorghum crops, with regards to overall per acre income, existing varieties were doing slightly better than ICMV 05777 and ICSSH-28 respectively and far better than other HBV varieties used in MLTs in farmers field during *kharif* 2015–16 (Tables 7 and 8). When it comes to biomass yield, 2015-ICFPM-1 of pearl millet crop and ICSSH-28 and ICSV 93046 were performing much better than existing varieties in 2015–16 *kharif*. In a base line survey done by CESS, Hyderabad in 2012–13 on impact of promoting food crops for biofuel cultivation, it was found that 38.44% of the households agreed that it will result in shortage of food grains while 61.56% did not perceive a reduction in the food supply. Majority of the respondents felt that there would not be any impact on food security, citing the reason that they would supplement sorghum/pearl millet either by procuring from fair price shops or from retail markets. Out of the 128 households which felt that there will be a reduction in food grains, 66.40%

TABLE 7 Details of grain and fodder yields of high biomass varieties vis-à-vis existing varieties during the year 2015–16 Kharif.

Particulars	Pearl millet						Sorghum		
	2015-ICFPM-7	2015-ICFPM-1	ICMV 05222	ICMV 05777	IP 6107	Existing varieties	ICSSH 28	ICSV 93046	Existing varieties
Grain yield in Qtls	4.00	–	4.66	6.66	4.00	12	7.82	–	10
Fodder yield in kgs	1260	3000	1740	1460	1410	2000	2924	2550	2000
Fodder income in Rs.	3150	7500	4350	3650	3525	6000	5263	4590	3000
Grain value in Rs.	4800	–	5592	7992	4800	14,440	9384	–	12,000
Cost of cultivation in Rs.	1180	1610	2095	1816	1498	10,000	6046	1986	6000
Gross income in Rs.	7950	7500	9942	11,642	8325	20,400	14,647	4590	15,000
Net income in Rs.	6770	5890	7847	9826	6827	10,400	8601	2604	9000

Source: Field survey 2016.

TABLE 8 Village-wise response of farmers regarding the impact of use of sorghum/pearl millet for biofuel production on fodder security in Madhya Pradesh.

Village name	Yes	No	Total
Nagada	47.06 (8)	52.94(9)	100.0(17)
Chinvani	55.6(10)	44.4(8)	100.0(18)
Nagaziri	61.1(33)	38.9(21)	100.0(54)
Rupkheda	46.7(7)	53.3(8)	100.0(15)
Baraha	37.03(20)	62.96(34)	100.0(54)
Bijoli	31.8(7)	68.2(15)	100.0(22)
Dahel	52.9(9)	47.1(8)	100.0(17)
Jakara	19.2(10)	80.8(42)	100.0(52)
Nahar Donki	100.0(22)	0.0(0)	100.0(22)
Ummed Garh	75.8(47)	24.2(15)	100.0(62)
Total	51.96(173)	48.04(160)	100.0(333)

Source: Field survey.

felt that such reduction in grains will impact the household food security, while 33.60% did not agree. The development of biofuels to meet the requirements of the transport sector can bring about changes in the land use pattern of the country and could threaten food security and other agrarian supplies. The potential diversion or displacement of food crops is also considered to be a serious problem for livestock sector. Though the analysis of CESS study shows that the impact might not be much regarding food grain security, there is a considerable amount of apprehension on its potential impact on fodder security. It is evident that even before the cultivation of these crops for biofuels production, a majority of the households (51.96%) believe that the use of these crops will affect the fodder security of their animals. On the other hand, 48.04% of the sample households perceived that there won't be any impact on fodder security. It was very interesting to see that across all study villages of the five districts, there were a few households that did perceive that there would be fodder insecurity in the event of cultivation of these crops for biofuels production. A further investigation was conducted to understand whether the diversion of fodder/biomass for biofuel production will affect the milk economy of the region. Nearly 33.9% of the samples households perceived that it will affect the milk economy, whereas 66.1% responded negatively.

6 Sustainable supply chain management

Sustainable supply chain management concept would necessarily consider the interdependence between the economic, the environmental, and the social performances of a biomass to energy plant (Chaabane et al., 2012). Economic sustainability would aim to optimization and scheduling of processes to maximize the net profits through

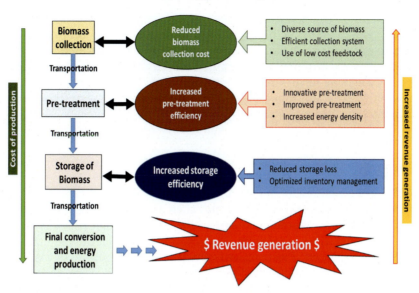

FIG. 5 Sustainable supply chain management.

maximizing revenue generation with minimal raw materials, inventory and production costs. Similarly, social sustainability, would ensure that the process meets the expectation of employees and local stakeholders. Environmental sustainability is generally linked to reduction in carbon footprint as well as environmental pollution. It also includes a reduced dependency on nonrenewable resources, increased energy efficiency, absence or decrease in the consumption of hazardous materials and lastly the frequency of environmental accidents (Gimenez et al., 2012). The complexity of logistics is the main challenge for large-scale biomass to energy production. The main components of the whole process is given in Fig. 5. For the sake of understanding let's consider biomass collection, pretreatment, storage and conversion of biomass to energy are the four basic building blocks of the overall supply chain. Each of these four steps is linked via means of transportation. Major tradeoff comes between transportation cost and capitol cost for establishing the production facilities. Let's consider each block an attempt their optimization in an attempt to understand the macropicture (Meixell and Gargeya, 2005).

6.1 Biomass collection

Steady supply of biomass is prerequisite of biomass energy production. For any given location for the establishment of a biomass energy unit, understanding the existing biomass generation and their usage needs to be critically analyzed. The assessment must attempt to visualize futuristic scenarios. For example, an area predominantly producing huge volume of cotton stock, the viability of cotton cultivation in terms of agro-climatic factors should be assessed. A diverse plethora of biomass stream invariably reduces the vulnerability of such a plant from unforeseeable situations such as advent of new pest or disease for a particular feedstock or competition with other usage of the same feedstock, cost of collection and segregation of a more widely distributed biomass for example kitchen waste from a cluster of villages etc. As highlighted in the previous section escalation of grain price may adversely effect agro-waste generation. Overall, diverse source of biomass which are sustainable for the foreseeable future and minimal transportation as well as collection cost leads to optimized biomass supply in the long-term. Financial attractiveness of the whole process is key as often over optimization with an aim to maximize profit for the top of the pyramid of value chain may be detrimental for the long-term viability of the plant.

6.2 Biomass pretreatment

Typically biomass contains a lot of moisture and are low in energy density. Lignocellulosic biomass comprising of structural carbohydrates such as cellulose and hemicellulose can be subjected to mechanical pretreatment, steam and steam explosion pretreatment, hot water pretreatment, ammonia fiber explosion and chemical pretreatments viz., acid, alkali, organosolv, CO_2 explosion etc. Pretreatment (say drying of rice straw) of the harvested biomass close to the source leads to reduced transportation cost and storage space requirement and increases and conversion efficiencies. Challenge here would be that if price is fixed at weight basis tendancy of keeping moisture or dust may look remuneratively attractive to the members of the value chain at the bottom of the pyramid. Hence, selection of pricing criteria guided by scientific unambiguous methods increases transparency and pretreatment efficiency. A comparable analogy would be the quality check for milk in terms of fat and protein content at the grass root level procurement process. Often 'what gets measured gets done' is the mode of operation which leads to efficient pretreatment process. For example, not harvesting a biomass that gets procured for biological valorization processes such as biogas generation immediately after spraying pesticide or herbicide would increase the safety in handling as well as augment the efficiency of biodegradation. Thus proper awareness built-up about what improves the process and technical back-stopping is of immense value in the long-term. Often such new practices need to be sincerely followed for multiple seasons to allow stakeholders to assimilate and adapt to the newer processes.

6.3 Storage

Often agro-waste gets generated seasonally with bulk volume of biomass generated over a handful few weeks of a season. Hence, the planning must be realistic and adopted through a consultative process with all the different stakeholders. A farmer hiring a tractor would surely harvest as much as possible in a single day in order to minimize the cost of harvesting. As such storage capacity is needed to ensure year-long availability of biomass. However, a good blend of short-duration crop residue, long-duration crop residue and perennial biomass waste stream leads to optimized storage space utilization over the temporal scale (Sharma et al., 2013). Please note here diversification of in-coming waste stream needs to be optimized and not the optimization of the storage of a single biomass stream. Scientific inventory management leads to minimal loss of biomass during storage. Adherence to safety guidelines is always beneficial in the long term.

6.4 Transport

Both biomass collection and delivery require extensive efforts in equipment selection, shift arrangement, vehicle routing, and fleet scheduling (DOE/EERE, 2013). Road transport is often preferred, due to the limited accessibility of some production sites. However, other modes of transport like rail can be used. In many cases, the fleet of vehicles is limited and the number of travels per period is restricted by various constraints like vehicle range or driving time regulations.

6.5 Conversion of biomass to energy

The conversion process for biomass to energy must be efficient, however, several factors need to be considered before the selection of any process. The most important criteria would be the per unit production cost of the process. Robustness of the process in a given location considering access to skill sets, quality of electricity and chemicals if any should be analyzed in detail. For example, a highly efficient and sophisticated process which need uninterrupted three-phase electricity may not be viable for certain locations in a rural area. However, shifting the unit closer to the urban clusters for the want of improved facilities and infrastructure may increase the transportation and/or storage cost. Different shortlisted suitable technologies must be evaluated based on their operational costs, biofuel productivity, and biomass requirement to make an optimal selection. Flexible variables can be used for the selection of biomass to energy conversion pathways at the supply chain level and the selection of catalysts, equipment, operation protocol, scheduling, and processing methods at the process level. Evaluation of the economic objectives viz. net present value, annualized total cost, etc. apart from conventional costs for construction, materials and labors, we should also take into account the government subsidies and revenue generation potential form selling the by-products such as bio-oil (DOE/EERE, 2013). Policy support in terms of subsidized electricity, low-cost land availability, tax holidays and accommodative flexible labor laws all play important role in establishing such a unit. Policy support can also lead to improved infrastructure and planning. For example, often biomass fuel lead to the generation of low-quality steam in terms of power generation. Setting up of fossil fueled based power plant which can buy this steam as input for power generation or an energy-intensive industry like cement factory may lead to a win-win situation for both. Here establishing the industrial ecosystem and infrastructure would lead to increased efficiency.

6.6 Focus on innovation and flexibility to changing local scenarios

Irrespective of meticulous planning at the time of the establishment of the plant, continuous adaptation, technology upgradation and improvement of the whole value chain with changing local scenario would ensure long-term viability of the plant. Hence, the optimized configuration of a sustainable biofuel supply chain may not be static and would rather evolve over time. In particular, application of multiperiod planning models proposed for generic supply chains to biofuel supply chains may be the preferred approach. Equitable distribution of the revenue generated across the pyramid of value chain would make the system sustainable in the long-term. Providing local farmers who supply a given biomass input say, pigeon pea stock with modern know-how to modern pigeon pea cultivation would be wise use of some fraction of the revenue in the long-term. Improving water use efficiency of the local farmers by irrigation scheduling or reuse of treated wastewater would similarly augment the long-term viability of biomass based energy plants.

6.7 Awareness generation and stakeholder meetings

The importance of awareness generation cannot be overrated. Yet, many a times the focus on establishing the perfect processing unit undermines the time and effort required to establishment rapport among different stakeholders. The biomass collection team, the supplier of various biomass, the pretreament units, the storage personnel and the people involved in biomass to energy generation process must discuss each other's requirement through numerous stakeholders' meetings. Setting-up of a dispute resolution committee with representation from each stakeholders would help to resolve potential conflicts. The processes involved must look for innovative methods to improve efficiency as a continual process. Overall, sustainable supply chain management should focus on both human and machine aspect of it as ignoring either would lead to failure.

7 Policy support for advanced biofuels

The eleventh five-year plan (2007–2012) highlighted the severe shortages of energy, the dominance of coal and the need to expand resources through exploration, energy efficiency, renewables, and research and development (Planning Commission, 2007). Subsequent, policy initiatives led to the development of National Action Plan on Climate Change, launched in June 2008. Though India does not have any binding emissions targets, the policies reflect a response to global concerns to address climate change. The National Mission for a 'Green India' aims to achieve afforestation of 6 million hectares of degraded forest lands and to expand forest cover from 23% to 33% of India's territory by 2022. The term 'biofuels' means liquid fuels that are derived from biomass, such as biodegradable agricultural, forestry or fishery products, wastes or residues, or biodegradable industrial or municipal waste. Biofuels are derived from biomass and use photo-synthetically fixed C, thus, facilitating recycling of atmospheric CO_2. Based on feedstock type, conversion process, technical specification of the fuel and its application, biofuels are categorized as first generation (1G), second generation (2G) and third generation (3G). Out of 83 billion liters biofuels which contribute about 1.5% of the global transport fuel consumption, 40% of global production of biofuel is in Brazil, China and Thailand outside the OECD region. Biofuel is expected to provide about 9% of the total transport fuel demand by 2030 with the production expected to rise to 159 billion liters in 5 years' time globally (IEA, 2018). In India, ethanol is primarily produced from sugarcane molasses and used as a biofuel for blending with petrol. In January 2003, Government of India (GOI) mandated 5% blending of ethanol with gasoline through its ambitious Ethanol Blended Petrol Programme (EBPP) which faced shortage of ethanol. Since then, petroleum with an ethanol blend has been developed and used in nine states and four union territories: Andhra Pradesh, Daman, Diu, Goa, Dadra, Nagar Haveli, Gujarat, Chandigarh, Haryana,

Pondicherry, Karnataka, Maharashtra, Punjab, Tamil Nadu and Uttar Pradesh (Ethanol Producer Magazine). In 2005, the country became the world's fourth largest producer of ethanol at 1.6 billion liters and at the same time the world's largest consumer of sugar. To promote biofuels as an alternative energy source, the GOI in December 2009 announced a comprehensive National Policy on Biofuels formulated by the Ministry of New and Renewable Energy (MNRE), calling for blending at least 20% of biofuels with diesel (biodiesel) and petrol (bioethanol) by 2030. However, greater push for biofuels is received through the National Biofuel Policy, 2018 that envisages 40% reduction in carbon emissions by 2030 (MNRE, 2018). In an attempt to curb the carbon footprint of biofuel new policy shifts the focus from first generation (1G) biofuels which are made from molasses and vegetable oils to 2G biofuels made from cellulosic and lingo-cellulosic biomass/woody crops, agricultural residues and municipal waste feed stocks. Such policies necessarily is an attempt to move toward a circular economy through encouraging "Waste to Wealth" initiatives. Also utilization of these wastes for ethanol generation will eliminate the problem of stress to arable land and water resources and other issues related to food security associated with 1G biofuels. In fact, in Union Budget for the year 2018–19, central govt. has also focused on Waste to Wealth conversion projects e.g. Gobar Dhan Scheme to produce bio CNG. New National Biofuel Policy 2018 will ensure cost-effective and pollution free import substitute of polluting fossil fuels. Moreover, govt. will authorize OMCs to sell EBP with ethanol percentage up to 10% under Ethanol Policy India. Furthermore, govt. will implement this National Biofuel Policy 2018 in a mission mode to make our environment pollution free. First generation biofuels such as bioethanol are produced mainly from starch derived from food or fodder crops like sugarcane, sugar beet, sweet sorghum stalks, corn, wheat etc. Biodiesel is produced by transesterification, whereby lipids (oils and fats) in edible/nonedible oil, such as palm, soybean, rapeseed, Jatropha, Pongamia, etc. are reacted with alcohols (ethanol or methanol) (Hileman et al., 2009). Bioethanol from molasses have competing uses and unfavorable policy acceptable by producers and GoI has led to only 1.7% blending, contrary to the national aspiration and commitment of reaching 5% blending by 2020. However, as indicated above the current level of ethanol blending is 4.7% in 2017–18 and new Biofuel Policy released by the Government of India has enabled the sugar factories to produce ethanol from molasses or sugar directly and signed the contracts for purchasing 2.37 billion liters of ethanol for blending (MNRE, 2018). The first-generation ethanol is the largest source of biofuel at present and search for alternative crops to food crops such as sugarcane for ethanol production is relevant in a country like India explore issues of water scarcity and food security. Central government of India in its 2018 New Biofuel Policy has indicated provision of incentives to all state-run oil marketing companies. These OMCs has made an agreement of long-term offtake of 2G ethanol under Biofuel Policy India. For this reason, OMCs are assuring suppliers for 15-year offtake contracts. Indian Oil Corporation (IOC) has recently signed an agreement with Punjab government. Under this agreement, IOC will establish various CNG plants in Punjab in upcoming 5 years. In addition to this, OMCs are going to set up 12 advanced biofuel refineries in several states. IOC is currently operating three biofuel plants and plans to increase its capacity from 100 to 1200 t per day in next 2 years. The most prominent driving forces for advanced biofuels on a global scale are political instruments, agreements, and regulations to reduce reliance on nonrenewable, imported fuels and to meet GHG reduction targets. The demand for biofuels, heat and electricity is increasing steadily around the globe. Major policy-related interventions for adoption and promotion of bioenergy have also been realized by several countries over the past few decades. Policy drivers such as blending targets, renewable portfolio standards have been more critical in influencing bioenergy expansion at local to global scales than market factors. Government commitment and support and financial incentives therefore continue to be important for significant, large-scale mobilization of the bioenergy supply chains.

8 Reduction of water foot-print of biofuel production

8.1 Wastewater and algal biofuel

Use of wastewater such as urban wastewater for algal cultivation could offer potential benefits serving a dual purpose of treating the wastewater as well as producing lipids-rich biomass, which could be used for biodiesel production (Chisti, 2013; Úbeda et al., 2017; Datta et al., 2019). Wastewater provides macro and micronutrients essential for algae growth. Algae assimilate nutrients by bio-sorption and utilize it for its metabolic activities and store excess energy in the form of lipids, carbohydrates and proteins. Among the various strategies possible for economical large-scale production of microalgal biomass, a coupling of wastewater treatment with algal farming is possibly the most sensible due to the similar scale and production facilities that both industries rely on (Delrue et al., 2016). The additional benefit from such coupling is the promotion of on-site local industries and more importantly, the elimination of a large negative environmental footprint that would otherwise arise from the pollution associated with nutrient manufacturing, transportation and change in land use. Despite these two opportunities, many research and development challenges have still to be overcome in order to benefit from the full potential of the combination of microalgae production and wastewater treatment.

8.2 Indian scenario on wastewater treatment and reuse in agriculture

At present of the 62,000 MLD (million liter per day) the total wastewater generated in major Indian cities only 23,277 MLD gets treated (CPCB, 2000). About 70% of people in India live in villages and rural wastewater management remains a challenge in India. The link between health and hygiene of the villagers and good wastewater management practice needs no further elaboration. Energy and chemical-intensive conventional wastewater treatment technologies such as activated sludge process, sequential batch reactors are neither feasible nor sustainable in rural setting with limited resources (Datta et al., 2016). Often in water scarce semiarid villages the raw wastewater from these sumps are utilized for salad crop or vegetable cultivation. The fitness of such agro-produce for human consumption is suspect, moreover, raw wastewater irrigation causes excessive weed growth, nutrient-rich run-off and eutrophication of freshwater sources nearby. The suspended solid particles get accumulated in the soil and over a long period can significantly deteriorate the physical property of the soil.

8.3 Potential of constructed wetlands

Despite their apparent simplicity constructed wetland (CW), a proven age-old wastewater treatment system, are complex ecosystems involving biogeochemical processes such as filtration, sedimentation, plant uptake or phytoremediation and microbial degradation. The recently concluded Indo-EU project titled "Water4Crops" funded by Department of Biotechnology (Govt. of India) and the European Commission under the seventh framework has established the potential of constructed wetland. The joint India-EU project review held in New Delhi (15th–16th June 2016) identified the decentralized wastewater treatment using constructed/engineered wetlands using filtration, phytoremediation and microbial transformations as a suitable and ready technology for scaling-up in India, as a business model (Datta et al., 2016; Tilak et al., 2016). The technology can be integrated in rural development schemes and could become part of the Swatch Bharat initiative of the Government of India. The various types of constructed wetlands used over the last four decades can be grouped into two broad categories viz. free water surface (FWS) wetlands or subsurface flow (SSF) wetlands. In short, the former involves a pond whereas the latter involves a dry surface (as their names suggest). One major advantage of SSF CWs (though being slightly expensive than FWS CWs owing to the filtering media cost) is the better control of mosquito menace. The CWs may also be used for growing algae for biofuel production and provide additional income source for the villagers during the construction, operation as well as maintenance activities.

9 Challenges

Producing advanced biofuels from biomass feedstocks is even more challenging than producing first generation biofuels. The major challenges are discussed hereunder.

9.1 Seasonal availability of biomass

Most of the biomass materials are seasonal and are required to be available in huge quantities to be qualified as feedstocks for biofuel production in biorefineries. Supply of biomass is critical to the reliable and efficient operation of any biomass-based biorefinery. The cost of feedstocks will significantly influence the cost of biofuel production. About one-third of biofuel production cost is associated with biomass cost and the cost of biomass ($ per ton) is directly proportional to the yield (ton per ha) (Duffy and Nanhou, 2002), which is influenced by soil fertility, location, and genetics. Another challenge would be to influence food grain growing farmers to cultivate biomass feedstocks assuring them of a guaranteed buy-back. The willingness of stakeholders to invest in infrastructure and technology is challenged by uncertainties surrounding long-term feedstock supply of both crops and value chain residues.

9.2 Biomass harvesting

Harvesting of different types of biomass requires different types of machinery which would influence the cost of harvesting making it an energy intensive process.

9.3 Moisture content in biomass

The presence of high moisture content in biomass causes biological degradation, mold formation and losses in the organic contents during storage (Johansson et al., 2006), that could reduce the yield of the fuel produced from these materials. Storing biomass at <10% can extend the conservation time of the materials and reduce major losses (sugars) in the biomass during the storage period (Balan, 2014). High oxygen contents of biomass materials can also negatively affect their conversion to various products such as fuels.

9.4 Density of biomass and transportation cost

Biomass has a relatively low energy density and hence requires more quantities of biomass to supply the same amount of energy as a traditional hydrocarbon fuel. The low density of biomass is reported to influence the transportation cost. Transportation cost is also influenced by the moisture content, distance from the field to biorefinery, available infrastructure, available on-site technology, and the mode of transportation (rail or road) (Balan, 2014; Kumar et al., 2006). Biomass material should be used in densified forms to overcome moisture, storage and handling problems.

9.5 Biomass supply chain and logistics

The supply chain steps pertaining to biomass production, harvesting, pretreating, storage and transporting biomass to centralized biorefineries will directly impact the cost of feedstock delivery. Apart from the above, the other constraints are nonavailability of standards for biomass classification, grading and quality. Besides this, there is a lack of established market pricing mechanism. In addition to these logistic challenges, efficient and commercially viable conversion technologies are also lacking for a number of supply chains and regions; and the valuation of by-products and co-products such as CO_2, ash, lignin is often lacking.

10 Conclusions

Rapid depletion of limited fossil fuels coupled with detrimental effects on environment has occurred due to human reliance. Production of bioenergy through utilization of dedicated, rapidly growing high-biomass feedstocks on nonarable lands and exploitation of agroindustrial waste materials can offer a solution to this issue. Utilization of Biomass feedstocks for advanced biofuel production depends on factors like availability, characteristics as fuel, and most importantly opportunity cost. The emerging concept of advanced biofuels would require a careful and judicious design of the biomass supply chain which holistically integrates different components of the supply chain to enhance the quantum of energy return, improves the greenhouse gas balance, reduces the water footprint of the bioenergy production facility and achieves environmental sustainability. At the global level, success in the commercial development and deployment of advanced biofuel technologies would require a significant amount of technological interventions through increased amounts of Research and Development to overcome the current cost barriers.

References

An, H.J., Wilhelm, W.E., Searcy, S.W., 2011. A mathematical model to design alignocellulosic biofuel supply chain system with a case study based on a regionin Central Texas. Bioresour. Technol. 102, 7860–7870.

Balan, V., 2014. Current challenges in commercially producing biofuels from lignocellulosic biomass. ISRN Biotechnol. 2014, 1–31. https://doi.org/10.1155/2014/463074.

Bartzialis, D., Giannoulis, K.D., Skoufogianni, E., Lavdis, A., Zalaoras, G., Charvalas, G., Danalatos, N.G., 2020. Sorghum dry biomass yield for solid bio-fuel production affected by different N-fertilization rates. Agron. Res. 18 (S2), 1147–1153.

Batog, J., Frankowski, J., Wawro, A., Łacka, A., 2020. Bioethanol production from biomass of selected sorghum varieties cultivated as main and second crop. Energies 13, 6291. https://doi.org/10.3390/en13236291.

Battaglia, M., Fike, J., Fike, W., Sadeghpour, A., Diatta, A., 2019. Miscanthus × giganteus biomass yield and quality in the Virginia piedmont. Grassl. Sci. 65 (4), 1–10.

Belal, E.B., 2013. Bioethanol production from rice straw residues. Braz. J. Microbiol. 44 (1), 225–234.

Binod, P., Sindhu, R., Singhania, R.R., Vikram, S., Devi, L., Nagalakshmi, S., Kurien, N., Sukumaran, R.K., Pandey, A., 2010. Bioethanol production from rice straw: an overview. Bioresour. Technol. 101 (13), 4767–4774.

Bischoff, K.P., Gravois, K.A., Reagan, T.E., et al., 2008. Registration of "L79-1002" sugarcane. J. Plant Reg. 2, 211–217.

Boschiero, B.N., de Castro, S.G.Q., da Rocha, A.E.Q., Franco, H.C.J., Carvalho, J.L.N., Soriano, H.L., Kolln, O.T., 2019. Biomass production and nutrient removal of energy.cane genotypes in northeastern Brazil. Crop Sci. 59 (1), 379–391. https://doi.org/10.2135/cropsci2018.07.0458.

Brandenburg, J., Poppele, I., Blomqvist, J., Puke, M., Pickova, J., Sandgren, M., Rapoport, A., Vedernikovs, N., Passoth, V., 2018. Bioethanol and lipid production from the enzymatic hydrolysate of wheat straw after furfural extraction. Appl. Microbiol. Biotechnol. 102, 6269–6277. https://doi.org/10.1007/s00253-018-9081-7.

Burner, D.M., Legendre, B.L., 2000. 2000. Phenotypic variation of biomass yield components in F1 hybrids of elite sugarcane crossed with Saccharum officinarum and S. spontaneum. J. Am. Soc. Sugar Cane Technol. 20, 81–87.

Center for strategic studies and management - CGEE, 2017. Second-generation Sugarcane Bioenergy & Biochemicals: Advanced Low-carbon Fuels for Transport and Industry. CGEE, Brasília, DF, p. 124.

Chaabane, A., Ramudhin, A., Paquet, M., 2012. Design of sustainable supply chains under the emission trading scheme. Int. J. Prod. Econ. 135 (1), 37–49.

Chisti, Y., 2013. Constraints to commercialization of algal fuels. J. Biotechnol. 167, 201–214.

CPCB, 2000. Manual on Hospital Waste Management. Central Pollution Control Board, Delhi.

da Silva, M.J., Carneiro, P.C.S., Carneiro, J.E.S., Damasceno, C.M.B., Parrella, N.N.L.D., Pastina, M.M., et al., 2018. Evaluation of the potential of lines and hybrids of biomass sorghum. Ind. Crop. Prod. 125, 379–385.

Datta, A., Wani, S.P., Patil, M.D., Tilak, A.S., 2016. Field scale evaluation of seasonal wastewater treatment efficiencies of free surface-constructed wetlands in ICRISAT, India. Curr. Sci. 110 (9), 1756–1763.

Datta, A., Thomas, K.M., Tiwari, A., Wani, S.P., 2019. The diatoms: from eutrophic indicators to mitigators. In: Gupta, S.K., Bux, F. (Eds.), Application of Microalgae in Wastewater Treatment. Vol. 1. Domestic and Industrial Wastewater Treatment. Springer Nature Switzerland, pp. 19–40.

de Abreu, L.G.F., Grassi, M.C.B., de Carvalho, L.M., da Silva, J.J.B., Oliveira, J.V.C., Bressiani, J.A., Pereira, G.A.G., 2020. Energy cane vs sugarcane: watching the race in plant development. Ind. Crop. Prod. 156, 112868.

Delrue, F., Álvarez-Díaz, P., Fon-Sing, S., Fleury, G., Sassi, J.-F., 2016. The environmental biorefinery: using microalgae to remediate wastewater, a win-win paradigm. Energies 9, 132–151.

Dias, M.O.S., Cunha, M.P., Jesus, C.D.F., Rocha, G.J.M., Pradella, J.G.C., Rossell, C.E.V., Maciel Filho, R., Bonomi, A., 2011. Second generation ethanol in Brazil: can it compete with electricity production? Bioresour. Technol. 102, 8964–8971. https://doi.org/10.1016/j.biortech.2011.06.098.

Diniz, A.L., Ferreira, S.S., ten Caten, F., Margarido, G.R.A., dos Santos, J.M., de Barbosa, G.V.S., et al., 2019. Genomic resources for energy cane breeding in the post genomics era. Comput. Struct. Biotechnol. J. 17, 1404–1414. https://doi.org/10.1016/j.csbj.2019.10.006.

DOE/EERE, 2013. Federal and State Laws and Incentives. Department of Energy, Office of Energy Efficiency & Renewable Energy. Available at: http://www.afdc.energy.gov/laws/. (Accessed 19 July 2013).

Duffy, M.D., Nanhou, V.Y., 2002. Switchgrass Production in Iowa: Economic Analysis. Special Publication for Cahriton Valley Resource Conservation District, Iowa State University Extension Publication, Iowa State University.

Engler, B., Schoenherr, S., Zhong, Z., Becker, G., 2012. Suitability of bamboo as an energy resource: analysis of bamboo combustion values dependent on the culm's age. Int. J. For. Eng. 23 (2), 114–121.

FAOSTAT, 2019. http://www.fao.org/faostat/en/#data/QC/visualize.

Giamalva, M., Clark, S., Stein, J., 1985. Conventional vs high fiber sugarcane. J. Am. Soc. Sugar Cane Technol. 4, 106–109.

Gimenez, C., Sierra, V., Rodon, J., 2012. Sustainable operations: their impact on the triple bottom line. Int. J. Prod. Econ. 140 (1), 149–159.

Gordon, J.C., 2001. Poplars: trees of the people, trees of the future. For. Chron. 77, 217–219.

Graham, R.L., Nelson, R., Sheehan, J., Perlack, R.D., Wright, L.L., 2007. Current and potential U.S. corn Stover supplies. Agron. J. 99, 1–11.

Han, M., Choi, G.-W., Kim, Y., Koo, B.-C., 2011. Bioethanol production by miscanthus as a lignocellulosic biomass: focus on high efficiency conversion to glucose and ethanol. BioResources 6 (2), 1939–1953.

Heaton, E.A., Dohleman, F.G., Long, S.P., 2008. Meeting US biofuel goals with less land: the potential of miscanthus. Glob. Chang. Biol. 14, 2000–2014. https://doi.org/10.1111/j.1365-2486.2008.01662.x.

Hileman, J.I., Ortiz, D.S., Bartis, J.T., Wong, H.M., Donohoo, P.E., Weiss, M.A., Waitz, I.A., 2009. Near-Term Feasibility of Alternative Jet Fuels. RAND Corporation.

Holfbauer, H., Knoef, H.A.M., 2005. Success stories on biomass gasification. In: Knoef, H.A.M. (Ed.), Handbook of Biomass Gasification. vol. 2005. BTG Biomass Technology Group BV, The Netherlands, pp. 115–161.

IEA, 2009. Bioenergy, 2009—A Sustainable and Reliable Energy Source. IEA, Paris, France.

IEA, 2018. Renewables 2018. IEA, Paris. https://www.iea.org/reports/renewables-2018.

Johansson, J., Liss, J., Gullberg, T., Bjorheden, R., 2006. Transport and handling of forest energy bundles-advantages and problems. Biomass Bioenergy 30, 334–341. https://doi.org/10.1016/j. biombioe.2005.07.012.

Kalinina, O., Nunn, C., Sanderson, R., Hastings, A.F.S., van der Weijde, T., Ozguven, M., Tarakanov, I., Schule, H., Trindade, L.M., Dolstra, O., et al., 2017. Extending miscanthus cultivation with novel germplasm at six contrasting sites. Front. Plant Sci. 8.

Kaushik, N., Biswas, S., 2007. Biochemical conversion of biomass – challenges and opportunities. In: Indian Engineering Congress (IEC-07), Udaipur.

Keshav, P.K., Naseeruddin, S., Rao, L.V., 2016. Improved enzymatic saccharification of steam exploded cotton stalk using alkaline extraction and fermentation of cellulosic sugars into ethanol. Bioresour. Technol. 214, 363–370.

Kim, S., Dale, B.E., 2004. Global potential bioethanol production from wasted crops and crop residues. Biomass Bioenergy 26 (4), 361–375.

Kumar, A., Sokhansanj, S., Flynn, P.C., 2006. Development of a multicriteria assessment model for ranking biomass feedstock collection and transportation systems. Appl. Biochem. Biotechnol. 129 (1–3), 71–87.

Kuttiraja, M., Sindhu, R., Varghese, P.E., Sandhya, S.V., Binod, P., Vani, S., Pandey, A., Rajeev, K.S., 2013. Bioethanol production from bamboo (Dendrocalamus sp.) process waste. Biomass Bioenergy 59, 142–150.

Lobovikov, M., Ball, L., Guardia, M., Russo, L., 2007. World Bamboo Resources: A Thematic Study Prepared in the Framework of the Global Forest Resources Assessment 2005. Food & Agriculture Organization, Rome, Italy.

Long, S.P., Karp, A., Buckeridge, M.S., Davis, S.C., Jaiswal, D., Moore, P.H., Moose, S.P., Murphy, D.J., Onwona-Agyeman, S., Vonshak, A., 2015. Feedstocks for biofuels and bioenergy. In: Souza, G.M., Victoria, R.L., Joly, C.A., Verdade, L.M. (Eds.), Bioenergy & Sustainabilty: Bridging the Gaps. SCOPE, Paris, pp. 302–346.

Mahmood, A., Honermeier, B., 2012. Chemical composition and methane yield of sorghum cultivars with contrasting row spacing. Field Crop Res. 128, 27–33.

Matsuoka, S., Kennedy, A.J., dos Santos, E.G.D., Tomazela, A.L., Rubio, L.C.S., 2014. Energy cane: its concept, development, characteristics, and prospects. Adv. Bot. 2014, 1–13. https://doi.org/10.1155/2014/597275.

McLaughlin, S.B., Kszos, L.A., 2005. Development of switchgrass (Panicum virgatum) as a bioenergy feedstock in the United States. Biomass Bioenergy 28, 515–535.

Meixell, M.J., Gargeya, V.B., 2005. Global supply chain design: a literature review and critique. Transport. Res. E-Log. 41, 531–550. https://doi.org/10.1016/j.tre.2005.06.003.

Mitchell, R.B., Vogel, K.P., Berdahl, J., Masters, R., 2010. Herbicides for establishing switchgrass in the central and northern Great Plains. Bioenergy Res. 3, 321–327.

Mitchell, R., Vogel, K.P., Uden, D.R., 2012. The Feasibility of Switchgrass for Biofuel Production. Nebraska Cooperative Fish & Wildlife Research Unit – Staff Publications 169. https://digitalcommons.unl.edu/ncfwrustaff/169.

MNRE, 2018. https://mnre.gov.in/img/documents/uploads/file_f-1597797108502.pdf.

Narra, M., James, J.P., Balasubramanian, V., 2015. Simultaneous saccharification and fermentation of delignified lignocellulosic biomass at high solid loadings by a newly isolated thermotolerant Kluyveromyces sp. for ethanol production. Bioresour. Technol. 179, 331–338.

Ohrnberger, D., 1999. The Bamboos of the World: Annotated Nomenclature and Literature of the Species and the Higher and Lower Taxa. Elsevier, Amsterdam, The Netherlands.

Otero, J., Panagiotou, G., Olsson, L., 2007. Fueling industrial biotechnology growth with bioethanol. Biofuels 108, 1–40.

Pang, S., 2016. Fuel flexible gas production: biomass, coal and bio-solid wastes. In: Oakey, J. (Ed.), Fuel Flexible Energy Generation. Woodhead Publishing, Sawston, Cambridge.

Panoutsou, C., Perakis, C., Elbersen, B., Zheliezna, T., Staritsky, I., 2017. Assessing potentials for agricultural residues. In: Panoutsou, C. (Ed.), Modeling and Optimization of Biomass Supply Chains. Academic Press, pp. 169–197, https://doi.org/10.1016/B978-0-12-812303-4.00007-0 (Chapter 7).

Parikka, M., 2004. Global biomass fuel resources. Biomass Bioenergy 27, 613–620.

Planning Commission, 2007. https://niti.gov.in/planningcommission.gov.in/docs/reports/genrep/cmtt_bio.pdf.

Prasad, S., Singh, A., Joshi, H.C., 2007. Ethanol as an alternative fuel from agricultural, industrial and urban residues. Resour. Conserv. Recycl. 50, 1–39.

Purohit, P., Dhar, S., 2015. Biofuel Roadmap for India – (November 2015). UNEP (United Nations Environment Program) DTU (Denmark Technical University) Partnership.

Purohit, P., Fischer, G., 2014. Second Generation Biofuel Potential in India: Sustainability and Cost Considerations, UNEP Risø Centre on Energy, Climate and Sustainable Development. Technical University of Denmark, Copenhagen.

Rivera-Burgos, L.A., Volenec, J.J., Ejeta, G., 2019. Biomass and bioenergy potential of Brown midrib sweet Sorghum germplasm. Front. Plant Sci. 10, 1142. https://doi.org/10.3389/fpls.2019.01142.

Rooney, W.L., Blumenthal, J., Bean, B., Mullet, J.E., 2007. Designing sorghum as a dedicated bioenergy feedstock. Biofuels Bioprod. Biorefin. 1, 147–157. https://doi.org/10.1002/bbb.15.

Roozeboom, K.L., Wang, D., McGowan, A.R., Propheter, J.L., Staggenborg, S.A., Rice, C.W., 2018. Long-term biomass and potential ethanol yields of annual and perennial biofuel crops. Agron. J. https://doi.org/10.2134/agronj2018.03.0172.

Sadiku, N.A., Oluyege, A.O., Sadiku, I.B., 2016. Analysis of the calorific and fuel value index of bamboo as a source of renewable biomass feedstock for energy generation in Nigeria. Lignocellulose 5 (1), 34–49.

Sanderson, M.A., Adler, P.R., Boateng, A.A., Casler, M.D., Sarath, G., 2007. Switchgrass as a biofuels feedstock in the USA. Can. J. Plant Sci. 86, 1315–1325.

Sandersson, J., 1999. Passing value to customers: on the power of regulation in the industrial electricity supply chain. Supply Chain Manag. 41, 199–208.

Sannigrahi, P., Ragauskas, A.J., Tuskan, G.A., 2010. Poplar as a feedstock for biofuels: a review of compositional characteristics. Biofuels Bioprod. Biorefin. 4, 209–226.

Satlewal, A., Agrawal, R., Bhagia, S., Das, P., Ragauskas, A.J., 2018. Rice straw as a feedstock for biofuels: availability, recalcitrance, and chemical properties. Biofuels Bioprod. Biorefin. 12 (1), 83–107.

Scagline-Mellor, S., Griggs, T., Skousen, J., et al., 2018. Switchgrass and giant miscanthus biomass and theoretical ethanol production from reclaimed mine lands. Bioenergy Res. 11, 562–573. https://doi.org/10.1007/s12155-018-9915-2.

Schmer, M.R., Vogel, K.P., Mitchell, R.B., Perrin, R.K., 2008. Net energy of cellulosic ethanol from switchgrass. Proc. Natl. Acad. Sci. 105, 464–469.

Scurlock, J., 2000. Bamboo: an overlooked biomass resource? Biomass Bioenergy 19, 229–244.

Shahzad, K., Sohail, M., Hamid, A., 2019. Green ethanol production from cotton stalk. IOP Conf. Ser. Earth Environ. Sci. 257. https://doi.org/10.1088/1755-1315/257/1/012025, 012025.

Sharma, B., Ingalls, R.G., Jones, C.L., Khanchi, A., 2013. Biomass supply chain design and analysis: basis, overview, modeling, challenges, and future. Renew. Sust. Energ. Rev. 24, 608–627.

Sharma, R., Wahono, J., Baral, H., 2018. Bamboo as an alternative bioenergy crop and powerful ally for land restoration in Indonesia. Sustainability 10 (12), 4367. https://doi.org/10.3390/su10124367.

Shimizu, T., Han, J., Choi, S., Kim, L., Kim, H., 2006. Fluidized-bed combustion characteristics of cedar pellets by using an alternative bed material. Energy Fuel 20, 2737–2742.

Silverstein, R.A., Chen, Y., Sharma-Shivappa, R.R., Boyette, M.D., Osborne, J., 2007. A comparison of chemical pretreatment methods for improving saccharification of cotton stalks. Bioresour. Technol. 98, 3000–3011.

Singh, S., Adak, A., Saritha, M., Sharma, S., Tiwari, R., Rana, S., Arora, A., Nain, L., 2017. Bioethanol production scenario in India: potential and policy perspective. In: Chandel, A.K., Sukumaran, R.K. (Eds.), Sustainable Biofuels Development in India. 2017. Springer International Publishing, Cham, Switzerland, pp. 21–37.

Sordi, R.A., Manechini, C., 2013. Utilization of trash: a view from the agronomic and industrial perspective. Sci. Agric. 70, 1–2. https://doi.org/10.1590/S0103-90162013000500002.

Tang, C., Li, S., Li, M., Xie, G.H., 2018. Bioethanol potential of energy sorghum grown on marginal and arable lands. Front. Plant Sci. 9, 440. https://doi.org/10.3389/fpls.2018.00440.

Tao, L., Aden, A., 2009. The economics of current and future biofuels. In Vitro Cell. Dev. Biol. 45 (3), 199–217.

Tilak, A.S., Wani, S.P., Patil, M.D., Datta, A., 2016. Evaluating wastewater treatment efficiency of two field scale subsurface flow constructed wetlands. Curr. Sci. 110 (9), 1764–1772.

Tishler, Y., Samach, A., Rogachev, I., Elbaum, R., Levy, A.A., 2015. Analysis of wheat straw biodiversity for use as a feedstock for biofuel pro- duction. BioEnergy Res. 8, 1831–1839. https://doi.org/10.1007/s12155-015-9631-0.

Townsend, T.J., Sparkes, D.L., Wilson, P., 2017. Food and bioenergy: reviewing the potential of dual-purpose wheat crops. GCB Bioenergy 9, 525–540. https://doi.org/10.1111/gcbb.12302.

Tripathi, Y.C., Khawlhring, L., 2010. Bamboo resource and its role in ecological security. Indian For. 136, 641–651.

Úbeda, B., Gálvez, J.Á., Michel, M., Bartual, A., 2017. Microalgae cultivation in urban wastewater: Coelastrum cf. pseudomicroporum as a novel carotenoid source and a potential microalgae harvesting tool. Bioresour. Technol. 228, 210–217.

Uppugundla, N., da Costa Sousa, L., Chundawat, S.P., et al., 2014. A comparative study of ethanol production using dilute acid, ionic liquid and AFEX™ pretreated corn stover. Biotechnol. Biofuels 7, 72. https://doi.org/10.1186/1754-6834-7-72.

Venuto, B., Kindiger, B., 2008. Forage and biomass feedstock production from hybrid forage sorghum and sorghum–sudangrass hybrids. Grassl. Sci. 54, 189–196.

Vermerris, W., Saballos, A., Ejeta, G., Mosier, N.S., Ladisch, M.R., Carpita, N.C., 2007. Molecular breeding to enhance ethanol production from corn and sorghum stover. Crop Sci. 47 (S3), S142–S153. https://doi.org/10.2135/cropsci2007.04.0013IPBS.

Virmond, E., Rocha, J.D., Moreira, R.F.P.M., José, H.J., 2013. Valorization of agroindustrial solid residues and residues from biofuel production chains by thermochemical conversion: a review, citing Brazil as a case study. Braz. J. Chem. Eng. 30 (2), 197–229.

Wang, M., Zhou, D., Wang, Y., Wei, S., 2016. Bioethanol production from cotton stalk: a comparative study of various pretreatments. Fuel 184, 527–532. https://doi.org/10.1016/j.fuel.2016.07.061.

Wu, J., Elliston, A., Le Gall, G., et al., 2018. Optimising conditions for bioethanol production from rice husk and rice straw: effects of pre-treatment on liquor composition and fermentation inhibitors. Biotechnol. Biofuels 11, 62. https://doi.org/10.1186/s13068-018-1062-7.

Wullschleger, S.D., Davis, E.B., Borsuk, M.E., Gunderson, C.A., Lynd, L.R., 2010. Biomass production in switchgrass across the United States: database description and determinants of yield. Agron. J. 102, 1158–1168.

Xie, G.H., 2012. Progress and direction of non-food biomass feedstock supply research and development in China. J. Chin. Agric. Univ. 17, 1–19 (in Chinses with English abstract).

Yamashita, Y., Shono, M., Sasaki, C., Nakamura, Y., 2010. Alkaline peroxide pretreatment for efficient enzymatic saccharification of bamboo. Carbohydr. Polym. 79, 914–920.

Zhang, J., Osmani, A., Awudu, I., Gonela, V., 2013. An integrated optimization model for switchgrass-based bioethanol supply chain. Appl. Energy 102 (2), 1205–1217. https://doi.org/10.1016/j.apenergy.2012.06.054.

Zhang, J., Zou, W., Li, Y., Feng, Y., Zhang, H., Wu, Z., Tu, Y., Wang, Y., Cai, X., Peng, L., 2015. Silica distinctively affects cell wall features and lig- nocellulosic saccharification with large enhancement on biomass production in rice. Plant Sci. 239, 84–91. https://doi.org/10.1016/j.plantsci.2015.07.014.

Zhao, D., Momotaz, A., LaBorde, C., Irey, M., 2020. Biomass yield and carbohydrate composition in sugarcane and energy cane grown on mineral soils. Sugar Tech 22 (4), 630–640.

CHAPTER

4

Microalgal biofuels: Challenges, status and scope

Dheeban Chakravarthi Kannan and Chaitanya Sampat Magar

The Energy and Resources Institute (TERI), Navi Mumbai, India

1 Introduction

Microalgae are one of the promising options for renewable fuel production. Microalgae are microscopic unicellular organisms (1–10 µm size) that are grown as suspensions in water. They are among the fastest growing organisms in the world and they produce lipids as they grow. They fix the abundant sunlight into fuel energy. Seawater and wastewater could be used as a water source. They can be grown in non-arable land; thus, they do not compete against food production. The oil yield of microalgae has been reported to be 10–50 times that of conventional oil crops (Chisti, 2007). Microalgae are one of the renewable fuel options that qualify by available resources to supply to the magnitude of world fuel needs.

Chisti (2007) presents the potential of microalgal biofuel in comparison with oil yield from other oil crops as below in Table 1. It illustrates how less the land requirement is for microalgae, compared to other crops.

The present annual transport fuel consumption of the world and India are 2627 mtoe (International Energy Agency (IEA), 2016; Le Fevre, 2019) and 119 mtoe (MoSPI, 2018). Individual fuel markets that microalgal biofuels are likely to cater to are that of diesel and jet fuel. The annual diesel fuel consumption of the world and India are 938 and 75–85 mtoe. The annual jet fuel consumption of the world and India are 390 mtoe (Reuters, 2020) and 8.7 mtoe (Mohanty, 2020). The transport fuel demand of India is expected to increase to 2–3 times the present demand by 2050 (IESS 2047, 2015).

Microalgae fix sunlight into biomass through the following photosynthesis reaction. Other products are formed by the subsequent reactions of metabolism.

$$6CO_2 + 6H_2O \xrightarrow{\text{Light energy}} C_6H_{12}O_6 + 6O_2$$

Carbon dioxide   Water   Glucose   Oxygen

TABLE 1   Comparison of biofuel potential from various oil feedstocks (Chisti, 2007).

Crop	Oil yield (L/ha)	Land area needed[a] (Mha)	Percent of existing US cropping area[a]
Corn	172	1540	846
Soybean	446	594	326
Canola	1190	223	122
Jatropha	1892	140	77
Coconut	2689	99	54
Oil palm	5950	45	24
Microalgae[b]	58,700	4.5	2.5

[a] For meeting 50% of all transport fuel needs of the United States.
[b] 30% oil (by weight) in biomass.

The various growth parameters that are involved in the production of microalgae are light, carbon dioxide, water, nutrients, oxygen removal, temperature, pH, salinity, and invasive contamination. Sunlight is usually the viable source of light energy. Carbon dioxide is either taken up from atmosphere (412 ppm; NASA, 2019) or supplied as dilute or concentrated gas from a commercial source. Water is a key growth component that serves both as the growth medium and a photosynthetic reactant. Nutrients refer mainly to major nutrients, nitrogen, phosphorus, and potassium which make up 6%–10%, 1%, and 1% of the dry microalgal biomass (Oswald, 1988; Larsdotter, 2006). Other nutrients such as iron, zinc, and copper are provided in minor quantities. Oxygen needs to be removed as in the case of a by-product of any forward-reverse equilibrium of any chemical reaction. In open raceway systems, this is addressed by the relatively shallow depth of open ponds and the mixing provided by the raceway paddlewheel. The temperature in the range of 20–35 °C is considered to be effective for outdoor algal cultivation. pH range of 7.5–8.5 is usually considered to be optimal for algal growth. Marine algae are usually cultivated at the representative seawater salinity of 35 g/L and salinities from 25 to 45 g/L have also been found to be effective. Some halotolerant algae grow at even higher salinities (Abubakar, 2016). After the basic photosynthesis reaction that results in the production of glucose sugar as platform molecule, the production of other cellular components such as carbohydrates, proteins and lipids through other metabolic reaction pathways is further influenced by various degrees by the aforementioned growth parameters (Juneja et al., 2013).

Microalgae are among the most efficient photosynthetic organisms on earth in utilizing incident light energy. They offer one of the best choices when it comes to harnessing the sunlight energy into useable fuel form through purpose-grown biomass (Darzins et al., 2010).

Microalgae are different from macroalgae (seaweed) (Cai et al., 2013). Macroalgae are multicellular, macroscopic marine organisms. They are also fast-growing organisms like microalgae. But they do not produce as much lipids as microalgae, thus they are not regarded as high oil-yielding crops. However, macroalgae also are considered for biofuel development as a biomass feedstock. They are easier to harvest and have been cultivated for food and nutrition applications so far. This chapter covers microalgal biofuels.

In this chapter, the various challenges in the field of microalgal biofuels in the context of the present background are discussed first, given that the field is still in development. Following this, the status of the microalgal biofuel field in addressing some of the challenges is covered. This order of discussion was adopted because there are a number of unresolved challenges in the field and the status is still a partial progress towards the challenges. Then the realistic scope of the microalgal biofuels in the near future and the way forward are discussed. The chapter aims to cover the prospects of the microalgal biofuels in the present term as detailed background and basic research of future prospects have been covered elsewhere.

2 Challenges

2.1 Outdoor algal growth systems

Though high yields have been obtained in laboratory conditions, translating those high yields to outdoor productivity to has been a major challenge. Improving light conversion efficiency has been reported as a major step needed in this regard (Ooms et al., 2016). Typical outdoor sunlight intensity is about $1000 \, W/m^2$, out of which only 45%–48% is photosynthetically active radiation (PAR) (1500–2000 μmol photons (PAR)/m^2/s) (Zhu et al., 2008). The theoretical maximum photosynthetic efficiency of PAR by microalgae is 27% (11%–12% of total sunlight) (Hay and Porter, 2006). But in reality, only 2%–5% of average photosynthetic efficiency (PAR) has been observed (Moheimani and Borowitzka, 2011). The saturation light intensity of microalgae is 200–400 μmol photons (PAR)/m^2/s (Chisti, 2007; Fabregas et al., 2004). Photosynthetic pigments cannot absorb all the light they receive (Parlevliet and Moheimani, 2014). There are 4–5 times excess light in outdoor sunlight and this often causes photoinhibition and lowers the growth rate (usually by half). The excess light is expelled out in the form of heat or fluorescence. The light and dark respirations further reduce the photosynthetic efficiency. Other factors that lower light utilization efficiency are a self-shading effect of algae in culture systems and seasonal variations in sunlight intensity that cause light limitation. Short-term bursts of high productivities ($50 \, g/m^2$/day) have been observed for natural blooms as well as in open pond systems, but extended productivities in reality are much lower (Field et al., 1998; Lee, 2001). Any development in outdoor algal growth system that can utilize the sunlight more effectively has the potential to improve the areal productivity by about two times.

Outdoor algal growth systems can either be open or closed (photobioreactors). The open raceway ponds (Fig. 1) are one of the most common outdoor systems employed presently for algal production (Chisti, 2007). They are much lower in cost than the closed photobioreactors. They are deemed to be the appropriate choice for algal biofuel production as the cost of closed photobioreactors is too high (Lundquist et al., 2010; Beal et al., 2015). The productivity of the open systems is typically in the range 12–22 g/m^2/day with 25–35 g/m^2/day in select instances (Lundquist et al., 2010; Quinn and Davis, 2015; Beal et al., 2015; Brennan and Owende, 2010). Indoor laboratory yields are typically in the range 0.1–0.4 g/L/day and this should translate to 30–60 g/m^2/day outdoor productivity if proportionally extrapolated. Hence the presently reported outdoor productivities are much lower than their inherent potential.

FIG. 1   Standard algal raceway pond set up at Seambiotic Ltd., Israel (Kloosterman, 2013).

Open systems are prone to contamination by invasive wild algal species and attacks by predators such as rotifers, paramecium, amoeba, worms and mosquitoes (Hannon et al., 2010). This tends to affect the consistency of the daily productivity and contributes to the lowering of long-term yields compared to the rated productivity. The open outdoor algal growth systems are prone to temperature variations due to diurnal and seasonal changes (Lundquist et al., 2010; Brennan and Owende, 2010). They are prone to water loss due to evaporation. In cases where carbon dioxide is supplied, it is utilized less efficiently due to loss by escape into atmosphere.

Closed system photobioreactor may be an option in limited cases such as treatment of wastewater that entails algal remediation of appreciable content of sugar or other organic constituents and production of high-value food and health supplements. Biofuel is likely to be a by-product in such cases. Photobioreactors are much more expensive than open systems (Lundquist et al., 2010; Brennan and Owende, 2010). They offer good control over growth parameters. Algae (2021) can be grown as pure species with high lipid content. They can be grown at higher concentrations. Volumetric productivities are higher. Gas exchange mechanisms need to be provided. Their cost of maintenance is high (Brennan and Owende, 2010). They require temperature regulation. Tubular photobioreactors (stacked-up, horizontal or vertical) are the largest and the most extensively employed among the photobioreactors (Fig. 2) (Brennan and Owende, 2010; Chisti, 2007). Vertical flat panel photobioreactors are reported to have lower oxygen accumulation and higher photosynthetic efficiency compared to the tubular photobioreactors (Fig. 3) (Brennan and Owende, 2010). Column photobioreactors are reported to have high mass transfer rates and less shear stress, but their

FIG. 2   A horizontal tubular algal photobioeactor set up at the Wageningen University, Netherlands (Bosma, 2014).

FIG. 3   A vertical flat panel photobioreactor set up at the Arizona Center for Algae Technology and Innovation, Arizona State University, USA (Energy.gov, 2021).

illumination areas are smaller and they need more sophisticated construction (Brennan and Owende, 2010). Column photobioreactors include bubble column photobioreactor (air bubbled through the liquid column) and airlift photobioreactor (liquid circulation induced by airflow in a demarcated draft channel) (Fig. 4) (Carvalho et al., 2006; Sánchez Mirón et al., 2000). Flexible plastic bag type systems (e.g., polyethylene) have also been employed as photobioreactors (Fig. 5) (Huang et al., 2017; Zhu et al., 2017).

Volumetric productivities of photobioreactors have been reported in the range 0.4–3.8 g/L/day (minimum photobioreactor size: 350 L) which corresponds to areal productivities of 23–38 g/m^2/day (Brennan and Owende, 2010; De Vree et al., 2015; Tang et al., 2012). It can be seen that the overall areal productivities of closed photobioreactors are not

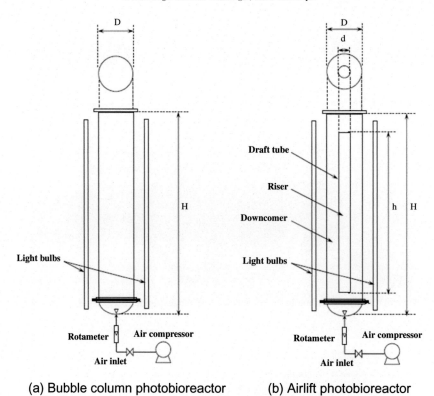

(a) Bubble column photobioreactor   (b) Airlift photobioreactor

FIG. 4   Schematic representation of column photobioreactors (A) bubble column reactor and (B) airlift reactor (Monkonsit et al., 2011).

FIG. 5   A representative plastic bag photobioreactor (Huang et al., 2017).

appreciably higher than that of the open systems. This is partly because the growth elements of closed photobioreactors (tubes, panels, columns) have interstitial vacant spaces where sunlight is wasted or ineffectively utilized. Photobioreactors facilitate a lower harvesting cost than open systems because of the higher algal density (Chisti, 2007). Closed systems are sometimes employed to develop inoculum or maintain inoculum in good health with appropriate provision of light and nutrients. The inoculum is then transferred to the open systems where actual production occurs (Narala et al., 2016; Stammbach et al., 2010).

2.2 Water

Water is a huge requirement as resource for microalgal production. Vast quantities of water are required for production of microalgae as it is the growth medium. Freshwater is a precious resource and would not be a viable water source in most parts of the world. Free abundant seawater is the most viable source of water for microalgal biofuel production. It is available in plenty around the coastlines of many countries and is an easily accessible water source for the cultivation of potential halophilic microalgal strains. Seawater also contains many trace metal nutrients which reduce the need for special supplementation of some metal ions, making it furthermore cost-effective (Jung et al., 2015).

It remains to be seen how far seawater can be brought inland. Areas with relatively flat terrains may have better range and ease of transportation of seawater than high terrain areas. Areas such as the deserts in the Middle East, Sonora Desert in the United States, Kutch region in Gujarat, India, select stretches of the Australian coast, stretches of west African Sahara Desert and far south Tamilnadu, India may facilitate bringing in of seawater to a considerable distance inside. In other regions, seawater drawing may have to be limited to about 5 km inland from the seashore. Off-shore cultivation of microalgae in the sea is another idea that has been mooted around, but the viability of this option is yet to be properly ascertained. The resource potential is enormous for off-shore cultivation, but infrastructure, rough weather, and environmental sustainability pose significant challenges that need to be addressed.

The open outdoor algal systems are associated with heavy water loss due to evaporation. Typical evaporation water loss is about 0.5 cm/day which translates to a water loss of about 50,000 L/ha/day in an outdoor open algal system. In the case of marine algal growth systems which typically operate at a salinity of 35 g/L, evaporation water loss relates to a salinity increase of 0.6 g/L per day. It remains to be seen if loss of water due to evaporation could be compensated by any form of make-up water addition or if the growth medium water should be periodically recycled with fresh seawater. In the latter case, it must be made sure that the spent media seawater discharged back into the sea meets the appropriate discharge standards. Microalgae must be completely removed and phosphorus, nitrogen and minor nutrients must either be brought to natural seawater concentration levels or undetectable levels. Then growth and lipid productivities must be made sure they do not drop to unviable levels at the low nutrient levels at the end before water is discharged back into sea. This shows that though microalgae are touted to be a feedstock that is not water-intensive for its cultivation, they could become heavily water-intensive if this challenge of evaporation water loss is not addressed for the marine systems.

2.2.1 *Wastewater as water source*

Wastewater is also a viable source of water for microalgal growth at appropriate conditions (Dalrymple et al., 2013). Typical municipal wastewater and various types of industrial wastewaters (a variety of organic pollutants from industries such as textile, tannery, distillery, paper and pulp industry; various inorganic pollutants) are wastewaters that could be treated by algae. Wastewaters may not only act as nutrient source for microalgae, but microalgae may provide a treatment option for wastewaters.

Concentrations of the major nutrients, nitrogen, phosphorus, and potassium in municipal wastewater are 20–35 mg/L (Li et al., 2017), 5–20 mg/L (Lenntech, 2021), and 10–30 mg/L (Arienzo et al., 2008), respectively. Biological Oxygen Demand (BOD) of municipal wastewater is about 300 mg/L (BMS, 2018). These concentrations are similar to that of typical growth media compositions defined for algal culturing. This shows that municipal wastewater is a good source of nutrients for microalgal growth.

For the treatment of wastewater such as municipal wastewater that involves removal of inorganic nutrients such as nitrogen, large areas of land would be required.

Consider a town like Indore in Madhya Pradesh India (population: 1,990,000 (2011 census)).

Wastewater generation of Indore: 450 million liters/day (Statista, 2020).

Total N nitrogen in domestic wastewater: 20–35 mg/L (Li et al., 2017).

Let us assume an average of 27.5 mg/L in the wastewater.

Total N that needs to be treated: 12,375 kg/day.

Let us assume nitrogen content in algae that would be grown in wastewater to be 10%.

Biomass that needs to be generated to treat wastewater: 123,750 kg/day.

Let us assume biomass productivity of 20 g/m^2/day.

Thus, the area of land required to produce the aforementioned biomass is 619 ha.

Assuming 15% lipid content in dry algae grown in wastewater, this relates to a production of 3063 tons of lipids/year for biofuel production.

A similar exercise for a city like Delhi would need 6049 ha of land.

In cases where such land is available near cities/towns/villages for municipal wastewater treatment and where land is available near industrial complexes that predominantly produce inorganic constituent-based effluent, algal wastewater treatment is a distinct possibility. Treatment of organic constituent-based wastewater that requires heterotrophic growth of algae may not have such a scale of land requirement. Feasibility of treatment of organic constituent-based wastewater treatment that requires mixotrophic growth of algae would depend on the amount of sunlight needed for the mixotrophic growth and that would dictate the extent of land requirement. For remediation of organic effluents, algal strains that are suitably acclimatized to assimilate the relevant organic compounds would need to be developed and they are often required to be isolated at the local discharge environments.

For microalgal treatment of municipal wastewater, the microalgal treatment step could either be introduced at the tertiary step where only inorganic nutrients are left in the wastewater to be treated, or, along with the secondary treatment step where it could be combined with the organic content removal. The algal wastewater treatment pilot units set up so far have not been able to demonstrate effective heterotrophic/mixotrophic assimilation of organic matter along with inorganic constituents for algal biomass production. If the latter scenario is realized, algal treatment of municipal wastewater could be achieved at a much smaller area than the area-scale stated above, since inorganic nutrients would be assimilated into algal biomass through the waste organic content instead of relying only on sunlight.

Demonstration of viable large-scale microalgal wastewater treatment is required and considerable R&D needs to be undertaken (Christenson and Sims, 2011). Value addition streams for the viability of biofuel production from microalgal wastewater treatment must be developed. Food and feed are unlikely to be value added products from microalgal wastewater treatment. Fuel production is likely to be a by-product and wastewater treatment may have to be the main purpose. Otherwise, other routes of value chain development must be pursued.

2.3 Land

Fixing sunlight as fuel energy means land is a resource that is required in vast stretches. Such vast requirement of land as a resource has always posed the question of whether it would compromise the land used for agriculture and thus interfere with food production. Microalgae can be cultivated in non-arable land. Wastelands, marginal lands and desert regions located near sea coast would be the suitable locations for marine algal production. Similar lands located near towns/cities and industries would be suitable locations for microalgal treatment of municipal wastewater and domestic wastewater. If microalgae are considered as a source of future fuel, it would need millions of acres of land under algae cultivation to cater to the magnitude of the fuel requirement of the countries around the world (Hannon et al., 2010). It would be challenging to find such vast stretches of land that is not under agricultural production. Identification of the kind of aforementioned non-arable land would need detailed area-specific survey. Such identified non-arable land must have adequate sunlight exposure and suitable climatic conditions throughout the year for the desired algal production.

Following are representative regions around the world that have suitable land areas near the sea coast: Sonora Desert in the United States; Kutch region in Gujarat, India; desert regions in the Middle East (Pakistan, Iran, Iraq, Kuwait, Saudi Arabia and other Gulf countries, Israel, Syria, Egypt, Sudan); coastal Sudan; Somalia; stretches of west African Sahara Desert, southern Africa (western South Africa, Namibia, western coast of Angola); stretches of the Australian coast (Western Australia, South Australia, Northern Territories, Queensland); Baja California, Mexico; south Tamilnadu, India; south-eastern Spain. In other regions, select suitable stretches of land may have to be found within about 5 km inland from the seashore.

2.4 Carbon dioxide

Carbon dioxide is a main rate-limiting factor in outdoor algal production. The concentration of carbon dioxide is low in the atmosphere (412 ppm) (NASA, 2019). Compared to other growth resources as reflected by the photosynthesis stoichiometric equation, carbon dioxide is the lowest in outdoor availability. Supply of carbon dioxide from emission sources is rated to be a key requirement for large-scale algal production. These emission sources could be power plants, cement industries and streel industries. Long distance piping transport of carbon dioxide from carbon dioxide reservoir has been reported to be a viable option (1.5–3 EUR/ton of CO_2/100 km for 180–750 km range) (ZEP, 2011; Gao et al., 2011; Mallon et al., 2013) and may need to be adopted for applications such as outdoor algal production if carbon dioxide is available from a distant location. Any provision of increased surface area in the growth system to enhance mass transfer of carbon dioxide from the atmosphere to growth system would also help. Carbon dioxide supplied to outdoor growth systems is utilized at low efficiency due to loss by escape into the atmosphere. Carbon dioxide in gaseous form has a limited and low solubility in water.

2.5 Low-cost nutrients

Commercial low-cost fertilizers must be adopted as nutrients effectively for large-scale microalgal biofuel production to be viable. Urea is the most widely employed nitrogen fertilizer in agriculture. Single superphosphate is a widely employed phosphorus fertilizer. Diammonium phosphate (DAP) is another popular commercial fertilizer that is a source of both nitrogen and phosphorus. Potash is the most commonly used potassium fertilizer. Various forms of mixed N:P:K fertilizers with varying ratios of N, P, and K have also been employed in agriculture. Some of the commercial fertilizers have fillers in them that would not be suitable for mixing in water. Adopted commercial fertilizers must be soluble in water and result in a fairly clear solution. Some of the phosphorus fertilizers such as single superphosphate have an appreciable content of water-insoluble phosphorus and this needs to be addressed. Though select instances of low-cost fertilizers have been reported for outdoor algal growth, extensive adoption of low-cost fertilizers and good algal productivity and lipid productivity from it need to be demonstrated.

2.6 Enhanced lipid production

Lipid content of microalgae typically tends to be 20%–25%, though higher contents up to 65% have been reported (Chisti, 2007). However, when lipid content tend to be high, algae are not in active growth mode. Lipid production pathway in algae is an energy storage mechanism (Vitova et al., 2015). Algae tend to shift to active lipid accumulation mode when there is a provision of light energy but growth in the form of its biomass platform expansion is hindered. Nutrient depletion (Zhu et al., 2016), high pH (Gardner et al., 2011) and increased carbon dioxide supply (Eloka-Eboka and Inambao, 2017) have been reported as factors that favor lipid production pathways (Zhu et al., 2016). Any scenario that limits nutrient supply and favors carbohydrate synthesis tend to favor fatty acid production pathways. The growth of algae is associated with the expansion of biomass protein platform that is based on nitrogen and other nutrients and cell division. Nutrient depletion and increased carbon dioxide supply are associated with the former while high pH is associated with the hindrance of cell division of the latter, as stress factors for lipid accumulation. These have been well-studied and widely reported in laboratory studies (Valenzuela et al., 2012, 2013; Mus et al., 2013; Lohman et al., 2015; Gardner et al., 2011). Thus, when algae are in such high lipid accumulation mode, the growth of the algae is hindered and this results in lower biomass productivity and lower lipid productivity overall. The 1998 US Department of Energy Aquatic Special Program report remarked that cultivation conditions that focus on high biomass productivity and lipid content in range of 20%–30% may be more beneficial as opposed to focusing on considerably higher lipid content (John et al., 1998). An ideal method would result in active growth phase initially followed by a short and rapid lipid enhancement-mode at the end (Qiao and Wang, 2009). Though such lipid enhancement methods have been reported at lab-scale (Xia et al., 2013; Yun et al., 2018; Nagappan et al., 2019; Aziz et al., 2020; Ho et al., 2014), the instances of such lipid enhancement in outdoor algal growth systems are limited (Shokravi et al., 2020). Growth parameter optimization and growth-lipid accumulation stage lipid enhancement methods tested successfully at lab-scale need to be demonstrated in open outdoor systems at large scale consistently and extensively.

2.7 Algal strain selection

Any algal strain adopted for open outdoor cultivation must be sturdy enough to thrive in the local conditions. High growth rate and high lipid content alone are not enough. Season changes and its effect on the algal culture sustenance must be taken into account. Mixed culture consortia (individual strains in a consortium may grow well at various points during the annual season) may prove to be effective and may need to be adopted for year-long balanced productivity. Screening and selecting optimal local strains may be effective for a given region. Any shortlisted foreign strain may need to be acclimatized to the local conditions and checked if it is able to sustain and thrive in the local conditions. Bacterial symbiosis with microalgae has been reported to be beneficial (Yao et al., 2019; Cole, 1982; Buchan et al., 2014; Croft et al., 2005; Amin et al., 2009) and could be explored for durable and better sustaining outdoor microalgal growth. Another key consideration relating to algal strain selection is that many of the challenges that are faced at the various processing stages need to be addressed at the strain selection stage itself. Strains that show better harvesting features may need to be identified at the algal strain screening stage. Strains that show better invasion resistance may need to be preferred over strains with high growth rates. Strains that facilitate easier lipid recovery due to their cell structure may need to be preferred.

2.8 Algal strain improvement through genetic studies

Direct genetic modifications or omics studies could help to improve the algal culture for good yield of biomass and their co-products. A risk is that the stability of such modifications may not be always long-lasting. Modified cultures could revert back to their original form or change to a less effective form. Many of the algal growth and the cell biosynthesis challenges have been attempted to be addressed through genetic tools. These include improving the light assimilation efficiency of microalgae, favoring lipid biosynthesis pathways and favoring value addition metabolite synthesis pathways. Genetically modified algae may face hurdles to be cultivated in open outdoor systems since many countries have a legal framework that discourages it. Random mutagenesis that does not involve genetic modification of organisms has been used to good effect to improve crop traits in conventional agricultural crops. Many of the growth physiology and metabolic pathway challenges may be attempted to be addressed through random mutagenesis studies in microalgae.

2.9 Harvest of microalgae

Harvesting of microalgae has been a decisive barrier in the field of microalgal biofuels. The stable suspension of unicellular microalgae that enables their fast growth and superior photosynthetic fixing of sunlight proves to be a hindrance when it comes to their removal from water for further processing. Microalgae remain a stable suspension in water due to the surface negative charge of the microalgal cells that makes the cells repel each other. Algal concentration in its growth suspension is 0.05%–0.10% (dry mass basis). The final harvested wet algal paste that is desired for further processing may contain 10%–20% solid content. This means 100–200 tons of water needs to be removed for each ton of wet algal biomass paste produced and this is highly energy intensive and presents a huge challenge for economic viability (Darzins et al., 2010).

Flocculation, filtration, centrifuging, and dissolved air floatation have been evaluated as the dewatering harvest methods, but they have largely proved to be unviable. Flocculation is usually achieved by the addition of chemical flocculants (e.g., alum). But this poses challenges in downstream processing because of the high solids ash content relating to the flocculant chemical. Value addition prospects are compromised because deoiled algae cannot be utilized as meal for food/animal feed purpose. Biodegradable flocculants (e.g., chitosan) would resolve this challenge, but they are expensive presently. The micro size of microalgae means that filtering algae is challenging because of their sheer small size and they clog the pores of filters quickly and form a filtration cake that presents a significant barrier even at slim thickness (Drexler and Yeh, 2014; Kumar et al., 2004; Qu et al., 2012; Singh and Patidar, 2018; Show and Lee, 2014; Huang et al., 2012; Japar et al., 2017). Traditional filtration methods have been found to pose laborious operational challenges as a result. Centrifuging microalgae from such dilute solids content to wet algal paste solids content is prohibitively energy-intensive. Dissolved air floatation (DAF) entails rigorous pumping of air into the liquid algal growth medium to form fine bubbles on which algal cells float and separate the biomass from liquid. Cationic chemical flocculants have also been used along with DAF to flocculate algae (Uduman et al., 2010; Liu et al., 1999). Utilizing the natural self-aggregating and settling features of select algae has been reported to be a method that has potential for viable microalgal harvesting in future (Lundquist et al., 2010).

2.10 Lipid recovery

Typical vegetable oil is interspersed copiously inside the oilseed kernel. However, in microalgae, lipids are well-encased inside sturdy cell walls. Harvested algae contain significant content of water. A non-polar organic solvent such as hexane is considered to be a representative industrial solvent that is well-suited to extract lipids (Lundquist et al., 2010; Kannan and Pattarkine, 2014). But the presence of water in microalgae presents a significant barrier for such a non-polar organic solvent to extract lipids. Thus, microalgae are typically required to be dried prior to lipid extraction. Drying of algae is resource-intensive in terms of land and energy. Hexane extraction of dry algae is considered to be the most economical option presently to recover lipids from algae (Lundquist et al., 2010; Kannan and Pattarkine, 2014). Sun drying of algae is considered to be the cheapest method to dry algae for biofuel production purpose (Lundquist et al., 2010). But sun drying of algae requires vast tracts of land. The land area required to sun dry algae is reported to be about 11% of the algal production area. Sun drying has also been reported to pose a risk of lipid loss (Patel and Kannan, 2021; Brennan and Owende, 2010). Drying using equipment such as spray dryer and rotary dryer is highly energy intensive (Show et al., 2015; Patel and Kannan, 2021). Thus, there is a need to recover lipids from wet microalgae without drying them.

The presence of cell wall presents an appreciable barrier for solvents to extract lipids. Pretreatment methods that disrupt the cell wall are required to facilitate extraction of lipids from algae (Kannan and Pattarkine, 2014).

If recovery of lipid from algae is changed from a solvent extraction route to hydrothermal liquefaction (HTL) route, the routes for development of the respective value-addition co-products must also be changed accordingly. For example, nitrogen may still be recovered but it may not be in the protein or amino acid form that would have been recovered in the solvent extraction route. Since microalgae are not seen as viable feedstock that can be pursued only for biofuel generation in the near future, development of suitable value chains is necessary for any method of biofuel development from them.

Microalgae may contain polar lipids as high as 30% in the total lipids (Patel and Kannan, 2021). Polar lipids are present inherently in microalgae because they are part of the cell membrane and offer appreciable value for lipid recovery. Any method of lipid recovery and drop-in fuel conversion such as solvent extraction and transesterification would need to evaluate the effective utilization of fatty acid component of the polar lipids for fuel conversion.

2.11 Value-addition biocommodities

Biofuels by themselves are unlikely to make the overall microalgal biofuel production economically viable, at least in the foreseeable future. Value addition from rest of the algal biomass after oil recovery is necessary. Useful value-addition co-products must be developed as biocommodities. The deoiled algae after extraction of lipids is similar to that soybean deoiled meal (post extraction of soybean oil) and should be able to find value as animal feed as good as that of soy meal. Deoiled algae contain appreciable content of proteins (30%–45%) and carbohydrates (30%–45%) and may contain other valuable pigments and nutrients that are not found in other oil crop-based meals. Deoiled algae could be used as various forms of animal feed—cattle feed, aquafeed (fish, shrimps), poultry feed. Aquafeed may yield even more value than cattle feed. The market size of animal feed is similar to that of transport fuel. It presently has a considerable scope for supplies from new sources given the demand (Feednavigator, 2012). Animal feed market is projected to have appreciable growth in future. Market price of animal feed is in the range $350–1000/ton (Agriwatch, 2021; Reus, 2020; EurofishMagazine, 2020). This presents a scope for competitive microalgal fuel price.

Apart from lipids for biodiesel, microalgae contain numerous cellular components that present good market value potential. These include proteins, amino acids, polysaccharides, a variety of color pigments and omega fatty acids (Mirón et al., 2003). Glycerol by-product from biodiesel conversion step has been reported to be a valuable platform chemical with potential for development of many novel value-addition commodities (Luo et al., 2016). Value addition commodity development is possible through other biomass conversion processes such as pyrolysis and hydrothermal liquefaction (Djandja et al., 2020). Material science-based development of novel products from biomass feedstocks is a promising ever-growing R&D area. The potential of value addition commodities from microalgae is still mostly at the concept level and lab-scale testing level. This needs extensive testing and demonstration in reasonably sized (1 ha) outdoor algal production and processing systems for convincing techno-economic viability.

2.12 Economic analysis

100–400 ha is referred to as the scale of land area need for economically viable algal production (Lundquist et al., 2010; Beal et al., 2015). The size of individual open raceway ponds may need to be of 1–4 ha size (Lundquist et al., 2010; Beal et al., 2015). No pilot or commercial algal production facility of this size exists presently. Hence the algal fuel production costs reported in the literature are estimated values based on models and utilize data where they are available and assumptions and projections otherwise. The present algal fuel production cost has normally been estimated in the range US$ 2–4/L (Lundquist et al., 2010; Beal et al., 2015; Davis et al., 2011, 2014) with some models reporting even higher costs US$ 4–6/L (Davis et al., 2011, 2012). Land and construction cost of the algal raceway pond systems have been reported to be in the range 23%–43% of the total capital cost, by far the biggest component in the capital cost (Lundquist et al., 2010; Beal et al., 2015). Construction cost of drying beds considered for drying of algae before lipid extraction relates to 10% of the capital cost (Lundquist et al., 2010). It requires 11% of land of the total algal production facility (Lundquist et al., 2010), thus it has a direct bearing on 11% of the final fuel cost. Manpower has been reported to be the highest operating cost component (34%–40%) (Lundquist et al., 2010; Beal et al., 2015). This suggests that regions with lower living costs, thus lower wage rates may be able to make a difference in the final fuel cost. Electricity is another notable operating cost component (17%–25%) (Lundquist et al., 2010; Beal et al., 2015). This indicates that any improvement in energy efficiency of the process operations would have an appreciable effect on the economics of the overall process. Algal culture mixing energy has been reported to be the key energy input cost, amounting to 49% of the total energy requirement at an algal facility (Lundquist et al., 2010). Lipid extraction of algae represented by the conventional hexane extraction of dry algae appears to require an even larger scale than the algal growth system to be economically viable. It has been reported that algae from four clusters of 100–400 ha size production sites would have to be brought to a single extraction plant for economically viable extraction (Lundquist et al., 2010). The cost of hauling biomass to centralized extraction plants located away from the production sites has been estimated to be 11% of the operating cost. Thus, a new extraction technology that is conducive for on-site decentralized processing of biomass may have a significant effect on the final production cost of the fuel. Value addition from co-product biocommodities is a critical intervention needed for the overall economic viability.

3 Present status

3.1 Outdoor algal growth systems

Open raceway ponds have been set up at their listed viable size of 1–2 ha (Lundquist et al., 2010; Beal et al., 2015). The raceway systems of this size require a paddlewheel mixing power of about 2 kW (Lundquist et al., 2010; Beal et al., 2015). Raceway pond system has been reported to be the most efficient open outdoor algal growth system presently. The layout of the raceway pond with channel width of 1.5–3.0 m and length of 1000–2000 m has been reported to be the most efficient in terms of the paddlewheel mixing energy (2 kW) (Lundquist et al., 2010). A surface velocity of 0.15–0.30 m/s in the growth medium of raceway pond powered by paddlewheel has been reported to be optimal (Cunha et al., 2020; Lundquist et al., 2010). This layout offers the least surface area resistance to water flow given the relatively shallow depth requirement of an open system for sunlight penetration and oxygen removal. Plastic pond liners are the most optimal raceway pond construction material presently (Cunha et al., 2020; Lundquist et al., 2010). These are typically polyethylene or polyvinyl chloride-based. They cost $ 0.5–0.8/m^2 ($ 1.0–1.2/kg) (IndiaMart, 2021; Lundquist et al., 2010). The productivity of the open systems has been reported in the range 12–22 g/m^2/day typically and productivities of 25–35 g/m^2/day have been reported in select instances (Lundquist et al., 2010; Quinn and Davis, 2015; Beal et al., 2015; Brennan and Owende, 2010).

Large-scale raceway ponds of individual size 0.8–1.25 ha (combined size of multiple individual ponds at a site: up to 5 ha) have been set up and tested extensively (Rogers et al., 2014; Craggs et al., 2012, 2014; Huntley et al., 2015; Beal et al., 2015). An open outdoor algal growth system based on systematic sunlight distribution has been reported to yield 1.5 times the areal productivity of the raceway pond standard (Kannan and Venkat, 2019). The method is aimed at segregating the direct and diffuse components of sunlight and consists of a horizontal bottom growth section similar to that of a raceway pond and vertical top panels (Fig. 6). The extended surface area is also likely to facilitate improved carbon dioxide transfer from the atmosphere to algal cell in cases where there is a dedicated carbon dioxide supply from emission sources. An outdoor open pond system based on side flow axial entry impeller for mixing instead of paddlewheel has been reported as an alternative growth system with better features (Lali et al., 2013; Sawant et al., 2018). The system was of 4500 L volume and dimensions (3.25 m length × 1 m depth × 1.5 m width).

In the case of any open outdoor system design that aims at sunlight distribution through transparent surface materials, polycarbonate and polyethylene are two representative materials that could be adopted (Kannan and Venkat, 2019). Polycarbonate is rated to be much longer lasting (7–14 years) than polyethylene (2–4 years). But polycarbonate is much more expensive than polyethylene. Polyethylene could be made as thin sheets and could be replaced and recycled. The cost associated with polyethylene turns out to be cheaper than polycarbonate overall.

Microalgal production focusing only on maximization of biomass should be pursued and demonstrated first. Maximized and sustained production of algal biomass over an extended period of time (annual) is required before attempting enhancement of lipid production. The biomass thus produced may be processed by alternate routes such as hydrothermal liquefaction and wet algal pyrolysis if lipid content is deemed to be low.

The cost of biodiesel production from soybean oil and rapeseed oil is reported to be $ 1.0–1.3/L (IRENA, 2021; Karmee et al., 2015) and $ 1/L from palm oil in Malaysia and Indonesia (IRENA, 2021; El-Gharbawy, 2017). The cost of biodiesel production from used cooking oil is about $ 0.4–0.5/L (El-Gharbawy, 2017; Karmee et al., 2015). These costs are lower than the present estimated cost of algal biofuels, but the scope of algal biofuels in terms of production scale is larger and there is scope for cost reduction with improvement in technology.

FIG. 6 A 220 m^2 (100,000 L) algal growth system based on sunlight distribution (separation of direct and diffuse components of sunlight) set up at The Energy and Resources Institute, Navi Mumbai, India.

Algal biofuels are third-generation biofuels in comparison to the other mentioned biodiesel forms which are mostly first-generation biofuels.

3.2 Water

Seawater and saline aquifer water have been utilized in a number of outdoor algal production systems (Nuwer, 2019; Awal and Christie, 2015). With regard to evaporation water loss and regular seawater replacement challenge, Huntley et al. (2015) in their hectare-scale integrated marine algal production and processing demonstration have reported that they have been able to replace seawater every 2–3 days. The active production cycle in the work entails bringing fresh seawater every 2–3 days and the spent seawater is discharged back into sea after sand filtration. They have reported that the nutrients are depleted to undetectable levels in the 2–3 days of production cycle and the marine strains (*Desmodesmus* sp., *Staurosira* sp. (diatom)) they employed settle in about half an hour leaving clear water at the top. This has proved to be a challenge in the past and it remains to be seen if this can be demonstrated extensively and consistently.

Viable provision of make-up water to balance the evaporation water loss apart from seawater replacement cycles has not yet been demonstrated.

3.2.1 Wastewater as water source

Though wastewater is an attractive resource for microalgae production, research and industrial trials on wastewater treatment-based algae production, harvest and subsequent processing for products recovery are lacking (Christenson and Sims, 2011).

Craggs et al. (2012) reported a high rate algal pond (HRAP) system-based municipal wastewater treatment plant in Christchurch, New Zealand. The system consisted of four 1.25 ha raceway ponds that spread over 5 ha. Carbon dioxide from the wastewater treatment generators was fed to the ponds through a sump. Biological Oxygen Demand (BOD) was reduced by 50%. Ammoniacal nitrogen removal was 65%. Dissolved reactive phosphorus removal was 19%. *E. coli* load reduced by two orders of magnitude. The algal/bacterial biomass productivity was about $8 \text{g/m}^2/\text{day}$. The algal biomass production yield was about 0.2 ton/ML of wastewater. Microalgae were harvested by natural settling of large flocs that were formed as a result of algae that formed colonies and combined with bacteria as bioflocculants. The 1% solid content harvested algal slurry was further concentrated to 30% solid content wet paste by centrifuging. This was then converted to bio-crude using a supercritical water reactor ($374\,°C$; 22.1 MPa) (30% conversion efficiency (dry basis)).

A US-based company, Gross-Wen Technologies uses a patented a revolving algal biofilm system (RAB) to recover nutrients from wastewater (algae.com). The RAB system uses vertically oriented conveyor belts that grow algae on their surface. The belts dip in wastewater that needs to be treated and utilizes surface-growing algae on their mats to remove the nutrients from the wastewater. They intend to sell the produced algal biomass as fertilizers, bioplastics, and biofuels.

The performance of algae-based sewage treatment plant (STP) in Mysore, Karnataka, India was covered by Mahapatra et al. (2013). The plant had 67.65 million liter water holding and treatment capacity per day. This STP plant consists of anaerobic lagoons and facultative ponds followed by maturation ponds. Their findings revealed that, at the entry of the treatment plant feed, 83% algal population observed was of Cyanophyceae (cyanobacteria). When the flow continued to facultative ponds, the Euglenophyceae population (green microalgae) was observed at majority. 97% and 91% of Euglenophyceae were observed at the beginning and at the end of facultative ponds, respectively. When the water flowed out from maturation pond, 78% of algae were Chlorophyceae (green microalgae). The results showed a reduction in total COD by 60% and reduction in by BOD by 82%. Nitrogen removal was not found to be efficient during the 14 days of residence time, but reduction in total suspended solids and carbon capture by euglenoids were found to be significant.

Algal treatment of municipal wastewater is typically proposed to be deployed at the tertiary stage of wastewater treatment (after the secondary stage of organic constituent treatment by oxidation). The tertiary stage deals with the removal of inorganic waste constituents such as nitrates, phosphates, and potassium salts. However, combining algal treatment with the secondary stage treatment of municipal wastewater has also been reported. Advantage here is that algae could utilize the carbon dioxide produced in the stage from of the oxidation of the organic waste constituents (Molazadeh et al., 2019). But this is a comparatively more complex treatment option than the tertiary stage option and is relatively less proven. Heterotrophic or mixotrophic treatment of the organic constituents by algae in the secondary stage itself has not been prominently reported in outdoor municipal wastewater treatment systems.

3.3 Carbon dioxide

Outdoor raceway pond systems with carbon dioxide provision in the form of flue gases from nearby emission sources have been set up (Cheng et al., 2015). Raceway pond size of 1191m^2 and productivity of $40.7 \text{g/m}^2/\text{day}$ has been reported (Cheng et al., 2015). A section of increased channel depth preceding the paddlewheel has been recommended as a sink in raceway pond for better mixing of carbon dioxide/flue gas since the depth is otherwise less for efficient gas mixing in water. Carbon dioxide capture efficiency of 33%–75% has been reported in outdoor raceway ponds (Cheng et al., 2015; Lundquist et al., 2010).

Any increase in surface area of open outdoor growth system is likely to improve carbon dioxide transfer from atmosphere to growth system even in cases where dedicated carbon dioxide supply from emission sources is possible.

In order to address the limited and low solubility of gaseous carbon dioxide in water, carbon dioxide can be absorbed in alkaline growth medium water by addition of alkali such as sodium hydroxide and potassium hydroxide at 0.01–1.00 mM concentrations. Thus, carbon dioxide can be supplied to algae in growth medium through carbonate-bicarbonate-gas equilibrium. The bicarbonate equilibrium facilitates steady and instantaneous release of gaseous carbon dioxide for algal cell uptake (Jones and Lu, 2003; Magar et al., 2019). This is a much more effective way of carbon dioxide utilization. There are some microalgae (*Nannochloropsis oculata*) that have been reported to uptake the higher concentrated bicarbonate ions directly (Munoz and Merrett, 1989). For other microalgae, the bicarbonate equilibrium would guarantee that carbon dioxide is always available at the least at the constant maximum water solubility concentration (Cai et al., 2018; Shim et al., 2016; Salmón et al., 2018; Rambhiya et al., 2021).

Carbon dioxide capture from large emission sources was accomplished mostly by employing organic amine-based absorbents (e.g., monoethanol amine). These are toxic chemicals and posed handling challenges. The cost of carbon capture was relatively high (Luis, 2016; Kiani et al., 2020). A company by name, Carbon Clean Solutions has developed a novel non-toxic chemical-based carbon dioxide capture as an alternative to the conventional amine-based carbon dioxide capture (Carbon Clean, 2021). They report that the cost of carbon capture has thus come down to a viable US$ 30/ton (Carbon Clean, 2021). They have set up a commercial carbon capture plant partnering with Tuticorin Alkali Chemical and Fertilizers Ltd. in Tuticorin, Tamilnadu, India in 2016.

As discussed in the background in the Challenges section, long distance piping transport of carbon dioxide looks promising in terms of viability. This should facilitate large-scale carbon dioxide supply for outdoor algal production whenever it is needed.

3.4 Low-cost nutrients

A variety of low-cost commercial fertilizers have been tested successfully for microalgal growth (Ashraf et al., 2011; El Nabris, 2012; Brito et al., 2013; Detchanamurthy et al., 2018; Castilla Casadiego et al., 2016; Sipaúba-Tavares et al., 2017; Mtaki et al., 2021). The low-cost commercial fertilizers include urea, ammonium sulfate, calcium superphosphate, diammonium phosphate, N:P:K fertilizers of various combinations, drip irrigation liquid fertilizer formulations and organic fertilizers. A variety of microalgae such as *Chlorella* sp., *Nannochloropsis* sp., *Dunaliella salina*, *Sinecosyfis* sp. (marine), *Chroomonas* sp. (marine), and *Ankistrodesmus gracilis* have been tested. Ashraf et al. (2011) tested commercial fertilizers (urea, Nutri-calcium, ammonium sulfate, Phosphorus Plus (P+), Potash-Plus (K+), Nitro-20 and diammonium phosphate (DAP)) employing microalga, *Chlorella vulgaris* in 1000 L fiber glass tanks and polyethylene bags and found that the commercial fertilizers yielded better results than the conventional media standard. Other aforementioned low-cost fertilizer studies were conducted at lab-scale 20–1000 mL batches. Castilla Casadiego et al. (2016) employed the drip irrigation liquid fertilizer formulation, Nutrifoliar effectively for culturing of saline microalgae, *D. salina*, *Sinecosyfis* sp. and *Chroomonas* sp. Detchanamurthy et al. (2018) employed organic fertilizer DRIPTM effectively for culturing of *D. salina*. Mahmood and Khudhair (2017) reported that when N:P:K 20:20:20 fertilizer was used for the culturing of *C. vulgaris*, the fertilizer resulted in 57% better growth than the conventional Chu10 medium. Commercial fertilizers such as urea, superphosphate, NPK (20:26:26), diammonium phosphate, potash and calcium nitrate have been employed satisfactorily in open raceway ponds of size, 50–1000 m^2 (Koley et al., 2019; Herrera et al., 2021; Baldev et al., 2018).

3.5 Enhancement of lipid production

Nutrient depletion (Zhu et al., 2016), high pH (Gardner et al., 2011), increased carbon dioxide supply (Eloka-Eboka and Inambao, 2017), adequate light throughout the culture volume/depth (Wen et al., 2016) and higher salinity (Takagi and Yoshida, 2006) have been reported at lab-scale to enhance lipid production in microalgae (Minhas et al., 2016). However, these are yet to be demonstrated reliably at larger-scale. Wen et al. (2016) have reported lipid enhancement in a 200 m^2 (40,000 L) open raceway pond with limited success, but they state that further field studies are required to maximize lipid enhancement at large-scale. The challenges they have reported include adequate light through the culture depth and culture competitiveness.

3.6 Microalgae-bacteria microbial consortium

In nature, photosynthetic producers and non-photosynthetic organisms are seen to have some form of association - symbiotic, opportunistic or parasitic, based on their growth demands and surrounding growth environment. In industrial treatment processes as well, microalgae are seen to be in mainly symbiotic association with many bacterial communities. This relationship is established due to the released dissolved organic carbon (DOC) or exopolysaccharides (EPS) by microalgae which attract many bacterial species and serve them as their food. In return the bacterial species utilize dead organic matter to reconstitute and remineralize the algal growth environment with sulfur, nitrogen, and phosphorus. In some cases, heterotrophic bacteria produce special components such as vitamin B and siderophores which act as growth promoter and ion binder, respectively, to offer more nutritionally rich environment for microalgae (Yao et al., 2019; Cole, 1982; Buchan et al., 2014; Croft et al., 2005; Amin et al., 2009). But the limited knowledge about this complex mechanism between algal and bacterial associations keeps this concept underutilized.

The examples presented by Watanabe et al. (2005) and Sandhya and Vijayan (2019) help to understand the microbial interactions between algae and other microorganisms. Watanabe et al. (2005) isolated and identified four species of bacteria and one species of fungus from a microbial consortium of *Chlorella sorokiniana* IAM C-212 under heterotrophic conditions. When axenic culture and consortium of *C. sorokiniana* were compared, it was found that the consortium was stable for almost 7 months. On the other hand, the growth of the axenic culture deteriorated after a few days and the chlorophyll content was seen to be declining. The study conducted by Sandhya and Vijayan (2019) on *Isochrysis galbana* MBTDCMFRI

S002 and its symbiotic association with two bacterial strains, *Alteromonas* sp. (MBTDCMFRI Mab 25) and *Labrenzia* sp. (MBTDCMFRI Mab 26) showed that algal growth increased with the growth-enhancing components released by these bacterial species. The growth promoters traced from the study includes antioxidants, siderophores and indole-3-acetic acid. Deep research to understand the potential mechanisms in these algae-bacterial associations will help to explore the potential of this biotechnological concept on commercial scale for cost effective algae production (Watanabe et al., 2005; Fuentes et al., 2016; Guo and Tong, 2014; Ramanan et al., 2016).

Various applications of algae-bacterial consortia in wastewater treatment have been reported. Weinberger et al. (2012) reported an algae and bacteria (ALBA)-based activated sludge development for paper mill wastewater treatment at lab-scale. To develop stable ALBA biocoenosis, they collected activated sludge from a nearby paper mill with total solids of 2–3 g/L and inoculated it with a selected algae suspension of 2 g/L algal concentration. The paper mill wastewater used in the study had a COD of 2400 mg/L. Crystal-clear treated water was obtained with settled sludge after 2 weeks. ALBA sludge displayed higher rate of waste degradation in comparison with conventional activated sludge. Wang et al. (2016) reviewed treatment of four industrial wastewaters (textile dye, pharmaceutical, agro-industrial and metal-containing) by microalgae-bacteria consortia and made a case for the feasibility and potential. They enlisted and gathered comprehensive data on various individual microalgal species, consortia of microalgal strains and bacterial strains and their reported efficiencies in wastewater treatment. Molazadeh et al. (2019) reported that employing bacteria and algae together in municipal wastewater treatment facilitated in provision of oxygen from algae to aerobic bacteria to digest organic matter and provision of carbon dioxide and nutrients from the bacteria-digested organic matter to algae for photosynthetic growth.

3.7 Algal strain improvement by genetic modification

Nuclear transformation methods in microalgae began to be developed from 1980s. The chlorophyte, *Chlamydomonas reinhardtii*, being haploid in the vegetative stage has been a model organism in green microalgae. *C. reinhardtii* has been exploited for the genetic basis studies in photosynthesis, ciliary pathologies and lipid production (Li et al., 2010; Li and Jonikas, 2016; Sharma et al., 2015). Polyethylene glycol and poly-L-ornithine have been employed to produce cell wall-deficient mutant of *C. reinhardtii* using yeast ARG4 locus. Then the stable variant was produced by performing a biolistic transformation to introduce the native gene for auxotrophic growth in mutant (Rochaix and Van Dillewijn, 1982; Debuchy et al., 1989; Kindle et al., 1989; Mayfield and Kindle, 1990). Nuclear genome transformation in *C. reinhardtii* was performed with more efficiency based on glass bead agitation, electroporation (droplet electroporation on microfluidic chip), *Agrobacterium tumefaciens*-mediation and nanoparticle-mediation. Kumar et al. (2020a) have given a comprehensive list of various transformation techniques and selectable markers - antibiotic features, herbicide resistance, auxotrophic selective markers—for transformant screening in many microalgal species, viz. *Chlamydomonas*, *Phaeodactylum*, *Nannochloropsis*, *Dunaliella* and *Haematococcus*. The conjugation-based vector delivery method using *E. coli* was found to be efficient for diatom species. Overall, low success rate of microalgal transformations is a persistent challenge in many microalgal species except *Chlamydomonas* due to limited exploitation of advanced transformation technologies (Karas et al., 2015; Kumar et al., 2004, 2020a; Kim et al., 2014).

For selective genetic transformation, specially-devised mechanisms are essential to edit the genome before insertion into the living cells. The zinc-finger nucleases (ZFNs), meganucleases (MNs), transcription activator-like effector nucleases (TALEN) were the first tools utilized in genome editing (Razzaq et al., 2019). But clustered regularly interspaced short palindromic repeats (CRISPR/Cas9) mediated genome editing in many organisms are now considered as an efficient method due to its easy and flexible utility (Kumar et al., 2020a).

High light exposure in mass algal production systems and self-shading effect by microalgal cells means that both photoinhibition and photo-limitation can simultaneously exist in the growth systems. Such conditions tend to reduce the overall biomass productivity of the culture system due to compromised photosynthetic efficiency. The photosynthetic efficiency of the culture can be enhanced by modifying the light harvesting complex of the cell or by enhanced light utilization, with a change in light composition and reduced non-photochemical quenching (NPQ). A random mutagenesis or RNAi mediated genetic modification of the light-harvesting antennae (chlorophyll *a* and *b*) to reduce their size in *C. reinhardtii* and *C. vulgaris* showed improved photosynthetic efficiency (Perrine et al., 2012; Shin et al., 2016a, b; Smith et al., 2017). Another method involves modification of intracellular spectral recompositioning of light (ISR) in diatom species, *Phaeodactylum tricornutum* to express green fluorescent protein. This absorbs blue spectra of the light and shifts it to green light to transmit green light which could then be absorbed by other accessory photosynthetic pigments and this enhances the overall productivity (Vinyard et al., 2014; Fu et al., 2017). This resolves the challenge of non-photochemical quenching. In dense cultures, this avoids photo-limitation by making the green light available for photosynthesis. Demonstration of these developments in large-scale applications is important and further laboratory studies are needed to achieve applicable solutions (Fu et al., 2019; Kumar et al., 2020a).

RuBiSco enzyme modification and enhancement of CO_2 assimilation process for increased biomass productivity by improving the selectivity and catalytic rate of the enzyme by genomic modification has shown limited success rate so far (Du et al., 2003; Spreitzer et al., 2005). Other target sites on RuBiSco genome to replace the targeted amino acids from functional protein chain of the enzyme could offer more potential variant of the enzyme instead of incorporation entire genome of other RuBiSco enzyme originated from different microalgae or plants (Genkov et al., 2010). Other enzymes like fructose 1.6-bisphosphatase (FBPase), fructose 1,6-bisphosphate aldolase (FBA), and sedoheptulose 1,7-bisphosphatase (SBPase) which are less abundant but actively take part in regeneration phase of Calvin cycle, have an effect on photosynthetic efficiency of microalgae. These enzymes could be altered to achieve the increased biomass productivity, though sometimes their overexpression could be detrimental for algae and reduce the biomass productivity.

With regard to lipid production, along with acetyl-CoA carboxylase (ACCase) enzyme which is the primary enzyme in lipogenesis, over expression of many other enzymes by genetic modification offers increased lipid accumulation in microalgal cells (Kumar et al., 2020a). The examples of such enzymes are malic enzyme, diacyl glycerol acyl transferase, pyruvate dehydrogenase, acetyl-CoA synthase, phosphoenolpyruvate carboxylase, NAD(H) kinase, glycerol kinase and RuBisCO activase.

Genetic modification in transcription factors that upregulate the lipid biosynthesis genes could also increase the higher lipid production in microalgae. With multiples models and predictions many of the options have been explored by scientific community to enhance the lipid production. However actual laboratory trials and stable microalgal variant generation are challenges that remain to be resolved. The novel research and advancement in gene editing tools such as CRISPR/Cas9 may allow development of robust genetically modified strains for high photosynthetic efficiency, high CO_2 fixation and high biomass productivities (Kumar et al., 2020a).

3.8 Harvesting of microalgae

Various microalgae harvesting methods are listed below (Mathur et al., 2017; Show et al., 2019). They are broadly classified as physical, chemical and biological methods of algae harvesting (Chen et al., 2011).

3.8.1 Physical methods

Screening: This is a technique that is commonly used in wastewater treatment plants and also has been introduced in microalgae harvesting. The concentrated algae culture of 7%–8% solid content can be effectively separated by the methods used in screening. Microstraining and vibrating screen are two well-known techniques used in screening. Microstrainer is a horizontal drum made up of a straining element made up of fabric, stainless steel or polyester and partially submerged into water from which algae is harvested while it rotates continuously. Due to the rotation, the big algal flocs are trapped inside the straining fabric and then separated. Similarly, vibrating screen has a sieving material placed on metallic frames and the screen vibrates continuously when algae culture is passed through. These methods have limited applications in harvesting the biomass of microalgae (Show et al., 2019). The method requires a preconcentration method to concentrate algae to the aforementioned 7%–8% solid content.

Filtration: The pore size of the filtration material required to harvest many microalgae is so minute that it cannot be achieved by conventional filtration materials. Therefore, recent studies focus on utilizing microfilters (0.2–0.8 μm) and ultrafilters (molecular weight cut-off: 10–100 kDa) for efficient algal biomass recovery. The materials utilized in constructing microfilters and ultrafilters are a combination of polyvinylidene fluoride (PVDF), *N,N*-dimethylformamide (DMF), mixed cellulose ester (MCE), polyether sulfone (PES), and polyacrylonitrile (PAN) and their proportion in construct determines the pore size of the final combined material (Rickman et al., 2012; Sun et al., 2013; Zhang et al., 2010). Other variations in filtration include the pressure filtration, vacuum filtration, deep bed filtration, crossflow filtrations. The physical microalgae coagulation techniques which use magnetic particles can also be clubbed with filtration and this is the combination of coagulation and filtration technique (Show et al., 2019). Filtration could be employed either as a preconcentration method or as the final step of harvesting depending on the conditions. The major limitation of this method is fouling of the filtration material. The flux which determines the filtration efficiency of the filter reduces within a few hours of biomass filtration. The microfilter and ultrafilters fail to sustain long duration due to fouling. The recent studies prove that it is essential for the filtration material to be hydrophilic in nature to increase the permeation flux and further advancements are needed in this direction (Sun et al., 2013; Sun et al., 2013b). A company, AlgaeVenture Systems in the US reported an algal filtration technology that is based on the capillary action seen in nature in plants for water distribution (ARPA-E, 2010). Preconcentrated algal slurry was filtered on flat laid-out sheets. The method is claimed to be less energy intensive and less expensive than the conventional methods. VITO, Belgium has developed a Membrane Algae Filtration (MAF) technology for continuous algae harvesting (VITO, 2019). An Integrated Permeate Channel (IPC) ultra-filtration backwashable membrane made of poly vinylidene fluoride (PVDF) has been reported to yield four times higher flux and less fouling compared to a benchmark membrane and result in 1/3rd of overall harvesting cost compared to a centrifuge (algae input concentration: 2 g/L) (De Baerdemaeker et al., 2013). The technology has been developed for value product applications and may be applicable as pre-concentration method for biofuel development.

Centrifuging: Centrifuging is another widely reported microalgae harvesting method in which algae sedimentation is achieved by gravitational force and centrifugal force with increasing rotational speed of the centrifuge drum (Molina Grima et al., 2003). The major drawback of this method is the high power consumption. Some design modifications have been proposed to overcome the high energy consumption. Increasing the flow of the biomass and lower biomass capture efficiency has been reported to reduce the energy consumption by 82% (Dassey and Theegala, 2013). The method is presently promising only after a pre-concentration step. Membrane microfiltration as a preconcentration step prior to centrifuging step has been reported to reduce energy consumption by up to 30%–34% (Dassey and Theegala, 2013; De Baerdemaeker et al., 2013). Dutch company, Evodos Inc. has developed an improved centrifuge technology, spiral plate technology that it claims to reduce the energy consumption of microalgal harvesting significantly (Evodos, 2020, https://www.evodos.eu/).

Ultrasound: In this method, the ultrasound chamber consisting of transducer and reflectors creates an ultrasonic field with high and low potential called bellies and nodes, respectively. In this chamber algal cells get attracted to each other and agglomerate at nodal knot of the sonic waves. These agglomerated biomass clumps settle down quickly after discontinuation of ultrasonic sound and this helps to recover algal biomass at 93% harvesting efficiency (Bosma et al., 2003).

Magnetic particles: Magnetic particles that exert electrostatic force on algal cells and absorb algal cells on their surface are used in this method. Earlier the magnetic particles used were precipitated magnetite, particles or beads either coated or modified by silica, diethyl aminoethyl (DEAE) and polyethylenimine (PEI) (Cerff et al., 2012; Prochazkova et al., 2013). Separation of algae from these particles required careful preparation of particle modifications, concentration of the particles and pH conditions and this makes the recovery of algae and the particles tedious. Recent advancements with naked ferric nanoparticles (Fe_3O_4) showed that electrostatic attraction between negatively charged algal cells and positively charged ferric nanoparticles aggregates algal cells at neutral to alkaline pH (Xu et al., 2011; Hu et al., 2013). This eliminates the coating and modifications of particles with other materials and makes the process easy. The biomass can then be quickly recovered with 95%–97% efficiency and the detached nanoparticles could be reused several times. The medium clarified after algae harvest and nanoparticles can be reused up to 5 cycles of repeated algal cultivation (Xu et al., 2011; Hu et al., 2013).

Flotation: Air is pumped into algal culture medium to form fine bubbles on which algal cells float and this separates the biomass from the liquid. This is called dispersed air floatation (DAF). Cationic chemical flocculants have also been used along with DAF to flocculate algae (Uduman et al., 2010; Liu et al., 1999). Pretreatment of the liquid with ozone prior to DAF lyse open the cells due to oxidizing effect of the ozone and increases the cell flotation (Widjaja et al., 2009).

3.8.2 Chemical methods

Chemical flocculation: Chemical flocculants like some metal salts and polyelectrolytes form effective aggregation of algal cells and allow them to settle rapidly due to gravity sedimentation. The separation of biomass after chemical flocculation is easy through filtration or centrifugation procedures (Brennan and Owende, 2010). The examples of chemical flocculants are aluminum sulfate (alum) and polyaluminum and polyferric salts. The organic synthetic polymers such as acrylamide, ethyleneimine and polyamine polymer are other options but the scale of requirement and low degradability make them inappropriate for large-scale application. Natural organic polymers such as chitosan, guar gum and cationic starch are some potential bio-based options. In a more recent and novel approach ammonia was utilized as a flocculant with 99% efficiency of biomass harvesting and it was reported to increase the fertilizer value of the harvested biomass due to ammonia being a nitrogen source (Wyatt et al., 2012; Beach et al., 2012; Vandamme et al., 2010; Zheng et al., 2012; Chen et al., 2012).

pH: pH-dependent algae flocculation is a method where a strong alkali like NaOH increases the pH of the water significantly to 9.3–11 and results in fast aggregate formation in suspended algae. The principle behind this phenomenon is: Bivalent ions in the algal culture medium such as Mg^{2+} and Ca^{2+} form precipitates of hydroxide and other salts at high pH which entrap algal cells, forming aggregates and making them settle down (Wu et al., 2012; Vandamme et al., 2012). Knuckey et al. (2006) have reported an advanced flocculation technique where alkali induced flocculation could be combined with a non-ionic polymer, Magnafloc-LT-25 for increased efficiency. Thus, the pH-based harvesting method is most likely a chemical-based flocculant method.

3.8.3 Biological methods

The biological methods of algae harvesting involves the use of biological flocculants such as bacteria, fungi and other algal species where extracellular polymeric substances (EPS) released by these agents mediate the flocculation procedure (Alam et al., 2016; Mathur et al., 2017). One of the speculated phenomena in biological interactions during algal flocculation procedures include bacteria which are positively charged with calcium ions (Ca^{2+}) on their cell membrane hold the negatively charged (COO^-, NH^{2-}, PO^{2-}) algal cells to form bioflocs. In case of fungi, the hyphae growth of fungus entraps the algal cells and forms the cellular aggregates to sediment the algal cells faster and separate from liquid. Apart from this, some genetic transformations conducted in algal cells to overexpress or silent some exogenous or endogenous genes, respectively, helps to produce secondary metabolites as well as facilitates cellular flocculation triggered by some externally added bioflocculants. The actual mechanisms have not been understood well and therefore detailed studies are required to explore the profitable benefits of bioflocculants.

The more widely used methods are filtration, centrifuging, dissolved air floatation, sedimentation, and chemical flocculation. Methods based on ultrasound, magnetic particles, electrochemical techniques, pH dependent-biomass separation and bio-flocculants are not widely employed yet.

Most harvesting methods prefer microalgae with better settling tendency to be economically viable for biofuel production. There are select microalgal species that are naturally aggregating (self-aggregating) and thus have better harvest features. Among freshwater strains, *Scenedesmus* species, *Monoraphidium* species and *Chlorella* species are naturally aggregating. Among marine strains, *Dunaliella* species, *Tetraselmis* species, *Desmodesmus* sp. and *Staurosira* sp. (diatom) are naturally aggregating (Patel and Kannan, 2021; Huntley et al., 2015).

The effectiveness of the harvesting methods would also dictate how viable and extensively deployable microalgae-based wastewater treatment would be since algae would need to be removed from the treated water before being discharged to environment.

3.9 Recovery of lipids

Mechanical pressing, bead grinding, wet milling, sonication, cavitation, enzymatic treatment, microwave irradiation, electro-mechanical pulsing, osmotic shock and acid/alkali treatment are various cell disruption pretreatment methods that have been pursued (Kannan and Pattarkine, 2014; Kapoore et al., 2018). Wet milling is a preferred pretreatment method presently for scaled-up operations. Microwave-assisted extraction is deemed to have scale-up challenges that are yet to be resolved. Other pretreatment methods have been tested successfully at small scale, but they are yet to be tested and demonstrated successfully at larger scale treating algae from large outdoor systems.

Dunaliella species are a select species of microalgae that lack of primary cell wall and consist of only the secondary cell membrane. This is likely to facilitate recovery of lipids without the need for a pretreatment step for cell disruption. *Dunaliella tertiolecta* has been tested successfully for maximum lipid recovery without any cell wall rupturing pretreatment in a wet algal lipid recovery method at lab-scale (Patel and Kannan, 2021). Such algae hold potential for lipid recovery without pretreatment at large scale outdoor production.

The kinetics of extraction of lipids out of the cells was reported to follow first order kinetics, attributed to diffusion based on concentration gradient. The solvent: biomass ratio depends on lipid content and the respective solvent-cellular interaction (strain-specific) (Halim et al., 2012). Increase in temperature has been reported to result in increase in lipid yield in many cases. But in some cases, beyond a certain temperature oxidative degradation may lower the yield (Halim et al., 2012).

Elevated temperature and pressure is one of the ways pursued to recover lipids from wet algae. It typically employs temperatures of about 100–200°C and pressures of about 50–100 atm (Patel and Kannan, 2021). Derwenskus et al. (2019) reported a temperature of 150°C and subcritical solvent pressures (100 atm) employing ethanol and ethyl acetate as solvents. It has been reported that high temperatures increase the solubility and extraction efficiency of solvent in wet biomass and the high pressure retains the solvent in liquid state at the temperatures above the atmospheric pressure boiling point (Pieber et al., 2012; Richter et al., 1996). Typically, polar solvents are employed in these conditions to extract lipids from wet algae and the aim is reduce the polarity (by lowering dielectric constants at these high temperatures) of the overall medium for effective lipid extraction.

Wet algal biomass has also been treated by lipid hydrolysis with sub-critical water at 250 °C and above 200 atm to form fatty acids, followed by supercritical in situ transesterification with ethanol at 275–325 °C and above 63 atm (Barreiro et al., 2013). Total treatment time reported has been 75–240 min. Lipid recovery efficiency was 77%–90%.

Hydrothermal liquefaction (HTL) is a thermochemical conversion technique that processes the whole wet algal biomass by applying near-critical temperature (350 °C) and high pressure (100–250 atm) (Gautam and Vinu, 2020; Djandja et al., 2020; Elliott et al., 2013). The method is still in early development stage, but it is seen as a promising option to process wet algae. It involves complex reactions such as hydrolysis, dehydration, decarboxylation, hydrogenation, and polymerization (Djandja et al., 2020). During hydrothermal liquefaction, lipids lead to fatty acids and glycerol among other products, through hydrolysis. Fatty acids are converted to long-chain hydrocarbons. Glycerol is converted to gas products, ethanol, methanol, allyl alcohol, acrolein and aldehydes. Proteins hydrolyze to amino acids and undergo decarboxylation and deamination, generating carbonic and organic acids, ammonia, and amines. These products repolymerize to aromatic compounds and long-chain hydrocarbons. Carbohydrates are converted to polar water-soluble organic compounds such as organic acids, benzenes, aldehydes, alcohols, and cyclic ketones. Benzene and aldehyde compounds then form larger hydrocarbons. The method intends to produce bio-oil with low molecular weight (Djandja et al., 2020). The method may include steps with or without catalysts. The method is reported to yield higher oil recovery (bio-oil) compared to the lipid yield of solvent extraction because part of proteins and carbohydrates also contribute to bio-oil formation (Djandja et al., 2020). It has been reported that basic reaction mechanisms need to be understood better and challenges relating to scale-up need to be addressed (Gollakota et al., 2018; Guo, 2019). Several parameters such as temperature, type of catalyst, reaction time, heating rate and biochemical composition must be optimized for better product yield (Kumar et al., 2020b; Djandja et al., 2020). The method would change the way value-addition commodities such as animal feed and feed supplements are envisioned for development from deoiled algae presently for the overall economic viability of the biofuel production process. A different elaborate value chain development is pursued under this method.

Supercritical carbon dioxide extraction (Kannan and Pattarkine, 2014) is another wet algal lipid extraction method, where pressures above 74 atm are applied above the critical temperature of carbon dioxide, 31 °C. It combines solvation contact property of liquids with high diffusivity of gases. But this is deemed to be expensive presently for a product like biofuels, owing to the high capital costs and the high operating pressures (Kannan and Pattarkine, 2014; Kapoore et al., 2018). The moisture in wet algae has been reported to reduce the effectiveness of supercritical fluid extraction of lipids (Kannan and Pattarkine, 2014; Dunford and Temelli, 1997).

Methods to recover lipids from wet algae at normal temperature and pressure conditions typically require long extraction times or result in lower extraction efficiency (Yoo et al., 2012; Lai et al., 2016; Balasubramanian et al., 2013; Silve et al., 2018). A method of lipid recovery from wet algae that involves gradual introduction of a polar solvent to wet algae followed by treatments of polar solvent: non-polar solvent mixture and non-polar solvent has been developed at lab-scale (Patel and Kannan, 2021). Solvent system was 2-propanol/hexane.

Lipid recovery as good as that of analytical lipid extraction method was obtained in 5–10 min of mixing time. The method is presently being tested at bench-scale in 100 L batch unit.

Microalgae need to be processed immediately after harvest as facilitated by wet algal lipid extraction because sun drying and any similar prolonged storage of algae in wet state before processing may lead to significant loss of lipids (Patel and Kannan, 2021).

Milking, electro-mechanical pulsing and electroporation methods that recover oil from algal cells directly into culture medium for recovery have been tested at lab-scale and promising results have been obtained (Kannan and Pattarkine, 2014; Kapoore et al., 2018). However, these methods are yet to be tested at larger scale and remain promising concepts as of now.

Pyrolytic recovery of oil from wet algae is also being attempted (pyrolysis: 400–800 °C in absence of oxygen) (Das, 2021). Pyrolysis of dried algae has already been reported (Vargas e Silva and Monteggia, 2015; Azizi et al., 2018; Pourkarimi et al., 2019). The bio-oil produced is likely to have a yield higher than that of the original lipid content of algae since part of the bio-oil is recovered from the non-lipid carbohydrate-based biomass (Das, 2021). The bio-oil could then be reformed and upgraded to drop-in fuels. Pyrolysis of deoiled algae is likely to recover an additional 10%–20% bio-oil after the recovery of lipids in a preceding solvent extraction step (Das, 2021; Maurya et al., 2016a, b).

Lipids from microalgae could be utilized as transport fuels in a variety of ways. One option is to convert the fatty acid glyceride esters of microalgae to biodiesel fatty acid methyl or ethyl esters through transesterification. Another option is to process the recovered crude lipids/oil in a petroleum refinery similar to the processing of petroleum crude oil. Drop-in fuel conversion costs are similar for both of the aforementioned conversion processes (Karatzos et al., 2017).

Almost all the lipid recovery methods except the hexane extraction of dry algae are yet to be reliably demonstrated at larger-scale at a cost range that would be viable for biofuel processing. The methods are still in development, but some show promise for viable scale-up for biofuel development.

3.10 Value-addition biocommodities

Deoiled algae after extraction of lipids contain 30%–45% proteins, 30%–45% carbohydrates, and 20%–25% salt and nutrient-related ash content in the case of marine microalgae. They have been reported to have promising potential as animal feed in future (Lum et al., 2013; Madeira et al., 2017; Benemann, 2013; Becker, 2004). Animal feed include cattle feed, aquafeed (fish, shrimps, other aquaculture), poultry feed and pig feed. Microalgae are a high-class source of food and technically retain good value as food and health supplement even after extraction of lipids because of the protein, pigment, carbohydrate, and nutrient value left (Madeira et al., 2017; Navarro et al., 2016). Aquafeed fetch a higher market price than that of cattle feed. At a market size of 1.3 billion tons per annum (IFIF, 2021; Alltech, 2019) and a market price of $ 350–1000/ton (Agriwatch, 2021; Reus, 2020; EurofishMagazine, 2020) (depending on the type/quality of animal feed), animal feed offer promising potential for commercialization of microalgal biofuels.

Microalgae are a naturally prominent source of health supplements, nutraceutical products, cosmetic products and pharmaceuticals (Hemantkumar and Rahimbhai, 2019; Barkia et al., 2019; Islam et al., 2017; Ariede et al., 2017; Joshi et al., 2018; Gujar et al., 2019). These are high-value products and their co-production may bring down the cost of biofuel lipids. Nutraceutical products include carotenoids, omega fatty acids (EPA, DHA), pigments and amino acids as functional food. Carotenoids include β-carotene, lutein, astaxanthin, cantha-xanthin, and zeaxanthin (Lum et al., 2013; Balakrishnan et al., 2020; Kaur, 2020; Nethravathy et al., 2019; Matos et al., 2017). Carotenoid market is ever growing. A considerable portion of the global carotenoid market is still catered to by synthetically produced carotenoids. There is a good market scope for naturally produced carotenoids. In addition to carotenoids, phycobiliproteins such as phycocyanin and phycoerythrin and chlorophylls (a,b,c,d,f) are prominent pigments from microalgae that offer scope for a variety of applications—cosmetic products, food colourants, skin care products, anti-oxidants, anticancer products, and anti-inflammatory products (Silva et al., 2020). An extensive variety of microalgae have been reported in literature for the production of specific products of interest, e.g., *C. vulgaris* (chlo-rophylls), *Spirulina platensis* (phycocyanin), *Haematococcus pluvialis* (astaxanthin), *D. salina* (β-carotene), Nannochloropsis spp., *Chaetoceros brevis* and *Thalassiosira weissflogii* (EPA, DHA), and, *Thraustrochytrium aureum*, *Schizochytrium* sp., and *Ulkenia* sp. (DHA) (Silva et al., 2020; Peltomaa et al., 2018; Oliver et al., 2020). *Spirulina* sp., *Chlorella* sp., *Haematococcus* sp., *Dunaliella* sp., *Phaeodactylum* sp., *Porphyridium* sp., *Chaetoceros* sp., *Crypthecodinium* sp., *Isochrysis* sp., *Nannochloris* sp., *Nitzschia* sp., *Schizochytrium* sp., *Tetraselmis* sp., and *Skeletonema* sp. are representative examples of microalgae that are pursued for commercial production of such products (Silva et al., 2020). The market prices of these products can vary from $ 300–10,000/kg depending on the type and quality of the products (Yaqoob et al., 2021). The aforementioned high value metabolites are produced by microalgae under specific growth parameters that promote them in their culturing conditions. Combined production of these metabolites with biofuel lipids needs specific culture condition optimization. In the case of recovery of value from nutraceutical products and cosmetic products, these prod-ucts of interest may have to be first recovered from extracted crude algal lipids after which the remaining lipids could be processed for biofuel production.

Glycerol by-product from biodiesel conversion process is a valuable platform chemical (Luo et al., 2016; Sun et al., 2017; Parate et al., 2018). Some of the products that could be pro-duced from glycerol are: acrolein (industrial chemical); acrolein polymers; 2,3-butanediol (fuel additive); glycerol oxidation products (dihydroxyacetone, hydroxypyruvic acid, formic acid, glycolic acid); 1,3-propanediol (solvent, adhesive); lactic acid; citric acid; succinic acid; polyglycerols (non-ionic surfactants used in the food, detergents and cosmetics industries); ethers (Luo et al., 2016; Sun et al., 2017; Parate et al., 2018).

Deoiled algae are being explored for the production of biodegradable plastics such as those of food-packaging plastics (Kalita et al., 2021). The nitrogen content in deoiled algae is suspected to result in better biodegradability of deoiled algae-based biocomposites compared to other biological feedstocks. Deoiled algae has been tested to be a suitable feedstock source for the production of good quality cellulose nanocrystals (Mokhena et al., 2018; Sucaldito and Camacho, 2017; Katiyar, 2021). Cellulose nanocrystals are a promising platform chemical that are pursued for the development of a novel material science-based products such as proton exchange membrane in fuel cell (PEMFC), electrodes in fuel cell and biodegradable plastics. Microalgae have also been tested as potential biofertilizers (Ronga et al., 2019; Gonçalves, 2021; Dineshkumar et al., 2019; Guo et al., 2020).

Polyurethane foam is being developed from the protein fraction of algae as a promising and valuable product (Pavlik et al., 2017). Characteristics of polyurethane produced from algal proteins have been reported to be of appreciable quality. Hydroythermal liquefaction (HTL) offers a different and an elaborate set of value-addition co-products which could be equally or even more valuable than those of solvent extraction co-products. The method offers an extensive network of products that are representative of biorefinery. These include alcohols, polymers, nitriles, ketones, cyclopentanones, aldehydes, furanone and furfural derivatives, monoaromatics, polycyclics, organic acids, alkanes, alkenes, tocopherols, amino acids, amines, amides and nutrients (Djandja et al., 2020; Barreiro et al., 2013; Yu et al., 2011; Usami et al., 2020; Biller, 2013; Wang et al., 2017). The market scope and value of such biorefinery products may be high if such a model comes to fruition.

The potential of value addition commodities from microalgae is still mostly at the concept level and lab-scale testing level. It needs to be demonstrated at a larger-scale in the context of viable biofuel production.

3.11 Commercial companies worldwide involved in microalgal biofuels

There is no commercial production of microalgal biofuels presently in the world. Whatever quantity of microalgal biomass commercially produced is mostly utilized for the production of food, nutraceuticals and cosmetics and these are high value, low volume market products. The three major commercial algae producers in United States are Earthrise Nutritionals LLC in California, Cyanotech Corporation in Hawaii, and Martek Biosciences Corporation in Maryland (Darzins et al., 2010). The first two use raceway ponds whereas the last one uses closed fermentation-based technology with sugar-based medium for microalgae production. The extensive hypersaline natural algal pond systems of Hutt Lagoons (520 ha) and Whyalla (440 ha) in Australia are managed and operated by Cognis Australia Pty Ltd. for cultivation of *D. salina* for β-carotene production (Darzins et al., 2010). These are among the largest algal production systems in the world. Parry Nutraceuticals (a division of EID Parry) is a major company involved in the production of health supplements and nutraceuticals from algae in India. Some representative microalgal companies with activities relating to biofuels are listed below.

3.11.1 Representative microalgal companies

Sapphire Energy, San Diego, California, USA

The company was established in 2007 for accelerated development of microalgal biofuels and co-products based on the biorefinery concept with huge investments (Wikipedia, 2021; sapphireenergy.com, 2021). They had set up large-scale microalgal cultivation called Green Crude Farms spread over 300 acres in 2012. It was set up in the desert outside of Columbus, New Mexico. They had developed an extensive team of about 1500 people for active research in all areas, top-to-bottom for microalgal biofuel production. They had employed freshwater and brackish water microalgae while stating that their operations do not use potable water. They had used genetically modified microalgae. They employed photobioreactors as inoculum development system and open raceway ponds for the active growth of the genetically modified microalgae. In 2014, they announced their diversification to include the production of nutritional supplements as well as biofuels. They were producing barrels of crude oil year-round and had planned to produce 100 of barrels per day (6 million liters/year) by year 2015. However, the company ceased operations in 2017 and was sold to a group of farmers who have listed that their mission is to produce microalgae for food, feed, nutritional and energy applications.

Seambiotic, Israel

Seambiotic is a marine microalga producing company primarily focused on carbon dioxide capture from emission source and assimilation through microalgae (seambiotic.com; Start-Up Nation Finder, 2021). In collaboration with Israel Electric Corporation, carbon dioxide from the latter's power plant in Ashkelon, Israel was channeled through pipelines to Seambiotic's open marine microalgae cultivation ponds. The company has been engaged in research methods for cultivation of various marine microalgal species for carbon dioxide sequestration and biofuel production.

Algenol, Florida, USA

Algenol (2021) is engaged in the development of a variety of biofuels (ethanol, green crude for gasoline, diesel and jet fuel), food products and personal care products from microalgae (algenol.com; LaMonica, 2008; Algenol, 2014). They reported that they employed genetically modified cyanobacteria in photobioreactor while utilizing carbon dioxide from emission sources. They had developed ethanol fuel in their registered DIRECT TO ETHANOL process. Algenol's system is reported to have algae produce ethanol inside the cell in gas form which is then siphoned off from the photobioreactor tubes.

Synthetic Genomics and ExxonMobil

Synthetic Genomics is a company that is involved in synthetic biology that entered into collaboration with Exxon Mobil for biofuel production from microalgae and other microorganisms through genetic modification and genetic tools (Synthetic Genomics, 2020; Anderson, 2018). In 2018, they announced that the next phase in their research that includes an outdoor field study that will grow naturally occurring algae in several contained ponds in California and aims at producing 10,000 barrels of algae biofuel per day by 2025.

Reliance Industries Ltd, India

Reliance Industries Ltd., India have set up extensive outdoor algal production systems and downstream processing systems in Jamnagar, Gujarat, India (GreenCarCongress, 2015; Lane, 2016; Bhujade et al., 2016; Posewitz, 2019). They have produced algae in large raceway pond systems for biofuel development. They have engaged a dedicated team of researchers to focus on the biology and screening aspects of microalgae as well. They have partnered with Algenol and set up demonstration module of Algenol's production platform in Jamnagar. The demonstration completed several production cycles of Algenol's wildtype host algae. Reliance Industries Ltd. has also been involved in extensive deployment of hydrothermal liquefaction (HTL) conversion of microalgae produced in their outdoor production systems to biofuels. This was aimed at simultaneous development of fuels, value-addition co-products and fertilizers.

PetroAlgae, Saratoga Springs, New York, USA

This is a company with stated focus on microalgal fuels (Petroalgae.com, 2019). The batch cultivation systems utilize specialized closed photobioreactors with controlled environment for temperature and light conditions. To harvest the biomass, acoustic focusing technology is practised for concentrating the growing microalgal cells and then oil is recovered from the biomass by chemical and solvents-based extraction methods.

Other representations of microalgal companies engaged in development of biofuels along with other products include AlgaEnergy, Spain, Aurora Algae, USA, Aquatic Energy, Green Sea, USA, Live Fuels, USA, AlgaeTech, Malaysia, Phycal, USA, Cellana, Hawaii, USA, General Atomics, San Diego, USA and Blue Marble Energy, USA.

Green Star

This is a US-based company involved in marine water-based algae cultivation for food and nutraceutical purpose (FinancialExpress, 2008). They are known for developing an efficient and viable commercial liquid fertilizer in partnership with Biotech Research, Inc. The liquid fertilizer they developed was called "Montana Micronutrient Booster (MMB)," nutritional formula for improved algal cultivation (34% improvement in daily growth rate). Testing has shown that MMB can be used at the dilution rate of 1:10,000 or even 1:20,000 over hectare-scale area with significant effectiveness on biomass productivity.

3.11.2 Representative companies using mcroalgae-based carbon dioxide capture

In addition to Seambiotic, Israel mentioned earlier, the following companies are involved in microalgal production utilizing carbon dioxide from large emission sources: MBD Energy, Australia, Linc Energy & BioCleanCoal (partnership) in Australia, Scottish Bioenergy, UK and RWE Energy, Germany, Carbon Capture Corporation, USA, and Arizona Public Service Co., USA. These companies produce algae biomass and utilize it mostly for biodiesel production (Apartments and Road, 2013).

4 Scope

Though algal productivities above $50\,g/m^2/day$ are theoretically possible, productivities in the range of $20–25\,g/m^2/day$ are the realistic targets in the near term. Similarly, lipid content that can be aimed for practical realization in the open outdoor algal systems in the near term is 15%–25%. Higher lipid content in open outdoor systems needs more studies and development. Open growth systems are rated to be the most viable growth system presently for biofuel production. Closed photobioreactors are cost-prohibitive both in terms of capital costs and operation/maintenance costs for a medium-value product such as a transport fuel. They are more suited for production of high value products. Also, daily productivity does not automatically translate to annual yields. Open systems are prone to productivity fluctuations and inconsistent culture heath status due to factors such as seasonal changes and contamination. Reliable annual yields of large-scale open algal systems representing data in years are yet to be demonstrated. Most global locations suited for algal production may face a downtime of 1–3 months due to winter, rainfall and maintenance requirements. Coastal regions of India receive monsoon rainfall for 2–3 months in year. Algal production is likely to be suspended during this time. Considering the aforementioned algal productivity and lipid content and a productive time period of 270 days on average in a year translates to an annual biomass yield of 54–68 tons/ha and an oil yield of 11–14 tons/ha. India has a coastal stretch of 7517 km (MHA, 2020). States such as Gujarat and Tamilnadu offer good prospects for marginal lands near the sea coast.

The present estimated cost of algal biofuels is high. There are viable scenarios that offer scope in future. Development of animal feed as a value-addition co-product is one scenario where the estimated cost of algal biofuels has been seen to approach commercially viable range. Aquafeed in particular could bring the cost of algal biofuels to Rs. 50–70/L. Animal feed in general could bring the cost of biofuels to Rs. 50–90/L. The market size of animal feed matches up with that of biofuels. There are other alternative value-addition co-products that show promise for similar viability potential and are being explored. Efforts are on around the world to evaluate the economic viability of the overall process through integrated production of fuels and value-addition commodities.

Since algae grows in a liquid medium, any algal production scenario gives a chance for carbon dioxide sequestration through deliberate supply. Carbon dioxide is a main rate limiting factor for outdoor algal production. Carbon dioxide can be supplied through long distance piping as done with transporting of carbon dioxide from carbon dioxide reservoirs. Carbon dioxide emissions from vehicles may be attempted to be captured in the vehicle itself and routed to algal production sites following similar reverse distribution logistics of fuel from production site to vehicles. At the fuel pumping stations, captured carbon dioxide can be recovered from vehicles at the same time the fuel is pumped into the vehicle.

If algae prove to be effective in wastewater treatment, biofuel from their lipids may be seen as a viable by-product since the algal production cost would be borne by wastewater treatment and only the fuel processing cost need to be borne by algal biofuel production stream.

It is likely that most of the light duty automobiles such as two-wheelers and cars would shift to electric drive in future. Heavier automobiles such as buses and trucks may also shift to electric mode partially. Aviation is one sector that looks likely to set to use liquid fuels for the foreseeable future. The present annual global jet fuel demand is 390 mtoe (Reuters, 2020). This is projected to increase to more than two times by 2050 (EIA, 2019). This is the sort of market size that algal biofuel could cater to in future.

5 Way forward

An integrated biorefinery platform is a widely recommended and compelling framework to pursue development of biofuels and biocommodities from microalgae. Bioeconomy development is seen as one of the key ways to realize sustainable development around the world in future. Microalgae are a natural bioeconomy feedstock given the range of products that could be developed from them. A number of possible biocommodity products that can be part of biorefinery network have been listed in the chapter.

Development of lipids, deoiled meal proteins and high value metabolites in an integrated way for applications such as fuel, feed and health supplements and algal wastewater treatment can be pursued as near-term prospects. These show potential for commercial viability in the near future. These could be pursued with an eye on long term possibilities. Long term possibilities include alternate fuel processing technologies, development of co-products based on novel material science, thermo-chemical processes and hydrothermal processes and development of biodegradable plastics and novel bio-based polymers.

One of the key challenges in the field of microalgal biofuels is the improvement of outdoor productivity. Utilizing the abundant and excessive sunlight more effectively through distribution or improved reception of sunlight in open systems is a recommended way to address this. Carbon dioxide is a main rate-limiting factor in outdoor algal production. Supply of carbon dioxide from emission sources has a critical role in improving the outdoor algal productivity. The demonstration systems set up to evaluate outdoor production should aim to achieve sustained long-term productivities in the form of annual yields. Robust strains and mixed cultures that can yield dependable productivity year-long through the seasonal variations are to be prioritized. Long term yields of just plain biomass production must be demonstrated first before pursuing enhancement in lipid productivity. Hydrothermal liquefaction and wet algal pyrolysis are options to process such biomass. After this, lipid enhancement triggers such as nutrient depletion and pH increase may have to be integrated seamlessly to follow the growth phase for demonstration in outdoor systems. To counter the salinity increase in the open outdoor systems due to evaporation water loss, regular intake and discharge cycles of fresh seawater must be optimized or water loss must be compensated by some form of make-up water addition. Employing algae with self-aggregating features, biocompatible flocculants, viable bioflocculation, and efficient filtration and centrifuging dewatering methods are promising ways to improve microalgal harvesting. To realize viable lipid/oil recovery from microalgae, wet algae must be processed without the need for drying, be it solvent extraction, pyrolysis, or hydrothermal liquefaction.

Wastewater is an option as water source in cases where land is available near towns for municipal wastewater treatment and where land is available near industrial complexes for wastewater treatment.

Since microalgae are a naturally protein-laden feedstock with carbohydrates, development of animal feed (aquafeed, cattle feed, poultry feed) from deoiled algae is an indispensable value-addition prospect to pursue for near-term commercial viability. Cost analyses indicate development of market saleable animal feed offers the most immediate chance of any commercial viability of algal fuels and commodities. Aquafeed in particular is seen to bring down the cost of algal biofuels by $ 1.35–2.45/L. Microalgae are natural food for fishes and other aquatic animals. It is only to be expected that deoiled algae whose protein, carbohydrates and other cellular contents are natural food ingredients for aquatic animals form a potential commercial aquafeed feedstock in the future. It is recommended that the algal biofuel initiatives strongly pursue this. Deoiled algae, in theory, are similar to (or perhaps even better than) soya meal which is widely used as cattle feed. Thus deoiled algae are a promising candidate to be used as cattle feed and poultry feed. Commercialization efforts should focus on development of cattle feed and poultry feed. These have the potential to bring down the cost of algal biofuels by $ 0.80–1.50/L. The global market size of animal feed is about 1.3 billion tons per annum (IFIF, 2021; Alltech, 2019) (aquafeed: 57 million tons per annum; (IMARC, 2021)) and these markets are reported to have scope for much larger growth (2–4 times by 2050) (IFIF, 2021). This market size is comparable to the scope of aforementioned algal fuel market size.

This should facilitate setting up of integrated algal fuel and commodity production in the initial phase before other commodities such as biodegradable plastics and platform chemicals become ready.

Microalgae are a promising source of nutraceutical products such as a variety of carotenoids, omega fatty acids and amino acids. There is a large untapped market for these products. Biofuels are likely to be a co-product here after the recovery of the nutraceutical component of interest. However, for this to happen, methods whereby the biofuel lipids/oil can be effectively recovered along with the nutraceutical products need to be developed. If nutraceutical components are selectively recovered from microalgae, fuel fraction must be recovered from the nutraceutical-recovered algae. If nutraceutical components and biofuel lipids are recovered together as crude extracts, biofuel lipids would need to be separated further. Facilitating commercial production of microalgae for such health supplement applications may set the platform to develop microalgae as feedstock for other applications such as biofuels and biocommodities. Open systems may be more appropriate for combined production of nutraceuticals and biofuels because the intensity of energy and costs and replication at a scale appropriate for biofuels may be challenges in closed photobioreactors. Value addition from nutraceuticals per liter of algal fuel would be high, but the scope of fuel production would be lower than that of animal feed value-addition, because of the corresponding market sizes. Cosmetic and skincare products, pigments, and food colors from microalgae also offer similar scope for commercial applications and value-addition. Microalgae are worth pursuing as a feedstock even just for food and food supplements. There is a considerable scope for human consumption. Starting with this could pave the way for other commercial applications from microalgae.

Among other value-addition commodities, polyurethane is rated to be a promising commodity that can add appreciable value to the commercial production of microalgae and biofuels. With a market price of $5–12/kg (PlasticsInsight, 2021) (market size: $75 billion (PlasticsInsight, 2021); 24 million tons (Statista, 2021)), polyurethane can fetch a value of about $1–3/L of algal fuel depending on processing and applications.

Commodities that can be developed from the promising hydrothermal liquefaction route such as aromatics, aldehydes, polycyclics, organic acids, alkanes, alkenes, amides, and polymers offer potential for commercial viability through alternate routes. Even if this may take time, it is worth pursuing them as a long-term commercial viability undertaking.

The potential for such a wide array of products from microalgae makes a case for their sustained pursuit and development. In the near future, the overall production process shall be broken down to individual process steps and thorough R&D efforts on the aforementioned individual challenges shall be undertaken. Then the integrated production process must be evaluated with adequate techno-economic analysis and life cycle analysis. Weighing up the potential of the technology and the many challenges that are yet to be addressed, it is worth pursuing algal biofuels with a measured investment. Caution shall be exercised in expectations and the projection of its realization.

Acknowledgments

Work referred to in Kannan and Venkat (2019), Patel and Kannan (2021) and Fig. 6 were supported by the Department of Biotechnology, Government of India.

References

Abubakar, A.L., 2016. Effect of salinity on the growth parameters of halotolerant microalgae, Dunaliella spp. Niger. J. Basic Appl. Sci. 24 (2), 85–91.

Anon., 2021. Agriwatch. http://www.agriwatch.com/animal-feed/feed-ingredients/.

Alam, M.A., Vandamme, D., Chun, W., Zhao, X., Foubert, I., Wang, Z., Yuan, Z., 2016. Bioflocculation as an innovative harvesting strategy for microalgae. Rev. Environ. Sci. Biotechnol. 15 (4), 573–583.

Algae, 2021. Green Wastewater Treatment System Gross-Wen Technologies. Retrieved 24 June 2021, from https://algae.com/.

Algenol, 2014. Direct to Ethanol. Algenol. Retrieved 23 June 2021, from https://web.archive.org/wcb/20140329011522/http://algenol.com/direct-to-ethanol/environmental-benefits.

Algenol, 2021. https://www.algenol.com/.

Alltech, 2019. Einstein-Curtis A, Alltech (2019, January 30). Broilers, Swine Production Support Global Feed Growth Past 1.1bn Metric Tons. Feednavigator.Com. https://www.feednavigator.com/Article/2019/01/30/Alltech-Global-feed-growth-passes-1.1bn-metric-tons.

Amin, S.A., Green, D.H., Hart, M.C., Kupper, F.C., Sunda, W.G., Carrano, C.J., 2009. Photolysis of ironsiderophore chelates promotes bacterial-algal mutualism. PNAS 106, 17071–17076.

Anderson, D., 2018, May 4. ExxonMobil and Synthetic Genomics Algae Biofuels Program Targets 10,000 Barrels per Day by 2025. Synthetic Genomics. https://syntheticgenomics.com/exxonmobil-and-synthetic-genomics-algae-biofuels-program-targets-10000-barrels-per-day-by-2025/.

Apartments, A., Road, N.H., 2013. Comprehensive Oilgae Report. Energy from Algae: Products, Market, Prosses & Strategies. A Preview. Oilgae, Home Algal Energy, p. 79.

Ariede, M.B., Candido, T.M., Jacome, A.L.M., Velasco, M.V.R., de Carvalho, J.C.M., Baby, A.R., 2017. Cosmetic attributes of algae—a review. Algal Res. 25, 483–487.

Arienzo, M., Christen, E., Quayle, W., Kumar, A., 2008. A review of the fate of potassium in the soil-plant system after land application of wastewaters. J. Hazard. Mater. 164, 415–422. https://doi.org/10.1016/j.jhazmat.2008.08.095.

ARPA-E, 2010. Algaeventure Systems. Advanced Research Projects Agency-Energy, US Department of Energy. https://arpa-e.energy.gov/technologies/projects/fuel-algae.

Ashraf, M., Javaid, M., Rashid, T., Ayub, M., Zafar, A., Ali, S., Naeem, M., 2011. Replacement of expensive pure nutritive media with low cost commercial fertilizers for mass culture of freshwater algae, Chlorella vulgaris. Int. J. Agric. Biol. 13 (4), 484–490.

Awal, S., Christie, A., 2015. Suitability of inland saline ground water for the growth of marine microalgae for industrial purposes. J. Aquac. Mar. Biol. 3 (2). https://doi.org/10.15406/jamb.2015.03.00063, 00063.

Aziz, M.M.A., Kassim, K.A., Shokravi, Z., Jakarni, F.M., Liu, H.Y., Zaini, N., Shokravi, H., 2020. Two-stage cultivation strategy for simultaneous increases in growth rate and lipid content of microalgae: a review. Renew. Sust. Energ. Rev. 119, 109621.

Azizi, K., Moraveji, M.K., Najafabadi, H.A., 2018. A review on bio-fuel production from microalgal biomass by using pyrolysis method. Renew. Sust. Energ. Rev. 82, 3046–3059.

Balakrishnan, J., Sekar, T., Shanmugam, K., 2020. Marine-microalgae as a potential reservoir of high value nutraceuticals. In: Marine Niche: Applications in Pharmaceutical Sciences. Springer, Singapore, pp. 221–236.

Balasubramanian, R.K., Yen Doan, T.T., Obbard, J.P., 2013. Factors affecting cellular lipid extraction from marine microalgae. Chem. Eng. J. 215–216, 929–936. https://doi.org/10.1016/j.cej.2012.11.063.

Baldev, E., Mubarakali, D., Saravanakumar, K., Arutselvan, C., Alharbi, N.S., Alharbi, S.A., Sivasubramanian, V., Thajuddin, N., 2018. Unveiling algal cultivation using raceway ponds for biodiesel production and its quality assessment. Renew. Energy 123, 486–498. https://doi.org/10.1016/j.renene.2018.02.032.

Barkia, I., Saari, N., Manning, S.R., 2019. Microalgae for high-value products towards human health and nutrition. Mar. Drugs 17 (5), 304.

Barreiro, D.L., Prins, W., Ronsse, F., Brilman, W., 2013. Hydrothermal liquefaction (HTL) of microalgae for biofuel production: state of the art review and future prospects. Biomass Bioenergy 53, 113–127.

Beach, E.S., Eckelman, M.J., Cui, Z., Brentner, L., Zimmerman, J.B., 2012. Preferential technological and life cycle environmental performance of chitosan flocculation for harvesting of the green algae Neochloris oleoabundans. Bioresour. Technol. https://doi.org/10.1016/j.biortech.2012.06.012.

Beal, C.M., Gerber, L.N., Sills, D.L., Huntley, M.E., Machesky, S.C., Walsh, M.J., Tester, J.W., Archibald, I., Granados, J., Greene, C.H., 2015. Algal biofuel production for fuels and feed in a 100-ha facility: a comprehensive techno-economic analysis and life cycle assessment. Algal Res. 10, 266–279.

Becker, W., 2004. 18 microalgae in human and animal nutrition. In: Handbook of Microalgal Culture: Biotechnology and Applied Phycology. vol. 312. Wiley Online Library.

Benemann, J., 2013. Microalgae for biofuels and animal feeds. Energies 6 (11), 5869–5886.

Bhujade, R., Ghadge, R., Konakandla, P., Sapre, A., 2016, October 26. Experience with Hydrothermal Processing to Convert Algae to Oil [Slides]. Algaebiomass.org. http://algaebiomass.org/wp-content/gallery/2012-algae-biomass-summit/2010/06/Bhujade-Ramesh-Reliance-Industries-Ltd-Experience-with-Hydrothermal-Processing-to-Convert-Algae-to-Oil.pdf.

Biller, P., 2013. Hydrothermal processing of microalgae (Ph.D. thesis). University of Leeds. https://etheses.whiterose.ac.uk/4687/1/PhD%20THESIS%20P%20BILLER%202013.pdf.

BMS, 2018. Sewage Parameters 1: BOD. Butler Manufacturing Services. https://www.butlerms.com/sewage-parameters-1-bod/. (Accessed 23 May 2018).

Bosma, D.R.R., 2014, November 21. Unusual Research Paper About Design and Construction Algae Pilot Plant. WUR. https://www.wur.nl/en/newsarticle/Unusual-research-paper-about-design-and-construction-algae-pilot-plant.htm.

Bosma, R., Van Spronsen, W.A., Tramper, J., Wijffels, R.H., 2003. Ultrasound, a new separation technique to harvest microalgae. J. Appl. Phycol. 15 (2), 143–153.

Brennan, L., Owende, P., 2010. Biofuels from microalgae—a review of technologies for production, processing, and extractions of biofuels and co-products. Renew. Sust. Energ. Rev. 14, 557–577.

Brito, D., Castro, A., Guevara, M., Gómez, E., Ramos-Villarroel, A., Aron, N.M., 2013. Biomass and pigments production of the mixed culture of microalgae (Hyaloraphidium contortum and Chlorella vulgaris) by cultivation in media based on commercial fertilizer. Ann. Univ. Dunarea Jos Galati. Fascicle VI: Food Technol. 37 (1), 85.

Buchan, A., LeCleir, G.R., Gulvik, C.A., Gonz_alez, J.M., 2014. Master recyclers: features and functions of bacteria associated with phytoplankton blooms. Nat. Rev. Microbiol. 12, 686–698.

Cai, T., Park, S.Y., Li, Y., 2013. Nutrient recovery from wastewater streams by microalgae: status and prospects. Renew. Sust. Energ. Rev. 19, 360–369. https://doi.org/10.1016/j.rser.2012.11.030.

Cai, Y., Wang, W., Li, L., Wang, Z., Wang, S., Ding, H., Zhang, Z., Sun, L., Wang, W., 2018. Effective capture of carbon dioxide using hydrated sodium carbonate powders. J. Mater. 11, 183.

Carbon Clean, 2021. Carbon Capture and Storage to Reach Net Zero. Carbon Clean. Retrieved 24 June 2021, from https://www.carbonclean.com/.

Carvalho, A.P., Meireles, L.A., Malcata, F.X., 2006. Microalgal reactors: a review of enclosed system designs and performances. Biotechnol. Prog. 22, 1490–1506.

Castilla Casadiego, D.A., Albis Arrieta, A.R., Angulo Mercado, E.R., Cervera Cahuana, S.J., Baquero Noriega, K.S., Suárez Escobar, A.F., Morales Avendaño, E.D., 2016. Evaluation of culture conditions to obtain fatty acids from saline microalgae species: Dunaliella salina, Sinecosyfis sp., and chroomonas sp. Biomed. Res. Int. 2016. https://www.ncbi.nlm.nih.gov/pmc/articles/PMC4916267/.

Cerff, M., Morweiser, M., Dillschneider, R., Michel, A., Menzel, K., Posten, C., 2012. Harvesting fresh water and marine algae by magnetic separation: screening of separation parameters and high gradient magnetic filtration. Bioresour. Technol. 118, 289–295.

Chen, C.Y., Yeh, K.L., Aisyah, R., Lee, D.J., Chang, J.S., 2011. Cultivation, photobioreactor design and harvesting of microalgae for biodiesel production: a critical review. Bioresour. Technol. 102 (1), 71–81.

Chen, F., Liu, Z., Li, D., Liu, C., Zheng, P., Chen, S., 2012. Using ammonia for algae harvesting and as nutrient in subsequent cultures. Bioresour. Technol. 121, 298–303. https://doi.org/10.1016/j.biortech.2012.06.076.

Cheng, J., Yang, Z., Huang, Y., Huang, L., Hu, L., Xu, D., Cen, K., 2015. Improving growth rate of microalgae in a 1191 m^2 raceway pond to fix CO_2 from flue gas in a coal-fired power plant. Bioresour. Technol. 190, 235–241. https://doi.org/10.1016/j.biortech.2015.04.085.

Chisti, Y., 2007. Biodiesel from microalgae. Biotechnol. Adv. 25 (3), 294–306.

Christenson, L., Sims, R., 2011. Production and harvesting of microalgae for wastewater treatment, biofuels, and bioproducts. Biotechnol. Adv. 29 (6), 686–702.

Cole, J.J., 1982. Interactions between bacteria and algae in aquatic ecosystems. Annu. Rev. Ecol. Evol. Syst. 13, 291–314.

Craggs, R., Sutherland, D., Campbell, H., 2012. Hectare-scale demonstration of high rate algal ponds for enhanced wastewater treatment and biofuel production. J. Appl. Phycol. 24 (3), 329–337. https://link.springer.com/article/10.1007/s10811-012-9810-8.

Craggs, R., Park, J., Heubeck, S., Sutherland, D., 2014. High rate algal pond systems for low-energy wastewater treatment, nutrient recovery and energy production. N. Z. J. Bot. 52 (1), 60–73.

Croft, M.T., Lawrence, A.D., Raux-Deery, E., Warren, M.J., Smith, A.G., 2005. Algae acquire vitamin B12 through a symbiotic relationship with bacteria. Nature 438, 90–93.

Cunha, P., Pereira, H., Costa, M., Pereira, J., Silva, J.T., Fernandes, N., Simões, M., 2020. Nannochloropsis oceanica cultivation in pilot-scale raceway ponds—from design to cultivation. Appl. Sci. 10 (5), 1725. https://www. mdpi.com/2076-3417/10/5/1725/pdf.

Dalrymple, O.K., Halfhide, T., Udom, I., Gilles, B., Wolan, J., Zhang, Q., Ergas, S., 2013. Wastewater use in algae production for generation of renewable resources: a review and preliminary results. Aquat. Biosyst. 9 (1), 1–11.

Darzins, A., Pienkos, P., Edye, L., 2010. Current status and potential for algal biofuels production. Rep. IEA Bioenergy Task 39 (13), 403–412.

Das, P., 2021. Personal Communication. Advanced Biofuels Division, The Energy and Resources Institute (TERI), New Delhi, India (relating to ongoing and yet to be published research).

Dassey, A.J., Theegala, C.S., 2013. Harvesting economics and strategies using centrifugation for cost effective separation of microalgae cells for biodiesel applications. Bioresour. Technol. 128, 241–245.

Davis, R., Aden, A., Pienkos, P.T., 2011. Techno-economic analysis of autotrophic microalgae for fuel production. Appl. Energy 88 (10), 3524–3531.

Davis, et al., ANL, NREL, PNNL, June 2012. Renewable Diesel From Algal Lipids: An Integrated Baseline for Cost, Emissions, and Resource Potential from a Harmonized Model. US Department of Energy Biomass Program. http://www.nrel.gov/docs/fy12osti/55431.pdf.

Davis, R.E., Fishman, D.B., Frank, E.D., Johnson, M.C., Jones, S.B., Kinchin, C.M., Skaggs, R.L., Venteris, E.R., Wigmosta, M.S., 2014. Integrated evaluation of cost, emissions, and resource potential for algal biofuels at the national scale. Environ. Sci. Technol. 48 (10), 6035–6042.

De Baerdemaeker, T., Lemmens, B., Dotremont, C., Fret, J., Roef, L., Goiris, K., Diels, L., 2013. Benchmark study on algae harvesting with backwashable submerged flat panel membranes. Bioresour. Technol. 129, 582–591.

De Vree, J.H., Bosma, R., Janssen, M., Barbosa, M.J., Wijffels, R.H., 2015. Comparison of four outdoor pilot-scale photobioreactors. Biotechnol. Biofuels 8, 215.

Debuchy, R., Purton, S., Rochaix, J.D., 1989. The argininosuccinate lyase gene of Chlamydomonas reinhardtii: an important tool for nuclear transformation and for correlating the genetic and molecular maps of the ARG7 locus. EMBO J. 8 (10), 2803–2809.

Derwenskus, F., Metz, F., Gille, A., Schmid-Staiger, U., Briviba, K., Schließmann, U., Hirth, T., 2019. Pressurized extraction of unsaturated fatty acids and carotenoids from wet *Chlorella vulgaris* and *Phaeodactylum tricornutum* biomass using subcritical liquids. GCB Bioenergy 11, 335–344. https://doi.org/10.1111/gcbb.12563.

Detchanamurthy, S., Kumar, D., Janardhanan, Y., Durairaj, P., Karunamoorthy, M., 2018. Effect of organic fertilizer on the growth, development and carotenoid production in marine micro algae *Dunaliella salina*. Asian J. Res. Pharm. Sci. Biotechnol. 6 (2), 20–28. https://seagrasstech.com/TechnicalPaper.pdf.

Dineshkumar, R., Subramanian, J., Gopalsamy, J., Jayasingam, P., Arumugam, A., Kannadasan, S., Sampathkumar, P., 2019. The impact of using microalgae as biofertilizer in maize (Zea mays L.). Waste Biomass Valoriz. 10 (5), 1101–1110.

Djandja, O.S., Wang, Z., Chen, L., Qin, L., Wang, F., Xu, Y., Duan, P., 2020. Progress in hydrothermal liquefaction of algal biomass and hydrothermal upgrading of the subsequent crude bio-oil: a mini review. Energy Fuel 34 (10), 11723–11751. https://pubs.acs.org/doi/abs/10.1021/acs.energyfuels.0c01973.

Drexler, I.L., Yeh, D.H., 2014. Membrane applications for microalgae cultivation and harvesting: a review. Rev. Environ. Sci. Biotechnol. 13 (4), 487–504.

Du, Y.C., Peddi, S.R., Spreitzer, R.J., 2003. Assessment of structural and functional divergence far from the large subunit active site of ribulose-1, 5-bisphosphate carboxylase/oxygenase. J. Biol. Chem. 278 (49), 49401–49405.

Dunford, N.T., Temelli, F., 1997. Extraction conditions and moisture content of canola flakes as related to lipid composition of supercritical CO_2 extracts. J. Food Sci. 62, 155–159. https://doi.org/10.1111/j.1365-2621.1997. tb04389.x.

EIA, 2019. Hanson, S. (2019, November 6). EIA Projects Energy Consumption in Air Transportation to Increase through 2050, Today in Energy. U.S. Energy Information Administration. https://www.eia.gov/todayinenergy/detail.php?id=41913.

El Nabris, K.J.A., 2012. Development of cheap and simple culture medium for the microalgae Nannochloropsis sp. based on agricultural grade fertilizers available in the local market of Gaza strip (Palestine). J. Al Azhar Univ. Gaza 14, 61–76.

El-Gharbawy, A.S.A.A., 2017. Cost analysis for biodiesel production from waste cooking oil plant in Egypt. Int. J. Smart Grid 1 (1), 16–25.

Elliott, D.C., Hart, T.R., Schmidt, A.J., Neuenschwander, G.G., Rotness, L.J., Olarte, M.V., Zacher, A.H., Albrecht, K.O., Hallen, R.T., Holladay, J.E., 2013. Process development for hydrothermal liquefaction of algae feedstocks in a continuous-flow reactor. Algal Res. 2, 445–454. https://doi.org/10.1016/j.algal.2013.08.005.

Eloka-Eboka, A.C., Inambao, F.L., 2017. Effects of CO_2 sequestration on lipid and biomass productivity in microalgal biomass production. Appl. Energy 195, 1100–1111. https://www.sciencedirect.com/science/article/pii/S0306261917303094.

Energy.gov, 2021. Algal Production. Office of Energy Efficiency & Renewable Energy, US Department of Energy. Retrieved 23 June 2021, from https://www.energy.gov/eere/bioenergy/algal-production.

EurofishMagazine, 2020. Importance of Fishmeal for Aquafeed Continues to Decline – More and More Vegetable Feed Components Are Being Used. Eurofish Magazine. https://eurofishmagazine.com/sections/aquaculture/item/717-importance-of-fishmeal-for-aquafeed-continues-to-decline-more-and-more-vegetable-feed-components-are-being-used. (Accessed 11 May 2020).

Evodos, 2020. Solving Algae Challenges. Evodos Dynamic Settlers. https://www.evodos.eu/. (Accessed 15 June 2020).

Fabregas, J., Maseda, A., Dominguez, A., Otero, A., 2004. The cell composition of Nannochloropsis sp. changes under different irradiances in semicontinuous culture. World J. Microbiol. Biotechnol. 20, 31–35.

Feednavigator, 2012. Gray, N. (2012, July 23). Meat Protein: Can Supply Match Growing Demand? Feednavigator. Com. https://www.feednavigator.com/Article/2012/07/23/Meat-protein-Can-supply-match-growing-demand.

Field, C.B., Behrenfeld, M.J., Randerson, J.T., Falkowski, P., 1998. Primary production of the biosphere: integrating terrestrial and oceanic components. Science 281 (5374), 237–240.

FinancialExpress, 2008. Green Star Announces Algae Breakthrough. The Financial Express. https://www.financialexpress.com/archive/green-star-announces-algae-breakthrough/314739/. (Accessed 26 May 2008).

Fu, W., Chaiboonchoe, A., Khraiwesh, B., Sultana, M., Jaiswal, A., Jijakli, K., Salehi-Ashtiani, K., 2017. Intracellular spectral recompositioning of light enhances algal photosynthetic efficiency. Sci. Adv. 3 (9), e1603096.

Fu, W., Nelson, D.R., Mystikou, A., Daakour, S., Salehi-Ashtiani, K., 2019. Advances in microalgal research and engineering development. Curr. Opin. Biotechnol. 59, 157–164.

Fuentes, J.L., Garbayo, I., Cuaresma, M., Montero, Z., González-Del-Valle, M., Vílchez, C., 2016. Impact of microalgae-bacteria interactions on the production of algal biomass and associated compounds. Mar. Drugs 14, 100.

Gao, L., Fang, M., Li, H., Hetland, J., 2011. Cost analysis of CO_2 transportation: case study in China. Energy Procedia 4, 5974–5981. https://doi.org/10.1016/j.egypro.2011.02.600.

Gardner, R., Peters, P., Peyton, B., Cooksey, K.E., 2011. Medium pH and nitrate concentration effects on accumulation of triacylglycerol in two members of the chlorophyta. J. Appl. Phycol. 23 (6), 1005–1016.

Gautam, R., Vinu, R., 2020. Reaction engineering and kinetics of algae conversion to biofuels and chemicals via pyrolysis and hydrothermal liquefaction. React. Chem. Eng. 5 (8), 1320–1373. https://doi.org/10.1039/D0RE00084A.

Genkov, T., Meyer, M., Griffiths, H., Spreitzer, R.J., 2010. Functional hybrid rubisco enzymes with plant small subunits and algal large subunits: engineered rbcS cDNA for expression in Chlamydomonas. J. Biol. Chem. 285 (26), 19833–19841.

Gollakota, A.R.K., Kishore, N., Gu, S., 2018. A review on hydrothermal liquefaction of biomass. Renew. Sust. Energ. Rev. 81, 1378–1392. https://doi.org/10.1016/j.rser.2017.05.178.

Gonçalves, A.L., 2021. The use of microalgae and cyanobacteria in the improvement of agricultural practices: a review on their biofertilising, biostimulating and biopesticide roles. Appl. Sci. 11 (2), 871.

GreenCarCongress, 2015. Algenol and Reliance Launch Algae Fuels Demonstration Project in India. Green Car Congress. Retrieved 23 June 2021, from https://www.greencarcongress.com/2015/01/20150121-algenol.html.

Gujar, A., Cui, H., Ji, C., Kubar, S., Li, R., 2019. Development, production and market value of microalgae products. Appl. Microbiol. 5, 162.

Guo, B., 2019. Hydrothermal liquefaction within a microalgae biorefinery (Doctoral dissertation). Karlsruhe Institute of Technology, Karlsruhe, Germany.

Guo, Z., Tong, Y.W., 2014. The interactions between Chlorella vulgaris and algal symbiotic bacteria under photoautotrophic and photoheterotrophic conditions. J. Appl. Phycol. 26, 1483–1492.

Guo, S., Wang, P., Wang, X., Zou, M., Liu, C., Hao, J., 2020. Microalgae as biofertilizer in modern agriculture. In: Microalgae Biotechnology for Food, Health and High Value Products. Springer, Singapore, pp. 397–411.

Halim, R., Danquah, M.K., Webley, P.A., 2012. Extraction of oil from microalgae for biodiesel production: a review. Biotechnol. Adv. 30 (3), 709–732.

Hannon, M., Gimpel, J., Tran, M., Rasala, B., Mayfield, S., 2010. Biofuels from algae: challenges and potential. Biofuels 1 (5), 763–784.

Hay, R.K.M., Porter, J.R., 2006. The Physiology of Crop Yield. Blackwell Publishing, Oxford.

Hemantkumar, J.N., Rahimbhai, M.I., 2019. Microalgae and its use in nutraceuticals and food supplements. In: Microalgae—From Physiology to Application. IntechOpen, p. 10.

Herrera, A., D'Imporzano, G., Fernandez, F.G.A., Adani, F., 2021. Sustainable production of microalgae in raceways: nutrients and water management as key factors influencing environmental impacts. J. Clean. Prod. 287, 125005. https://doi.org/10.1016/j.jclepro.2020.125005.

Ho, S.H., Ye, X., Hasunuma, T., Chang, J.S., Kondo, A., 2014. Perspectives on engineering strategies for improving biofuel production from microalgae—a critical review. Biotechnol. Adv. 32 (8), 1448–1459.

Hu, Y.R., Wang, F., Wang, S.K., Liu, C.Z., Guo, C., 2013. Efficient harvesting of marine microalgae nannochloropsis maritima using magnetic nanoparticles. Bioresour. Technol. 138, 387–390. https://doi.org/10.1016/j.biortech.2013.04.016.

Huang, C., Chen, X., Liu, T., Yang, Z., Xiao, Y., Zeng, G., Sun, X., 2012. Harvesting of chlorella sp. using hollow fiber ultrafiltration. Environ. Sci. Pollut. Res. 19 (5), 1416–1421.

Huang, Q., Jiang, F., Wang, L., Yang, C., 2017. Design of photobioreactors for mass cultivation of photosynthetic organisms. Engineering 3 (3), 318–329.

Huntley, M.E., Johnson, Z.I., Brown, S.L., Sills, D.L., Gerber, L., Archibald, I., Greene, C.H., 2015. Demonstrated large-scale production of marine microalgae for fuels and feed. Algal Res. 10, 249–265. https://doi.org/10.1016/j.algal.2015.04.016.

IESS 2047, 2015. India Energy Security Scenarios 2047. NITI Aayog, Government of India. http://iess2047.gov.in/.

IFIF, 2021. The Global Feed Industry. The International Feed Industry Federation. Retrieved 24 June 2021, from https://ifif.org/global-feed/industry/.

IMARC, 2021. Aquafeed Market Size, Share, Growth, Trends and Outlook 2021–2026. IMARC Group. Retrieved 24 June 2021, from https://www.imarcgroup.com/aquafeed-market.

IndiaMart, 2021. https://dir.indiamart.com/mumbai/polyvinyl-chloride.html.

International Energy Agency (IEA), 2016. Key World Energy Statistics. https://www.ourenergypolicy.org/wp-content/uploads/2016/09/KeyWorld2016.pdf.

IRENA, 2021. Biodiesel, Transportation, Costs. International Renewable Energy Agency. https://www.irena.org/costs/Transportation/Biodiesel.

Islam, M.N., Alsenani, F., Schenk, P.M., 2017. Microalgae as a sustainable source of nutraceuticals. In: Microbial Functional Foods and Nutraceuticals. Wiley, p. 1.

Japar, A.S., Takriff, M.S., Yasin, N.H.M., 2017. Harvesting microalgal biomass and lipid extraction for potential biofuel production: a review. J. Environ. Chem. Eng. 5 (1), 555–563.

John, S., Terri, D., John, B., Paul, R., 1998. A Look Back at the U.S. Department of Energy's Aquatic Species Program—Biodiesel From Algae, A National Laboratory of the U.S. Department of Energy Operated by Midwest Research Institute Under Contract No. DE-AC36-83CH10093.

Jones, I.S.F., Lu, C.H., 2003. Engineering carbon sequestration in the ocean. In: Proceedings of the 2nd Annual Conference on Carbon Sequestration, May 6, 2003, Washington, DC.

Joshi, S., Kumari, R., Upasani, V.N., 2018. Applications of algae in cosmetics: an overview. Int. J. Innov. Res. Sci. Eng. Technol. 7 (2), 1269.

Juneja, A., Ceballos, R.M., Murthy, G.S., 2013. Effects of environmental factors and nutrient availability on the biochemical composition of algae for biofuels production: a review. Energies 6 (9), 4607–4638.

Jung, J.Y., Lee, H., Shin, W.S., Sung, M.G., Kwon, J.H., Yang, J.W., 2015. Utilization of seawater for cost-effective cultivation and harvesting of Scenedesmus obliquus. Bioprocess Biosyst. Eng. 38 (3), 449–455.

Kalita, N.K., Damare, N.A., Hazarika, D., Bhagabati, P., Kalamdhad, A., Katiyar, V., 2021. Biodegradation and characterization study of compostable PLA bioplastic containing algae biomass as potential degradation accelerator. Environ. Chall. 3, 100067. https://doi.org/10.1016/j.envc.2021.100067.

Kannan, D.C., Pattarkine, V.M., 2014. Recovery of lipids from algae. In: Algal Biorefineries. Springer, Dordrecht, pp. 297–310. http://www.springer.com/de/book/9789400774933.

Kannan, D.C., Venkat, D., 2019. An open outdoor algal growth system of improved productivity for biofuel production. J. Chem. Technol. Biotechnol. 94 (1), 222–235. https://doi.org/10.1002/jctb.5768.

Kapoore, R., Butler, T., Pandhal, J., Vaidyanathan, S., 2018. Microwave-assisted extraction for microalgae: from biofuels to biorefinery. Biology (Basel) 7, 18. https://doi.org/10.3390/biology7010018.

Karas, B.J., Diner, R.E., Lefebvre, S.C., McQuaid, J., Phillips, A.P., Noddings, C.M., Weyman, P.D., 2015. Designer diatom episomes delivered by bacterial conjugation. Nat. Commun. 6 (1), 1–10.

Karatzos, S., van Dyk, J.S., McMillan, J.D., Saddler, J., 2017. Drop-in biofuel production via conventional (lipid/fatty acid) and advanced (biomass) routes. Part I. Biofuels Bioprod. Biorefin. 11 (2), 344–362.

Karmee, S.K., Patria, R.D., Lin, C.S.K., 2015. Techno-economic evaluation of biodiesel production from waste cooking oil—a case study of Hong Kong. Int. J. Mol. Sci. 16 (3), 4362–4371. https://doi.org/10.3390/ijms16034362.

Katiyar, V., 2021. Personal Communication. Department of Chemical Engineering, IIT Guwahati, India (relating to ongoing and yet to be published research).

Kaur, P., 2020. Microalgae as nutraceutical for achieving sustainable food solution in future. In: Microbial Biotechnology: Basic Research and Applications. Springer, Singapore, pp. 91–125.

Kiani, A., Jiang, K., Feron, P., 2020. Techno-economic assessment for CO_2 capture from air using a conventional liquid-based absorption process. Front. Energy Res. 8, 92.

Kim, S., Lee, Y.C., Cho, D.H., Lee, H.U., Huh, Y.S., Kim, G.J., Kim, H.S., 2014. A simple and non-invasive method for nuclear transformation of intact-walled *Chlamydomonas reinhardtii*. PLoS One 9 (7), e101018.

Kindle, K.L., Schnell, R.A., Fernández, E., Lefebvre, P.A., 1989. Stable nuclear transformation of Chlamydomonas using the Chlamydomonas gene for nitrate reductase. J. Cell Biol. 109 (6), 2589–2601.

Kloosterman, K., 2013. Seambiotic Makes Algae for Food and Biofuel. Green Prophet. Retrieved from 11 April 2013 www.greenprophet.com/2011/09/seabiotic-biofuel-algae/. (Accessed 23 June 2021).

Knuckey, R.M., Brown, M.R., Robert, R., Frampton, D.M., 2006. Production of microalgal concentrates by flocculation and their assessment as aquaculture feeds. Aquac. Eng. 35 (3), 300–313.

Koley, S., Mathimani, T., Bagchi, S.K., Sonkar, S., Mallick, N., 2019. Microalgal biodiesel production at outdoor open and polyhouse raceway pond cultivations: a case study with Scenedesmus accuminatus using low-cost farm fertilizer medium. Biomass Bioenergy 120, 156–165. https://doi.org/10.1016/j.biombioe.2018.11.002.

Kumar, S.V., Misquitta, R.W., Reddy, V.S., Rao, B.J., Rajam, M.V., 2004. Genetic transformation of the green alga—*Chlamydomonas reinhardtii* by *Agrobacterium tumefaciens*. Plant Sci. 166 (3), 731–738.

Kumar, G., Shekh, A., Jakhu, S., Sharma, Y., Kapoor, R., Sharma, T.R., 2020a. Bioengineering of microalgae: recent advances, perspectives, and regulatory challenges for industrial application. Front. Bioeng. Biotechnol. 8, 914. https://www.frontiersin.org/articles/10.3389/fbioe.2020.00914/full.

Kumar, M., Sun, Y., Rathour, R., Pandey, A., Thakur, I.S., Tsang, D.C., 2020b. Algae as potential feedstock for the production of biofuels and value-added products: opportunities and challenges. Sci. Total Environ. 716, 137116. https://doi.org/10.1016/j.scitotenv.2020.137116.

Lai, Y.J.S., De Francesco, F., Aguinaga, A., Parameswaran, P., Rittmann, B.E., 2016. Improving lipid recovery from Scenedesmus wet biomass by surfactant-assisted disruption. Green Chem. 18, 1319–1326. https://doi.org/10.1039/c5gc02159f.

Lali, A.M., Pandit, R.A., Gunjan, P., Mathpati, C.S., Gangal, S.P., Vira, C.P., Palkar, J.A., Patil, S.D., Gaikwad, S.P., 2013. Raceway Pond System for Increased Biomass Productivity. Pub. No: WO/2013/186626, International Application No: PCT/IB2013/001224, India.

LaMonica, M., 2008, June 13. Algae Farm in Mexico to Produce Ethanol in '09. CNET. https://www.cnet.com/news/algae-farm-in-mexico-to-produce-ethanol-in-09/.

Lane, J., 2016, November 1. Hydrothermal Processing and Algae to Oil: The Digest's 2016 Multi-Slide Guide to Reliance Industries: Biofuels Digest. The Digest. https://www.biofuelsdigest.com/bdigest/2016/11/01/hydrothermal-processing-and-algae-to-oil-the-digests-2016-multi-slide-guide-to-reliance-industries/.

Larsdotter, K., 2006. Wastewater treatment with microalgae—a literature review. Vatten 62 (1), 31.

Le Fevre, C.N., 2019. A Review of Prospects for Natural Gas as a Fuel in Road Transport. Oxford Institute for Energy Studies, Oxford, UK. https://www.oxfordenergy.org/wpcms/wp-content/uploads/2019/04/A-review-of-prospects-for-natural-gas-as-a-fuel-in-road-transport-Insight-50.pdf.

Lee, Y.K., 2001. Microalgal mass culture systems and methods: their limitation and potential. J. Appl. Phycol. 13 (4), 307–315.

Lenntech, 2021. Phosphorous Removal from Wastewater. Retrieved 24 June 2021, from https://www.lenntech.com/phosphorous-removal.htm.

Li, X., Jonikas, M.C., 2016. High-throughput genetics strategies for identifying new components of lipid metabolism in the green alga Chlamydomonas reinhardtii. In: Lipids in Plant and Algae Development. Springer, Cham, pp. 223–247.

Li, Y., Han, D., Hu, G., Dauvillee, D., Sommerfeld, M., Ball, S., Hu, Q., 2010. Chlamydomonas starchless mutant defective in ADP-glucose pyrophosphorylase hyper-accumulates triacylglycerol. Metab. Eng. 12 (4), 387–391.

Li, Y.H., Li, H.B., Xu, X.Y., Xiao, S.Y., Wang, S.Q., Xu, S.C., 2017. Fate of nitrogen in subsurface infiltration system for treating secondary effluent. Water Sci. Eng. 10 (3), 217–224. https://doi.org/10.1016/j.wse.2017.10.002.

Liu, J.C., Chen, Y.M., Ju, Y.H., 1999. Separation of algal cells from water by column flotation. Sep. Sci. Technol. 34 (11), 2259–2272.

Lohman, E.J., Gardner, R.D., Pedersen, T., Peyton, B.M., Cooksey, K.E., Gerlach, R., 2015. Optimized inorganic carbon regime for enhanced growth and lipid accumulation in *Chlorella vulgaris*. Biotechnol. Biofuels 8 (1), 1–13.

Luis, P., 2016. Use of monoethanolamine (MEA) for CO_2 capture in a global scenario: consequences and alternatives. Desalination 380, 93–99.

Lum, K.K., Kim, J., Lei, X.G., 2013. Dual potential of microalgae as a sustainable biofuel feedstock and animal feed. J. Anim. Sci. Biotechnol. 4 (1), 1–7.

Lundquist, T.J., Woertz, I.C., Quinn, N.W.T., Benemann, J.R., 2010 October. A Realistic Technology and Engineering Assessment of Algae Biofuel Production. Energy Biosciences Institute, University of California, Berkeley, CA.

Luo, X., Ge, X., Cui, S., Li, Y., 2016. Value-added processing of crude glycerol into chemicals and polymers. Bioresour. Technol. 215, 144–154. https://doi.org/10.1016/j.biortech.2016.03.042.

Madeira, M.S., Cardoso, C., Lopes, P.A., Coelho, D., Afonso, C., Bandarra, N.M., Prates, J.A., 2017. Microalgae as feed ingredients for livestock production and meat quality: a review. Livest. Sci. 205, 111–121.

Magar, C.S., Rambhiya, S.J., Deodhar, M.A., 2019. Evaluation of CO_2 removal efficiency of Pseudanabaena limnetica (Lemm.) Komárek grown in Na_2CO_3 enriched seawater medium in 60 L airlift flat panel photobioreactor. J. Environ. Sci. Technol. 12 (4), 186–196.

Mahapatra, D.M., Chanakya, H.N., Ramachandra, T.V., 2013. Treatment efficacy of algae-based sewage treatment plants. Environ. Monit. Assess. 185 (9), 7145–7164.

Mahmood, K.A., Khudhair, E.M., 2017. Experimental study for commercial fertilizer NPK (20: 20: 20+ te N: P: K) in microalgae cultivation at different aeration periods. Iraqi J. Chem. Pet. Eng. 18 (1), 99–110.

Mallon, W., Buit, L., van Wingerden, J., Lemmens, H., Eldrup, N.H., 2013. Costs of CO_2 transportation infrastructures. Energy Procedia 37, 2969–2980. https://doi.org/10.1016/j.egypro.2013.06.183.

Mathur, M., Bhattacharya, A., Malik, A., 2017. Advancements in algal harvesting techniques for biofuel production. Algal Biofuels, 227–245.

Matos, J., Cardoso, C., Bandarra, N.M., Afonso, C., 2017. Microalgae as healthy ingredients for functional food: a review. Food Funct. 8 (8), 2672–2685.

Maurya, R., Ghosh, T., Saravaia, H., Paliwal, C., Ghosh, A., Mishra, S., 2016a. Non-isothermal pyrolysis of de-oiled microalgal biomass: kinetics and evolved gas analysis. Bioresour. Technol. 221, 251–261.

Maurya, R., Paliwal, C., Ghosh, T., Pancha, I., Chokshi, K., Mitra, M., Mishra, S., 2016b. Applications of de-oiled microalgal biomass towards development of sustainable biorefinery. Bioresour. Technol. 214, 787–796.

Mayfield, S.P., Kindle, K.L., 1990. Stable nuclear transformation of *Chlamydomonas reinhardtii* by using a *C. reinhardtii* gene as the selectable marker. Proc. Natl. Acad. Sci. 87 (6), 2087–2091.

MHA, 2020. Annual Report 2019–20. Ministry of Home Affairs, Government of India. https://www.mha.gov.in/documents/annual-reports. (Accessed 30 June 2021).

Minhas, A.K., Hodgson, P., Barrow, C.J., Adholeya, A., 2016. A review on the assessment of stress condition for simultaneous production of microalgal lipids and carotenoids. Front. Microbiol. 7, 1–19.

Mirón, A.S., Garcıa, M.C.C., Gómez, A.C., Camacho, F.G., Grima, E.M., Chisti, Y., 2003. Shear stress tolerance and biochemical characterization of *Phaeodactylum tricornutum* in quasi steady-state continuous culture in outdoor photobioreactors. Biochem. Eng. J. 16 (3), 287–297.

Mohanty, S.J.N., 2020, January 22. Analysis: India's Jet Fuel Demand to Emerge From the Red as Turbulence Subsides. S&P Global Platts. https://www.spglobal.com/platts/en/market-insights/latest-news/oil/012220-india-jet-fuel-demand-to-emerge-from-red.

Moheimani, N.R., Borowitzka, M.A., 2011. Increased CO_2 and the effect of pH on growth and calcification of Pleurochrysis carterae and Emiliania huxleyi (Haptophyta) in semicontinuous cultures. Appl. Microbiol. Biotechnol. 90 (4), 1399–1407.

Mokhena, T.C., Sefadi, J.S., Sadiku, E.R., John, M.J., Mochane, M.J., Mtibe, A., 2018. Thermoplastic processing of PLA/cellulose nanomaterials composites. Polymers 10 (12), 1363. https://doi.org/10.3390/polym10121363.

Molazadeh, M., Ahmadzadeh, H., Pourianfar, H.R., Lyon, S., Rampelotto, P.H., 2019. The use of microalgae for coupling wastewater treatment with CO_2 biofixation. Front. Bioeng. Biotechnol. 7, 42.

Molina Grima, E., Belarbi, E.H., Acién Fernández, F.G., Robles Medina, A., Chisti, Y., 2003. Recovery of microalgal biomass and metabolites: process options and economics. Biotechnol. Adv. 20, 491–515. https://doi.org/10.1016/S0734-9750(02)00050-2.

Monkonsit, S., Powtongsook, S., Pavasant, P., 2011. Comparison between airlift photobioreactor and bubble column for *Skeletonema costatum* cultivation. Eng. J. 15, 53–64. https://doi.org/10.4186/ej.2011.15.4.53.

MoSPI, 2018. Energy Statistics. Ministry of Statistics and Programme Implementation, Government of India. http://mospi.nic.in/publication/energy-statistics-2018.

Mtaki, K., Kyewalyanga, M.S., Mtolera, M.S., 2021. Supplementing wastewater with NPK fertilizer as a cheap source of nutrients in cultivating live food (*Chlorella vulgaris*). Ann. Microbiol. 71 (1), 1–13.

Munoz, J., Merrett, M.J., 1989. Inorganic-carbon transport in some marine eukaryotic microalgae. Planta 178, 450–455.

Mus, F., Toussaint, J.P., Cooksey, K.E., Fields, M.W., Gerlach, R., Peyton, B.M., Carlson, R.P., 2013. Physiological and molecular analysis of carbon source supplementation and pH stress-induced lipid accumulation in the marine diatom *Phaeodactylum tricornutum*. Appl. Microbiol. Biotechnol. 97 (8), 3625–3642.

Nagappan, S., Devendran, S., Tsai, P.C., Dahms, H.U., Ponnusamy, V.K., 2019. Potential of two-stage cultivation in microalgae biofuel production. Fuel 252, 339–349.

Narala, R.R., Garg, S., Sharma, K.K., Thomas-Hall, S.R., Deme, M., Li, Y., Schenk, P.M., 2016. Comparison of microalgae cultivation in photobioreactor, open raceway pond, and a two-stage hybrid system. Front. Energy Res. 4, 29.

NASA, 2019. The Atmosphere: Getting a Handle on Carbon Dioxide, Part Two, Sizing Up Humanity's Impacts on Earth's Changing Atmosphere: A Five-Part Series by Alan Buis. NASA's Jet Propulsion Laboratory. https://climate.nasa.gov/news/2915/the-atmosphere-getting-a-handle-on-carbon-dioxide/. (Accessed 9 October 2019).

Navarro, F., Forján, E., Vázquez, M., Montero, Z., Bermejo, E., Castaño, M.Á., Vega, J.M., 2016. Microalgae as a safe food source for animals: nutritional characteristics of the acidophilic microalga Coccomyxa onubensis. Food Nutr. Res. 60 (1), 30472.

Nethravathy, M.U., Mehar, J.G., Mudliar, S.N., Shekh, A.Y., 2019. Recent advances in microalgal bioactives for food, feed, and healthcare products: commercial potential, market space, and sustainability. Compr. Rev. Food Sci. Food Saf. 18 (6), 1882–1897.

Nuwer, R., 2019, September 30. How Algae Could Help Save us from the Impending Freshwater Crisis. Massive Science. https://massivesci.com/articles/algae-salt-water-ocean-climate-change-agriculture/.

Oliver, L., Dietrich, T., Marañón, I., Villarán, M.C., Barrio, R.J., 2020. Producing omega-3 polyunsaturated fatty acids: a review of sustainable sources and future trends for the EPA and DHA market. Resources 9 (12), 148. https://doi.org/10.3390/resources9120148.

Ooms, M.D., Dinh, C.T., Sargent, E.H., Sinton, D., 2016. Photon management for augmented photosynthesis. Nat. Commun. 7, 12699.

Oswald, W.J., 1988. Micro-algae and waste-water treatment. In: Borowitzka, M.A., Borowitzka, L.J. (Eds.), Micro-Algal Biotechnology. Cambridge University Press, Cambridge, pp. 305–328.

Parate, R.D., Rode, C.V., Dharne, M.S., 2018. 2,3-Butanediol production from biodiesel derived glycerol. Curr. Environ. Eng. 5 (1), 4–12. https://doi.org/10.2174/2212717805666180112162517.

Parlevliet, D., Moheimani, N.R., 2014. Efficient conversion of solar energy to biomass and electricity. Aquat. Biosyst. 10 (1), 1–9.

Patel, S., Kannan, D.C., 2021. A method of wet algal lipid recovery for biofuel production. Algal Res. 55, 102237. https://doi.org/10.1016/j.algal.2021.102237.

Pavlik, D., Zhong, Y., Daiek, C., Liao, W., Morgan, R., Clary, W., Liu, Y., 2017. Microalgae cultivation for carbon dioxide sequestration and protein production using a high-efficiency photobioreactor system. Algal Res. 25, 413–420.

Peltomaa, E., Johnson, M.D., Taipale, S.J., 2018. Marine cryptophytes are great sources of EPA and DHA. Mar. Drugs 16 (1), 3. https://doi.org/10.3390/md16010003.

Perrine, Z., Negi, S., Sayre, R.T., 2012. Optimization of photosynthetic light energy utilization by microalgae. Algal Res. 1 (2), 134–142.

Petroalgae.com, 2019, June 3. Petro Algae—Providing a Renewable, Sustainable, Carbon-Neutral Source of Oil. Petroalgae. http://www.petroalgae.com.

Pieber, S., Schober, S., Mittelbach, M., 2012. Pressurized fluid extraction of polyunsaturated fatty acids from the microalga *Nannochloropsis oculata*. Biomass Bioenergy 47, 474–482. https://doi.org/10.1016/j.biombioe.2012.10.019.

PlasticsInsight, 2021. Polyurethane Production, Pricing and Market Demand. Plastics Insight. https://www. plasticsinsight.com/resin-intelligence/resin-prices/polyurethane/.

Posewitz, M., 2019, March. DOE Bioenergy Technologies Office (BETO) 2019 Project Peer Review PACE: Producing Algae for Coproducts and Energy [Slides]. Energy.Gov. https://www.energy.gov/sites/prod/files/2019/03/f61/PACE%20-%20Producing%20Algae%20for%20Coproducts%20and%20Energy_EE0007089.pdf.

Pourkarimi, S., Hallajisani, A., Alizadehdakhel, A., Nouralishahi, A., 2019. Biofuel production through micro-and macroalgae pyrolysis—a review of pyrolysis methods and process parameters. J. Anal. Appl. Pyrolysis 142, 104599.

Prochazkova, G., Podolova, N., Safarik, I., Zachleder, V., Branyik, T., 2013. Physicochemical approach to freshwater microalgae harvesting with magnetic particles. Colloids Surf. B: Biointerfaces 112, 213–218.

Qiao, H., Wang, G., 2009. Effect of carbon source on growth and lipid accumulation in *Chlorella sorokiniana* GXNN01. Chin. J. Oceanol. Limnol. 27 (4), 762.

Qu, F., Liang, H., Wang, Z., Wang, H., Yu, H., Li, G., 2012. Ultrafiltration membrane fouling by extracellular organic matters (EOM) of *Microcystis aeruginosa* in stationary phase: influences of interfacial characteristics of foulants and fouling mechanisms. Water Res. 46 (5), 1490–1500.

Quinn, J.C., Davis, R., 2015. The potentials and challenges of algae based biofuels: a review of the techno-economic, life cycle, and resource assessment modelling. Bioresour. Technol. 184, 444–452.

Ramanan, R., Kim, B.H., Cho, D.H., Oh, H.M., Kim, H.S., 2016. Algae-bacteria interactions: evolution, ecology and emerging applications. Biotechnol. Adv. 34, 14–29.

Rambhiya, S.J., Magar, C.S., Deodhar, M.A., 2021. Using seawater-based Na_2CO_3 medium for scrubbing the CO_2 released from bio-CNG plant for enhanced biomass production of *Pseudanabaena limnetica*. SN Appl. Sci. 3 (2), 1–17.

Razzaq, A., Saleem, F., Kanwal, M., Mustafa, G., Yousaf, S., Imran Arshad, H.M., Joyia, F.A., 2019. Modern trends in plant genome editing: an inclusive review of the CRISPR/Cas9 toolbox. Int. J. Mol. Sci. 20 (16), 4045.

Reus, A., 2020, December 11. US Animal Feed Costs Expected to Rise 12% in 2021. Feed Strategy. https://www.feedstrategy.com/business-markets/us-animal-feed-costs-expected-to-rise-12-in-2021/.

Reuters, 2020. Sharafedin, B. T. H. (2020, April 15) Jet Fuel Demand to Remain Low as Airlines Buckle up for Tough Ride. https://www.reuters.com/article/us-global-oil-jet-fuel-idUSKCN21X1DS.

Richter, B.E., Jones, B.A., Ezzell, J.L., Porter, N.L., Avdalovic, N., Pohl, C., 1996. Accelerated solvent extraction: a technique for sample preparation. Anal. Chem. 68, 1033–1039. https://doi.org/10.1021/ac9508199.

Rickman, M., Pellegrino, J., Davis, R., 2012. Fouling phenomena during membrane filtration of microalgae. J. Membr. Sci. 423, 33–42.

Rochaix, J.D., Van Dillewijn, J., 1982. Transformation of the green alga Chlamydomonas reinhardii with yeast DNA. Nature 296 (5852), 70–72.

Rogers, J.N., Rosenberg, J.N., Guzman, B.J., Oh, V.H., Mimbela, L.E., Ghassemi, A., Donohue, M.D., 2014. A critical analysis of paddle wheel-driven raceway ponds for algal biofuel production at commercial scales. Algal Res. 4, 76–88. https://doi.org/10.1016/j.algal.2013.11.007.

Ronga, D., Biazzi, E., Parati, K., Carminati, D., Carminati, E., Tava, A., 2019. Microalgal biostimulants and biofertilisers in crop productions. Agronomy 9 (4), 192.

Salmón, I.R., Cambier, N., Luis, P., 2018. CO_2 capture by alkaline solution for carbonate production: a comparison between a packed column and a membrane contactor. Appl. Sci. 2018 (8), 996.

Sánchez Mirón, A., García Camacho, F., Contreras Gómez, A., Grima, E.M., Chisti, Y., 2000. Bubble-column and airlift photobioreactors for algal culture. AIChE J. 46, 1872–1887.

Sandhya, S.V., Vijayan, K.K., 2019. Symbiotic association among marine microalgae and bacterial flora: a study with special reference to commercially important Isochrysis galbana culture. J. Appl. Phycol. 31 (4), 2259–2266.

sapphireenergy.com, 2021. Sapphire Energy. http://sapphireenergy.com/. (Accessed 13 March 2021).

Sawant, S.S., Gajbhiye, B.D., Mathpati, C.S., Pandit, R., Lali, A.M., 2018, September. Microalgae as sustainable energy and its cultivation. IOP Conf. Ser. Mater. Sci. Eng. 360 (1), 012025. https://iopscience.iop.org/article/10.1088/1757-899X/360/1/012025/pdf.

Sharma, S.K., Nelson, D.R., Abdrabu, R., Khraiwesh, B., Jijakli, K., Arnoux, M., Salehi-Ashtiani, K., 2015. An integrative Raman microscopy-based workflow for rapid in situ analysis of microalgal lipid bodies. Biotechnol. Biofuels 8 (1), 1–14.

Shim, J.-G., Lee, D.W., Lee, J.H., Kwak, N.-S., 2016. Experimental study on capture of carbon dioxide and production of sodium bicarbonate from sodium hydroxide. Environ. Eng. Res. 21 (3), 297–303.

Shin, S.E., Lim, J.M., Koh, H.G., Kim, E.K., Kang, N.K., Jeon, S., Jeong, B.R., 2016a. CRISPR/Cas9-induced knockout and knock-in mutations in *Chlamydomonas reinhardtii*. Sci. Rep. 6 (1), 1–15.

Shin, W.S., Lee, B., Jeong, B.R., Chang, Y.K., Kwon, J.H., 2016b. Truncated light-harvesting chlorophyll antenna size in *Chlorella vulgaris* improves biomass productivity. J. Appl. Phycol. 28 (6), 3193–3202.

Shokravi, Z., Shokravi, H., Chyuan, O.H., Lau, W.J., Koloor, S.S.R., Petrů, M., Ismail, A.F., 2020. Improving 'lipid productivity' in microalgae by bilateral enhancement of biomass and lipid contents: a review. Sustainability 12 (21), 9083.

Show, Y., Lee, D.J., 2014. Microalgal biomass harvesting. In: Pandey, A., Lee, D.J., Chisti, Y., Soccol, C.R. (Eds.), Biofuels From Microalgae. Elsevier.

Show, K.Y., Lee, D.J., Mujumdar, A.S., 2015. Advances and challenges on algae harvesting and drying. Dry. Technol. 33, 386–394. https://doi.org/10.1080/07373937.2014.948554.

Show, K.Y., Yan, Y.G., Lee, D.J., 2019. Algal biomass harvesting and drying. In: Biofuels From Algae. Elsevier, pp. 135–166.

Silva, S.C., Ferreira, I.C., Dias, M.M., Barreiro, M.F., 2020. Microalgae-derived pigments: a 10-year bibliometric review and industry and market trend analysis. Molecules 25 (15), 3406. https://doi.org/10.3390/molecules25153406.

Silve, A., Papachristou, I., Wüstner, R., Sträßner, R., Schirmer, M., Leber, K., Guo, B., Interrante, L., Posten, C., Frey, W., 2018. Extraction of lipids from wet microalga Auxenochlorella protothecoides using pulsed electric field treatment and ethanol-hexane blends. Algal Res. 29, 212–222. https://doi.org/10.1016/j.algal.2017.11.016.

Singh, G., Patidar, S.K., 2018. Microalgae harvesting techniques: a review. J. Environ. Manag. 217, 499–508.

Sipaúba-Tavares, L.H., Segali, A.M.D.L., Berchielli-Morais, F.A., Scardoeli-Truzzi, B., 2017. Development of low-cost culture media for Ankistrodesmus gracilis based on inorganic fertilizer and macrophyte. Acta Limnol. Bras., 29. http://www.scielo.br/pdf/alb/v29/2179-975X-alb-29-e5.pdf.

Smith, E.G., D'angelo, C., Sharon, Y., Tchernov, D., Wiedenmann, J., 2017. Acclimatization of symbiotic corals to mesophotic light environments through wavelength transformation by fluorescent protein pigments. Proc. R. Soc. B Biol. Sci. 284 (1858), 20170320.

Spreitzer, R.J., Peddi, S.R., Satagopan, S., 2005. Phylogenetic engineering at an interface between large and small subunits imparts land-plant kinetic properties to algal rubisco. Proc. Natl. Acad. Sci. U. S. A. 102, 17225–17230. https://doi.org/10.1073/pnas.0508042102.

Stammbach, M.R., De Nys, P., Heimann, K., Rogers, A. (2010). Method of culturing photosynthetic organisms. PCT Patent Application WO2010/132917 (25 November 2010).

Start-Up Nation Finder, 2021. Seambiotic. Start-Up Nation Finder. Retrieved 23 June 2021, from https://finder.startupnationcentral.org/company_page/seambiotic.

Statista, 2020. Sewage Generation and Treatment Capacity India 2017 by City (2020, November 16), Jaganmohan M. Statista. https://www.statista.com/statistics/1168389/india-sewage-generation-and-treatment-capacity-by-city/.

Statista, 2021. Global Polyurethane Market Volume 2015–2021. Statista. https://www.statista.com/statistics/720341/global-polyurethane-market-size-forecast/. (Accessed 23 June 2021).

Sucaldito, M.R., Camacho, D.H., 2017. Characteristics of unique HBr-hydrolyzed cellulose nanocrystals from freshwater green algae (*Cladophora rupestris*) and its reinforcement in starch-based film. Carbohydr. Polym. 169, 315–323. https://doi.org/10.1016/j.carbpol.2017.04.031.

Sun, X., Wang, C., Tong, Y., Wang, W., Wei, J., 2013. A comparative study of microfiltration and ultrafiltration for algae harvesting. Algal Res. 2 (4), 437–444.

Sun, W., Liu, J., Chu, H., Dong, B., 2013b. Pretreatment and membrane hydrophilic modification to reduce membrane fouling. Membranes 3 (3), 226–241. https://doi.org/10.3390/membranes3030226.

Sun, D., Yamada, Y., Sato, S., Ueda, W., 2017. Glycerol as a potential renewable raw material for acrylic acid production. Green Chem. 19 (14), 3186–3213. https://doi.org/10.1039/C7GC00358G.

Synthetic Genomics, 2020, October 14. Synthetic Genomics, Inc. Synthetic Genomics, Inc. https://syntheticgenomics.com/.

Takagi, M., Yoshida, T., 2006. Effect of salt concentration on intracellular accumulation of lipids and triacylglyceride in marine microalgae Dunaliella cells. J. Biosci. Bioeng. 101 (3), 223–226. https://doi.org/10.1263/jbb.101.223.

Tang, H., Chen, M., Ng, K.Y., Salley, S.O., 2012. Continuous microalgae cultivation in a photobioreactor. Biotechnol. Bioeng. 109, 2468–2474.

Uduman, N., Qi, Y., Danquah, M.K., Forde, G.M., Hoadley, A., 2010. Dewatering of microalgal cultures: a major bottleneck to algae-based fuels. J. Renew. Sust. Energy 2 (1), 012701.

Usami, R., Fujii, K., Fushimi, C., 2020. Improvement of bio-oil and nitrogen recovery from microalgae using two-stage hydrothermal liquefaction with solid carbon and HCl acid catalysis. ACS Omega 5 (12), 6684–6696.

Valenzuela, J., Mazurie, A., Carlson, R.P., Gerlach, R., Cooksey, K.E., Peyton, B.M., Fields, M.W., 2012. Potential role of multiple carbon fixation pathways during lipid accumulation in *Phaeodactylum tricornutum*. Biotechnol. Biofuels 5 (1), 1–17.

Valenzuela, J., Carlson, R.P., Gerlach, R., Cooksey, K., Peyton, B.M., Bothner, B., Fields, M.W., 2013. Nutrient resupplementation arrests bio-oil accumulation in *Phaeodactylum tricornutum*. Appl. Microbiol. Biotechnol. 97 (15), 7049–7059.

Vandamme, D., Foubert, I., Meesschaert, B., Muylaert, K., 2010. Flocculation of microalgae using cationic starch. J. Appl. Phycol. 22 (4), 525–530.

Vandamme, D., Foubert, I., Fraeye, I., Meesschaert, B., Muylaert, K., 2012. Flocculation of *Chlorella vulgaris* induced by high pH: role of magnesium and calcium and practical implications. Bioresour. Technol. 105, 114 119.

Vargas e Silva, F., Monteggia, L.O., 2015. Pyrolysis of algal biomass obtained from high-rate algae ponds applied to wastewater treatment. Front. Energy Res. 3, 31.

Vinyard, D.J., Gimpel, J., Ananyev, G.M., Mayfield, S.P., Dismukes, G.C., 2014. Engineered photosystem II reaction centers optimize photochemistry versus photoprotection at different solar intensities. J. Am. Chem. Soc. 136 (10), 4048–4055.

VITO, 2019. From Micro to Macro: Algae Are Full of Useful Components. Retrieved June 30 from https://vito.be/en/news/micro-macro-algae-are-full-useful-components.

Vitova, M., Bisova, K., Kawano, S., Zachleder, V., 2015. Accumulation of energy reserves in algae: from cell cycles to biotechnological applications. Biotechnol. Adv. 33 (6), 1204–1218.

Wang, Y., Ho, S.-H., Cheng, C.-L., Guo, W.-Q., Nagarajan, D., Ren, N.-Q., Lee, D.-J., Chang, J.-S., 2016. Perspectives on the feasibility of using microalgae for industrial wastewater treatment. Bioresour. Technol. 222, 485–497. https://doi.org/10.1016/j.biortech.2016.09.106.

Wang, W., Zhang, S., Yu, Q., Lin, Y., Yang, N., Han, W., Zhang, J., 2017. Hydrothermal liquefaction of high protein microalgae via clay material catalysts. RSC Adv. 7 (80), 50794–50801.

Watanabe, K., Takihana, N., Aoyagi, H., Hanada, S., Watanabe, Y., Ohmura, N., Tanaka, H., 2005. Symbiotic association in chlorella culture. FEMS Microbiol. Ecol. 51 (2), 187–196.

Weinberger, G., Hentschke, C., Neis, U., Ergünsel, A., Pereira, R., Gere, P., Thiebeaut, Q., 2012. Combined ALgal and BActerial Waste Water Treatment for High Environmental QUAlity Effluents (ALBAQUA). Project Report. PTS, Germany. Available from: http://www.ptspaper.de/fileadmin/PTS/Dokumente/Forschung/Forschungsprojekte/EU_ALBAQUA.pdf.

Wen, X., Du, K., Wang, Z., et al., 2016. Effective cultivation of microalgae for biofuel production: a pilot-scale evaluation of a novel oleaginous microalga *Graesiella* sp. WBG-1. Biotechnol. Biofuels 9 (123). https://doi.org/10.1186/s13068-016-0541-y.

Widjaja, A., Chien, C.C., Ju, Y.H., 2009. Study of increasing lipid production from fresh water microalgae Chlorella vulgaris. J. Taiwan Inst. Chem. Eng. 40 (1), 13–20.

Wikipedia, 2021. Sapphire Energy. Wikipedia. https://en.wikipedia.org/wiki/Sapphire_Energy.

Wu, Z., Zhu, Y., Huang, W., Zhang, C., Li, T., Zhang, Y., Li, A., 2012. Evaluation of flocculation induced by pH increase for harvesting microalgae and reuse of flocculated medium. Bioresour. Technol. 110, 496–502.

Wyatt, N.B., Gloe, L.M., Brady, P.V., Hewson, J.C., Grillet, A.M., Hankins, M.G., Pohl, P.I., 2012. Critical conditions for ferric chloride-induced flocculation of freshwater algae. Biotechnol. Bioeng. 109 (2), 493–501.

Xia, L., Ge, H., Zhou, X., Zhang, D., Hu, C., 2013. Photoautotrophic outdoor two-stage cultivation for oleaginous microalgae *Scenedesmus obtusus* XJ-15. Bioresour. Technol. 144, 261–267.

Xu, L., Guo, C., Wang, F., Zheng, S., Liu, C.Z., 2011. A simple and rapid harvesting method for microalgae by in situ magnetic separation. Bioresour. Technol. 102 (21), 10047–10051.

Yao, S., Lyu, S., An, Y., Lu, J., Gjermansen, C., Schramm, A., 2019. Microalgae–bacteria symbiosis in microalgal growth and biofuel production: a review. J. Appl. Microbiol. 126 (2), 359–368.

Yaqoob, S., Riaz, M., Shabbir, A., Zia-Ul-Haq, M., Alwakeel, S.S., Bin-Jumah, M., 2021. Commercialization and marketing potential of carotenoids. In: Carotenoids: Structure and Function in the Human Body. Springer Nature Switzerland, p. 799, https://doi.org/10.1007/978-3-030-46459-2_27.

Yoo, G., Park, W.K., Kim, C.W., Choi, Y.E., Yang, J.W., 2012. Direct lipid extraction from wet Chlamydomonas reinhardtii biomass using osmotic shock. Bioresour. Technol. 123, 717–722. https://doi.org/10.1016/j.biortech.2012.07.102.

Yu, G., Zhang, Y., Schideman, L., Funk, T., Wang, Z., 2011. Distributions of carbon and nitrogen in the products from hydrothermal liquefaction of low-lipid microalgae. Energy Environ. Sci. 4 (11), 4587–4595.

Yun, J.H., Cho, D.H., Lee, S., Heo, J., Tran, Q.G., Chang, Y.K., Kim, H.S., 2018. Hybrid operation of photobioreactor and wastewater-fed open raceway ponds enhances the dominance of target algal species and algal biomass production. Algal Res. 29, 319–329.

ZEP, 2011. European Technology Platform for Zero Emission Fossil Fuel Power Plants (2011, July). The Costs of CO_2 Transport, Post-Demonstration CCS in the EU (No. 119811-Cover-82451). Global CCS Institute. https://www.globalccsinstitute.com/archive/hub/publications/119811/costs-co2-transport-post-demonstration-ccs-eu.pdf.

Zhang, X., Hu, Q., Sommerfeld, M., Puruhito, E., Chen, Y., 2010. Harvesting algal biomass for biofuels using ultra-filtration membranes. Bioresour. Technol. 101, 5297–5304. https://doi.org/10.1016/j.biortech.2010.02.007.

Zheng, H., Gao, Z., Yin, J., Tang, X., Ji, X., Huang, H., 2012. Harvesting of microalgae by flocculation with poly (γ-glutamic acid). Bioresour. Technol. 112, 212–220. https://doi.org/10.1016/j.biortech.2012.02.086.

Zhu, X.G., Long, S.P., Ort, D.R., 2008. What is the maximum efficiency with which photosynthesis can convert solar energy into biomass? Curr. Opin. Biotechnol. 19 (2), 153–159.

Zhu, L.D., Li, Z.H., Hiltunen, E., 2016. Strategies for lipid production improvement in microalgae as a biodiesel feedstock. Biomed. Res. Int. 2016, 8792548.

Zhu, H., Zhu, C., Cheng, L., Chi, Z., 2017. Plastic bag as horizontal photobioreactor on rocking platform driven by water power for culture of alkalihalophilic cyanobacterium. Bioresour. Bioprocess. 4 (1), 1–10.

Biodiesel and green diesel

Deepak Tuli[a] and Sangita Kasture[b]

[a]Bioenergy-Biosciences, DBT-IOC Bioenergy Centre, Faridabad, India [b]Department of
Biotechnology, Ministry of Science and Technology, New Delhi, India

1 Introduction

The production and use of renewable fuels has been intensified to reduce the dependence on fossil fuels and ensure the availability of energy, mainly for countries like India which are dependent upon imported oil. Green House Gas Emissions (GHG) from the transport sector is currently about 26% of the total GHG emissions, and road transport is the major emitter. Alternate methods of powering the internal combustion engines have been researched and electric mobility is being favored for light duty transport, passenger cars, however heavy duty trucking, rail and public transport shall remain on diesel for a foreseeable future. While western countries and US are more of gasoline dependant, developing countries with robust public transport require more of diesel e.g., India consumes diesel which is about 3.5 times gasoline consumption. The search for either blending or substitution of conventional fuels prompted the look for biofuel alternatives that are not only technically feasible but are also economically viable with ensured sustainable availability of raw materials. Biodiesel is one such option that can be blended with petro diesel in different proportions depending upon its type. The feedstock widely used is triglyceride lipids which are obtained from vegetable oils, animal fats or can be extracted from algae.

Though the biofuels ideally should not compete with food chain and should not be produced as a result of change of land use, however presently in several instances it is not the case. Ethanol for blending in gasoline is made in US from corn and in Brazil from sugarcane. Globally, biodiesel is mostly produced from palm oil (31%), soybean (27%), oilseed rape (20%), and used cooking oil (10%). In the EU biodiesel is produced, as per global market survey report of 2019, from oilseed rape (44%), palm oil (29%), used cooking oil (15%), and soybean oil (5%). The rests represent sunflower, coconut, peanuts, hemp, jatropha, corn, and algae (UFOP Report on Global Market Supply, 2018/2019). However, now biofuels are being produced from, though presently in small quantities, from lignocellulosic materials

(Agricultural/forestry residues) for ethanol and increasingly used cooking oil/animal fats is for biodiesel.

Biodiesel production from lipid containing oils is primarily from two routes. The first route, and most commonly used in industry, is by a process of transesterification of oil using a lower alcohol, mostly methanol, and in the presence of a catalyst. The biodiesel so obtained is chemically fatty acid methyl ester or FAME. The second route to produce biodiesel from lipids is by cracking and hydro-processing in the presence of catalysts, and the biodiesel is called HVO (hydro processed vegetable oil) and is also called green diesel. These two types have mutual advantages and some issues and the general scheme as shown in Fig. 1.

2 Green diesel

Green diesel is a mixture of straight chain and branched saturated hydrocarbons which are typically C15 to C18 molecules. This composition resembles petroleum diesel and allows green diesel utilization in internal combustion engines. Unlike biodiesel FAME which can be blended to a certain extent, green diesel can be used in pure form or as a blend with any blend ratio, without engine modifications. Like biodiesel, green diesel is also a product of biological origin and its use leads to GHG reduction. Studies have shown that life cycle greenhouse gas emissions of green diesel are slightly lower than those of FAME, if both are made from the same feedstock.

Green diesel can be produced from biomass through four technologies: (a) hydro-processing, (b) catalytic upgrading of sugars, starches and alcohols, (c) thermal conversion (pyrolysis) and upgrading of bio-oil and (d) biomass to liquid (BTL) thermochemical processes. However presently green diesel is almost exclusively produced by adopting the hydro-processing route in which the feed stock is lipids like vegetable oils and animal fats. This places a severe constraint on the availability on green diesel as these feedstock are not available in large quantities. In view of this technology platforms like pyrolysis, biomass to liquid and sugar upgradation are being developed as these can utilize biomass, agricultural/forest residue, algae which are sustainably available in extremely large amounts.

Hydro-processing aims at the conversion of the triglycerides of the biomass oils and fats into saturated hydrocarbons through catalytic treatment in the presence of hydrogen.

FIG. 1 General schemes for FAME and Green Diesel production from lipids.

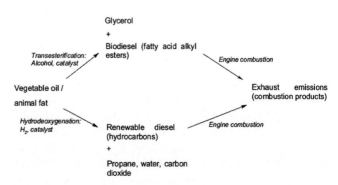

The catalytic upgrading of sugars and alcohols involves liquid phase technologies called aqueous phase reforming (APR). Thermal conversion involves the pyrolysis of biomass and the production of bio-oil which is then refined into green diesel. Finally, BTL processes involve the high temperature gasification of the biomass for the production of syngas which is rich in H2 and CO and the subsequent chemical synthesis of liquid green diesel through the well known Fischer-Tropsch (FT) process. The green diesel produced by the Fischer-Tropsch method is sometimes referred as FT green diesel.

According to a global market survey report of 2018, the global market of green diesel is growing, from 0.3 billion gallons in 2011 to 2.1 billion gallons per year in 2017. Given that green diesel can be mixed with conventional petroleum diesel and it meets the automotive fuel specifications, some oil refineries have developed methods for the simultaneous co-processing of triglyceride feedstocks with petroleum intermediates such as straight run gas oil and/or vacuum gas oil. This leads to process economics as no new infrastructure is required. There are a number of companies co-processing vegetable oils with petroleum distillates such as Petrobras (Brazil), Cepsa (with several refineries in Spain), Repsol (with several refineries in Spain) and British Petroleum (Australia). Indian Oil developed a technology for hydro-processing of Jatropha oil and co-processed this with VGO in one of the refinery. A report in biorrefineria.blogspot looked at existing large green diesel HVO refineries. Standalone biorefineries of 1 million ton per year capacities based on vegetable oils and animal fats were in operation and some of these are listed below (Table 1).

It is observed that most of the producers have developed proprietary technologies (Neste NExBTL, UOP/Eni EcofiningT) and standalone plants comprised by the biomass clean up and pretreatment section, the deoxygenation (hydrotreatment) section, a hydro-isomerisation reactor and a separation column. A typical process diagram is as shown in Fig. 2.

TABLE 1   Some large stand-alone HVO plants.

Company/location	Feedstock	Capacity	Technology
Neste/Netherland	Vegetable oil, waste animal fat	1,000,000 TPA	NExBTL
Neste/Singapore	Vegetable oil, waste animal fat	1,000,000 TPA	NExBTL
UOP/Italy	Used cooking oil/animal fat	780,000 TPA	Ecofining
Diamond Green/USA	Non edible oil/animal fat	900,000 TPA	Ecofining

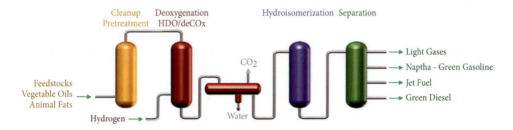

FIG. 2   A typical stand-alone biorefinery for production of green diesel.

3 Green diesel by hydro-processing of lipids

Green diesel is produced by a hydrotreating triglycerides of vegetable oils with hydrogen. The hydrotreating process consists of three main reactions: hydrodeoxygenation (HDO), decarbonylation (DCO) and decarboxylation (DCO2), is operated to remove oxygen, carbon monoxide and water, and carbon dioxide respectively. Green diesel is a straight-chain hydrocarbon fraction similar to that produced by the hydrogenation of triglycerides from vegetable oils (He and Wang, 2012). The injection of hydrogen win the presence of a bimetallic catalyst for the hydrotreating reaction of the triglyceride compound is capable of breaking the ester or carboxylic group bond of the glycerol group (Chew and Bathia, 2008; Hilten et al., 2011). Then followed the breaking of carboxylic groups from the existing fatty acid bonds as well as the saturation reaction of the double chain carbon bonds into straight chain hydrocarbon compounds (Lima et al., 2004; Maher and Bressler, 2007).

The main reactions are as depicted in Fig. 3.

Hydrotreated vegetable oil (HVO) or green diesel is a biofuel produced from fatty matters and obtained from the hydrogenation process, has similar physical–chemical properties to conventional diesel and can be produced in the existing conventional diesel production process in petroleum refineries (Knothe, 2010).

Biodiesel is mostly used as a blend component in petro-diesel and seldom as pure because of some inferior properties like oxidation stability and gum forming tendency. However, the green diesel (HVO) has advantages that it can be added to petro-diesel in very high percentage, as it is pure hydrocarbon like diesel and provides superior properties like higher cetane number. Lower quality lipids like having free fatty acids are an issue in transesterification but these can be handled in hydro processing with some pretreatment (Stumborg et al., 1996).

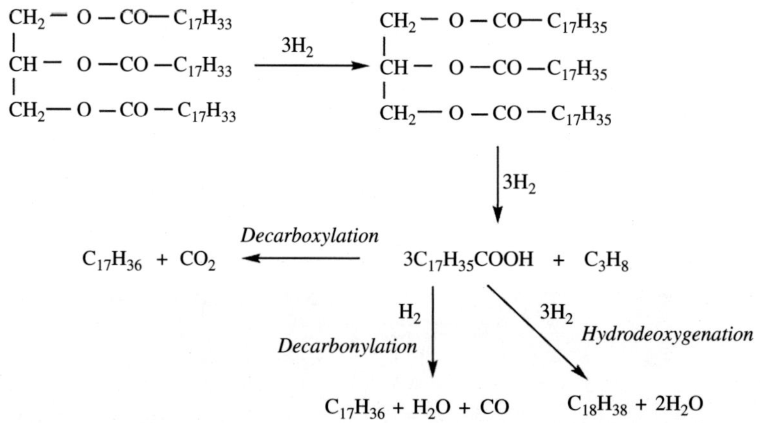

FIG. 3   Reaction pathway of hydrotreatment.

4 Hydrogenation and hydrodeoxygenation

Hydro-treatment or hydro-processing is series of catalytic reactions using hydrogen to eliminate the hetero-atoms such as sulfur, nitrogen, and oxygen, each of which is ejected as hydrogen sulfide (H2S), ammonia (NH3) and water (H2O), respectively during the production process. Hydrogenation process is usually followed by hydrodeoxygenation reaction (HDO), decarboxylation (DCO), and decarbonylation (DCN) reactions (Arun et al., 2015). Hydrodeoxygenation (HDO) is a process for eliminating oxygen hydrogenolysis of a material by cutting the carbon-oxygen bond using hydrogen gas (Mohammad et al., 2013).

The process economics largely depends upon the reaction conditions and the way heat integration is managed in the process. In a study of production of HVO from used palm oil, the reaction conditions and their effect on the yield of HVO has been discussed (Amin, 2019). The procedure in this study employed NiMo/ -Al2O3 catalyst and waste cooking oil into a reactor. Then HDO processes were varied on time, temperature, pressure, and number of reaction stage. Some conclusions indicate that the main factors affecting yield of HVO were pressure, temperature, number of passes, and time. Optimizing these conditions, the quantitative conversion could be achieved (Amin, 2019).

In the preparation of green diesel, the hydrogenation (or hydrodeoxygenation) process removes oxygen from triglyceride molecules and convert it into water. The hydrotreatment occurs, in general, under temperatures of 240–360°C and pressures that can reach up to 80 bar (Nikolopoulos et al., 2020).

The main factors that affect the green diesel production process by hydrotreatment are pressure, temperature, hydrogen/fat material ratio, catalyst type and reaction speed. Three parallel reactions occur during the process of transformation of oils and fats are hydrogenation, hydrodeoxygenation, and decarboxylation. Water, and n-paraffins, mainly n-C16 and n-C18, are formed during the hydrogenation and hydrodeoxygenation reactions. The products generated in the hydrogenation and decarboxylation are n-paraffins, mainly n-C15 and n-C17. Hydrogenation leads to saturation of double bonds (Amin, 2019). Hydrodeoxygenation processes generally results, in alkanes which are one carbon shorter than lipid fatty acid carbon chain.

Due to the high similarity to diesel, the green diesel use has several advantages. The main advantages of green diesel over biodiesel is the possibility of using it at a higher proportion in mixtures with conventional diesel and its production, using the infrastructure used in the production of this fossil fuel. Green diesel is comparable to ultra-low sulfur diesel (ULSD) and as such has no sulfur. Since it has higher density, green diesel has higher net energy balance than petroleum or biodiesel and it considerably reduced GHG emissions. Catalytic hydro-processing is a common process in oil refineries and is used to reduce sulfur and nitrogen in the fuels and the same units can be used to make green diesel.

5 Green diesel from conversion of olefins

Alkene oligomerization describes the conversion of lower alkenes such as propylene, butylene and pentene to heavier alkenes in the boiling range of gasoline and petroleum diesel.

The reaction may proceed over acid catalysts such as zeolites or over supported Ni catalysts. Two commercial processes of alkene oligomerization are the Mobil Olefins to Gasoline and Distillates (MOGD) and Conversion of Olefins to Diesel (COD) of Lurgi. The COD process gives olefins of C3-C23 range and which upon hydrogenation and distillation gives green diesel.

6 Green diesel through pyrolysis and hydro-thermal liquefaction

Flash pyrolysis of powdered biomass having 10% moisture at about 500°C in less than 1 s residence time, is carried out in a fluidized bed reactor. Any solid matter and solid char is removed through cyclones and the bio-oil vapor exits the reactor. After removing solids/ char, bio-oil vapor is then quickly liquefied in a condenser providing the brown liquid bio-oil with an approximately 75% yield on a dry weight basis.

Two-stage catalytic hydrotreatment is carried out, first stage the mild hydrotreating of bio-oil takes place at about 240°C and 170 bar in excess of hydrogen using Co—Mo type catalysts. At the second stage, heavy hydrotreating takes place at about 370°C and 140 bar in excess of hydrogen in which deoxygenation of bio-oil takes place producing mixtures of saturated hydrocarbons. Fractional distillation gives green diesel. Presently there are no commercial scale plants on this technology though large pilots have been taken. Catalyst deactivation is one of the major issue.

Hydrothermal liquefaction (HTL), also known as hydrous pyrolysis or direct liquefaction, is another thermal technology used for the conversion of biomass into a liquid product similar to bio-oil which is usually called bio-crude. The HTL process usually takes place at temperatures of 200–400°C and at elevated pressures of 50–200 bar with relative long residence times of 10–60 min. HTL biocrude can be upgraded by catalytic hydrotreatment, just like pyrolysis liquid. The advantage of HTL process is that there is no need to dry the feed and it is being considered best option for algae. HTL process involves very complex reactions whose kinetic is still being studied.

7 Biodiesel from lipids by transesterification

Biodiesel is alkyl ester of fatty acids and has oxygen and this content of oxygen has a very important role in estimating the characteristics of the fuel to be generated. Oxygen levels in vegetable oils is very high and therefore the FAME produced suffers from the low calorific value, low thermal stability, and therefore cannot be mixed with fossil fuels in high present as there is an increasing tendency toward polymerization reaction (Mohammad et al., 2013).

Transesterification is a mature technology and is a better choice for biodiesel production compared with other existing methods. Transesterification reaction is added by catalyst and two kinds of major catalytic systems, chemical and biological are employed. Further the chemical catalyst can be either solid heterogenous type or homogenous. Homogenous catalysis involving sodium or potassium hydroxides is most popular and most widely used in commercial plants. In heterogeneous catalysts, external-surface active species of porous solid

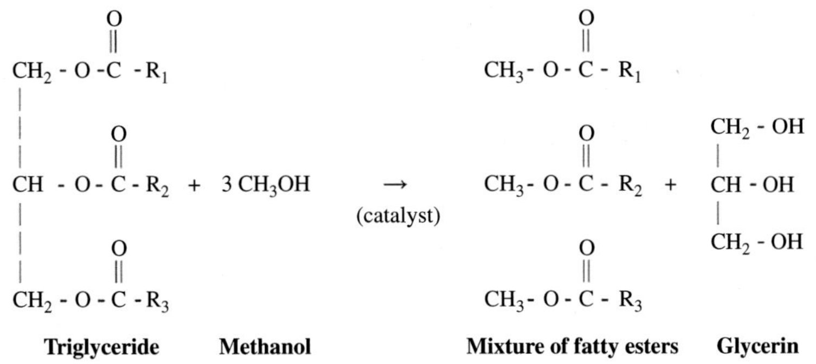

FIG. 4   Transesterification reaction.

support only is involved and catalyst leaching is a serious issue (Thangaraj et al., 2019). Lipase type of biocatalyst were also examined but the cost is high and their application on commercial scale has not been possible.

Transesterification reaction is an equilibrium reaction and use of excess methanol favors the product formation. Though stoichiometrically only 3 mol of methanol is required for a mole of lipid but in practice six to 9 mol excess of methanol is used and recovered after reaction for recycle. Though the reaction, as shown in Fig. 4 is simple but there are several product purification steps so as to meet stringent quality parameters as determined by ASTM test method. The final product cannot have more than 0.1% of methanol, low ppb level of alkali metals, and free from glycerol. This is achieved by water washing and at times vacuum distillation. Though alkali catalyzed transesterification is widely used, it involves several water washing steps and the resultant water washes need treatment before recycling the water. Lipids having enhanced level of free fatty acids are troublesome for base catalyzed reactions as these lead to soap formation and product loss. The development of effective and inexpensive catalysts with an environmentally benign process is a continued research interest essential to overcome the present issues.

8 Transesterification by homogeneous catalysts

Generally, homogeneous chemical catalysts have several advantages, including high conversion, lower catalyst costs, higher reaction rate, low side products and easy optimization of activity (Ma and Hanna, 1999). The common alkali catalysts being employed are sodium hydroxide, sodium methoxide and potassium hydroxide. NaOH is preferred over KOH because it dissolves quickly in methanol though the reaction rate is much faster in KOH. NaOH is preferred in transesterification due to its high purity, easy availability, low cost and a relatively smaller quantity is needed as compared to KOH, on a stoichiometric basis (Jitputti et al., 2006).

However, alkali metal alkoxides are found to be more active than hydroxides (Saydut et al., 2016).

In alkali-catalysed transesterification, even if water-free vegetable oils and alcohol are used, a certain amount of water is formed by the sodium methoxide solution because of the interaction between NaOH and methanol. The presence of water and FFA leads to soap formation by hydrolysis of the triglycerides, which reduces the biodiesel yield and affects the quality of the product (Veljkovic et al., 2006). There are strict limits of water in diesel.

Product recovery becomes complicated in an alkali-catalysed transesterification reaction if the oil has >1% Free fatty acids (FFA); in such an event, part of the process may take the saponification route, thereby forming soap. The soap thus formed inhibits the separation of biodiesel and glycerine. Used cooking oils or lower grade rancid oils generally have higher amount of free fatty acids (3%–6%) and hence the simple base catalyzed transesterification cannot be used as most of the product is lost in soap emulsions. Because of this, two-step transesterification—acid first and alkali next—is recommended (Thangaraj et al., 2014).

Lower alcohols are much more efficient and reaction is much faster, methanol is better than ethanol for overall recovery and is easy to get as water free (Sanli and Canakci, 2008).

9 Transesterification by heterogeneous catalysts

Heterogeneous catalysts have been explored extensively for their ease of separation, better tolerance of moisture and use in cheap oils having free fatty acids, leading to lower costs. (Refaat, 2010; Endalew et al., 2011). These catalysts tolerate high FFA and moisture content of the vegetable oils (Refaat, 2010).

Solid base catalysts like CaO, MgO, KNO3/Al2O3 have been explored and these have higher catalytic activity than solid acid catalysts (Ranganathan et al., 2008).

Heterogeneous catalysts include noble metals (with no sulfur), nonnoble metals, phosphide metal, carbide metal, and nitrate metals (Li et al., 2018). Lipases have been used for transesterification lipases immobilized on inert surfaces are preferred. However commercial application of Lipases is very limited as these give lower reaction rate and a two phase system is difficult to handle on industrial scale.

10 Biodiesel from algae

Algae has been explored for last two decades as a source which can give lipids and biomass and there were high expectations for using algae biodiesel for many years, but the quantities of algal biodiesel currently available are too little and its contribution is almost negligible in the diesel pool.

The photosynthetic microorganisms like microalgae require mainly light, carbon dioxide, and some nutrients (nitrogen, phosphorus, and potassium) for its growth, and to produce large amount of lipids and carbohydrates, which can be further processed into different biofuels and other valuable co-products (Brennan and Owende, 2010; Nigam and Singh, 2011).

Algae can either be autotrophic or heterotrophic. The autotrophic algae require only inorganic compounds such as CO_2, salts, and a light energy source for their growth, while the heterotrophs are nonphotosynthetic, which require an external source of organic compounds as well as nutrients as energy sources (Brennan and Owende, 2010). Microalgae contain more lipids than macroalgae and have the faster growth in nature (Lee et al., 2014).

The cycle of growing algae is very short (1–10 days), which allows for several harvests. This means that the yield of algal biomass is much higher compared to conventional crops, which further allows the production of a several times more oil than crops such as oilseed rape and soybeans (Khan et al., 2018).

Some popular algae species like *Schizochytrium* sp., *Nitzschia* sp., and *Botyococcus braunii* contain over 50% of the oil reduced to dry biomass, which can be extracted and processed into fuel (Adeniyi et al., 2018; Alaswad et al., 2015).

As the production of algal lipids is an energy-intensive process, so the price of algal-derived biodiesel is high. Opportunities for more economical production of algal biodiesel are seen in integration with other processes, such as wastewater treatment, but this does not translate into large-scale production. Behera et al. reviewed the production of algae, lipid extraction and biodiesel production (Behera et al., 2015).

Essential strengths/advantages of algae for biodiesel are discussed in details (Bošnjakovic and Sinaga, 2020). Algae productivity is higher compared to most effective crops and a large number of algae species can be genetically modified to increase growth profile and also lipid content. The CO2 footprint of algae biodiesel is smaller than conventional diesel. Ability to grow throughout the year, therefore, algal oil productivity is higher in comparison to the conventional oil seed crops. There is no requirement of herbicides or pesticides in algal cultivation and the ability to grow under harsh conditions like saline, brackish water, coastal seawater, which does not affect any conventional agriculture (Dragone et al., 2010).

From algae to produce biodiesel, there are several steps such as drying, cell disruption, oils extraction, transesterification, and biodiesel refining. There are some factors that algal biofuels are not yet commercial on large scale. The main drawback is that biodiesel production from algae is an extremely energy-intensive process that, in some cases, results in a negative energy balance. Production costs are significantly higher compared to the production of conventional diesel. Harvesting of algae require high energy input, which is approximately about 20%–30% of the total cost of production. Though several techniques such as centrifugation, flocculation, floatation, sedimentation, and filtration are usually used for harvesting and concentrating the algal biomass, but still these are expensive (Ho et al., 2011). Also the water footprint is large and fresh water algae is almost ruled out in countries/regions which experience water scarcity. The main problems are related with the high water content of the harvested algal biomass (over 80%), which overall increases the cost of whole process.

There are reports of different methods of oil extraction from wet algae, such as mechanical and solvent extraction (Li et al., 2014). In situ transesterification of wet algae skips the oil extraction step. The alcohol acts as an extraction solvent and an esterification reagent as well, which enhances the porosity of the cell membrane. Yields found are higher than via the conventional route, and waste is also reduced.

A lot of research is still taking place to address some of the issues which hamper the commercial exploitation of algal based biodiesel. Linking biodiesel production and wastewater treatment, different sources of wastewater containing nutrients like nitrogen and phosphorus can be utilized for algal cultivation apart from providing any additional nutrient Algal biomass-based co-products can provide the necessary revenue to reduce the net cost of biodiesel production. Developing and applying less energy-intensive technologies for the biodiesel production process,

Optimization of the biodiesel production process by introducing less energy-intensive technologies and application of postharvest water recycling can really make a difference.

The key challenges in mass production of biodiesel from algae appear to be high infrastructure, operational, and maintenance costs. Progress in minimizing/reducing the energy, water, and land-use footprints needs to be a primary objective of enhancing larger-scale algal biodiesel production. Private sector is therefore not coming forward to invest in algal systems for biodiesel. It is not to be expected that large quantities of algae derived biodiesel will be used in transport soon.

11 Comparison of green diesel and FAME

Green diesel is also known as hydro-processed renewable diesel hydrotreated vegetable oil (HVO), can be used as such in CI engines as it resembles normal diesel in carbon chain length. The engine performance is not affected when diesel blended with green diesel is used and very large number of engine tests have been reported (No, 2014; Singh et al., 2018). Green diesel has no issues with: stability, water separation, microbiological growth, impurities causing precipitation. Biodiesel is a mixture of fatty acid alkyl esters, whereas green diesel is a paraffinic that contain, C12 − C18 which is what petroleum diesel has. Both the conventional biodiesel, also called FAME (Fatty acid methyl esters) and the green diesel reduce the GHG emissions. The extent of reduction of GHG emissions depend upon the feedstock used and the technology of production (Arvidsson et al., 2011).

The ignition properties of green diesel in internal combustion engines are similar to those of conventional diesel. It presents a high cetane number, good oxidative stability when stored for long periods, and no sulfur, aromatic acids, or ashes. It presents a high calorific power (44 MJ kg^{-1}), low specific gravity. (Kalnes et al., 2009; Yoon, 2011). In the HVO production process, hydrogen is used to remove the oxygen from the triglycerides and does not produce any glycerol as a side product. Additional chemicals, like methanol for FAME production, are not needed. Hydrogenation removes all oxygen from the vegetable oils while esterification does not.

Green diesel has up to 40% lower particulate matter in the engine exhaust, without changes in the NOx emissions, when compared to conventional diesel (Dimitriadis et al., 2020). HVO, which is generally produced in conventional refineries, is hydrocarbon like petro diesel and thus can be added in any proportion without any engine modifications and without the loss of engine efficiency. Biodiesel is fatty acid methyl esters which has about 10%–12% oxygen and thus is chemically different from petro diesel. It has lower oxidation stability and makes polymeric gums specially at higher temperatures and thus can be added up to some amounts, less than 20% in blends. Since biodiesel has oxygen, its fuel efficiency is lower than petro diesel.

Used cooking oil (UCO) is now a major feedstock both for FAME and for green diesel. Repeated use of cooking oil leads to generation polymeric chemical formation which can be harmful to human health. The first sign of cooking oil damage is the formation of acrolein in cooking oil. This acrolein causes itching in the throat when eating fried foods using cooking oil repeatedly. Recently in India specifications have been enforced for use of vegetable oils for cooking and this can lead to higher availability of used cooking oil for conversion to green diesel or biodiesel.

12 Biodiesel-global status

According to REN21, Renewables Now report, the world's biodiesel supply grew from 3.9 billion liters in 2005 to 18.1 billion liters in 2010 was about 33 billion liters in 2016. In 2019 US, Indonesia, Brazil and Germany were first largest producers of biodiesel (IFPEN, 2021) which together more or less exceeds the EU production. In Europe, the largest producer countries are Germany, France and the Netherlands. In Germany and France production consist of smaller-scale distributed FAME production units, while in the Netherlands the majority of the biodiesel production takes place in the Neste large-scale HVO plant (Nyström et al., 2019). As shown in Fig. 5, there has been a sharp growth in biodiesel production beginning 2005 and the production of biodiesel increased substantially in Europe.

Out of the total biodiesel production, FAME biodiesel is about 82% and HVO at 18%. However the growth of HVO is much more than FAME biodiesel. Some large HVO plants are recently announced by US and EU companies which will substantially increase the supply of HVO. In 2018, Neste (Finland), the world's largest HVO producer, announced an investment of USD 1.6 billion to more than double its renewable diesel production capacity in Singapore by adding a further 1.7 billion liters of annual capacity (Kotrba, 2018). There is increased emphasis on using nonfood feedstocks to produce HVO fuels. For example, Neste now produces

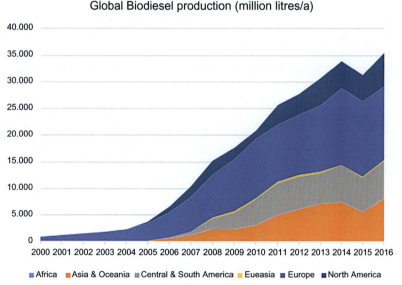

FIG. 5 Global Biodiesel (FAME) production. *Adopted from Nyström, I., Bokinge, P., Franck, P.A., 2019. Production of Liquid Advanced Biofuels-Global Status. CIT Industriell Energi AB.*

its HVO from 80% waste vegetable oils and residual materials rather than from virgin feedstock (https://www.neste.com/en/companies/products/renewable-fuels, n.d.). In the US, where large expansion HVO production occurred, Renewable Energy Group increased the combined capacity at its 13 biomass-based diesel refineries to more than 2 billion liters per year. In Europe, Eni (Italy) ramped up HVO production at its Venice refinery to 250,000 t (320 million liters) in 2018 and aims to expand the facility's capacity to 770 million liters; the company also expected its Sicily plant to come online in 2019 (*Italy–Eni Produces More HVO*, n.d.).

The most dramatic development in renewable biofuel production since 2010 has been the evolvement of commercial HVO production. Globally, there are currently about 15 large-scale, commercial plants for the production of HVO (including co-processing with fossil feedstock in traditional refineries), which is used as a renewable drop-in diesel fuel mostly blended in fossil diesel (Landälv et al., 2017; Greenea, 2017). The global HVO market is dominated by Neste (Neste, 2018).

13 Biodiesel-Indian status

Though India in its first biofuel policy of 2009, planned 20% biofuel blending in gasoline and diesel, the actual blending was only a very small fraction specially in diesel (https://mnre.gov.in/biofuels, n.d.).

The policy focused on research and development on cultivation, processing and production of biofuels, biodiesel through mainly Jatropha to ensure a minimum level of biofuels which are readily available in the market. An indicative blending mandate of 20% was for both bio-ethanol and bio-diesel by 2017 (Biodiesel later reduced to 5%). Jatropha was considered as the main feedstock for conversion of its oil into biodiesel. Very large estimates were made based on very few jatropha plantations. However the projection of jatropha yield per tree and need for watering and care was not encouraging. Thus, using jatropha proved untenable due to a host of agronomic and economic constraints. Large scale jatropha plantations were not undertaken. The learnings of 2009 policy led to declaration of biofuel policy of 2018- duly adopted by Indian government.

India's National Biofuel Policy 2018 (National Biofuel Policy, 2018) seeks to achieve a national average of E20 for gasoline and B5 for diesel by 2030. The ethanol blending has seen a very quantum jump and we may achieve the 20% target even much before 2030. In the current ethanol supply year, which started in October 2020, India plans to have 8.5% ethanol-blending with gasoline. This would be raised to 10% by mixing in 4 billion liters of ethanol by 2022 (Times of India, 2020).

In contrast, biodiesel market penetration has remain stuck at 0.14% due to limited supply, insufficient feedstocks, supply chain constraints, and restrictions on imports (GAIN, 2019). The majority of the biodiesel produced goes mostly to informal sectors and no, or very negligible, organized blending by oil companies is done.

The new biofuel policy encourages setting up of supply chain mechanisms for biodiesel production from nonedible oilseeds, Used Cooking Oil, short gestation crops animal tallow, acid oils, and algal feedstock (National Biofuel Policy, 2018). To date, biodiesel is manufactured from imported palm stearin, and small volumes of nonedible oils, UCO and domestically sourced animal fats. Domestically sourced UCO has been identified as a feedstock with large potential for biodiesel production. From 2018 regulations are framed for monitor the quality of oil during frying and maximum permissible limit of total polar compound in edible oils is 25% (Business Standard, 2019).

Oil Marketing Companies will provide complete off-take guarantee for bio-diesel produced by UCO in 100 select cities initially. The oil companies have fixed the price of buying bio-diesel at Rs. 51/L for the first year, and thereafter increase in subsequent years. In order to achieve the 5% blending target by 2030, 5 billion liters of biodiesel is required a year. Based on estimates by oil-marketers, approximately 22.7 million tons per annum or 27 billion liters of cooking oil is used. Of that 1.2 million tons (1.40 billion liters) UCO can be collected from bulk consumers for conversion, which will give approximately 1.10 billion liters of biodiesel in a year (Business Standard, 2019).

Biofuel policy 2018 also says that farmers will be encouraged to grow a variety of different biomass as well as oilseeds on their marginal lands as intercrops, and as a second crop wherever only one crop is raised under rain-fed conditions. Suitable supply chain mechanisms, feedstock collection centers, and fair price mechanisms.

Presently, India has six plants with combined annual capacity of 650 million liters of biodiesel per year. The production capacity of existing plants ranges from 11 million liters to 280 million liters. (GAIN, 2019; Economic Times, 2019).

Biodiesel producers use nonedible industrial oil (palm stearin), UCO, animal fats, tallows and 'other oils' (sludge, acidic oils, and tree-borne oils etc.) to produce biodiesel, thereby utilizing 29% of the total installed capacity. While the use of animal fats and tallows has remained constant, the remaining feedstock use has shown steady growth, namely nonedible industrial oil and UCO.

Presently HVO is not produced in India. HVO in India has been limited to mostly R&D and in one case refinery trial In this process de-gummed and de-metalled nonedible oil is mixed with diesel feed and fed into the DHDS/DHDT reactor along with recycled hydrogen. Indian Oil R&D has developed a demetallation process which is carried outside refinery. The salient features of this technology include utilization of existing refinery infrastructure with minor modifications. The trial run of the technology was successfully carried out on a commercial scale in April 2013 at the DHDT unit of Manali Refinery218 of Chennai Petroleum Corporation with 6.5% of Jatropha oil with diesel feed (Advance Biofuels in India, 2018).

There are reports of production of sustainable bio aviation fuels from used vegetable oil and nonedible oil seeds and therefore a competing use.

References

Adeniyi, O.M., Azimov, U., Burluka, A., 2018. Algae biofuel: current status and future applications. Renew. Sust. Energ. Rev. 90, 316–335.

Advance Biofuels in India, 2018. A Comparative Analysis Between India and the EU for Cooperation and Investment. Artfuel Final Report Oct and see www.iocl.com/.

Alaswad, A., Dassisti, M., Prescott, T., Olabi, A.G., 2015. Technologies and developments of third generation biofuel production. Renew. Sust. Energ. Rev. 51, 1446–1460.

Amin, A., 2019. Review of diesel production from renewable resources: catalysis, process kinetics and technologies. Ain Shams Eng. J. 10, 821–839.

Anon. https://mnre.gov.in/biofuels.

Anon. https://www.neste.com/en/companies/products/renewable-fuels.

Anon., Italy–Eni Produces More HVO. https://www.agra-net.com/agra/world-ethanol-and-biofuels-report/biofuel-news/biodiesel/italy-eni-produces571843.htm.

Arun, N., Sharma, R.V., Dalai, A.K., 2015. Green diesel synthesis by hydrodeoxygenation of bio-based feedstocks: strategies for catalyst design and development. Renew. Sust. Energ. Rev. 48, 240–255.

Arvidsson, R., Persson, S., Froling, M., Svanstrom, M., 2011. Life cycle assessment of hydrotreated vegetable oil from rape, oil palm and Jatropha. J. Clean. Prod. 19, 129–137.

Behera, S., Singh, R., Arora, R., Sharma, N.K., Shukla, M., Kumar, S., 2015. Scope of algae as third generation biofuels. Front. Bioeng. Biotechnol. 2, 90.

Bošnjakovic, M., Sinaga, N., 2020. The perspective of large-scale production of algae biodiesel. Appl. Sci. 10, 8181.

Brennan, L., Owende, P., 2010. Biofuels from microalgae—a review of technologies for production, processing, and extractions of biofuels and o-products. Renew. Sust. Energ. Rev. 14, 557–577.

Business Standard, 2019. Used Cooking Oil May Fuel Cars as OMCs Look to Procure Biodiesel. Aug 10, and also see fssai.gov.in.

Chew, T.L., Bathia, S., 2008. Catalytic processes toward the production of biofuels in a palm oil and oil palm biomass-based biorefinery. Bioresour. Technol. 99, 7911–7922.

Dimitriadis, A., Seljak, T., Baskovic, U.Z., Dimaratos, A., Bezergianni, S., Samaras, Z., Katrašnik, T., 2020. Improving PM-NOx trade-off with paraffinic fuels: a study toward diesel engine optimization with HVO. Fuel 265, 116921.

Dragone, G., Fernandes, B., Vicente, A.A., Teixeira, J.A., 2010. Third generation biofuels from microalgae. In: Mendez-Vilas, A. (Ed.), Current Research, Technology and Education Topics in Applied Microbiology and Microbial Biotechnology. Elsevier, pp. 1315–1366.

Economic Times, 2019. India's Bio-Diesel Blending for Road Transport to Remain Muted in 2019. US Dept of Agriculture; Economic Times. Aug 28.

Endalew, A.K., Kiros, Y., Zanzi, R., 2011. Inorganic heterogeneous catalysts for biodiesel production from vegetable oils. Biomass Bioenergy 35, 3787–3809.

GAIN, 2019. Report Number: IN9069 India Biofuels Annual Biofuels Annual.

Greenea, 2017. New Players Join the HVO Game. [Online]. Available https://www.greenea.com/publication/new-players-join-the-hvo-game/. Retrieved: 2018-11-14.

He, Z., Wang, X., 2012. Hydrodeoxygenation of model compounds and catalytic systems for pyrolysis bio-oils upgrading. Catal. Sust. Energy 1, 28–52.

Hilten, R., Speir, R., Kastner, J., Das, K.C., 2011. Production of aromatic green gasoline additives via catalytic pyrolysis of acidulated peanut oil soap stock. Bioresour. Technol. 102, 8288–8294.

Ho, S.H., Chen, C.Y., Lee, D.J., Chang, J.S., 2011. Perspectives on microalgal CO_2-emission mitigation systems—a review. Biotech. Adv. 29, 189–198.

IFPEN, 2021. FO Licht Report.

Jitputti, J., Kitiyanan, B., Rangsunvigit, P., Bunyakiat, K., Attanatho, L., Jenvanitpanjakul, P., 2006. Transesterification of crude palm kernel oil and crude coconut oil by different solid catalysts. Chem. Eng. J. 11, 61–66.

Kalnes, T.N., Koers, K.P., Marker, T., Shonnard, D.R., 2009. A technoeconomic and environmental life cycle comparison of green diesel to biodiesel and syndiesel. Environ. Prog. Sust. Energy 28, 111–120.

Khan, M.I., Shin, J.H., Kim, J.D., 2018. The promising future of microalgae: current status, challenges, and optimization of a sustainable and renewable industry for biofuels, feed, and other products. Microb. Cell Factories 17, 1–21.

Knothe, G., 2010. Biodiesel and renewable diesel: a comparison. Prog. Energy Combust. Sci. 36, 364–373.

Kotrba, R., 2018. Neste to more than Double Renewable Diesel Capacity in Singapore. Biodiesel Magazine.

Landälv, I., Maniatis, K., Waldheim, L., Vanden, H.E., Kalligeros, S., 2017. Building up the future—technology status and reliability of the value chains, European Commission. In: Sub Group on Advanced Biofuels. Sustainable Transport Forum.

Lee, K., Eisterhold, M.L., Rindi, F., Palanisami, S., Nam, P.K., 2014. Isolation and screening of microalgae from natural habitats in the midwestern United States of America for biomass and biodiesel sources. J. Nat. Sci. Biol. Med. 5, 333.

Li, Y., Naghdi, F.G., Garg, S., Adarme-Vega, T.C., Thurecht, K.J., Ghafor, W.A., Tannock, S., Schenk, P.M., 2014. A comparative study: the impact of different lipid extraction methods on current microalgal lipid research. Microb. Cell Fact. 13, 1–9.

Li, X., Luo, X., Jin, Y., Li, J., Zhang, H., Zhang, A., Xie, J., 2018. Heterogeneous sulphur-free hydrodeoxygenation catalysts for selectively upgrading the renewable bio-oils to second generation biofuels. Renew. Sust. Energ. Rev. 82, 3762.

Lima, D.G., Soares, V.C.D., Ribeiro, E.B., Carvalho, D.A., Cardoso, E.C.V., Rassi, F.C., Mundim, K.C., Rubim, J.C., Suarez, P.A.Z., 2004. Diesel-like fuel obtained by pyrolysis of vegetable oils. J. Anal. Appl. Pyrol. 71, 987–996.

Ma, F., Hanna, M.A., 1999. Biodiesel production: a review. Bioresour. Technol. 70, 1–15.

Maher, K.D., Bressler, D.C., 2007. Pyrolysis of triglyceride materials for the production of renewable fuels and chemicals. Bioresour. Technol. 98, 2351–2368.

Mohammad, M., Hari, T.K., Yaakob, Z., Sharma, Y.C., Sopian, K., 2013. Renewable and sustainable energy reviews. Appl. Therm. Eng. 22, 121–132.

National Biofuel Policy, 2018. MoPNG.gov.in.

Neste, 2018. Neste Annual Report 2017. [Online]. Available https://www.neste.com/corporate-info/news-inspiration/material-uploads/annual-reports.

Nigam, P.S., Singh, A., 2011. Production of liquid biofuels from renewable resources. Prog. Energy Combust. Sci. 37, 52–68.

Nikolopoulos, I., Kogkos, G., Kordouli, E., Bourikas, K., Kordulis, C., Lycourghiotis, A., 2020. Waste cooking oil transformation into third generation green diesel catalyzed by nickel–alumina catalysts. Mol. Catal. 482, 110697.

No, S.Y., 2014. Application of hydrotreated vegetable oil from triglyceride-based biomass to CI engines—a review. Fuel 115, 88–96.

Nyström, I., Bokinge, P., Franck, P.A., 2019. Production of Liquid Advanced Biofuels-Global Status. CIT Industriell Energi AB.

Ranganathan, S.V., Narasimhan, S.L., Muthukumar, K., 2008. An overview of enzymatic production of biodiesel. Bioresour. Technol. 99, 3975–3981.

Refaat, A.A., 2010. Different techniques for the production of biodiesel from waste vegetable oil. Int. J. Environ. Sci. Technol. 7, 183–213.

Sanli, H., Canakci, M., 2008. Effects of different alcohol and catalyst usage on biodiesel production from different vegetable oils. Energy Fuels 22, 2713–2719.

Saydut, A., Kafadar, A.B., Aydin, F., Erdogan, S., Kaya, C., Hamamci, C., 2016. Effect of homogeneous alkaline catalyst type on biodiesel production from soybean [Glycine max (L.) Merrill] oil. Ind. J. Biotechnol. 15, 596–600.

Singh, D., Subramanian, K.A., Bal, R., Singh, S.P., Badola, R., 2018. Combustion and emission characteristics of a light duty diesel engine fueled with hydro-processed renewable diesel. Energy 154, 498–507.

Stumborg, M., Wong, A., Hogan, E., 1996. Hydroprocessed vegetable oils for diesel fuel improvement. Bioresour. Technol. 56, 13–18.

Thangaraj, B., Ramachandran, K.B., Raj, S.P., 2014. Homogeneous catalytic transesterification of renewable *Azadirachta indica* (neem) oil and its derivatives to biodiesel fuel via acid/alkaline esterification processes. Int. J. Renew. Energy Biofuel. 2014, 1–11.

Thangaraj, B., Solomon, P.R., Muniyandi, B., Ranganathan, S., Lin, L., 2019. Catalysis in biodiesel production—a review. Clean Energy 2–23.

Times of India, 2020. Govt Looks to Advance 20% Ethanol Blending With Petrol. Dec 21st http://timesofindia.indiatimes.com/articleshow/79838255.cms?ut.

UFOP Report on Global Market Supply, 2018/2019. Union zur Förderung von Oel-und Protein Pflanzen. https://www.ufop.de/files/4815/4695/8891/WEB_UFOP_Report_on_Global_Market_Supply_18-19.pdf.

Veljkovic, V.B., Lakicevic, S.H., Stamenkovic, O.S., Todorovic, Z.B., Lazic, M.L., 2006. Biodiesel production from tobacco (*Nicotiana tabacum* L.) seed oil with a high content of free fatty acids. Fuel 85, 2671–2675.

Yoon, J.J., 2011. What's the Difference Between Biodiesel and Renewable (Green) Diesel. Advanced BioFuels USA, Frederick, MD.

6

Development of second-generation ethanol technologies in India: Current status of commercialization

Harshad Ravindra Velankar, Chiranjeevi Thulluri, and Anu Jose Mattam

Bioprocess Division, Hindustan Petroleum Corporation Limited, HP Green R&D Centre, KIADB Industrial Area, Tarabahalli, Devanagundi, Hoskote, Bengaluru, India

1 Introduction

The British Socialist R.H Tawney once mentioned that the situation of the rural population is that of a man standing permanently up to the neck in water so that even a ripple is sufficient to drown him. Although, this remark was made in a different context, it applies to India, a country whose greater dependence on agriculture makes her vulnerable to climate changes. Moreover, diminishing natural resources and the ever-growing demand for agricultural products by an ever-increasing population makes it necessary for India to rapidly increase farmland productivity. While doing this, the farmers' incomes will have to be increased such that the differential between farm and nonfarm occupations is minimal and the availability of subsequent generations of food-providers are safeguarded. Multiple initiatives taken by the government of India (GOI) such as the distribution of soil health cards, efficient water management measures, promoting organic farming, creation of a unified national agriculture market, etc. are suitable for increasing farm productivity and farmers' incomes (Bhuvaneshwari et al., 2019). In addition, the utilization of farm-generated agro-waste for the production of value added products and energy would prove to be a shot in the arm for the farmers and the country (Rajkhowa et al., 2019).

At this stage, there is a growing consensus about climate change and the necessity to save humanity from catastrophic environmental events. The foremost cause for the deteriorating climate is the increased emission of greenhouse gases, resulting from the widespread use of fossil fuels as the primary source of energy (Perera, 2018). Therefore, the replacement of conventional fuels with alternative cleaner energy sources such as renewable fuels has become very necessary.

Renewable energy is produced from a variety of natural sources such as water (hydro), wind, solar, geothermal, marine or biomass (Panwar et al., 2011). All these resources are abundantly available on earth and can be converted into usable forms of energy. While most of these forms of energy are considered cleaner, the direct combustion of biomass for generating energy causes air pollution. However, instead of direct burning, specific components of biomass can be converted into usable forms of energy such as ethanol/butanol (biofuels), which can then replace conventional fossil fuels to burn in a cleaner manner (Luque et al., 2008). Technologies for producing biofuels from lignocellulosic biomass have been under development for at least the last three decades, albeit only a few processes have attained technocommercial success. The subsequent sections in this chapter discuss the commercialization status of lignocellulosic biofuels in India, the challenges involved and the way forward.

2 Cellulosic ethanol: Commercialization status

The production of bio-ethanol from sugar-bearing crops has been widely commercialized in India (known as first-generation ethanol plants) and most of the currently blended ethanol in gasoline comes from these plants. Currently, around 195 crore liters of ethanol is being blended with gasoline (9%) in India. These figures indicate that the ethanol blending percentage achieved till now is still very low and needs to be augmented for achieving the higher blending targets of 20% by 2025. Any substantial increase in bioethanol production can happen only when the feedstock range is expanded beyond the existing sugar or starch-bearing substrates. Naturally, lignocellulosic biomass is the most abundant and relatively cheaper feedstock that can be used to produce bioethanol. With the availability of a technology that can convert biomass into ethanol in a cost-effective manner, India can set itself on the path of energy independence. In India, the current prices of ethanol produced from biomass is at least twice that of 1st generation ethanol (produced from sugar-based feedstock). From December 2020, the price of ethanol produced from sugarcane juice and C-heavy molasses has been increased to Rs. 62.65/L and Rs. 45.69/L. The rate of ethanol produced from B-heavy molasses is now Rs. 54.27/L. Although the Indian government has approved variable pricing of ethanol depending upon the source of production, the production of lignocellulosic ethanol is still not considered economically viable. For the widespread commercialization of 2G ethanol technologies, the cost of ethanol production from biomass will have to be competitive with 1G ethanol.

The complexities involved in converting biomass into biofuel have been known for a long time. In 1981, a report authored by Prof. T K Ghosh of the Indian Institute of Technology, New Delhi foresaw the benefits of exploring biotechnology in the Indian context for the "displacement of the processes based on petro chemistry by renewable resources such as lignocellulose, sugar or starch". The article further mentioned that biofuels are "less likely to be affected by the same kind of cost spirals which are inherent in the conventional technologies based on nonrenewables as long as the substrates for these processes do not compete with food". Prof. Ghosh also developed a protocol for estimating cellulase activities and even after four decades, the same is still being followed across the world (Ghose, 1987). Despite having the knowledge and expertise about biomass conversion , the progress made towards the development of a viable biomass conversion technology did not result in notable successes in India. At the same time, scientists and engineers in other parts of the world could successfully demonstrate lignocellulosic ethanol technologies at larger scales. Many of these cellulosic ethanol plants were shut down due to issues related to higher operating expenses, sustainable feedstock supply, environmental factors, etc. In several cases, geopolitical or economic factors such as fluctuating oil prices, political instability, R&D policies or the availability of alternative forms of energy such as shale gas, which also reduced the investments required for the development of biofuels. Despite these setbacks, some progress continued to be made, the most notable being the 2G biorefinery set up by Beta Renewables in Italy, which was operated for considerable time, before it had to be closed down due to reasons such as feedstock logistics and financing problems (Cardona Alzate et al., 2020).

As of today, only Raízen Energia S/A (JV between Cosan and Royal Dutch Shell) operates a 10 MMgy second-generation ethanol plant in Sao Paulo, Brazil. As per the available information, the plant uses technology developed by Iogen Energy to convert sugarcane bagasse and cane straw into ethanol.

In India, biomass conversion technologies developed by PRAJ Industries-Pune , Institute of Chemical Technology, Mumbai (funded by the Department of Biotechnology) were demonstrated at large scales (10 T/day biomass) with varying degrees of success whereas technologies developed by others such as Nagarjuna Fertilizers-Hyderabad, NIIST-Trivandrum, IIT-Kharagpur were demonstrated only at the pilot scales. Lately, the oil marketing companies (OMC's) of India have shown keen interest in the development and scale-up of biomass conversion processes. The Indian Oil Corporation Limited (IOCL) established the DBT-IOC Centre for Advanced Bioenergy Research for conducting research on the development of new and economical pre-treatment processes for the conversion of cotton stalk and rice straw into ethanol and developing robust strains for fermentations. Likewise, Hindustan Petroleum Corporation Limited has set up a state-of-the-art R&D center in Bangalore to carry out research activities on different aspects of energy including renewable energy. The HP Green R&D center has developed a single-step pretreatment process (HP-ASAP) for biomass fractionation and has been actively working in different areas related to the development of enzymes and pentose converting strains.

Being an agricultural country, India has a rich experience of handling different types of agricultural residues that are potential feedstock for biofuel production. Indian farmers grow a wide range of crops and generate ~600 million tons (MT) of crop residues annually (TIFAC report) (TIFAC, 2018). Some of these agricultural residues such as rice and wheat straw (~ 140 MT) are burnt on farms every year to clear the field from straw and stubble of the preceding crop for sowing of the succeeding crop (TIFAC report). This wastage crops and the resultant air pollution caused by burning can be avoided by using it as a substrate for producing biofuels.

Energy feedstock (such as rice straw and wheat straw) is a low-density material and its collection from farms and transport to the nearest biorefinery is a labor-intensive and an expensive process. For example, the bulk density of chopped straw (bone dry) is

$\sim$50–120 kg m^{-3} and that of rice straw pellets is $\sim$600–700 kg m^{-3} (Migo, 2019). This means that a 4-ton truck can carry either 2400–2800 kg of rice straw pellets or 200–480 kg of chopped straw (bone dry) which is clearly uneconomical. An alternative would be to increase the bulk density of rice straw by densifying the material into bales (density 100–200 kg m^{-3}), prior to its transportation. An earlier study indicates that the process of raking can also be used for biomass densification (Kadam et al., 2000). Usually, round bales can shed off rainwater during storage whereas square shaped bales are easy to handle, store and transport. The Indian government provides subsidies to Indian farmers for purchasing farm equipment (such as balers) through newly created digital platforms such as Kisan Suvidha, Pusa Krishi, and MKisan. Several start-up companies in India are also creating businesses around biomass supply chain management.

The cost of transporting biomass from the farms to the nearest biorefinery has to be kept minimal and several studies have indicated that a distance of $\sim$100–200 km is suitable for collection of biomass. In India, a wide network for transporting sugarcane to sugar factories already exists and a 'bolt-on approach' to transport agricultural residues can minimize the costs of transportation. The location of the biorefinery is also very important since availability and access to basic facilities and infrastructure (roads, electricity, water, and land) have a major impact on the overall project feasibility. Moreover, any new technology that is appended to an existing facility can benefit significantly by the sharing of utilities (power, steam, water and labor).

When sugarcane is transported to the factory, it undergoes water washing for the removal of bound soil particles. Similarly, for lignocellulosic biomass, the de-baling, stone/metal removal is followed by a cleaning step. Although complete soil removal cannot be attained during washing of biomass, reducing the content of soil can have a significant impact on increasing the efficiencies of subsequent acid/enzyme-catalyzed hydrolysis processes. The removal of stone or metal is necessary to protect the mechanical parts of equipment from getting damaged during subsequent operations. One of the most energy-intensive operations is the sizing of biomass which is influenced by the moisture content in the biomass. The effectiveness of subsequent processing steps (pretreatment, hydrolysis) can get affected if suitable sized biomass is not generated. Technology providers are expected to consider all these aspects about biomass sizing before claiming their technology to be feedstock agnostic For a technology to be truly agnostic, multiple feedstocks should be processable. However, this may still require equipment modifications due to compositional variations of feedstock. In a recent incident, when cotton stalk was processed in an equipment designed for rice straw sizing, the material caught fire due to heat generated by friction during sizing. The lint fibers in cotton stalk have a lower ignition temperature than rice straw and are therefore easily combustible.

After sizing, the biomass is subjected to thermochemical pretreatment wherein a catalyst, in presence of heat and pressure removes lignin and improves the accessibility of cellulose to enzymes. Several single or multi-step pretreatments such as acid and/or alkali, steam explosion, ammonia (AFEX), water (hydrothermal), organosolv, etc., have been developed and each of these come with their own advantages and disadvantages (Kumar and Sharma, 2017). Depending on its effectiveness, the choice of pretreatment can have a significant impact on the process economics. Pretreatment processes using acids, alkalis, organic solvents can be carried out in conventional reactors (CSTR), whereas advanced processes (steam explosion and

AFEX treatments) may require specialized equipment, thereby increasing the capital cost (CAPEX). In some pretreatments, despite higher CAPEX, the use of conventional solvents (water for steam explosion and hydrothermal treatments) or advanced catalysts (organic solvents) can reduce the operating costs. While majority of the pretreatment processes use acids and alkalis for biomass hydrolysis, enzymatic pretreatments have been used for delignification (Bhuvaneshwari et al., 2019). Compared to thermochemical pretreatment processes, enzymatic methods are much slower, less efficient and moreover, the cost of producing enzymes for large scale applications could become a detriment to their application. Although the degree of delignification is an indicator of the effectiveness of pretreatment, it is necessary to estimate the carbohydrate losses during pretreatment. Another important aspect to increase the profitability of a biorefinery is to utilize waste streams (lignin) for different applications such as electricity generation. However, in some cases, the co-gen of electricity by the combustion of lignin-rich streams may not be possible due to economic/space constraints. In such cases, suitable alternatives for lignin valorization will have to be developed.

Amongst all the unit operations involved in biomass conversion, the enzymatic hydrolysis process is considered as the most important step that can tip the overall economics. This is because the enzymatic hydrolysis step requires the use of cellulolytic enzymes that are still expensive. Although substantial investments were made to develop cellulolytic enzymes for second-generation ethanol production, there are only a few companies such as Novozymes, who have the capacity to supply commercial quantities of these enzymes. An alternative to using commercial enzymes is to produce these "on-site". However, the production of cellulolytic enzymes is still very expensive due to low enzyme yields and productivities. Some reports also indicate that cellulases developed for industrial applications such as textiles could be used, however, their effectiveness for biomass hydrolysis remains to be proven.

In order to achieve techno-economic success with a 2G ethanol process, it is necessary to convert all the released sugars into ethanol . Although the yeast *Saccharomyces cerevisiae* is widely used for producing ethanol from C6 sugars, it cannot convert pentose (C5) sugars into ethanol. Other yeast such as *Candida* and *Pichia* can partially convert both, C5 and C6 sugars, however, the yields are lower and the production of undesired products (s.a xylitol) can be problematic. Strain improvement via genetic engineering can further improve the C5 and C6 conversion to near-theoretical yields. Currently, the Indian government permits the use of genetically engineered organisms at large scales only after obtaining the necessary approvals from regulatory authorities (IBSC, RCGM & GEAC).

Fractional distillation is the most commonly used process for the separation of ethanol from the fermented broth . In case of lignocellulosic ethanol, the broth contains less ethanol (~5%) compared to that obtained during conventional sugar fermentations (~12%) and therefore higher energy input is required for its separation. Several engineering companies in India are already in the business of supplying technologies to distill ethanol and their expertise can be leveraged for the separation and purification of 2G ethanol.

The success of second-generation ethanol technology depends on collaboration between scientists and engineers. While newer and efficient catalysts, biocatalysts, and processes are being developed by scientists, it is important for the engineers to focus on designing rapid and efficient processes for converting lignocellulosic biomass into ethanol. Overall, the 2G ethanol process being a multi-step operation generates large data sets which have to be processed for building predictive simulation models.

3 National Biofuel Policy (NBP)

The Indian government has earlier formulated several policies and guidelines for increasing the use of ethanol as an alternative source of energy. The Power Alcohol Act of 1948 directed the use of surplus molasses for the production of "power alcohol" for blending in petrol. Although the Act mandated blending of 20% (v/v) ethanol in petrol, the target was never met due to several reasons such as limited ethanol availability, oil price fluctuations, poor implementation, etc., and therefore the Act was finally repealed in 2000. In the year 2002, the Ethanol Blending Programme (EBP) was notified and a seemingly realistic target of 5% blending in selected states by 2003 was set. Unfortunately, this target could also not be met due to similar reasons (mentioned earlier) and therefore in 2004, the Indian government announced ethanol blending to be an optional activity. However, due to rising oil prices by 2007, ethanol blending (5%) was made mandatory across the country. Again, the implementation of EBP was only partially successful due to the unavailability of feedstock, delay in ethanol procurement by the oil marketing companies (OMCs) and an unfavorable pricing policy.

While several policies were formulated for ethanol blending in petrol, similar efforts were also made for replacing diesel with biodiesel. In 2003, the National Mission on Biodiesel (NMB) encouraged the cultivation of oil crops such as Jatropha on wastelands to produce biodiesel for blending in diesel up to 5% by 2007 and 20% by 2011–12. The OMCs of India assured the offtake of biodiesel thereby establishing a ready market. Unfortunately, the biodiesel blending program met with limited success due to low oil yields.

Based on the learnings of EBP and NMB programs, a new National Policy on Biofuels of 2009 was formulated by the government of India which aimed to achieve a blending of 20% (v/v) ethanol and biodiesel in petrol and diesel, respectively, by 2017. For the first time, the policy clearly acknowledged the roles of financial institutions in achieving the overall objectives. The policy further declared bioethanol and biodiesel as "declared goods," thus enabling their free movement across the country and encouraging foreign investment in this sector (up to 100%).

In 2018, the growing concerns about the environment and the urgency of reducing oil imports to save foreign exchange led to the formulation of a National Biofuel Policy which also considered other aspects such as could increasing farmers' income, generating rural employment, optimal usage of drylands, etc. The National Biofuel Policy (NBP) of 2018 set an ambitious target of 20% (v/v) bioethanol and 5% (v/v) biodiesel blending by 2030 (Refer http://petroleum.nic.in/national-policy-biofuel-2018-0.) and also provided financial incentives for 2G ethanol production. In addition, for developing the necessary infrastructure, Viability Gap Funding (VGF) of Rs. 5000 crores over a period of 6 years was introduced for setting up of both, commercial and demonstration scale 1G/2G biorefineries with access to subsidized electricity and water. The Ministry of Petroleum & Natural gas (MoPNG) has announced the setting up of 12 2G ethanol plants using various feedstock (Rice straw, wheat straw, sugarcane bagasse, cotton stalk, etc.) across the country.

To attain the objectives of the NBP 2018, ~3136 million liters of ethanol would be required annually to achieve 10% blending target, while only ~1400 million liters of ethanol could be produced in 2017–18. As per preliminary estimates, the ethanol deficit is expected to build up to over 2604 million liters by 2022. Therefore, the inclusion of other feedstock (such as sugarcane juice and B-molasses) with potential to produce ethanol was felt necessary and included in the policy. As a result of this change, the sugar mill owners stand benefited since the excess cane juice can be diverted for ethanol production during years of surplus sugar production.

In addition to the sugar-based feedstock, starchy substrates such as damaged foodgrains, rotten potatoes, broken rice, sugar beet, corn, etc., and lignocellulosic biomass such as farm waste and agri-residues (rice and wheat straw, sugarcane bagasse, etc.) have also been included as a potential feedstock for the production of ethanol under NBP 2018 (http://petroleum.nic.in/national-policy-biofuel-2018-0.)

The policy allows 100% Foreign Direct Investment (FDI) in the biofuel sector but restricts the import (for encouraging domestic production capabilities and ensuring price stability) and the export (to ensure internal utilization) of ethanol. The policy has identified OMCs as the nodal agencies for the storage, distribution, and marketing of biofuels. On the sustainability front, the NBP of 2018 promotes the use of indigenous feedstock grown on agricultural land or wasteland, so as to not disturb the flora and fauna of the country. The biofuel policy further aims to reduce greenhouse gas emissions by around 33%–35% by 2030.

4 Second-generation ethanol processes and major challenges—A brief overview

Lignocellulosic biomass (LCB) is abundantly available in nature and is a potential substrate for producing ethanol. However, the route for lignocellulosic biomass conversion into bio-ethanol or butanol involves a multi-step biochemical process (Fig. 1) wherein lignin

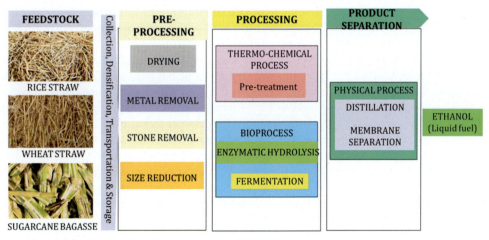

FIG. 1 The various steps involved in the conventional 2G ethanol process.

present in the biomass is first removed and the cellulosic components are hydrolyzed into fermentable sugars for producing bio-alcohols (Al-Zuhair et al., 2013). The biochemical route for lignocellulosic ethanol production is still not considered to be economically viable for various reasons such as the requirement of expensive catalysts, low efficiency of C5 utilizing strains, sugar losses, higher energy and water requirement as well as other process-related challenges. Each of these problems will have to be carefully studied and suitable solutions need to be found for making the process viable. The following sections discuss various problems associated with the commercialization of 2G ethanol technology.

4.1 Unit operations and associated challenges in 2G ethanol production

Although severe process conditions can disintegrate the biomass, it can cause the degradation of the polymeric carbohydrates, thereby making them unavailable for ethanol production. Therefore, it is necessary to select a suitable thermochemical pretreatment process that is highly effective and less severe.

4.1.1 Thermochemical pretreatment of biomass

Over the years, researchers have developed several methods for biomass deconstruction using catalysts such as acids, alkalis, ionic liquids, solvents, steam, etc., in a single or a two-step process (Kumar and Sharma, 2017). Generally, the use of chemical catalysts during biomass pretreatment causes the partial removal of lignin from biomass and the hydrolysis of the hemicellulosic component. Under low process severity, the cellulose component of biomass is maximally retained and hydrolyzed in subsequent unit operations. The choice of a pretreatment process depends upon its effectiveness in removing lignin and hydrolyzing hemicellulose minimizing inhibitor formation while ensuring the generation of cellulose that is highly amenable to cellulases. Although commonly used acids (H_2SO_4, HCl, HNO_3, H_3PO_4 etc.) are relatively cheaper, their use may generate sugar dehydration products (fermentation inhibitors like furfural and 5-hydroxymethyl furfural) which may require an additional neutralization step in the process (Jönsson and Martín, 2016). Similarly, the use of alkaline chemicals such as NaOH, KOH, NH_4OH, etc. for delignification of biomass can also lead to loss of the hemicellulose component. The treatment of biomass with steam or ammonia requires specialized equipment leading to higher CAPEX, whereas, the use of advanced catalysts such as ionic liquids could lead to lower fermentation efficiencies as even their trace quantities in the sugar stream may be toxic to the biocatalyst . Moreover, in order to increase the effectiveness of a pretreatment process, higher mass transfer via uniform mixing needs to be ensured which may be difficult due to the heterogeneous nature of the reaction mixture. Ideally, lower the biomass particle size, greater is the mass transfer during mixing operations. To obtain finer biomass particles ($\sim$0.5–1 mm), specialized or additional (cutting followed by grinding) equipment might be required resulting in higher energy inputs and increased capital expenditure. However, the generation of finer size biomass may lead to greater biomass losses during subsequent unit operations.

4.1.2 Enzymatic hydrolysis of pretreated biomass

Worldwide, considerable research has gone into developing highly efficient and cost-effective cellulase enzymes but has met with limited success. Due to this, there are only a handful of companies that have commercialized the production of these enzymes. Cellulases comprise of *exo*-glucanase, *endo*-glucanase, beta-glucosidase and other accessory proteins (Payne et al., 2015). The expensive nature of cellulases can be attributed to the inability of any single organism to produce a mixture of all of the above-mentioned enzymes and accessory proteins at higher concentrations (Payne et al., 2015). Moreover, some of the components used in the fungal cultivation media (Cellulsoe, peptone, yeast extract, vitamins and surfactants) are costly and the protein secretion levels are relatively lower. Some 2G ethanol processes have been able to demonstrate the recycling of enzymes using membrane technologies; however, issues related to membrane life and fouling, loss of enzyme activity need to be addressed. To reduce the cost of enzymes, in-situ enzyme production has been suggested as an alternative for commercial enzymes. The on-site production of enzymes may seem to be an attractive option since, other than biomass separation, no other processing (enzyme purification or concentration) is needed prior to biomass hydrolysis (Li et al., 2017). However, unless the enzyme expression levels improve significantly, this may not become a viable process for enzyme production.

Some pretreatment processes may result in only the partial removal of lignin leading to the nonspecific adsorption of cellulases on to the residual lignin resulting in reduced hydrolysis performance.

4.1.3 Ethanol fermentation

To maximize ethanol production all sugars released from biomass need to be converted. The fermentative conversion of sugars into ethanol occurs in hydrolysates that are nutritionally deficient and contain inhibitory compounds such as furans, phenols, organic acids, etc., generated during pretreatment (Jönsson and Martín, 2016). The main challenge therefore is to develop a robust biocatalyst that can tolerate these inhibitors and convert sugars into ethanol.

4.1.4 Waste management

The production of lignocellulosic ethanol is a water-intensive process. Depending upon the type of the technology, the nature of liquid and solid effluents generated during the processing of biomass will vary. Suitable effluent treatment systems need to be designed and carefully evaluated for treating the waste streams of a 2G ethanol process.

For scaling up bioprocesses wherein multiple variables are involved, large volume of lab/pilot scale data collected over longer periods needs to be processed and analyzed for building simulation models. These models are needed to predict the process performance at larger scales. In most cases, reliable data (in terms of quality and quantity) for estimating the mass and energy balance of all the unit operations involved in biomass conversion is not available to build accurate predictive models.

4.1.5 *Feedstock availability and supply chain management*

One of the most important aspects of commercial 2G ethanol process is to ensure the continuous supply of feedstock at a stable and lower price. The prices of agricultural residues can increase if the feedstock supply becomes limited due to environmental factors or in case alternative applications are allowed within the vicinity of a biorefinery. Ideally, any biorefinery would seek legal guarantees for ensuring the supply of biomass at fixed prices (with annual marginal increments) and for longer durations of 5 years. In addition, the volatility in fuel prices could affect the biomass transportation costs and therefore mechanisms have to be found to minimize the impact. To ensure process continuity, mega-scale plants would require a facility for storing biomass to suffice their short-term (from 1 week to 1 month) requirements.

5 Analytical framework in 2G ethanol process development

The composition of biomass may vary depending on the source and type of biomass, environmental factors, and even cultivation practices. To determine the exact potential of particular biomass for producing biofuels, it is necessary to know the exact composition of biomass that also includes the fermentable carbohydrates. The procedures developed by the National Renewable Energy Laboratory (NREL), USA are globally recognized as standard protocols for estimating biomass components (Sluiter et al., 2011). Likewise, for the estimation of cellulase enzyme activities, the protocols developed by IUPAC and T.K. Ghose are widely accepted to be accurate (Ghose, 1987). However, despite the availability of these standard analytical procedures, wide differences are observed in values published by different laboratories w.r.t the content of cellulose and hemicellulose in biomass, the moisture content, fungal enzyme activities, etc. thereby making their comparative evaluation difficult.

Some biofuel research groups are known to practice easier methods of biomass compositional analysis (TAPPI methods) which may not be adequate to determine the potential biomass for biofuel production. The TAPPI methods can approximately estimate the cellulose and hemicellulose components whereas NREL procedures are more comprehensive and can accurately estimate glucan, xylan, arabinan, mannan, and galactan content in biomass. Further, TAPPI methods are used to determine lignin content in paper pulp samples (with <5% w/w) and may not be suitable for estimating the higher lignin content in biomass (~20 % wt.). Moreover, NREL procedures can determine both, acid soluble (ASL) and acid insoluble (AISL) lignin content which may help to indicate the effectiveness of a delignification process. In comparison with TAPPI methods, NREL procedures are complex and sensitive. For example, during biomass sample preparation for NREL estimations, any deviations from the recommended particle size of 180–850 μm can result in erroneous lignin estimation (for particles <180 μm) or under-estimation of carbohydrates (for particles >850 μm). Similarly, biomass samples with higher moisture content (>10%) may cause the dilution of sulfuric acid, leading to incomplete hydrolysis and incorrect carbohydrate estimations. Other aspects of the methods such as proper mixing of the reactants during stage 1 hydrolysis, careful washing of the solid biomass residue post hydrolysis, using the recommended apparatus (pressure tubes) and determining the sugar recovery factors are critical for accurate estimation of biomass composition by NREL methods.

6 Technology readiness level (TRL) and commercialization prospects:

The unit operations (biomass pretreatment, enzymatic hydrolysis, fermentation, distillation) of a typical 2G ethanol process are at various stages of Technology Readiness Levels (TRL) which are described below;

6.1 Biomass pretreatment

Any pretreatment process aims to maximize access to fermentable carbohydrates in biomass. However, due to the compositional variations in biomass, a single pretreatment process may not be suitable for processing of different feedstock. At the same time, it would be impractical to have separate pretreatments for various feedstock. In such a scenario, it would be advisable to develop a pretreatment method that can hydrolyze at least two to three different types of biomass. In addition, the pretreatment process should also be able to absorb minor variations in physical characteristics (moisture content, size, soil content, etc.) of the feedstock. The use of certain feedstock such as rice straw can increase the wear and tear of the sizing and pretreatment equipment due to the presence of abrasive components such as silica and therefore the MOC (material of construction) of the equipment is to be suitably selected. In addition, the pretreatment reactor design should ensure proper mass transfer, avoid creation of dead zones as well as, enable uninterrupted material transfer. Some of the expectations from a pretreatment process are as below;

- Low severity process to minimize sugar losses
- High amenability of cellulose to enzymes
- Minimal chemical and water requirement
- Low energy requirement and carbon-footprint
- Low CAPEX and OPEX

Of the several pretreatment methods developed, conventional processes like ball milling, steam explosion, hydrothermal, dilute acid, and alkali have been demonstrated at larger scales (TRL 5–7) (Tables 1 and 2) whereas advanced methods that use ionic liquids, deep eutectic solvents, supercritical CO_2, biocatalysts, etc., are still at the laboratory scale (TRL 2–4) (Tables 1 and 2).

6.2 Enzymatic hydrolysis

Enzymatic hydrolysis of pretreated biomass causes depolymerization of cellulose into fermentable sugars. Most of the laboratory studies are carried out at low biomass solid concentrations (2%–5% w/v) in accordance with NREL saccharification protocols and report >80% hydrolysis. However, for enzymatic hydrolysis to become feasible higher solid loadings of at least 10% w/v are essential and therefore, the experimental values generated at the lab scale may not be directly applicable at the pilot/demo scales. Moreover, for handling hydrolysis reactions with higher solids different reactor designs may be required. Further, several approaches such as the use of higher enzyme concentrations, additives (BSA, Tweens & PEG) (Obeng et al., 2017) or accessory proteins (esterases, laccases, LPMOs) (Banerjee et al., 2010; Silva et al., 2016), modification of the reactor internals for overcoming mass transfer problems, etc., (Zhang et al., 2010) have also been reported for improving cellulose

TABLE 1   Technology readiness level (TRL) scale.

TRL	Definition	Explanation	Stage
0	Idea	Unproven concept, no testing has been performed	Research
1	Preliminary research	Postulating principles and observed but no considerable experimental proof available	
2	Technology formulation	Conceptualization and application have been formulated	
3	Applied research	Laboratory studies and establishing proof of concept	
4	Small scale prototype	Built in laboratory environment	Development
5	Large scale prototype	Tested in intended environment	
6	Prototype system	Tested in intended environment close to expected performance (Pilot-scale studies)	
7	Demonstration system	Operating in operational environment at pre-commercial scale	Deployment
8	First-of-a-kind commercial system	Manufacturing issues solved	
9	Ready for commercialization	Technology available for consumers	

TABLE 2   Technology readiness level (TRL) and future prospects of biomass pretreatment during 2G ethanol production.

Pretreatment technology	TRL	Advantages	Disadvantages	Scope for improvement	Suitable for
Steam explosion	6–8	• Cost-effective • High glucose yields • Lignin and hemicellulose removal • Low environmental impact	• Steam often needed to optimized • Formation of inhibitors and toxic compounds	• Development of new catalysts • Developing micro-organisms more tolerant to inhibitors	• A variety of herbaceous and woody biomass
Hydrothermal/ autocatalysis	4–6	• No chemical use • High hemicelluloses removal	• Higher operating temperature • Inhibitor formation	• Introduction of new catalysts for removal of lignin component	• Low lignin feedstocks
Dilute-acid	5–7	• High hydrolysis of hemicelluloses	• Degradation by-products (salts) and inhibitors • Pseudo-lignin formation • Equipment corrosion	• Developing micro-organisms more tolerant to inhibitors • Reducing intensity of pre-treatment • New enzyme developments	• Low lignin feedstocks

TABLE 2 Technology readiness level (TRL) and future prospects of biomass pretreatment during 2G ethanol production—cont'd

Pretreatment technology	TRL	Advantages	Disadvantages	Scope for improvement	Suitable for
Dilute-alkali	5–7	• Low CAPEX • Low inhibitor formation • High cellulose conversions	• Residue formation • Need to recycle chemicals • Enzyme adjustment needed	• New enzyme development • Recovery and reuse of chemicals	• Smaller-scale plants
AFEX	3–5	• High size biomass can b treated • Low inhibitor formation • High accessible surface area	• High cost due to solvent • Not effective for high % lignin	• Recovery and reuse of chemicals	• Smaller decentralized plants
Organosolv	4–6	• Causes lignin and hemicellulose hydrolysis or dissolution	• High CAPEX & OPEX • Solvent may inhibit cell growth	• Recovery and reuse of chemicals • Develop methods to add value to lignin	• High quality lignin co-product
Supercritical CO_2 cooking	2–4	• Increases accessible surface area • Low inhibitors or residues	• Does not affect lignin and hemicellulose • Very high pressure, high capex	• Improve process technology • Develop methods to add value to lignin	• Continuous technology • Suitable for smaller-scale plants
Ionic liquids and DES	2–3	• Effective & selective dissolution of all Lignocellulose components • Low degradation products	• Expensive technology and recovery required	• Recovery and reuse of chemicals • Develop process technology	• A variety of herbaceous and woody biomass
Biological (fungal/ laccase)	3–4	• Low energy requirement • No corrosion Suitable for lignin and hemicellulose removal	• Time-consuming • Some carbohydrate losses	• Development of robust micro-organisms	• Composting the biomass
Ball milling	5–6	• Reduces cellulose crystallinity • No inhibitors or residues	• Energy intensive process • Poor sugar yields	• Process integration • Combine with mild chemical treatments	• High value products

hydrolysis at laboratory scale. However, unless the cost of cellulolytic enzymes is reduced, the inclusion of external additives may not be feasible. Considering the above, the TRL of enzymatic hydrolysis could be around 7.

6.3 Fungal cellulase production

There are only a few enzyme suppliers with the capacity to produce cost-effective and highly active cellulase enzymes. Enzymes are still very expensive and contribute to about 25%–30% of the product cost. Based on the available literature, only a few enzyme-producing strains (*Trichoderma* sp., *Penicillium*, and *Aspergillus* sp.) have been developed for cellulase production and enzyme titers of these strains still remain low. In addition to the main cellulase enzymes, the role of accessory proteins (expansins, swollenin, LPMOs, etc.) in enhancing the enzyme-substrate interaction has been reported, however, these efforts are limited to the laboratory scale. Although the production of cellulases for other applications (textile, animal-feed processing, etc.) has been commercialized, the TRL for cellulolytic enzymes required for biofuel production may be 7.

6.4 Ethanol fermentation

Although C6 sugar fermentation is carried out by distilleries using molasses and sugarcane juice as the feedstock, the conversion of pentose sugars in biomass into ethanol remains a challenge as only a few fermenting organisms have this ability. In addition, biomass hydrolysates are typically nutrient deficient and contain inhibitory compounds which may further reduce fermentation efficiencies leading to longer process times. Although the TRL for yeast mediated C6 fermentation have achieved commercial success (TRL 9), the TRL for C5 fermentation is around demonstration scale (TRL 7).

7 Future perspectives and conclusion

In order to develop a techno-economically feasible process for producing lignocellulosic ethanol, major scientific advancements are needed with respect to developing cellulolytic enzymes and biocatalysts. For unit operations such as size reduction, existing equipment will have to be customized and fine-tuned for generating suitable feed for subsequent processes. Similarly, for pretreatment and enzymatic hydrolysis, the best processes developed at laboratory scales will have to be optimized at the larger scales and evaluated using the established protocols. In case of fermentation, efforts will have to be focused on overcoming the nutritional deficiencies of hydrolysates and optimizing the bioprocess conditions at large scale. At the same time, several non-technical issues such as biomass supply chain management, ethanol distribution, pricing, etc. also need to be resolved for making 2G ethanol processes techno-commercially viable. Finally, although the development of biofuels would be extremely advantageous for India, the replacement of fossil fuels with renewables cannot be realized without addressing each of the above mentioned challenges. Already, with the implementation of the Government of India, Ministry of New & Renewable Energy (2018), a definitive step in the right direction has been taken and concerted efforts by the various stakeholders (feedstock suppliers, technologists, regulatory authorities, etc.) will result in the development of a sustainable second-generation ethanol production technology.

References

Al-Zuhair, S., Al-Hosany, M., Zooba, Y., Al-Hammadi, A., Al-Kaabi, S., 2013. Development of a membrane bioreactor for enzymatic hydrolysis of cellulose. Renew. Energy 56, 85–89.

Banerjee, G., Car, S., Scott-Craig, J.S., Borrusch, M.S., Bongers, M., Walton, J.D., 2010. Synthetic multi-component enzyme mixtures for deconstruction of lignocellulosic biomass. Bioresour. Technol. 101 (23), 9097–9105.

Bhuvaneshwari, S., Hettiarachchi, H., Meegoda, J.N., 2019. Crop residue burning in India: policy challenges and potential solutions. Int. J. Environ. Res. Public Health 16, 832.

Cardona Alzate, C.A., Serna-Loaiza, S., Ortiz-Sanchez, M., 2020. Sustainable biorefineries: what was learned from the design, analysis and implementation. J. Sustain. Dev. Energy Water Environ. Systems 8 (1), 88–117.

Ghose, T.K., 1987. Measurement of cellulase activities. Pure Appl. Chem. 59, 257–268.

Government of India, Ministry of New & Renewable Energy, 2018. National Policy on Biofuels 2018. Government of India, Ministry of New & Renewable Energy. http://petroleum.nic.in/national-policy-biofuel-2018-0.

Jönsson, L.J., Martín, C., 2016. Pretreatment of lignocellulose: formation of inhibitory by-products and strategies for minimizing their effects. Bioresour. Technol. 199, 103–112.

Kadam, K.L., Forrest, L.H., Alan Jacobson, W., 2000. Rice straw as a lignocellulosic resource: collection, processing, transportation, and environmental aspects. Biomass Bioenergy 18 (5), 369–389.

Kumar, A.K., Sharma, S., 2017. Recent updates on different methods of pretreatment of lignocellulosic feedstocks: a review. Bioresour. Bioproces. 4, 7.

Li, Y., Zhang, X., Xiong, L., Mehmood, M.A., Zhao, X., Baia, F., 2017. On-site cellulase production and efficient saccharification of corn Stover employing cbh2 overexpressing *Trichoderma reesei* with novel induction system. Bioresour. Technol. 238, 643–649.

Luque, R., Herrero-Davila, L., Campelo, J.M., Clark, J.H., Hidalgo, J.M., Luna, D., Marinas, J.M., Romero, A.A., 2008. Biofuels: a technological perspective. Energ. Environ. Sci. 1, 542–564.

Migo, M.V.P., 2019. Optimization and Life Cycle Assessment of the Direct Combustion of Rice Straw Using a Small Scale, Stationary Grate Furnace for Heat Generation. Unpublished Master's thesis. University of the Philippines Los Baños.

Obeng, E.M., Budiman, C., Ongkudon, C.M., 2017. Identifying additives for cellulase enhancement—a systematic approach. Biocatal. Agric. Biotechnol. 11, 67–74.

Panwar, N.L., Kaushik, S.C., Kothari, S., 2011. Role of renewable energy sources in environmental protection: a review. Renew. Sustain. Energy Rev. 15 (3), 1513–1524.

Payne, C.M., Knott, B.C., Mayes, H.B., Hansson, H., Himmel, M.E., Sandgren, M., Ståhlberg, J., Beckham, G.T., 2015. Fungal cellulases. Chem. Rev. 115 (3), 1308–1448.

Perera, F., 2018. Pollution from fossil-fuel combustion is the leading environmental threat to global pediatric health and equity: solutions exist. Int. J. Environ. Res. Public Health 15 (1), 16.

Rajkhowa, D.J., Sarma, A.K., Bhattacharyya, P.N., Mahanta, K., 2019. Bioconversion of agricultural waste and its efficient utilization in the hilly ecosystem of Northeast India. Int. J. Recycl. Org. Waste Agric. 8, 11–20.

Silva, A.S., Souza, M.F., Ballesteros, I., Manzanares, P., Ballesteros, M., Bon, E.P.S., 2016. High-solids content enzymatic hydrolysis of hydrothermally pretreated sugarcane bagasse using a laboratory-made enzyme blend and commercial preparations. Process Biochem. 51, 1561–1567.

Sluiter, A., et al., 2011. Determination of Structural Carbohydrates and Lignin in Biomass. National Renewable Energy Laboratory (NREL), Analytical Procedure (LAP). http://www.nrel.gov/biomass/analytical_procedures.html.

TIFAC, 2018. Estimation of Surplus Crop Residues in India for Biofuel Production. Technology Information, Forecasting & Assessment Council (TIFAC). https://tifac.org.in/images/pdf/pub/TIFACReports/newreports/biomass_w(1).pdf.

Zhang, J., Chu, D., Huang, J., Yu, Z., Dai, G., Bao, J., 2010. Simultaneous saccharification and ethanol fermentation at high corn Stover solids loading in a helical stirring bioreactor. Biotechnol. Bioeng. 105, 718–728.

Biomass characterization

Bijoy Biswas[a,b], Bhavya B. Krishna[a,b], M. Kiran Kumar[c], Rajeev K. Sukumaran[c], and Thallada Bhaskar[a,b]

[a]Material Resource Efficiency Division (MRED), CSIR-Indian Institute of Petroleum (IIP), Dehradun, India [b]Academy of Scientific and Innovative Research (AcSIR), Ghaziabad, Uttar Pradesh, India [c]Microbial Processes and Technology, CSIR-National Institute for Interdisciplinary Science and Technology, Trivandrum, India

1 Introduction

Human impact on global warming and climate change are increasingly evident and are a serious concern that can affect our very existence on the planet. The burning of fossil fuels is a major contributor to greenhouse gas emissions, and this warrants urgent attention and action transitioning towards renewable and green alternatives. Geopolitical issues related to fossil fuels across the globe are the reasons behind every country's drive to become self-reliant in terms of energy requirements. The recent bold commitments made at COP-26 by global leaders has increased the drive to safeguard the planet by slowly transitioning toward renewable sources of energy.

In this scenario, researchers are developing renewable energy-based processes for electricity production. Huge amount of research and development is being carried out on electric mobility which can help in decarbonizing the world economy. Light-duty vehicles can run on batteries for limited distance, but alternate options need to be identified for heavy-duty transport and long-haul travel. Bio-derived fuels could be a transitional arrangement till electric vehicles start to cover long distances with single charge or heavy-duty transportation starts relying on other fuels like hydrogen, ammonia derived fuels, etc.

When it comes to materials and chemicals, renewable carbon is still required to make the process carbon neutral. Biomass is the only source of such renewable organic carbon within its macromolecular structure that can be exploited to produce bio-materials/bio-chemicals/bio-fuels (Wu et al., 2020).

Several kinds of biomass can be utilized for the production of different end-products. Currently, first-generation biofuels have been deployed in many countries across the globe. First-generation biofuels are derived from edible biomass such as edible oil(s), sugars, etc. The most widely used bio-fuels from edible sources are bio-ethanol and bio-diesel. However, in certain countries they cause the dilemma of food vs fuel leading to the movement toward utilization of nonedible feedstocks. Agricultural crop residues, forestry wastes, defatted (oil removed) seed cakes, invasive terrestrial species belong to the category of lignocellulosic biomass. Bio-fuels derived from these feedstocks are termed as second-generation bio-fuels. Third-generation bio-fuels are derived from algal sources, be it microalgae or macroalgae (seaweeds). Both the second-generation and third-generation bio-fuels are yet to increase their market visibility and are at different levels of development in various parts of the globe. Aquatic biomass feedstocks are another group of plants that grow on water and most of the times cause issues like pollution, eutrophication, etc. Some of these aquatic weeds are water hyacinth, paragrass, etc., which need to be removed from the water bodies to save them. The efficiency of conversion of biomass to biofuels/chemicals/materials depends on the biomass chemical composition and its structure. Therefore, it is important to first understand the biomass composition and its characteristics before it is used in any biorefinery applications (Raj et al., 2015; Gundekari et al., 2020).

The composition of biomass will help in identifying the possible products that can be derived from it. The other characteristics of biomass will help in understanding its structure, bonds present, functional groups available, etc. This structural information about the feedstock may provide an idea of the conversion step that maybe necessary for its valorization. The following chapter aims to discuss in detail the different methods of characterization available for understanding biomass feedstocks. The variations in the information provided by different techniques for a particular characterization will be discussed. In addition, the manuscript also aims to provide information on how data from different characterization techniques can be corroborated to obtain a holistic picture of the feedstock under study. This will provide confidence to the researchers to use it in any conversion process for obtaining the desired product.

2 Major components in biomass

2.1 Lignocellulosic biomass

2.1.1 Cellulose

Cellulose is one of the major components of lignocellulosic biomass, and it constitutes about 30%–50% of the biomass. The earliest extraction of cellulose from wood was in 1838 by HNO_3 treatment. It is usually present as microfibrils in the plant cell ranging in length between 3 and 35 nm depending on its source (Rajinipriya et al., 2018). Cellulose is a linear polymer made of unbranched anhydro-β-glucose rings linked through oxygen covalently bonded to C1 of one glucose ring and the C4 of adjacent glucose which is rotated 180° axially. This bonding structure leads to the nomenclature of β-1-4 glycosidic bond (Chatterjee et al., 2015) (Fig. 1). High stability of cellulose is because of the intra- and intermolecular hydrogen bonds between the units. The intramolecular hydrogen bonds are found in-between C3 hydroxyl group with adjacent in-ring oxygen and C2 hydroxy group with hydroxymethyl group oxygen on C6. The intermolecular hydrogen bonds exist in between monomeric units connected through C6 hydroxymethyl with C3 hydroxy groups of the adjacent chain. As it is a polymer, cellulose can be denoted by its degree of polymerization and it varies depending on the plant type (Lee et al., 2014).

Meyer–Misch model in the form of Cellulose I was used to explain the crystalline form of native cellulose (Mark and Meyer, 1929). Here, it is assumed that cellulose I is a monoclinic unit cell. Few researchers differed from this view and suggested that native cellulose is a composite of two different crystal units, cellulose Ia (with one-chain triclinic structure) and cellulose Ib (a two-chain monoclinic structure) (Atalla and Vanderhart, 1984; Karimi et al., 2013). Different methods are used by industries specific to their requirement due to their preference for a particular fraction of lignocellulosic biomass. Technical Association of Pulp and Paper Institute's (TAPPI) standards are generally preferred by the pulp and paper industries using woody biomass as feedstocks as they are more interested in the cellulose fraction. The 2G ethanol industries usually prefer the NREL method of biomass compositional analysis since they require both the cellulose and the hemicellulose fractions for further processing. The food and forage industries prefer the Association of Official and Analytical Chemists International (AOAC) standards since they need to understand the digestibility of feed and hence for them, the digestible fiber is of utmost importance (Agblevor and Pereira, 2013; Theander et al., 1995).

Understandably, there are stark compositional differences between various categories of lignocellulosic biomass but even within the same category minor variations are observed. Like for example, forest residues have more lignin than crop residues and even within forest residues, softwood has more lignin content than hardwood. In aquatic biomass, the percentage of cellulose is generally higher as compared to the lignocellulosic biomass. Due to the low lignin content, cellulose is majorly observed alongside lipid and protein content. The carbohydrate content in algal feedstocks depends on the species under study and the conditions in which it is cultivated (Markou et al., 2012).

2.1.2 Hemicellulose

Hemicellulose is an amorphous and heterogeneous polysaccharide present in lignocellulosic biomass constituting about 20–50 wt% of its weight. It has a lower degree of polymerization than cellulose. It is a highly branched group of pentose sugars (β-D-xylose,

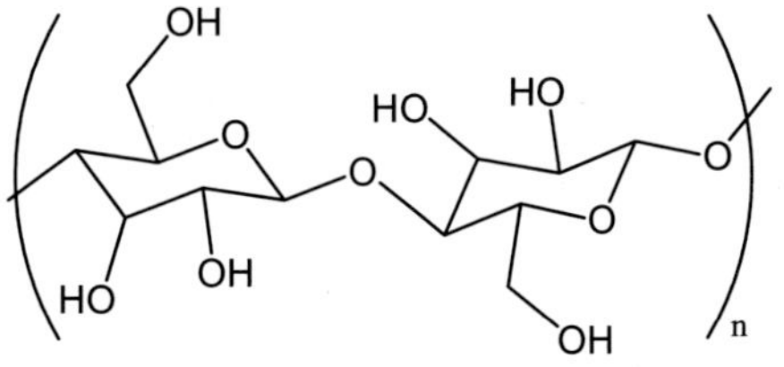

FIG. 1 Structure of cellulose.

α-L-arabinose), hexose sugars (β-D-mannose, β-D-glucose, α-D-galactose), hexuronic acids (4-*O*-methyl-D-glucuronic acid, galacturonic acid, and glucuronic acid), small amounts of rhamnose, fucose, and acetyl groups (Lee et al., 2014) (Fig. 2). Like cellulose, the composition of hemicelluloses also varies for forest and crop residues. Woody biomass usually is comprised of xylan and agro residues usually contain mannan, glucan and galactan (Lee et al., 2014). The sugars and acids which comprise the hemicellulose also vary from one plant to another. Hardwoods are made up of 4-*O*-methylglucuronoxylan with acetyl substituents. The xylopyranose (around 10 wt%) which is present in hardwood is substituted by 4-*O*-methyl glucuronic acid. Softwood hemicelluloses majorly contain *O*-acetyl-galactoglucomannan units (combination of mannose with glucose) and xylan contains nonacetyl substituents (Luo et al., 2019).

Usually xylose, mannose, and galactose are present in the backbone of hemicellulose chain while arabinose, galactose, 4-*O*-methyl-D-glucuronic acid are found in the side chains (Liu et al., 2019; Zhou et al., 2017).

From the open literature, it can be seen that the hemicellulose content is different in different biomasses as expected. It was observed to be around 25% in rice straw and nearly 48% in some macroalgae (Wang et al., 2020).

2.1.3 *Lignin*

Lignin is the third major component of lignocellulose biomass constituting 10%–35% by weight and 40% by energy. It is the second largest heterogeneous natural organic biopolymer. Lignin is a three-dimensional cross-linked amorphous polymer whose monomers are coumaryl alcohol, coniferyl alcohol and sinapyl alcohol (Biswas et al., 2021) (Fig. 3). These monomers are connected by strong C—O (~60%–70%) and C—C (~25%–35%) linkages (Chio et al., 2019). Softwood contains 46 wt% β-*O*-4-aryl ethers, 6 wt% of α-*O*-4-aryl ethers, 3.5 to 4 wt% of 4-*O*-5-diaryl ether, 9–12% of β-5-phenylcoumaran units, 9.5%–11% of 5-5-biphenyl units, 7% of β-1-(1,2-diarylpropane) unit, 2% of β-β-(Resinol) and rest 13% of others.

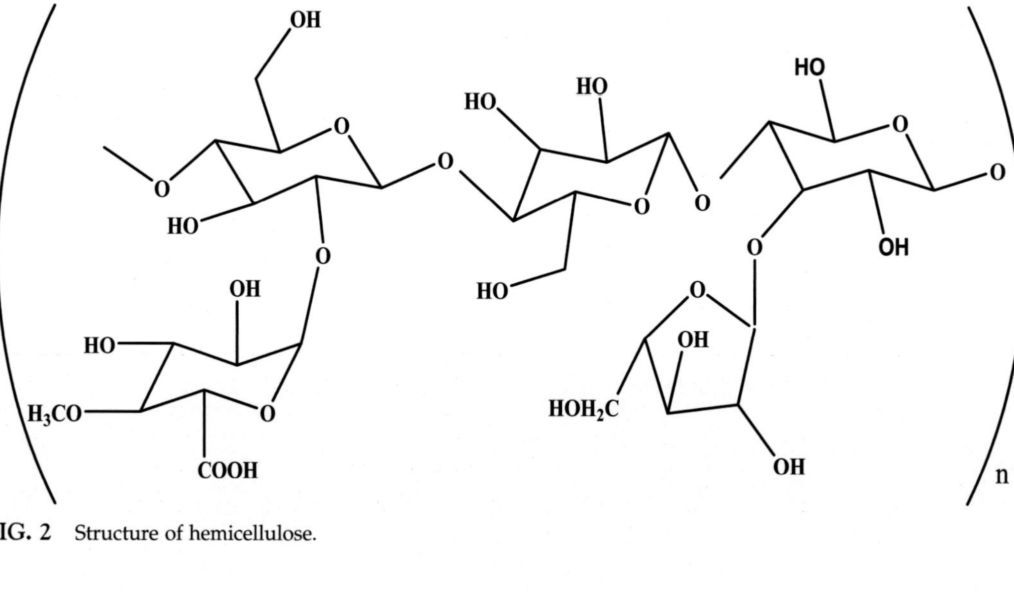

FIG. 2   Structure of hemicellulose.

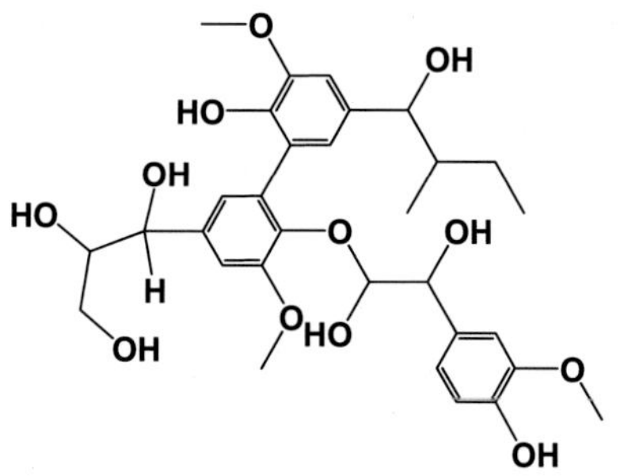

FIG. 3 Structure of lignin.

On the other hand, hardwoods comprise of 60 wt% β-*O*-4-aryl ethers, 8% of α-*O*-4-aryl ethers, 6.5% of 4-*O*-5-diaryl ethers, 6% of β-5-phenylcoumaran units, 4.5% of 5-5-biphenyl units, 7% of β-1-(1,2-diarylpropane, 3% of β-β-(Resinol) and 5% of others (Pandey and Kim, 2011).

It is very well known that the different linkage such as β-*O*-4-aryl ethers, α-*O*-4-aryl ethers, 4-*O*-5-diaryl ethers, β-5, β-β-(Resinol), and 5-5-biphenyl which are a part of lignin vary from plant to plant leading to also the differences in the types of linkages between them. Usually, more amounts of lignin are present in softwoods than in hardwoods and least in grasses. Lignin is produced as a waste/by-product by the paper, sugar, ethanol and other bio-based industries. During lignin depolymerization, the aryl ether linkages (e.g., β-*O*-4) are easier to break than any condensed linkages (e.g., C—C) (Chio et al., 2019). Several depolymerization methods are being studied to add value to lignin by producing valuable products such as oxidative/reductive depolymerization, solvolytic/hydrothermal liquefaction, thermal/catalytic pyrolysis. The products are a mixture of monomers such as phenols, dihydroxybenzenes, alkylphenols, methoxy phenols, alkylbenzenes, and alkyl-methoxy substituted phenols (Sun et al., 2018; Xu et al., 2014).

2.1.4 Ash

Ash is another component of lignocellulosic biomass which, even though is a minor component has a lot of effect during any conversion. Ash is mainly composed of certain mineral components and trace elements. It also has some alkali and alkaline earth metals. Some of the ash components are K, Na, Ca, Si, Mg, P, etc. Ash is present in biomass usually through two routes: either from soil contamination during the harvest and collection practices or it is structural ash found within the plant cell wall formed naturally through photosynthetic process. The variation in ash content is usually the highest among all the other components of biomass. Other than variations within categories of biomass, ash content also varies for the same feedstock also depending on its location. The cultivation conditions, harvest and practices, handling operations, etc., lead to the differences in the ash percentage in any biomass. In some cases, it is observed that late harvesting periods may lead to reduction in the ash content of biomass. Ash content varies from less than 1% (wood) to as high as 25% in certain crop residues. Grasses also usually contain more ash than woody biomass (Tortosamasia et al., 2007). Some of the problems that are encountered while using high ash containing biomass are slagging and fouling. During combustion, ash may melt leading to deposits on the combustor surfaces (fouling) or remain as hard chunks of material at the base of the combustion chamber (slagging/clinkering) (Bajpai, 2016).

2.1.5 Extractives

Extractives are organic and/or inorganic compounds present in biomass which can be extracted using polar and nonpolar solvents before it is sent for further processing. Some of the extractives in biomass are proteins, fats, fatty acids, terpenes, resins, etc. Extractives are usually less than 2% of biomass weight but they are characteristic to the nature of the biomass. Some extractives are water-soluble; some are ether soluble and some are hexane soluble. Extractives in herbaceous crops/plants may be different from the extractives found in forest residues (Tekin et al., 2014).

2.2 Aquatic biomasses

In the past decade, algal biomass has been gaining attention due to several advantages over terrestrial biomass (Benson et al., 2014). It is touted to have high potential due to various reasons such as high rate of atmospheric CO_2 absorption, ability to be cultivated in waste waters, no requirement of land, etc. There are many different types of algae observed which are either made up of a single cell like in microalgae or are multicellular in nature like macroalgae or seaweeds. They utilize sunlight to convert carbon dioxide in the atmosphere and nutrients in the growing media to produce organic molecules. Macroalgae or seaweeds form multicellular thalli, like Rhodophyta, Chlorophyta, and Phaeophyta (Andersen, 2005).

Microalgae are microscopic in nature and their size usually ranges from 5 to 50 μm. They can be cultivated in seawater, fresh water and even in waste waters (Spolaore et al., 2006). There are several algal species that are being cultivated across the globe and the composition of the strain also varies with the environment in which they are grown. The lipid, protein and carbohydrate content of algae varies depending on the species, and conditions of growth namely intensity of light, temperature, pH and the nutrient media. They usually have more protein in the range of 10%–60% DM, lipids around 2%–90% DM, carbohydrates (starch, sugar, and other polysaccharides) in the range of 5%–50% DM (Chisti, 2007). Algae may also have some pigments, and essential vitamins like nicotinate, A, B_1, B_2, B_6, B_{12}, folic and pantothenic acid besides sometimes having essential fatty acids like eicosapentaenoic acid (EPA), decosahexaenoicacid (DHA) in their lipids (Spolaore et al., 2006). Algae can be used as such to produce nutraceuticals, etc. or be extracted to recover useful content, in which case defatted algal cakes are generated. Algae are now being cultivated to produce bio-diesel using their lipids and here as well, defatted cakes would be generated. During pyrolysis or liquefaction, large amounts of nitrogen heterocycles, pyrroles and indoles are observed to be produced from their protein content, cyclic ketones and phenols are produced from carbohydrates, and lipids give rise to fatty acids (Biswas and Bhaskar, 2019). These may also be used to produce functional hydrocarbons/chemicals (Foley et al., 2011). Though several processes are being developed for algal growth and further utilization, concerns over their economic viability and sustainability still need to be addressed. This can only be done by clearly understanding the composition of algae and relating it to the conversion processes.

Once this is understood, the type of algae required for a particular product can be grown or identified from the existing species.

Aquatic weeds usually grow in water absorbing the nutrients present and lead to eutrophication. The marine species present in the water suffer due to loss of dissolved oxygen and loss of valuable nutrients required for their growth/survival. This leads to water pollution and the entire water body becomes useless due to the rapid growth of weeds. Some of the aquatic weeds commonly found are water hyacinth, paragrass, etc. Their composition is like that of lignocellulosic biomass but there are variations in the ratios of the three polymers present. These biomasses have higher carbohydrate composition with different ash content. It has been showed that aquatic floating plant such as water hyacinth has higher carbohydrates (54.4%) with low amount of lignin (6%), however, other aquatic plants such as paragrass and phumdi biomass has higher lignin content (13.7% and 25%). In case of lignocellulosic biomass, higher lignin content was observed with low ash content (Kumar et al., 2021). Lignocellulosic biomass generally does not have lipids and proteins unlike aquatic biomass and their hemicellulose content is also low. Protein is made up of different amino acids containing amine functional group ($-NH_2$) leading to the high nitrogen content of algal biomass. For example, the carbohydrate content in chlorella is around 26% and protein and lipid content are around 55% and 12% respectively (Li et al., 2017).

3 Physico-chemical characterization of biomass

3.1 Compositional analysis

The three major components in lignocellulosic biomass are cellulose, hemicellulose, and lignin. The differences in the ratios of these components lead to the differences in the products obtained during biochemical or thermochemical conversion processes. It is extremely important to carry out the compositional analysis of lignocellulosic biomass to know the amount of cellulose, hemicellulose and lignin present in the feedstock. This will help in designing the best possible process for its valorization. For example, feedstock with low lignin and high cellulose content are preferred for bio-ethanol production, whereas lignin-rich feedstocks are preferred for thermochemical conversion to high calorific value bio-oils or for the production of high value functional/specialty chemicals. Sample preparation to maintain uniformity is the main step while using the ASTM standard E1757-01 method (Cai et al., 2017). This can be used for biomass within a specific particle size range and moisture content. The steps involve the drying of sample at 45 ± 3°C for 24–48 h followed by grinding the dry sample to ensure that the particles pass through a 2 mm screen. The ground sample is sieved and the −20/+80 mesh fraction is used for further analysis. Once the fraction is obtained, any of the three methods mentioned below can be used for further analysis: sulfuric acid hydrolysis method, near infrared spectroscopy (NIRS) method or kinetic analysis method (Kaurase and Singh, 2020).

3.1.1 Sulfuric acid hydrolysis method

The two-step sulfuric acid hydrolysis is one of the oldest and most common methods used for biomass compositional analysis. The National Renewable Energy Laboratory (NREL) has standardized a method for the estimation of structural carbohydrates and lignin biomass and it is widely used (Sluiter et al., 2008a; Marrugo et al., 2016).

Biomass is first subjected to an ethanol extraction step and the extracted components undergo a strong sulfuric acid hydrolysis step at room temperature. This is followed by a dilute sulfuric acid hydrolysis step at high temperature for further breakdown to produce monomers. For proper quantification of polysaccharide sugars, Klason lignin and acid-soluble lignin can be separated and measured. The acid-soluble lignin samples are quantified using ultraviolet spectroscopy. Klason lignin content is obtained from the acid-insoluble residue gravimetrically. The reported cellulose, hemicellulose, and lignin values are a combination of glucose (representing the cellulose fraction) and xylose, galactose, and arabinose (representing the hemicellulose) values measured using NREL's LAP determination of structural carbohydrates and lignin in biomass (Sluiter et al., 2008b), glucose, and xylose values predicted using NIR-based predictive models developed at NREL. The most used procedure for the determination of carbohydrates, acetate, lignin, and ash in biomass is the one provided by the National Renewable Energy Laboratory (NREL). As mentioned earlier, it is preferred for applications in the second-generation biofuels and chemicals where holocellulose is the required fraction (Karimi and Taherzadeh, 2016).

3.1.2 NIRS method

Though the sulfuric acid hydrolysis method is reliable, it is labor-intensive, time-consuming, costly and requires preconditioning to remove extractives (Pandey et al., 2014).

Instead, the NIRS method which is relatively less complicated, fast, and precise can be used. Here, hazardous chemicals are not used in the analysis. The various functional groups present in the macromolecular backbone of biomass give rise to peaks in the infrared spectra, they can be used to identify the corresponding bonds. Statistical models which can connect the spectra to the chemical bonds and provide an insight into the chemical composition are however required. Principle component analysis (PCA), partial least squares (PLS), artificial neural networks (ANN), and support vector machines (SVM) can be used for qualitative analysis (Roggo et al., 2007).

Another set of samples are prepared for external validation of the predicted values from the calibration model and reference method. Successful predictions of the composition of rice straw have been carried out so far using NIRS (Jin and Chen, 2007).

3.1.3 Analysis using Fibertec analysis unit

Another method to carry out the compositional analysis has been developed by the IBERS Analytical Chemistry Laboratory at the University of Aberystwyth. Gravimetric measurements of neutral detergent fiber (NDF), acid detergent fiber (ADF), and acid detergent lignin (ADL) are made using Gerhardt fibrecap system that is an improvised Van Soest method (Soest et al., 1991). NDF, the total cell wall content is calculated by weighing the residue left after refluxing for 1 h in a neutral buffered detergent solution and then corrected for ash. Similarly, ADF (measure of cellulose and lignin) is the weight of solid remaining after refluxing the samples in a solution of cetylammoniumbromide (CTAB) in 2 M sulfuric acid followed by correction for ash. ADL is obtained by treating ADF with 72% sulfuric acid to solubilize the cellulose to determine crude lignin. When the sample is heated at 600°C in a muffle furnace for at least 4 h, ash content is obtained. Hemicellulose and cellulose quantities are obtained using the Eqs. (1) and (2), respectively (Akinrinola et al., 2014).

$$\text{Hemicellulose}\% = \text{NDF}\% - \text{ADF}\% \tag{1}$$

$$\text{Cellulose}\% = \text{ADF}\% - \text{ADL}\% \tag{2}$$

3.2 Proximate analysis

Proximate analysis is used to estimate the amount of moisture, ash, volatile matter, and fixed carbon of the biomass. Moisture content is observed due to the presence of water in biomass, and it is expressed as a percentage of the material weight (Karunanithy et al., 2013). There are two types of moisture in biomass: external and internal, external is usually present outside the cell walls, whereas internal moisture is absorbed within the cell walls. Moisture content is usually an important parameter studied during biomass characterization as its quantity affects the product calorific value and process energy consumption.

Moisture content of biomass can be found using a standard protocol developed by NREL (NREL/TP-51042620) using an infrared drier or by ASTM standard E1756-08 [ASTM E1756-08]. The weight of the feedstock is initially noted down and then, it is heated at $105 \pm 3°C$ in an oven for a minimum of 3 h (no longer than 72 h). After cooling, the sample weight is noted, and the percentage of moisture is calculated using the weight loss observed.

The amount of ash present in the biomass is calculated by weighing the solid residue, left over after the biomass sample is completely burnt (Cai et al., 2017). Volatile matter content of biomass is the amount of condensable vapors and permanent gases (other than water vapor) released upon heating. Fixed carbon is estimated by weighing the solid residue left over after the removal of volatile matter from biomass.

Volatile matter can be estimated for biomass samples using ASTM D317 (Milne et al., 1990). In this procedure, around 1 g of biomass (kept in quartz crucible) is kept in a muffle furnace at 950°C for 7 min. This is followed by cooling the crucible to room temperature by placing it in a desiccator. The difference in weight represents the volatile matter of the biomass feedstock and representative values of some biomasses are provided in Table 3.

There exists a standard NREL protocol (NREL/TP-510-42,622) for ash estimation as well (Sluiter, 2008). In this procedure, biomass (~1 g) is heated to $575 \pm 25°C$ in muffle furnace for 4–6 h. The crucible is allowed to cool down to room temperature in a desiccator. Finally, fixed carbon in biomass is calculated by using the formula: Fixed carbon = 100—(Moisture + Ash + Volatile matter).

3.3 Ultimate analysis

Ultimate analysis is carried out to estimate the amount of carbon, hydrogen, nitrogen, sulfur and oxygen and reported on dry biomass basis or dry ash free basis. Elemental analyzers are used for the study where a weighed quantity of biomass is combusted in a controlled atmosphere and the gaseous products are identified by analysers (Kirmse, 2012). The principle of the same is that the carbon present in the biomass is converted to CO_2 while the hydrogen content is converted to H_2O. Upon combustion, nitrogen content comes out as NOx and sulfur content leads to SO_2 production. Heated high purity copper is used to convert NOx to N_2. The gas analyzers provide the amount of individual gases observed which in turn provides the information of presence of each element in the biomass sample (Jong and Ommen, 2014; Garcia et al., 2012). Heating value of biomass or its calorific value is necessary to understand how much heat can be generated once the biomass is burnt. Higher heating values may be estimated using different calorimeters. Once the ultimate analysis is known, higher heating values of biomass can also be calculated theoretically.

3.3.1 Carbon

Carbon is one of the most important elements present in biomass. The carbon content of the biomass dictates its end use and contributes to its calorific value. The presence of this carbon makes the biomass sustainable and renewable in nature. Upon complete combustion, this carbon is left into the atmosphere as CO_2 and in case of incomplete combustion, it is released as poisonous CO gas (Wang et al., 2011; Vallero, 2008). Carbon is present in all the three components of lignocellulosic biomass namely cellulose, hemicellulose and lignin. Carbon is also left behind as solid bio-char residue after pyrolysis or liquefaction processes (Karampinis et al., 2012).

3.3.2 Hydrogen

Hydrogen is the second most important constituent of biomass and is also present in both the carbohydrate and phenolic fraction of biomass. On burning, hydrogen is released as H_2O and this contributes to the calorific value of biomass (Kreith, 2007). Hydrogen content is usually more in woody biomasses than herbaceous biomasses (Karampinis et al., 2012).

3.3.3 Nitrogen

Nitrogen is one of the important constituents of biomass primarily due to its nutrient value. The manures and fertilizers used during the growth of biomass along with the soil quality and type of plant determine its nitrogen content. In the recent times, crop residues are showing more nitrogen content than forestry biomass (Hirel et al., 2011). The amount of nitrogen present in the biomass sample affects the biochemical or biotechnological conversion processes such as anaerobic digestion and fermentation (Karampinis et al., 2012).

3.3.4 Sulfur

Sulfur is another constituent of biomass affecting certain conversion processes. It is typically low and at times, undetectable in biomass as compared to the fossil fuels currently being used in huge quantities. The lower content or absence of sulfur has resulted in preference of biomass-derived fuel in certain heavy-duty applications as SOx production would be minimal. The current liquid fuels in use are continually being upgraded to reduce the sulfur content to meet the specifications and, in this regard, bio-fuels have an edge over fossil fuels. Herbaceous biomasses are known to have sulfur content of around 2–5wt%.

(Robbins et al., 2012). When present, it is released at high temperatures leading to extra gas cleaning steps before it can be used further for other applications to avoid corrosion (Karampinis et al., 2012).

3.3.5 Oxygen

Oxygen is the last major element of the lignocellulosic biomass discussed here. The amount of oxygen content in biomass dictates its end use, like carbon, but the implications are opposite to that of carbon. It is usually not analyzed but estimated by difference using the values of C, H, N, S obtained by ultimate analysis. Higher oxygen content in biomass leads to an increased oxygen content of the products, which reduces the chances of its utilization as fuel (Elliott, 2011). The removal of oxygen from the thermal conversion products requires another energy-consuming process to be used for fuel applications (Karampinis et al., 2012).

3.3.6 Calorific value of biomass

Calorific value of biomass provides information on the amount of energy that can be obtained when it is combusted completely/burnt. The higher the calorific value, the better is its suitability for usage as fuel (Dahlquist, 2013). Calorific value may also be termed as heating value and there are two types of heating values. Higher heating value (HHV) is the total amount of energy that can be obtained from biomass including the latent heat of vaporization of water. The lower heating value (LHV) is the actual amount of energy available after removing the moisture from the feed or products (Miller et al., 2010). HHV of biomass can be estimated using a bomb calorimeter following the ASTM standard D5865–13 (Ohliger et al., 2013).

In this method, a small amount of biomass is placed in a sealed container and heated in the presence of oxygen. The heat released from combustion is measured as the higher heating value. The hydrogen present in the biomass also leads to the production of steam upon burning. The latent heat of condensation of steam can be subtracted from the HHV to obtain LHV (Demirbas, 2007). The heating values of lignocellulosic biomass are dependent on various factors. They vary with the type of biomass, soil in which they are grown, etc. and hence, the calorific value is usually presented in range for different types of biomasses (Williams et al., 2016). The calorific value of forest biomass (18.5–22.5 kJ/mol) is generally higher than those of crop residues (15.5–19.5 kJ/mol).

Typical compositional analysis data of representative biomass feedstocks in each of the categories are shown in Table 1. Tables 2 and 3 show the proximate and ultimate analyses of the representative biomasses, respectively. Thus, it can be observed that there are variations in the presence of each component. The wide variations within each component in the same category are due to the differences in the conditions in which biomass is grown namely region of cultivation, soil/media, temperature, nutrient availability, etc. This reiterates the necessity for understanding the biomass nature before it can be used for any conversion process.

3.4 Elemental analysis (ICP-AES/OES)

Ash is a component of lignocellulosic biomass which needs more attention in understanding due to the challenges it creates during processing. This inorganic content could be introduced from the soil or through other forms of contamination. The other kind of ash is the inherent macro and micronutrients present within the plant structure formed naturally through photosynthesis and bound to the polymers (Li et al., 2016; Agblevor and Besler, 1996). The ash content and its composition vary with the type of lignocellulosic biomass, soil

TABLE 1   Compositional analysis data of lignocellulosic and aquatic biomass (Cai et al., 2017; Azizi et al., 2018).

Biomass	Cellulose, wt%	Hemicellulose, wt%	Lignin, wt%
Lignocellulosic biomass			
Harwood (Poplar)	50.8–53.3	26.2–28.7	15.5–16.3
Softwood (Pine)	45.0–50.0	25.0–35.0	25.0–35.0
Rice straw	29.2–34.7	23.0–25.0	17.0–19.0
Wheat straw	35.0–39.0	23.0–30.0	12.0–16.0
Grasses	25.0–40.0	25.0–50.0	15.0–20.0
Water hyacinth	32.0	22.4	6.0
Aquatic biomass			
Biomass	Protein, wt%	Carbohydrates, wt%	Lipid, wt%
Chlorella sp.	29.6	15.5	17.7
N. gaditana	30.7	21.6	12.7
S. platensis	48.3	30.2	13.3
D. tertiolecta	61.3	21.6	2.8
C. vulgaris	41.5	20.9	15.6

TABLE 2   Proximate analysis of lignocellulosic and algal biomass (Biswas et al., 2017a; Biswas et al., 2017b; Kumar et al., 2021; Biswas et al., 2016).

Biomass	Moisture, wt%	Volatile matter, wt%	Fixed carbon, wt%	Ash, wt%
Lignocelluloses biomass				
Corn cob	12.77	91.16	6.54	2.30
Wheat straw	12.81	83.08	10.29	6.63
Rice straw	11.69	78.07	6.93	15.00
Rice husk	10.89	73.41	11.44	15.14
Shimbal	3.44	89.32	5.34	5.34
Coked	3.94	87.83	3.68	8.48
Aquatic biomass				
Azolla	11.0	88.3	4.4	7.3
Sargassum tenerrium	13.5	61.5	11.9	23.2
Water hyacinth	9.5	77.8	0.4	21.8
Para grass	6.1	80.2	11.4	8.5
Phumdi	5.9	73.8	14.6	11.5

TABLE 3 Elemental analysis of lignocellulosic and algal biomass (Biswas et al., 2017a; Azizi et al., 2018; Kumar et al., 2021).

Biomass	C, wt%	H, wt%	N, wt%	S, wt%	O, wt%	HHV, MJ/kg
Lignocelluloses biomass						
Corn cob	42.10	5.90	0.50	0.48	51.02	16.00
Wheat straw	38.34	5.47	0.60	0.37	55.22	14.68
Rice straw	36.07	5.20	0.64	0.26	57.83	14.87
Rice husk	41.92	6.34	1.85	0.47	49.52	12.87
Shimbal	45.84	7.47	0.42	0.02	46.25	17.43
Coked	44.34	7.10	0.40	0.01	48.15	17.12
Aquatic biomass						
Chlorella sp.	46.1	6.1	6.7	0.4	40.7	20.4
D. tertiolecta	38.2	6.1	11.1	–	44.6–0	11.6
Sargassum tenerrimum	32.1	4.7	0.9	1.55	62.55	11.9
S. platensis	46.1	7.1	10.5	0.7	35.60	20.5
Water hyacinth	35	6.5	0.8	1.4	56.3	14.4

in which it is grown, other environmental conditions, harvest practices, etc. (Lacey et al., 2018). The NREL also has standardized a method for ash analysis where a weighed amount of sample is kept in a TGA unit and heated until750°C and the weight loss percentage is considered as the ash quantity (Sluiter et al., 2005). The ASTM E1755-01 method provides information on total ash and inorganic species. The other methods that can be used to estimate the inorganic content is using the inductively coupled plasma-optical emission spectrometry (ICP-OES), inductively coupled plasma-mass spectroscopy (ICP-MS) and flame atomic absorption spectrometry (FAAS). All these methods have a standard sample preparation method mostly using the acid digestion method. Another method called Laser-induced breakdown spectroscopy (LIBS) is being developed recently which may provide more accurate results than ICP-OES methods and the sample preparation steps are also not as elaborate (Westover and Emerson, 2015).

   Other analytical methods that can be used for real-time observation of inorganic species to improvise online process control are near-infrared (NIR) spectroscopy (Xiao et al., 2014) and X-ray fluorescence (XRF) spectrometry (Morgan et al., 2015). The recent advanced approach can differentiate exogenous and naturally embedded ash by combining energy-dispersive XRF and NIR spectroscopy. These new methods will help in understanding the ash content in biomass better and help in designing better downstream processes (Thyrel et al., 2019). NIR though powerful has limitations in terms of measuring only inorganics that are organically bonded and hence, it must be combined with other analytical tools to corroborate the results (Lestander and Rhén, 2005; Lestander et al., 2009).

3.5 Thermogravimetric analysis (TGA, DTG, DSC)

Thermogravimetry is a robust tool that can provide a lot of information about the ligno-cellulosic biomass feedstock. This analytical tool is used by all researchers during the preliminary characterization of biomass itself. It is specifically used for understanding the decomposition profile of a sample in the absence of oxygen. Here, a small quantity of biomass (usually 4–6 mg) is placed in the sample holder and heated up to 900°C and the weight loss is measured automatically. Thermogravimetric data is also used recently for evaluating the kinetic triplet, understanding the pyrolysis decomposition mechanism and estimating the thermodynamic parameters (Awasthi et al., 2020).

Upon plotting the temperature vs weight data, three distinct regions are usually observed for most lignocellulosic biomasses. The first region usually belongs to the moisture removal, and second region is due to organic matter decomposition and last region is due to secondary reactions or bonded carbon related reactions. In terms of the components of lignocellulosic biomass, hemicellulose usually decomposes in the range of 100°C–250°C and cellulose decomposition is observed from 300°C–500°C (Biswas et al., 2020). The last component lignin decomposes over a wide range at higher temperatures of 500°C–750°C. Complex decomposition patterns are also observed in cases where there is high ash content and this leads to the observance of multiple peaks in DTG curves leading to different "T_{max}" temperatures where maximum weight loss occurs (Guimar et al., 2009; Paiva et al., 2004).

Unlike lignocellulosic biomass, aquatic biomass degradation temperature is different due to it different composition contents. In most cases, it has been observed that there is more than 1 hump in the DTG curves of aquatic biomass. This is indicative of the complex degradation mechanism of the feedstock. This is mainly also due to the high inorganic content as observed earlier leading to multiple reactions during thermal degradation (Fig. 4).

3.6 Functional groups analysis by FT-IR

Fourier Transform-Infra red spectroscopy can be used to understand the functional groups present in the backbone of macromolecular structure of biomass. Depending on the specific functional groups, vibrations are observed in typical wavenumbers. Usually, the solid sample is prepared by mixing it with KBr and then placed for analysis in an FT-IR unit. The scanning range usually starts from 400 cm^{-1} and goes up to 4000 cm^{-1}. The main peaks from the resulting spectrums are attributed to bonds such as C-Ph, C=C, O–CH$_3$, C=O and others and they later can be used to identify the components such as lignin, hemicellulose, and cellulose (Yang et al., 2007; Marrugo et al., 2016). The wide band at around 3365 cm^{-1} belongs to the O—H bond that is characteristic of lignin and carbohydrate phenolic (pH), alcoholic groups, carboxylic functional groups and hydrogen links. Usually, the peak at around 2942 cm^{-1} is attributed to the vibration of C—H functional groups and elongations –CH$_2$ and –CH$_3$. The band at 2870 cm^{-1} is characteristic of the O-CH$_3$ vibration mostly present in lignin. The peak around 1730 cm^{-1} corresponds to C=O in conjugate or nonconjugate systems (carbonyl/carboxyl) and acetyl groups present in hemicellulose. These peak intensities can be used to identify the abundance of the three components of biomass (Marrugo et al., 2016) (Fig. 5).

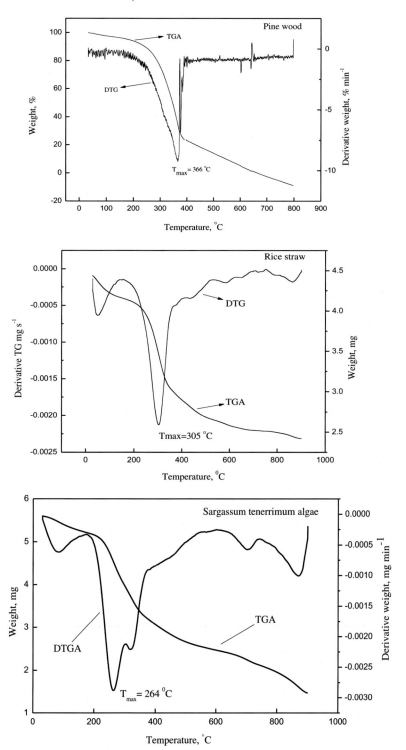

FIG. 4 TGA/DTGA analysis results of different biomass.

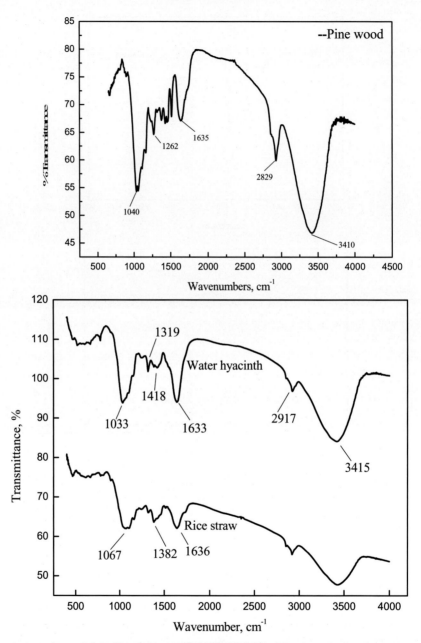

FIG. 5   Fourier transform—infrared spectroscopy (FT-IR) of different biomass.

The bands at 1604 and 1509 cm^{-1} are observed due to C-Ph and C=C, respectively. This is usually due to the aromatic structure of lignin. The peaks at 1456 and 1426 cm^{-1} are due to the presence of polysaccharides and the deformations of aliphatic C—H. C—H flexing due to lignin shows peak at 1375 cm^{-1} and 1230 cm^{-1}. The C-O-C bonds in all components of lignocellulosic biomass lead to peak at1170 cm^{-1}. The OH primary and secondary groups show bands at 1051 and 1165 cm^{-1}. Monosaccharaides in biomass have β-glycosidic links between them and this can be identified at 897 cm^{-1} (Marrugo et al., 2016; Biswas et al., 2017a, 2017b; Biswas et al., 2016). In case of feedstocks with high nitrogen content (mainly due to proteins), bands are observed in the region 1590–1484 cm^{-1}. This peak is due to C—O, N—H and C—N stretching vibrations found in amide complexes. Lipid peak is identified by strong vibrations of the C—O observed in the region 1778–1706 cm^{-1} (Biswas et al., 2017b). The major point to be remembered while inferring from the peak intensities is that the peak intensity is dependent on the concentration of the sample and this should be majorly considered as a qualitative analysis unless standardized/measured values are used (Marrugo et al., 2016).

3.7 Crystallinity by XRD

Cellulose is a crystalline component of lignocellulosic biomass, and the loss of crystallinity can be considered as an indication that cellulose has been broken down or converted into products after processing. Several methods are available to calculate the crystallinity using XRD data. Some of them are Segal method (Segal et al., 1959). Ruland–Vonk method, Rietveld refinement method, and Debye calculation method (Ruland, 1961; Rietveld, 1967). XRD scanning is carried out from 10 to 80° 2θ values using copper as the radiant source. The background intensity without the sample is usually subtracted from the final intensity to eliminate noise. The crystallinity index (CrI) can be calculated as $CrI = (I_{002} - I_{am})/I_{002} * 100$, where I_{002} is the total intensity and $(I_{002} - I_{am})$ is the peak height. It is measured by the ratio of the intensity of the main crystalline plane (002) at 22.4° and the amorphous at 16.3° of 2θ. The peak at I_{002} indicates the presence of crystalline material in the biomass and the higher CrI is mainly due to lower amount of amorphous material, i.e., hemicellulose and lignin. The XRD pattern of biomass is often characterized by an intensive amorphous halo with a major maximum at 20–23° and a minor maximum at 13–17° which signifies the occurrence of cellulose (Segal et al., 1959).

Not all celluloses are crystalline and amorphous cellulose due to unorganized chains is also present. Highly crystalline cellulose is difficult to break and strong chemical conditions are necessary. Hence, an understanding of the biomass crystallinity will help in designing the down stream processing accordingly. Similarly, the effectiveness of the process can be understood by observing the crystallinity of the product and the amorphous nature may indicate effective processing. Hemicellulose and lignin are amorphous polymers and hence, no inference about their nature can be obtained from XRD (Xu et al., 2013).

Lignocellulosic biomasses usually show XRD peaks at 22° 2θ associated with the crystalline region of cellulose and generally attributed to carbon solids with a long-terms structural order (Biswas et al., 2017a). The XRD spectrum of forest biomass such as pine wood showed a peak of greatest intensity at 22° 2θ and small peak was observed at 16° 2θ due to the presence of amorphous lignin that is present in higher amounts than other agricultural and aquatic

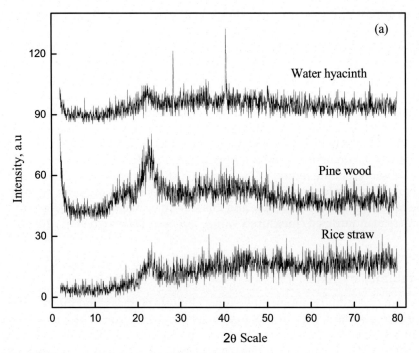

FIG. 6 Powder XRD of water hyacinth, pinewood and rice straw.

biomass. Moreover, in case of aquatic biomass several other high intensity peaks were found at 28 and 43°, which is due to the inorganic metal content. This is well in agreement with the ash content of the biomass; higher ash content is observed in case of aquatic biomass as compared to forest and agricultural biomass (Fig. 6).

3.7.1 Solid state NMR

NMR characterization technique is nondestructive and can be used to understand the nature of different biomasses. Biomass is a complex material and hence, detailed understanding may not be possible due to several overlapping peaks and poor resolution. In this state, advanced two-dimensional solid-state NMR maybe used for better resolution and understanding (Yann et al., 2015). Peaks are observed at specific values depending on their characteristics (Table 4).

In the area of 105–64 ppm, ^{13}C signal due to aliphatic lignin is observed (Liitia et al., 2000). Correlation between these carbons with O-alkyl protons and aromatic protons are observed at $\delta C/\delta_H$ 50.0–110/3–7 ppm. At the ^{13}C chemical shift of 74 and 104 ppm and ^{1}H shift at 6–7 ppm due to aromatic protons, peaks may be due to carbohydrates carbon and lignin aromatic proton interaction or within the lignin structure itself (Yann et al., 2015).

These peaks can thus be used to identify the groups present in the biomass feedstock and correlated with the results of other analytical tools to understand the macromolecular structure of biomass.

TABLE 4 2D solid state NMR peaks and their attributed structures (Mao et al., 2010; Liitia et al., 2000; Yann et al., 2015).

$\delta C/\delta_H$	Attributed structures
64.4/3.6	C_6 (CH_2OH in glucose unit)
74.9–72.3/3.6	$C_{2,3,5}$
83.6/3.6	C_4 disordered structure
87.9/3.6	C_4 ordered structure
104.8/3.6	C_1 carbons of polysaccharides
56.2/3.6	O-alkyl protons from CH_3O
56.2/6.35–7	aromatic protons from lignin aromatic rings
110–112/6.7–7.5	G_2 and its protons
113–115/6.5–7.5	G_5 and its protons
118–120/6.3–7.0	G_6 and its protons
111.5/3.6	G_2 and methoxyl protons
116.9/6.9 ppm	Para-coumarate group (p-CA_3 and p-CA_5)
132–136/3.0–4.0	S_1 and G_1 units (aromatic carbons) with Aliphatic CH–O protons
138.0–140.0/3.2–4.0	Aromatic C—O carbons with aliphatic CH–O protons (S_4)
146.0–149.5/3.1–4.0	Aromatic C—O carbons with aliphatic CH–O protons (G_{3-4})
150.0–154.0/3.2–4.0	Aromatic C—O carbons with aliphatic CH–O protons (S_{3-5} etherified)
145–147/6.8–7.2	Aromatic C—O carbons with aromatic protons (G_4)
147–149/7.0–7.5	Aromatic C—O carbons with aromatic protons (G_3)
168–174/1.5–7.0	Carbonyl groups
172–173/1.9–2.1	Acetate group (CH_3–CO)
167–168/3.6–4.0	Carbonyl group with ferulate (lignin–hemicellulose cross-linking)

3.8 Raman spectroscopy

Vibrational spectroscopic techniques can also be used to understand the structure of biomass. In Raman spectroscopic technique, scattered photons generated during the interaction between light and matter are measured. The study can be done using ultraviolet (UV), visible, or near-infrared (NIR) lasers. Produced scattering maybe identical (elastic), higher (inelastic), or lower (inelastic) frequency than that of the excitation source called as Rayleigh, Stokes, and anti-Stokes, respectively (Smith and Dent, 2005; Popp, 2006). Intense Rayleigh scattering needs to be removed completely from the optical beam path using specialized optics such as holographic notch filters (HNFs) (Dao, 2006). In case of its presence, Rayleigh scattering leads to detector saturation and obscures Raman signal from Stokes scattering which is a much weaker phenomenon. Most commonly used Stokes scattering leads to an energy shift toward higher vibrational levels. Anti-Stokes scattering occurs when there is a shift from a higher to lower vibrational level and is less common as the probability of molecules populating higher vibrational levels at ambient conditions is low.

In this method, molecules are promoted to short-lived, virtual vibrational levels. It is not sufficient to match the excitation frequency to that necessary to promote molecules from the ground state to the first excited vibrational level. It is necessary to look out for a change in the polarizability of the electron cloud during the interaction of the molecule with light. Vibrational modes for C—C, C=C, C—H, C—O, H–C–C, C–O–H, H–C–H are usually observed (Landsberg and Mandelstam, 1928). Generally, symmetric bonds show major changes in polarizability leading to strong Raman signals. Raman spectra are observed usually in the range of 800–$1800\,cm^{-1}$. Vibrations due to lignin are observed from 1595 to $1650\,cm^{-1}$ and peaks in the range of 973–$1183\,cm^{-1}$ are due to cellulose and hemicellulose (Nanda et al., 2013). As hemicellulose is not ordered as cellulose and more complex, their Raman bands are broader in nature. Even in case of cellulose, the band type depends on the crystallinity and fiber orientation making it possible to differentiate the cellulose polymorphs. The dominant lignin vibrational mode near $1600\,cm^{-1}$ is due to ring breathing indicating the presence of any phenyl-containing molecules, like flavonoids. The presence of high amount of extractives also leads to peak at $1600\,cm^{-1}$ in addition to lignin. Hence, when lignin is required to be characterized, it is highly essential to remove the extractives before analysis. The peak at $1600\,cm^{-1}$ contains overlapping signals from S, G, and H lignin monomers further complicating the studies. The presence of amorphous cellulose leads to peak at $900\,cm^{-1}$ while that of crystalline cellulose is $1098\,cm^{-1}$. Cellulose and hemicelluloses show C—O stretching vibration and C—H deformation at 1056 and $1375\,cm^{-1}$. Characteristic broad peaks in the range of 1580–1610 and 1325–$1380\,cm^{-1}$ are due to graphite (G band) and defect (D band) structures in sample respectively (Nanda et al., 2013).

3.9 Surface properties by scanning electron microscopy (SEM) and transmission electron microscopy (TEM)

The surface morphology of any biomass sample can be understood using the SEM and TEM analytical tools. Electron beams are used to understand the size and shape of sample morphology. The resultant details after processing provide a picture of the sample along with the size, shape and particle dimensions. Using the EDS technique, the elemental composition of the sample can also be identified. Fibrous structure, granular structure, etc. can be very well identified and the way they are distributed can also be understood using these techniques (Natarajan et al., 1998; Lin et al., 2003).

In case of biomass, it is usually composed of small amounts of K, Na, Mg, Ca, P, etc. which can be identified using EDS. These results can be corroborated along with the ash composition using ICP techniques discussed earlier (Yusuf and Inambao, 2020). The physical structure of rice straw, pine wood, and water hyacinth are shown in Fig. 7. Three types of biomass are different in physical structures, especially water hyacinth biomass (layered structure) being obviously different from those of the other two biomass rice straw (plate-like layer structure) and pine wood (uniform granular structure) and these structures are dependent on the biomass sources (Biswas et al., 2017a, b; Biswas et al., 2016).

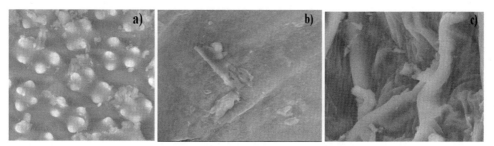

FIG. 7 SEM of rice straw, pine wood and water hyacinth biomass (a, b and c).

3.9.1 Challenges and opportunities

In general, biomass feedstocks are complex in nature as it is a natural bio-polymer. The macromolecular backbone structure is made up of several monomeric units. There is elaborate cross-linking between the different monomeric units, which could be intraunit or interunit linkages. Over the years, it has become a separate area of research to understand the complex structure of biomass using different characterization techniques. It is highly imperative to understand the structure of biomass to correlate the products formed, and to understand the decomposition mechanism of biomass. Though the structures of cellulose and hemicellulose have been understood to a higher extent, the structure of lignin still needs more detailed analyses. An in-depth knowledge of the different linkages in lignin will help in developing processes to produce value-added chemicals/materials from the same. Advanced characterization techniques need to be developed and at times used even in combination to be able to better elucidate the biomass structure. Other challenges in this area are also standardization of the characterization methods. Due to the inherent differences in the biomass nature, methodologies developed for any kind of biomass may not be directly employed for other biomass. Similarly, the methods may need changes for utilization in different parts of the world, due to the differences in the biomass grown in different soil and climatic conditions. Overall, the first step of biomass characterization area is of utmost importance and needs continual development for the design of optimal methods.

References

Agblevor, F.A., Besler, S., 1996. Inorganic compounds in biomass feedstocks. 1. Effect on the quality of fast pyrolysis oils. Energy Fuel 10, 293–298.

Agblevor, F.A., Pereira, J., 2013. Progress in the Summative Analysis of Biomass Feedstocks for Biofuels Production Aqueous Pretreatment of plant Biomass For Biological and Chemical Conversion to Fuels and Chemicals. John Wiley & Sons Ltd, pp. 335–354.

Akinrinola, F.S., Darvell, L.I., Jones, J.M., Williams, A., Fuwape, J.A., 2014. Characterization of selected Nigerian biomass for combustion and pyrolysis applications. Energy Fuel 28, 3821–3832.

Andersen, R.A., 2005. Algal Culturing Techniques. Phycological Society of America, Elsevier, Academic, Amsterdam, p. 578.

Atalla, R.H., Vanderhart, D.L., 1984. Native cellulose—a composite of 2 distinct crystalline forms. Science 223, 283–285.

Awasthi, A., Dhyani, V., Biswas, B., Kumar, J., Bhaskar, T., 2020. Production of phenolic compounds using waste coir pith: estimation of kinetic and thermodynamic parameters. Bioresour. Technol. 274, 1730179.

Azizi, K., Moraveji, M.K., Najafabadi, H.A., 2018. A review on bio-fuel production from microalgal biomass by using pyrolysis method. Renew. Sustain. Energy Rev. 82, 3046–3059.

Bajpai, P., 2016. Pretreatment of Lignocellulosic Biomass for Biofuel Production. Springer, Singapore, pp. 71–75.

Benson, D., Kerry, K., Malin, G., 2014. Algal biofuels: impact significance and implications for EU multi-level governance. J. Clean. Prod. 72, 4–13.

Biswas, B., Bhaskar, T., 2019. Hydrothermal upgradation of algae into value-added hydrocarbons. In: Biofuels from Algae, second ed. Elsevier, pp. 435–459.

Biswas, B., Krishna, B.B., Singh, R., Kumar, J., Bhaskar, T., 2016. Slow pyrolysis of pine wood: effect of CO2 and N2 atmospher. J. Energy Environ. Sustain. 2, 7–12.

Biswas, B., Mundada, G., Krishna, B.B., Singh, R., Kumar, J., Bhaskar, T., 2020. Pyrolysis of forest biomass residues: comparative study of shimbal, cokad, cheed and cerus. J. Energy Environ. Sustain. 10, 38–46.

Biswas, B., Pandey, N., Bisht, Y., Singh, R., Kumar, J., Bhaskar, T., 2017a. Pyrolysis of agricultural biomass residues: comparative study of corn cob, wheat straw, rice straw and rice husk. Bioresour. Technol. 237, 57–63.

Biswas, B., Singh, R., Krishna, B.B., Kumar, J., Bhaskar, T., 2017b. Pyrolysis of azolla, sargassum tenerrimum and water hyacinth for production of bio-oil. Bioresour. Technol. 242, 139–145.

Biswas, B., Kumar, A., Krishna, B.B., Bhaskar, T., 2021. Effects of solid base catalysts on depolymerization of alkali lignin for the production of phenolic monomer compounds. Renew. Energy 175, 270–280.

Cai, J., He, Y., Yu, X., Banks, S.W., Yang, Y., Zhang, X., Yu, Y., Liu, R., Bridgwater, A.V., 2017. Review of physicochemical properties and analytical characterization of lignocellulosic biomass. Renew. Sustain. Energy Rev. 76, 309–322.

Chatterjee, C., Pong, F., Sen, A., 2015. Chemical conversion pathways for carbohydrates. Green Chem. 17, 40–71.

Chio, C., Sain, M., Qin, W., 2019. A review of lignin depolymerization from various aspects. Renew. Sustain. Energy Rev. 107, 232–249.

Chisti, Y., 2007. Biodiesel from microalgae. Biotechnol. Adv. 25, 294–306.

Dahlquist, E., 2013. Biomass as Energy Source: Resources, Systems and Applications. CRC Press.

Dao, N.Q., 2006. Dispersive Raman spectroscopy, current instrumental designs. In: Meyers, R.A. (Ed.), Encyclopedia of Analytical Chemistry. John Wiley & Sons, Hoboken, NJ, pp. 13024–13058.

Demirbas, A., 2007. Effects of moisture and hydrogen content on the heating value of fuels. Energy Sources, Part A 29, 649–655.

Elliott, D., 2011. Thermochemical Processing of Biomass. John Wiley & Sons Ltd., Chichester, UK.

Foley, P.M., Beach, E.S., Zimmerman, J.B., 2011. Algae as a source of renewable chemicals: opportunities and challenges. Green Chem. 13, 1399–1405.

Garcia, R., Pizarro, C., Lavín, A.G., Bueno, J.L., 2012. Characterization of Spanish biomass wastes for energy use. Bioresour. Technol. 103, 249–258.

Guimar, J.L., Frollini, E., da Silva, C.G., Wypych, F., Satyanarayana, K.G., 2009. Characterization of banana, sugarcane bagasse and sponge gourd fibers of Brazil. Ind. Crop Prod. 30, 407–415.

Gundekari, S., Mitra, J., Varkolu, M., 2020. Classification, characterization, and properties ofedible and non-edible biomass feedstocks. In: Advanced Functional Solid Catalysts for Biomass Valorization. Elsevier, pp. 89–120.

Hirel, B., Tétu, T., Lea, P.J., Dubois, F., 2011. Improving nitrogen use efficiency in crops for sustainable agriculture. Sustainability. 3, 1452–1485.

Jin, S., Chen, H., 2007. Near-infrared analysis of the chemical composition of rice straw. Ind. Crop Prod. 26, 207–211.

Jong, D.W., Ommen, V.J.R., 2014. Biomass as a Sustainable Energy Source for the Future: Fundamentals of Conversion Processes. Vol. 2014 Wiley.

Karampinis, E., Grammelis, P., Zethraeus, B., Andrijvskaja, J., Kask, U., Hoyne, S., et al., 2012. Proceedings of the 20th European biomass conference and exhibition. In: Bioenergy System Planners Handbook Bisyplan, Milan.

Karimi, K., Taherzadeh, M.J., 2016. A critical review of analytical methods in pretreatment of lignocelluloses: composition, imaging, and crystallinity. Bioresour. Technol. 200, 1008–1018.

Karimi, K., Shafiei, M., Kumar, R., 2013. Progress in physical and chemical pretreatment of lignocellulosic biomass. In: Gupta, V.K., Tuohy, M.G. (Eds.), Biofuel Technologies. Springer, Berlin, Heidelberg, pp. 53–96.

Karunanithy, C., Muthukumarappan, K., Donepudi, A., 2013. Moisture sorption characteristics of corn Stover and big bluestem. J. Renew. Energy. 939504.

Kaurase, K.P., Singh, D., 2020. Delonix Regia fruit fibers: a new potential source of cellulosic fibers. Mater. Sci. Forum 979, 185–196.

Kirmse, W., 2012. Organic Elemental Analysis: Ultramicro, micro, and Trace Methods. Elsevier Science.

Kreith, F.G.Y., 2007. Handbook of Energy Efficiency and Renewable Energy. Taylor and Francis, Boca Raton, Florida.

Kumar, A., Biswas, B., Kaur, R., Krishna, B.B., Bhaskar, T., 2021. Predicting the decomposition mechanism of Loktak biomass using Py-GC/MS. Environ. Technol. Innov. 23, 101735.

Lacey, J.A., Aston, J.E., Thompson, V.S., 2018. Wear properties of ash minerals in biomass. Front. Energy Res., 1–6.

Landsberg, G., Mandelstam, L., 1928. Eine neueErscheinungbei der Lichtzerstreuung in Kristallen. Naturwissenschaften 16, 557–558.

Lee, H.V., Hamid, S.B., Zain, S.K., 2014. Conversion of lignocellulosic biomass to nanocellulose: structure and chemical process. Sci. World J. 631013.

Lestander, T.A., Rhén, C., 2005. Multivariate NIR spectroscopy models for moisture, ash and calorific content in biofuels using bi-orthogonal partial least squares regression. Analyst 130 (8), 1182–1189.

Lestander, T.A., Johnsson, B., Grothage, M., 2009. NIR techniques create added values for the pellet and biofuel industry. Bioresour. Technol. 100 (4), 1589–1594.

Li, C., Aston, J.E., Lacey, J.A., Thompson, V.S., Thompson, D.N., 2016. Impact of feedstock quality and variation on biochemical and thermochemical conversion. Renew. Sustain. Energy Rev. 65, 525–536.

Li, K., Zhang, L., Zhu, L., Zhu, X., 2017. Comparative study on pyrolysis of lignocellulosic and algal biomass using pyrolysis-gas chromatography/mass spectrometry. Bioresour. Technol. 234, 48–52.

Liitia, T., Maunu, S.L., Hortling, B., 2000. Solid-state NMR studies of residual lignin and its association with carbohydrates. J. Pulp Pap. Sci. 26, 323–330.

Lin, W., Dam-Johansen, K., Frandsen, F., 2003. Agglomeration in bio-fuel fired fluidized bed combustors. Chem. Eng. J. 95, 171–185.

Liu, Y., Nie, Y., Lu, X., Zhang, X., He, H., Pan, F., Zhou, L., Liu, X., Ji, X., Zhang, S., 2019. Cascade utilization of lignocellulosic biomass to high-value products. Green Chem. 21, 3499–3535.

Luo, Y., Li, Z., Li, X., Liu, X., Fan, J., Clark, J.H., Hu, C., 2019. The production of furfural directly from hemicellulose in lignocellulosic biomass: a review. Catal. Today 319, 14–24.

Mao, J.D., Holtman, K.M., Franqui-Villanueva, D., 2010. Chemical structures of corn Stover and its residue after dilute acid Prehydrolysis and enzymatic hydrolysis: insight into factors limiting enzymatic hydrolysis. J. Agric. Food Chem. 58, 11680–11687.

Mark, H., Meyer, K.H., 1929. About the formation of crystalized proportions in the cellulose II. Z. phys. Chem., B Chem. Elem.proz. Aufbau Mater. 2, 115–145.

Markou, G., Irini, A., Georgakakis, D., 2012. Microalgal carbohydrates: an overview of the factorsinfluencing carbohydrates production, and of mainbioconversion technologies for production of biofuels. Appl. Microbiol. Biotechnol. 96, 631–645.

Marrugo, G., Valdés, C.F., Chejne, F., 2016. Characterization of Colombian agroindustrial biomass residues as energy resources. Energy Fuel 30, 8386–8398.

Miller, F.P., Vandome, A.F., John, M.B., 2010. Higher Heating Value. VDM Publishing.

Milne, T., Brennan, A., Glenn, B.H., 1990. Sourcebook of Methods of Analysis for Biomass Andbiomass Conversion Processes. Springer, pp. 300–371.

Morgan, T.J., George, A., Boulamanti, A.K., Álvarez, P., Adanouj, I., Dean, C., Vassilev, S.V., Baxter, D., Andersen, L.-K., 2015. Quantitative X-ray fluorescence analysis of biomass (switchgrass, corn Stover, Eucalyptus, beech, and pine wood) with a typical commercial multi-element method on a WDXRF spectrometer. Energy Fuel 29 (3), 1669–1685.

Nanda, S., Mohanty, P., Pant, K.K., Naik, S., Kozinski, J.A., Dalai, A.K., 2013. Characterization of north American lignocellulosic biomass and biochars in terms of their candidacy for alternate renewable fuels. Bioenergy Res. 6, 663–677.

Natarajan, E., Nordin, A., Rao, A.N., 1998. Overview of combustion and gasification of rice husk in fluidized bed reactors. Biomass Bioenergy 14, 533–546.

Ohliger, A., Förster, M., Kneer, R., 2013. Torrefaction of beech wood: a parametric study including heat of reaction and grindability. Fuel 104, 607–613.

Paiva, J.M.F., Trindade, W.G., Frollini, E., Pardini, L.C., 2004. Carbon fiber reinforced carbon composites from renewable sources. Polym. Technol. Eng. 43, 1187–1211.

Pandey, M.P., Kim, C.S., 2011. Lignin Depolymerization and conversion: a review of thermochemical methods. Chem. Eng. Technol. 34, 29–41.

Pandey, A., Negi, S., Binod, P., Larroche, C., 2014. Pretreatment of Biomass: Processes and Technologies. Elsevier Science.

Popp, J.A.K.W., 2006. Raman scattering, fundamentals. In: Meyers, R.A. (Ed.), Encyclopedia of Analytical Chemistry. John Wiley & Sons, Hoboken, NJ, pp. 13104–13139.

Raj, T., Kapoor, M., Gaur, R., Christopher, J., Lamba, B., Tuli, D.K., Kumar, R., 2015. Physical and chemical characterization of various Indian agriculture residues for biofuels production. Energy Fuel 29, 3111–3118.

Rajinipriya, M., Nagalakshmaiah, M., Robert, M., Elkoun, S., 2018. Importance of agricultural and industrial waste in the field of nanocellulose and recent industrial developments of wood based nanocellulose: a review. ACS Sustain. Chem. Eng. 6, 2807–2828.

Rietveld, H., 1967. Line profiles of neutron powder-diffraction peaks for structure refinement. Acta Crystallogr. 22, 151–152.

Robbins, M.P., Evans, G., Valentine, J., Donnison, I.S., Allison, G.G., 2012. New opportunities for the exploitation of energy crops by thermochemical conversion in northern Europe and the UK. Prog. Energy Combust. Sci. 38, 138–155.

Roggo, Y., Chalus, P., Maurer, L., Lema-Martinez, C., Edmond, A., Jent, N.A., 2007. Review of near infrared spectroscopy and chemometrics in pharmaceutical technologies. J. Pharm. Biomed. Anal. 44, 683–700.

Ruland, W., 1961. X-ray determination of crystallinity and diffuse disorder scattering. Acta Crystallogr. 14, 1180–1185.

Segal, L., Creely, J.J., Martin, A.E., Conrad, C.M., 1959. An empirical method for estimating the degree of crystallinity of native cellulose using the X-ray diffractometer. Text. Res. J. 29, 786–794.

Sluiter, A., 2008. Determination of Ash in Biomass: Laboratory Analytical Procedure (LAP). National Renewable Energy Laboratory: Golden, Co, pp. 1–5.

Sluiter, A., Hames, B., Ruiz, R., Scarlata, C., Sluiter, J., Templeton, D., 2005. Determination of Ash in Biomass. National Renewable Energy Laboratory.

Sluiter, A., Hames, B., Ruiz, R., Scarlata, C., Sluiter, J., 2008a. Determination of Structural Carbohydrates and Lignin in Biomass. Laboratory Analytical Procedures (LAP), National Renewable Energy Laboratory (NREL), Golden, Co, pp. 1–15.

Sluiter, A., Hames, B., Ruiz, R., Scarlata, C., Sluiter, J., Templeton, D., Crocker, D., 2008b. Determination of Structural Carbohydrates and Lignin in Biomass, Laboratory Analytical Procedure, NREL/TP-510-42618.

Smith, W., Dent, G., 2005. Modern Raman Spectroscopy. John Wiley & Sons, Chichester.

Soest, P.J.V., Robertson, J.B.B., Lewis, A., 1991. Methods for dietary fiber, neutral detergent fiber, and nonstarchpolysaccharides in relation to animal nutrition. J. Dairy Sci. 74, 3583–3597.

Spolaore, P., Cassan, C.J., Duran, E., Isambert, A., 2006. Commercial appliations of microalgae. J. Biosci. Bioeng. 101, 87–96.

Sun, Z., Fridrich, B., de Santi, A., Elangovan, S., Barta, K., 2018. Bright side of lignin depolymerization: toward new platform chemicals. Chem. Rev. 118, 614–678.

Tekin, K., Karagoz, S., Bektas, S., 2014. A review of hydrothermal biomass processing. Renew. Sustain. Energy Rev. 40, 673–687.

Theander, O., Aman, P., Westerlund, E., Andersson, R., Petersson, D., 1995. Total dietary fiber determined as neutral sugar residues, uronic acid residues, and Klason lignin (the Uppsala method): collaborative study. J. AOAC Int. 78, 1030–1044.

Thyrel, M., Aulin, R., Lestander, T.A., 2019. A method for differentiating between exogenous and naturally embedded ash in bio-based feedstock by combining ED-XRF and NIR spectroscopy. Biomass Bioenergy 122, 84–89.

Tortosamasia, A.A., Buhre, B.J.P., Gupta, R.P., Wall, T.F., 2007. Characterizing ash of biomass and waste. Fuel Process. Technol. 88, 1071–1081.

Vallero, D., 2008. Fundamentals of Air Pollution, fourth ed. Elsevier Inc., San Diego.

Wang, W., Kuang, Y., Huang, N., 2011. Study on the decomposition of factors affecting energy-related carbon emissions in Guangdong province. China. Energies. 4, 2249–2272.

Wang, G., Fan, B., Chen, H., Li, Y., 2020. Understanding the pyrolysis behavior of agriculture, forest and aquatic biomass: products distribution and characterization. J. Energy Inst. 93, 1892–1900.

Westover, T.L., Emerson, R.M., 2015. Rapid analysis of inorganic species in herbaceous materials using laser-induced breakdown spectroscopy. Ind. Biotechnol. 11, 322–330.

Williams, C.L., Westover, T.L., Emerson, R.M., Tumuluru, J.S., Li, C., 2016. Sources of biomass feedstock variability and the potential impact on biofuels production. Bio Energy Res. 9, 1–14.

Wu, X., Luo, N., Xie, S., Zhang, H., Zhang, Q., Wang, F., Wang, Y., 2020. Photocatalytic transformations of lignocellulosic biomass into chemicals. Chem. Soc. Rev. 49, 6198–6223.

Xiao, L., Wei, H., Himmel, M.E., Jameel, H., Elley, S.S., 2014. NIR and Py-mbms coupled with multivariate data analysis as a high-throughput biomass characterization technique: a review. Front. Plant Sci. 5 (388).

Xu, F., Shi, Y.C., Wang, D., 2013. X-ray scattering studies of lignocellulosic biomass: a review. Carbohydr. Polym. 94, 904–917.

Xu, C., Arancon, R.A., Labidi, J., Luque, R., 2014. Lignin depolymerisation strategies: towards valuable chemicals and fuels. Chem. Soc. Rev. 43, 7485–7500.

Yang, H., Yan, R., Chen, H., Lee, D.H., Zheng, C., 2007. Characteristics of hemicellulose, cellulose and lignin pyrolysis. Fuel 86, 1781–1788.

Yann, L.B., Luc, D., Jesus, R., Nicolas, B., Roger, G., Anthony, D., 2015. High resolution solid state 2D NMR analysis of biomass and biochar. Anal. Chem. 87, 843–847.

Yusuf, A.A., Inambao, F.L., 2020. Characterization of Ugandan biomass wastes as the potential candidates towards bioenergy production. Renew. Sustain. Energy Rev. 117, 109477.

Zhou, X., Li, W., Mabon, R., Broadbelt, L.J., 2017. A critical review on hemicellulose pyrolysis. Energ. Technol. 5, 52–79.

8

Pretreatment of lignocellulosic biomass for bioethanol production

Shailja Pant, Ritika, and Arindam Kuila

Department of Bioscience & Biotechnology, Banasthali Vidyapith, Banasthali, Rajasthan, India

1 Introduction

Rising concerns over declining fossil fuels, global climate change and recent advancements in mandates to meet the global renewable energy targets has augmented the need to search for alternative transportation fuels. The production of bioethanol represents one of the most promising materials which can be blended with fossil gasoline and hence has been explored globally. Currently, nearly all of the bioethanol (>99%) is produced as first-generation biofuel utilizing either sugar (sugarcane) or starch rich food crops (maize) as feedstocks with Brazil and U.S.A being the lead producers from these feedstocks respectively. Yet the sustainability of this first-generation bioethanol has been a subject of much scrutiny due to food versus fuel debate and for diversion of arable land. In lieu of these concerns, second-generation bioethanol production has gained much attention over the past few years and has shifted the interest of researchers toward the development of an efficient lignocellulose-based process for bioethanol production (Cooper et al., 2020).

Biomass to bioethanol conversion process is carried out in four major steps: (a) pretreatment (b) hydrolysis (c) fermentation (d) distillation and dehydration. The primary role of pretreatment is to efficiently alter the lignin part while preserving cellulose and hemicellulose, reducing the cellulose crystallinity and eventually increasing the porosity of the material (Chiaramonti et al., 2012). This chapter discusses all the pretreatment methods employed specifically for bioethanol production and how these treatments affect the downstream processing of bioethanol.

2 Lignocellulosic biomass and its conversion to bioethanol

Lignocellulose is a term used for a heterogeneous group of plant-based material (agricultural and forest wastes, municipal solid waste, energy crops, etc.) that consists of three major components: cellulose $(C_6H_{10}O_5)_n$, hemicellulose $(C_5H_8O_4)_m$, and lignin $[C_9H_{10}O_3(OCH_3)_{0.9-1.7}]_x$. However, the ratio of these components varies from one species to another as shown in Table 1. Due to large-scale availability of lignocellulosic biomass, limited competition with food crops and its cellulose-rich nature, it has been researched as a

TABLE 1   Composition of commonly used lignocellulosic biomass.

Lignocellulosic biomass	Composition of lignocellulosic biomass			References
	Cellulose (%)	Hemicellulose (%)	Lignin (%)	
Wheat straw	35–45	23–30	8–15	Sarkar et al. (2012)
Rice straw	32–47	19–27	5–24	Sarkar et al. (2012)
Corn cob	39	35	15	Machado et al. (2016)
Coco peat	21–36	15	48–53.5	Israel et al. (2011)
Coconut coir	48	26	18	Raveendran et al. (1995)
Bamboo	45	24	20	Li et al. (2015)
Banana	13.2	14.8	14	Sánchez (2009)
Mustard stalk and straw	39.5	18.7	22.5	Kapoor et al. (2015)
Millet husk	33	27	14	Raveendran et al. (1995)
Sugarcane bagasse	50	25	18.4–25	Pandey et al. (2000)
Switch grass	31–37	20–29	17–19	Sánchez (2009)
Timothy grass	34	30	18	Nanda et al. (2013)
Elephant grass	36	24	28	Scholl et al. (2015)
Napier Grass	47	31	22	Reddy et al. (2018)
Reed	49.40	31.50	8.74	Tutt and Olt (2011)
Hemp	53.86	10.60	8.76	Tutt and Olt (2011)
Newspaper waste	40–55	25–40	18–30	Prasad et al. (2007)
Poplar	44	20	29	Kim et al. (2009)
Rye straw	33–35	27–30	16–19	Rowell (1992)
Hardwood stem	40–55	24–40	18–25	Malherbe and Cloete (2002) Prasad et al. (2007)
Softwood stem	45–50	25–30	25–35	Dale (1999) Sánchez (2009)
Paper waste	60–70	10–20	5–10	Chen et al. (2004)

promising feedstock for the production of second-generation bioethanol. Unfortunately, the interaction between different components in lignocellulosic materials causes recalcitrance and leaves the biomass resistant to any kind of enzymatic or chemical degradation.

The different components in lignocellulosic biomass are connected to each other via interpolymer linkages such as ether bond (cellulose-lignin, hemicellulose-lignin), ester bond (hemicellulose-lignin) and hydrogen bond (cellulose-hemicellulose, hemicellulose-lignin, cellulose-lignin). Among the three types of linkages, ether bond holds the utmost significance in context to conversion of biomass into bioethanol. It is the most dominant bond present in lignin polymer and holds the glucose molecules together in a cellulose chain. Presence of lignin in biomass acts as a physical barrier and prevents the efficient digestion of biomass for cellulose extraction. Therefore, for efficient utilization of lignocellulosic biomass, it is essential to cleave the ether bond for separation of lignin from cellulose and hemicellulose before subjection to enzymatic hydrolysis (Harmsen et al., 2010).

3 Pretreatment

Pretreatment is an indispensable step in the conversion of biomass to biofuel and its primary role is to separate the different components of biomass by breaking down the lignin seal and making the carbohydrate part more amenable for subsequent bioprocessing (Chen et al., 2017). Pretreatment changes the structure and chemical composition of biomass thus making it more amenable for degradation. Various pretreatment methods including physical, chemical, biological and combinational have been extensively studied and compared over the past few decades (Fig. 1); however, no method can be concluded as 'the best' for all types of biomass. Since pretreatment is an interaction-based reaction, depending on the ratio of different components in lignocellulosic material, it results in different outcomes. Therefore, the choice

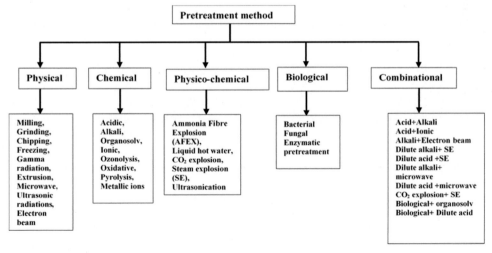

FIG. 1 Different pretreatment strategies utilized for bioconversion of lignocellulosic biomass.

of pretreatment method is crucial and should take into consideration the distribution of components in the biomass, their interaction among themselves and the desired outcomes.

3.1 Goals of pretreatment

Pretreatment leading to maximum cellulose enrichment accompanied by maximum lignin removal was being considered as ideal for bioethanol production. However, over the past few decades interest of researchers has shifted more toward the development of an economical and environment-friendly bioprocess rather than just a production process with good bioethanol production efficiency. In general, an ideal pretreatment method should meet some important criteria such as

1. Method should require either no or minimum particle size reduction
2. Since lignin forms an undesirable fraction in the bioethanol production process, the method should offer its maximum removal while preserving the cellulose and hemicellulose part in solids.
3. There should be minimum formation of inhibitors during the process
4. Method must be energy and cost-effective
5. Should make use of non-corrosive chemicals to minimize bioreactor cost
6. Should facilitate high recovery of individual polymers like lignin or ligno-xylan to increase the economic efficiency of the process

3.2 Physical treatment

Physical treatments are the most conventional method for biomass treatment that work on the principle of size reduction. Reduced particle size of biomass leads to increase in surface area eventually making the biomass more accessible for pretreatment. Physical pretreatment includes methods like mechanical extrusion, microwave irradiation, ultrasonic treatment and pyrolysis (Behera et al., 2014).

Mechanical extrusion is one of the commonly used physical method for size reduction of biomass. A general extrusion machine consists of one or two rotating screw extruders that spin in a tight barrel and a temperature-controlled system (Zheng and Rehmann, 2014). The method makes use of high temperature (>300°C) and shear forces generated by the spinning action of screw blades in the barrel to disrupt the recalcitrant structure of biomass (Baruah et al., 2018). Screw configuration, screw speed and the temperature of barrel are the major factors that affects the efficiency of mechanical extrusion. Milling is another conventional method applied to decrease cellulose crystallinity and particle size (up to 0.2 mm) of biomass. Depending on the type of mill used for the process, it could be classified as ball milling, rod milling, two roll milling, wet desk milling, colloid milling and hammer milling (Kumar and Sharma, 2017). Efficacy of method depends on the adopted milling process, processing time and the type of biomass. Ultrasonication is relatively a new technique used for the pretreatment of biomass. In the method, the employment of ultrasonic radiations forms small cavitation bubbles throughout the biomass and generates shear forces that alter the morphology of biomass by opening the crystalline fractions of cellulose. The severity of biomass decomposition depends on irradiation frequency, sonication time, sonication power and the process temperature. Due to the low vibrating energy of ultrasonic waves that fails to bring out a conformational change in biomass structure, ultrasonication technique cannot be employed as a stand-alone method and therefore is generally performed as ultrasound assisted alkali treatment. Addition of alkali aids in cleavage of interpolymer hydrogen bonds consequently increasing the reaction kinetics. Pyrolysis pretreatment is a thermal degradation process that includes subjection of biomass to high temperature generally ranging from 500 to 800°Cin absence of any catalyzing agent. The treatment is generally applied for production of bio-oil and is unpopular in case of bioethanol production.

Physical treatment methods reduce the crystallinity of cellulose up to some extent but are not efficient enough to separate the different components of biomass and are therefore generally accompanied by other treatment methods to increase the efficiency. Depending on the type of feedstock and the final particle size required for hydrolysis, the size reduction of biomass costs from low to a huge amount of energy that makes it an energy extensive step and difficult to use for commercial purpose. Hence, research nowadays is being emphasized on the development of methods that avoids or minimize the need for size reduction of biomass.

3.3 Chemical method

3.3.1 Acid and alkaline pretreatment

pH has a large effect on the outcome of conversion efficiency of any pretreatment process. The main purpose of pretreatment is to remove lignin to a maximum extent while preserving the cellulose and hemicellulose fraction. Therefore, the selection of a suitable pH for pretreatment process is of utmost importance. Pretreatment of feedstock at low pH or with acids results in hydrolysis of mainly hemicellulosic part while keeping the cellulose part intact whereas a high pH treatment or alkali treatment results in dissolution of lignin fraction while preserving the cellulose and hemicellulose fraction. Hence selection of the pretreatment method depends on the composition of feedstock and the desired outcomes. Depending on the concentration of acid or alkali used for treatment process, degradation of biomass can vary up to a high extent (Galbe and Wallberg, 2019).

Acid hydrolysis is the most widely used pretreatment method for lignocellulosic biomass where the feedstock is treated with a defined concentration of either some inorganic acid (sulfuric acid, nitric acid, phosphoric acid, hydrochloric acid, nitric acid) or in combination with organic acids (propionic acid, formic acid, acetic acid) under high temperature, low pressure or low temperature, high pressure depending on the concentration of acid used. During acid treatment, hydronium ions generated from acid leads to cleavage of inter and intra polymer glucosidic linkages between hemicellulose-cellulose and results in release of monomeric sugars.

Two types of acid treatments are being reported for hydrolysis of biomass:

- Dilute acid treatment: (0.2–2.5)% acid concentration, high temperature (120–210°C) and pressure for a few minutes to few hours. (Bensah and Mensah, 2013)
- Concentrated acid treatment: more than 40% acid concentration, at low temperature (30–60°C) and pressure for a few seconds to few minutes. (Bensah and Mensah, 2013)

Although concentrated acid treatment can result in conversion efficiency of more than 90% for both cellulose and hemicellulose, it suffers from some severe disadvantages such as production of high amount of inhibitors (furfurals, 5 HMF, phenolic acids) (Wang et al., 2018) and corrosion of bioreactor. Considering the disadvantages associated with the use of concentrated acids, dilute acid pretreatments are preferred nowadays. Studies have reported that among the acids, dilute sulfuric acid is the most efficient and extensively used acid for pretreatment of lignocellulosic biomass because of easy large scale availability and lower cost. Acid pretreatment of biomass requires subsequent detoxification for removal of inhibitors from the reaction mixture either by chemical or biological processes that either neutralizes the levels of inhibitor present in reaction mixture or convert them into inert substances. However, the most ideal way to keep the inhibitor level in check is to choose a feedstock that are more susceptible to degrade under dilute conditions and generate lesser amount of inhibitors.

In contrary to acid pretreatments, alkali pretreatment method require lesser temperature and time and therefore can be performed under milder conditions (Jędrzejczyk et al., 2019). The method relies on dissolution of lignin in alkali while having negligible to very less effect on solubilization of cellulosic fractions. Alkali reagents bring out saponification reaction of ester bonds and degrades the side chain of glycosides leading to alternation in lignin conformation, cellulose selling and partial solvation of hemicellulose ions (by removal of acetyl groups and uronic acids). The most commonly employed alkalis for pretreatment of lignocellulosic biomass include NaOH, Ca (OH)$_2$, KOH, and ammonia. Sodium hydroxide has been reported as the most effective in terms of lignin removal from softwoods. Alkali pretreatment is advantageous to acid pretreatments as the process could be carried out under milder conditions therefore require lesser energy and avoids the need for detoxification. The only disadvantage associated with alkali pretreatment lies in the fact that a large amount of water is utilized in removing the alkali salts adsorbed onto the surface of biomass during the process and much higher cost of alkali.

3.3.2 *Oxidation pretreatment*

Oxidation pretreatment includes degradation of lignocellulosic biomass in the presence of an oxidizing agent like ozone, the process being called as c or oxygen also known as wet oxidation. Ozone is a powerful oxidizing agent that is used for disintegration of lignin and hemicellulose. At 60% removal of lignin, the method can increase the efficiency of saccharification process by five folds. Moreover, the process can be carried out under mild conditions at room temperature and pressure. The main drawback in utilization of ozone for large scale is the high cost of ozone and environmental considerations making the process less suitable for commercial purpose. Wet oxidation is another approach followed for pretreatment of biomass utilizing oxygen. The chemical oxidation of oxygen occurs in the presence of water and therefore the process is carried out under high moisture conditions. Reaction rate of wet oxidation increases with increase in temperature following Arrhenius dependence (Palonen et al., 2004). At high severity of oxygen, conversion of liberated sugars into organic acids is initiated therefore optimization of oxygen concentration is needed in the pretreatment of specific biomass for desired downstream process (Chen et al., 2017).

3.3.3 Green solvents

Most of the chemical based pretreatment strategies employed in bioethanol production requires either the addition of an acid or base for efficient degradation of biomass which in turn leads to generation of toxic wastes possessing detrimental environmental effects. In the same context, ionic solvents also considered as green solvents due to their non-corrosive nature and minimal waste generation are being promoted as sustainable substitutes. Ionic liquids are basically salts that are ionic in nature and consist of an anionic (both inorganic and organic) and a cationic (organic) species. These liquids remain in fluid state at ambient temperature and have melting point below 100°C (Hassan et al., 2018). Some of the common anions and cations used to design ionic solvents are shown in Table 2.

Process of pretreatment with ionic liquids involves soaking of biomass in ionic solvent at ambient pressure with constant heat (temperature of 90–130°C) and stirring for few hours (1–24 h.) (Brodeur et al., 2011). IL-biomass solution is then precipitated with the addition of an antisolvent which commonly a mixture of acetone and water or deionized water. Acids and alcohols are also utilized for the same purpose. Anti-solvents are recovered from the solution by evaporation and lignin is filtered out.

3.3.3.1 Dissolution mechanism of cellulose

Solubility of cellulose in an ionic solvent depends on the type of cation and anion used for its construction. The most effective cations employed for dissolution of cellulose are imidazole based and this effectivity can be attributed to high hydrogen bond basicity as well as high bond dipolarity (values near to 1). Usmani et al. (2020) in their review mined the data from various researches and concluded that the solubility of lignin in ionic solvents depends on various solvatichromic parameters like polarizability (π^*), acidity (α) (depends on the nature of cation), and basicity (β) (depends on the nature of anion) of hydrogen bond. Basicity of hydrogen bond is an indirect measure of biomass dissolution in ionic liquids. Therefore,

TABLE 2   Common ionic species in ionic solvents.

Cations	(a) **Imidazolium based ($[(C_3N_2)X_n]^+$):** e.g., 1-Alkyl-2,3-dimethylimidazolium, 1-Alkyl-pyridinium, 1-Alkyl-3-methylimidazolium, 1-butylimidazolium, 1-propyl-3-methylimidazolium
	(b) **Ammonium-based $[NX_4]^+$:** e.g., Tetraalkylammonium, 2-hydroxyethylammonium, triethyl-(2-(2-methoxyethoxy)ethoxy)ethylammonium,
	(c) **Pyrrolidinium based ($[C_4N)X_n]^+$):** e.g., 1-Alkyl-1-methyl-pyrrolidinium
	(d) **Pyridinium based ($[(C_5N)X_n]^+$):** e.g., 1-alkyl-1-methyl-piperidinium, 1-ethyl-3-(hydroxymethyl) pyridine, 1-butyl-3-methylpyridinium
	(e) **Sulfonium-based $[SO_3]^+$:** e.g., trialkylsolfonium
	(f) **phosphonium-based $[PX_4]^+$:** e.g., trialkylsolfonium and
	(g) Choline
Anions	Hexafluorophosphate, tetrafluoroborate, acetate, halides (chloride, fluoride, bromide), bis(trifluoromethylsulfonyl)imide, methyl sulfate, dimethyl phosphate, trifluoromethanesulfonate, dicyanamide

Source: da Costa Lopes, A.M., João, K.G., Morais, A.R.C., Bogel-Łukasik, E., Bogel-Łukasik, R., 2013. Ionic liquids as a tool for lignocellulosic biomass fractionation. Sustainable Chem. Processes. 1(1), 1–31; Yoo, C.G., Pu, Y., Ragauskas, A.J., 2017. Ionic liquids: promising green solvents for lignocellulosic biomass utilization. Curr. Opinion Green Sustain. Chem. 5, 5–11.

solvents with high hydrogen bond basicity (more than 0.8) are preferred for pretreatment reactions.

Ionic liquids work by competing with cellulosic components for hydrogen bonding. Interaction between the hydrogen atom of cellulose hydroxyl moieties present in biomass with the anion present in ionic liquids lead to formation of an electron donor-electron acceptor (EDA) complex which further leads to cleavage of cellulose-cellulose interactions thus disrupting the structural composition of biomass (Usmani et al., 2020). Cations present in ionic liquids also leads to cleavage of hydrogen bond present between lignin moieties and alter the aromatic character of lignin by π-π and n-π interactions. Various IL systems like [1-butyl-3-methylimidazolium hydroxide][Cl], [1-alkyl-3-methylimidazolium hydroxide] [Cl], [1-alkyl-3-methylimidazolium hydroxide][OAc] [1-ethyl-3-methylimidazolium][OAc], [1-butyl-3-methylimidazolium hidroxide][OAc], [1-butyl-3-methylimidazolium hidroxide] [HSO$_4$] [1-ethyl-3-methylimidazolium][MeO(H)PO$_2$], [1-alkyl-3-methylimidazolium hydroxide][OFo],], [1-butyl-3-methylimidazolium hidroxide][Cl] have been investigated for dissolution of lignin in ionic liquids (Halder et al., 2019). The tunability of ionic solvents in terms of endless combinations possible with different types of anions and cations has positioned them as a strong promising alternative for large-scale biorefineries. However, recovery of ionic liquid from pretreatment hydrolysate and high cost associated with the treatment process are still some obstacles that need to be overcome in order to make the treatment feasible for large-scale biorefineries.

3.4 Biological pretreatment

Biological pretreatment makes use of microbes or their products for the degradation of lignocellulosic biomass. The method relies on the theory that certain microbes in absence of simple sugars like glucose, xylose initiates the transcription of enzymatic genes and thus release a set of extracellular enzymes that aid in the conversion of complex polysaccharides present in biomass into simple sugars. Most of the microbes explored for bioconversion belong to the kingdom of fungi among which white, brown and soft rot fungi are the most explored ones. Although biological pretreatment offer several advantages like mild reaction conditions, less energy requirement, absence of toxic waste production, absence of need to neutralize and detoxify the biomass after treatment still the slow rate of degradation makes biological pretreatments impractical for use in commercial processes.

Biological treatments can be carried out in two ways: As whole cell pretreatment where a microbe is allowed to grow on biomass directly and at the same time degrades it or as enzymatic degradation. For enzymatic degradation, not microbes but their product (enzymes) is utilized for degradation of biomass.

One of the major disadvantages of whole cell pretreatment is that the optimal conditions required for the growth of microbe and efficient enzymatic activity differs from each other by a high degree. As a result of which enzyme released by microbe is not efficiently utilized for the degradation process and remains unused in the production media. Therefore, enzymatic degradation of biomass is preferred over whole cell pretreatment for which the enzyme is harvested from the production media and biomass is subjected to enzymatic hydrolysis under suitable conditions but the life cycle becomes too long to be feasible for large scale pretreatment of biomass. However, with the advancements in genetic engineering technologies there is still a hope for biological treatments to be recognized for pretreatment purpose in the near future.

3.4.1 Enzymes involved in biological pretreatment

Biomass degrading enzymes include three classes of enzyme, i.e., cellulases, hemicellulases, and lignolytic enzymes based on the fraction of biomass hydrolyzed.

Cellulases: Cellulases represent a group of hydrolytic enzymes (endoglucanase, beta-glucosidase, and cellobiohydrolase) that act in synergism for saccharification of cellulose into monomeric sugars by cleavage of beta-1,4-glycosidic bond connecting the glucose molecules. Cellulases are synthesized in nature by a number of micro-organisms commonly fungi, however, these cellulase enzymes are only active on soluble cellulose such as carboxymethylcellulose and less active on native cellulose. Some of the common cellulolytic fungi are *Trichoderma reesei, Trichoderma viridie, Aspergillus, Penicillium funiculosum, Phanerochaete chrysosporium, Chaetomium thermophile* and *Thermoascus aurentiacus*. Relatively few aerobic cellulolytic bacteria have been recorded to produce cellulases for, e.g., *Aerobacter, Achromobacte, Acetobacter xylinum, Agrobacterium, Alacaligenes, Azotobacter, Pseudomonas* and *Sarcina* (Sharma et al., 2019).

Hemicellulases: Hemicellulases are the group of enzymes that catalyzes the degradation of hemicellulose in biomass. The enzymes can be categorized as either glycoside hydrolase that hydrolyze the glycosisdic bond present between xylose molecules or carbohydrate esterase that cleave the ester bonds of acetyl or ferulic groups attached to hemicellulose moieties (Ummalyma et al., 2019).

Lignolytic enzymes: These are the oxidative enzymes that oxidize the phenolic and non-phenolic components of lignin fraction. Based on the mode of action lignolytic enzymes falls under one of the two categories: laccase (Lac) or peroxidases. Laccases are structurally a group of multicopper enzyme that comprises of four copper atoms in their catalytic domain. Large molecular weight (more than 70 kDa) of laccase molecules restricts their entry into the biomass and therefore the enzymes could only act on the surface of the biomass. These enzymes work by radical catalyzed mechanism and can oxidize only phenolic components due to their low redox potential (0.5–0.8 V). However, studies have reported the oxidation of non-phenolic components by laccases in the presence of a strong oxidation mediator system also known as LMS (Laccase Mediator System). ABTS (2,2'-azinobis-(3-ethylbenzthiazoline-6-sulfonate) is the most widely used LMS. These LMS system works by acting as an electron shuttle between laccase and the substrate. Firstly, the mediator gets oxidized by laccases and then gets reduced into the parent form by substrate (Zabed et al., 2019).

Peroxidases are the group of extracellular oxidoreductases that are classified as lignin peroxidase (LiP) and manganese peroxidase (MnP). Both enzymes are heme containing glycoprotein and oxidizes the phenolic and non-phenolic components of lignin in the presence of H_2O_2. Oxidation by LiP involves formation of two reaction intermediates, LP-I and LP-II (through the oxidation of ferrous ion) which further get reduced into phenoxy radicals in the presence of H_2O_2. These phenoxy radicals aids in the cleavage of side chains in non-phenolic components of lignin and cause their rearrangements. In a similar way, MnP works by oxidizing the Mn^{2+} to Mn^{3+} that gets chelated into a complex form in the subsequent steps.

The chelated complex exhibits a strong oxidizing capacity and thus oxidize the various non-phenolic components by bond cleavages and their rearrangements.

Fungal enzymes hold more ecological and commercial importance as bacteria seldom produces complete microbial secretome (set of biomass degrading enzymes). Despite the higher activity and specificity of enzymes produced from fungi, the cost constraint, longer production cycle time and low yield of enzyme present major drawbacks in utilization of fungi for enzyme production. In the same context, cloning and expression of fungal genes into bacterial system could combine the best properties of two species in a single organism resulting in increased rates of hydrolysis, fermentation, and product recovery. This would have significant impact in decreasing the cost of ethanol production from agricultural and cellulosic biomass feedstocks.

3.5 Combinational pretreatment strategies

3.5.1 Sequential acid and alkali treatment

Pretreatment of lignocellulosic biomass with acid or base provides efficiency either in removal of hemicellulose or lignin but never for both. In context to bioethanol production, both hemicellulose and lignin complicates the bioconversion of lignocellulosic biomass and hence are desired to remove before saccharification and fermentation. A way to achieve the removal of both components (hemicellulose and lignin) is using a sequential acid and alkali treatment. The method involves impregnation of biomass with dilute acid (usually sulfuric or nitric acid) followed by treatment with subsequent dilute alkali treatment (usually NaOH). A combination of acid and alkali makes the removal of hemicellulose and lignin possible thereby leaving the only cellulosic fraction in biomass (Gupta et al., 2020). The method increases the enzymatic susceptibility and fermentability of biomass. However due to removal of hemicellulosic and lignin fractions from the biomass, the crystallinity index is increased. The method gives better efficiency with lignocellulosic materials possessing low levels of hemicellulose, lignin and high levels of cellulose (Ascencio et al., 2020).

3.5.2 Dilute acid + steam explosion

Due to no use of chemicals and less energy consumed, steam explosion is considered as an effective and economical method of pretreatment. The method includes subjection of feedstock to a saturated steam with a temperature of 160–260°C at 0.69–4.83 MPa (Duque et al., 2016). Sudden drop in pressure at the end of pretreatment reaction results in decompression of biomass and lignin depolymerization through homolytic cleavage of β-O-4′interlinkages. Steam pretreatment is not that efficient for degradation of lignocellulosic biomass having high content of hemicellulose and lignin and is therefore generally accompanied by external catalyst (acid or base) addition. Combining the steam explosion method with acid or alkali treatment increases the bioconversion efficiency by reducing the reaction time and temperature. Moreover, addition of acid or base minimizes the formation of inhibitors during the process ultimately leading to better hydrolysis efficiency in subsequent bioprocessing.

Among the various acidic alternatives, sulfuric acid and sulfur dioxide gas are the most popular catalysts. Gaseous catalysts offer even distribution of catalyst all over the biomass, hence in application SO_2 is preferred over H_2SO_4. Pretreatment of biomass with a combination of dilute acid and steam explosion method has been tested widely at pilot scales for different types of lignocellulosic biomass and studies have concluded that irrespective of nature of biomass, the method has been successful in complete removal of hemicellulose with significant modifications in lignin structure (Auxenfans et al., 2017). In light of all the advantages that method offers in terms of efficient degradation (removal of hemicellulose or lignin by addition of acid or base respectively), the process came closest to upscaling for commercial production of bioethanol. The method is currently being used at pilot scale at Iogen demonstration plant, Canada (Aftab et al., 2019).

3.5.3 Alkali + steam explosion/ammonia fiber explosion (AFEX)

Ammonia fiber explosion or AFEX is a method that combines steam explosion with alkali (Ammonia) treatment. As discussed under alkaline pretreatment, alkali aids in dissolution of lignin part from biomass while keeping the cellulose and hemicellulose fraction as solids. Steam explosion causes the swelling and rupturing of lignocellulosic fibers that further results in increase of surface area and subsequently better impregnation of biomass with ammonia. Catalyst used in the process is liquid anhydrous ammonia (1–2 kg ammonia/Kg biomass) under moderate temperature range (40–140°C) and high pressure (250–300 psi). Method is considered beneficial for herbaceous material and not softwoods since the content of hemicellulose in softwoods is more as compare to lignin. Some of the advantages of AFEX over dilute acid-steam explosion method includes complete recovery of ammonia, minimal formation of inhibiting compounds (because of very less degradation of hemicellulose whose degradation products get converted to inhibitory compounds like furfural, HMF etc.) during the process and no need of neutralization (Balan et al., 2009).

3.5.4 CO_2 explosion

A key drawback in implication of ammonia fiber explosion method is highly corrosive nature of ammonia with serious environmental effects and its high cost. In the same context, supercritical CO_2 appears to be a safer and greener alternative for pretreatment. Supercritical CO_2 is a gas with liquid characteristics which means it enters the spaces between biomass components like a gas and dissolve the components like a liquid (Serna et al., 2016). In the process, CO_2 is passed through the biomass under high pressure and temperature in a tightly closed vessel. CO_2 molecules interact with the water molecules present in biomass and forms carbonic acids which further catalyzes the hydrolysis of hemicellulosic fractions. Therefore, the technique is not considered suitable for biomass with low moisture content (Rosero-Henao et al., 2019; Baruah et al., 2018).

3.5.5 Wet explosion (wet oxidation + steam explosion) (WEx)

Wet explosion combines the use of an oxidizing agent and steam explosion for better dissolution of lignin as compared to conventional steam explosion. In general, Wet explosion process was first reported in 2004 by Ahring et al. Method includes exposure of biomass to a temperature of 140–210°C at 0–3.5 MPa pressure for a time of 5–120 min (Biswas et al., 2015). The extent of biomass degradation depends on the temperature and severity of oxygen employed in the method. Sudden decompression occurs in the pretreated material on termination of reaction by opening the pressure valve leading to swelling of biomass fibers and increase in porosity. The option of sudden decompression is not available in case of wet oxidation. The most popular oxidizing agents used in wet explosion are compressed atmospheric air, O_2 and H_2O_2 among which oxygen has been considered as the most effective for lignocellulose bioconversion. Wet explosion eliminates the need for addition of external chemical (acid or base) and is therefore advantageous in manner that no degradation products are formed that could inhibit the subsequent saccharification and fermentation of pretreated biomass. Advantages and disadvantages of each type of pre-treatment has been shown in Table 3.

TABLE 3   Comparison of different pretreatment methods.

Method	Type	Advantages	Disadvantages	References
Physical	Milling Ultrasonic Microwave Electron beam	Increase in surface area, Reduction in crystallinity and degree of polymerization, Short processing time, less energy consumption; uniform heat distribution (microwave) avoids degradation of lignocellulosic material into humic acid and furfural	Incapability of removing lignin makes it a less suitable option as enzyme accessibility to the substrate is reduced in the presence of lignin	Bhatt and Shilpa (2014), Aftab et al. (2019)
Chemical	Acidic	Efficient removal of lignin and hemicellulose, greater accessibility of cellulose for enzymatic hydrolysis, efficiency of sugar recovery (glucose and xylose) is high	Corrosive and dangerous, formation of chemicals at low pH such as furfurals during the degradation of hemicelluloses that inhibit the following enzymatic hydrolysis and microbial fermentation	Chiaramonti et al. (2012), Chen et al. (2017), Galbe and Wallberg (2019)
	Alkali	Decrease in crystallinity of cellulose and degree of sugar polymerization, greater removal of lignin and partly hemicellulose	Longer pretreatment time, formation of inhibitory compounds, large amount of water is required for washing, high cost of alkali catalyst.	
	Ionic	Disruption of lignin and hemicellulose, decrease in crystallinity of internal and external accessible cellulose	High chemical cost, industrially not feasible, toxicity of some ionic liquids, separation of sugars from ionic liquids is tedious	
	Organosolv	Useful for pure lignin, cellulose and hemicellulose recovery	High chemical cost requires recovery of chemical, not suitable for industrial purpose, toxic effects on environment and fermentation process	
	Oxidation Ozonolysis	Efficient removal of lignin, no toxic compound generation	High cost	

TABLE 3 Comparison of different pretreatment methods—cont'd

Method	Type	Advantages	Disadvantages	References
Physico-chemical	AFEX	Increases the porosity of the materials by breaking the chemical bonds between lignin and cellulose and hemicelluloses, increases cellulose accessibility to enzyme, less inhibitors formation	High cost of ammonia, not efficient for the biomass with high lignin content	Bhatt and Shilpa (2014), Chen et al. (2017), Galbe and Wallberg (2019)
	SE	Cost effective method for complete hemicellulose removal, causes swelling of biomass hence causes hemicellulose solubilization and lignin removal,	High temperature and pressure is required Lignin is not efficiently removed; toxic products are generated	
	CO$_2$ Explosion	Cost effective, increases the accessible surface area for enzymatic hydrolysis, no toxic compound generation	Very high pressure is required, not efficient for the raw material with high lignin content.	
Biological		Degradation of lignin and hemicellulose at normal temperature, low energy consumption, no inhibitor generations, environment friendly	Slow rate of hydrolysis, microorganisms efficiently consume cellulose and free sugars instead of lignin and hemicellulose	Chiaramonti et al. (2012) Kumar and Sharma (2017)
Combinatorial		Efficient removal of lignin and hemicellulose Improved enzymatic saccharification process	High energy requirement, specific equipment is required, accumulation of toxic chemicals effects downstream processing	Chen et al. (2017), Kumar and Sharma (2017)

3.6 Inhibitor formation during pretreatment

One of the major issues in the pretreatment of lignocellulosic biomass is generation of inhibitors. Bioconversion of lignocellulosic materials give rise to degradation products that may act as inhibitor of microbes in the subsequent steps of saccharification and sugar fermentation including aliphatic and aromatic acids, furanaldehydes, alcohol derivates, phenolic compounds and other pretreatment products. These pretreatment products affect ethanol production process negatively by creating unfriendly conditions for fermentative microbes and thus leading to loss of cell density along-with increase in the length of lag phase (Zhang et al., 2019).

3.6.1 Cellulose and its degradation products

However, the primary purpose of pretreatment step is to separate lignin and hemicellulose to maximum extent keeping the cellulose fibers intact, cellulose is still degraded to hexoses to different extent depending on the method employed. In early 1819, Braconnot et al., in his study demonstrated that the treatment of linen with concentrated sulfuric acid in the presence of heat lead to production of fermentation sugars. Acids by breaking intra, interchain hydrogen and glycosidic bonds present in cellulose disrupt its integral structure leading to cellulose decrystallization and its disintegration into glucose monomers. Hydrolysis of cellulose during pretreatment leads to formation of hexose monomers which are further converted to HMF (5-hydroxymethyl furfural) by dehydration process. Chemically, 5-hydroxymethyl furfural consists of a furan ring with aldehydic group. Due to more stability of furan ring as compared to attached aldehydic group, chemical transformation takes place at aldehydic group and furan ring is dehydrated to form levulinic acid and formic acid. Various studies conducted in the past have addressed 5-HMF as toxic and inhibitory to the growth of ethanalogens. In as early as 1923, Heuser and Schott identified formation of HMF as a by-product in acid hydrolysis of lignocellulosic biomass. In 1938, Mashevitskaya and Plevako in their study confirmed the inhibitory impact of HMF on growth of *Monilia murmanica*. Numerous literature studies conclude the toxic effect of HMF on ethanogens. HMF and furfural is assimilated by yeast during fermentation and is converted into less toxic compounds, however conversion of HMF is initiated only after whole of furfural has been converted. Although immediate effect of HMF is less than that of furfural on growth of ethanolegens, however due to slow conversion rate HMF resides for longer time in medium and possess more problem. Xiao et al. (2004) showed that cellobiose, xylose, mannose and galactose inhibit activity of cellulases and b-glucosidases, cellulases being more sensitive than b-glucosidase.

3.6.2 Hemicellulose and its degradation products

Hemicellulose accounts for 20%–40% of total biomass composition and is the second most abundant polymer found in LCB. Hemicellulose is composed of different monosaccharides like xylose and arabinose, mannose, glucose, galactose and sugar acids. To make cellulose more susceptible to biological treatment for ethanol production in subsequent steps, it is important to solubilize hemicellulose for which lignocellulosic feedstock is first treated and then incorporated into ethanol production process. Similar to cellulose degradation, hydrolysis of hemicellulose during pretreatment process also leads to formation of a variety of products such as xylose, arabinose, mannose, galactose, aliphatic acids like acetic acid, formic acid and levulinic acid, and furan aldehydes [5-hydroxymethylfurfural (HMF) and furfural]. These degradation products act as inhibitors for subsequent bioprocessing. Furfural is the most abundant compound found in pretreatment hydrolysate of hemicellulose formed from dehydration of pentoses, uronic acids and is investigated as the most toxic product for subsequent bioprocessing. Without any exception, degradation of hemicellulose in all types of lignocellulosic biomass leads to the formation of acetic acid (due to acetyl group dehydration). Since formation of acetic acid cannot be blocked, pretreated biomasses are generally detoxified before fermentation.

Along with furfural, weak acids formed during degradation of hemicellulose are also reported to possess inhibitory effect on ethanol yield. Wang et al. (2018) in his study identified formation of nine compounds in acid treated hydrolysate from xylan. The compounds were furfural, acetic acid, formic acid, 5-methylfurfural, acrylic acid, 2(5H) furanone, glutanoic anhydride, levulinic acid and 2-furancarboxylic acid. Among the various acids, studies have confirmed acetic acid as the most potent inhibitor among all acids. Acids lead to drop in intracellular pH of ethanogens. Acids enter into the microbial cell in an undissociated form and gets dissociated under neutral cell conditions. To neutralize the cell environment, ATPase pump starts pumping protons out of the cell. Due to exhaustion of proton pumping capacity of cell, cytoplasm acidification occurs resulting in inhibition of cell activity. Fu et al. (2014) pointed out negative impact of formic acid, guaiacol and vanillin on ethanol production. Addition of formic acid at conc. of $2.5\,g\,L^{-1}$ decreased ethanol yield by 8% whereas 50% and 20% decrease in ethanol yield was observed for guaiacol and vanillin at a conc. of $3\,g\,L^{-1}$ and $2.5\,g\,L^{-1}$ respectively. Eva Palmquist.et.al in his studies showed weak acids at low conc. $<100\,nmol\,L^{-1}$ has stimulating effect and increased the yield at Ph 5.5 as compared to the fermentation with no aliphatic acids however at concentrations higher than $>200\,nmol\,L^{-1}$ inhibition of fermentation occurs.

3.6.3 *Lignin and its degradation products*

Lignin is the third most abundant polymer found in lignocellulosic biomass after cellulose and hemicellulose. The highest levels of lignin is found in softwoods (lignocellulosic biomass) while lowest being in herbaceous plants. To increase kinetics of saccharification reaction or cellulose degradation, lignin is generally removed from the biomass in the pretreatment process. Separation of lignin from lignocellulosic biomass relies on its solubilization under alkaline conditions. Alkaline pretreatment of lignin leads to formation of several inhibitory products among which vanillic acid methyl ester and vanillic acid ethyl ester has been reported as the dominant ones (Jönsson et al., 1998).

3.7 Conclusion

High cellulose crystallinity and wrapping of cellulosic fibers with hemicellulose-lignin complex renders a recalcitrance nature to lignocellulosic biomass and limits its biodegradability. Therefore, pretreatment is an indispensable step for bioconversion of lignocellulosic biomass into bioethanol. Although over the past few decades various physical, chemical and biological pretreatments have been investigated (Table 4) for a variety of lignocellulosic biomass, no pretreatment can be considered ideal for all types of biomasses. Since lignocellulosic biomass is highly heterogeneous what works for one type of biomass might not translate into an efficient process for another type of biomass. Analysis of different pretreatment methods brings us to the conclusion that each method has its own advantages and disadvantages and huge variation in composition of lignocellulosic biomass makes it difficult to design a single general process. However, an ideal pretreatment method must provide a balance between the digestion of biomass into separate components and minimum formation of inhibitory compounds. In order to establish a pretreatment technology that is suitable for commercial production of bioethanol, current researches are oriented toward the identification of optimal process parameters as well as pretreatments that are efficient at even high biomass loading. Combinational approaches could combat the disadvantages associated with stand-alone pretreatment strategies in terms of lower time and energy required and provides a holistic approach toward development of economical and environment friendly technologies by reducing the number of operational steps. Physio-dilute acid treatment followed by steam explosion has already reached the commercial phase for pretreatment of lignocellulosic biomass

TABLE 4   Different types of pretreatment of lignocellulosic biomass.

Title	Method	Reference
Process for production of pure glucose from cellulose	Two step acid treatment **Conditions**: Acid Conc.: 1%–3% Temp.: 100–130°C Time: 5–20 min	Lali et al. (2019)
Preliminary treatment method for improved enzymatic hydrolysis	Sequential pretreatment of biomass with aqueous solutions of hydrazine hydrate and sodium sulfite **Conditions**: Di-amine conc.: 1%–99% (w/w) Sulfur Conc.: 0.1%–10% (w/w) Temp.: 80–20°C Time: 5 min to 4 h	Satlewal et al. (2017)
A process for recovering higher sugar from biomass	Hot water extraction followed with subsequent soaking of biomass in acid and steam pretreatment. **Conditions**: Hot water extraction.: 70–90°C, 5–120 min Acid soaking: 1%–2% acid solution (w/w), 5–120 min Steam pretreatment: 150–200°C, 1–3 min, 5–30 bar pressure, pH 3–5	Kumar et al. (2019)

and serves as reference for other pretreatment methods. However, there is a still a room for improvement in terms of better bioreactor designs, exploring catalysts in different combinations for better sugar yield.

Acknowledgment

Authors are thankful to Department of Biotechnology (DBT), New Delhi, India for the financial support (Grant No. BT/PR31154/PBD/26/762/2019) to complete this manuscript.

References

Aftab, M.N., Iqbal, I., Riaz, F., Karadag, A., Tabatabaei, M., 2019. Different pretreatment methods of lignocellulosic biomass for use in biofuel production. In: Biomass for Bioenergy-Recent Trends and Future Challenges. IntechOpen, London, UK.

Ascencio, J.J., Chandel, A.K., Philippini, R.R., da Silva, S.S., 2020. Comparative study of cellulosic sugars production from sugarcane bagasse after dilute nitric acid, dilute sodium hydroxide and sequential nitric acid-sodium hydroxide pretreatment. Biomass Conv. Bioref. 10 (4), 813–822.

Auxenfans, T., Crônier, D., Chabbert, B., Paës, G., 2017. Understanding the structural and chemical changes of plant biomass following steam explosion pretreatment. Biotechnol. Biofuels 10 (1), 1–16.

Balan, V., Bals, B., Chundawat, S.P., Marshall, D., Dale, B.E., 2009. Lignocellulosic biomass pretreatment using AFEX. Biofuels, 61–77.

Baruah, J., Nath, B.K., Sharma, R., Kumar, S., Deka, R.C., Baruah, D.C., Kalita, E., 2018. Recent trends in the pretreatment of lignocellulosic biomass for value-added products. Front. Energy Res. 6, 141.

Behera, S., Arora, R., Nandhagopal, N., Kumar, S., 2014. Importance of chemical pretreatment for bioconversion of lignocellulosic biomass. Renew. Sustain. Energy Rev. 36, 91–106.

Bensah, E.C., Mensah, M., 2013. Chemical pretreatment methods for the production of cellulosic ethanol: technologies and innovations. Int. J. Chem. Eng. 2013, 1–22.

Bhatt, S.M., Shilpa, 2014. Bioethanol production from economical agro waste (groundnut shell) in SSF mode. Res. J. Pharm. Biol. Chem. 5, 1210–1218.

Biswas, R., Uellendahl, H., Ahring, B.K., 2015. Wet explosion: a universal and efficient pretreatment process for lignocellulosic biorefineries. Bioenergy Res. 8 (3), 1101–1116.

Brodeur, G., Yau, E., Badal, K., Collier, J., Ramachandran, K.B., Ramakrishnan, S., 2011. Chemical and physicochemical pretreatment of lignocellulosic biomass: a review. Enzyme Res. 2011.

Chen, Y., Knappe, D.R., Barlaz, M.A., 2004. Effect of cellulose/hemicellulose and lignin on the bioavailability of toluene sorbed to waste paper. Environ. Sci. Technol. 38 (13), 3731–3736.

Chen, H., Liu, J., Chang, X., Chen, D., Xue, Y., Liu, P., Lin, H., Han, S., 2017. A review on the pretreatment of lignocellulose for high-value chemicals. Fuel Process. Technol. 160, 196–206.

Chiaramonti, D., Prussi, M., Ferrero, S., Oriani, L., Ottonello, P., Torre, P., Cherchi, F., 2012. Review of pretreatment processes for lignocellulosic ethanol production, and development of an innovative method. Biomass Bioenergy 46, 25–35.

Cooper, J., Kavanagh, J., Razmjou, A., Chen, V., Leslie, G., 2020. Treatment and resource recovery options for first and second generation bioethanol spentwash–a review. Chemosphere 241, 124975.

Dale, B.E., 1999. Biobased industrial products: bioprocess engineering when cost reality counts. Biotechnol. Prog. 15 (5), 775–776.

Duque, A., Manzanares, P., Ballesteros, I., Ballesteros, M., 2016. Steam explosion as lignocellulosic biomass pretreatment. In: Biomass Fractionation Technologies for a Lignocellulosic Feedstock Based Biorefinery, pp. 349–368.

Fu, S., Hu, J., Liu, H., 2014. Inhibitory effects of biomass degradation products on ethanol fermentation and a strategy to overcome them. Bioresources 9 (3), 4323–4335.

Galbe, M., Wallberg, O., 2019. Pretreatment for biorefineries: a review of common methods for efficient utilisation of lignocellulosic materials. Biotechnol. Biofuels 12 (1), 1–26.

Gupta, R., Aswal, V.K., Saini, J.K., 2020. Sequential dilute acid and alkali deconstruction of sugarcane bagasse for improved hydrolysis: insight from small angle neutron scattering (SANS). Renew. Energy 147, 2091–2101.

Halder, P., Kundu, S., Patel, S., Setiawan, A., Atkin, R., Parthasarthy, R., Shah, K., 2019. Progress on the pre-treatment of lignocellulosic biomass employing ionic liquids. Renew. Sustain. Energy Rev. 105, 268–292.

Harmsen, P.F.H., Huijgen, W., Bermudez, L., Bakker, R., 2010. Literature Review of Physical and Chemical Pretreatment Processes for Lignocellulosic Biomass.

Hassan, S.S., Williams, G.A., Jaiswal, A.K., 2018. Emerging technologies for the pretreatment of lignocellulosic biomass. Bioresour. Technol. 262, 310–318.

Israel, A.U., Ogali, R.E., Akaranta, O., Obot, I.B., 2011. Extraction and characterization of coconut (Cocos nucifera L.) coir dust. J. Sci. Technol. 33 (6).

Jędrzejczyk, M., Soszka, E., Czapnik, M., Ruppert, A.M., Grams, J., 2019. Physical and chemical pretreatment of lignocellulosic biomass. In: Second and Third Generation of Feedstocks, pp. 143–196.

Jönsson, L.J., Palmqvist, E., Nilvebrant, N.O., Hahn-Hägerdal, B., 1998. Detoxification of wood hydrolysates with laccase and peroxidase from the white-rot fungus Trametes versicolor. Appl. Microbiol. Biotechnol. 49 (6), 691–697.

Kapoor, M., Raj, T., Vijayaraj, M., Chopra, A., Gupta, R.P., Tuli, D.K., Kumar, R., 2015. Structural features of dilute acid, steam exploded, and alkali pretreated mustard stalk and their impact on enzymatic hydrolysis. Carbohydr. Polym. 124, 265–273.

Kim, Y., Mosier, N.S., Ladisch, M.R., 2009. Enzymatic digestion of liquid hot water pretreated hybrid poplar. Biotechnol. Prog. 25 (2), 340–348.

Kumar, A.K., Sharma, S., 2017. Recent updates on different methods of pretreatment of lignocellulosic feedstocks: a review. Bioresour. Bioprocess 4 (1), 1–19.

Kumar, R., Gaur, R., Raj, T., Kakkar, M.K., Satlewal, A., Gupta, R.P., Tuli, D.K., 2019. Method of Pretreatment for Enhanced Enzymatic Hydrolysis. U.S. Patent No. 10,487,347.

Lali, A.M., Odaneth, A.A., Victoria, J.J., Choudhari, V.G., Wadekar, P.C., Patil, M.L., Huang, X., 2019. Process for Production of Pure Glucose From Cellulose. U.S. Patent No. 10,465,257. Washington, DC: U.S. Patent and Trademark Office.

Li, X., Sun, C., Zhou, B., He, Y., 2015. Determination of hemicellulose, cellulose and lignin in moso bamboo by near infrared spectroscopy. Sci. Rep. 5 (1), 1–11.

Machado, G., Leon, S., Santos, F., Lourega, R., Dullius, J., Mollmann, M.E., Eichler, P., 2016. Literature review on furfural production from lignocellulosic biomass. Nat. Resour 7 (3), 115–129.

Malherbe, S., Cloete, T.E., 2002. Lignocellulose biodegradation: fundamentals and applications. Rev. Environ. Sci. Biotechnol. 1 (2), 105–114.

Nanda, S., Mohanty, P., Pant, K.K., Naik, S., Kozinski, J.A., Dalai, A.K., 2013. Characterization of north American lignocellulosic biomass and biochars in terms of their candidacy for alternate renewable fuels. Bioenergy Res. 6 (2), 663–677.

Palonen, H., Thomsen, A.B., Tenkanen, M., Schmidt, A.S., Viikari, L., 2004. Evaluation of wet oxidation pretreatment for enzymatic hydrolysis of softwood. Appl. Biochem. Biotechnol. 117 (1), 1–17.

Pandey, A., Soccol, C.R., Nigam, P., Soccol, V.T., 2000. Biotechnological potential of agro-industrial residues. I: sugarcane bagasse. Bioresour. Technol. 74 (1), 69–80.

Prasad, S., Singh, A., Joshi, H.C., 2007. Ethanol as an alternative fuel from agricultural, industrial and urban residues. Resour. Conserv. Recycl. 50 (1), 1–39.

Raveendran, K., Ganesh, A., Khilar, K.C., 1995. Influence of mineral matter on biomass pyrolysis characteristics. Fuel 74 (12), 1812–1822.

Reddy, K.O., Maheswari, C.U., Dhlamini, M.S., Mothudi, B.M., Kommula, V.P., Zhang, J., Zhang, J., Rajulu, A.V., 2018. Extraction and characterization of cellulose single fibers from native african napier grass. Carbohydr. Polym. 188, 85–91.

Rosero-Henao, J.C., Bueno, B.E., de Souza, R., Ribeiro, R., Lopes de Oliveira, A., Gomide, C.A., Tommaso, G., 2019. Potential benefits of near critical and supercritical pre-treatment of lignocellulosic biomass toward anaerobic digestion. Waste Manag. Res. 37 (1), 74–82.

Rowell, R.M., 1992. Opportunities for Lignocellulosic Materials and Composites.

Sánchez, C., 2009. Lignocellulosic residues: biodegradation and bioconversion by fungi. Biotechnol. Adv. 27 (2), 185–194.

Sarkar, N., Ghosh, S.K., Bannerjee, S., Aikat, K., 2012. Bioethanol production from agricultural wastes: an overview. Renew. Energy 37 (1), 19–27.

Satlewal, A., Tuli, D.K., Kakkar, M.K., Gupta, R.P., Ravindra, K., Ruchi, G., Tirath, R., 2017. Gupta, R. P. Preliminary Treatment Method for Improved Enzymatic Hydrolysis. Patent No. BR102016028978A2.

Scholl, A.L., Menegol, D., Pitarelo, A.P., Fontana, R.C., Zandoná Filho, A., Ramos, L.P., Dillon, A.J.P., Camassola, M., 2015. Elephant grass pretreated by steam explosion for inducing secretion of cellulases and xylanases by *Penicillium echinulatum* S1M29 solid-state cultivation. Ind. Crop Prod. 77, 97–107.

Serna, L.D., Alzate, C.O., Alzate, C.C., 2016. Supercritical fluids as a green technology for the pretreatment of lignocellulosic biomass. Bioresour. Technol. 199, 113–120.

Sharma, H.K., Xu, C., Qin, W., 2019. Biological pretreatment of lignocellulosic biomass for biofuels and bioproducts: an overview. Waste Biomass Valoriz. 10 (2), 235–251.

Tutt, M., Olt, J., 2011. Suitability of various plant species for bioethanol production. Agron. Res. 9 (1), 261–267.

Ummalyma, S.B., Supriya, R.D., Sindhu, R., Binod, P., Nair, R.B., Pandey, A., Gnansounou, E., 2019. Biological pretreatment of lignocellulosic biomass—current trends and future perspectives. In: Second and Third Generation of Feedstocks. Elsevier, Netherland, pp. 197–212.

Usmani, Z., Sharma, M., Gupta, P., Karpichev, Y., Gathergood, N., Bhat, R., Gupta, V.K., 2020. Ionic liquid based pretreatment of lignocellulosic biomass for enhanced bioconversion. Bioresour. Technol. 304, 123003.

Wang, Q., Tian, D., Hu, J., Shen, F., Yang, G., Zhang, Y., Deng, S., Zhang, J., Zeng, Y., Hu, Y., 2018. Fates of hemicellulose, lignin and cellulose in concentrated phosphoric acid with hydrogen peroxide (PHP) pretreatment. RSC Adv. 8 (23), 12714–12723.

Xiao, Z., Zhang, X., Gregg, D.J., Saddler, J.N., 2004. Effects of sugar inhibition on cellulases and β-glucosidase during enzymatic hydrolysis of softwood substrates. In: Proceedings of the Twenty-Fifth Symposium on Biotechnology for Fuels and Chemicals Held May 4–7, 2003, in Breckenridge, CO. Humana Press, Totowa, NJ, pp. 1115–1126.

Zabed, H.M., Akter, S., Yun, J., Zhang, G., Awad, F.N., Qi, X., Sahu, J.N., 2019. Recent advances in biological pretreatment of microalgae and lignocellulosic biomass for biofuel production. Renew. Sustain. Energy Rev. 105, 105–128.

Zhang, X., Zhu, J., Sun, L., Yuan, Q., Cheng, G., Argyropoulos, D.S., 2019. Extraction and characterization of lignin from corncob residue after acid-catalyzed steam explosion pretreatment. Ind. Crop Prod. 133, 241–249.

Zheng, J., Rehmann, L., 2014. Extrusion pretreatment of lignocellulosic biomass: a review. Int. J. Mol. Sci. 15 (10), 18967–18984.

Recent developments in cellulolytic enzymes for ethanol production

Jitendra Kumar Saini

Department of Microbiology, Central University of Haryana, Mahendergarh, Haryana, India

1 Introduction

Globally, the search for an environmentally benign alternatives to fossil fuels are being looked upon and a priority research area during the recent times. Biofuels are products of biomass usually in the form of bioalcohols, biodiesel, biogas, and other chemicals and among them bioethanol production from lignocellulosic biomass is a promising technology (Saini et al., 2015). Bioconversion of the pretreated or untreated lignocellulosic materials requires the application of cellulases, for carrying out its hydrolysis to glucose and then its subsequent fermentation to ethanol. Lignocellulose-to-bioethanol conversion by cellulases and utilizing the sugar released on hydrolysis is foreseen as its largest application and the demands for cellulases are expected to increase many fold as more biofuel plants are set up globally (Hemansi et al., 2019).

Bioethanol is the an extensively used biofuel globally, due to its better fuel properties in comparison to gasoline, such as lesser carbon dioxide emissions, release of air pollutants in comparison to gasoline combustion, improved oxidation of hydrocarbons, and higher octane number (Saini et al., 2015). Based upon substrate types used, bioethanol can be categorized as first-generation (1G) bio-ethanol if it is produced from food crops or as second-generation bio-ethanol (2G) produced from nonfood resources, such as lignocellulose-rich agricultural residues. Due to several drawbacks, including food versus fuel nexus, 2G ethanol is more favored. It uses complex lignocellulosic biomass as substrate which provides abundant quantity of fermentable monosaccharides, such as glucose and xylose. Lignocellulosic biomass is the most abundantly present organic mass in biosphere. For bioethanol production, lignocellulosic substrate has to be first treated with various physical and chemical methods. Lignin acts as a major physical barrier and plays a negative role in converting biomass to biofuel, affecting pretreatment as well as enzymatic hydrolysis process due to its resistant nature and cross-linked networks. Thus, lignin removal is the main target during pretreatment involving alkaline, alcohol-based organosolv, ionic liquid pretreatment, and biological methods such as enzymes and fungi (Yoo et al., 2020). The next step involves the release of sugars (majorly hexose and some pentoses) from the pretreated biomass, with the help of cellulose (and hemicellulose) hydrolyzing enzymes. This is followed by the fermentation of the sugars to produce ethanol. The lignocellulolytic enzymes (LCEs) involved in saccharification of biomass include cellulases, hemicellulases, and accessory enzyme including oxidative enzymes. Ligninolytic enzymes are produced by certain fungi and bacterial strains in large amounts. The efficient and beneficial characteristics of these strains are highly preferable for biocatalyst development for biofuel production, biopulping, and textile industries, and platform chemicals (Gaur, 2018). The on-site enzyme production and tailor-made enzyme cocktail formulation can be used to hydrolyze lignocellulosic biomass effectively. However, the cost-effective production of 2G bioethanol is majorly hampered due to the several factors. One of the major bottlenecks to cost-effective production of cellulosic bioethanol is the enzyme cost. These added enzymes are of high-cost commodity due to the production cost, including nutrient costs, operational and capital cost, formulation, transport cost, and enzyme activity. Moreover, the performance of the enzyme is another limitation that usually differs based on the lignocellulosic substrates (Dragone et al., 2020). Many industries are nowadays producing robust and superior LCEs and their costs have been significantly reduced. However, the enzyme cost still contributes a significant portion of the total costs of biomass to finished bioethanol product. Therefore, the present chapter discusses concepts, challenges, and prospects of the development of efficient LC, including their commercialization aspects.

2 Lignocellulosic biomass and its composition

LCB is a highly reliable and sustainable feedstock for biofuels and bioproduct synthesis. The lignocellulosic residues comprise an abundance of complex carbon components derived from plant sources after harvesting or processing. Based on the source of origin, the LCB feedstocks can be categorized into agricultural residues, forestry residues, energy crops (Miscanthus, Porospis, etc.), **industrial and municipal wastes,** and **algal biomass** (Raud et al., 2019). Agro-residues include rice and wheat straw, sugarcane bagasse, corn stover, stalks, sweet sorghum bagasse sugarcane, cassava, and plant seeds, etc. These plant-based feedstocks are sustainable and have the potential to be generated under harsh conditions like saline, drought, and hot climates (Dar et al., 2018). Agricultural biomass produced as a waste from the agricultural practice is a cheaper source of LCB which can also be used as renewable feedstock for sustainable production bioethanol (Adewuyi, 2020). Industrial and municipal waste can also be used for the generation of renewable fuels and chemicals. Municipal wastes, including animal waste, rotten vegetables and fruits, and tubers, have been used for bioethanol production (Adewuyi, 2020). Paper mill sludge from paper and milling industries can also be used as renewable feedstocks for biofuel production using feasible biological conversion approaches, as it contains higher cellulose and hemicelluloses, with lower content of lignin (Tawalbeh et al., 2021). Algal biomass, including microalgae, seaweeds, and natural algae is a large-scale and renewable feedstock used for biofuel production. Macro-algae, such as *Ulva* sp., predominantly known as seaweeds, are rich in sugars (at least 50%), and can be utilized as a bioethanol feedstock (Nagarajan et al., 2020; Margareta et al., 2020). Microalgae produces several different kinds of renewable biofuel, such as (a) anaerobic digestion of the algal biomass produces methane, (b) biodiesel from micro algal oil, and (c) biohydrogen through photobiological mechanism (Rajkumar et al., 2014).

Plant-based lignocellulosics are the main feedstock for bioethanol production due to their high abundance and cheaper availability. Plant cell walls consist of: (i) middle lamella, (ii) primary cell wall (iii) secondary cell wall structures. In general, the plant cell wall composition varies considerably among monocots, dicots, softwood, and hardwood, and their lignin content also varies considerably (Vogel, 2008; Rytioja et al., 2014). The major plant polysaccharides are cellulose, hemicelluloses, lignin, pectins, proteins, and other extractives (Yoo et al., 2020). In general, LCB has a cellulose content of 30%–50%, hemicellulose content of 20%–40% and lignin is found at 15%–40% of the dry weight of LCB. The cellulose component of LCB is a polymer of glucose, hemicellulose is a heterogeneous component made of both pentoses and hexose sugars, and lignin is a complex macromolecule composed of monomeric units of *para*-coumaryl, coniferyl, and sinapyl alcohols that are cross-linked via stable covalent bonds between the polysaccharides and lignin polymer. Being a polymer of monomeric glucose units linked by β-1-4 glycosidic bonds, cellulose is linear crystalline structure and contains cellobiose, a disaccharide of β-1-4-linked glucose units. Hemicellulose is located between the cellulose and lignin. It is a complex polysaccharide mainly composed of arabinoxylan with branched hetero polymers of D-sugars: galactose, glucose, mannose, and xylan. Xylan and lignin are covalently bound together by cinnamic acids. The lignin-based cross-linked matrix complicates the degradation of biomass using microorganisms (Dragone et al., 2020). Lignin forms the protective layer for the hemicellulose and cellulose matrix. Lignin is the most difficult to degrade among the LCB constituents, as it is a complex polymer of phenolic, amorphous, and hydrophobic nature due to its varied precursor components (Pereira et al., 2015).

3 Lignocellulolytic enzymes (LCEs)

Biomass hydrolyzing enzymes require synergistic action of many enzymes, which are functionally different and belong to different classes. Since, the complexity of biomass varies enormously, the exact amount of each type of enzyme for its efficient hydrolysis may also vary considerably. The depolymerization of LCB needs the synchronized functioning of many enzymes, which may be oxidizing, hydrolyzing, and nonhydrolyzing in nature (Sista Kameshwar and Qin, 2018) and may be classified, as per CAZy, under: (i) glycoside hydrolase-family (GHs), (ii) glycosyl-transferase family (GTs), (iii) polysaccharide lyase family (PLs), (iv) carbohydrate esterase family (CEs), (v) carbohydrate binding modules (CBMs), and (vi) enzyme families possessing auxiliary activities (AAs), based on structural or sequence similarities (Lombard et al., 2013).

3.1 Cellulases

Efficient degradation of cellulose requires three types of enzymes: cellobiohydrolases (CBHs) to cleave the cellobiose; endo-β-glucanases, to cleave the glycosidic bonds; and β-glucosidases, for hydrolysis of the cellobiose and cellodextrin fractions Hydrolysis of the xylan component of hemicellulose requires endoxylanases and accessory enzymes like β-xylosidases, α-L-arabinofuranosidases, 4-O-methyl-D-glucuronidases, and acetyl xylan esterases (Cintra et al., 2020). Endoglucanase, 1,4-β-D-glucan-4-glucanohydrolase, or carboxymethylcellulases (CMCases) (EC 3.2.1.4) cut randomly at the β-1,4-bonds of cellulose chains, generating new ends and hydrolyze the amorphous cellulose for generation of more accessible free chain-ends as substrate for CBH. Exoglucanases, 1,4-β-D-glucan glucohydrolases (EC. 3.2.1.74), or CBHs acts processively on the reducing/nonreducing cellulose chain ends, releasing cellobiose, or glucose as major products. CBHs acts on cellulose microfibril ends, cleaving the cellobiose from it in a processive manner. For its processivity, active sites similar to tunnel are necessary, so that the substrate chain is received only through the ends (Yeoman et al., 2010; Kurasin and Väljamäe, 2011). β-glucosidase, having EC number 3.2.1.21, (also known as "cellobiase"), are necessary for complete cellulose hydrolysis, by lysing cellobiose through glucose removal from the sugar's nonreducing terminal. The β-glucosidases (BGL) hydrolyze β-glucosidic in disaccharides, oligosaccharides, or conjugated glucosides. Based on substrate specificity, β-glucosidases are divided into three groups: aryl-β-glucosidases, cellobiases, and broad-specificity β-glucosidases. Aryl-β-glucosidases hydrolyse aryl-β-glucoside-containing substrates. In contrast, cello-oligosaccharides and cellobiose are only the substrates hydrolyzed by the enzyme cellobiase. Some β-glucosidases have broad-specificity, i.e., acting on either of the substrate types. Mostly, microbial BGL belong to this category of enzymes (Bhatia et al., 2002). β-glucosidase is the rate-limiting enzyme because it hydrolyzes the final step of lignocellulose breakdown in which cellobiose and short cellodextrins are converted into glucose.

3.2 Hemicellulases

Based on the mode of action on the substrate, hemicellulose is majorly hydrolyzed by endo-1,4-β-xylanase or endoxylanases and exo-1,4-β-xylosidase or β-xylosidase or xylobiase, having EC nos. 3.2.1.8 and 3.2.1.37, respectively. The efficient hydrolysis of xylan demands the use of endo-β-1,4-xylanases having random cleaving action on xylan to release a diverse range of products, including xylobiose, xylotriose, xylotetraose, and xylooligomeric products (Collins et al., 2005). β-xylosidase hydrolyzes the nonreducing ends of xylose chains, xylobiose, and xylooligomers to release xylose, without hydrolytic action on xylan (Yan et al., 2008; Knob et al., 2010; Huy et al., 2015). Several supplementary enzymes, including α-L-arabinofuranosidase, α-D-glucuronidase, α-D-galactosidase, acetyl xylan esterase, and ferulic acid esterase, participate in xylan biomass hydrolysis.

β-Mannanases or endo-β-1,4-mannanase hydrolyze mannan linkages by cleaving β-1,4 bonds and producing new reducing and nonreducing ends; whereas, β-mannosidase or exo-β-1,4-mannosidases hydrolyse terminal, nonreducing β-D-mannose residues (Andlar et al., 2018). Additionally, removal of side-chain acetyl substituents attached at various points on the mannan structure requires the action of acetyl mannan esterase (AME). α-L-arabinofuranosidases (ARA) removes the arabinose residues from its substrates (arabinan, arabinoxylan or pectin), thereby, debranching and degrading xylan and disrupting the lignin-carbohydrate complex. α-glucuronidases removes glucuronic acid or its derivative 4-O-methylglucuronic acid from xylan, and is synergistic to endoxylanases. Whereas, α-D-galactosidases cleave terminal α-1,6-linked galactose residues of galactomannans, galactoglucomannans, and oligosaccharides (Lei et al., 2016; Ademark et al., 2001).

Carbohydrate esterases act synergistically with hemicellulases and the accessory enzymes are acetyl xylan esterase (AX), feruloyl esterase (FAE), *para*-coumaroyl esterase (CAE), exo acting α-L-arabinofuranosidase, endo acting arabinofuranosidase, xylan α-1,2-glucuronosidase, and α-glucuronidase (Andlar et al., 2018; Malgas et al., 2019; Zhang et al., 2011). AXEs release acetic acid from acetylated polysaccharides by hydrolysis of ester bonds which increases the accessibility of the chain toward GH enzymes. FAEs cleave ester linkages of hydroxyl-cinnamate and acetyl xylan, releasing derivatives of phenolic acids (ferulic, FA, or *para*-coumaric acids, CA) (Wong et al., 2013). Glucuronoyl esterases break the ester-linkages of lignin-aliphatic alcohols with 4-O-methyl-D-glucuronic acid derivatives of glucuronoxylans (Baath et al., 2016). FAE and CAE cleave linkages of hydroxycinnamic acids with sugars, thereby releasing FA and CA, respectively. α-Glucuronidase catalyzes the hydrolysis of xylan into glucuronic acid or 4-O-methyl-glucuronic acid. The action of esterases can also enhance the accessibility of the cellulose fibers and be used to produce bioactive chemicals and biofuels (Polizeli et al., 2005). Pectinases cause the depolymerization and de-esterification of pectin components of cell walls.

3.3 Lignin depolymerizing enzymes

Lignin depolymerizing enzymes are termed "ligninases", and are categorized into CAZy auxiliary activities (AA) family, AA1-AA3, which can be divided into two main families, phenol oxidases (laccases) and peroxidases (Dashtban et al., 2010). Lignin degradation is achieved by the cooperative actions of fungal and bacterial enzyme systems. Degradation of native lignin is carried out by lignin depolymerizing enzymes such as lignin peroxidase (LiP) manganese peroxidase (MnP) (EC nos. 1.11.1.14 and 1.11.1.13, respectively), and laccase (EC 1.10.3.2) secreted by white-rot fungi (Gold and Alic, 1993). Recombinant DypB (dye decolorizing) peroxidases have a significant role in lignin degradation, and oxidize polymeric lignin and β-aryl ether lignin compounds. This was the first recombinant bacterial lignin peroxidase characterized (Ahmad et al., 2011). Versatile peroxidases possess the characteristics of both LiP and MnP, carrying out the oxidation of both phenol containing as well as nonphenolic substrates. Aryl alcohol oxidases oxidizes many aromatic ring containing primary alcohol. Fungal aryl-alcohol dehydrogenases and quinone reductases tyrosinases, and catechol oxidases also assist in degrading lignin (Andlar et al., 2018).

Bacteria produce ligninolytic multi-copper oxidase (MCO) superfamily enzymes including laccases, ascorbate oxidases ferroxidases, bilirubin oxidases, and other enzyme subfamilies. A novel lignin-oxidizing manganese superoxide dismutase, solubilizing organosolv and kraft lignin to generate a mixture of polymeric and monocyclic aromatic products, has been reported in *Sphingobacterium* sp. (Rashid et al., 2015). Glutathione-dependent β-etherase cata-lyzes the cleavage of the β-aryl ether linkage by targeting glutathione at the β position of an oxidized aryl unit containing a ketone group (Marinović et al., 2018).

3.4 Accessory and nonhydrolytic proteins involved in cellulose hydrolysis

Apart from major cell-wall-degrading enzymes (cellulases, hemicellulases, and ligninases), efficient conversion of plant biomass to commodity chemicals requires other en-zymes, which help in achieving the complete hydrolysis of the biomass. Redox enzymes carry out the auxiliary activities (AAs), assisting and working simultaneously with other GHs to saccharify LCB efficiently. Cellulose decomposition was thought to be mediated primarily through the hydrolytic action of cellulases. Later, polysaccharide degradation was discovered to be mediated by oxidative reactions catalyzed by CBM33s (chitin-binding proteins in bac-teria) and GH61s (EGs in fungi) (Vaaje-Kolstad et al., 2010). These enzymes are referred to as lytic polysaccharide monooxygenases (LPMOs) and are reclassified as AA families 10 and 9, respectively, in the CAZy database (Levasseur et al., 2013). LPMOs are a new class of oxidative cellulases and their catalytic activities have not yet fully characterized (Østby et al., 2020). LPMOs act on soluble cello-oligosaccharides, hemicelluloses, and starch, in ad-dition to oxidatively cleaving chitin and cellulose. In CAZy database, LPMOs are grouped into six CAZy AA families (AA9, AA10, AA11, AA13, AA14, AA15, and AA16) (Filiatrault-Chastel et al. 2019; Levasseur et al., 2013; El-Gebali et al., 2019). LPMOs act on recalcitrant polysaccharides by oxidatively cleaving β-1,4-glycosidic bonds, either in a monooxygenase reaction using molecular O_2 and a reductant (Vaaje-Kolstad et al., 2010) or in a peroxygenase reaction using H_2O_2 (Bissaro et al., 2017, 2020). LPMOs make the cellulose more accessible for hydrolytic enzymes (Beckham et al., 2014).

The nonhydrolytic proteins, termed "swollenin (SWO1)" are similar to the expansins, and have no catalytic activity. These disrupt cellulose microfibers by breaking H-bonds (Saloheimo et al., 2002). SWO1 enhances endoxylanase in synergistic manner, thereafter pro-moting endoglucanase or cellobiohydrolase activities by rendering the xylan portion of LCB more accessible to hemicellulases during saccharification, thereby indirectly promoting cel-lulases action (Gourlay et al., 2013). Cellulose-induced protein-1 and -2 (termed "CIP1 and CIP2") effectively cleave LCB (Foreman et al., 2003; Banerjee et al., 2010). CIP1 has synergistic activity with swollenins, while CIP2 cleaves hemicellulose-lignin cross-links (Jacobson et al., 2013). CIP2 is a glucuronoyl esterase of the carbohydrate esterase family 15. The glucuronoyl esterase activity might play a significant role in plant biomass degradation by separating lig-nin and hemicellulose through cleavage of ester linkage of 4-O-methyl-D-glucuronic acid moi-eties of glucuronoxylans with aromatic alcohols of lignin (Pokkuluri et al., 2011).

3.5 Cellulase mechanism of action and synergism

Cellulose is converted to glucose monomers through action of exoglucanases and endoglucanases in the initial step and β-glucosidase in the next step. Initial depolymerization of cellulose during saccharification released cellobiose, which in the final stage is cleaved to cellobiose. Endo- and exo-β-glucanases as well as exo-β-glucanases that act from the reducing and nonreducing ends have been found to be synergistic in their action. Four different types of synergism exist among these enzymes as proposed by Teeri (1997): (i) endo–exo type synerism among endo- and exoglucanases, (ii) exo–exo type among exoglucanases acting on reducing and the nonreducing sugar ends, (iii) among exoglucanases and BGL, and (iv) intramolecular synergism among CBMs and catalytic modules. CBMs aid in disrupting the cellulose and at the same time also assist in cellulases binding with cellulose (Zhang and Zhang, 2013).

4 Cellulase producing microorganisms

Cellulolytic microorganisms have developed two major cellulase strategies: discrete noncomplexed cellulases and complexed cellulases (Lynd et al., 2002; Zhang and Lynd, 2004). Many cellulose-degrading microbes function by producing cellulases having carbohydrate-binding module (CBM) connected to linker. The linker is a flexible peptide which on the other side connects to the catalytic module (CM). The CBM is located in the N- or the C-terminal end of the CM. In contrast, most anaerobic microorganisms produce large (>1 million Da molecular mass) multienzyme complexes adhered to surface of microbial cells, known as "cellulosomes" (Bayer et al., 2004). Few anaerobic bacteria are able to secrete both cellulosomes as well as free cellulases.

4.1 Fungi

The mold *Trichoderma reesei* was discovered on the Solomon Islands owing to its growth on and degradation of cellulosic cloths. It was the first-known cellulase-producing microorganism. *T. reesei* enzymes saccharify lignocellulose of plant biomass for producing bio-based fuels and chemicals (Bischof et al., 2016). The mutant *Trichoderma* sp. QM6a was developed to degrade native crystalline cellulose more efficiently than the parent strain. This strain was regarded as the *T. reesei* reference strain, and most of the mutants used in industry today have been derived from this strain. Subsequently, significant enhancements (20-fold) in the extracellular protein production capability of parent strain QM6a was achieved through random mutagenesis approach, which practically improved industrial applications of *T. reesei* cellulases (Bischof et al., 2016). By the end of the 1990s, *Hypocrea jecorina*, the sexual form of *T. reesei*, was discovered. Since then, numerous cellulolytic microorganisms have been discovered, and their cellulases have been characterized. In general, fungal EG's CM may or may not be associated with CBM. Contrastingly, bacterial EG has been found to be associated with many CMs, CBMs and other modules, whose functions are yet not defined.

One of the major obstacles in efficient biomass hydrolysis by cellulases of *T. reesei* is the lack of BGL activities, despite having excellent endoglucanase content. Due to this, the sugar released during hydrolysis inhibit the overall saccharification by feedback inhibition of the cellulolytic enzymes. Therefore, to alleviate this challenge, a number of researchers have searched for the alternative fungal sources of the cellulolytic enzymes. Aspergillus and Penicillium are some of the efficient fungal genera, which have been reported to produce higher quantities of cellulases having better efficiencies than those produced by *T. reesei*. In comparison to large number of reports of synergistic cellulolytic blends of *Aspergillus* sp. and *Trichoderma* sp. through mixed/co-culture or fermentation broth addition, fewer studies have used *Penicillium* spp., which are now established for their excellent cellulolytic potential (Saini et al., 2016). Nowadays, increasing a greater number of commercial enzymes are being made using cellulases from recombinant *Aspergillus* sp. and *Penicillium* sp.

4.2 Bacterial cellulases

Cellulases of bacteria differ from their fungal counterparts in their rapid production due to better growth rate of bacteria, their diversity, and ease in changing their genetic makeup (Chandel et al., 2010). Many bacteria produce endo-glucanases and carry out hydrolysis of noncrystalline cellulose, e.g., carboxymethylcellulose, but can be limited in the efficient hydrolysis of crystalline cellulose (Wilson, 2011). A few bacterial synthesize EGs are capable of degrading AVICEL (microcrystalline type of cellulose) (Han et al., 1995). Furthermore, thermo-tolerant bacteria-derived cellulases are not feedback inhibited due to better BGL secretion, thereby, their application enhances saccharification performance (Bhalla et al., 2013). No doubt, efficient saccharification as well as better enzyme synergy demands best pretreatment of the biomass.

4.3 Cellulosomes

Cellulosomes are nanostructured multicomponent extensive enzyme systems that can overcome the recalcitrant and complex nature of lignocellulosic biomass for efficient degradation. A cellulosome complex was first discovered from the thermophilic anaerobe *Clostridium thermocellum* (Bayer et al., 1983; Lamed et al., 1983). Later, several other microbial species, mainly anaerobes, were reported to produce such complexes (Bayer et al., 2004). Cellulosomes can overcome the energy limitation for enzyme production by the anaerobic bacteria (Artzi et al., 2017). These complexes also facilitate the separation of individual cellulose microfibrils from larger particles, thereby, increasing the cellulose accessibility to enzyme active sites (Ding et al., 2012; Resch et al., 2013). Cellulosome activity have been identified in several bacterial species from a variety of anaerobic and extreme ecosystems.

5 Cellulase production

Despite continuous research efforts in lignocellulose conversion by cellulases, cellulases production is still in its infancy and needs more precise efforts. The production of cellulase in low titers is the major concern for large-scale production and industrial applications, which needs multiple solutions, including bioprocess development. Microbial cellulases, can be produced either by submerged fermentation (SmF) and solid-state fermentation (SSF) configurations. Both these technologies find applications in cellulase production and have their own advantages and disadvantages, differing significantly in the choice of microorganisms, suitable combination of production parameters, and bioreactor design and operation (Hemansi et al., 2019).

5.1 Submerged fermentation (SmF)

Submerged fermentation is carried out in the presence of excess free water. Generally, small-scale production employs shake-flask conditions and large-scale production employs Stirred-tank type bioreactors (STRs). Due to better monitoring and ease of handling, almost most of the industries employ this methodology for large-scale enzyme production. Along with fungi, bacteria (including filamentous actinomycetes) are also employed for cellulase production, but bacterial cellulase production is commercially nonfeasible due to low activities (Singhania et al., 2010). The major commercial cellulase is produced by filamentous fungi *Trichoderma* (Neagu et al., 2012; Oszust et al., 2017) and *Aspergillus* (Gomathi et al., 2012; Mrudula and Murugammal, 2011; Reddy et al., 2015a, b; Hemansi et al., 2018). The production of cellulase can be controlled by optimal performance of process variables directly or indirectly affecting enzyme synthesis, such as substrate type and amount, initial pH, nutrients in media, inducing compounds, temperature and aeration level, etc. Thus, submerged fermentation requires prior knowledge of microbial physiology and proper monitoring/control of microbial enzyme production (Singhania et al., 2010).

The medium formulation and optimization of process parameters are often necessary for attaining optimum enzyme levels using either native or improved microbial strain (Sajith et al., 2016). For better productivity, various synthetic or natural carbon sources such as Avicel (Legodi et al., 2019) and Solka Floc, wheat bran (Reddy et al., 2015a, b; Sharma et al., 2016), wheat straw (Khokhar et al., 2014; Toscano-Palomar et al., 2015), corn cob, rice straw (Mrudula and Murugammal, 2011; Reddy et al., 2015a, b), paper industry waste and more lignocellulosic waste are used in SmF (Reddy et al., 2015a, b). Smf is often affected by low productivity, which is primarily caused by carbon catabolite repression (CCR) or lack of inducing compounds. These limitations can be met by fed-batch or continuous production to prevent accumulation of sugars as reported by (Li et al., 2018) or by developing mutants against catabolite repression (Zhao et al., 2019) or by using soluble inducers which are mostly pure cellulose preparations, but use of cellulosic biomass is in majority as it does not promote CCR (Singhania et al., 2010). Optimum growth and enzyme production varies according to the microorganism. Moreover, numerous environmental, physiological, and nutritional factors affect cellulase production.

5.2 Solid state fermentation (SSF)

One of the main interests of the society in the last decades is the valorization of waste and for SSF refers to that mode of fermentation, in which microorganisms are cultivated in complete or near absence of free moisture. Such conditions closely resemble environmental conditions of the nature to which microbes are already adapted, thereby showing maximal enzyme production (Abdul Manan and Webb, 2017). The process involves use of solid organic substrates that include agricultural and forestry residues, which not only provide a solid support by staying inert and insoluble for nutrients absorption and biomass growth, but they are also a source of carbon and nutrients (Cerda et al., 2019; Krishna, 2005; Soccol et al., 2017).

Cellulolytic enzymes are the most prominent product being produced through SSF (Cerda et al., 2019). Fungal cellulase production via SSF generates high cellulase titers (Singhania et al., 2010). Several species of filamentous fungi, such as *Trichoderma* sp., *Aspergillus* sp., *Penicillium* sp. etc., have been used for cellulase production under SSF from agro-wastes (Hemansi et al., 2019). Among them maximum enzyme was reported to be produced (1–100 Units/g substrate) by *T. reesei* using agricultural residues as feedstock (Cerda et al., 2019; Singhania et al., 2010). Cellulose content in the substrates used for cellulase production is a crucial factor as cellulose below 30% leads to a lesser cellulase yield, proving cellulose as inducer in cellulase production (Cerda et al., 2019; Mejias et al., 2018). Cellulase production, when compared between SmF and SSF, showed a 10-fold cost reduction when SSF was employed due to more concentrated enzyme form compensating the recovery costs (Singhania et al., 2010).

The properties of cellulase produced under SSF are also found to be better in terms of stability toward temperature, pH, metal ions, etc., as cellulase produced by *Penicillium citrinum* under SSF was found to have alkali tolerance (Dutta et al., 2008; Singhania et al., 2010). SSF proves to be advantageous to get concentrated enzyme solutions for application purposes of cellulases where general purity is not a concern, and it was also found that the same substrate which is to be used for hydrolysis could be used for enzyme production and hence improves the overall efficiency (Lynd et al., 2002). The advantages SSF offers over other production technique are many such as, higher yield in terms of volumetric production along with high concentration of products, lesser effluents and lower CCR making it sustainable (Singhania et al., 2010).

SSF has one more unique advantage, that is use of mixed cultures that provides us opportunity to study the metabolic synergy in various organisms and also may result in better enzyme production (Krishna, 2005). Higher cellulase activities using mixed cultures are found, but the synergism between them needs to be explored further (Saini et al., 2016).

6 Challenging aspects of cellulolytic enzyme applications for bioethanol production

Biofuel production from lignocellulosic residues seems to be the most common alternatives, especially from an economic point of view (Beig et al., 2021). However, the production of biofuels from lignocellulosic biomass is challenging. It is necessary to increase energy density and reduce reactivity of the biomass for economic production of biofuels, such as bioethanol (Alonso et al., 2012). The conversion of lignocellulose into hexoses or pentoses as fuel precursors for ethanol, has several stages that start with a pretreatment, followed by enzymatic saccharification and, ends with the sugar fermentation (Yang et al., 2020). The pretreatment applied to biomass represents between one-third to one-fifth of the total bioethanol production costs (Axelsson et al., 2012; Beig et al., 2021). Hydrolysis of hemicellulose and cellulose into reducing sugars is a critical point in the conversion process (Alonso et al., 2012). So far, biofuel production from lignocellulosic material has been limited. This is mainly attributed to the hydrolysis stage, where some barriers are present, which affect the viability of the process; one of them is the cost associated with the enzymes in charge of hydrolyzing the biomass (Lin et al., 2019), the low hydrolysis efficiency, and the high production costs (Wang and Lü, 2021). On the other hand, there is a need for low-cost feedstocks that can be effectively digested by hydrolytic enzymes, with low processing costs (Sandesh and Ujwal, 2021).

Hydrolyzing lignocellulose into monosaccharides remains a technical challenge due to the indigestibility of the cellulose structure (Khaire et al., 2021). Pretreatment of lignocellulosic biomass is paramount to improve cellulose accessibility for enzymes to release fermentable sugars at the hydrolysis stage and reduce enzyme usage (Marulanda et al., 2019), so many studies have focused on determining process conditions that allow better yield at a lower cost. Pretreatments applied to lignocellulosic biomass before enzymatic hydrolysis, require high energy demands (Beig et al., 2021). These are applied singly or in combination, to improve enzyme efficacy during hydrolysis (Jamaldheen et al., 2018). The combined pretreatment strategy results in higher rate of biomass saccharification and better sugar yields. The main constraints in developing an efficient pretreatment process are: better energy efficiency without compromising production, least degradation or loss of sugars, least generation of compounds that may act as inhibitors of the enzyme or fermenting microbes, avoiding biomass washing or liquor neutralization to save water, energy and costs, process integration and synchronization of subsequent operations for achieving better process efficiency (Houghton, 2006; Axelsson et al., 2012; Beig et al., 2021). Large-scale enzymatic saccharification is also challenged by the biomass structure, costly enzymes, and poor enzymatic performances (Guo et al., 2018).

In the process of bioethanol production, the cost for LCB saccharification is still extraordinarily high. Many currently available industrial cellulases are not optimal for harsh conditions and lack sufficient enzymes for complete hydrolysis. LCB-dependent industries prefer cellulase cocktails having resilience toward heat, pH changes, changes in pH, ionic strength, and solvents. Additionally, the saccharification process to achieve higher sugar recovery needs to be optimized.

Genetic engineering has played a crucial role in the design of new enzymes or enzyme preparations. At present, improving the heat stability of the enzyme is a primary target for attaining improved enzymes, the temperature being a determining factor in the viability of the plant material. The enzyme production yield and its catalytic efficiency, as well as the reduction of protein production costs and inhibition of the final product, are some of the issues under study (Elleuche et al., 2014; Valdivia et al., 2016).

The production of technical enzymes is affected by Research & Development (R&D) activities and environmental policies and legislation in each country. The Paris Climate Agreement encourages the production of fuels from renewable sources, to reduce greenhouse gases, a situation that has favored the demand for saccharifying enzymes (Dewan, 2014). However, the cost of technical enzymes remains an important factor in the growth of the market.

Enzymatic hydrolysis is of great interest because it could overcome the disadvantages of acid and alkali catalyzed hydrolysis. However, there are still some downsides, such as a slow reaction rate and limited enzymatic accessibility to polysaccharides. Pretreatment is necessary to open the biomass's cell wall structure, which would increase the enzymatic accessibility during enzymatic hydrolysis (Saeed and Saleem, 2018).

An obstacle for large-scale enzymatic saccharification practice includes high enzyme costs and low enzyme activity for practical use (Chen and Fu, 2016). Genetic engineering has been one of the solutions tools for enzyme technology. However, it suffers from various drawbacks, such as posttranslational modifications, inclusion bodies, costly, tedious, time-consuming, and requisite expertise. Immobilization has been the foremost enzyme technology being used due to its simplicity, decrease labor, and cost-efficacy. It leads to physical confinement or localization of enzymes in a specifically defined region of space with retention of their catalytic activities, less sensitivity toward their environment with insistent usability (Dwevedi and Kayastha, 2011).

7 Newer trends in enzymatic saccharification technologies

As mentioned above, the enzymatic saccharification processes have been accompanied by genetic recombination processes of microorganisms to obtain higher enzyme titers and enzyme recovery for reuse in search of an increase in production yields (Guo et al., 2018). Nowadays, research on the production of biofuels continues to make use of enzymatic saccharification processes. The main difficulties derive from the crystallinity and degree of polymerization of cellulose, the accessibility to the substrate surface, and mainly from the presence of lignin. The last one prevents the fibers' swelling and produces nonproductive adsorption of cellulases (Sheng et al., 2021). Novel pretreatments accompany the enzymatic saccharification processes to contribute to the delignification of the lignocellulosic biomass.

The hybrid pretreatment of ultrasound and organic solvents consists of solvents' synergistic action with free radicals' production. These radicals are produced through the sonochemical effect to exert an attack on the biomass components and reduce the cellulose's crystallinity through a rearrangement of molecules through mechanoacoustic development (Lee et al., 2020). Research has been carried out on the use of lytic polysaccharide monooxygenases (LPMOs) in conjunction with cellulase enzymes to improve saccharification yields through the oxidation of substrates' surface, facilitating the access of hydrolytic enzymes (Velasco et al., 2021).

In recent years, the use of deep eutectic solvents (DES) as a pretreatment in enzymatic saccharification has been reported. DES has properties similar to ionic liquids, although they stand out for being simple to synthesize, biodegradable, and have a low cost (Ling et al., 2020). As lignin is an essential component of lignocellulosic materials, its revaluation in saccharification processes could contribute to biofuel production's profitability (Huang et al., 2021). Research has recently been conducted using DES with lignin derivatives. In 2020, the first report of DES prepared with p-hydroxybenzoic acid (derived from lignin) and colin chloride for the pretreatment of woody biomass improved the percentage of delignification enzymatic hydrolysis, also achieving a sustainable process by recycling the DES used (Wang et al., 2020). Similarly, Huang et al. (2021) reported that using a DES pretreatment consisting of choline chloride/guaicol (derived from lignin) with traces of $AlCl_3$ contributed significantly to the degradation of hemicellulose and lignin, resulting in complete enzymatic hydrolysis from wheat Straw.

The use of alkaline hydrogen peroxide has been reported, increasing the enzymatic digestibility of corn stubble thanks to the breaking of the hydrogen bonds of cellulose and hemicellulose and the elimination of lignin, reducing the nonproductive adsorption of cellulases to the biomass (Yang et al., 2021). The use of a novel hybrid pretreatment has recently been reported by Tang et al. (2021). They combined an organic surfactant (humic acid) with dilute sulfuric acid, achieving an increase in the percentage of lignin and hemicellulose removal and the ES of wheat straw, reaching a saccharification percentage of 92.9% (Tang et al., 2021). Similarly, a pretreatment effect based on ozonolysis with a subsequent washing with sulfuric acid under mild conditions before enzymatic saccharification has been evaluated with good results working with cane bagasse (Perrone et al., 2021).

There is currently a trend toward the use of enzymatic saccharification using macroalgal biomass. The use of macroalgal biomass is because, in contrast to the terrestrial lignocellulosic biomass sources, they do not deserve land for agriculture or fertilizers, water, or pesticides. In some cases, they respond better to thermal pretreatments, increasing the percentage of saccharification compared to some terrestrial biomass (Thygesen et al., 2020).

The addition of surface-active additives and proteins to the reaction mixture has become a practical solution to improve the hydrolysis of LCB, by mitigating the cellulase-lignin nonproductive binding. Nonionic surfactants are found particularly effective for stimulating LCB hydrolysis (Eriksson et al., 2002; Kristensen et al., 2007). A role for surfactants is found in pretreatment, enzymatic hydrolysis, and enzyme recycling stages of LCB bioconversion. Surfactants like Tween-80, dodecylbenzene sulfonic acid, and PEG-4000 remove lignin effectively from biomass and reduce the nonproductive binding of enzymes to its surface (Qing et al., 2010). Surfactants can also stabilize the enzyme during hydrolysis by preventing denaturation and can alter the structure of the substrate, thereby making it more accessible to the enzymes (Kaar and Holtzapple, 1998; Helle et al., 1993; Kim et al., 1982). Hence the addition of surfactants significantly enhances the saccharification efficiency (Kumar and Wyman, 2009).

8 Commercialization of cellulolytic enzymes

New enzymatic technologies have made it possible to overcome the problems of converting recalcitrant biomasses or lignocellulosic materials. Commercially available enzymes for biofuel production from different feedstocks can be grouped into cellulases, amylases, β-glucosidases, xylanases, proteases, lipases, keratinases, laccases, lignin peroxidase, and manganese peroxidase (Pereira Scarpa et al., 2019). The applications of these enzymes can vary according to the type of fermentation, such as solid-state (SSF) or submerged (SmF) fermentation (de Castro and de Castro, 2012). Among the saccharifying enzymes that dominate the market are cellulases, due to the wide range of industrial applications, other enzymes such as lipases, catalase, and xylanase are being investigated based on catalytic activity (Chapman et al., 2018). In the market, these enzymes can be prepared as cocktails, which contain different enzymes with specific properties and other substances such as secondary metabolites produced by microbial strains (Álvarez et al., 2016). Currently, saccharolytic enzymes are produced in the market by blends of enzymes (Lange, 2017; Lange et al., 2021). Many enzymes are available on the market mainly by Novozymes (Denmark), Danisco/Dupont (US), BASF (Germany), DSM (Netherlands), and Abengoa, representing an important segment of total enzyme production (Dewan, 2014). Other important companies are Denykem (UK), Megazyme (Ireland), Advanced Enzymes Technologies (India), and MetGen (Finland). Modern cellulolytic blends have LPMO activities added to enhance saccharification rate and yields. The most productive major source of cellulases comes from the filamentous fungi and mutant strains of *Trichoderma* (*T. viride*, *T. reesei*, and *T. longibrachiatum*).

The two leading companies that supply commercial cellulases are Novozymes and Genencor. The latter has developed four new cocktail variants of its popular brand Accelerase, each of which includes two or more enzymes. Accelerase1500 includes exoglucanase, endoglucanase, hemicellulase, and β-glucosidase. XP and XC variants enhance both xylan and glucan conversion. Accellerase Duet has exoglucanase, endoglucanase, β-glucosidase, and xylanase enzymes and effectively saccharify LCB into monomeric hexose and pentose sugars. In contrast, Accelerase BG includes only β-glucosidase enzyme designed as an accessory product to supplement whole cellulases deficient in beta-glucosidase. Similarly, Novozymes has produced CellicCTec, which can be supplemented with HTec variant rich in hemicellulases. It effectively hydrolyses many types of LCB biomasses. All the commercial cellulases have an optimal condition at 50°C and pH of 4.0–5.0. More recently, enzyme mixtures produced from other companies (i.e., Biocellulase A from Quest Intl. (Sarasota, Fl) and Cellulase AP 30 K from Amano Enzyme Inc.) can work at higher temperatures from 50 to 60°C (Verardi et al., 2012). Among the main options in the biofuel enzymes market, the most prominent enzymes used globally are mentioned in Table 1.

Novozymes released an annual report in 2019, estimating that it comprises approximately 48% of the global enzyme market, also reported that sales during the same year increased moderately, with predominant growth in India and a weakening of the market in China and emerging markets. Currently, Novozymes has launched the product Fortiva, composed of alpha-amylase, to increase ethanol production yields by 1%. It should also be noted that this company has focused its efforts on the production of yeasts to produce first-generation biofuels, under the name Innova yeast technology. According to its annual report, sales in the bioenergy sector are expected to grow between 1% and 5%.

Market studies have shown that the price of the enzyme should stabilize at $0.4/gal, however, this cost may increase in a commercial presentation. Global market of enzymes used in biofuel sector is expected to be nearly 1.3 billion USD by the end of 2026, with an yearly growth of (CAGR) 7.3% during 2021–26 (Global Biofuel Enzymes Industry, 2021). Previous studies report enzyme costs higher than 30% in bioethanol production (Solarte-Toro et al., 2019). Stabilization of enzyme cocktail costs requires increasing demand and competition. For this, the design of new enzymes or preparations from lignocellulosic biomass is necessary to ensure the stability of this market segment (Valdivia et al., 2016).

TABLE 1 Commercially available enzymes in the biofuel market.

Industry	Trade Names	Enzyme Type	Reference
Novozymes	Celluclast, Cellic CTec2 Cellic CTec3	Cellulase,	Khare et al. (2015), Scott et al. (2016), Brar et al. (2019)
	HTec3	Hemicellulase, Cellulase	Sharma et al. (2016)
	Viscozyme L	Multienzyme	Gama et al. (2015)
	Novozyme 188	glucosidase	Khare et al. (2015)
Danisco	BrewZymeLP	β-Glucanase	Sharma et al. (2016)
Genencor	Spezyme, Accelarase 1500	Cellulase	Khare et al. (2015)

9 Conclusion and perspective

Many challenges are still affecting cost-effective enzyme development, both at production and hydrolysis application levels. Combination of more than one pretreatment method is helpful in effective delignification and deconstruction, resulting in maximal saccharification efficiency. Another approach for effective and economical conversion of biomass is to find suitable multifunctional thermophilic GHs. In general, the stability of the cellulase over a range of pH, temperature, metal ions, and organic solvents offers a broad scope for its applications in biorefineries to produce sugars and concomitant fermentation products. Therefore, methods to enhance saccharification efficiency are urgently needed and may be solved by staggered enzyme loading, making an enzyme cocktail, and optimizing the conditions for maximal sugar recovery. Additionally, candidate microbial strains that produce multifunctional GHs should possess cellulase activity in the presence of hydrophobic solvents at thermo-alkali conditions. These microbial features are desirable for optimal cellulase production and subsequent bioconversion of LCB into fuels/platform chemicals. Such strains are more potent in terms of activity and stability and, thereby, make the strain a cost-efficient resource. High-value commodity chemicals from biomass can be produced using the optimized process of in-house thermophilic GHs production through submerged and solid-state fermentations. More availability of scientific knowledge on enzymes in general and cellulases in particular, is based on the recent advancements in biotechnology, molecular biology, and genetics-based tools. Based upon the recent development in the enzyme technology, combining saccharification technology with the microbial strain improvements, it seems obvious that in coming times, more robust enzymes with high activities, better performances, and lesser cost will be available in the market for cost-effective bioethanol production.

References

Abdul Manan, M., Webb, C., 2017. Modern microbial solid state fermentation technology for future biorefineries for the production of added-value products. Biofuel Res. J. 4 (4), 730–740.

Ademark, P., de Vries, R.P., Hagglund, P., et al., 2001. Cloning and characterization of *Aspergillus niger* genes encoding an α-galactosidase and a β-mannosidase involved in galactomannan degradation. Eur. J. Biochem. 268 (10), 2982–2990.

Adewuyi, A., 2020. Challenges and prospects of renewable energy in Nigeria: a case of bioethanol and biodiesel production. Energy Rep. 6, 77–88.

Ahmad, M., Roberts, J.N., Hardiman, E.M., et al., 2011. Identification of DypB from *Rhodococcus jostii* RHA1 as a lignin peroxidase. Biochemistry 50 (23), 5096–5107.

Alonso, D.M., Wettstein, S.G., Dumesic, J.A., 2012. Bimetallic catalysts for upgrading of biomass to fuels and chemicals. Chem. Soc. Rev. 41 (24), 8075–8098. https://doi.org/10.1039/c2cs35188a.

Álvarez, C., Reyes-Sosa, F.M., Díez, B., 2016. Enzymatic hydrolysis of biomass from wood. Microb. Biotechnol. 9 (2), 149–156. https://doi.org/10.1111/1751-7915.12346.

Andlar, M., Rezić, T., Marđetko, N., et al., 2018. Lignocellulose degradation: an overview of fungi and fungal enzymes involved in lignocellulose degradation. Eng. Life Sci. 18 (11), 768–778.

Artzi, L., Bayer, E., Moraïs, S., 2017. Cellulosomes: bacterial nanomachines for dismantling plant polysaccharides. Nat. Rev. Microbiol. 15, 83–95.

Axelsson, L., Franzén, M., Ostwald, M., Berndes, G., Lakshmi, G., Ravindranath, N.H., 2012. Perspective: Jatropha cultivation in southern India: assessing farmers' experiences. Biofuels Bioprod. Biorefin. 6 (3), 246–256. https://doi.org/10.1002/bbb.

Baath, J., Giummarella, N., Klaubauf, S., et al., 2016. A glucuronoyl esterase from *Acremonium alcalophilum* cleaves native lignin-carbohydrate ester bonds. FEBS Lett. 590, 2611–2618.

Banerjee, G., Car, S., Scott-Craig, J.S., et al., 2010. Synthetic enzyme mixtures for biomass deconstruction: production and optimization of a core set. Biotechnol. Bioeng. 106, 707–720.

Bayer, E.A., Kenig, R., Lamed, R., 1983. Adherence of *clostridium thermocellum* to cellulose. J. Bacteriol. 156, 818–827.

Bayer, E.A., Belaich, J.P., Shoham, Y., Lamed, R., 2004. The cellulosomes: multienzyme machines for degradation of plant cell wall polysaccharides. Annu. Rev. Microbiol. 58, 521–554.

Beckham, G.T., Ståhlberg, J., Knott, B.C., et al., 2014. Toward a molecular-level theory of carbohydrate processivity in glycoside hydrolases. Curr. Opin. Biotechnol. 27, 96–106.

Beig, B., Riaz, M., Raza, S., Hassan, M., Zheng, Z., Karimi, K., Million, M., 2021. Current challenges and innovative developments in pretreatment of lignocellulosic residues for biofuel production: a review original equipment manufacturer. Fuel 287. https://doi.org/10.1016/j.fuel.2020.119670, 119670.

Bhalla, A., Bansal, N., Kumar, S., Bischoff, K.M., Sani, R.K., 2013. Improved lignocellulose conversion to biofuels with thermophilic bacteria and thermostable enzymes. Bioresour. Technol. 128, 751–759.

Bhatia, Y., Mishra, S., Bisaria, V.-S., 2002. Microbial ß-glucosidases, cloning, properties, and applications. Crit. Rev. Biotechnol. 22, 375–407.

Bischof, R.H., Ramoni, J., Seiboth, B., 2016. Cellulases and beyond: the first 70 years of the enzyme producer *Trichoderma reesei*. Microb. Cell Factories 15, 106.

Bissaro, B., Røhr, Å.-K., Müller, G., et al., 2017. Oxidative cleavage of polysaccharides by monocopper enzymes depends on H_2O_2. Nat. Chem. Biol. 13 (10), 1123–1128.

Bissaro, B., Streit, B., Isaksen, I., Eijsink, V.G.H., Beckham, G.T., DuBois, J.L., Røhr, Å.K., 2020. Molecular mechanism of the chitinolytic peroxygenase reaction. Proc. Natl. Acad. Sci. U. S. A. 117 (3), 1504–1513.

Brar, K.K., et al., 2019. Evaluating novel fungal secretomes for efficient saccharification and fermentation of composite sugars derived from hydrolysate and molasses into ethanol. Bioresour. Technol. 273, 114–121. https://doi.org/10.1016/j.biortech.2018.11.004.

Cerda, A., Artola, A., Barrena, R., Font, X., Gea, T., Sánchez, A., 2019. Innovative production of bioproducts from organic waste through solid-state fermentation. Front. Sustain. Food Syst. 3, 63.

Chandel, A.K., Singh, O.V., Chandrasekhar, G., et al., 2010. Key drivers influencing the commercialization of ethanol-based biorefineries. J. Commer. Biotechnol. 16, 239–257.

Chapman, J., Ismail, A., Dinu, C., 2018. Industrial applications of enzymes: recent advances, techniques, and outlooks. Catalysts 8 (6), 238. https://doi.org/10.3390/catal8060238.

Chen, H., Fu, X., 2016. Industrial technologies for bioethanol production from lignocellulosic biomass. Renew. Sust. Energ. Rev. 57, 468–478. https://doi.org/10.1016/j.rser.2015.12.069.

Cintra, L.C., da Costa, I.C., de Oliveira, I.C.M., et al., 2020. The boosting effect of recombinant hemicellulases on the enzymatic hydrolysis of steam-treated sugarcane bagasse. Enzym. Microb. Technol. 133, 109447.

Collins, T., Gerday, C., Feller, G., 2005. Xylanases, xylanase families and extremophilic xylanases. FEMS Microbiol. Rev. 29, 3–23.

Dar, R.A., Dar, E.A., Kaur, A., Phutela, U.G., 2018. Sweet sorghum-a promising alternative feedstock for biofuel production. Renew. Sust. Energ. Rev. 82, 4070–4090.

Dashtban, M., Schraft, H., Syed, T.A., et al., 2010. Fungal biodegradation and enzymatic modification of lignin. Int J Biochem Mol Biol 1, 36–50.

de Castro, S.M., de Castro, A.M., 2012. Assessment of the Brazilian potential for the production of enzymes for biofuels from agroindustrial materials. Biomass Conversion and Biorefinery 2 (1), 87–107. https://doi.org/10.1007/s13399-012-0031-9.

Dewan, S.S., 2014. Global Markets for Enzymes in Industrial Applications. BCC Research, Wellesley.

Ding, S.Y., Liu, Y.S., Zeng, Y., et al., 2012. How does plant cell wall nanoscale architecture correlate with enzymatic digestibility? Science 338, 1055–1060.

Dragone, G., Kerssemakers, A.A.J., Driessen, J.L.S.P., et al., 2020. Innovation and strategic orientations for the development of advanced biorefineries. Bioresour. Technol. 302, 122847.

Dutta, T., Sahoo, R., Sengupta, R., Ray, S.S., Bhattacharjee, A., Ghosh, S., 2008. Novel cellulases from an extremophilic filamentous fungi *Penicillium citrinum*: production and characterization. J. Ind. Microbiol. Biotechnol. 35 (4), 275–282.

Dwevedi, A., Kayastha, A.M., 2011. Enzyme immobilization: a breakthrough in enzyme technology and boon to enzyme based industries. Proteomics Res. J. 3 (April), 31–50.

El-Gebali, S., Mistry, J., Bateman, A., et al., 2019. The Pfam protein families database in 2019. Nucleic Acids Res. 47, D427–D432.

Elleuche, S., Schroeder, C., Sahm, K., Antranikian, G., 2014. Extremozymes-biocatalysts with unique properties from extremophilic microorganisms. Curr. Opin. Biotechnol. 29, 116–123.

Eriksson, T., Börjesson, J., Tjerneld, F., 2002. Mechanism of surfactant effect in enzymatic hydrolysis of lignocellulose. Enzym. Microb. Technol. 31 (3), 353–364.

Foreman, P.K., Brown, D., Dankmeyer, L., et al., 2003. Transcriptional regulation of biomass-degrading enzymes in the filamentous fungus *Trichoderma reesei*. J. Biol. Chem. 278, 31988–31997.

Gama, R., Van Dyk, J.S., Pletschke, B.I., 2015. Optimisation of enzymatic hydrolysis of apple pomace for production of biofuel and biorefinery chemicals using commercial enzymes. 3 Biotech 5 (6), 1075–1087. https://doi.org/10.1007/s13205-015-0312-7.

Gaur, N., 2018. Biochemical and kinetic characterization of laccase and manganese peroxidase from novel *Klebsiella pneumoniae* strains and their application in bioethanol production. RSC Adv. 8, 15044–15055.

Global Biofuel Enzymes Industry, 2021. Global Biofuel Enzymes Market to Reach $1.3 Billion by 2026. Available from: https://www.reportlinker.com/p04707128/?utm_source=GNW (Accessed 10 November 2021).

Gold, M.H., Alic, M., 1993. Molecular biology of the lignin-degrading basidiomycete *Phanerochaete chrysosporium*. Microbiol. Rev. 57 (3), 605–622.

Gomathi, S., Ambikapathy, V., Pannerselvam, A., 2012. Potential strain of Trichoderma spp. to control damping off of disease in chilli. Indian J. Appl. Pure Biol. 27 (1), 133–140.

Gourlay, K., Hu, J., Arantes, V., et al., 2013. Swollenin aids in the amorphogenesis step during the enzymatic hydrolysis of pretreated biomass. Bioresour. Technol. 142, 498–503.

Guo, H., Chang, Y., Lee, D.J., 2018. Enzymatic saccharification of lignocellulosic biorefinery: research focuses. Bioresour. Technol. 252, 198–215. https://doi.org/10.1016/j.biortech.2017.12.062.

Han, S.J., Yoo, Y.J., Kang, H.S., 1995. Characterization of a bifunctional cellulase and its structural gene the cel gene of *Bacillus* sp. D04 has exo-and endoglucanase activity. J. Biol. Chem. 270, 26012–26019.

Helle, S.S., Duff, S.J., Cooper, D.G., 1993. Effect of surfactants on cellulose hydrolysis. Biotechnol. Bioeng. 42, 611–617.

Hemansi, Gupta, R., Kuhad, R.C., Saini, J.K., 2018. Cost effective production of complete cellulase system by newly isolated *Aspergillus niger* RCKH-3 for efficient enzymatic saccharification: medium engineering by overall evaluation criteria approach (OEC). Biochem. Eng. J. 132, 182–190.

Hemansi, Chakraborty, S., Yadav, G., Saini, J.K., Kuhad, R.C., 2019. Comparative study of cellulase production using submerged and solid-state fermentation. In: Srivastava, N., Srivastava, M., Mishra, P.K., Ramteke, P.W., Singh, R.-L. (Eds.), New and Future Developments in Microbial Biotechnology and Bioengineering. Elsevier, USA, pp. 99–113.

Houghton, J., 2006. Breaking the Biological Barriers to Cellulosic Ethanol: A Joint Research Agenda. EERE Publication and Product.

Huang, C., Zhan, Y., Cheng, J., Wang, J., Meng, X., Zhou, X., Fang, G., Ragauskas, A.J., 2021. Facilitating enzymatic hydrolysis with a novel guaiacol-based deep eutectic solvent pretreatment. Bioresour. Technol. 326, 124696.

Huy, N.D., Le, N.C., Seo, J.W., et al., 2015. Putative endoglucanase PcGH5 from *Phanerochaete chrysosporium* is a beta-xylosidase that cleaves xylans in synergistic action with endo-xylanase. J. Biosci. Bioeng. 119, 416–420.

Jacobson, F., Karkehabadi, S., Hansson, H., et al., 2013. The crystal structure of the core domain of a cellulose induced protein (Cip1) from *Hypocrea jecorina*, at 1.5 Å resolution. PLoS One 8, e70562.

Jamaldheen, S.B., Sharma, K., Rani, A., Moholkar, V.S., Goyal, A., 2018. Comparative analysis of pretreatment methods on Sorghum (*Sorghum durra*) stalk agrowaste for holocellulose content. Prep. Biochem. Biotechnol. https://doi.org/10.1080/10826068.2018.1466148.

Kaar, W.E., Holtzapple, M.T., 1998. Benefits from tween during enzymic hydrolysis of corn Stover. Biotechnol. Bioeng. 59, 419–427.

Khaire, K.C., Moholkar, V.S., Goyal, A., 2021. Bioconversion of sugarcane tops to bioethanol and other value added products: an overview. Mater. Sci. Energy Technol. 4, 54–68. https://doi.org/10.1016/j.mset.2020.12.004.

Khare, S.K., Pandey, A., Larroche, C., 2015. Current perspectives in enzymatic saccharification of lignocellulosic biomass. Biochem. Eng. J. 102, 38–44. https://doi.org/10.1016/j.bej.2015.02.033.

Khokhar, Z.U., Qurat-ul-Ain Syed, J.W., Athar, M.A., 2014. On-site cellulase production by Trichoderma reesei 3EMS35 mutant and same vessel saccharification and fermentation of acid treated wheat straw for ethanol production. EXCLI J. 13, 82.

Kim, M.H., Lee, S.B., Ryu, D.D.Y., Reese, E.T., 1982. Surface deactivation of cellulase and its prevention. Enzym. Microb. Technol. 4, 99–103.

Knob, A., Terrasan, C.R.F., Carmona, E.C., 2010. β-Xylosidases from filamentous fungi: an overview. World J. Microbiol. Biotechnol. 26, 389–407.

Krishna, C., 2005. Solid-state fermentation systems—an overview. Crit. Rev. Biotechnol. 25 (1–2), 1–30.

Kristensen, J.B., Borjesson, J., Bruun, M.H., Tjerneld, F., Jorgensen, H., 2007. Use of surface-active additives in enzymatic hydrolysis of wheat straw lignocelluloses. Enzym. Microb. Technol. 40, 888–895.

Kumar, R., Wyman, C.E., 2009. Effects of cellulase and xylanase enzymes on the deconstruction of solids from pretreatment of poplar by leading technologies. Biotechnol. Prog. 25, 302–314.

Kurasin, M., Väljamäe, P., 2011. Processivity of cellobiohydrolases is limited by the substrate. J. Biol. Chem. 286, 169–177.

Lamed, R., Setter, E., Bayer, E.A., 1983. Characterization of a cellulose-binding, cellulase-containing complex in clostridium thermocellum. J. Bacteriol. 156, 828–836.

Lange, L., 2017. Fungal enzymes and yeasts for conversion of plant biomass to bioenergy and high-value products. In: The Fungal Kingdom. ASM Press, Washington, DC, USA, pp. 1027–1048, https://doi.org/10.1128/9781555819583.ch51.

Lange, L., Connor, K.O., Arason, S., Bundgård-Jørgensen, U., Canalis, A., Carrez, D., Gallagher, J., Gøtke, N., Huyghe, C., Jarry, B., Llorente, P., 2021. Developing a sustainable and circular bio-based economy in EU: by partnering across sectors, upscaling and using new knowledge faster, and for the benefit of climate, environment & biodiversity, and people & business. Front. Bioeng. Biotechnol. 8, 1456.

Lee, K.M., et al., 2020. Synergistic ultrasound-assisted organosolv pretreatment of oil palm empty fruit bunches for enhanced enzymatic saccharification: an optimization study using artificial neural networks. Biomass Bioenergy 139 (May). https://doi.org/10.1016/j.biombioe.2020.105621, 105621.

Legodi, L.M., La Grange, D., Van Rensburg, E.L., Ncube, I., 2019. Isolation of cellulose degrading fungi from decaying banana pseudostem and Strelitzia alba. Enzyme Res. 2019, 1390890.

Lei, Z., Shao, Y., Yin, X., et al., 2016. Combination of xylanase and debranching enzymes specific to wheat arabinoxylan improve the growth performance and gut health of broilers. J. Agric. Food Chem. 64, 4932–4942.

Levasseur, A., Drula, E., Lombard, V., et al., 2013. Expansion of the enzymatic repertoire of the CAZy database to integrate auxiliary redox enzymes. Biotechnol. Biofuels 6, 41.

Li, Y.H., Zhang, X.Y., Zhang, F., Peng, L.C., Zhang, D.B., Kondo, A., Bai, F.W., Zhao, X.Q., 2018. Optimization of cellulolytic enzyme components through engineering Trichoderma reesei and on-site fermentation using the soluble inducer for cellulosic ethanol production from corn Stover. Biotechnol. Biofuels 11 (1), 1–14.

Lin, W., Chen, D., Yong, Q., Huang, C., Huang, S., 2019. Improving enzymatic hydrolysis of acid-pretreated bamboo residues using amphiphilic surfactant derived from dehydroabietic acid. Bioresour. Technol. 293 (August). https://doi.org/10.1016/j.biortech.2019.122055, 122055.

Ling, Z., Guo, Z., Huang, C., Yao, L., Xu, F., 2020. Deconstruction of oriented crystalline cellulose by novel levulinic acid based deep eutectic solvents pretreatment for improved enzymatic accessibility. Bioresour. Technol. 305, 123025.

Lombard, V., Golaconda Ramulu, H., Drula, E., et al., 2013. The carbohydrate-active enzymes database (CAZy) in 2013. Nucleic Acids Res. 42 (D1), D490–D495.

Lynd, L.R., Weimer, P.J., van Zyl, W.H., Pretorius, I.S., 2002. Microbial cellulose utilization: fundamentals and biotechnology. Microbiol. Mol. Biol. Rev. 66, 506–577.

Malgas, S., Mafa, M.S., Mkabayi, L., et al., 2019. A mini-review of xylanolytic enzymes with regards to their synergistic interactions during hetero-xylan degradation. World J. Microbiol. Biotechnol. 35, 187.

Margareta, W., Nagarajan, D., Chang, J., Lee, D., 2020. Dark fermentative hydrogen production using macroalgae (Ulva sp.) as the renewable feedstock. Appl. Energy 262, 11457.

Marinović, M., Nousiainen, P., Dilokpimol, A., et al., 2018. ACS Sustain. Chem. Eng. 6 (3), 2878–2882.

Marulanda, V.A., Gutiérrez, C.D.B., Álzate, C.A.C., 2019. Thermochemical, biological, biochemical, and hybrid conversion methods of bio-derived molecules into renewable fuels. In: Advanced Bioprocessing for Alternative Fuels, Biobased Chemicals, and Bioproducts. Woodhead Publishing, pp. 59–81.

Mejias, L., Cerda, A., Barrena, R., Gea, T., Sánchez, A., 2018. Microbial strategies for cellulase and xylanase production through solid-state fermentation of digestate from biowaste. Sustainability 10 (7), 2433.

Mrudula, S., Murugammal, R., 2011. Production of cellulase by *Aspergillus niger* under submerged and solid state fermentation using coir waste as a substrate. Braz. J. Microbiol. 42 (3), 1119–1127.

Nagarajan, D., Nandini, A., Dong, C., et al., 2020. Bioengineering lactic acid production from renewable feedstocks using poly-vinyl alcohol immobilized *Lactobacillus plantarum* 23. Ind. Eng. Chem. Res. 59 (39), 17156–17164.

Neagu, D.A., Destain, J., Thonart, P., Socaciu, C., 2012. Trichoderma reesei cellulase produced by submerged versus solid state fermentations. Bull. UASVM Agric. 69 (2), 2012.

Østby, H., Hansen, L.D., Horn, S.J., et al., 2020. Enzymatic processing of lignocellulosic biomass: principles, recent advances and perspectives. J. Ind. Microbiol. Biotechnol. 47 (9–10), 623–657.

Oszust, K., Pawlik, A., Siczek, A., Janusz, G., Gryta, A., Bilińska-Wielgus, N., Frąc, M., 2017. Efficient cellulases production by Trichoderma atroviride G79/11 in submerged culture based on soy flour-cellulose-lactose. Bioresources 12 (4), 8468–8489.

Pereira Scarpa, J.d.C., et al., 2019. Saccharification of pretreated sugarcane bagasse using enzymes solution from Pycnoporus sanguineus MCA 16 and cellulosic ethanol production. Ind. Crop. Prod. 141. https://doi.org/10.1016/j.indcrop.2019.111795, 111795.

Pereira, J.D.C., Marques, N.P., Rodrigues, A., et al., 2015. Thermophilic fungi as new sources for production of cellulases and xylanases with potential use in sugarcane bagasse saccharification. J. Appl. Microbiol. 118 (4), 928–939.

Perrone, O.M., de Souza Moretti, M.M., Bordignon, S.E., de Cassia Pereira, J., da Silva, R., Gomes, E., Boscolo, M., 2021. Improving cellulosic ethanol production using ozonolysis and acid as a sugarcane biomass pretreatment in mild conditions. Bioresour. Technol. Rep. 13, 100628.

Pokkuluri, P.R., Duke, N.E.C., Wood, S.J., et al., 2011. Structure of the catalytic domain of glucuronoyl esterase Cip2 from *Hypocrea jecorina*. Proteins 79, 2588–2592.

Polizeli, M.L.T.M., Rizzatti, A.C.S., Monti, R., et al., 2005. Xylanases from fungi: properties and industrial applications. Appl. Microbiol. Biotechnol. 67, 577–591.

Qing, Q., Yang, B., Wyman, C.-E., 2010. Impact of surfactants on pretreatment of corn Stover. Bioresour. Technol. 101, 5941–5951.

Rajkumar, R., et al., 2014. Potential of the micro and macro algae for biofuel production: a brief review. Bioresources 9 (1), 1606–1633.

Rashid, G.M.M., Taylor, C.R., Liu, Y., et al., 2015. Identification of manganese superoxide dismutase from *Sphingobacterium* sp. T2 as a novel bacterial enzyme for lignin oxidation. ACS Chem. Biol. 10, 2286–2294.

Raud, M., Kikas, T., Sippula, O., Shurpali, N.J., 2019. Potentials and challenges in lignocellulosic biofuel production technology. Renew. Sust. Energ. Rev. 111, 44–56.

Reddy, G.P.K., Narasimha, G., Kumar, K.D., Ramanjaneyulu, G., Ramya, A., Kumari, B.S., Reddy, B.R., 2015. Cellulase production by aspergillus Niger on different natural lignocellulosic substrates. Int. J. Curr. Microbiol. App. Sci. 4 (4), 835–845.

Resch, M.G., Donohoe, B.S., Baker, J.O., et al., 2013. Fungal cellulases and complexed cellulosomal enzymes exhibit synergistic mechanisms in cellulose deconstruction. Energy Environ. Sci. 6, 1858–1867.

Rytioja, J., Hildén, K., Yuzon, J., et al., 2014. Plant-polysaccharide-degrading enzymes from basidiomycetes. Microbiol. Mol. Biol. Rev. 78 (4), 614–649.

Saeed, S., Saleem, M., 2018. Novel pretreatment methods to improve enzymatic saccharification of sugarcane bagasse: a report. Iran. J. Chem. Chem. Eng. 37 (5), 225–234.

Saini, J.K., Saini, R., Tewari, L., 2015. Lignocellulosic agriculture wastes as biomass feedstocks for second-generation bioethanol production: concepts and recent developments. 3 Biotech 5 (4), 337–353.

Saini, J.K., Singhania, R.R., Satlewal, A., Saini, R., Gupta, R., Tuli, D., Mathur, A., Adsul, M., 2016. Improvement of wheat straw hydrolysis by cellulolytic blends of two penicillium spp. Renew. Energy 98, 43–50.

Sajith, S., Priji, P., Sreedevi, S., Benjamin, S., 2016. An overview on fungal cellulases with an industrial perspective. J. Nutr. Food Sci. 6, 1.

Saloheimo, M., Paloheimo, M., Hakola, S., et al., 2002. Swollenin, a *Trichoderma reesei* protein with sequence with sequence similary to the plant expansins, exhibits disruption activity on cellulosic materials. Eur. J. Biochem. 269, 4202–4211.

Sandesh, K., Ujwal, P., 2021. Trends and perspectives of liquid biofuel – process and industrial viability. Energy Convers. Manag. 10. https://doi.org/10.1016/j.ecmx.2020.100075, 100075.

Scott, B.R., et al., 2016. Catalase improves saccharification of lignocellulose by reducing lytic polysaccharide monooxygenase-associated enzyme inactivation. Biotechnol. Lett. 38 (3), 425–434. https://doi.org/10.1007/s10529-015-1989-8.

Sharma, A., Parashar, D., Satyanarayana, T., 2016. Acidophilic Microbes: Biology and Applications. Springer, Cham, pp. 215–241, https://doi.org/10.1007/978-3-319-13521-2_7.

Sheng, Y., et al., 2021. Enzymatic conversion of pretreated lignocellulosic biomass: a review on influence of structural changes of lignin. Bioresour. Technol. 324 (October 2020). https://doi.org/10.1016/j.biortech.2020.124631, 124631.

Singhania, R.R., Sukumaran, R.K., Patel, A.K., Larroche, C., Pandey, A., 2010. Advancement and comparative profiles in the production technologies using solid-state and submerged fermentation for microbial cellulases. Enzym. Microb. Technol. 46 (7), 541–549.

Sista Kameshwar, A.K., Qin, W., 2018. Comparative study of genome-wide plant biomass-degrading CAZymes in white rot, brown rot and soft rot fungi. Mycology 9, 93–105.

Soccol, C.R., da Costa, E.S.F., Letti, L.A.J., Karp, S.G., Woiciechowski, A.L., de Souza Vandenberghe, L.P., 2017. Recent developments and innovations in solid state fermentation. Biotechnol. Res. Innov. 1 (1), 52–71.

Solarte-Toro, J.C., Romero-García, J.M., Martínez-Patiño, J.C., Ruiz-Ramos, E., Castro-Galiano, E., Cardona-Alzate, C.A., 2019. Acid pretreatment of lignocellulosic biomass for energy vectors production: a review focused on operational conditions and techno-economic assessment for bioethanol production. Renew. Sust. Energ. Rev. 107, 587–601.

Tang, W., Wu, X., Huang, C., Ling, Z., Lai, C., Yong, Q., 2021. Natural surfactant-aided dilute sulfuric acid pretreatment of waste wheat straw to enhance enzymatic hydrolysis efficiency. Bioresour. Technol. 324, 124651.

Tawalbeh, M., Rajangam, A.S., Salameh, T., 2021. Science direct characterization of paper mill sludge as a renewable feedstock for sustainable hydrogen and biofuels production. Int. J. Hydrog. Energy 46, 4761–4775.

Teeri, T.T., 1997. Crystalline cellulose degradation: new insights into the function of cellobiohydrolases. Trends Biotechnol. 15, 160–167.

Thygesen, A., Ami, J., Fernando, D., Bentil, J., Daniel, G., Mensah, M., Meyer, A.S., 2020. Microstructural and carbohydrate compositional changes induced by enzymatic saccharification of green seaweed from West Africa. Algal Res. 47, 101894.

Toscano-Palomar, L., Montero-Alpirez, G., Stilianova-Stoytcheva, M., Vertiz-Pelaez, E., 2015. Cellulase production from filamentous fungi for its application in the hydrolysis of wheat straw. MRS Online Proc. Libr. 1763, 25–30.

Vaaje-Kolstad, G., Westereng, B., Horn, S.J., et al., 2010. An oxidative enzyme boosting the enzymatic conversion of recalcitrant polysaccharides. Science 330, 219–222.

Valdivia, M., Galan, J.L., Laffarga, J., Ramos, J.L., 2016. Biofuels 2020: biorefineries based on lignocellulosic materials. Microb. Biotechnol. 9 (5), 585–594.

Velasco, J., Pellegrini, V.D.O.A., Sepulchro, A.G.V., Kadowaki, M.A.S., Santo, M.C.E., Polikarpov, I., Segato, F., 2021. Comparative analysis of two recombinant LPMOs from Aspergillus fumigatus and their effects on sugarcane bagasse saccharification. Enzym. Microb. Technol. 144, 109746.

Verardi, A., De Bari, I., Ricca, E., Calabrò, V., 2012. Hydrolysis of lignocellulosic biomass: current status of processes and technologies and future perspectives. In: Bioethanol. IntechOpen, pp. 95–122.

Vogel, J., 2008. Unique aspects of the grass cell wall. Curr. Opin. Plant Biol. 11 (3), 301–307.

Wang, T., Lü, X., 2021. Overcome saccharification barrier: advances in hydrolysis technology. In: Lü, X. (Ed.), Woodhead Publishing Series in Energy, Advances in 2nd Generation of Bioethanol Production. Woodhead Publishing, ISBN: 9780128188620, pp. 137–159, https://doi.org/10.1016/B978-0-12-818862-0.00005-4.

Wang, Y., Meng, X., Jeong, K., Li, S., Leem, G., Kim, K.H., Pu, Y., Ragauskas, A.J., Yoo, C.G., 2020. Investigation of a lignin-based deep eutectic solvent using p-hydroxybenzoic acid for efficient woody biomass conversion. ACS Sustain. Chem. Eng. 8 (33), 12542–12553.

Wilson, D.B., 2011. Microbial diversity of cellulose hydrolysis. Curr. Opin. Microbiol. 14, 259–263.

Wong, W.S.W., Chan, V.J., Liao, H., Zidwick, M.J., 2013. Cloning of a novel feruloyl esterase gene from rumen microbial metagenome and enzyme characterization in synergism with endoxylanases. J. Ind. Microbiol. Biotechnol. 40, 287–295.

Yan, Q.J., Wang, L., Jiang, Z.Q., et al., 2008. A xylose-tolerant β-xylosidase from *Paecilomyces thermophila*: characterization and its co-action with the endogenous xylanase. Bioresour. Technol. 99, 5402–5410.

Yang, H., Jin, Y., Shi, Z., Wang, D., Zhao, P., Yang, J., 2020. Effect of hydrothermal pretreated bamboo lignin on cellulose saccharification for bioethanol production. Ind. Crop. Prod. 156 (June). https://doi.org/10.1016/j.indcrop.2020.112865.

Yang, L., Ru, Y., Xu, S., Liu, T., Tan, L., 2021. Features correlated to improved enzymatic digestibility of corn Stover subjected to alkaline hydrogen peroxide pretreatment. Bioresour. Technol. 325, 124688.

Yeoman, C.-J., Han, Y., Dodd, D., et al., 2010. Thermostable enzymes as biocatalysts in the biofuel industry. Adv. Appl. Microbiol. 70, 1–55. ISSN 0065-2164, ISBN 9780123809919.

Yoo, C.G., Meng, X., Pu, Y., Ragauskas, A.J., 2020. The critical role of lignin in lignocellulosic biomass conversion and recent pretreatment strategies: a comprehensive review. Bioresour. Technol. 301, 122784.

Zhang, Y.-H.-P., Lynd, L.-R., 2004. Toward an aggregated understanding of enzymatic hydrolysis of cellulose: non-complexed cellulase systems. Biotechnol. Bioeng. 88, 797–824.

Zhang, X.Z., Zhang, Y.H.P., 2013. Cellulases: characteristics, sources, production, and applications. In: Yang, S., El-Enshasy, H.A., Thongchul, N. (Eds.), Bioprocessing Technologies in Biorefinery for Sustainable Production of Fuels, Chemicals, and Polymers, first ed. John Wiley & Sons, Inc., Hoboken, NJ, pp. 131–146.

Zhang, J., Siika-Aho, M., Tenkanen, M., Viikari, L., 2011. The role of acetyl xylan esterase in the solubilization of xylan and enzymatic hydrolysis of wheat straw and giant reed. Biotechnol. Biofuels 4, 60.

Zhao, S., Liao, X.Z., Wang, J.X., Ning, Y.N., Li, C.X., Liao, L.S., Liu, Q., Jiang, Q., Gu, L.S., Fu, L.H., Yan, Y.S., 2019. Transcription factor Atf1 regulates expression of cellulase and xylanase genes during solid-state fermentation of ascomycetes. Appl. Environ. Microbiol. 85 (24), e01226-19.

C H A P T E R

10

Yeast-mediated ethanol fermentation from lignocellulosic pentosan

Abhilek K. Nautiyal[a,b], Tripti Sharma[a,b], Diptarka Dasgupta[a,b], Thallada Bhaskar[a,b], and Debashish Ghosh[a,b]

[a]Material Resource Efficiency Division (MRED), CSIR-Indian Institute of Petroleum (IIP), Dehradun, India [b]Academy of Scientific and Innovative Research (AcSIR), Ghaziabad, Uttar Pradesh, India

1 Introduction

The decline in petroleum products, their economic challenges, and the environmental problems of gas burning have led to research into new renewable energy sources (Zabed et al., 2016). Lignocellulosic biomass is a permanent cradle of fossil fuels for bioethanol production and value-added products. Lignocellulose is among the utmost widely accessible renewable biomass cradles on the planet, and its utilization may not race with land usage in food production. (Budzianowski, 2017; Lian et al. (2018)). The composite and repetitive structure of lignocellulosic biomass consist of cellulose (crystalline, glucose compatible homopolymer), hemicellulose (branched heteropolymer, amorphous hexoses, and pentoses) (Zabed et al., 2016). Inhibitor compounds, including 5-hydroxymethylfurfural (HMF), furfural, and acetic acid, are released between the phases of hydrolysis and pretreatment to produce burning sugars from lignocellulosic Biomass (Kim et al., 2015). These compounds significantly impact bacterial growth and ethanol fermentation (Jönsson and Martín, 2016; Cunha et al., 2019). Lignocellulosic hydrolysates contain the most monosaccharides, namely glucose and xylose, which make up 60%–70% and 30%–40% of their sugar, respectively (Kwak and Jin, 2017). Among the many ways to hydrolyze these polysaccharides into boiling sugar, enzymatic hydrolysis is the most widely used method within the current bioethanol business. However, enzymatic hydrolysis (cellulose saccharification) cannot be performed immediately in biomass because the cellulose in biomass is lined with lignin and hemicellulose (Cunha et al., 2019).

For this reason, Biomass reuptake facilitates the availability of enzymes in cellulose, exploits cellulose regeneration, and, simultaneously, it reducing the amount of enzyme inhibitors that appear in the previous treatment. In recent ages, significant progress has been made in the conception of xylose metabolism, and identifying and developing issues with augmented xylose fermentation. Examining many microorganisms that can synthesize xylose in ethanol directly shows that wild-type yeast and regenerative bacteria have a complete release depending on the final concentration of ethanol, high yield, and volumetric production. Thus, progress in the conversion of xylose is progressing rapidly, but there are still several questions about the critical signs of xylose metabolism in bacteria and yeast. There is already a substantial body of works on the use of xylose yeast. In addition, details on the production of synthetic viruses are increasing rapidly.

The preferred microorganism *Saccharomyces cerevisiae* is not naturally consuming xylose for the production of large-scale ethanol. Therefore, the hemicellulosic portion composed primarily of xylose must be transformed into ethanol capably by constructing *S. cerevisiae* strain to absorb xylose and have high inhibitory compounds resistance. Xylose assimilation is obtained by converting xylose into xylulose and ultimately phosphorylated to xylulose-5-phosphate, additionally metabolized into a pentose phosphate pathway. The most important recent advances in xylose conversion research are the assembly of recombinant bacterial and yeast strains with improved xylose fermentation efficiency. Genetic engineering, to generate a recombinant strain of *S. cerevisiae*, is proficient in high-yield xylose fermentation approaching the end of its execution. Furthermore, progress in the formation of higher quality and recombinant bacterial strains is being made. In light of the rapid developments in the metabolic engineering of high-end xylose-fermenting organisms and the improvement of recombinant organisms in a state of abundance, cutting-edge enactment appears to be one in the near future.

2 Selection of lignocellulosic biomass

2.1 Biomass availability

Biomass assets that can be acquired on a renewable base and recycled without delay, such as fuel or improved to an alternative type of energy product, are recognized as "feedstocks." Biomass feedstocks contain reliable green energy, agricultural crop residues, forest residues, algae, municipal, and wet waste.

Energy crops donated by nonedible crops can be mature on the sides of the earth (land now no longer suitable for traditional crops such as sorghum and beans), primarily to provide biomass. This is analyzed by standard categories: herbaceous and woody. For herbaceous plants that have a long life cycle (for more than 2 years of age), the grass is harvested each year and reaches occupied fruiting bodies after taking 2–3 years. These embrace switchgrass, bamboo, sweet cereals, wheatgrass, and more. Woody plants alternate with fast-growing hardwoods that are garnered within 5–8 years of planting. These contain hybrid poplar, hybrid willow, oriental cottonwood, black walnut, and sweetgum. Numerous plants have the prospective to develop water and soil health, enhance the habitat of plants and animals associated with annual plants, diversify profitable assets, and improve the average productivity of the farm.

The ethanol business uses starch-based maize as its primary feedstock. In the high capture of the local distribution of inconsistent ethanol production, the energy sources of the process lignocellulosic biomass are considered. It was revealed that maize production sites already control large amounts of ethanol production and that few unused maize products continue to be commercially exploitative. Recovery of various sugar products, including other types of biomass, including lignocellulosic substances, becomes an essential factor in the industry as it expands, in moderation, and the need to satisfy the authority of the authorities. The ability of bioconversion and thermochemical modification to produce biofuels from lignocellulosic biomass is being studied.

2.2 Physicochemical characterization of biomass

Lignocellulosic biomass enhances dry plantation, clearly marked with cellulose, hemicellulose, and lignin (Lian et al., 2018). Lignocellulosic biomass feedstocks are accessible for energy found mainly in the following sectors: agriculture, woodland, and industry. Residues and wood residues are promising high-quality biomass feedstocks for their efficiency and low cost (Kim et al., 2015). Traditional lignocellulosic biomass is prohibited from heating and cooking, leading to significant environmental and degradation and desertification by thermochemical or chemical modification pathways. Instead, lignocellulosic biomass can be transformed into electrical or energy conductors. The thermochemical modification uses thermal and chemical processes to generate energy goods from biomass and combustion, pyrolysis, gasification, and liquefaction (Cunha et al., 2019). Biomass chemical modification involves bacteria, microorganisms, or enzymes that break down biomass into gas or liquid fuels containing biogas or bioethanol (Kwak and Jin, 2017). The traditional technology for biomass conversion and its essential products is shown in Fig. 1.

The entire biomass-to-biofuel system consists of asset management methods, pretreatment, and modification of lignocellulosic Biomass (Kumar and Sharma, 2017). The procedure of asset

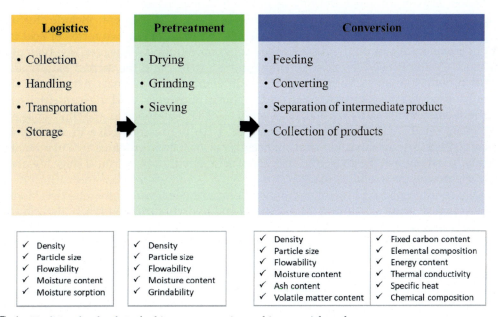

FIG. 1   Traditional technology for biomass conversion and its essential products.

management includes the collection, handling, and distribution of biomass feedstocks. Pretreatment methods include drying, digestion, and reduction of feedstocks. The conversion process consists of feeds, conversions, product separation, collection and improvement, and sales. The physicochemical properties of lignocellulosic biomass are important figures aimed at the design and implementation of these techniques.

3 Xylose recovery from biomass

Many preexisting therapies have been industrialized to expand cellulose efficiency and escalate the yield of boiling sugar. Traditional purposes of early treatment comprise of (1) the manufacture of high-quality and effective soluble substances that enrich the sugar crop with enzyme hydrolysis, (2) preventing the sugar decomposition (especially pentoses) and those resultant from hemicellulose, and (3) reducing the production of inhibitors by following steps for fermentation, (4) the return of lignin to be converted into valuable byproducts, and (5) by operating in small-sized reactors and reducing heat and energy requirements.

The main box of treatment is to improve the other areas to eliminate the cellulose's ability and the exclusion of hemicellulose and lignin. Ace-and-outs of the various techniques is defined in the following segments. Over-exposure to the biomass of the species does not preclude using a single type of preprocessing of various types of raw materials.

3.1 Chemical methods

3.1.1 Pretreatment with acid

Acids are employed to treat lignocellulosic biomass in these procedures. Pretreatment preferences heavily influence the mix of products that prevent acid treatment. Inhibitory compounds created by substantial amounts of acid pretreatment include furfurals, 5-hydroxymethylfurfural, aldehydes, and phenolic acid. There are two forms of acid treatment in advance, reliant on the nature of the program required. However, one type of fast time, i.e., 1–5 min, is used in extreme temperatures $>180°C$, and the second long type, i.e., 30–90 min, with low temperatures $<120°C$. A different step of biomass hydrolysis can be omitted due to hydrolysis employing an acid treatment. However, washing is essential before sugar fermentation to neutralize acid (Sassner et al., 2008; Salerno et al. (2009); Kumar et al., 2011; Luengo et al., 2015; Kumar and Sharma, 2017). For pretreatment of acid, reactors must display conflict to destructive, harmful, and acidic poisons; thus, previous acid treatment can be costly. To strengthen the economic viability of acid rehydration, a critical stage toward the end of the procedure is acid recovery.

Highly concentrated acids are commonly employed to deal with lignocellulosic biomass. The most typical acids are sulfuric acid (H_2SO_4) and hydrochloric acid (HCl). Preacid treatment can be used to extend the hydrolysis process to extract boiling sugar from lignocellulosic biomass. Sulfuric acid pretreatment is traditionally employed for poplar, switchgrass, spruce, and corn stover. A decrease in the sugar from 19.71% and 22.93% was produced due to Bermuda grass acidity, respectively (Kim et al., 2011). In the percolation reactor, retreatment of rice straw was completed in two phases consuming aqueous ammonia and dilute sulfuric acid. When ammonia was used, 96.9% sugar intake was acquired; simultaneously, a 98%

yield was obtained when dilute acid was used. In the concentration of 4% weight of sulfuric acid, pretreatment is desired due to the procedure's meagre cost and efficiency. Combining sulfuric acid causes biomass hydrolysis, after which the addition of xylose to furfural is completed. High temperatures favor hydrolysis by reducing sulfuric acid (Mosier et al., 2005). Elimination of hemicellulose is essential for the growth of sugar from cellulose, and reducing sulfuric acid can have a powerful effect on this goal. Low-cost biomass conversion needs to obtain high levels of xylan-to-xylose. Xylan represents 33% of available carbohydrates in high lignocellulosic substances.

For the realignment of lignocellulosic biomass, in addition to H_2SO_4 and HCl, different acids such as maleic and oxalic acid are also employed. After using oxalic and maleic acid, it would attain a higher pK_a and pH solution than sulfuric acid (Esteghlalian et al., 1997). Due to the two values of pK_a, dicarboxylic acid hydrolyze biomass is more effective than H_2SO_4 and HCl (Brennan et al., 1986). Various benefits embrace lower toxicity to yeast, no odor, increased pH range and hydrolysis temperature, and no glycolysis disruption. Maleic acid contains higher khyd/kdeg; which favors cellulose hydrolysis to glucose over breakdown of glucose (Lee and Jeffries, 2011). Therefore, compared to oxalic acid, the use of maleic acid results in an overabundance of xylose and glucose.

3.1.2 Pretreatment with alkali

The local form of lignocellulosic biomass is not permitted for enzymatic degradation. There is, therefore, a need for a well-designed treatment that can eliminate Biomass enzymatic hydrolysis with an inexpensive amount of processing. To date, some specialized forms of advanced medical technology have been established with the same objective of building biomass more vulnerable to enzymatic anointing. An advanced treatment using alkaline reagent has emerged as one of the most effective methods of the program due to its preinitiation effect and more straightforward method. In addition to acids, a few bases are used in biomass treatment. The main functions of alkaline regeneration are selectively eliminating lignin without causing carbohydrate degradation and will enhance porosity and surface area, improving enzymatic hydrolysis. The effect of alkaline treatment was greatly influenced by lignin content. In comparison to earlier treatments, alkaline treatment necessitates less pressure, temperature, and environment. However, previous alkaline treatment requires time in days and hours. In addition, the reduction of sugar in alkali medicine is much lower than in acid treatment.

In addition, the elimination and regulation of dominant salt are possible and smooth in the event of an alkaline solution. For alkaline treatment, ammonium, calcium, sodium, and potassium hydroxides are utilized, but sodium hydroxide is the most frequently utilized agent of alkaline softening. In contrast, calcium hydroxide ($Ca(OH)_2$) is the least expensive but the strongest among all other alkaline treatments. With the help of calcium dioxide reduction, calcium can be obtained without difficulty in insoluble calcium carbonate using a generation of lime fire; $Ca(OH)_2$ can be rejuvenated. Equipment needed for alkali production is primarily temperature control, tank, CO_2 scarf, water jacket, water and air repetition, pump, tray, body, temperature sensor, and temperature. The initial stage in pretreatment is to create a lime slurry with water, followed by spraying this loose lime on biomass; after spraying, keep biomass for many hours or, in some circumstances, days. Temperature increases might reduce the contact phase (MacDonald et al., 1983; Elshafei et al., 1991; Lee et al., 1994; Kim and

Holtzapple, 2006). Crystallinity index rises to pretreatment due to lignin and hemicellulose elimination. The structural factors caused by previous lemon treatment affect premade biomass hydrolysis. Chang and Holtzapple reported a combination of three structural elements: lignin, acetyl content and crystallinity, and enzymatic digestion (Chang and Holtzapple, 2000). It is established that (1) to achieve excess digestion, a good enough formation, whichever is crystallinity and acetyl content (2) The parameters associated with hydrolysis are eliminated by dehydration and blood sales. (3) Although crystallinity has little effect on eventual sugar generation, it does contribute to earlier hydrolysis. It is indicated on those points that the lignin content should be decreased to 10% and that all acetyl groups should be eliminated with a robust pretreatment process. As a result of exposing cellulose to enzymes, alkaline regeneration performs a critical function. With growing enzymes entering the cellulose and hemicellulose penetration and eradicating nonproductive advertising spots, lignin deduction can increase enzyme activity.

3.1.3 Organosolv

Ethanol generation via bioconversion of lignocellulosic biomass has piqued the interest of researchers in the last couple of years. However, as the first therapeutic process to increase cellulose degradation by this enzyme, it has become an essential factor in the commercial production of cellulosic ethanol. Many of the existing treatments are designed to shrink the exhaustion of biomass over the past few years, but none of them seems to be very hopeful. From the standpoint of the combined use of lignocellulosic biomass, organosolv reconstruction offers a mechanism for the decomposition of biomass. Organosolv pulp is the process of extracting lignin from lignocellulosic raw materials and with natural cleaning solvents or strong detergents. Since the mid-1970s, organosolv production has a great deal of interest. The traditional methods of exploration, the kraft and the sulfite techniques, have various drawbacks, leading to severe health problems, such as airborne and waterborne pollutions. Before, the administration of the organosolv is similar to that of the management of the organosolv; however, the rate of reactivation is not always necessary, as unusual as that of the digestive system.

Correspondingly, organosolv-based therapies have several rewards as trails: (1) solvents derived from organic matter remain easier to improve by using filtered beverages and used for recycling; (2) Chemical rejuvenation in organosolv pulping techniques can differentiate lignin as solids and carbohydrates as syrup; together they display potency as chemical feedstocks (Aziz and Sarkanen, 1989; Johansson et al., 1987; Lora and Aziz, 1985). It looks that organosolv treatment apparently could be more effective in the lignocellulosic biomass treatment, which monitors the use of biomass substances entirely. Nevertheless, there is a flaw in the previous preparation of organosolv. Preformed solids should be cleaned with organic solvent prewashed in water to avoid reducing dissolved lignin, leading to more extensive washing processes. Natural solvents are always costly, so they should be recycled as much as possible, increasing energy consumption.

Similarly, pretreatment of organosolv needs to be eliminated under vital and decomposing compounds, compensating for the instability of organic solvents. Grinding leaks will not be permitted due to the risk of fires and explosions (Aziz and Sarkanen, 1989). As a result, previous organosolv treatment is more expensive than current biomass treatment (Table 1).

TABLE 1 The benefits and drawbacks of various lignocellulosic biomass pretreatment methods.

Pretreatment method	Benefits	Drawbacks
Acid	• High ethanol yield. • Hemicellulose is dissolved.	• Expensive acid costs and the need for recovery • Expensive corrosive-resistant equipment. • The production of inhibitors.
Alkali	• Effective lignin removal. • Inhibitor formation is minimal.	• Expensive alkaline catalyst. • Changes in the structure of lignin.
Ionic liquid	• Lignin and hemicellulose hydrolysis. • Capability to dissolve heavy loadings of various biomass types. • Low temperatures settings.	• Expensive solvents • The requirement for solvent recovery and recycling.
Steam	• Affordability. • Transformation of lignin and hemicellulose solubilization • High glucose and hemicellulose yield in a two-pace method.	• Toxic compound generation. • Partial hemicellulose degradation. • A high lignin content material necessitates the use of an acid catalyst.
LHW	• Parting of practically unadulterated hemicellulose from the remainder of the feedstock. • There is no requirement for a catalyst. • Hydrolysis of hemicellulose.	• Any solid substance (cellulose/lignin) that remains must be dealt with. • Extensive use of energy and water.

3.1.4 Ionic liquids

The scientist was intrigued by the usage of an ionic liquid in the processing of lignocellulosic biomass. Ionic liquids encompassing cations or anions have been a new family of solvents for decades, with a high level of thermal stability, polarity, much less at the melting point, and insignificant vapor pressure (Behera et al., 2014; Zavrel et al., 2009). Ionic liquids are typically formed by critical organic cations and tiny inorganic anions. Factors such as the amount of the anion charge delocalization and cation structure significantly influence ionic liquid's physical, chemical, and biological properties. The interaction between the ionic liquids and biomass gain was plagued by temperature, cations and anions nature, and the pretreatment period. The science behind ionic liquids is like they contest for hydrogen bonding with lignocellulosic constituents, and in this antagonism, breakdown of the linkage arises. Moderate modifications in biomass composition happened after ionic liquid pretreatments, even though massive adjustments have been found in the biomass structure. Due to excessive thermal and chemical balance, much less risky processing conditions, the low vapor pressure of solvents, and sustaining liquid state over a wide temperature range, ionic liquid pretreatment is considerably less preferred over other procedures. Ionic liquids are easily recyclable and nonderivatizing. The downside of ionic liquid pretreatment is that cellulase unfolding and inactivation occur due to cellulase incompatibility with ionic liquids. Because cellulose solubilizes at low temperatures at significantly lower viscosities, viscosity is a significant aspect to consider when employing ionic liquids concerning the overall power depletion of the technique. High temperatures will lead to more side effects and poor performance, such as a decrease in the ionic liquid balance (Behera et al., 2014).

3.1.5 Ozonolysis

Preozone depletion is another excellent way to reduce lignin content in the lignocellulosic biomass content. The use of ozone pretreatment improves biomass in vitro digestibility. In this situation, no inhibitors are created, which is a massive benefit because alternative chemical treatments produce harmful residues. Ozone works as a catalyst to lignin degradation in this process. Ozone gas dissolves in water, and a dominant oxidant emits minute amounts of molecules and soluble compounds by breaking down lignin. To minimize lignin and hemicellulose, wheatgrass, bagasse, cotton grass, unripe grass, poplar sawdust, pineapple, and pine can be mixed with ozone. Hemicellulose undergoes only minor correction, whereas cellulose undergoes essentially no alteration. Ozonolysis equipment includes an ozone catalytic destroyer, an iodine trap used to check the performance of the catalyst, an oxygen cylinder, an ozone generator, a three-way valve, an ozone UV spectrophotometer, pressure regulation, a process gas humidifier, a vent, and automatic control gas flow (Kumar et al., 2009). The moisture content significantly impacts lignin oxidation via ozone pretreatment because lignin oxidation reduces with increasing intensity within the biomass moisture content. Ozone depletion is restricted to low concentrations of water, which is ultimately due to its reuse with biomass. Long-term ozone depletion results from the closure of pores through water film (Mamleeva et al., 2009). The pH of water decreases during Ozonolysis due to the creation of natural acids. Because it eliminates lignins attached to carbs, alkaline media induces retardation. The synthesis of inhibitory compounds is linked to biomass separation. As a result of regeneration, several aromatic and polyaromatic compounds are formed (Travaini et al., 2013).

3.2 Physicochemical pretreatment

3.2.1 Explosion by steam

Steam-Explosion pretreatment is one of the most extensively recycled preferences worldwide, as it practices chemical and physical methods to disrupt the formation of lignocellulosic substances. This process of hydrothermal pretreatment puts the matter at high pressures and short-term temperatures and then quickly compresses the system, disturbing the formation of the fiber. Fiber interruption will increase cellulose access to enzymes during the period of hydrolysis. Particle size is a crucial factor subsidizing the method's performance, and it appears that the larger particles have been able to produce more sugar concentrations. This is a good acquisition, as reducing particle size entails treating the raw fabric machine using production costs (Ballesteros et al., 2002). Temperatures from 190°C to 270°C were used in indoor conditions for 10 and 1 min, respectively. The initial factors and the size of the particles to be processed can significantly impact the relationship between time and temperature (Duff and Murray, 1996; Viola et al., 2008).

It has also been explored with a two-step vapor barrier option that allows dissolving the structure in distances that develop the highest quality component during a period of hydrolysis. The first place involves temperatures of 180°C to dissolve and eliminate the hemicellulose fraction. The succeeding pace applied a treatment suppressed with high temperatures up to 210°C, not surpassing 240°C, in which the cellular portion turned out to be a violation of its carbohydrate interactions (Söderström et al., 2002; Tengborg et al., 1998). This two-step approach increases the yield of low ethanol by increasing the approachability of the cellulose structure due to the reduction of the hemicellulose fraction. Operating costs are also reduced as a small amount of the enzyme is required due to the high availability of cellular fractions (Söderström et al., 2002). However, the extended cost of the system required to process and use additional power for the second method of smoke extraction is requisite. Acid catalysts have been cast off within the steam explosion method to reduce the components that allow you to enhance hemicellulose hydrolysis during all cellulose treatment and treatment within this system. Combine strong acids to reduce the storage time and temperature of the current active structures or allow softwood in this previous treatment, where it has now been shown to be less expensive at first. With a reduction in storage time and temperature by incorporating this acidic compound, the reduction of compound blockers is observed. Complete deletion of hemicellulose is imminent, and temporary; temporary hydrolysis in production is available (Ballesteros et al., 2006).

3.2.2 CO_2 explosion

In the same way, carbon dioxide is charged to the reactor in some form. Depending on the pretreatment level, CO_2 has been applied to pressures flanked by 30 bars (such as steam-based explosions) and 275 bars (top notch-important CO_2) (Kumar and Sharma, 2017; Zheng et al., 1998; Cha et al., 2014). Temperatures are almost lower than steam blasting, operating between 35°C and 175°C (Zheng et al., 1998; Kim and Hong, 2001; Cha et al., 2014; Kumar and Sharma, 2017). CO_2 processes are performed without aggregates throughout the house, while with external CO_2 pressures, Cha et al. (2014), took advantage of the feedstock impeller as it is highly compressed under pressure. The reactor base can be allied to a second outlet box, and the pressure variance between the two vessels conducts the reduction function. The usage of this type of reactor enables CO_2 to be reused. Previous treatment methods for lignocellulosic waste hydrolysis used a complete explosion-based CO_2 process. It has been found that decompression eruptions induced with the help of CO_2 to penetrate hollow networks are beneficial in increasing the available waste area (Zheng et al., 1998). This is especially true when CO_2 is used separately because the type of reactant contains liquid and air. As a liquid, CO_2 can pass through feedstock and form carbonic acid, although it is considered a weak acid origin to undergo hydrolysis (Zheng et al., 1998; Sun and Cheng, 2002; Mood et al., 2013; Kumar and Sharma, 2017). It is unnecessary to dry feedstock material to obtain two-edged acid-designed hydrolysis and shear-based force based on the explosion. CO_2 obtains its acidic qualities when dissolved in water (Mood et al., 2013; Kumar and Sharma, 2017). Also, the energy consumption of fibrous bursts caused by pressure appears to be very low. However, overpressure equipment's initial cost and potential protection costs, especially when making alarmingly effective solutions, can be of very high prices. In the context of the environment, that is a raw form of pretreatment as the CO_2 has just been used, as is the case with CO_2 consumed during thermochemical processing to be recycled (Cha et al., 2014). The practical solution of carbon dioxide explosion originated as non-economical while using the high pressures were specified (Brodeur et al., 2011).

3.2.3 Liquid hot water

Hydrothermal methods are innovative because of the affordable prices of the principal reagent, water. In addition, this refers to the pretreatment physicochemical biomass treatment as the simplicity of the comparison of this method can be done from the lab scale to the driving and sales scale; pretreatment of hot water treatment is frequently compared to a smoke explosion. This process, however, is carried out in the fluid stage at a high temperature of 160–250°C. Also, it works at a distance of less than miles ($\sim$5bar); which means that average passing time reduces chemical and mechanical fatigue (Agbor et al., 2011; Kim et al., 2016; Zhuang et al., 2016; Kumar and Sharma, 2017; Li et al., 2017). Another process value can be prevented compared to the different chemical processes necessary to reduce or, after the treatment, wash the biomass waste that was generated. This pretreatment factor roots hemicellulose hydrolysis and reduces lignin content, whereas only 4–22 wt% of accessible cellulose is removed (Agbor et al., 2011; Kumar and Sharma, 2017). Except for not using an external printing source while using the specified high temperature, the H$^+$ water content is determined to be 20 times better than during room temperature (Kim et al., 2016). However, hot prewater treatment techniques will serve as a double-edged sword, and even minor acid hydrolysis is also possible. This previous treatment was conducted without immediately removing the reactor but managing the stay period (15–70 min) as another effective strategy to boost cellulosic content (Zhuang et al., 2016; Li et al., 2017). Hot water preheat treatment results in infractions. The first element is a liquid containing dissolved hemicellulose. The second component is a previously treated solid component that can be dehydrated and thermochemically changed to generate energy and carbonaceous compounds (Zhuang et al., 2016).

3.2.4 Wet oxidation

Using an oxidizing agent in a liquid environment is a beneficial prerupture treatment for lignocellulose in its components (An et al., 2019). An oxidative method has been observed to improve the mix of hemicellulose and breakdown lignin. It is usually utilized as a part of a complete pretreatment method and a variety of digestive and digesting procedures. This is why this process was previously mentioned as oxidative delignification (Zhuang et al., 2016; Li et al., 2017). Examples of previously used oxidizers are compressed gases containing air and oxygen (5–30bar) or liquid-based peroxides in particular (McGinnis et al., 1983; An et al., 2019). The effective temperature of this scheme within the presence of the oxygen zone has exceeded between 120°C and 350°C (Munir et al., 2018). The duration of stay is usually between 0.5 and 4h, while pre-hydrogen peroxide has passed through temperatures for miles (30°C) with a very long shelf life of $\sim$8h (Munir et al., 2018). Pretreatment interventions influencing wheatgrass and hemicellulose solubility have been temperature-dependent, increasing the process temperature from 150–185°C, tripling the concentration of the solution (Schmidt and Thomsen, 1998). Where complete oxidation happens, the oxidative process produces carbon dioxide and water as products.

Furthermore, the hydrolysis of polymeric chains can create molecular weight carboxylic acids, aldehydes, and alcohol (Palonen et al., 2004). Similar techniques of free interaction are evident in early sonochemical treatments (mentioned earlier); This can result in hydrogen peroxide, a powerful bonding agent (Munir et al., 2018). Liquid-liquids are often combined with a soluble water base, even if they provide additional costs in the whole process. They are beneficial when separating excess cellulose yields from lignocellulosic biomass waste (Schmidt and Thomsen, 1998).

3.2.5 *Sulfite*

Cell walls of lignocellulosic biomass are natively resistant to microbial and enzymatic breakdown, referred to as "recalcitrance." Lignocellulose's recalcitrance is a significant impediment to the commercially viable growth of biobased fuels and commodities (Himmel et al., 2007). Unfortunately, no methodologies or platforms exist to convert woody biomass, particularly softwood, to fuel ethanol efficiently and cost-effectively. Existing alkaline, dilute acid, hot water/steam(explosion), ammonia, and organosolv pretreatment processes have impacted (Mosier et al., 2005; Pan et al., 2005). Some fundamental challenges related to woody biomass bioconversion, however, remained unaddressed. A pretreatment approach based on Sulfite, specifically sodium bisulfite, has been created to resolve this problem. The raw material, woody chips, is processed in an aqueous sulfite solution before being mechanically reduced by disk refining.

Impregnation of wood chips allows pretreatment chemicals to infiltrate these chips uniformly before reactions at the ultimate pretreatment temperature. Wood chips were immediately processed with $NaHSO_3$ with or without H_2SO_4 preceding size reduction. At 90°C, the wood chips were first saturated with the pretreatment liquid. The time required for impregnation ranged from 0 to 3 h. The temperature was raised to 180°C in about 30 min after impregnation and sustained for another 30 min. To achieve uniform delignification in acid sulfite and bisulfite chemical pulping, impregnation is critical (Fig. 2).

Impregnation is often performed by immersing the wood chips in the pretreatment solution for the final few minutes to several hours at temperatures lower than the pretreatment or pulping temperature. Impregnating the wood chips in acid sulfite pulping takes 3–6 h. Although essential in sulfite pulping, a long impregnation time affects productivity and energy. All pretreatment processes were carried out for 30 min at 180°C. The solid was acquired after pretreatment and delivered straight to disk refiners for size reduction without washing.

3.3 Biological pretreatment

Since the 1950s, enzymatic hydrolysis has been utilized. Thanks to the renewed interest in cellulose hydrolysis in biology, it has gotten a lot of attention lately. This is a route in which enzymes aid in the breakdown of bonds in molecules by introducing water components. Enzymatic hydrolysis is based on the chemical interaction of an enzyme, such as cellulose or trypsin, with the cell wall to loosen the connections and permit for easier extraction of lipids after cell breaking. The procedure entails the inclusion of cellulases to hydrolyze pretreated lignocellulosic feedstock into fermentable sugars in the present background. The procedure includes several critical steps: (1) allocation of enzymes from the aqueous medium to the skin of the cellulose (2) adsorption of enzymes and establishment of enzyme–substrate complexes (3) hydrolysis of the cellulose (4) allocation of hydrolysis products from the cellulose particle surface to the bulk aqueous phase (5) hydrolysis of sugar-type products to glucose. Enzymatic

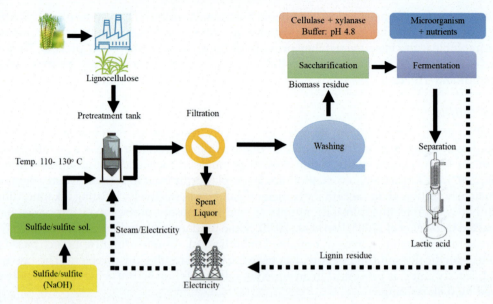

FIG. 2   Flow diagram of the sulfite process.

hydrolysis transforms cellulose to reducing sugars, which can then be fermented to ethanol by yeasts or bacteria (Sun and Cheng, 2002). Cellulase may degrade cellulose to sugar at 45–50°C and pH 4.8. This stepwise response emerges in a heterogeneous channel in which indecipherable cellulose is firstly decomposed at the solid–liquid interface by the synergetic effect of endoglucanases (EC 3.2.1.4) and exoglucanases/cellobiohydrolases (EC 3.2.1.91). Following that, liquid-phase hydrolysis of aqueous transitional products, especially short cellulo-oligosaccharides and cellobiose, occurs, which is enzymatically split to release glucose by the acting of glucosidase (EC 3.2.1.21) (Andrić et al., 2010. The total pace of the procedure is prejudiced by the structural topographies of the lignocellulosic feedstock and the configuration and availability of the cellulases. Despite the advantages of enzymatic lysis, the disintegration of cell walls with enzymes is not lucrative due to its high cost and low efficacy. Because it is usually performed under favorable conditions and does not generate corrosion concerns, enzymatic hydrolysis is low-priced than acid or alkaline hydrolysis (Duff and Murray, 1996). Both bacteria and fungi can produce cellulases for the hydrolysis of lignocellulosic materials.

Glucose is fermented in fermentation methods to make a thinned ethanol–water mixture. Although it is challenging to transform xylose to ethanol, it does happen. Lignin, which cannot be transformed biologically, can be improved by chemicals or, more commonly, merely incinerated as boiler fuel. Simultaneous saccharification and fermentation (SSF) was created to ferment sugars generated from lignocellulose (Han and Chen, 2010). To eliminate end-product inhibition, the method incorporates hydrolysis and fermentation. The combination of hydrolysis and fermentation removes glucose quickly before it can hinder further hydrolysis. The input material is processed and hydrolyzed to create a cocktail pentose which includes both (xylose and arabinose) and fiber. Before combining cellulose and hemicellulose enzymes, the solution is neutralized with limestone. The leftover hemicellulose and cellulose are digested to produce hexose (glucose) and pentoses (xylose and arabinose), which are promptly fermented to produce ethanol. The ideal temperature for the hydrolysis/fermentation response is somewhere between the ideal temperature for cellulase action and the yeast. Lignin is extracted from the mixture and utilized as boiler fuel, or it is processed into high-value octane enhancers blended with gasoline.

One of the most important elements influencing the yield and starting rate of cellulose enzymatic hydrolysis is substrate concentration. At low substrate concentrations, increasing the substrate concentration usually results in a higher yield and hydrolysis reaction rate (Cheung and Anderson, 1997). Therefore, several strategies for reducing hydrolysis inhibition have been devised: high enzyme concentrations, including glucosidases and sugar removal throughout hydrolysis, via ultrafiltration or (SSF).

4 Strategies for fermentation of xylose to ethanol

Many yeasts have been assessed for their capacity to ferment xylose to ethanol for high yields directly. More information on the cultivation of yeast is accessible, notably for bacteria and fungus yeast. Before the growth of recombinant *E. coli* strains, xylose-fermenting yeast was thought to be the most efficient way to produce high-yield ethanol. The benefits of adopting a yeast-based (e.g., *S. cerevisiae*) route for xylose translation arises from how yeast operates sound at truncated pH and is tolerant of ethanol. These two variables decrease the prospect of bacterial contamination. An additional benefit is that antibiotics can eradicate contamination when pH and nutritional excursions are insufficient. Furthermore, yeasts are essential as a byproduct that can be utilized as a food or protein source to improve animal diets. Many evidence is available of large-scale yeast farming.

4.1 Separate hydrolysis and fermentation

Separate hydrolysis and fermentation (SHF) is a procedure that sequentially performs enzymatic hydrolysis and fermentation. Enzymatic saccharification of starchy biomass or pretreated lignocellulosic biomass is performed initially in this method at the ideal temperature of the saccharifying enzyme. Following that, suitable microorganisms are introduced to ferment the saccharified solution. The temperatures of enzymatic hydrolysis and fermentation can be optimized independently in the SHF process. Enzymatic hydrolyses require fewer saccharifying enzymes than simultaneous saccharification and fermentation since they are accomplished at the ideal temperature.

Furthermore, the risk of contamination is lowered since saccharified solutions containing fermentable sugar can be sterilized. On the other hand, the SHF method is carried out in two separate processes that need two distinct reactors for the saccharification and fermentation processes; thus, the capital costs are much higher than that of the simultaneous process.

4.2 Separate hydrolysis and co-fermentation

Separate hydrolysis and co-fermentation of hexoses and pentoses is a procedure that is identical to separate hydrolysis and fermentation, but that hexose and pentose fermentation happen concurrently. SHF can simplify processing while lowering capital costs. The marketing of bioethanol for the second generation has been hampered by many glitches, including inefficient biomass usage and high production costs. The ethanol output and ethanol concentration in the fermentation broth are widely acknowledged as the most critical metrics in lowering production costs. Agricultural leftovers include a high concentration of hemicellulose; using xylose to boost ethanol concentration and production during fermentation is a viable option. Many ethanol-fermented microorganisms that naturally arise are not xylose but can also co-ferment glucose and xylose using genetically engineered yeast strains. The rate of absorption of xylose is raised, however, only if the glucose levels are low.

For example, xylose and glucose are competitively carried by the matching transport protein in *S. cerevisiae* (Kilian and Van Uden, 1988; Meinander et al., 1999); however, xylose has a 200-fold lower affinity (Kötter and Ciriacy, 1993). The content of glucose in the medium must therefore be minimal to prevent glucose inhibition. A low but not nil glucose level was found to increase the use of xylose using a gradual release or feeding glucose (Meinander et al., 1999). Since glucose is released during hydrolysis, this is the sole reason why SSCF has become an appealing alternative process.

4.3 Engineered yeasts for ethanol fermentation from pentose sugar

Over the last few years, remarkable progress has been achieved in the knowledge of xylose metabolism and the recognition and production of strains with superior xylose fermentation attributes. The efficient development of recombinant bacterial and yeast strains with enriched xylose fermentation functionality is possibly the single most impactful achievement in recent years in xylose transformation research. The present rate of advancement in the field of xylose conversion is significant. There is already a substantial form of literature available on the enactment of xylose-fermenting yeast. Statistics on the efficacy of recombinant *E. coli* are also rapidly piling up. Attempts are being made through genetic engineering to generate a recombinant *S. cerevisiae* strain susceptible to high-yield xylose fermentation, and efforts to create superior bacterial strains are also underway. The numerous advancements being made in the realization of the effective expansion of large-scale technologies based on efficient high-yield xylose fermentation should be possible within the next decade.

Bioethanol from lignocellulosic biomass is a potential substitute energy source. *S. cerevisiae* is the best microorganism for producing ethanol due to its extraordinary skill to ferment glucose and its high acceptance of ethanol and inhibitors prevalent in lignocellulosic hydrolysates (Stambuk et al., 2008). Unfortunately, it cannot ferment xylose, which is prevalent in biomass hydrolysates. In *S. cerevisiae*, xylose is metabolized into xylulose by two enzymes that utilize diverse cofactors, prominent in a redox imbalance and the inability to ferment xylose. (Stambuk et al., 2008). Two principal tactics have been used to tackle this challenge: Cloning of xylose reductase and xylitol dehydrogenase that are both allied to almost the same coenzyme or cloning of xylose isomerase that translates xylose openly into its isomer xylulose. Unfortunately, yeasts engineered to ferment xylose digest it sluggishly, and store xylitol (Gao et al., 2019). As a result, considerable genetic modifications were performed to enhance xylose-specific intake, rate, and yield of ethanol production: (i) overexpression of the enzymes required to translate xylulose into intermediate glycolytic; (ii) obliteration of the endogenous aldose reductase, which transforms xylose to xylitol. Aside from the metabolic engineering technique, evolutionary engineering has been applied to develop the cell function and durability of the recombinant strain (Cai et al., 2012).

Another bottleneck of economic ethanol synthesis from hydrolysates of biomass is the parallel conversion of xylose and glucose (Kang et al., 2014; Eiteman et al., 2008). Xylose fermented yeast cannot take xylose until the glucose is fully consumed. One plausible cause for this occurrence is that glucose curbs the expression of genes required for xylose catabolism via Mig1, a crucial transcription factor in the catabolic repression process. Mig1 speedily travels from the cytoplasm into the nucleus in the presence of high amounts of glucose and fixes to the promoters of glucose-repressible genes. Mig1 is transported back to the cytoplasm when cells are starved of glucose, relieving glucose repression (Bisson et al., 2016). To overwhelm the inhibitory impact of glucose on xylose usage, a recent publication reported a genetically modified yeast strain deliberate to accomplish intracellular hydrolysis of cellobiose, permitting cellobiose and xylose consumption (Kang et al., 2014). Notably, xylose's successful utilization after glucose depletion could be attributable to xylose–glucose competition during uptake. This pentose is transferred in *S. cerevisiae* by active transport regulated by hexose permeases, which also transport xylose with shallow empathy compared to glucose absorption (Subtil and Boles, 2012). The development of a xylose-specific transporter that is not repressed by glucose and has a high similarity and transport capability may increase cellular enactment in xylose fermentation in biomass hydrolysates (Ruchala et al., 2020).

5 Influence of several xylose metabolic pathways on xylose fermentation

Several yeasts were observed to be proficient in manufacturing ethanol from xylose. Among these, *Pachysolen tannophilus*, *Candida shehatae*, and *Pichia stipitis* were the top-notch of his ability to provide, especially the excessive ethanol yield from xylose. Significant studies have been conducted on the physiology and biochemistry of that yeast (Ko et al., 2016). The results confirmed that the degree and yield of xylose-derived ethanol using this yeast was lower than that of glucose fermentation (Lee et al., 2014). This low ethanol yield was converted into a simultaneous use of ethanol before high xylose concentrations (Winkelhausen and Kuzmanova, 1998). It produced xylitol, ribitol, arabitol, and acetic acid (Baptista et al., 2018). Yeast that converts xylose well into defined media is often detrimental to used biomass hydrolysates or alcohol waste products of lignocellulosic substances. The occurrence of hexoses is a crucial concern (especially glucose and mannose) that contend or block the use of xylose (Almeida et al., 2008; Ruiz et al., 2012). Pentose-fermenting yeast can translate glucose or xylose discretely into ethanol. The fermentation of the xylose found in combination with glucose, however, is no longer progressing well. This is because glucose fermentation preceded by xylose and pentose-fermenting yeast usually does not tolerate enough ethanol to complete a one-second process. A new delinquent is the existence of cofactor imbalances within the primary enzymes of xylose metabolism, XR with a better attraction for NADPH and xylitol dehydrogenase (XDH) most actively acting on NAD (Ishii et al., 2013).

In some cases of xylose metabolism, certain coenzymes secreted by XR and XDH lead to the accretion of NADP and NADH from primary and secondary reactions. This cofactor mismatch limits metabolism along these pathways, and oxygen regeneration of cofactors is expected to maintain xylose metabolism. This prompted a proposal that having XR selective NADH would obviate the requisite for cofactor renewal, allowing xylose to be boiled improperly (Bruinenberg et al., 1984). The oxygen demand for the use of xylose by *P. tannophilus* was discovered, and this has been identified as one of the causes of ineffective xylose fermentation (Neirinck et al., 1984).

Physicochemical regeneration reveals a hybrid combination of inhibitors that negatively distress yeast growth, function, and fermentation (Aguilera and Prieto, 2001). Furthermore, the availability of oxygen throughout the xylose fermentation subsidizes the simultaneous production and use of ethanol simultaneously as the mass of xylose remains in between (Wahlbom and Hahn-Hägerdal, 2002). Any other disadvantage is the manifestation of many preexisting inhibitors that undesirably mark yeast growth and fermentation. Different settings are required for dissimilar substrates, and those elements can lead to the production of different concentrations of dissimilar inhibitors. The overhead encounters have steered struggles to report these concerns to enhance the pentose fermenting general enactment of lignocellulosic hydrolysates.

The preferred microorganism *S. cerevisiae* does not naturally absorb xylose to produce significant ethanol. The hemicellulosic section composed mainly of xylose must be converted to ethanol by forming an S-type. However, *cerevisiae* can absorb xylose and have a high concentration of chemical compounds. Xylose synthesis is obtained by converting xylose into xylulose and eventually phosphorylated into xylulose-5-phosphate, also synthesized into a pentose phosphate method (Fig. 3). Two previously modified mechanisms have been proposed in *S. cerevisiae*; two distinct pathways are utilized to convert xylose to xylulose: oxidoreductase and isomerase. Several forms of xylose yeast use the oxidoreductase technique, which comprises two enzymatic processes catalyzed by xylose reductase (XR) and xylitol dehydrogenase (XDH) (Liu et al., 2008). Instead of NADPH over NADH as a cofactor, XR first converts xylose to xylitol; xylitol is then linked to xylulose by XDH, which utilizes just NAD + as a cofactor. The XR/XDH line exhibits a cofactor mismatch between XR-dependent XR and XDH–XDH-dependent XDH, causing xylitol accumulation and reducing ethanol synthesis (Salerno et al., 2009).

Furthermore, *S. cerevisiae* introduces the natural gene GRE3, which incorporates the unadulterated aldose reductase of NADPH to convert xylose into xylitol (Kumar et al., 2011). The expression of this enzyme, which is very effective in using NADPH as a cofactor, can irritate redox imbalances, thus leading to the accumulation of excess xylitol and improper maturation of Xylose (Luengo et al., 2015). In this sense, the removal of GRE3 (Mosier et al., 2005, Sassner et al., 2008) and genetic mutations aid cofactor regeneration, (Esteghlalian et al., 1997) decreases xylitol, and thus then ethanol production increases.

5.1 Genetic improvement for ethanol production in native xylose-fermenting yeasts

The most important source of bioconversion of sugar for biofuel and other vital chemicals is lignocellulosic substrates. To expand the economics of biomass conversion, most of the sugars in hydrolysate must be transformed into desired products. While hexoses are added

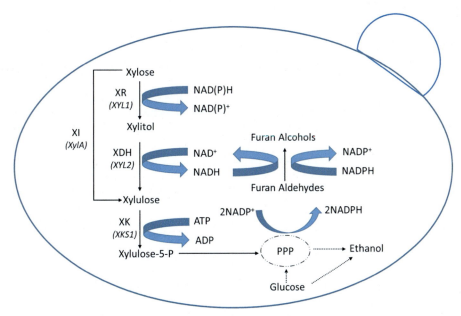

FIG. 3 Schematic picture of diverse xylose utilizing pathways, glucose metabolic pathway, and furan detoxification in *S. cerevisiae*. The fungal route employs xylose reductase (XR) and xylitol dehydrogenase (XDH), though the bacterial process employs xylose isomerase (XI) (XI). Together, these pathways yield D-xylulose, which is transformed to D-xylulose-5P by endogenous xyulokinase (XK). Then, under suitable conditions, D-xylulose-5P enters the pentose phosphate pathway (PPP), which is converted to produce ethanol. Arrows indicate the chemical steps.

to ethanol and other high-value compounds, pentoses are not converted to hydrolysates. As a result, this is still one of the essential chemicals in the evolution of lignocellulosic biomass. Indigenous pentose yeast can boil both glucose and xylose in lignocellulosic biomass in ethanol. They act below hydrolysate inhibitors, show less tolerance to ethanol and glucose despotism, and fermentation breathes less efficiently than primary glucose hexoses and mannose.

The process of adaptability is perhaps lazy and time-consuming but somewhat beneficial. The development of industrial resilience often includes randomly modified organisms where most high competitors are chosen for mutagenesis and testing. A sensitive and powerful screen is needed to select the desired changes. Despite the effectiveness of random mutagenesis, only one study has used this method to detect xylose genes by increasing acceptance to inhibitors in lignocellulosic hydrolysates.

Genome shuffling includes conventional mutagenesis and the opportunity of reunification to allow for the growth of the same pressure. This approach uses repeated repetition to create the beauty of genetic variants with continuous phenotypes, taking into account the advent of new mutant compounds and problems with advanced features (Biot-Pelletier and Martin, 2016). Genome regeneration quickens the formation of many favorable mutations in simultaneous pathways and the ability to reverse the dreaded genetic mutations, thus recalling the engineering of composite phenotypic signals with minimal time and effort (Johannsen et al., 1985; Kavanagh et al., 2003). The key to successful genetic shuffling is an operative and subtle screen to pick continuous mutations as with random mutagenesis. In genetic mutations, regeneration can be performed using protoplast fusion replication or cross-linking.

Major accounts on genome shuffling enhance accepting an inhibitor of yeast ethanol product based on protoplast fusion (Kavanagh et al., 2003; Khattab et al., 2013). However, completed genome mixing with the help of protoplast fusion can also produce hybrid species (Giudici et al., 2005). Old-fashioned genetic modification techniques can also lead to random mutations during the genome, a few of which may be unfavorable. On the other hand, metabolic engineering processes permit the defined modification of particular pathways (Jeffries and Jin, 2004; Jeffries, 2006). The challenge with the latter techniques is that the details of what alterations are required to achieve the desired phenotypic effect are unknown. The lignocellulosic biomass plant embodies one of the leading striking food sources for ethanol fuel production and chemicals. To enhance the financial viability of this type of approach, potential biocatalysts need to be able to successfully convert hexoses and pentoses cut into industrial lignocellulosic hydrolysate into the anticipated products.

Indigenous xylose yeast has the potential to convert essential sugars into lignocellulose hydrolysates into ethanol, and it has been a significant source of interest since its unearthing in the initial 1980s. On the other hand, this yeast ferments xylose efficiently and produces a significantly lower output than glucose fermentation. Furthermore, they have glucose intolerance, poor ethanol acceptance, and are vulnerable to several therapeutic-based inhibitors in industry-appropriate hydrolysates. These issues have impeded their biotechnological capacity, and they have worked for the past three decades to enhance their structures utilizing a combination of traditional and molecular techniques. Despite the improved pressures, ethanol production and xylose synthesis no longer matched the overall performance of typical fermented yeast *S. cerevisiae* in glucose.

In addition, the overall enactment of this yeast in preindustrial hydrolysate works remains poor. The genetic development of microorganisms and traditional fermentation yeast will continue to improve (powerful) forms of lignocellulosic biomass mutations. With the introduction of advanced testing and several "omics" techniques, weight gain is geared toward moderate improvement. Essentially, new methods used for pentose-fermenting yeast can find new vistas in detail in biological and metabolic barriers that inhibit the proper fermentation of pentose sugar in lignocellulosic hydrolysates. Those new details keep their assurance regarding the creation of solid strains capable of fermenting most of the sugars in lignocellulosic hydrolysates into ethanol.

6 Ethanol from xylose: Techno-economic feasibility

Early days economic estimation for cellulosic ethanol production indicated toward the cost of conversion as the primary factor influencing the overall process economics (Lynd et al., 51,991), while other TEA studies cited the optimization of operational expenditure (OPEX) (Nguyen and Saddler, 1991; Von Sivers and Zacchi, 1995; Wyman, 1994). However, a better understanding of process design and economics brought a bright side to cellulosic ethanol production. The typical substantial factors contributing toward high economic estimations of lignocellulosic ethanol involve the feedstock prices, product yield, and the cost of associated enzymes for enzymatic hydrolysis (Chovau et al., 2013). However, most publications' main critical factor is the capital cost of cellulosic ethanol production (Galbe et al., 2007).

The National Renewable Energy Laboratory (NREL) first and second reports were published in 1999 and 2002 for lignocellulosic ethanol production. The cost of ethanol cited in the first report was 0.38 US$/L (Wooley et al., 1999). In contrast, the second report with revised strategies suggested improvements in the technology to achieve a minimum estimated selling price of 0.28 US$/L in 2010. The further published report in 2011 by NREL presented an MESP of 0.57 US$/L (Gubicza et al., 2016).

Zhao et al. (2015) carried out techno-economic and sensitivity analyses to identify the key elements responsible for influencing the plant-gate price (PGP). The considered process scheme included dilute-acid pretreatment of corn stover, enzymatic hydrolysis to extract glucose, and subsequent fermentation of glucose and xylose into ethanol. The estimated PGP was in the range of $4.68–$6.05/gal. He concluded that ethanol production from lignocellulosic biomass is not feasible at the time of this study compared to fossil gasoline prices in China. Several factors resulting in higher PGP were biomass cost, conversion efficiency and rate of enzymatic hydrolysis, the ratio of pentose sugars converted to bioethanol, and amount of enzyme loaded, etc.

Li et al., 2018, compared the biomass sugars extraction efficiency of three low boiling point polar solvents, viz., tetrahydrofuran (THF), acetone, and 1,4-dioxane. Enzymatic hydrolysis was performed separately to extract out glucose, followed by co-fermentation of C_5 and C_6 by a recombinant strain of *Zymomonas mobilis*. The biomass solvent liquefaction carried out at 2000 metric tons per day scale had ethanol minimum fuel selling price in the range of $2.98–$4.06 per gallon. THF was found to have the lowest MFSP. However, to improve process economics, further process optimization was suggested.

7 Future prospect and way forward

Several developments have been done in the recent past to produce lignocellulosic bioethanol from xylose by yeast fermentation. Also, numerous bottlenecks like biomass conversion, inhibitors generation, rate of xylose conversion, and lower ethanol yield have been tackled via different strategies to make 2G ethanol production economically feasible. Although several biomass hydrolysis methods are well portrayed in the literature, compelling production of lignocellulosic sugars without compromising yields and efficient conversion into ethanol is the prerequisite to achieve sustainable ethanol productivity and recovery (Girio et al., 2010).

Yang et al., 2009, reviewed the studies performed with genetically modified strains of *S. Cerevisiae* capable of assimilating xylose for ethanol production by considering three aspects: xylose transporters and their importance xylose metabolic pathway and tolerance toward inhibitors incorporated in the hydrolysate. The use of conventional yeast strains for lignocellulosic ethanol production has not been considered a favorable option. Sharma and Arora (2020) have discussed almost all the metabolic engineering strategies for ethanol fermentation from lignocellulosic biomass. Selim et al., 2020 highlighted the recent developments focused on native xylose assimilating yeasts by emphasizing the potential of a newly isolated strain S. Passalidarum to overcome the bottlenecks of xylose consumption through genetic engineering. In attribution to biological alteration through the approach of inborn or foreign metabolic pathways, adaptive engineering plays an integral part in this process. Because of the metabolic difficulty associated with the growth of xylose-fermenting yeast, it is critical to apply metabolic engineering in conjunction with other current approaches, for instance, global transcription machinery engineering (gTME), to speed up physiological adaptation in xylose. In addition, a future challenge will be the vigilant optimization of the altered yeast through the application of a mixture of diverse technologies.

8 Conclusion

In line with the increasing worldwide emphasis on sustainable development through the use of lignocellulosic wastes to make biofuels and other value-added compounds in the paradigm of biorefineries, boosting *S. cerevisiae's* ability to metabolize xylose is crucial. Many metabolic obstacles are to be addressed, including xylose absorption, the catalytic proficiency of xylose isomerases, xylitol generation, and enhanced unambiguous evolution in xylose. Research of process techniques using a wide range of newly generated xylose/xylose & glucose fermenting wild yeasts should be promoted to develop an economically sustainable commercial technique for generating ethanol from lignocellulosic hydrolysates expressing xylose. However, when revisiting 2G ethanol process carefully in terms of different unit operations starting from field to wheel, and carbon recovery in each step, the xylose-to-ethanol's techno-economics may not be favorable with ethanol as a sole product. Rather, it is recommended that even if a robust xylose fermenting wild yeast variety is available, conversion of xylose into high-value biochemicals other than ethanol, could suffice the ethanol cost to compete in the biofuel market.

Acknowledgment

Authors thankfully acknowledge Dr. Anjan Ray, Director CSIR-Indian Institute of Petroleum, for his constant encouragement and support. A.N. and T.S. thankfully acknowledge the Academy of Scientific and Innovative Research (AcSIR) for academic support.

Funding sources

None.

Conflict of interest

The authors declare that they have no competing interests.

References

Agbor, V.B., Cicek, N., Sparling, R., Berlin, A., Levin, D.B., 2011. Biomass pretreatment: fundamentals toward application. Biotechnol. Adv. 29 (6), 675–685.

Aguilera, J., Prieto, J., 2001. The *Saccharomyces cerevisiae* aldose reductase is implied in the metabolism of methylglyoxal in response to stress conditions. Curr. Genet. 39 (5), 273–283.

Almeida, J.R., Modig, T., Röder, A., Lidén, G., Gorwa-Grauslund, M.F., 2008. Pichia stipitis xylose reductase helps detoxifying lignocellulosic hydrolysate by reducing 5-hydroxymethyl-furfural (HMF). Biotechnol. Biofuels 1 (1), 1–9.

An, S., Li, W., Liu, Q., Xia, Y., Zhang, T., Huang, F., Chen, L., 2019. Combined dilute hydrochloric acid and alkaline wet oxidation pretreatment to improve sugar recovery of corn Stover. Bioresour. Technol. 271, 283–288.

Andrić, P., Meyer, A.S., Jensen, P.A., Dam-Johansen, K., 2010. Reactor design for minimizing product inhibition during enzymatic lignocellulose hydrolysis: I. Significance and mechanism of cellobiose and glucose inhibition on cellulolytic enzymes. Biotechnol. Adv. 28 (3), 308–324.

Ballesteros, I., Oliva, J.M., Negro, M.J., Manzanares, P., Ballesteros, M., 2002. Enzymic hydrolysis of steam-exploded herbaceous agricultural waste (*Brassica carinata*) at different particular sizes. Process Biochem. 38 (2), 187–192.

Aziz, S., Sarkanen, K., 1989. Organosolv pulping—a review. Tappi J. 72, 162–175.

Ballesteros, I., Negro, M.J., Oliva, J.M., Cabañas, A., Manzanares, P., Ballesteros, M., 2006. Ethanol production from steam-explosion pretreated wheat straw. In: Twenty-Seventh Symposium on Biotechnology for Fuels and Chemicals. Humana Press, pp. 496–508.

Baptista, S.L., Cunha, J.T., Romaní, A., Domingues, L., 2018. Xylitol production from lignocellulosic whole slurry corn cob by engineered industrial *Saccharomyces cerevisiae* PE-2. Bioresour. Technol. 267, 481–491.

Behera, S., Arora, R., Nandhagopal, N., Kumar, S., 2014. Importance of chemical pretreatment for bioconversion of lignocellulosic biomass. Renew. Sust. Energ. Rev. 36, 91–106.

Biot-Pelletier, D., Martin, V.J.J., 2016. Seamless site-directed mutagenesis of the Saccharomyces cerevisiae genome using CRISPR-Cas9. J. Biol. Eng. 10 (1), 1–5.

Bisson, L.F., Fan, Q., Walker, G.A., 2016. Sugar and glycerol transport in *Saccharomyces cerevisiae*. In: Yeast Membrane Transport. Springer, pp. 125–168.

Brennan, A.H., Hoagland, W., Schell, D.J., Scott, C.D., 1986, January. High-temperature acid hydrolysis of biomass using an engineering-scale plug flow reactor. Results of low testing solids. In: Biotechnology and Bioengineering Symposium (United States) no. CONF-860508. vol. 17. Solar Energy Research Institute, Golden, CO, USA.

Brodeur, G., Yau, E., Badal, K., Collier, J., Ramachandran, K.B., Ramakrishnan, S., 2011. Chemical and physicochemical pretreatment of lignocellulosic biomass: a review. Enzyme Res., 1–17. https://doi.org/10.4061/2011/787532.

Bruinenberg, P.M., de Bot, P.H., van Dijken, J.P., Scheffers, W.A., 1984. NADH-linked aldose reductase: the key to anaerobic alcoholic fermentation of xylose by yeasts. Appl. Microbiol. Biotechnol. 19 (4), 256–260.

Budzianowski, W.M., 2017. High-value, low-volume bioproducts coupled to bioenergies with the potential to enhance the business development of sustainable biorefineries. Renew. Sust. Energ. Rev. 70, 793–804.

Cai, Z., Zhang, B., Li, Y., 2012. Engineering *Saccharomyces cerevisiae* for efficient anaerobic xylose fermentation: reflections and perspectives. Biotechnol. J. 7 (1), 34–46.

Cha, Y.L., Yang, J., Ahn, J.W., et al., 2014. The optimized CO_2-added ammonia explosion pretreatment for bioethanol production from rice straw. Bioprocess Biosyst. Eng. 37, 1907–1915. https://doi.org/10.1007/s00449-014-1165-x.

Chang, V.S., Holtzapple, M.T., 2000. Fundamental factors affecting biomass enzymatic reactivity. In: Twenty-First Symposium on Biotechnology for Fuels and Chemicals. Humana Press, Totowa, NJ, pp. 5–37.

Cheung, S.W., Anderson, B.C., 1997. Laboratory investigation of ethanol production from municipal primary wastewater solids. Bioresour. Technol. 59 (1), 81–96.

Chovau, S., Degrauwe, D., van der Bruggen, B., 2013. Critical 12 analysis of techno-economic estimates for the production cost of lignocellulosic bio-13 ethanol. Renew. Sust. Energ. Rev. 26, 307–321.

Cunha, J.T., Romaní, A., Costa, C.E., Sá-Correia, I., Domingues, L., 2019. Molecular and physiological basis of *Saccharomyces cerevisiae* tolerance to adverse lignocellulose-based process conditions. Appl. Microbiol. Biotechnol. 103 (1), 159–175.

Duff, S.J., Murray, W.D., 1996. Bioconversion of forest products industry waste cellulosics to fuel ethanol: a review. Bioresour. Technol. 55 (1), 1–33.

Eiteman, M.A., Lee, S.A., Altman, E., 2008. A co-fermentation strategy to consume sugar mixtures effectively. J. Biol. Eng. 2 (1), 1–8.

Elshafei, A.M., Vega, J.L., Klasson, K.T., Clausen, E.C., Gaddy, J.L., 1991. The saccharification of corn Stover by cellulase from *Penicillium funiculosum*. Bioresour. Technol. 35 (1), 73–80.

Esteghlalian, A., Hashimoto, A.G., Fenske, J.J., Penner, M.H., 1997. Modeling and optimization of the dilute-sulfuric-acid pretreatment of corn Stover, poplar, and switchgrass. Bioresour. Technol. 59 (2–3), 129–136.

Galbe, M., Sassner, P., Wingren, A., Zacchi, G., 2007. Process engineering economics of 20 bioethanol production. Adv. Biochem. Eng. Biotechnol. 108, 303–327.

Gao, M., Ploessl, D., Shao, Z., 2019. Enhancing the co-utilization of biomass-derived mixed sugars by yeasts. Front. Microbiol. 9, 3264.

Girio, F.M., Fonseca, C., Carvalheiro, F., Duarte, L.C., Marques, S., Bogel Łukasik, R., 2010. Hemicelluloses for fuel ethanol: a review. Bioresour. Technol. 101, 4775–4800.

Giudici, P., Solieri, L., Pulvirenti, A.M., Cassanelli, S., 2005. Strategies and perspectives for genetic improvement of wine yeasts. Appl. Microbiol. Biotechnol. 66, 622–628.

Gubicza, K., Nieves, I.U., Sagues, W.J., Barta, Z., Shanmugam, K.T., Ingram, L.O., 2016. Techno-economic analysis of ethanol production from sugarcane bagasse using a liquefaction plus simultaneous Saccharification and co-fermentation process. Bioresour. Technol. 208, 42–48.

Han, Y., Chen, H., 2010. Biochemical characterization of a maize stover β-exoglucanase and its use in lignocellulose conversion. Bioresour. Technol. 101 (15), 6111–6117. https://doi.org/10.1016/j.biortech.2010.02.108.

Himmel, M.E., Ding, S.Y., Johnson, D.K., Adney, W.S., Nimlos, M.R., Brady, J.W., Foust, T.D., 2007. Biomass recalcitrance: engineering plants and enzymes for biofuels production. Science 315 (5813), 804–807.

Ishii, J., Yoshimura, K., Hasunuma, T., Kondo, A., 2013. Reduction of furan derivatives by overexpressing NADH-dependent Adh1 improves ethanol fermentation using xylose as sole carbon source with Saccharomyces cerevisiae harboring XR–XDH pathway. Appl. Microbiol. Biotechnol. 97 (6), 2597–2607.

Jeffries, T.W., 2006. Engineering yeasts for xylose metabolism. Curr. Opin. Biotechnol. 17, 320–326.

Jeffries, T.W., Jin, Y.S., 2004. Metabolic engineering for improved fermentation of pentoses by yeasts. Appl. Microbiol. Biotechnol. 63, 495–509.

Johannsen, E., Eagle, L., Bredenhann, G., 1985. Protoplast fusion used for the construction of presumptive polyploids of the D-xylose-fermenting yeast Candida shehatae. Curr. Genet. 9, 313–319.

Johansson, A., Aaltonen, O., Ylinen, P., 1987. Organosolv pulping—methods and pulp properties. Biomass 13 (1), 45–65. https://doi.org/10.1016/0144-4565(87)90071-0.

Jönsson, L.J., Martín, C., 2016. Pretreatment of lignocellulose: formation of inhibitory byproducts and strategies for minimizing their effects. Bioresour. Technol. 199, 103–112.

Kang, Q., Appels, L., Tan, T., Dewil, R., 2014. Bioethanol from lignocellulosic biomass: current findings determine research priorities. Sci. World J. 2014, 298153.

Kavanagh, K.L., Klimacek, M., Nidetzky, B., Wilson, D.K., 2003. Structure of xylose reductase bound to NAD+ and the basis for single and dual co-substrate specificity in family 2 aldo-keto reductases. Biochem. J. 373 (2), 319–326.

Khattab, S.M.R., Saimura, M., Kodaki, T., 2013. Boost in bioethanol production using recombinant Saccharomyces cerevisiae with mutated strictly NADPH-dependent xylose reductase and NADP(+)-dependent xylitol dehydrogenase. J. Biotechnol. 165, 153–156.

Kilian, S.G., Van Uden, N., 1988. Transport of xylose and glucose in the xylose-fermenting yeast Pichia stipitis. Appl. Microbiol. Biotechnol. 27 (5), 545–548.

Kim, S., Holtzapple, M.T., 2006. Effect of structural features on enzyme digestibility of corn Stover. Bioresour. Technol. 97 (4), 583–591.

Kim, Y., Yu, A., Han, M., Choi, G.W., Chung, B., 2011. Enhanced enzymatic saccharification of barley straw pretreated by ethanosolv technology. Appl. Biochem. Biotechnol. 163 (1), 143–152.

Kim, K.H., Hong, J., 2001. Supercritical CO_2 pretreatment of lignocellulose enhances enzymatic cellulose hydrolysis. Bioresour. Technol. 77 (2), 139–144. https://doi.org/10.1016/S0960-8524(00)00147-4.

Kim, S.K., Jin, Y.S., Choi, I.G., Park, Y.C., Seo, J.H., 2015. Enhanced tolerance of Saccharomyces cerevisiae to multiple lignocellulose-derived inhibitors through modulation of spermidine contents. Metab. Eng. 29, 46–55.

Kim, S.M., Dien, B.S., Singh, V., 2016. Promise of combined hydrothermal/chemical and mechanical refining for pretreatment of woody and herbaceous biomass. Biotechnol. Biofuels 9 (1), 1–15.

Ko, J.K., Um, Y., Woo, H.M., Kim, K.H., Lee, S.M., 2016. Ethanol production from lignocellulosic hydrolysates using engineered Saccharomyces cerevisiae harboring xylose isomerase-based pathway. Bioresour. Technol. 209, 290–296.

Kötter, P., Ciriacy, M., 1993. Xylose fermentation by Saccharomyces cerevisiae. Appl. Microbiol. Biotechnol. 38 (6), 776–783.

Kumar, A.K., Sharma, S., 2017. Recent updates on different methods of pretreatment of lignocellulosic feedstocks: a review. Bioresour. Bioprocess. 4 (1), 1–19.

Kumar, P., Barrett, D.M., Delwiche, M.J., Stroeve, P., 2009. Methods for pretreatment of lignocellulosic biomass for efficient hydrolysis and biofuel production. Ind. Eng. Chem. Res. 48 (8), 3713–3729.

Kumar, P., Barrett, D.M., Delwiche, M.J., Stroeve, P., 2011. Pulsed electric field pretreatment of switchgrass and wood chip species for biofuel production. Ind. Eng. Chem. Res. 50 (19), 10996–11001.

Kwak, S., Jin, Y.S., 2017. Production of fuels and chemicals from xylose by engineered *Saccharomyces cerevisiae*: a review and perspective. Microb. Cell Factories 16 (1), 1–15.

Lee, J.W., Jeffries, T.W., 2011. Efficiencies of acid catalysts in the hydrolysis of lignocellulosic biomass over a range of combined severity factors. Bioresour. Technol. 102 (10), 5884–5890.

Lee, D., Alex, H.C., Wong, K.K., Saddler, J.N., 1994. Evaluation of the enzymatic susceptibility of cellulosic substrates using specific hydrolysis rates and enzyme adsorption. Appl. Biochem. Biotechnol. 45 (1), 407–415.

Lee, S.M., Jellison, T., Alper, H.S., 2014. Systematic and evolutionary engineering of a xylose isomerase-based pathway in *Saccharomyces cerevisiae* for efficient conversion yields. Biotechnol. Biofuels 7 (1), 1 8.

Li, M., Cao, S., Meng, X., Studer, M., Wyman, C.E., Ragauskas, A.J., Pu, Y., 2017. The effect of liquid hot water pretreatment on the chemical–structural alteration and the reduced recalcitrance in poplar. Biotechnol. Biofuels 10 (1), 1–13.

Li, W., Ghosh, A., Bbosa, D., Brown, R., Wright, M.M., 2018. Comparative techno-economic, uncertainty and life cycle analysis of lignocellulosic biomass solvent liquefaction and sugar fermentation to ethanol. ACS Sustain. Chem. Eng. 6 (12), 16515–16552.

Lian, J., Mishra, S., Zhao, H., 2018. Recent advances in metabolic engineering of *Saccharomyces cerevisiae*: new tools and their applications. Metab. Eng. 50, 85–108. https://doi.org/10.1016/j.ymben .2018.04.011.

Liu, Z.L., Moon, J., Andersh, B.J., Slininger, P.J., Weber, S., 2008. Multiple gene-mediated NAD (P) H-dependent aldehyde reduction is a mechanism of in situ detoxification of furfural and 5-hydroxymethylfurfural by *Saccharomyces cerevisiae*. Appl. Microbiol. Biotechnol. 81 (4), 743–753.

Lora, J.H., Aziz, S., 1985. Organosolv pulping: a versatile approach to wood refining. Tappi J. 68 (8), 94–97.

Luengo, E., Martínez, J.M., Coustets, M., Álvarez, I., Teissié, J., Rols, M.P., Raso, J., 2015. A comparative study on the effects of millisecond-and microsecond-pulsed electric field treatments on the permeabilization and extraction of pigments from *Chlorella vulgaris*. J. Membr. Biol. 248 (5), 883–891.

MacDonald, D.G., Bakhshi, N.N., Mathews, J.F., Roychowdhury, A., Bajpai, P., Moo-Young, M., 1983. Alkali treatment of corn stover to improve sugar production by enzymatic hydrolysis. Biotechnol. Bioeng. 25 (8), 2067–2076.

Mamleeva, N.A., Autlov, S.A., Natal'ya, G.B., Lunin, V.V., 2009. Delignification of softwood by ozonation. Pure Appl. Chem. 81 (11), 2081–2091.

McGinnis, G.D., Wilson, W.W., Mullen, C.E., 1983. Biomass pretreatment with water and high-pressure oxygen. The wet-oxidation process. Ind. Eng. Chem. Prod. Res. Dev. 22 (2), 352–357.

Meinander, N.Q., Boels, I., Hahn-Hägerdal, B., 1999. Fermentation of xylose/glucose mixtures by metabolically engineered *Saccharomyces cerevisiae* strains expressing XYL1 and XYL2 from *Pichia stipitis* with and without overexpression of TAL1. Bioresour. Technol. 68 (1), 79–87.

Mood, S.H., Golfeshan, A.H., Tabatabaei, M., Jouzani, G.S., Najafi, G.H., Gholami, M., Ardjmand, M., 2013. Lignocellulosic biomass to bioethanol, a comprehensive review with a focus on pretreatment. Renew. Sust. Energ. Rev. 27, 77–93.

Mosier, N., Wyman, C., Dale, B., Elander, R., Lee, Y.Y., Holtzapple, M., Ladisch, M., 2005. Features of promising technologies for pretreatment of lignocellulosic biomass. Bioresour. Technol. 96 (6), 673–686.

Munir, M.T., Mansouri, S.S., Udugama, I.A., Baroutian, S., Gernaey, K.V., Young, B.R., 2018. Resource recovery from organic solid waste using hydrothermal processing: opportunities and challenges. Renew. Sust. Energ. Rev. 96, 64–75.

Neirinck, L.G., Maleszka, R., Schneider, H., 1984. The requirement of oxygen for incorporation of carbon from D-xylose and D-glucose by *Pachysolen tannophilus*. Arch. Biochem. Biophys. 228 (1), 13–21.

Nguyen, Q.A., Saddler, J.N., 1991. An integrated model for the technical and economic evaluation of an enzymatic biomass conversion process. Bioresour. Technol. 35, 275–282.

Palonen, H., Thomsen, A.B., Tenkanen, M., Schmidt, A.S., Viikari, L., 2004. Evaluation of wet oxidation pretreatment for enzymatic hydrolysis of softwood. Appl. Biochem. Biotechnol. 117 (1), 1–17.

Pan, X., Arato, C., Gilkes, N., Gregg, D., Mabee, W., Pye, K., Saddler, J., 2005. Biorefining of softwoods using ethanol organosolv pulping: preliminary evaluation of process streams for manufacture of fuel-grade ethanol and co-products. Biotechnol. Bioeng. 90 (4), 473–481.

Ruchala, J., Kurylenko, O.O., Dmytruk, K.V., Sibirny, A.A., 2020. Construction of advanced producers of first-and second-generation ethanol in *Saccharomyces cerevisiae* and selected species of non-conventional yeasts (*Scheffersomyces stipitis, Ogataea polymorpha*). J. Ind. Microbiol. Biotechnol. 47 (1), 109–132.

Ruiz, H.A., Silva, D.P., Ruzene, D.S., Lima, L.F., Vicente, A.A., Teixeira, J.A., 2012. Bioethanol production from hydrothermal pretreated wheat straw by a flocculating Saccharomyces cerevisiae strain–effect of process conditions. Fuel 95, 528–536.

Salerno, M.B., Lee, H.-S., Parameswaran, P., Rittmann, B.E., 2009. Using a pulsed electric field as a pretreatment for improved biosolids digestion and methanogenesis. Water Environ. Res. Biol. Treat. 81 (8), 831–839.

Sassner, P., Mårtensson, C.G., Galbe, M., Zacchi, G., 2008. Steam pretreatment of H2SO4-impregnated Salix for the production of bioethanol. Bioresour. Technol. 99 (1), 137–145.

Schmidt, A.S., Thomsen, A.B., 1998. Optimization of wet oxidation pretreatment of wheat straw. Bioresour. Technol. 64 (2), 139–151.

Selim, A.K., Easa, M.S., Diwany, E.A., 2020. The xylose metabolizing yeast *Spathaspora passalidarum* is a promising genetic treasure for improving bioethanol production. Fermentation 6 (1), 33.

Sharma, S., Arora, A., 2020. Tracking strategic developments for conferring xylose utilization/fermentation by *Saccharomyces cerevisiae*. Ann. Microbiol. 70, 50.

Söderström, J., Pilcher, L., Galbe, M., Zacchi, G., 2002. Two-step steam pretreatment of softwood with SO2 impregnation for ethanol production. Biotechnology for Fuels and Chemicals. Applied Biochemistry and Biotechnology, 1st. Humana Press, Totowa, NJ, pp. 5–11.

Stambuk, B.U., Eleutherio, E.C.A., Florez-Pardo, L.M., Souto-Maior, A.M., Bon, E.P., 2008. Brazilian potential for biomass ethanol: challenge of using hexose and pentose cofermenting yeast strains. J. Sci. Ind. Res. 67 (11), 918–926.

Subtil, T., Boles, E., 2012. Competition between pentoses and glucose during uptake and catabolism in recombinant *Saccharomyces cerevisiae*. Biotechnol. Biofuels 5 (1), 1–12.

Sun, Y., Cheng, J., 2002. Hydrolysis of lignocellulosic materials for ethanol production: a review. Bioresour. Technol. 83 (1), 1–11.

Tengborg, C., Stenberg, K., Galbe, M., et al., 1998. Comparison of SO2 and H2SO4 impregnation of softwood prior to steam pretreatment on ethanol production. Appl. Biochem. Biotechnol. 70. https://doi.org/10.1007/BF02920119, 3.

Travaini, R., Otero, M.D.M., Coca, M., Da-Silva, R., Bolado, S., 2013. Sugarcane bagasse ozonolysis pretreatment: effect on enzymatic digestibility and inhibitory compound formation. Bioresour. Technol. 133, 332–339.

Viola, E., Cardinale, M., Santarcangelo, R., Villone, A., Zimbardi, F., 2008. Ethanol from eel grass via steam explosion and enzymatic hydrolysis. Biomass Bioenergy 32 (7), 613–618.

Von Sivers, M., Zacchi, G., 1995. A techno-economical comparison of three processes for the production of ethanol from pine. Bioresour. Technol. 51, 43–52.

Wahlbom, C.F., Hahn-Hägerdal, B., 2002. Furfural, 5-hydroxymethyl furfural, and acetoin act as external electron acceptors during anaerobic fermentation of xylose in recombinant *Saccharomyces cerevisiae*. Biotechnol. Bioeng. 78 (2), 172–178.

Winkelhausen, E., Kuzmanova, S., 1998. Microbial conversion of D-xylose to xylitol. J. Ferment. Bioeng. 86 (1), 1–14.

Wooley, R., Ruth, M., Sheehan, J., Ibsen, K., Majdeski, H., Galvez, A., 1999. Lignocellulosic biomass to ethanol process design and economics utilizing co-current dilute acid prehydrolysis and enzymatic hydrolysis current and futuristic scenarios. Report no. NREL/TP-580-21 26157. National Renewable Energy Laboratory, Golden, CO.

Wyman, C.E., 1994. Ethanol from lignocellulosic biomass: technology, economics, and opportunities. Bioresour. Technol. 50, 3–16.

Yang, Y.J., Lu, R.J., Dang, Y.H., Li, Y., Ge, S.B., 2009. Development on ethanol production from xylose by recombinant *Saccharomyces cerevisiae*. Nat. Sci. 1 (3), 210–215.

Zabed, H., Sahu, J.N., Boyce, A.N., Faruq, G., 2016. Fuel ethanol production from lignocellulosic biomass: an overview on feedstocks and technological approaches. Renew. Sust. Energ. Rev. 66, 751–774.

Zavrel, M., Bross, D., Funke, M., Büchs, J., Spiess, A.C., 2009. High-throughput screening for ionic liquids dissolving (ligno-) cellulose. Bioresour. Technol. 100 (9), 2580–2587.

Zhao, L., Zhang, X., Xu, J., Ou, X., Chang, S., Wu, M., 2015. Techno-economic analysis of bioethanol production from lignocellulosic biomass in China: dilute-acid pretreatment and enzymatic hydrolysis of corn Stover. Energies 8, 4096–4117.

Zheng, Y., Lin, H.M., Tsao, G.T., 1998. Pretreatment for cellulose hydrolysis by carbon dioxide explosion. Biotechnol. Prog. 14 (6), 890–896.

Zhuang, X., Wang, W., Yu, Q., Qi, W., Wang, Q., Tan, X., Yuan, Z., 2016. Liquid hot water pretreatment of lignocel-lulosic biomass for bioethanol production accompanying with high valuable products. Bioresour. Technol. 199, 68–75.

11

Pyrolytic bio-oil—Production and applications

Piyali Das

The Energy and Resources Institute (TERI), New Delhi, India

1 Introduction

Among the different thermochemical conversion routes, biomass pyrolysis has emerged as one of the most promising routes for making bio-oils/bio-crude, biochar, and non-condensable gases from wood and lignocellulosic biomass. Another unique advantage of pyrolysis technology is the manipulation of relative product distribution through variation of pyrolysis heating conditions especially the rate of heating and temperature of pyrolysis. Even though work on biomass pyrolysis has accelerated over the last 2 decades, it is only recently that the technology has proven to be of commercial value. With continuing techno-development and a large number of biorefineries being built in the United States for extraction of value-added chemicals alongside fuels from the bio-oils, this technology has finally gained its due importance.

Practically any form of biomass can be considered for pyrolysis. Though most of the work reported is on wood due to its consistency, and comparability between tests, several other biomass are increasingly been tested worldwide. It ranges from agricultural wastes and nut-shells to energy crops such as Miscanthus and Sorghum and solid wastes such as sewage sludge, leather wastes, MSW, etc., to name a few (Mohan et al., 2006).

Under fast pyrolysis conditions, bio-oil is a major product with biochar and non-condensable gases as minor products whereas, slow pyrolysis maximizes the biochar with bio-oil and non-condensable gases as minor products. In contrast to slow pyrolysis, the by-product, biochar obtained under fast pyrolysis conditions has much lower surface area, negligible porosity, and high heating values. Most of the fast pyrolysis technologies use this biochar as a fuel in the process, while in some cases value addition happens through downstream activation to high-grade carbon (Garg and Das, 2018; Garg and Das, 2019

Fast pyrolysis liquids or bio-oil already received due credence as a substitute for a variety of fuel oils. Combustion tests performed using different scale boilers, internal combustion engines and gas turbines have demonstrated efficient burning of bio-oil in standard and modified equipment. Inconsistent quality of bio-oil and lack of sustainable supply of biomass over the life of commercial pyrolysis plants are perceived as the major impediments towards the large-scale deployment of this technology. Apart from being energy-dense, lowest cost liquid biofuel, and positive CO_2 balance, bio-oil has several advantages over gaseous fuels. It can be easily stored and transported and is thus best suited for a decentralized mode of fuel production alike mini refineries with subsequent distribution and use in rural areas/industries and/or as a fuel oil substitute in furnaces and boilers.

Keeping in view of the global scenario, speedy advancement of pyrolysis technology, and soaring need for renewable fuels, it is high time to recognize the real value of bio-oil and oil-based renewable products and develop more commercial processes with indigenous biomass resources for decentralized oil production.

1.1 Reactors for fast pyrolysis

In recent years, the European Commission has been financing the Pyrolysis Network (PyNE), led by Aston University, UK. Presently, PyNE consists of over 15 states including Canada and the United States. PyNE allows its members to exchange experiences, to be well informed of the latest fast pyrolysis technologies developments, and publishes its own magazine. Under the current IEA (International Energy Agency, 2011), the PyRA (Pyrolysis Activity) has specified standards for pyrolysis liquids analogous to petroleum fuels. Petroleum-derived fuel oil specifications from six countries in North America and Europe were compared and used as the basis for analogous fuel oil specifications for bio-oils. The specifications for bio-oils differ mainly in density, heating value, water content, and corrosiveness.

The market attractiveness and the technology strength matrix for some of the well-developed pyrolysis technologies are reported by IEA's Pyrolysis Research Network (PyNE) which is shown in Fig. 1 (Butler et al., 2011). The relative merits of different

FIG. 1 Market attractiveness and technology strength of various competing pyrolysis technologies (Brown and Holmgren, 2006).

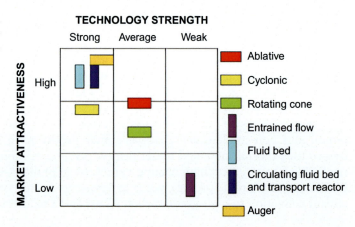

Property	Status	Bio oil yield on dry biomass	Complexity	Feed size specification	Inert gas requirements	Specific reactor size	Scale up	Gas quality
Fluid bed	Commercial	75 wt.%	Medium	High	High	Medium	Easy	Low
CFB & Transported bed	Commercial	75 wt.%	High	High	High	Medium	Easy	Low
Rotating Cone	Commercial	70wt.%	High	High	Low	Low	Medium	High
Entrained Flow	Laboratory	60 wt.%	Medium	High	High	Medium	Easy	Low
Ablative	Laboratory	75 wt.%	High	Low	Low	Low	Difficult	High
Screw or Auger	Demonstration	60 wt.%	Medium	Medium	Low	Low	Medium	High
Vacuum	None	60 wt.%	High	Low	Low	High	Difficult	Medium
The darker the cell color, the less desirable the process.				Lab: 1-20 kg h^{-1} Pilot:20-200 kg h^{-1} Demo: 200-2000 kg h^{-1}				

FIG. 2   Relative merits of different reactors. *Adapted from IEA-Task 34 Bioenergy Report https://task34.ieabioenergy.com/pyrolysis-reactors/.*

technologies are depicted in Fig. 2 (IEA-PyNE Bioenergy Report). It reveals that some technologies though score high in energy efficiency, suffer on account of feedstock size flexibility, whereas there are others with a good degree of scale-up flexibility. Hence there is still scope for further technological innovation with respect to the specific need for bio-oil applications and feedstock nature. Fluidized bed technology though undoubtedly offers the most desirable features for high-scale commercial applications, Auger and Rotating cone technologies appear as strong contenders especially for medium to small-scale applications.

Significant research progress is made especially in the domain of bio-oil and its applications which is evident through the appearance of plenty of publications and patents. A detailed account of the major technologies and processes developed over the past 20 years is reviewed (Bridgwater and Peacocke, 2000; Bridgwater et al., 1999a, c; Scott et al., 1999; Bridgwater, 2003). The technical and economic performances especially of thermal processes to generate electricity from woody feedstock are assessed. These reviews have largely covered fast pyrolysis process designs, reactor configurations, technology status, etc., in various countries including commercialization. Fast pyrolysis chemistry of a wide range of lignocellulosic feeds is critically reviewed in light of chemical point of view in another study (Mohan et al., 2006).

The 19th century witnessed the appearance of the "acid-wood industry," alternatively named as "wood distillation industry" to generate charcoal and liquid by-products such as acetic acid, methanol, and acetone (Canham, 2010) and is considered as the harbinger of the modern petrochemical industry. Later slow pyrolysis or carbonization industry appeared in the chronicle of Industrial Chemistry. Whereas, the 20th century saw the rise of the petroleum industry making cheaper products, and thus leading to the decline of the pyrolysis industry. However, the 1970s oil crisis reinforced biomass pyrolysis as a technology that could potentially contribute to reducing dependency on fossil fuel (Garcia-Nunez et al., 2017).

Pyrolysis reactors are classified differently based on dissimilar aspects. In one study, pyrolysis reactors are classified based on the pyrolysis vapors residence time (VRT) inside the reactor, the highest temperature of pyrolysis, the rate of heating, and the biomass particle size (Bridgwater, 2012). Fine powdery biomass is typically used in fast pyrolysis actors. In this approach, the fast, intermediate, and slow pyrolysis is defined as fast (pyrolysis temperature: 500 °C, particle diameter < 2 mm, VRT: 1 s), intermediate (pyrolysis temperature: 500 °C, small particles, VRT: 1 s), and slow (final temperature: 500 °C, logs or chips, VRT: in days).

On the contrary, slow pyrolysis commonly called carbonization reactors are sub-classified as kilns, retorts, and converters (Emrich, 1985). The term kiln is traditionally used for making char from wood logs where other by-products are not recovered. Industrial reactors where char is recovered alongside liquid condensate and syngas are referred to as retorts or converters. Retort reactors reportedly operate with pilewood or wood logs over 30 cm long and over 18 cm in diameter. Whereas, converters use relatively smaller biomass such as chipped or pelletized wood. The converters thus operate at conditions close to the biomass intermediate pyrolysis reactors as described by (Bridgwater, 2012). In another review, the aspect of the heating mechanism is considered as a ground for reactor grading. According to this study, the reactors can be grouped under slow pyrolysis (Kiln and retort), intermediate pyrolysis (converters), fast pyrolysis, and microwave pyrolysis (Emrich, 1985; Bridgwater, 2012; Motasemi and Afzal, 2013; Mushtaq et al., 2014). Other possible ways of defining the pyrolysis reactors are based on targeted final products (oil, char, heat, electricity, gases), mode of operation (batch or continuous), type of heating (direct or indirect heating, auto-thermal, microwave), nature of heat source (electric, gas heater, biomass combustion), method of loading the reactor (by hand, mechanical), the reactor pressure (vacuum, atmospheric, pressurized), the material of construction (soil, brick, concrete, steel), reactor portability (stationary, mobile), etc. Further information is given in another reference (Boateng et al. 2015).

In spite of the fact that the type of pyrolysis reactor and its operating conditions greatly determine the quality of the final products, there is a large literature gap on finding any correlation between them. Another void area in literature is the know-how of systematic methodologies for designing pyrolysis reactors. To address this gap, a useful stepwise strategy of reactor sizing is illustrated in a recent review (Garcia-Nunez et al. 2017).

In this study, only fast and intermediate pyrolysis reactors are briefly discussed.

1.1.1 *Fast pyrolysis*

In fast pyrolysis, biomass is rapidly heated in the absence of oxygen to generate vapors, aerosols, and biochar. A dark brown mobile liquid formed upon condensation of the vapors with heating value nearly half of that of conventional fuel oil is called bio-oil (http://www.pyne.co.uk). Fast pyrolysis processes yield 60–75 wt% of liquid bio-oil, 15–25 wt% of solid char (Garcia-Nunez et al. 2017), and 10–20 wt% of non-condensable gases, depending on the feedstock used. Four attributes of a fast pyrolysis process are very high heating and heat transfer rates which necessitates finely ground biomass feedstock, carefully controlled pyrolysis temperature typically between 425 and 500 °C, short vapor residence times less than 2 s, and rapid quenching and cooling of pyrolysis vapors and aerosols to give bio-oil (Bridgwater, 2003). Fig. 3 shows the schematic of the most common fast pyrolysis reactors (Garcia-Nunez et al. 2017).

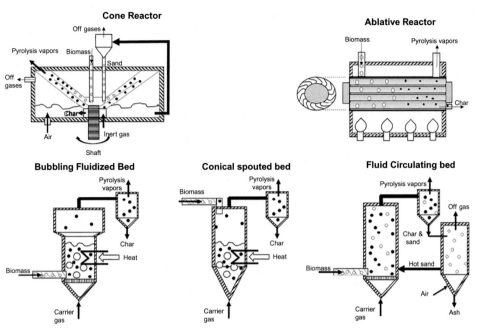

FIG. 3   Schemes of fast pyrolysis. *Adapted from Venderbosch and Prins, 2010.*

1.1.2 Bubbling fluidized bed reactors

In bubbling fluidized bed reactors, feed sizes are maintained typically below 0.5 mm, in order to achieve a high heat transfer rate and avoid the retention of aerosols inside the particle (Shen et al., 2009; Westerhof et al., 2012). Combustion of pyrolysis non-condensable gases and chars provide the heat of pyrolysis which is transferred to biomass by heating sand and the carrier. Given the low heat transfer rates between combustion gases and the bed (100–200 W/m^{2}K) at least 10–20 m^2 of the surface area is required to transfer the heat required to pyrolyze 1 t/h of biomass. These heat transfer surfaces are very susceptible to attrition from the sand (Venderbosch and Prins, 2010).

Prominent pyrolysis plants operating with fluidized bed technology are illustrated (Garcia-Nunez et al. 2017). One of the best-known examples of fluidized bed reactors is made by Dynamotive, Ltd., the technology of which is rooted in the University of Waterloo. Agritherm at the University of Western Ontario has developed a 610 t/day fluidized bed mobile pyrolysis unit with a unique design where the reactor is made using an annulus with a burner at the core supplying the pyrolysis energy (http://agri-therm.com) (Bridgwater, 2012; Meier et al. 2013).Avello Bioenergy in the State of Iowa (USA) commercializing bubbling fluidized bed technology where fractionation strategies are applied to obtain different products from bio-oils (http://www.avellobioenergy.com). Bioware is a Brazilian company commercializing auto-thermal fluidized bed reactors producing bio-oil, char and phenolic resins (https://www.bioware.com.br; Mesa-Pérez et al., 2014). Nettenergy BV in the Netherlands has commercialized a 100 kg/h mobile unit with a unique multi-stage separation design (http://www.nettenergy.com/index.php/en/).

1.1.3 Circulating fluidized beds reactors

In this type of reactors biomass, hot recirculated sand and hot fluidizing gas enter in an up-flowing transported bed and get pyrolyzed. The products pass through two cyclones that separate both solids from the pyro-vapors and subsequently undergo rapid cooling and quenching in multiple stages (Bridgwater and Peacocke, 2000). The solid and vapor residence time in these reactors is nearly the same (Bridgwater, 2012). The recirculation of gases from secondary char combustion is the main heat source. A second reactor is required for char combustion and sand reheating unless the combustion occurs in the bottom part of the reactor thus eliminating the requirement of the second reactor.

Rapid thermal processing (RTP)TM technology originated at the University of Western Ontario in the late 1970s and early 1980s and was further developed by Ensyn is the only pyrolysis technology that is commercial for long period. (http://www.ensyn.com; https://www.envergenttech.com). Examples of other larger scale units include ENEL plant built by Ensyn in Italy (15.6t/day), several 40t/day units at Red Arrow (USA) operating for the production of smoke aromas, and the Ensyn 50t/day unit at their R&D center in Renfrew Canada (Motasemi and Afzal, 2013). Some specific features worthwhile to mention are precise control of pyrolysis temperature, use of large-size particles, suitability for large throughputs, and well-understood technology (Bridgwater, 2012).

One of the major drawbacks of this technology is attributed to the high volumes of inert carrier gases which leads to a dilution of the pyrolytic gases resulting in difficulty of bio-oil recovery. Another challenge is a high degree of char and sand attrition due to high fluidizing velocities. Char and sand are separated from pyro vapor using cyclones. Catalysts in place of hot sand are used in the technology developed by KIOR, now Inaeris Tech in Houston, Texas (http://www.inaeristech.com/). In this process, catalytic cracking is used to deoxygenate bio-oil in a Fluid Catalytic Cracking (FCC) reactor. Metso, UPM, and Fortum have been operating a 400kg/h circulating bed pyrolysis reactor coupled with a condensation system in Joensuu, Finland. The bio-oil is used in a fluidized bed power boiler (Bridgwater, 2012).

1.1.4 Rotating cone reactors

In this technology, the heat carrier hot sand and the biomass are transported up in a conical bed by the centrifugal forces created by the rotation of the cone (Bridgwater, 2012). This technology was originally developed by the University of Twente and was subsequently commercialized by BTG-BTL (Biomass Technology Group-Biomass to Liquid, Netherlands) (http://www.btgworld.com/en/). Some successful examples are demonstration plant of 50t/day capacity. This technology has been also used by Empyro for the construction of a 5t/d plant that operates since 2015 in Hengelo, the Netherlands generating pyrolysis oil, process steam, and electricity. In both plants, gas and char are burned to heat the sand, which is recycled back to the pyrolysis reactor. BTG Netherland has built a Pilot plant in Malaysia using empty fruit bunches from palm oil trees as feedstocks.

1.1.5 Ablative pyrolysis reactors

The overall concept of ablative pyrolysis markedly differs from other fast pyrolysis techniques. The pyrolysis mechanism is best described as melting butter in a frying pan, where the melting rate can be significantly influenced by pressing down and moving the butter over the heated pan surface (Bridgwater et al., 2002a, b, Reprint, 2008). Pyrolysis heat is transferred from the hot surface of the reactor to melt wood which is made to touch the reactor surface under pressure. The oily layer thus formed on the pan evaporates and condenses to give bio-oil. A unique feature of this process is that it is limited to the rate of heat supply to the biomass in contrast to the rate of heat absorption by the biomass as in other reactors. Theoretically, there is no particle size limitation in this process.

The main downsides are that these reactors require a heated surface area control system, operate with moving parts at high temperatures increasing their complexity, and induce inevitable wear and tear on the moving components (Bridgwater et al., 1999a, c). The yields of char, oil, and gases are comparable with those obtained with similar feedstock with fluidized bed reactors.

Compared with the fluidized bed, the key advantages of ablative reactors are: large feed size, no requirement of inert gas, no milling efforts needed for biomass, energy and cost efficiency as no heating and cooling of the fluidized bed is required, condensation units with smaller volume requiring less space and lower cost (Meier et al., 2007). NREL (Golden, Colorado, USA) and CNRS laboratories (France) conducted most of the pioneering studies on ablative reactors (Bridgwater, 2012). The University of Hamburg has built three plants which include 20 kg/h; 250 kg/h and a demonstration unit with a capacity of 2 t/h (Bridgwater, 2012; Venderbosch and Prins, 2010). Aston University (UK), Institute of Engineering Thermophysics (Ukraine), Latvian State Institute (Latvia), and the Technical University of Denmark have active programs on this technology (Bridgwater, 1999).

1.1.6 Auger reactors

The auger reactor typically consists of a feeding screw that carries the biomass to the pyrolysis zone where subsequently the condensable and non-condensable vapors get released. A schematic of Auger reactor is presented in Fig. 4. The condensable vapors are condensed to make bio-oil. The residence time of pyrolysis vapors inside these types of reactor varies between average between 5–30 s (Motasemi and Afzal, 2013). Conventionally hot sand, steel, or ceramic balls are used as heat carriers to biomass (Meier et al., 2013; Brown and Brown 2012; Mura et al. 2013).Experimental studies with woody biomass show yields of char between 17 and 30 wt% and yields of oil between 48 and 62 wt% (Meier et al., 2013). The bio-oil yield is slightly lower than that of fluidized bed reactors and contains more water 30%–55%. As expected, the oil yield of agricultural residues is much lower due to the high ash content of this feedstock. Although difficult to compare, it seems that the yields obtained with sand heat carriers are slightly higher than those obtained without.

In one invention, a gas-fired auger reactor is reported (Das et al., 2020; Das, 2014). In the auger reactor, the biomass from the hopper is carried through a screw conveyor to have a free fall in the reactor. The biomass discharge rate is controlled by regulating the conveyor speed via a frequency controller. The movement of pyrolyzed biomass inside the reactor is adjusted by controlling the shaft speed with a variable speed motor. This unit is coupled with the thermal gasifier. The producer gas generated in the thermal gasifier is combusted outside the pyrolyzer and the flue gases thus generated come in direct contact with the pyrolysis feed material inside the pyrolyzer. This results in good heat transfer rates that are fairly comparable to those of fluidized bed systems. This design is also relatively more suitable for

FIG. 4   Schemes of auger reactor (Garcia-
Nunez et al., 2017).

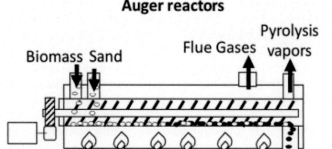

scale-up than an externally heated reactor. The pyrolysis vapors along with the flue gas are collected in the staged condensing train attached to the reactor. The condensing train consists of a series of condensers maintained at three different pre-set temperature ranges ranging between 0 and 90°C, with chilled coolant circulated through the condenser jackets (Garg and Das, 2018). The non-condensable gases are passed through a flow meter and a vacuum pump and finally flared.

Other prominent examples of auger reactors are developed by ABRI-Tech in Canada which has sold several small-scale testbeds 1 t/day unit to many academic universities carrying our active research on this technology (Meier et al., 2013). Auburn University (USA), KIT (FZK) (Germany), Mississippi State University (USA), Michigan State University (USA), Texas A&M (USA), and Washington State University (USA) 61 have active research programs on this technology (Bridgwater, 2012).

1.2   Pyrolysis technology world status

The United States, Canada, and Europe have spearheaded the development devoting a large amount of money and research into the development of this pyrolysis technology. As a result, many technologies have been developed, typically using fast pyrolysis towards maximizing liquid yields. Fluidized, entrained bed and vacuum technologies have been the most common.

A comprehensive list of technology status is presented by IEA Bioenergy Task 34 covering aspects like TRL status, technology types, state of operation, and raw material used (*IEA-Tas 34 Bioenergy Report*, 2015).

It is evident that most of the commercially available pyrolysis plants use fluidized bed technologies (large-scale plant of 100 tpd and above) with fast pyrolysis where bio-oil is the major product and the biochar produced is combusted for process heat. Nowadays, auger-based technologies (mostly below 10 TPD) with fast to intermediate pyrolysis are also gaining importance because of their small to medium scale operation, decentralized nature, and no energy-intensive feedstock pre-treatment steps. These units are also flexible to different biomass feedstock sizes where biochar is obtained as a by-product.

There are currently many commercial plants operational worldwide at the moment. Some of the significant commercial facilities under implementation are:

- Roseburg facility, Vienna, Georgia: 20 million gallons per year facility being developed by Ensyn, Renova Capital Partners, and Roseburg Forest Products. Feedstock is forest residues and thinning from local sources. Product is targeted for US refineries.
- Aracruz facility, Brazil: 22 million gallons per year facility being developed in partnership with Fibria Celulose. Located at Fibria's pulp mill in Aracruz, Espirito Santo. Feedstock is eucalyptus forest residues. Product supply is targeted for US refineries.
- Cote Nord, Quebec: 10 million gallons per year facilities being constructed by Ensyn and Arbec Forest Products. RTP equipment supplied by Envergent Technologies. RTP modules fabricated by Honeywell UOP's petrochemical suppliers. Product will be sold to heating and refining customers in the US Northeast.
- BTG Bio Liquids, Netherlands: BTG Bio Liquids has partnered with Technip and developed a Co FCC processing route. This process envisages co-processing of Pyrolysis oil produced from FPBO plant with Vacuum Gas oil in FCCs to produce Gasoline, Diesel, and LPG.
- Many other pyrolysis oil commercialization efforts are underway in the United States (e.g., Anellotech, GTI's new IH$_2$ technology licensed by Criterion with a demonstration unit being built in Bangalore) and Europe.

It is worth noticing that co-firing is increasingly being pursued globally and thus most of the commercial plant is also planned with bio-oil refining options.

A detailed list of the current demonstration, first-of-a-kind commercial and commercial pyrolysis plants are presented in Tables 1, 2, and 3.

2 Bio-oil composition and characterization

Pyrolysis liquid is known by many names including pyrolysis oil, bio-oil, bio-crude-oil, wood liquids, pyrolygnious tar, etc., to name a few. The crude bio-oil is brown in color and roughly resembles parent biomass in elemental composition (Bridgwater et al., 1999a, c). The bio-oil liquid has a distinctive odor that arises from low molecular weight aldehydes and ketones. The density of the oil is very high around 1.2 kg/L compared to light fuel oil around 0.85 kg/L. This means that the oil has about 42% of the energy content of fuel oil on a weight basis, and 61% on a volumetric basis. This has deep implications on the design and specifications of equipment like pumps. The viscosity of fresh bio-oil can vary widely from 25–1000 cst at 40 °C or more depending on the feedstock, water content, process of collection, etc. Once the pyrolytic vapors are quenched and condensed they cannot be distilled completely through heating. Over 100 °C it rapidly undergoes polymerization and produces solid residue of around 50 wt% of the original liquid along with distillate containing volatile organic compounds and water. Hence, through low-temperature vacuum distillation or solvent fractionation are resorted to fractionate the oil.

A special subject group under PyNe network carried out extensive work on analysis, characterization, and test method developments for bio-oil. This task force reviewed all the

TABLE 1 TRL 6–7 Demonstration-scale plants details.

S. no.	Technology provider (partner)	Technology	Status	Input: raw materials Output: products	City/country
1	Agri Therm (Partner: University of Western Ontario)	Fluidized bed	Under commissioning	**Input:** agricultural residues **Output:** different chemicals	London Ontario/ Canada
2	ALTACA Energy http://www.altacaenerji.com/projeler/catliq%2Ddemo%2Dtesisi/	Based on hydrothermal liquefaction technology, CatLiq, invented by SCF Technologies (Denmark)	Operational since 2016	**Input:** biomass sources including biogas plant digestate, forest waste, sewage sludge, agricultural waste, food plant waste, organic household waste **Output:** HTL bio-crude, with 70% recovery of process heat bio-oil (20,000 m³/y)	Gönen/Turkey
3	BIOGAS ENERGY USA (Partners: Thermophil International (Hamburg), Bio Energy Concept (Lüneberg) Joint BioEnergy Institute and California State University) **BIOGAS Energy Project**	Ablative	Under construction	**Input:** 500 kg/h forest residues, e.g., demolition wood, bark beetle infested trees, forest, and agriculture residues **Output:** pyrolysis oil (1300 m³/y)	Sacramento-California/ United States
4	Karlsruhe Institute of Technology (KIT) (Partner: Air Liquide) Project Name: bioliq	Auger	Operational	**Input:** various residues (500 kg/h) **Output1:** pyrolysis oil (300 kg/h) **Output2:** bioslurry (400 kg/h)	Eggenstein-Leopoldshafen/ Germany
5	Licella Pty Ltd., Australia (Partners: Armstrong Chemicals (UK), Canfor Canada)	Catalytic hydrothermal liquefaction technology trademarked as HTRTM	Operational	**Input 1:** wood, agricultural residue, pulp, and paper **Input 2:** End-of-life plastics **Output:** bio-oil (125 t/y) (Cat-demonstration-scale (10,000 t/y) Biofuels and chemicals Additionally, Licella partnered with Armstrong Chemicals (UK) to build the first commercial plant for chemical recycling of end-of-life plastics using the Cat-HTRTM technology	Somersby-New South Wales/ Australia

#	Company / Project (Partner)	Technology	Status	Input / Output	Location
6	Shanxi Yingjiliang Biomass Company and Shanghai Jiao Tong University (Partner: Shanxi Yingjiliang Biomass) **DALI COUNTY Facility Project**	Downdraft circulating fluidized bed fast pyrolysis	Operational	**Input:** Rice husk, 1–3 t/h **Output:** bio-oil (4500 m^3/y), biochar	Dali County / China
7	Shell Catalysts & Technologies Partner Gas Technology Institute (GTI), CRI Catalyst Company (CRI) https://www.shell.com/business-customers/catalysts-technologies/licensed-technologies/benefits-of-biofuels/ih2-technology.html IH2 Demonstration Facility Project GTI licensed the IH^2 technology to CRI for exclusive worldwide deploy	Integrated hydro pyrolysis and hydroconversion IH2	Operational	**Input:** agricultural residues, forestry, and urban waste 5 t/d **Output:** bio-LPG 500 m^3/y	Bangalore / India
8.	SILVA Green Fuel Partner" Steeper Energy Silva Green Fuel is a joint venture between Statkraft and Sødra for the development of advanced biofuels	Steeper Energy's hydrothermal liquefaction technology (HydrofactionTM)	Operational by 2021 commercial-scale by 2025	**Input:** forest residues **Output:** bio-oil (4 m^3/y), upgrade the bio-crude oil to produce renewable diesel, jet, or marine fuel	Tofte / Norway
9.	Valmet Partner-VTT Netherlands	Fluid bed	Operational	**Input:** forest residues Output: pyrolysis oil	Tampere / Finland
10	Versa Renewables LLC Project: Versa Pyrolysis	Not known	Operational	**Input:** lignocellulosics 42 kg/h **Output:** chemicals 25 kg/h	Albany (Georgia) / United States

Data taken from IEA Bioenergy 2020 Report.

TABLE 2　TRL 8 First-of-a-kind commercial-scale plants.

S. no.	Technology provider/ partner	Technology	Status	Input: **raw materials** Output: **products**	Country
1	BTG-BTLEMPYRO Netherlands **Project EMPYRO** Partners: BTG Bioliquids, Empyro	Fast pyrolysis based on rotating cone technology rotating cone	Operational	**Input**: organic residues and waste streams, e.g., wood pellet processing waste (5000 kg/h) **Output1**: pyrolysis oil 3200 kg/h **Output2**: steam **Output3**: power (electricity)	Hengelo/ Netherlands
2	Canfor (CPPI) Pulp Mills and Licella Pty Ltd. Joint Venture Canada	Licella's Cat-HTRTM technology	Under construction	**Input**: wood and pulp residues from kraft pulping **Output**: bio-oil 8,000,000 m^3/years	Prince George/ British Columbia/ Canada
3	Ensyn Brazil (Partner Fibria)	Circulating fluid bed	Planned	**Input**: forest residues, eucalyptus (16,667 kg/h) **Output**: pyrolysis oil 11,470 kg/h	Brazil
5	Fortum Finland (Partner: Valmet & VTT)	Fluidized bed	Operational	**Input**: forest residues pine 10,000 kg/h **Output**: pyrolysis oil 6313 kg/h	Joensuu/ Finland
6	Pyrocell AB Pyrocell Sweden (Partners: TechnipFMC and BTG BioLiquids (BTG-BTL) Pyrocell AB is a joint venture between Swedish wood processing company Setra and oil refinery Preem AB)	TechnipFMC and BTG BTG Bio Liq first pyrolysis	Ground-breaking at Setra's Kastet sawmill in March 2020 Plant is expected to be operational by the end of 2021	**Input**: forest residues, sawdust **Output**: pyrolysis oil 25,000 t/y	Kastet/ Sweden
7	Red Arrow United States	Circulating fluidized bed	Operational	**Inputs**: forestry **Output**: chemicals, food additives	Wisconsin/ United States

Data taken from IEA Bioenergy 2020 Report.

analytical methods and their appropriateness for bio-oil analysis and results are published in IEA Bioenergy Handbook (Bridgwater et al., 1999a, c). Additionally, alternative standardization methods for physical and chemical characterization of complex pyrolysis liquid/bio-oils are also developed and subsequently evaluated (Oasmaa et al. 1997). Several round-robin tests are carried out in 12 laboratories with bio-oils derived from varied feedstocks and produced through different technologies. An updated guide to physical property characterization of

TABLE 3 TRL 9 commercial-scale plants.

S. no.	Technology provider/ partner	Technology	Status	Input: **raw materials** Output: **products**	Country
1	BioLiquids Project GREEN FUEL NORDIC OY Funded by venture capital, the North Karelia ELY Centre and Green Fuel Nordic Oy shareholders	Rotating Cone BTG-BTL fast pyrolysis technology implemented by Technip, core unit being manufactured by Zeton	Operational	**Inputs**: forestry, sawdust **Output**: pyrolysis oil (2000 m^3/y/20 ML/y pyrolysis oil	Lieksa, Finland
2	Ensyn http://www.ensyn.com/quebec.html	Circulating fluidized bed	Commissioning	**Inputs**: forest residues **Output**: pyrolysis oil (40,000 m^3/y	Port Cartier, Quebec/ Canada
3	Ensyn, Arbec Forest Products Project: Cote Nord This project is developed by Ensyn, Arbec Forest Products, and Groupe Rémabec, and the technology is supplied by Envergent Ensyn owns a minority interest in the equity of the project	Ensyn's RTP fast pyrolysis technology	Operational since 2018	**Input**: 65,000 dry metric tons/y cellulosic woody biomass **Output**: 38 ML/year pyrolysis oil to customers in the Northeast US and in Eastern Canada for heating purposes and as a renewable feedstock for refinery co-processing	Port Cartier/ Quebec/ Canada
4	Ensyn, Suzano SA (Partners: Suzano (formerly FibriaCelulose S.A.) Project: ARACRUZ PROJECT	Circulating fluid bed	Late-stage development	Eucalyptus forest residues (17,000 kg/h) Pyrolysis oil (11,000 kg/h) an 83 ML/year pyrolysis oil production facility that is	Barra do Riacho/ Brazil

Data taken from IEA Bioenergy 2020 Report.

biomass-derived fast pyrolysis oil is published (Oasmaa and Peacocke, 2001; Oasmaa, 2003). The objective of the EU PyNe round-robin was to compare existing analytical methods without any restriction. The primary aim of the IEAA PYRA round-robin has been the determination of the interlaboratory precision and methods applied for elemental composition, water, pyrolytic lignin, and main compounds. The round-robin by 12 labs is carried out not to compare pyrolysis liquids but the compare the analyses in different laboratories.

Some critical learning from the round-robin are as follows:

- Liquid sample handling plays a very important role
- The precision of carbon and hydrogen is very good
- Oxygen by difference is viable and oxygen by direct determination is poor

- Water by Karl–Fishcher titration is accurate but should be calibrated by the water addition method
- Accuracy of density is good
- High variation is obtained for nitrogen viscosity, pH, and solids
- The method of determination of pyrolytic lignin and measuring the stability of the liquid needs improvement

2.1 Bio-oil characterization

Conventionally chemical characterization involves fractionation of bio-oil into different groups of chemical functionalities followed by the analysis of fractions through GC/MS. Later simpler characterization schemes including fractionation into water-soluble and insoluble fractions are developed (Mohan et al., 2006).

A chemical analysis scheme of the oils based on water fractionation is reported in the literature as shown in Fig. 5 (Venderbosch and Prins, 2010). The basic scheme is a relatively simple procedure consisting of the fractionation of oil with water, analyses of water-soluble and water-insoluble fractions, and of the further extraction of the water-soluble fraction with diethyl ether. The water-insoluble fraction comprises of high molecular mass lignin-derived material whereas, water-soluble fractions are dominated by water, volatile acids, alcohols, evaporation residues of diethyl ether insoluble and soluble compounds. The characterization scheme is suggested to be used as such or after modification appropriate to specific purposes. The most appropriate method of characterization of the fractions post-separation is graphically presented in Fig. 5 (Venderbosch and Prins, 2010; Oasmaa et al., 1997; Fig. 6).

Fig. 5 illustrates the preferential extraction of functional groups against the solvent used as

- Water solubles mainly contains acids, alcohols, and diethylethers.
- Ether solubles mainly contain aldehydes, ketones, lignin monomers, etc.
- Ether insoluble mainly contains anhydosugars, hydroxyl acids.
- *n*-Hexane solubles mainly contains fatty acids, extractives, etc.

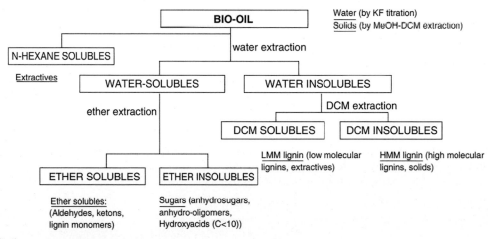

FIG. 5   Bio-oil fractionation scheme (Oasmaa and Kuoppala, 2003).

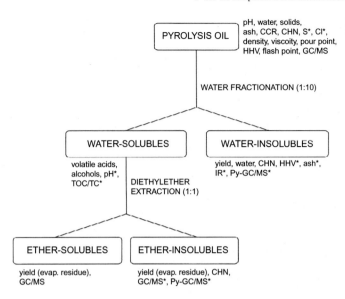

FIG. 6 Characterization of biomass-based fast pyrolysis oil (Oasmaa et al., 1997).

- DCM solubles mainly contains low molecular lignin fragment, extractives.
- DCM insoluble mainly contains degraded lignins, high molecular lignin fragments, including solids.

Different other bio-oil fractionation methods for a wide range of feedstocks and subsequent characterization schemes reported by several research groups are critically reviewed and compared in great detail (Mohan et al. 2006). These fractionation protocols are expected to be extremely beneficial in the future for chemical recovery from bio-oil as opposed to fuels. The suitability of standard protocols for analysis of the bio are listed in Table 4 (Oasmaa and Kuoppala, 2003; Oasmaa and Meier, 2002).

2.2 Bio-oil properties

Bio-oils differ significantly from petroleum-based liquid fuels. This difference in bio-oil physical properties from their petroleum counterpart is attributed to the difference in their chemical composition. The typical physicochemical properties of bio-oil and its comparison with petroleum fuels are presented in Table 5 (Oasmaa and Peacocke, 2010) and Table 6 (Oasmaa and Peacocke, 2010). Pyrolysis liquids are highly polar, containing about 35–40 wt% oxygen (dry basis), while mineral oils contain oxygen at ppm levels. The chemical composition of fast pyrolysis liquids is difficult to analyze with conventional methods like GC/MSD due to low volatility resulting from the polarity and high molecular mass of the compounds in the liquid. Solvent fractionation methods as mentioned in the earlier section are found extremely useful for the chemical characterization of whole pyrolysis liquids. The results of solvent fractionation and GC/MSD complete each other as shown in Table 6 (Oasmaa and Peacocke, 2010).

TABLE 4 Standards for bio-oil analysis (Oasmaa and Kuoppala, 2003).

S. no.	Analysis	Method
1	Water (wt%)	ASTM E 203
2	Solids (wt%)	Ethanol-insoluble
		MeOH-DCM-insoluble
3	Particle size distribution	
4	Ash (wt%)	EN 7
5	CHN (wt%)	ASTM D 5291
6	Sulfur and chlorine (wt%)	Capillary electrophoresis
7	Alkali metals (wt%)	AAS
8	Metals (wt%)	ICP, AAS
9	Conradson carbon (wt%)	ASTM D 189
10	Stability test	80 °C 24h
		80 °C 6h
		80 °C 1 week
11	Viscosity (20 and 40 °C) (cSt)	ASTM D 445
12	Viscosity (mPas)	Rotational viscometry
13	Density (15 °C, kg/dm^3)	ASTM D 4052
14	Flash point (°C)	ASTM D 93
15	Pour point (°C)	ASTM D 97
16	Heating value (MJ/kg)	
17	Calorimetric (HHV)	DIN 51900
18	Effective (LHV)	
19	pH	pH meter

Bio-oils are acidic, viscous, and thermally unstable. It is a complex mixture of over 200 chemical compounds comprising hydroxy aldehydes, hydroxy ketones, sugars, carboxylic acids, and phenolics in addition to the single largest compound water. High presence of oxygen up to 45–50wt% and other reactive polar compounds lead to an increase of viscosity over storage due to condensation and polymerization reactions catalyzed by acids and traces of inorganic elements in biochar.

A systematic study has been made to confirm the role of char in catalyzing the aging process (Agblevor et al., 1995). The study is conducted by adding char externally to the oil and comparing the aging process with respect to filtered bio-oil. It is reported that the presence of char enhanced the rate of increase in viscosity in comparison to the viscosity of the filtered fresh bio-oil. The char particles are understood to be forming catalytic condensation sites and thereby leading to "aging" due to apparent agglomeration of tar particles (French and Milne, 1994).

TABLE 5 Physical properties of fast pyrolysis bio-oils and mineral oil US #4 FO (Oasmaa and Peacocke, 2010).

Analysis	Typical bio-oil	US # 4 FO
Water (wt%)	20–30	0.5 (water and sediment)
Solids (wt%)	Below 0.5	0.5 (water and sediment)
Ash (wt%) 0.01–0.2	0.01–0.2[a]	0.1 max
Nitrogen (wt%)	Below 0.4	–
Sulfur (wt%)	Below 0.05	Varies
Stability	Unstable[b]	–
Viscosity (40 °C)	15–35[c]	5.5–24
Density (15 °C) (kg/dm³)	1.10–1.30[c]	–
Flash point (°C)	40–110	55 min
Pour point (°C)	−9 to −36	6 min
LHV (MJ/kg)	13–18[c]	–
pH	2–3	–
Distillability	No-distillable	–

[a] Note that metals form oxides during ashing and may yield ash values that are larger than the total solids in the liquid.
[b] Unstable at high temperatures and for prolonged periods of time.
[c] Depends on water content.

All the ash (i.e., the inorganic elements) in the biomass is retained in the char. The inorganic or mineral content of biomass is found in many forms; in aqueous solution in association with various counter ions, bound to organic acids or as deposits. The counterions in the solution include carbonates, oxalates, phosphates, silicates, chlorides, and sulfates (Shihadeh, 1998). Ash is suspected to take part in oil stability as some of the minerals present in ash are potential catalysts for reactions (e.g., Acetal formation), which are important in aging, e.g., chlorides of Sodium, Potassium, Calcium, Lithium, Iron, Magnesium, Manganese, and Zinc.

Under room temperature storage, aging takes place over the first 6-month period. The water content increases due to condensation reactions (Oasmaa and Kuoppala 2003). The decrease in volatile aldehydes and ketones leads to a rise of flash and pour points of the liquid during storage. In order to verify the minimum self-life of 6 months for customer user acceptance, stability tests and accelerated aging tests of pyrolysis oils are conducted. The increase of viscosity and water content is tested while the oil is maintained for 24 h at 80 °C, and 6 h at 80 °C. The chemical changes at 80 °C for 6/24 h are found to correlate roughly with the changes in 3–4 months/1 year storage at room temperature (Oasmaa and Kuoppala, 2003).

In-situ separation of acids and water during the collection of bio-oil through staged condensation is also found effective in making bio-oil highly stable (Das, 2014). The addition of alcohols is reported to largely arrest the thickening and aging of bio-oils (Oasmaa and Peacocke, 2001). The addition of up to 10 wt% of methanol in bio-oil is reportedly reducing the rate of increase of viscosity by nearly 20 times than for the oils without additives (Oasmaa et al., 2004a, b). With 10 wt% alcohol additions, the aging reactions are retarded for nearly

TABLE 6   Composition of a pine pyrolysis liquid (CHNO of dry matter) when combining solvent fractionation and GC-MSD[a] (Oasmaa and Peacocke, 2010).

Fast pyrolysis bio-oil (pine)		Wet	Dry	C	H	N	O	pK$_a$
Whole oil		23.9	0	53.3	6.5	0.08	40	
Water	wt%	23.9	0					15.7
Acids	wt%	4.3	5.6	40.0	6.7	0	53.3	3–5
Formic acid	wt%		1.5					3.8
Acetic acid	wt%		3.4					4.7
Propionic acid	wt%		0.2					
Glycolic acid	wt%		0.6					3.8
Alcohols	wt%	2.2	2.9	37.5	12.5	0	50.0	15–16
Ethylene glycol	wt%		0.3					15
Methanol	wt%		2.6					16
Aldehydes, ketones, furans, pyrans	wt%	15.4	20.3					16–36
Nonaromatic aldehydes	wt%		9.72	40.0	6.7	0.0	53.3	17
Aromatic aldehydes	wt%		0.009					17
Nonaromatic ketones	wt%		5.36	48.6	8.11	0.0	43.2	20
Furans	wt%		3.37					32–36
Pyrans	wt%		1.10					32–34
Sugars	wt%	34.4	45.3	44.1	6.6	0.1	49.2	3–16
Anhydro-ß-D-arabino-furanose, 1,5-	**wt%**		**0.27**					
Anhydro-ß-D-glucopyranose (levoglucosan)	**wt%**		**4.01**					
Dianhydro-α-D-glucopyranose, 1,4:3,6-	wt%		0.17					
Hydroxy, sugar acids	wt%							3–5
LMM lignin	wt%	13.4	17.7	68	6.7	0.1	25.2	9–10
Catechols	wt%		0.06					
Lignin-derived phenols	wt%		0.09					10
Guaiacols (methoxy phenols)	wt%		3.82					10
HMM lignin	wt%	1.95	2.6	63.5	5.9	0.3	30.3	
Extractives	wt%	4.35	5.7	75.4	9.0	0.2	15.4	7–16
Fatty acids	wt%							9–10
Triglycerides	wt%							
Resin acids	wt%							

[a] Analyzed at the vTI(Germany). LMM = dichloromethane soluble lower-molecular mass fraction of water-insoluble(WIS). HMM = dichloromethane insoluble higher molecular mass fraction of water-insoluble (WIS).

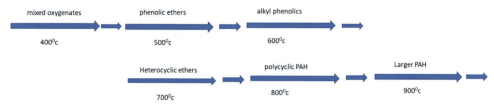

FIG. 7 Hypothesized schemes of chemical reaction pathways between alcohol and bio-oil.

1 year period. Interestingly it has been hypothesized that alcohol additives influence the liquid viscosity via three mechanisms as shown in Schemes 1–3.

  i. Physical dilution without affecting the chemical reaction rates

  ii. Reducing the reaction rate by molecular dilution or by changing the liquid microstructure

  iii. Chemical reactions between the solvent and the liquid components prevent further chain growth.

The underlying chemical reactions between bio-oil and alcohols are perceived to be esterification (Scheme 1) and acetalization (Schemes 2 and 3) (Diebold, 2002; Diebold and Czernik, 1997; McMurry, 1998).

Acetals serve as protecting groups for aldehydes and ketones. Alcohol addition is so far recommended as one of the most practical approaches bio-oil upgrading owing to its low cost and simplicity (Oasmaa and Czernik, 1999).

The pyrolysis reaction temperature and the presence of catalysts have an extremely significant role in manipulating the chemistry of bio-oils (Mohan et al., 2006). Increasing the cracking severity (time–temperature relationship) lowers the molecular weight distribution in the resulting oils yielding more gases. Dehydrogenation/aromatization reactions are enhanced at very high temperatures causing the formation of larger polynuclear aromatic hydrocarbons that eventually lead to increased carbonization. The relationship between the types of compounds in the products and the temperature at which the vapors are exposed prior to quenching is illustrated below (Fig. 7) (Elliot, 1986). As temperature increases, alkyl groups cleave/detach from aromatic compounds leading to the formation of polycyclic aromatic hydrocarbons (PAHs) at higher temperatures.

3 Pyrolysis oil specifications and standards

Although the pyrolysis liquid is called bio-oil, it is hardly miscible with liquid hydrocarbons due to its inherent high polarity and hydrophilic nature. In the recent industrial-scale bio-oil combustion tests, bio-oil has been found technically suitable in replacing heavy fuel

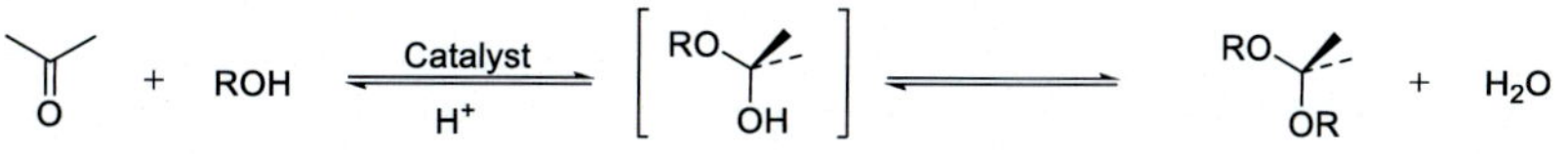

SCHEME 1 Aldehydes and ketones esterification.

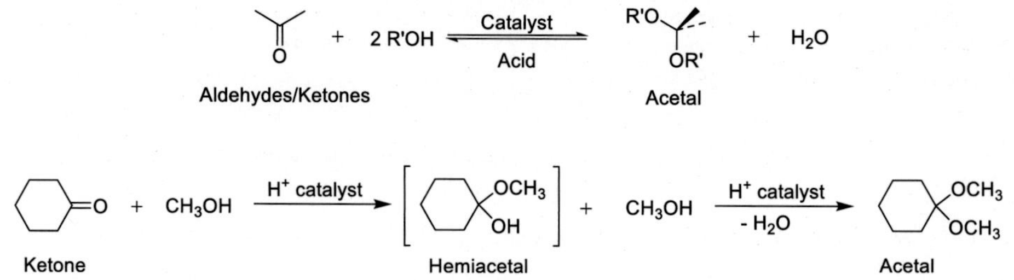

SCHEME 2 Aldehydes and ketones applications in synthesis of hemiacetal and acetal.

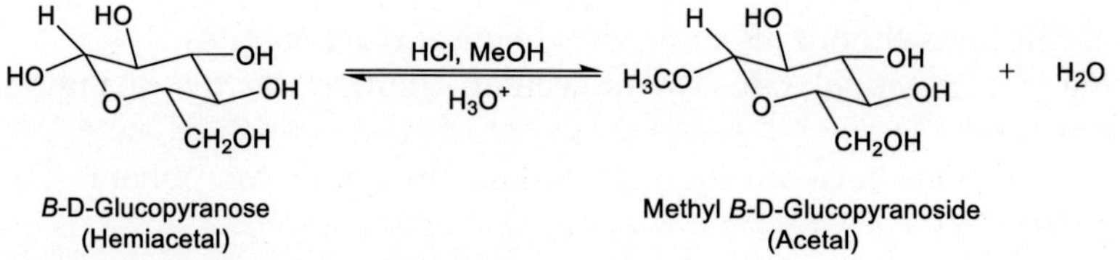

SCHEME 3 Acetal synthesis from monosaccharides.

oil for heating applications. This kind of replacement, however, demands more suitable metallurgy especially for the parts exposed to bio-oil.

In general, the emissions in bio-oil combustion largely depend on the solids, water, and nitrogen content of the oil. While the SOx emissions are literally nullified, the NOx-emission in bio-oil combustion primarily originates from fuel-bound nitrogen. Staged combustion is resorted to as a viable solution for NOx-reduction since successful air staging in natural gas, heavy and light fuel oil combustion has already been practiced. Current burner designs are quite sensitive to bio-oil properties, hence the quality consistency of the bio-oil is of supreme importance to avoid challenges in ignition, flame detection, and flame stabilization. Therefore, in order to create highly efficient, reliable bio-oil combustion systems, bio-oil grades necessitate appropriate standardization. Similar standards are also needed for other bio-oil applications as well.

The IEA Bioenergy has formed Task 34 for pyrolysis involving several countries such as Finland, Germany, Netherlands, Norway, Sweden, and the UK with the leadership of the United States with one of the objectives as standardization of bio-oils (*IEA-Tas 34 Bioenergy Report*, 2015). The task force supports CEN working group which is developing standards for bio-oil use in Europe. Towards this goal, five major bio-oil round robins are conducted on the standardization of specifications are 1988 (IEA), 1997 (EU PyNE,IEA), 2000 (EU PyNE& IEA), and 2012 (IEA). Working group (WG 41) established under European committee CEN/TC 19 (Gaseous and liquid fuels, lubricants, and related products of petroleum, synthetic and biological origin) for the task of bio-oil standardization is known as CEN/TC 19/WG 41. In response to EC mandate M/525 (2013), It is currently developing each of the following standards for fast pyrolysis oils (CEN Report).

   i. European standard for quality specification for pyrolysis oil replacing heavy fuel oil in boilers

ii. European standard for quality specification for pyrolysis oil replacing light fuel oil in boilers

iii. European standard for the quality specification for pyrolysis oil replacing fuel oils in stationary internal combustion engines

iv. Technical specifications for a quality specification for pyrolysis oil suitable for gasification feedstock for the production of syngas and synthetic biofuels

v. Technical specification for a quality specification for pyrolysis oil suitable for mineral oil refinery co-processing

Among the above five, three standards have already been developed for the replacement of heavy fuel oil, light fuel oil, and for use of bio-oils in stationary combustion engines. The remaining two technical specifications are in the process to be introduced for use of fast pyrolysis oils as gasification feedstocks and for mineral oil refinery co-processing.

CEN/TR 17103:2017 (publication date 2017-05-24), is the standard developed for fast pyrolysis bio-oil for stationary internal combustion engines and its quality determination (CEN Standards 2018 and 2019) This technical report describes the key properties of fast pyrolysis bio-oils and their importance to the fuel quality for use in stationary internal combustion engines. Internal combustion engine (ICE) in the scope of this document means a type of engine in which heat energy and mechanical energy are produced inside the engine. ICE includes compression ignition engines (diesel engines) and gas turbines. Attention is drawn to differences especially in those properties, which can have an effect on the required engine performance, such as ash, acidity, viscosity, combustion properties, and sulfur content. In addition to the quality requirements and related test methods for fast pyrolysis bio-oil (FPBO)s further instructions on storage, sampling, and materials compatibility are given.

EN 16900:2017 (publication date 2017-03-15), is the standard guidelines for fast pyrolysis bio-oils for industrial boilers, its requirements, and test methods (CEN Standard, 2018, 2019). This European Standard specifies requirements and test methods for fast pyrolysis bio-oils for boiler use at industrial-scale (>1 MW thermal capacity), not for domestic use. Two different grades are specified. It is recommended to draw attention to differences especially in those properties, which can have an effect on the required flue gas treatment system, such as ash, nitrogen, and sulfur content. National and local regulations determine the requirements for a flue gas treatment system. In addition to the quality requirements and test methods for fast pyrolysis bio-oils, further instructions on storage, sampling, and materials compatibility are given.

Bio-oil use in the boiler is the present focus due to its commercial readiness (Table 7, Oasmaa and Peacocke, 2010; Table 8, *IEA-Tas 34 Bioenergy Report*, 2015) provide the standard specification of Fast pyrolysis bio-oil (FPBO) as boiler fuel and the recommended test methods, respectively.

Presently two fuel oil grades are established by ASTM D 7544 for FPBO for bio-oil in the boiler. A major difference in these grades lies in the maximum solids and ash content in the bio-oil.

The recommended bio-oil as boiler fuel restricts the solid and ash content up to 0.15 wt% thus to ensure that inorganics in the form of ash and sand are present at as low a concentration as possible to reduce particulate matter. While the water content and kinematic viscosity limit

TABLE 7   ASTM Burner fuel standard D 7544 for fast pyrolysis bio-oil (Oasmaa et al., 2015; Oasmaa and Peacocke, 2010).

Property	Grade G	Grade D
Gross heat of combustion MJ/kg (min)	15	15
Water content, % mass (max)	30	30
Pyroysis solid content, % mass (max)	2.5	0.25
Kinematic viscosity at 40 °C, mm^2/s (max)	125	125
Density at 20 °C, kg/dm^3	1.1–1.3	1.1–1.3
Sulfur content, % mass (max)	0.05	0.05
Ash content, % mass (max)	0.25	0.15
Ph	Report	Report
Flash point, °C (max)	45	45
Flash point, °C (min)	−9	−9

TABLE 8   Properties of fast pyrolysis bio-oil and suitable test methods (d.b. refers to dry basis) (*IEA-Tas 34 Bioenergy Report*, 2015).

Property	Unit	Typical range	Applicable test methods
HHV	MJ/kg	14–19	DIN51900, ASTM D240
LHV	MJ/kg	13–18	DIN51900, ASTM D240, ASTM D5291 for H
Water	wt%	20–30	ASTM E203
pH	–	2–3	ASTM E70
TAN	Mg KOH/g	70–100	ASTM D664
Kinematic viscosity at 40 °C	mm^2/s	15–40	EN ISO3104, ASTM D445
Density at 150 °C	kg/dm^3	1.11–1.30	EN ISO 12185, ASTM D4052
Pour point	0C	−9 to −36	ENISO 3016, ASTM D97
Carbon	wt% on d.b.	50–60	ASTM D5291
Hydrogen	wt% on d.b.	7–8	ASTM D5291
Nitrogen	wt% on d.b.	<0.5	ASTMD5291
Sulfur	wt% on d.b.	<0.05	EN ISO 20846, ASTM D5453
Oxygen	wt% on d.b.	35–40	As difference
Solids	wt%	<1	ASTM D7579
MCR, CCR	wt%	17–23	ASTM D4530, ASTM D189
Ash	wt%	<0.3	ENISO6245
Flash point	°C	40–110	ENISO 2719, ASTM D 93B
Sustained combustibility	–	Does not sustain	EN ISO 9038
Na, K, Ca, Mg	wt% on d.b.	<0.06	EN ISO 16476
Chlorine	ppm	<75	Not specified

are easily achievable through mild upgrading of bio-oil, the threshold flash point (max) of 45 ° C leaves more room for research innovation in the near future.

Apart from this, several other standardization committees are formed under CEN and ISO for solid, liquid, and gaseous biofuel standardization by the Task Force such as CEN/TC 335 & CEN/TC 238 (solid biofuels), CEN/TC 019 (petroleum products, lubricants, and related products), CEN/ TC 343 & CEN/TC 300 (solid recovered fuels), etc.

3.1 Bio-oil conversion to drop-in fuels

Drop-in fuels refer to both transportation fuels as well as boiler fuels, in certain cases. Though there is no well-defined definition for drop-in fuels, there is a growing consensus that it refers to a blendstock that does not require substantial changes in refining or distribution infrastructure. Drop-in fuels are obtained from renewable sources such as biomass, vegetable oils, algae, animal fats, etc. (Araujo et al., 2017). Most of the drop-in fuels are produced through pyrolysis/ hydro pyrolysis/hydroprocessing should meet the finished diesel/jet fuel/gasoline specifications. Bio-oils need extensive upgradation to qualify drop-in fuels standards. No separate specification is recommended for drop-in renewable or green diesel. Nevertheless, the diesel blend needs to meet the regular specifications mandated in the respective country or region, e.g., ASTM D 975 in the United States, EN 590 in Europe, IS 1460:2017 in India. In case of upgraded bio-oil that is high in aromatics and olefins, or which is compatible enough to be blended with existing refinery streams thus could be a possible route producing green diesel.

Another aspect that might need critical consideration while making drop-in fuels from bio-oil or upgraded bio-oil is hydrogen to carbon ratio (H/C ratio). This ratio is conventionally used as an indicator to assess the hydrogen richness and energy density of fossil fuels. The production of drop-in biofuels targets to elevate the low H/C ratio of the biomass feedstock to a degree nearly compatible with that of diesel, jet, and gasoline fuels with H/C ratios close to 2 as shown in Fig. 8 (Vennestrøm et al., 2011). High O_2 or low H/C feedstocks require more processing and H_2 inputs During combustion, the oxygen within the biomass consumes hydrogen and thus reduces its effective H/C ratio. Thus, using a biomass feedstock where the main elemental components are hydrogen carbon and oxygen, the H/C ratio must account for the relatively high level of oxygen (in contrast to petroleum feedstock which contains practically no oxygen) as each oxygen atom consumes two hydrogen atoms to form a water molecule (H_2O) that contributes no energy to the combustion system (Karatzos et al., 2014). The "effective" H/C ratio for oxygenated biomass feedstocks, Heff/C, is calculated by Eq. (1).

$$\text{Heff/C} = \frac{n(H) - 2n(O)}{n(C)} \tag{1}$$

where n is the number of atom of each element.

Thermochemical bio-oil is well suited for long-term drop-in biofuels. The upgradation of bio-oil seems avoidable for its suitability for any applications. The low-cost availability of H_2 and innovation in catalyst thus appear extremely important for the sustainable production of upgraded bio-oil.

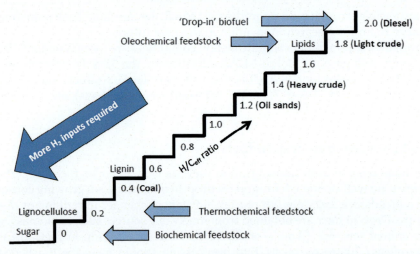

FIG. 8   "Staircase." Model For upgradation of stock based on hydrogen to carbon ratio (Karatzos et al., 2014).

4 Bio-oil for heat and power

The use of pyrolysis liquid in gas turbines and in a co-firing mode in large power stations is technically the most advanced. Recent work with diesel engines also appears to be quite promising. A brief account of the state of art in the area of combustion of bio-oil in boilers, gas turbines, and diesel engines is presented below.

Heat and Power are currently one of the most successful bio-oil applications worldwide. Commercialized applications are in Finland, the Netherlands, Canada, and the United States.

4.1 Combustion of bio-oil in the boiler

Bio-oil seems to be the most suitable boiler fuel, as long as it has uniform characteristics, as furnaces and boilers can accommodate a large variety of fuel.

The only commercial system, using bio-oil in a 5 MW swirl burner is at the Red Arrow Products pyrolysis plant in Wisconsin (Freel, et al. 1996), and is in operation for over 10 years. Extensive research on bio-oil combustion in boilers has been carried out in Finland. Tests have been performed at Neste Oy (Gust, 1997), in a 2.5 MW Danstoker boiler supplied with a dual-fuel burner and the boiler operation has been found to be satisfactory with varying fuel oil to bio-oil ratio. The operation on pyrolysis oil, without an auxiliary fuel, required only relatively minor modifications to improve combustion stability. VTT Energy in collaboration with Oilon Oy (Oasmaa et al., 2001a, b), performed a series of tests on a wide range of bio-oils in an 8 MW nominal capacity furnace operated at 4 MW output. Co-firing of bio-oil with conventional fuels has been identified as a promising option. Large-scale tests have been carried out at the Manitowac power station (Sturzl, n.d.), where pyrolysis liquids from the Red Arrow operation were co-fired with coal for the commercial production of electricity. Bio-oil was co-fired for about 370 h providing 5% of the thermal input to the 20 MWe boiler over a month period. The combustion of bio-oil was found to be clean and efficient without having any adverse changes on the boiler operation or on the emission levels. A successful co-firing test with 15 tons of bio-oil is conducted in 2002 in a 350 MWe natural-gas-fired power station in the Netherland (Khodier et al., 2009). Oil from the Malaysian BTG plant is routinely used to replace expensive diesel for the start-up of a fluid bed combustor near Kuala Lumpur International Airport although results are not available in the open literature. Since 2006, BTG is engaged in research on the oil in a standard 250 kW hot water generation unit, to replace diesel and/or natural gas. Towards this objective, appropriate quantities of palm-derived oil from Malaysia are transported to the Netherlands, and a dedicated oil lance is developed. Large-scale testing in larger, commercial boiler set-up, owned by Stork Thermeqare planned (Venderbosch and Prins, 2010).

In conclusion, a commercial large-scale application needs a constant supply of uniform quality oil available at an attractive price. Minor burner modifications are recommended for retrofitting bio-oils in existing boilers.

4.2 Combustion of bio-oil in gas turbines (https://task34.ieabioenergy.com/gas-turbines/)

Gas turbines are recently being explored as one of the most viable options for producing power and a combination of heat and power using low-grade bio-oils/bio-crude as a renewable substitute to conventional feedstock like petroleum distillates or natural gas. The efficient combustion of bio-oil in gas turbines needs specific modifications in the fuel supply system & burner combustion chamber to overcome the limitation of lower combustion speed associated with bio-oil burning. In addition compatibility of bio-oils with the materials of construction in fuel systems (acid corrosion) and in blades (erosion, alkali hot corrosion) is of critical importance.

Since 1995, Magellan-Orenda Aerospace Corporation of Canada has been actively working on using bio-oil as feedstock in gas turbines as the first of its kind in the world. 2.5 MWe class GT2500 engine originally designed and built by Mashproekt in Ukraine is tested as the most suitable turbine for low-grade fuels including bio-oils. The principal advantage of this engine is its "silo" type combustion chamber placed above the turbine thus enabling it to be easily modified and optimized for such fuels. Advanced coating of the whole hot section provides protection against contaminants (alkali). The use of bio-oil is reported to produce lesser NO_X and SO_X emissions and higher levels of particulates compared to operation with diesel fuel (IEA Bioenergy task 34 report). A complete gas turbine package including heat recovery unit and fuel pre-treatment skid installed at Dynamotive's commercial pyrolysis demonstration site in West-Lorne, Canada. In this plant, up to 48 tons of bio-oil/day are fed to fuel a 2.5 MWe direct-fired Orenda gas turbine. Unfortunately, this operation has become irregular due to a lack of supply of bio-oil. The economics of this application are found competitive with other gas turbine-powered bioenergy technologies and thus showing its commercial viability. There is a high degree of emission benefits of bio-oil use over conventional diesel on account of NOx and SO_2 with a moderate increase of CO as shown in Table 9.

OPRA Turbines is a Dutch company based in Hengelo, the Netherlands. It develops, manufactures, markets, and maintains generator sets in the 2 MW power range using the OP16 series of gas turbines. The OP16 gas turbine is of an all-radial design, which provides

TABLE 9 Gas turbine typical performance data for different fuels.

Oil	Fuel flow (L/h)	Electrical output (kW)	Inlet temp. (°C)	Outlet temp. (°C)	CO (ppm)	NOx (ppm)	SO$_2$ (ppm)
Diesel	1071	2510	−3	403	1	321	7
Ethanol	1800	2510	2	415	3	101	2
Biodiesel	1200	2550	11	467	4	321	1
Pyrolysis oil—Ensyn	1800	2650	−10	420	55	60	1
Pyrolysis oil—Dynamotive	1883	2510	−2	417	49	57	2

Reproduced from IEA Bioenergy task 34 report.

robustness, reliability, and the highest efficiency in its class. A key feature of the OP16 gas turbine is the ability to utilize a wide range of fuels.

The combustion of pyrolysis has been tested (*IEA-Tas 34 Bioenergy Report*, 2015; Beran and Axelsson, 2014). It is found that between 70% and 100% load it is possible to burn 100% pyrolysis oil without the need of mixing it with ethanol. Based on this research OPRA has been able to design a new combustor for burning pyrolysis oil and other low-calorific fuels. The new combustor is large enough to provide sufficient residence time for complete combustion of the pyrolysis oil. It is found that the maximum permissible droplet size of the pyrolysis oil spray should be restricted to 50%–70% of the droplet size for diesel fuel, to achieve efficient combustion. Similar investigations on a regenerated small turbine engine are also performed with the liquefied wood, a bio-crude produced via solvolysis of lignocellulosic biomass in acidified glycols, with fuel properties in close proximity to pyrolysis oil (Seljak and Katrašnik, 2016).

Several test campaigns using bio-oil in OP16 series of gas turbines at 2 MW power range are reported (Martin Beran and Axelsson, 2013; Lupandin et al., 2005). Non-pretreated pine wood-derived bio-oil provided by BTG Bioliquids is tested initially to generate baseline data. Under standard configuration, 25% pyrolysis oil and 75% ethanol mixture are found as optimum for sustaining stable combustion. Due to the differences in fuel composition, especially the large amount of dilutants associated with the pyrolysis oil, the residence time for pyrolysis oil is longer than for fossil fuels. The combustor geometry is redesigned to troubleshoot the formation of sediment on the flame tube inner wall and in the exhaust as a result of incomplete combustion.

The pressure loss of the combustor is an important parameter governing the mixing process and the function of the air blast atomizer. Therefore, it was important to decrease the effective combustor area and air mass flow to maintain the same pressure loss as the original combustor. Pintle air blast nozzle, capable of handling bio-oils of high kinematic viscosity, designed by OPRA, is used as the fuel injector in place of standard pressure nozzle. This specially designed nozzle eliminated the need for any external source of air or steam.

The overall aim of the R&D project was to develop a combustor suitable for the efficient combustion of pyrolysis oil. It was found that between 70% to 100% combustor loading it is possible to burn 100% pyrolysis oil without the need of mixing it with ethanol. This research insight led the OPRA to design a new combustor of larger size to provide sufficient residence time to pyrolysis oil ensuring its complete combustion.

A recent report shows the combustion of the high share of FPBO-ethanol blend fuel (1:1 vol ratio) in a test rig based on a modified micro gas turbine (MGT) (Buffi et al., 2018). The modification includes a redesigned silo combustor and a modified injection line. The modified configuration of the test rig allowed stable MGT operation at different loads, with blends of pyrolysis oil/ethanol at 20/80 and 50/50% (volume fractions). The most significant result of the present study revealed that the use of 50% (volume fraction) of fast pyrolysis bio-oil (blended with ethanol) allowed stable combustion at 20 kW power output in the revised configuration of the micro gas turbine. Test towards 100% FPBO feeding in this MGT configuration showed unstable operation, and the analysis of carbon deposits on the hot parts of the combustor confirmed this statement (Buffi et al., 2018). In order to achieve stable combustion with pure pyrolysis oil without the support of pilot injectors, further investigation on local heat transfer on the injection nozzle is required. A possible solution is envisaged as the modification of the atomization section along with the increase of temperature of the primary combustion air.

4.3 Combustion of bio-oil in diesel engines

While boilers are mostly used to produce heat, diesel engines offer high efficiency (up to 40%) in power generation and can also be adapted to the combined heat and power (CHP) process where overall efficiencies can be enhanced beyond 70%. Medium and slow-speed engines are known for better fuel flexibility and can operate on low-grade fuels. Some specific features of bio-oil that need special consideration before using in diesel engines are ignition difficulty (resulting from low heating value and high water content), corrosiveness (acids), and coking (thermally unstable compounds). However potential advantages of using bio-oils for power generation have led to important research activities in several countries.

In 1993, Solantausta et al., at VTT energy tested pyrolysis oil in a 500 cc (maximum power 4.8 kW) high-speed, single-cylinder, direct-injection petter diesel engine with a compression ratio of 15.3:1, (Solantausta et al., 1993). Bio-oil was seen to have very poor auto-ignitability, and even with 9% of additive, the ignition delay for bio-oil was 9 crank angle degrees (CAD) compared to 6 CAD for No. 2 Fuel Oil. In addition, coke formed during bio-oil combustion resulting in rapid clogging of injection nozzles. However, CO, NO_X, and hydrocarbon emissions (after the catalytic converter) from pyrolysis oil were comparable to those from diesel fuels. Further tests at VTT Energy (84 kWe engine) and Wartsila (1.5 MW_e electricity) showed the potential use of bio-oil in the pilot–ignited medium-speed diesel engines (Solantausta et al., 1994). The challenges identified were difficulty in adjusting the injection system (excess variability in the composition of bio-oil), wear and corrosion of certain injection and pump elements (acids, particulates), and high CO emissions. However, with better quality oil, proper selection of materials for injection nozzles, and catalytic converter for exhaust, the challenges seemed to be overcome easily.

The study conducted by Suppes in the University of Kansas with a blend of 72% pyrolysis oil, 24% methanol, and 4% cetane enhancer (tetramethylene glycol dinitrate) performed in a single-cylinder, air-cooled Lister Petter diesel engine, showed comparable performance as that with diesel fuel. It is recommended to restrict the use of pure (unblended) bio-oil to low-speed diesel engines with relatively high compression ratios, whereas, the bio-oil-methanol blend could be used in high-speed engines, especially with cetane improving additives.

Long hours, over 400 h, of testing on a modified dual-fuel slow-speed diesel engine have been achieved by Ormrod Diesels in the United Kingdom (Ormod and Webster, 2000). Three cylinders of the six-cylinder 250 kWe engine have been modified to run on bio-oil using 5% diesel as a pilot fuel to initiate combustion. The emission results are found promising though the CO being reported a little higher compared to diesel fuel. The engine has been successfully operated entirely on bio-oil by shutting the diesel supply to the unmodified cylinders (Leech, 1997). The minimum diesel contribution as a pilot fuel to provide satisfactory operation is 5% in energy terms. Even though deposits on pumps and injectors are reported, the overall performance is found unaffected.

The use of fast pyrolysis bio-oil in a modified diesel engine is reported (PyNE 41, 2017). The use of Fast Pyrolysis Bio-Oil (FPBO) in stationary diesel engines can be a valuable approach for small-scale, combined heat & power (CHP) applications. However, direct application (i.e., without chemical upgrading) is challenging due to the specific properties of FPBO. The pH of FPBO is below 3 and the water content is in the range of 20–25 wt%. As a result, standard diesel engine components are expected to quickly corrode when contacted with FPBO. To enable fast and complete combustion of FPBO, proper atomization is of utmost importance.

Compared to conventional diesel FPBO, has a higher viscosity and a higher density which will result in larger droplets. Additionally, the dynamic surface tension of FPBO is significantly higher than for diesel causing a further increase in droplet size (van de Beld et al., 2011, 2013, 2017). Adding some ethanol to the FPBO strongly improves the atomization properties. FPBO is difficult to ignite and special measures are required. Different approaches are proposed such as increasing the air inlet temperature, increasing the compression ratio, adding cetane improvers to the fuel, or pilot fuel injection. BTG has modified two compression ignition engines to develop this application, viz. a one-cylinder, and four-cylinder prototype development. Initial work is done with the 1-cylinder engine which has the advantage that only 1 fuel injection system needed reconstruction. More recently, the modification of the four-cylinder engine was initiated which can be seen as a prototype for a commercial size CHP system. The prototype is based on the Weichai 226B engine. Successful experiments were conducted with a modified four-cylinder Weichai 226B engine as a prototype for a commercial CHP system. and the complete Genset is assembled and supplied by ABATO Motoren BV, the Netherlands. Subsequently, the unit is modified by BTG, Netherland to enable FPBO fueling.

To overcome the poor ignition characteristics of FPBO, preheating of the incoming air is applied while keeping the compression ratio to the original value of 18. So far all tests are carried out with 20 wt% ethanol blended wood-based bio-oil produced in Empyro plant in Hengelo, Netherlands. The major engine modifications were carried out primarily in fuel feeding pumps and fuel injectors. The fully modified system is tested with different fuels including FPBO. Experimental results of fuel consumption as a function of the electrical load for the three configurations, e.g., conventional fuel system, modified fuel pumps, and modified pump and injectors are shown in Fig. 9. Hardly any difference is observed which indicates that the system is working reasonably well with FPBO. In Fig. 10 the NOx concentration in the flue gas is shown as a function of the flue gas temperature. These tests are carried out at different engine loads (up to 20 kWe) and different air inlet temperatures ranging between 50 and 200 °C. The new system leads to a lower NOx content probably due to some delay in the fuel injection caused by the indirect pump system. In Fig. 11, the specific fuel consumption is plotted in g/kWh as a function of the electrical load for diesel, butanol, ethanol, and FPBO. Though bio-oil's specific fuel consumption shows the highest owing to its lower heating value, the fuel consumption gradually decreases with increasing load. It is worthwhile to mention that 20 kWe corresponds to only 40% engine load. Nevertheless, different fuels show nearly identical overall efficiency.

5 Combustion of bio-oil in marine engines

The greenhouse gas (GHG) emissions of the marine sector were around 2.6% of world GHG emissions in 2015 and are expected to increase 50%–250% to 2050 under a "business as usual" scenario, making the decarbonization of this fossil fuel-intensive sector an urgent priority. Biofuels, which come in various forms, is one of the most promising options to replace existing marine fuels as marine engines are more tolerant of lower-quality fuel than the road and aviation sectors.

Marine distillate fuels are categorized into three major groups namely, DMA, DMB, and DMC based on their specifications. Among these DMA fuels, commercially denominated as marine gas oil or MGO are derived from the distillate fraction of crude oil and are much easier to handle than that derived from residual fuel oil. The relative properties of marine fuels are detailed in Table 10.

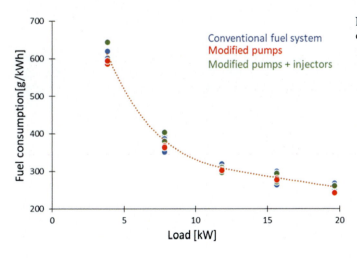

FIG. 9 Fuel consumption as a function of the load (PyNE 41, 2017).

FIG. 10   NOx concentration vs flue gas temperature.

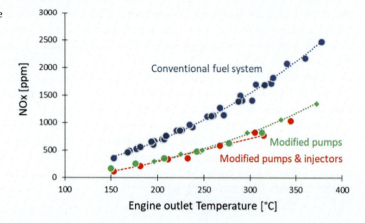

FIG. 11   Specific fuel consumption vs electrical load for different fuels (PyNE 41, 2017).

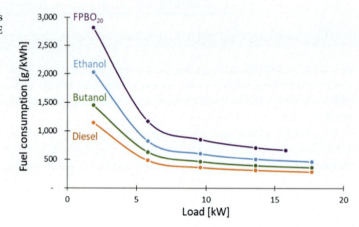

The EU directive/regulation 2012/33/EU to use marine fuel with reduced sulfur and carbon content has driven the quest for sustainable renewable fuel in the marine sector. Researchers have reported the compatibility of biodiesel with marine fuel and its miscibility at all proportions of bio-oil (Lin, 2013). Unlike the bio-oil and biodiesel blend fuel, and biodiesel-marine fuel blend fuel, the reports on bio-oil marine fuel blend are scarce (Alcala and Bridgwater, 2013; Garcia et al. 2007, 2010; Nguyen and Honnery, 2008; Chong and Bridgwater, 2017). As a part of ReShip project, led by Paper and Fiber Research Institute in Norway, Aston and Norwegian University of Sc. and Technology has carried out extensive research on developing/cost competitive pyrolysis oil-based fuel blends as potential replacements to marine fuels. A mixed fuel blend containing bio-oil, butanol, Rapeseed oil biodiesel (RME), and MGO is found to fall short of the acceptable specification limits of current marine fuel. Although few blends containing <40% bio-oil met the MGO viscosity specifications, the

TABLE 10 Requirement for marine distillate fuels (Vermeire, 2007).

Characteristic	Unit	Limit	DMX	DMA (MGO)	DMB	DMC	Test method reference
Density at 15°C	kg/m^3	Max.	–	890	900	920	ISO 3675 or ISO 12185
Viscosity at 40°C	mm^2/s	Min.	1.40	1.50	–	–	ISO 3104
		Max.	5.50	6.00	11.0	14.0	ISO 3104
Flashpoint	°C	Min.	–	60	60	60	ISO 2719
		Max.	43	–	–	–	
Pour point (upper)	°C						
Winter quality		Max.	–	−6	0	0	ISO 3016
Summer quality		Max.	–	0	6	6	ISO 3016
Cloud point	°C	Max.	−16	–	–	–	ISO 3015
Sulfur	% (m/m)	Max.	1.00	1.50	2.00	2.00	ISO 8754 or ISO 14596
Cetane index	–	Min.	45	40	35	–	ISO 4264
Carbon residue on 10% (V/V) distillation bottoms Carbon residue	% (m/m)		0.30	0.30	–	–	ISO 10370
		Max.	–	–	0.30	2.50	
Ash	% (m/m)	Max.	0.01	0.01	0.01	0.05	ISO 6245
Appearance	–	–	Clear and bright	–	–		ISO 10307-1
Total sediment, existent	% (m/m)	Max	–	–	0.10	0.10	
Water	% (V/V)	Max.	–	–	0.3	0.3	ISO 3733
Acid number	mgKOH/g	Max.	0.5	0.5	0.5	0.5	ASTM D664
Vanadium	mg/kg	Max	–	–	–	100	ISO 14597 or IP 501 or IP 470
Aluminum plus silicon	mg/kg	Max	–	–	–	25	ISO 10478 or IP 501
Used lubricating oil (ULO)						The fuel shall be free of ULO	
– Zinc	mg/kg	Max	–	–	–	15	IP 501 or IP 470
– Phosphorous	mg/kg	Max	–	–	–	15	IP 501 or IP 500
– Calcium	mg/kg	Max	–	–	–	30	IP 501 or IP 470

flashpoint of the solvent and the high acidity and moisture content of bio-oil are found to be critical influencers in noncompliance of target specifications. The overall inconclusive results of finding the most appropriate blend to comply with marine fuel specifications led to future investigations with upgraded bio-oils and solvents of higher flash points, e.g., 1-hexanol or ethylene glycol. It has also been reported that crude bio-oil cannot mix with current marine fuels in any proportion (Chong and Bridgwater, 2017). Mild upgrading of bio-oil through hydrodeoxygenation (HDO), wherein the oxygen content is reduced rather than being eliminated thus found essential to meet the standard (Boscagli et al., 2015) while also arresting the cost escalation and carbon footprint (Geraedts, 2017). Subsequently, multicomponent blend fuels made with mild catalytically hydrotreated stable pyrolysis oil, conventional diesel, and surfactants are explored to overcome the limitations posed by non-upgraded bio-oils. In these studies, the properties of MGO are benchmarked for blended marine fuels. Esterification coupled with azeotropic water removal for achieving the dual benefit of reduction of bio-oil acidity and moisture content to a level suitable enough for use in slow diesel engines has also been proposed (Sundqvist et al. 2015).

In a recent study, a comprehensive review is performed in order to identify the most suitable/ feasible biofuel for marine fuel (Mukherjee et al. 2020). This study is performed in order to identify the most suitable/feasible biofuel as a marine fuel. In this study, it is argued that the cheap marine fuel availability, massive refueling infrastructure, exclusion of the marine sector from the Paris climate agreement are major impediments of biofuel penetration in the shipping sector. In order to address this, four high-potential biofuel options, e.g., bio-methanol, bio-dimethyl ether, bio-liquefied natural gas, and bio-oil are assessed/evaluated on account of six criteria—cost, potential availability, present technology status, GHG mitigation potential, infrastructure compatibility, and carbon capture and storage (CCS) compatibility via both secondary literature review as well as stakeholder consultations (Mukherjee et al. 2020).

All the four biofuel's comparative scores nearly matched in spite of each possessing certain strengths and limitations that could favor that fuel under specific circumstances. As seen in Table 11 (Mukherjee et al., 2020), critical assessment is also made on the possibility of integrating deployment of these biofuels in combination with CCS for ambitious emission reduction in the marine sector. This awareness though presently is lacking among industrial stakeholders but is likely to increase enhance in near future.

In a comparison between HTL bio-crude and pyrolysis oil, it is argued that though the former is favorably considered as a drop-in fuel for heavy marine engines owing to its lower moisture content, higher calorific value, and higher H:C ratio (Hsieh and Felby, 2017) the later, being a near-commercial technology, with a higher TRL level deserves closer attention as well. Nevertheless, the simplicity, maturity, and low cost of pyrolysis bio-oil production must be balanced against the complexity and cost of its upgrading. The GHG emissions reductions also need careful estimation, with poorly devised upgrading scenarios potentially enhancing life cycle emissions beyond that of fossil diesel.

This work shows that, among the various biofuel options, bio-methanol, bio-DME, bio-LNG, and bio-oil are the most suited for the marine sector in the medium term owing to their potential for scale-up, advanced production status, and low costs. Although the options are close to each other in their overall suitability. The implementation of CCS is not presently very high on the list of priorities of the shipping industry, but this may change in line with a greater

TABLE 11 Scores for the fuels.

	Weight	Bio-methanol			Bio-DME			Bio-LNG			Bio-oil		
		A	B	C	A	B	C	A	B	C	A	B	C
Present technology status	0.08	3.5	3.8	1.8	3	3.2	1.5	4	3.7	1.8	3.5	3.3	1.6
Potential availability (EJ/y)	0.2	3	3.3	4.0	3	2.9	3.5	3	3.3	4.0	3.5	3.4	4.1
GHG mitigation potential (%)	0.2	4	3.9	4.7	4	3.7	4.4	3	3.4	4.1	3.5	3.7	4.4
Cost (€/Gj)	0.31	3.5	3.5	6.5	3.5	3	5.6	2	2.7	5.0	3	3.3	6.1
Infrastructure compatibility	0.16	3	3.4	3.3	3.5	3.6	3.5	3.5	3.3	3.2	3	3.7	3.6
CCS compatibility	0.05	3	3.2	1.0	3	2.9	0.9	4	2.7	0.8	2	2.7	0.8
Sum		20.0		21.2	20.0		19.4	19.5		18.8	18.5		20.6

A: Score allotted to fuel for criterion based on literature study; B: Score allotted to fuel criterion by stakeholders; C: Weighted score of fuel for criterion (C = Weight × B × 6, as 6 criteria used; rounded to one decimal place) (Mukherjee et al. 2020).

emphasis on CCS in the broader green energy sector and hence should be kept on the table as a deep decarbonization option (Mukherjee et al. 2020).

6 Bio-oil gasification to methanol and DME

Production of clean synthetic fuels in a decentralized manner from biomass at a scale to fit the economic collection radius of biomass resources makes considerable sense (Olah, 2013; *IEA Bioenergy Report, n.d.*; Bansal and Bandivadekar, 2013; Bruins and Sanders, 2012). Presently, direct biomass gasification and pyrolysis oil gasification are the two most prominent/synonym choices available for making methanol /or DME via syngas (Das and Bhatnagar, 2018). Over large-scale, direct biomass gasification process, bio-oil gasification with energy-dense bio-crude transported from smaller decentralized pyrolysis units operating at moderate 400–600 °C under atmospheric pressure could be proven more sustainable in near future. Although the technical feasibility of bio-oil to green synthetic fuel via syngas has already been proven up to pilot scale, however, a large-scale demonstration is yet far from reality. However, for correct evaluation of the relative performance of bio-oil and biomass-based synfuel processes, technical comparison in conjunction with life cycle and economic analysis hold critical importance.

Karlsruhe Institute of Technology (KIT), Germany has successfully demonstrated the unique Bioliqprocess for the production of syngas through bio-oil gasification. The syngas is finally converted to DME in a single-step process. In this process, fast pyrolysis of ligno-cellulosic biomass (e.g., wood and straw) is performed to produce energy-dense slurry containing bio-oil and char also called bio-syn crude. The slurry is comprised of 20% char and 50%–60% bio-oil (water and ash-free basis). The bio-syncrude, thus generated, is gasified in a pressurized entrained flow (PEF) gasifier with technical oxygen, at nearly 1200 °C. The gasifier operates at 8 MPa pressure and has a thermal fuel capacity of 5 MW (Dahmen et al., 2011).

Raw syngas generated in this process has a much lower H_2/CO ratio (0.5–0.7) while desired H_2/CO ratio is 2 for Methanol synthesis using Cu catalysts or FT synthesis with Co catalysts. After removal of particles, ammonia, HCN, CO_2, and other impurities, the syngas obtained is catalytically converted at 250 °C and 5 MPa pressure in presence of Cu/ ZnO/ Al_2O_3 for methanol synthesis and γ-Al_2O_3 for dehydration of methanol in a single-step process leveraging the low H2/CO ratio to directly produce DME (Dahmen et al., 2011). Air Liquide, in association with KIT completed the second stage of the Bioliq Process integrating all the steps, e.g., fast pyrolysis, entrained flow gasification, hot gas clean-up, and synthesis.

These bio-oil gasification processes are typically conducted at a higher pressure in order to reduce the number of syngas compression steps required for the methanol/DME production unit (5–10 MPa). Nevertheless, high-pressure and high-temperature gas clean-up systems, which are not as advanced or common as lower pressure–temperature processes are a major limiting factor for these routes. The complexity and safety aspects of such a high-pressure reactor are also practical challenges for the decentralized operation of such gasifiers in rural or remote areas for medium and small-scale units (Das and Bhatnagar, 2018).

ETC from Sweden and BTG from the Netherlands under the project entitled "Sustainable Products from Economic Processing of Biomass in Highly Integrated Bio Refineries" or SUPRABIO, also has successfully demonstrated a two-stage process for conversion of biomass to DME through bio-oil gasification in an entrained flow reactor. Among the many salient features, the ability to ensure a steady continuous flow of the acidic, pyrolysis oil into a pressurized environment at 1000 °C and 0.3 MPa pressure is worth mentioning (SUPRABIO Newsletter, 2013).

In recent years, steam gasification of bio-oil and bio-oil char mixture in atmospheric fluidized bed gasifier has been explored. High H_2/CO ratio apt for synthetic fuel production is obtained using nickel-based naphtha steam reforming catalyst and silica sand at bed material. High carbon deposition on catalysts being a common phenomenon, more work needs to be aimed at exploring the appropriate fractions of bio-oil and finding the optimum condition for steam reforming to achieve continuous syngas formation. Even if the formation of the carbon deposit cannot be completely avoided, efforts should be made to suppress it to the level so that there is no clogging in the reactor leading to the continuous formation of syngas. Further experimental investigations are needed to ascertain the optimal conditions for steam reforming of the bio-oil model mixtures.

7 Bio-oil upgrading

There have been several motives and approaches for upgrading bio-oil. Over the last few years, different upgradation techniques have been reported to elevate the bio-oil quality prior to its application. For example, to reduce char and ash content, to lower acidity and viscosity, to decrease the water content and increase its heating value, to render the oil more suitable to blend with petroleum fuels, to stabilize the oil, or to convert the bio-oil to transport fuel.

Upgrading is classified as physical and chemical upgrading. Physical upgrading involves the hot gas or cold oil filtration to separate char from the oil, addition of water or organic solvents to decrease the oil viscosity, and forming micro-emulsions to increase bio-oil miscibility in petroleum fuels by using surfactants and addition of solvents. Chemical upgrading occurs via hydrotreating, catalytic treating (with or without catalyst), esterification of bio-oil, olefination of esterified bio-oil, or simultaneous olefination/esterification of bio-oil.

7.1 Physical upgrading methods

Cyclones are traditionally being used to separate chars from the hot bio-oil vapor in most of the existing pyrolysis plants. There are some reports of using liquid filtration in which cases the oil agglomerates resulting in clogging of the filters. Hence hot gas filtration prior to condensation at around 400–420 °C has been preferred using sintered metal and flexible ceramic fabric elements (Scahill et al., 1997). The hot gas filtration reports a very very low level of char (<0.01%) and alkali metal (<10 ppm) retention, nevertheless, the flipside is decrease of oil yield by 10 wt% over the processes where only cyclones have been used to sequester the char particles. It is worthwhile to mention that thermal cracking of pyrolysis vapors during hot gas filtration is found to be highly beneficial with respect to bio-oil quality. The thermal cracking causes the size reduction of oligomers and hence filtered oil has shown a significant increase in burning rate and ignition delay over the unfiltered oil (Shihadeh, 1998).

The simplest use of bio-oil as a transport fuel seems to be in forming stable bio-oil-diesel emulsions. To enhance the miscibility of bio-oils in petro-fuels different surfactants are developed and stable micro-emulsions with 5%–30% bio-oil in diesel are reported by CANMET (Ikura et al., 1998). The resultant emulsions show promising ignition characteristics. However, high cost of surfactants and the energy-intensive emulsification process restricts the success of this approach. In addition, significantly higher levels of corrosion/erosion are observed in engine applications for such emulsions over unblended bio-oil or diesel.

There are studies both on the upgradation of bio-oil in terms of pH and viscosity. Since biofuels are being considered as alternative sources of fuel, it has been of interest to investigate its miscibility characteristics with both oxygenated (such as ethanol, methanol, and acetone) and conventional fuels like diesel (Bakhshi and Adjaye 1995). A potential upgrading process is being esterification of bio-oil. As mentioned in Section 2.2 the monofunctional alcohols react with the acidic compounds present in the bio-oils to form polyesters which in turn undergo esterification reactions with alcohols to form low molecular weight materials. While the esterification reaction has been shown to enhance most bio-oil properties except heating value, the flip side is the production of water from the reaction thereby lowering the overall energy content (Zhang et al. 2006). Moreover, the resulting flashpoint of the bio-oil solvent blend has not been reported, though a mention by Boucher et al. is critical as they have stated that "a limitation of methanol, however, is its low flash point in the blend" (Boucher et al. 2000).

In another upgrading method, organic acids in bio-oil have been esterified by application of a dicatopmoc ionic liquid (C6(mim)2-H_2SO_4) as an alternative to traditional acid (Xiong et al., 2009). Loss of around 51% of the total reaction products makes this process unlikely to be commercialized.

7.2 Chemical upgrading methods

Upgrading bio-oil to transport fuels like Diesel and Gasoline though has been proven technically but not yet commercialized. Upgrading of bio-oil into transport fuel requires full deoxygenating, which is conventionally accomplished via two main routes: hydrotreating and catalytic vapor cracking, and these methods are proposed in the 1980s.

Hydrotreating of bio-oil is carried out at a high temperature, high hydrogen pressure ($\sim$100–200 atm) in hydrogen, hydrogen, and carbon monoxide or hydrogen donor solvents in the presence of catalysts resulting in the elimination of oxygen as water and in hydrogenation-hydrocracking of large molecules. The catalysts that have been are typically used are sulfided CoMo or NiMo supported on alumina and the process conditions resemble those used in petroleum refining. To avoid polymerization, a low-temperature 250°C) pre-treatment of bio-oil is carried out prior to the standard hydrotreating process at 400°C. This process results in a naphtha-like product. Commercial viability of this process is restricted due to the requirement of high-pressure systems and the high consumption of hydrogen. Additionally, high levels of coke formation cause catalyst deactivation resulting in reduced hydrocarbon yield. The development of new catalysts and simultaneous use of catalyst regeneration technologies is expected to overcome these challenges. However, approximately 3 wt% of hydrogen is required and nearly 30% of the energy contained in the original raw bio-oil is lost in the process. Mild hydrogenation, instead of complete hydrogenation as described above, has also been tried and reported. Mild hydrogenation saturates the olefins and converts aldehydes to alcohols and is expected to be less expensive, but reportedly causes phase separation and thereby drastically increases the viscosity of the organic phase (Diebold, 2002).

Alternatively, catalytic vapor cracking using heterogenous acid catalysts has been extensively used to simultaneously dehydrate-decarboxylate bio-oil. In this approach, the bio-oil vapor is passed over acidic zeolite catalysts and at around 450°C and atmospheric pressure where oxygen is rejected as H_2O, CO_2, and CO making the resultant oil highly aromatic with a dominance of single-ring aromatic compounds (Chang and Silvestri, 1977). Partial deoxygenation also improves the stability of pyrolysis oils by reducing the concentration of reactive oxygenated functional groups which improves downstream hydroprocessing. This also reduces coke formation and improves catalyst life. Several catalysts have been studied including ZnO, phosphoric acid, and other salts, but zeolites (e.g., HZSM-5) and porous silicates have received maximum attention (Carlson et al., 2008; Williams and Nugranad, 2000; Adam et al., 2005; Jackson et al., 2009; Zhang et al., 2009).

Methods are reported where activated carbon-supported hydrotreating catalysts are used which results in reduced coke formation (Centeno et al., 1997). Vapor cracking using basic oxides instead of acidic zeolites and steam addition is also reported to have an enhanced hydrocarbon yield. Zeolite catalysts with micropores present good catalytic characteristics in the biomass catalytic pyrolysis process. However, large-molecule oxygenates produced from pyrolysis cannot enter their pores and would form coke on their surfaces, which decreases hydrocarbon yield and deactivates catalyst rapidly. An interesting result is reported on using some mesoporous and microporous catalysts, e.g., Gamma-Al_2O_3, CaO, and MCM41 mixed with a microporous catalyst (LOSA-1) for biomass catalytic pyrolysis (Zhang et al. 2013). While the added catalysts help in cracking the large-molecule oxygenates into small-molecule oxygenates, the LOSA-1 facilitates in converting these small-molecule oxygenates into olefins and aromatics. The results show that all the additives in LOSA-1 enhanced hydrocarbon yield significantly. The maximum aromatic and olefin yield of 25.3% was obtained with 10% Gamma-Al2O3/90% LOSA-1, boosted by 39.8% compared to that obtained with pure LOSA-1. Besides, all the additives in LOSA-1 improved the selectivities of low-carbon components in olefins and aromatics significantly.

A two-stage upgrading method is developed for a large number of agroresidue-based bio-oils produced at gas-fired auger pilot pyrolyzer. The bio-oil vapor is in-situ separated through staged condensation into organic oil fraction and acid-dominated pyrolytic aqueous fraction termed as "PAF." This oil is further upgraded through indigenously developed zeolite-supported bimetallic Ni, Co catalysts followed by mild hydrotreatment of nearly 200 °C at H_2 pressure up to 5 bar. The oil is upgraded to high gasoline fractions and aromatics with an increase of heating value in the range of 20%–40% (Das and Kalita, 2021). More experiments are in progress. The results will be published in near future.

In a recent patent disclosure, a water fractionation method has been proposed to fractionate the raw bio-oil to produce pyroligneous fraction prior to esterification or olefination which is then converted to gasoline and/or diesel hydrocarbons by two stages: hydrotreating followed by hydrocracking (Marker and Petri, 2009). In this process, the aqueous fraction is proposed to be reformed to make hydrogen which in turn can be used in subsequent hydrotreating and hydrocracking processes. The pyroligneous fraction produced by this process is, however, no process has been recommended for practical hydrotreating.

The olefination of bio-oil model compounds and bio-oil has been attempted with limited success due to the immiscibility of olefin hydrocarbons with the water emulsion that comprises bio-oil (Zhang et al., 2010). In another patent disclosure, an upgradation method has been produced on olefination of esterified bio-oil or simultaneous olefination/esterification of bio-oil (Steele et al. 2011). This process reports a dramatic reduction of water content producing fuel with a higher heating value of >20 Mg/kg. The upgraded product is proposed to be utilized as boiler fuel, a turbine fuel, or to power slow diesel engines. It has been proposed to produce hydrocarbons from the product by hydrodeoxygenation. One of the main advantages of olefination is reported to have reduced hydrogen requirement during hydrodeoxygenation resulting in a significant increment in hydrocarbon yield due to reduced water content.

In another patent disclosure, a process has been developed for producing a diesel boiling point range product and an aviation boiling point range product from renewable feedstocks such as plant and animal oils (Brady et al. 2009; Shen et al., 2009). This invention relates to a method and device to produce esterified, olefinated/esterified, or thermo-chemolytic reacted bio-oils to upgrade the bio-oil as a fuel. The bio-oil esterification reaction is catalyzed by the addition of alcohol and acid catalyst The water content of this olefinated product is reduced by this process, producing an upgraded product that could be utilized as a boiler fuel, a turbine fuel, or as a fuel to power a slow diesel, or as any combustion fuel for which a mildly upgraded bio-oil is adequate.

The feasibility of bio-oil upgradation to green transport fuel though has been proven technically and the first demonstration unit of integrated Pyrolysis based Biomass to Liquid of transport fuel grade (BtL) plant is currently under engineering design in Kapolei, Hawaii, by Envergent Technology (Joint venture between Ensyn and Honeywell's UOP). The development of this technology has also secured funding from DoE, USA. Dynamotive technologies (Canada) too have reported successful production of transport fuel from lignocellulosic bio-oil in their R&D facility in Waterloo, Canada through a hydrotreating process that is planned for commercialization in near future.

Another research group in GTI (Gas Technology Institute, USA) has completed proof of principle work on Integrated Hydropyrolysis and Hydroconversion (IH2) technology with DoE funding for direct production of Gasoline and Diesel from biomass. GTI has the lab-scale catalytic pyrolysis method to make upgraded bio-oil to be transported to oil refineries for processing into fungible fuels. IH2 is carried out in two integrated stages. The first stage is a catalytically assisted, fast hydro pyrolysis step completed in a fluid bed under moderate hydrogen pressure. Vapors from the first stage pass directly to a second stage hydroconversion step where a hydrodeoxygenation catalyst removes all remaining oxygen and produces gasoline and diesel boiling range material. An existing small-scale GTI gasifier was reconfigured for use in hydro pyrolysis. It has also been shown that catalytic pyrolysis technology is more economically attractive than lignocellulosic ethanol production through fermentation and thus could be the optimal technology choice for biomass conversion. GTI's present focus is testing more than 4000 h, to establish catalyst stability for different catalyst combinations and to produce enough oil for engine testing (Shell Report).

Co-pyrolysis of lignocellulosic biomass residues with waste plastics like PE, PP, PS, PET, and rubber wastes are also being explored to obtain upgraded bio-oil with high content of aliphatic and aromatic hydrocarbons and lower oxygenates as compared to biomass-based bio-oil. Co-pyrolysis also leads to the enhancement of other physicochemical properties like TAN, moisture, viscosity, flash point, and heating value (Nguyen et al. 2021).

The native oxygen content in biomass (40–50 wt%) is reported to be reduced significantly from 40 to 50 wt% to as low as 10 wt% in co-pyrolysis bio-oil. The exact amount of oxygen in co-pyrolysis bio-oil will eventually depend on the type of polymer and ratio of biomass: a polymer that is pyrolyzed as a mixture. For example, with hydrogen-rich polymers like PE and PP, a great extent of interactions between the intermediates in the form of hydrogen transfer reactions will aid in better deoxygenation. However, with oxygenated polymers or with polymers containing aromatics in the backbone, the bio-oil will be carbon-rich with oxygen content slightly on the higher side of ca. 20 wt%. Nonetheless, this will be significantly lower than the oxygen content in the bio-oil from individual biomass pyrolysis.

Upgrading of bio-oil into transport fuel requires full deoxygenating, which is convention-ally accomplished via two main routes: hydrotreating and catalytic vapor cracking. Hydrotreating of bio-oil is carried out at a high temperature, high hydrogen pressure (as high as 50 bar or higher), and in the presence of catalysts resulting in the elimination of oxygen as water. The catalysts (typically sulfided Co-Mo or Ni-Mo supported on alumina) and the pro-cess conditions are similar to those used in the refining of petroleum cuts. Catalytic vapor cracking makes deoxygenation possible through simultaneous dehydration-decarboxylation over acidic zeolite catalysts and at around 450 °C and atmospheric pressure, oxygen is rejected as H_2O, CO_2, and CO producing mostly aromatics. Upgrading bio-oil directly to a quality transport liquid fuel in stand-alone systems still is not currently economically attractive due to the use of hydrogen as an upgrading agent and rapid catalyst deactivation in most cases.

There are inherent merit and demerit for both processes. Hydrotreating route has high carbon retention but high H_2 requirement whereas, catalytic upgrading has comparatively much lower H_2 requirement but it sacrifices carbon. Both the upgrading processes thus need to be developed to a point and scale where the comparative Techno-economic evaluation will be possible.

A recent publication reviewed different catalytic hydrotreating (HDT) processes that are considered effective in removing oxygen from bio-oils to produce hydrocarbon-rich refinery intermediates or infrastructure-compatible finished biofuels (Zacher et al., 2014a, b). Exten-sive research is performed over the years to understand HDT of bio-oils. These studies are conducted with various catalysts, process conditions (temperature, pressure, and space velocity), and reactor configurations (Wang et al. 2013). Fundamental studies using model compounds are also pursued to elucidate the chemistry of upgrading bio-oils by HDT (Zacher et al., 2014a, b). However, HDT of bio-oil continues to be challenging due to the poor physicochemical properties of the feedstock and the lack of understanding of how each class of oxygenated compounds presents in the bio-oil impacts the upgrading process. Suitable cat-alysts that are stable, active, and selective for bio-oil HDT are still in the early stages of de-velopment (Pham et al. 2014). Typically, rapid catalyst coking as a result of condensation/thermal polymerization reactions is inevitable (Resaseo, 2011). Another major drawback of bio-oil HDT is that small oxygenated intermediates (<C4) are usually eliminated as C2–C3 gases and not retained in the liquid product, thus lowering the carbon efficiency of the pro-cess (Baker and Elliott, 1988). The two-stage hydroprocessing configuration reported by Elliot and Baker (Elliott, 2007) in 1988 has been used to overcome some of the challenges associated with hydrotreating raw bio-oils and to achieve long run times during continuous operation. In the two-stage continuous flow configuration, the bio-oil is stabilized in the first reactor at low temperatures (<300 °C) and low liquid hourly space velocity (LHSV); the second reactor is then used to perform deep deoxygenation at a relatively higher temperature (e.g., 350–400 °C) and LHSV. Studies have shown that this two-stage approach still suffers from coke formation while processing raw bio-oils (Olarte et al. 2016; http://www.btgworld.com/en/).

Ultimately, it is recognized that the physicochemical properties of bio-oil need to be improved to fundamentally address the challenges related to bio-oil hydrotreating. Preprocessing or stabilization of the raw bio-oil by mild hydrodeoxygenation (HDO) is now considered a prerequisite to achieving a successful deep HDT upgrading (Zacher et al., 2014a, b). Examples of preprocessing steps include physical modifications and chemical modifications like esterification, ion exchange, and azeotropic distillation to name a few (Thilakaratne et al. 2014). As an alternative to the traditional approach for the production of hydrocarbon-based biofuels from pyrolysis-derived bio-oil, RTI International is developing an advanced biofuels technology that integrates catalytic biomass pyrolysis and hydrotreating. The bio-crude from catalytic fast pyrolysis has more desirable physical and chemical properties than the raw pyrolysis bio-oil. Characteristically, the bio-crude from catalytic pyrolysis is relatively thermally stable, less acidic, and less oxygenated. Consequently, it is hypothesized that this bio-crude could be hydrotreated fairly effectively without a preprocessing or stabilization step. The major concern with catalytic pyrolysis is that the carbon yield is relatively low compared with conventional pyrolysis due to excessive coke and gas formation. Notwithstanding, the new process has the foreseeable advantage of potentially minimizing the overall hydrogen demand and increasing hydrocarbon product yield and quality. In fact, a recent techno-economic analysis found that the integration of mild catalytic biomass pyrolysis with hydroprocessing is a promising approach for the production of transportation fuels (Yang et al. 2012).

8 Bio-oil refinery integration/co-refining

Looking at the list of several upcoming commercial plants, it is realized that co-refining of bio-oil is increasingly being pursued globally.

One study demonstrates the technical feasibility of co-processing raw bio-oil with VGO, a fossil feedstock, in the fluid catalytic cracking (de Pinho et al., 2017). Different feeding strategies for integrating the bio-oil into the FCC unit are shown to be necessary due to its low miscibility with petroleum streams. Up to 10% of bio-oil having an oxygen content of approximately 50% is directly fed into a demonstration-scale FCC riser reactor that had multiple feed injection points. The tests conducted with the same bio-oil in two different experimental test series showed that while a 9-month-old bio-oil did not cause operating problems in the FCC unit, while a 21-month aged bio-oil affected operating conditions. The oxygen present in the bio-oil was almost completely removed through catalytic cracking, mostly as water and as CO or CO_2 to a lesser extent. However, the concentrations of the oxygenated compounds, especially alkyl phenols, increased in co-processed gasoline and diesel products compared to the amounts present in the processing of VGO alone. The presence of renewable carbon is confirmed in gasoline and diesel cuts through 14C isotopic analysis, showing that renewable carbon is not only being converted into coke, CO, and CO_2 but also into valuable refined liquid products. It is shown that gasoline and diesel could be produced from lignocellulosic raw materials through a conventional refining scheme, which uses the versatile catalytic cracking process. The bio-oil renewable carbon conversion into liquid products (carbon efficiency) is approximately 30%, well above the efficiency found in the literature for FCC bio-oil upgrading from laboratory-scale equipment, whose characteristics enhance coke formation in the reactor and deteriorate the yields profile. The successful demonstration of raw pyrolysis oil (50 wt% oxygen) co-processing suggests that it would also be feasible to co-process partially upgraded bio-oils as well. This also suggests there would be an optimal hand-off point between the biorefinery and the petroleum refinery. More data for co-processing partially upgraded pyrolysis oils are needed to further investigate this optimization. The subsequent analysis is underway to evaluate detailed economic feasibility and lifecycle analysis of the co-processing strategies.

The major questions for bio-crude refinery integration seem to be the correct assessment of the trade-offs for co-processing conventional petroleum feeds with biomass-derived crude, degree of upgradation required to co-process bio-crude into the refinery, the technical risks, and finally how the future economy looks like based on the pilot-scale techno-economic assessment studies.

8.1 Current status of research and development of bio-oil co-refining

Bio-oil upgrading processes are usually conducted in relatively complex facilities that require both high hydrogen inputs and capital costs. In most thermochemical processes there is a trade-off between capital costs, product yield, and the extent of hydrogen requirements.

Co-locating thermochemical processes at refineries can be used to leverage oil refinery assets, reduce capital costs and ensure a relatively low-cost source of hydrogen. Pyrolysis platforms appear particularly well suited to exploit co-location and synergy with existing refineries as pyrolysis oils can be processed using similar equipment to that currently used to upgrade crude oil. However, in practice, pyrolysis liquids contain relatively high levels of water and oxygenated species and thus are chemically quite distinct from crude oil and poorly suited to being "dropped into" existing petroleum processing units. However, it is likely that downstream refinery units such as FCCs and hydrocrackers could be configured to process thermochemical biofuel intermediates such as FT liquids and hydrotreated bio-oils to drop-in fuel blendstocks.

The majority of the processes and catalysts used to upgrade pyrolysis oils originate in the oil refining industry. It has also been suggested that pyrolysis oils or their derivatives could be "dropped into" existing refineries for final processing (Corma et al. 2007). The main benefit of this approach is capital cost savings by utilizing facilities and off-take infrastructure that has already been built. As mentioned earlier, neat pyrolysis oils cannot be readily co-processed with petroleum feeds as they typically contain up to 30% water and 40% oxygen and are thus not miscible with the apolar petroleum liquids (Venderbosch and Prins, 2011).

As the oxygen content of bio-oils also increases coking and deactivation of zeolite and HDO catalysts they cannot be readily inserted in oil refineries unless they have been partially deoxygenated (hydrotreated). However, it will be important to deoxygenate only enough to meet the minimum requirements of the refinery since deoxygenation gets disproportionately costlier when approaching oxygen-free bio-oils (Elliott, 2007). Once the oxygen content of the bio-oil has been reduced by hydrotreatment, it becomes a liquid hydrocarbon intermediate (such as hydro deoxygenated oil—HDO) that can potentially be inserted into an oil refinery. As HDO bio-oils, even when partially deoxygenated) they cannot be directly inserted with crude oil at an early process stage of the refinery. Thus bio-oil insertion is likely to occur at the refinery's hydroprocessing (hydrotreatment and hydrocracking) or fluid catalytic cracking reactors.

A simplified schematic showing HDO bio-oil insertion points (red arrows) within a typical refinery is outlined in the figure as shown (Fig. 12).

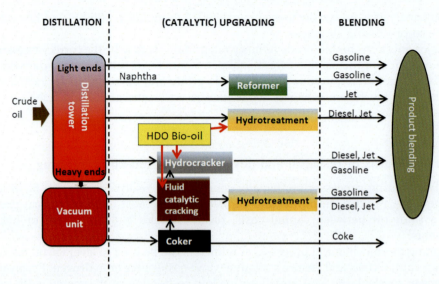

FIG. 12   Refinery insertion points (*red arrows*) for HDO bio-oils (Karatzos et al., 2014; *IEA-Tas 34 Bioenergy Report,* 2015).

There are several lab-scale as well as pilot-scale studies reported in the literature on bio-crude co-refining. In the domain of FCC co-refining around 12 publications are available on blends of VGO with raw, catalytic, and hydrotreated pyrolysis oil. Inspite of a variety of conditions, catalysts, scales are reported with modeling data the trends are found conflicting and difficult to discern. The major question still remained as to how to assess the trade-offs for co-processing conventional petroleum feeds with biomass-derived intermediates in a petroleum refinery? Towards this several pilot-scale studies on bio-crude co-refining are undertaken under DOE-BETO program. Most of the pilot-scale bio-crude refinery integration studies are carried out between 2012 and 2017 to identify costs, opportunities, technical risks, information gaps, research needs associated with co-processing.

Highlights of five significant pilot studies are mentioned below.

Study 1 (Zacher, 2015).
DOE Bioenergy Technologies Office (BETO) Project "Optimizing Co-Processing of Bio-oil in Refinery Unit Operations Using a Davison Circulating Riser (DCR) 2.4.2.402."
Objectives

- How much stabilization is required to co-process bio-oil into a refinery?
- What is the quality of fuel products produced after co-processing?

Partners

- W.R. Grace (FCC Catalyst).
- Tesoro (leading Refinery in the USA, presently known as Andeavor).
- VTT (Pyrolysis Technology/Oil provider).

- Oak Ridge National Laboratory (ORNL)
- PNNL (Pyrolysis Bio-oil provider)
- Aston University (Pyrolysis Technology Expert).

Project details

- In this project, four VGO samples were used for co-processing studies. Three samples were from TOSORO and one from Grace
- Two types of bio-oils from woody biomass (by VTT's 20 kg/h pilot system), and crop residues (Straw bio-oil from VTT and corn stover bio-oil from PNNL) are studied for co-processing.
- Crop residues were produced from two sources due to biomass availability and the need for both C3 (wood, straw) and C4 (corn stover) carbon fixation pathways.
- Partially as well as deep hydroprocessed oil with six different levels of oxygen content (6DCR Feed) is used for the co-processing study at pilot scale (400 mL/800 mL).

Key findings

- Raw bio-oil (38–40 wt% O) up to 5 wt% co-processed possible with difficulty. It produced increased coke with respect to non-blended VGO and reduced gasoline yield.
- Mild hydroprocessed (up to 20–22 wt% O) co-processed only up to 3 wt% with difficulty. It too produced increased coke with reduced gasoline yield.
- Medium treated (8–12 wt% O) co-processed successfully up to 10 wt% with yield exactly similar to VGO processing.
- Severe hydroprocessed (1–3 wt% O) co-processed successfully up to 10 wt% yield similar to VGO processing.
- Information on the performance of structural materials is being collected, in-situ corrosion coupons in DCR 2–5 co-processing tests, ~100 h, 1000 h ex situ corrosion analysis performed on available stabilized bio-oils, in situ DCR corrosion correlates with the degree of stabilization.
- Carbon accounting data shows that stabilized bio-oils may preferentially partition to the liquid products. Radioactive 14C analyses of products are a proven way to follow biogenic carbon, but it is expensive, time-consuming, and not readily available to the refiner. Stable isotopes (C, H, N, O) offer a potentially new way to trace biogenic carbon, it is inexpensive, fast, and performed by readily available equipment.

The results of this study are given in Tables 12 and 13.

TABLE 12 Co-processing results (Zacher, 2015).

Feed	Biomass	Stabilization	Dry wt% O	Conditions
DCR #2	Pine	Severe	1%–3%	3 bed, sulfided, 140–420 °C
DCR #3	Pine	Mild	20%–22%	2 bed, sulfided, 140–190 °C
DCR #4	Pine	Medium	8%–12%	3 bed, noble, 140–340 °C
DCR #5	Straw	Medium	8%–10%	3 bed, noble, 140–340 °C
DCR #6	Straw	Severe	1%–3%	3 bed, sulfided, 140–420 °C
DCR #6a	Corn stover	Severe	1%–3%	2 bed, sulfided, 140–420 °C

TABLE 13 Co-processing results (Zacher, 2015).

DCR	Feed bio-oil	Dry wt% O	Max. VGO feed preheat w/o nozzle plugging	% bio-oil co-processed	Yield observations
1	Pine, raw	38%–40%	~200 °F	Up to 5 wt% with difficulty	Increased coke, reduced gasoline
3	Pine, mild	20%–22%	~200 °F	Up to 3 wt% with difficulty	Increased coke, reduced gasoline
4	Pine, medium	8%–12%	Up to 700 °F	10 wt%	Yields similar to VGO
2	Pine, severe	1%–3%	Up to 700 °F	10 wt%	Yields similar to VGO
5	Straw, medium	8%–10%	Up to 700 °F	10 wt%	Pending
6	Straw/corn, severe	1%–3%	Pending	Pending	Pending

Study 2(Baldwin, 2016).

DOE/BETO sponsored-Project:

"Identification of optimal strategies of bio-crude co-processing with high carbon efficiency, low CAPEX and OPEX."

Project Timeline: 2012–2017.

Overall objective:

To identify optimal strategies of bio-crude co-processing with high carbon efficiency, low CAPEX, and OPEX.

Partners:

NREL (Overall analysis of chemical, technical, economic, and environmental/GHG emission).

Ensyn (Pyrolysis bio-oil provider).

LANL (biogenic C tracking tests).

ORNL (corrosion and fouling tests).

Project details

- Multiple liquefaction technologies and bio-crude have been examined.
- Different types of bio-crude oil including Fast pyrolysis oil (FPO), Catalytic pyrolysis oil (CPO, produced through both in-situ catalytic upgradation and ex-situ catalytic upgradation), hydropyrolysis (HYP) bio-oil, Hydrothermal (HTU) bio-oil are used.
- Impact of partial upgrading is assessed.
- Multiple refinery "insertion points" are evaluated.

- Experimental work coupled with rigorous TEA, LCA, and refinery modeling conducted through different consortia.
- Studies on biogenic C tracking (LANL) and corrosion and fouling (ORNL) undertaken.

Key findings on HYP bio-oil insertion at different insertion points

- Highly aromatic material possesses a significantly lower hydrogen to carbon ratio relative to typical refinery cracked naphthas (coker and FCC). The material would likely join heavily cracked naphtha refinery streams for hydroprocessing. Therefore possible insertion point recommended is at cracked naphtha hydroprocessing.
- Likely to be more highly aromatic than FCC light cycle oil (LCO), which suggests significant hydrogen addition would improve potential for diesel blending. Therefore possible Insertion point(s) suggested are high-pressure hydroprocessing or hydrocracking.
- Properties resemble FCC heavy cycle oil (CHO) or unconverted bottoms (slurry oil). Therefore possible insertion point(s) suggested are Hydrocracker, Coker, Asphalt or Bunker Fuel Blending.

Study 3 (Chum, L. H. 2015).

U.S. DOE Bioenergy Technologies Office (BETO): Brazil, USA, Canada Trilateral (Petrobras-NREL CRADA).

Credible commercial-scale results of co-processing raw fast pyrolysis oil in Fluid Catalytic Cracking (FCC) and techno-economic and life cycle analyses of FCC products with and without pyrolytic bio-oil.

Objectives

- To demonstrate technical and economic feasibility of co-processing raw fast pyrolysis oil in Fluid Catalytic Cracking (FCC) operation with credible commercial-scale and partners and also to support US and Brazil Strategic Energy Dialogue goals in advanced biofuels and industrial partnerships across these and other countries.
- Major outcome of this US-Brazil strategic project is the techno-economic and life cycle analyses of FCC products with and without pyrolytic bio-oil.

Partners

- NREL, USA (Chemical, technical, economic, and environmental/emission analyses)
- Petrobras USA (CENPES -Petrobras R&D Centre) (experiments in SIX Refinery, Brazil)
- Ensyn Corp. Canada (Pyrolysis Bio-oil provider)

Project details

In this project, 6 tons of pyrolytic bio-oil is produced from two different biomass, oak and bagasse in 10 kg/h pyrolysis unit at NREL and in Ensyn. The bio-oil is co-processed at kg/h scale in Petrobras CENPES center. Further, the bio-oil is co-processed at Petrobras SIX refinery at 200 kg/h scale for final products and overall heat and mass balance. 70 h uninterrupted production w/5% bio-oil/VGO. Products distilled into FCC gasoline and LCO.

Processing bio-oil produced from thermochemical process (pyrolysis) of wood residues to produce naphtha and Light Cycle Oil (LCO), stream having the same diesel distillation range, has been a major recent challenge of the Research and Production teams at six refineries in Petrobras. The sawdust from bio-oil, produced in Canada by Ensyn Corp., on the other hand has several undesirable characteristics such as high content of oxygen and water, high acidity, some chemical instability, and low heating value. One of the routes studied is catalytic cracking, a field of technology of great expertise at Petrobras. It was felt that it would not be possible to develop this route without a larger FCC unit as the U-144.

Other research groups had tried to process bio-oil in experiments conducted in small-scale reactors, but the results were very poor with feeding problems that cannot be resolved at the laboratory-scale. Bio-oil does not mix with fossil fuel processing streams. Therefore, at small scales, with a single dispersing point, it is necessary to use an emulsifier, which distorts the results and makes their interpretation difficult. The feed dispersion becomes greatly impaired as it is not possible to heat the feeding lines without coking them. In Petrobras's FCC unit U-144, this problem was solved by injection of fossil and renewable currents in different feed nozzle axial points, making it possible to heat the fossil fuel normally, while the bio-oil is kept at 40°C.

Thus, while laboratory-scale results show very high coke yields, indicating that bio-oil co-processing, even at low proportions of 5%–10% of the heavy gas oil is inviable, demonstration-scale results from the Petrobras SIX U-144 (scale of 200 kg/h = 32 bbl/day) processed up to 10% without any major problems.

Bio-oil is found to have no significant effects on operational factors and FCC yields, showing the potential for more profitable results, near-commercial reality. In summary, in this case, going to a larger scale proved to be especially important. The higher the scale of the tests, the better the bio-oil results became, rescuing a route initially ruled out nearly worldwide.

The processing is part of a cooperation agreement between Petrobras, an American oil company, and the National Renewable Energy Laboratory (NREL), a laboratory of the US Department of Energy.

Fibria shipped FCC intermediates (naphtha and LCO) to the United States for hydrogenation and testing of final gasoline and diesel. NREL evaluated the reduction in greenhouse gases achieved by this route.

The results of this study are presented in Table 14.

TABLE 14   Co-processing yields (Chum, 2015).

S. no.	Weight %	100%VGO	95% VGO+5% Bio-oil	90% VGO+10% bio-oil
1	Dry gas	3.5	2.8	2.8
2	LPG (C3–C4)	13.8	13.8	12.5
3	Gasoline (C5-220C)	39.9	40.6	38.8
4	Diesel (220-344C)	20.3	19.6	19.2
5	Bottoms (+344C)	16.1	14.4	14.4
6	Coke	6.4	6.0	6.5
7	CO	0.0	1.0	1.7
8	CO_2	0.0	0.4	0.6
9	Water	0.0	1.4	3.5

Key findings

- Literature states that it is not feasible to process raw bio-oil in the FCC due to the increase in coke and hydrotreating is necessary to improve the processing characteristics of bio-oil.
- Petrobras experience shows that lab-scale units (literature results) overestimate coke yields whereas in pilot and demo-scale studies much less coke deposition happened.
- Coke Lab g/min ≫ Coke Pilot1 kg/h > Coke Demo 200 kg/h ≥ Coke Commercial.
- Negligible catalyst deactivation caused by alkaline metals was found after 24 h with 5 wt% bio-oil.
- Co-processing up to 5% is economically feasible in the near term (with current biomass price in ≥88.2 US$/dry tons), available bio-oil production technology, and refinery technologies.
- Co-processing up to 10% is economically feasible with progress in Industry and pyrolysis technology Demo scale experiments with 5% and 10% showed improved oxygen mass balance.
- Low-cost biomass/bio-oil could accelerate the economical deployment of co-processing.
- Co-processing 5 and 10 wt% produces similar product yield and quality to that of petroleum feed.
- Co-processed fuels have one of the lowest carbon intensity values 21–25 GCO_2/MJ.

Study 4 (Biddy and Jones, 2017).
DOE Bioenergy Technologies Office (BETO) Project

Process model development of three key petroleum refining conversion systems converting mixtures of conventional and biomass-derived intermediates.

Objectives

- Develop a suite of models to understand impacts, opportunities, and gaps associated with co-processing.
- To develop detailed process models of three key petroleum refining conversion systems converting mixtures of conventional and biomass-derived intermediates to identify costs, opportunities, technical risks, information gaps, research needs associated with co-processing.

Partners

NREL (44%), PNNL (56%), Assistance from Aspen Tech.

Bio-crude considered

- Partially hydrotreated bio-oil
- HTL bio-crude
- Lipids

Petroleum-derived feed is chosen for study: Vacuum Gas Oil (VGO).

Project details

i. Technical model development.

- **FCC** Aspen Plus model development for pure compounds, stoichiometric reactor and with 0%, 10%, 20% partially hydrotreated (upgraded) bio-oil blend with VGO (Transferred developed models in other BETO projects).
- **Hydrocracker (HCK)** Aspen Plus model development using pure compounds, stoichiometric reactor and 0%, 10%, 20% partially hydrotreated bio-oil blend with VGO (transferred developed models in other BETO projects).

ii. **Economic model development**

- Pricing basis is a function of crude oil benchmark price (WTI) and at what crude price does co-processing become economically attractive to refiners (without policy incentives).
- Models developed in this project helped to support inclusion of refinery integration state of technology case in 2017 Multiyear programme plan (MYPP).

Key accomplishments

- Three refinery models and methods for co-refining (HCK, FCC, HT), first-of-a-kind, developed.
- Three types of biomass intermediates (pyrolysis bio-oil, HTL bio-crude, lipid feed) assessed.
- Economic assessment method reflecting a range of crude oil scenarios/prices developed.
- Resource assessment tool to estimate potential fuel volumes developed.
- Results under publication.
- Vetted assumptions with industry and adopted feedback to consider additional real-world constrained scenarios.
- Transferred models and methods to appropriate BETO projects for continued use/improvement toward understanding opportunities for renewable fuel cost reduction. This project Supports BETO Multiyear Program Plan critical conversion research area to "Work with petroleum refiners to address the integration of biofuels into refinery processes" and to reduce the cost of conversion to $3/gge by 2022.
- Publishing report and HYSYS models for use by stakeholders.

Study 5 (Dayton, 2017).

US Department of Energy (DOE) Bioenergy Technologies Office (BETO) Project:

Ex situ catalytic pyrolysis step to produce a low oxygen content, thermally stable bio-crude intermediate that can be upgraded in a hydroprocessing unit to produce infrastructure-compatible biofuels.

Objectives

Demonstration of an advanced biofuels technology at the pilot scale that integrates a catalytic biomass pyrolysis step to produce a low oxygen content, thermally stable bio-crude intermediate that can be upgraded in a hydroprocessing unit to produce infrastructure-compatible biofuels. The result of the study is presented in Table 15.

Partners

- RTI (project lead, I TPD CFP technology development for bio-crude, Hydroprocessing, process modeling)

TABLE 15 Hydrotreating of bio-crude: physicochemical properties (Dayton, 2017; Dayton et al., 2015).

Fractions	HDT product oxygen content (wt%)			
	0.73	4.32	4.94	5.16
Naphtha/gasoline IBP-165°C	30.1%	22.9%	24.8%	22.3%
Kerosene/jet fuel 165–250°C	34.8%	26.2%	25.2	25.5%
Heavy diesel/heating 250–340°C	23.6%	17.2%	17.8%	15.9%
Diesel (total)	58.4%	43.4%	43.0%	41.4%
Gas oils 340–550°C	6.5%	24.7%	19.3%	31.4%

~2:1 diesel/gasoline range products obtained.

- HaldorTopsøe A/S (HTAS) (optimized process condition for hydroprocessing of RTI bio-crude, bio-crude refinery intermediate co-processing, process modeling for 2000 TPD integrated process design and economics)
- Idaho National Laboratory (High impact feedstock providers)
- Iowa State University (High impact feedstock providers)

Key achievements

- 1 TPD unit operational for more than 3 years
- Four catalysts tested; five feedstocks—loblolly pine, hybrid poplar, corn stover, hardwood pellets, red oak, 12-h parametric studies
  - Temperature was the most influential factor.
  - Moderate temperatures ($450 \leq T < 500$) favored higher yields.
  - Anhydro sugars are cracked at higher temperature > 500°C.
  - Formation of simple phenols, catechols, and PAH increases.
  - Short residence times reduced biomass devolatilization.
- Overall material balances routinely 80%–100%
- Gas yield: 5–11C%; total liquid yield: 15–30C%; biochar yield: 48–70C%
- Steady-state yield varied between 38 and 50 gal/dry ton of biomass
- Over 200-gal of loblolly pine bio-crude and 20-gal of red oak bio-crude were produced for upgrading.
- Hydroprocessing evaluation and optimization
  - Developed strategy for upgrading RTI bio-crude samples (catalyst selection and initial process conditions)
  - Design, fabrication, and installation of hydroprocessing unit at RTI
  - Baseline testing to validate system performance
  - Three baseline experiments with vacuum gas oil (VGO)
  - More than 300 h time-on-stream and 30 gal of VGO hydrotreated
  - Mass closures between 99 and 100 wt%

- Hydrodesulfurization between 93.8% and 95.3% and hydrodenitrification between 84.1% and 89.7%
- Over 1500 cumulative hours of bio-crude upgrading since April 2015.
  - High carbon yields (55%–95%).
  - More stable hydrotreating catalyst activity (>100-h) with lower oxygen-containing bio-crudes (<15 wt%).
  - The HDT product contained mostly naphthenic hydrocarbons.
  - HDT products with greater than 5 wt% O contained more mono-phenols and aromatics hydrocarbons (Mono-, Di-, and Tri-).
  - HDT product carbon number distribution up to C27.
- Co-processed bio-crude and refinery intermediate blends
  - 1180 h total co-processing with 542-h 30%CPO/70%LGO
  - For feed oxygen in the range 0.5–2%wt, pressure (inlet pH_2) 70 bar, reactor temp. 330–345 °C, LHSV = $0.4 h^{-1}$, H_2/oil ratio: 625 Nl/L Product oxygen in the range 1–200 wt. ppm
  - 85% diesel-range product

The above-mentioned studies have produced several publications in the recent past. Study 3 results are shared in a recent publication (Zacher et al., 2014a, b). The Study 5 results have been reported in another publication (Dayton, 2017).

The key points from the overall literature review of bio-oil co-refining are as follows:

- High oxygen in bio-crude (due to the presence of water, acids, and oxygenates) is the major limiting factor in co-refining
- Bio-oil properties like O%, moisture content, and TAN No., C/H ratio are most significant for co-refining
- Most of the studies recommend bio-crude having ≤15 wt% O as the suitable feed for successful co-processing up to 10 wt%
- Bio-crude co-processing studies (with VGO and LGO) yield almost similar products to conventional crude
- Multiple insertion points in refinery should be adopted depending on the suitability of the bio-crude-type, e.g., CPO (catalytic pyrolysis oil), HDT (hydrotreated oil), HYP (hydropyrolysed oil) HTU (hydrothermally or catalytically upgraded oil)
- 5–10 wt% bio-crude integration economically viable with minimum modification in FCC and Hydrocracking units in refinery
- Co-processing 5 &10 wt% produces similar product yield and quality to that of petroleum feed
- Co-processed fuels have one of the lowest carbon intensity values
- Catalyst characterization beyond brief "proof of concept" needed
- Technologies to make pyrolytic bio-crude is mature internationally
- Technologies to make pyrolytic bio-crude in India reached up to pilot scale that needs to be upscaled
- Low-cost biomass/bio-oil could accelerate the economical deployment of refinery co-processing

(A)

(B)

FIG. 13 (A) TERI's 20 kg/h (max) PLC controlled automated pyrolysis unit and (B) pyrolysis vapor cracking and upgradation unit (integrated).

9 Pyrolysis technology status: Indian scenario

Research on pyrolysis for liquid fuel and chemicals in India has reached from bench-scale to pilot scale. However, no commercial plant has yet been built.

IIT Bombay over the last decade has carried out substantial work in the area of biomass characterization and thermochemical conversion especially in the area of vacuum pyrolysis. Work has been done with contributions in basic research on bio-oil production from different biomass on a lab-scale. The research has produced three Ph. Ds (Author being one of them) and a number of M. Tech projects and publications (Karumuri, 2002; Raveendran, 1995; Das, 2004; Das and Ganesh, 2003; Das et al., 2004a, b).

Recently Indian Institute of Petroleum (IIP), Dehradun reported significant work on biomass hydro pyrolysis (Greg Perkins et al., 2018; Kumar et al. 2020). IIP Dehradun has developed technology for producing bio-crude from biomass and lab trials are also conducted for co-processing of Jatropha-derived pyrolysis oil with petroleum-derived vacuum gas oil in advanced cracking evolution FCC unit. Furthermore, IIP has scaled up a 1tpd mobile plastic pyrolyzer and deployed it in the field.

Some work on Pyrolysis of waste agro biomass residues by a thermo-catalytic process using microwave in pebble and fluidized bed reactor are initiated in VIT University, Vellore under the financial support of Ministry of New and Renewable Energy (MNRE). With funding from GAIL (India) Ltd., Indian Institute of Technology (IIT) Madras has demonstrated the production of high-quality bio-oil with low oxygen content in a bench-scale microwave pyrolysis unit coupled with catalytic upgradation of vapors(Suriapparao et al., 2014, 2015, 2018, Xhou et al., 2016). The properties of the upgraded bio-oil are reported to be nearly equivalent to light fuel oil (LFO). Up to 1 kg of the combined feedstock of biomass and plastics are processed in batch mode in the reactor, and there are plans to upgrade the unit into a semi-batch or continuous reactor for better throughput.

In Shell Technology Centre Bangalore, a 5 tpd demo unit has been commissioned recently using CRI's IH2 technology that adopts Hydro pyrolysis and Hydro conversion to produce fuels such as Diesel and Petrol with an oil yield of 26% from MSW, Agro waste, etc. A commercial plant of capacity 500 dry tpd based on bagasse is being built by M/s. Sunlight fuels, UP. (Shell Report).

The Energy and Resources Institute (TERI) has worked extensively on pyrolysis technology development over a decade producing many publications and patents (Mukherjee et al., 2014; Das et al. 2013; Das, 2014; Das et al., 2016, 2020); TERI's technology is developed to suit Indian scenario for small to medium scale decentralized production of bio-crude from heterogeneous biomass. Indigenous pyrolysis technology is developed with joint funding from TERI, Ministry of New and Renewable Energy (MNRE) & Ministry of Petroleum and Natural Gases (MoP&NG) for making refinery-grade bio-oil and biochar from different agro-industrial biomass residues (MNRE Report; http: 164.100.94.214). TERI's Patented Pilot Scale Pyrolysis Test Units (PTU of capacity 20 kg/h) are uniquely designed fully automated Programmable Logic Control (PLC) based gas-fired auger pyrolyzer reactor systems as shown in Fig. 13. The advantage of having the gas-fired auger system is to avoid the energy-intensive feedstock pre-treatment steps and to make the unit extremely flexible to heterogeneous and lignocellulosic biomass feedstocks, a major source for biofuel production in

India. In this process, the combustion of flue gases from a coupled biomass gasifier and the pyrolysis non-condensable gases provide the pyrolysis heat. The added advantage of the gas-fired auger is its compactness and operation at lower process temperature (400–500 °C) without carrier gas. The biochar produced as a by-product is a potential precursor for fuels, catalysts, activated carbon, and soil improver.

This novel design is suitable to scale-up from pilot to modular commercial units. The Pilot Reactor has been tested extensively with agro-industrial crop residues and non-edible oil seed residues and shells like Cotton stalks, Paddy, Wheat, and Maize stalks, Mustard husk and stalk, Ground nut Shell, Cashew nut-shells, Bamboo wastes, Jatropha, Karanja oil seed cakes and Lignin residues from the lignocellulosic ethanol plant. The advantage of having the gas-fired auger system is to avoid the energy-intensive feedstock pre-treatment steps and to make the unit extremely flexible to heterogeneous and lignocellulosic biomass feedstocks, a major source for biofuel production in India for decentralized pyrolysis. Another unique feature of this technology is in situ separation of oil, aqueous & acidic fractions through staged vacuum condensation thereby avoiding post distillation of bio-oil.

On average, the pyrolysis product distribution obtained at 450–500 °C is total liquid ~50–60 wt%, biochar ~25–30 wt% and rest non-condensable gases. The total liquid comprises ~35–40 wt% combustible bio-oil fraction and 15–20 wt% in-situ separated Pyrolytic aqueous fraction called PAF. The overall moisture content in the combustible bio-oil fraction is in the range of 3%–5% whereas, for PAF fractions it is considerably high between 40 and 50 wt%. The higher heating values of bio-oils are in the range of 31–35 MJ/kg. The pyrolytic biochar has moderate heating values in the range of 25–26 MJ/kg. Activated carbon is successfully made from biomass and de-oiled algal biomass pyrolytic biochar through downstream activation. The biochar produced are having surface areas as high as $2400\,m^2/g$ and they are found highly suitable for CO_2 adsorption (Garg and Das, 2018, 2019).

TERI's approach to pyrolysis process development is twofold. Non-catalytic pyrolysis process and catalytic pyrolysis through the upgrading of pyrolytic vapor. The non-catalytic pyrolysis process is aimed at making hi-grade stable bio-oil with minimum moisture content (1–2 wt% achieved so far) free from acidic components which could be co-processed in a refinery with mild modification or directly used in industries for heat and power applications. The refinery-grade bio-oil fraction has a high heating value of 30–35 MJ/kg, very low moisture, acids (max up to 5 wt%), and O_2 content (3–12 wt%). This high-grade stable bio-oil that could be produced in decentralized units also has a great potential as a renewable and green substitute to conventional fuel oil in various heating applications.

The generated pyrolysis oil has been used by blending in Oil Corporation (IOC) refinery. In IOC, the bio-oil supplied by TERI is blended at 5–10 wt% with the heavy furnace oil (FO) originating from IOCL's refinery and co-processed in the furnace to see the effect of bio-oil in improving the overall combustion efficiency. Preliminary results have been found to be promising. The bio-oil blending experiments have also been performed in the delayed coker unit in IOC. All over the world, refineries have been looking for a reduction in coke yield in DCU to improve the process economics. In an MNRE-supported project, 5 wt% and 10 wt% blends of TERI's bio-oil with VR (obtained from IOCL's Panipat Refinery) were processed in the DCU furnace. The blending of bio-oil is found to result in a significant decrease in coke yield in DCU with a simultaneous increase in liquid yield, which can thus improve the overall process economics. All the blending experiments are done at the IOCL R&D facility in Faridabad.

Past few years' intense research on catalytic pyrolysis process is aimed at downstream catalytic upgradation of pyrolytic vapor produced by the same non-catalytic pyrolyzer to further superior quality bio-oil for refinery integration or possible applications as green transport fuels. TERI's catalytic upgradation process resulted in >16wt% upgraded bio-oil yield under mild conditions of <200 °C at 5bar (max) H_2 pressure for agro-straw residues. The catalysts are indigenously made bimetallic catalysts impregnated on Zeolite based matrix with Si/Al_2O_3 ratio between 5 and 50. Catalysts show significant improvement of in-situ separated bio-oil with majorly hydrocarbon fraction and aromatics (Das and Kalita, 2021). More experiments are in progress.

10 Conclusions

Pyrolysis technology has reached technical maturity mostly in developed countries like the United States, Canada, and Europe resulting in a large number of upcoming commercial and demonstration plants. Among the fast pyrolysis technologies aimed at maximizing liquid bio-oil, fluidized and entrained bed technologies are the most advanced in the commercial space. However, there is an emerging trend in the development of mobile and smaller pyrolysis units at a 1–5 tpd scale. These units are described as densification facilities to produce bio-oil/bio-crude as intermediate along with biochar that is intended to be either decentralized use of products or for centralized application of oil after being transported and subsequently upgraded at large centrally located facilities. To leverage the existing refinery infrastructure the upgradation plants are also increasingly being co-located within oil refineries.

Two sets of European standards for fuel use of bio-oil in stationary IC engines and industrial boilers have already been developed, while, the specifications for refinery co-processing and as a precursor for synthetic fuel through gasification are due in near future. Concurrently, recent progress in large-scale first-generation heat and power application of bio-oil are significant steps towards greater adoption of pyrolysis technologies. This will help in market penetration of pyrolysis products and large-scale availability of standard bio-oils which in turn will create opportunities for next-generation applications, such as a drop-in fuel blendstock, conversion to gasoline/diesel, or synfuels like methanol and DME through gasification.

The present research focus is more on bio-oil quality improvement to make it compliant for different applications. A large volume of data and know-how on pyrolysis technologies, bio-oil properties, characterization methods, and upgradation techniques are published PyNE newsletters and in other peer-reviewed journals.

Successful gas turbine operations with 100% pyrolysis oil without solvent (e.g., ethanol) mix is achieved at 2MW scale. However, more research is needed for operation at smaller turbines. Application in the marine sector looks promising too. Although there is room for technical innovation in low-cost metallurgical options for parts of engines and turbines, nevertheless, the quality of the upgraded bio-oil and its sustainable low-cost supply holds the key for pyrolysis technologies' large deployment in these sectors and accrue the real benefit of GHG emissions reductions over petroleum fuel.

In the area of synthetic fuel via oil gasification, the technology of biochar slurry gasification is restricted to high-pressure–temperature conditions. In view of the limitations and challenges of the high-pressure entrained bed slurry reactor and the positive aspects in the atmospheric steam gasification of bio-oil-char mixture, future research needs to be focused on innovation in bio-oil gasification at near or slightly above atmospheric pressure in oxygen and steam conditions for decentralized & sustainable production of methanol and DME.

In terms of providing energy security in an environmentally sustainable way, co-refining holds maximum promise. It will reduce dependence on crude imports. The major questions for bio-crude refinery integration seem to be the proper assessment of the trade-offs for co-processing petroleum crude with bio-oil, degree of upgradation required to co-process bio-crude into a refinery, the technical risks, and finally how the future economy looks like based on the pilot-scale techno-economic assessment studies.

Both the FCCs and hydroprocessing units within an oil refinery could potentially be used to upgrade bio-crude or bio-crude-petroleum blends. However, current petroleum catalysts are incompatible with highly oxygenated biomass-derived compounds and biomass feedstock increases the risk of contaminating downstream refinery units with oxygenated and inorganic species (originating from bio-feeds). With current technology, pyrolysis oil hydroprocessing in oil refineries requires the pyrolysis oil oxygen content to be lower than about 5 wt%.

Regarding strategies to process bio-oil in a refinery, the recent pilot-scale studies on co-refining give some clarity on viable insertion points. Two favorable insertion points suggested are either a refinery's fluid catalytic cracking (FCC) units or its hydroprocessing (hydrotreating and hydrocracking) units. The rationale is that FCC units have a similar configuration to the fluidized bed reactors used for pyrolysis. These units are at the heart of an oil refinery and are mainly used to crack heavy petroleum cuts and maximize the production of lighter cuts such as gasoline blendstocks. Considering the worldwide "dieselification" trend, many refineries FCC units are becoming underutilized. Hence, inserting minimally upgraded pyrolysis oils into FCCs (e.g., in blends with vacuum gas oil) offers a strategy to put these "stranded" assets into better use besides providing a green credit to a refinery's product slate. Additionally, the FCC insertion point for pyrolysis oils represents a complete trade-off approach where minimum hydrogen is used and the bio-oils are minimally pre-processed and then co-processed with heavy vacuum gas oils to produce a relatively low-value hydrocarbon intermediate and large amounts of renewable power. Recent studies under the European BIOCOUP program indicated that the insertion of bio-oil containing 20 wt% oxygen for co-processing with heavy (petroleum-based) oils in FCCs which has potential as one strategy for upgrading bio-oils within an oil refinery.

The second, insertion point proposed for bio-oils in a refinery is before the refinery's hydroprocessing unit. Hydroprocessing is more sensitive to oxygen and impurities than FCC units and unlike the FCC insertion, hydroprocessing insertion of pyrolysis oils relies on substantial hydrogen inputs (e.g., 800 L H_2 per kg of bio-oil processed, including pre-refinery hydrotreatment processing), much costlier catalysts (e.g., ruthenium-based vs ZSM5-based) and requires extensively pre-processed bio-oils (e.g., de-oxygenated to contain only 3–5 wt% oxygen).

While the strategy of inserting prior to hydroprocessing is clearly more expensive, it produces a larger amount of higher value middle distillates such as diesel and jet fuels than the FCC insertion. On the contrary, the cheaper FCC insertion strategy favors lower value light and especially heavy hydrocarbons such as gasoline and bunker fuels, respectively, as well as the generation of coke and gaseous by-products (potential green fuels for process heat and power). Both these refinery insertion strategies are synergistically beneficial but remain more technically challenging than is generally acknowledged by bio-oil commercialization companies and related stakeholders. Additionally, refineries can produce cheap hydrogen for pyrolysis and other drop-in biofuel facilities. Access to low-cost hydrogen is important to achieving favorable drop-in biofuel economics and makes co-locating drop-in biofuel facilities with oil refineries synergistically beneficial.

In India, pyrolysis technology has reached up to pilot scale and no commercial unit has yet been built. However, In India, there lies a huge potential in producing bio-oil from surplus agro and agro-industrial residues available at decentralized locations. Distributed pyrolysis technologies are gaining importance over the last few years because of their lower capital cost, small to medium scale, and decentralized nature. Upscaling of the existing technologies through a modeling simulation study and co-refining study of mildly upgraded bio-crude in Indian refinery set upholds critical importance for a correct assessment of techno-economic feasibility for sizeable replacement of petrocrude with bio-crude. Also, the expected cost reduction through upscaling and integration of a highly efficient upgrading process will help improve competitiveness with fossil fuels and drive the commercial deployment of this technology.

Needless to say that since the refineries application of agro residue-based bio-oil as a precursor for transport fuels via upgradation is yet untested in many countries including India, a strong R&D drive is required for making it viable. Clear R&D Strategy on thermochemical biomass conversion, with special thrust on pyrolysis technology upscaling, capability generation, more systematic experimental and modeling work for the development of techno-economically viable demonstration unit for both the heat & power and transport fuel through refinery upgradation of this the urgent need.

In fact, once the hydrogen sourcing and catalyst issues are resolved, the pyrolysis platform holds great potential since it can effectively utilize a range of biomass feedstocks and has relatively low capital costs compared to gasification, particularly leveraging the existing petroleum refining infrastructure. It is also interesting to note that, unlike HVOs and most other drop-in biofuel platforms, pyrolysis-derived biofuels contain aromatics produced from the conversion of the phenolics in the lignin component of biomass.

Aside from catalyst improvements, it is also suggested that a two-step hydrotreatment process might be a more cost-effective approach to bio-oil upgrading. The first step would stabilize the bio-oil and, a second step would be used to complete hydrotreatment. In this two-step approach, the first stabilization steps could be performed at decentralized locations with smaller-scale facilities near feedstock supplies, and the second final upgrading step in larger centralized facilities or perhaps petroleum refineries to attend where greater economies of scale.

Efforts should enable O_2 content of less than 15% to mitigate the metallurgical issue, coking on the catalyst surface, downstream contamination risk, and venting out oxygenated gases (CO, CO_2, and H_2O), and to explore low-cost hydrogen generation facilities like pyrolytic char gasification.

References

Adam, J., Blazsó, M., Mészáros, E., Stöcker, M., Nilsen, M.H., Bouzga, A., Hustad, J.E., Grønli, M., Øye, G., 2005. Pyrolysis of biomass in the presence of Al-MCM-41 type catalysts. Fuel 84, 1494–1502.

Agblevor, F.A., Besler, S., Evans, R.J., 1995. Influence of inorganic compounds on char formation and quality of fast pyrolysis oils. In: Abstracts of the ACS 209th National Meeting. Anaheim, CA, April 2–5.

Alcala, A., Bridgwater, A.V., 2013. Upgrading fast pyrolysis liquids: blends of biodiesel and pyrolysis oil. Fuel 109, 417–426.

Araujo, K., Mahajan, D., Ryan Kerr, R., da Silva, M., 2017. Global biofuels at the Crossroads: An overview of technical, policy, and investment complexities in the sustainability of biofuel development. Agriculture 7, 32. https://doi.org/10.3390/agriculture7040032. from www.mdpi.com/journal/agriculture.

Baker, E.G., Elliott, D.C., 1988. Upgrading Biomass Pyrolysis Oils. Springer.

Bakhshi, N.N., Adjaye, J.D., 1995. Characteristics of a fast pyrolysis bio-fuel and its miscibility with oxygenated and conventional fuels. In: Proc. of 2nd Biomass Conference of America-Energy, Environment, Agriculture and Industry, pp. 1079–1088. Portland, Oregan, August 21–24.

Baldwin, M.R., 2016. NREL "refinery integration of bio-oil". In: Presentation to California Air Resources Board.

Bansal, G., Bandivadekar, A., 2013. Overview of India's Vehicle Emissions Control Program. International Council on Clean Transportation. Available at: http://www.indiaenvironmentportal.org.in/content/383741/overview-of-indias-vehicle-emissions-control-program-past-successes-and-future-prospects/.

Beran, M., Axelsson, L.-U., 2014. Development and experimental investigation of a tubular combustor for pyrolysis oil burning. J. Eng. Gas Turbines Power 137, 31508. https://doi.org/10.1115/1.4028450.

Biddy, M., Jones, S., 2017. U.S. DOE BETO-2017 Project Peer Review Report, Refinery Integration- Analysis and Sustainability 4.1.1.31 NREL 4.1.1.51 PNNL, NREL.

Boateng, A.A., Garcia-Perez, M., Masek, O., Brown, R., del Campo, B., 2015. Chapter 4: Biochar production technology. In: Lehmann, J., Joseph, S. (Eds.), Biochar for Environmental Management. Science, Technology, and implementation, second ed, p. 63. 2015.

Boscagli, C., Raffelt, K., Zevaco, T.A., Olbrich, W., Otto, T.N., Sauer, J., Grunwaldt, J.-D., 2015. Mild hydrotreatment of the light fraction of fast pyrolysis oil produced from straw over nickel based catalysts. Biomass Bioenergy 83, 525–538.

Boucher, M.E., Chaala, A., Roy, C., 2000. Bio-oils obtained by vacuum pyrolysis of softwood bark as a liquid fuel for gas turbine. Part I: properties of bio-oil and its blends with methanol and a pyrolytic aqueous phase. Biomass Bioenergy 19, 337–350.

Brady, J.P., Kalnes, T.N., Marker, T.L., 2009. Production of Transportation Fuel From Renewable Feedstocks. US Patent No. 20090229172.

Bridgwater, A.V., 1999. Principles and practice of biomass fast pyrolysis processes for liquids. J. Anal. Appl. Pyrolysis 51, 3–22. 19, 351–361.

Bridgwater, A.V., 2003. Renewable fuels and chemicals by thermal processing of biomass. Chem. Eng. J. 91 (2–3), 87–102 (21).

Bridgwater, A.V., 2012. Review of fast pyrolysis of biomass and product upgrading. Biomass Bioenergy 38, 68–94.

Bridgwater, A.V., Peacocke, G.V.C., 2000. Fast pyrolysis processes for biomass. Renew. Sust. Energ. Rev. 4 (1), 1–73.

Bridgwater, A.V., Meier, D., Radlein, D., 1999a. An overview of fast pyrolysis of biomass. Org. Geochem. 30, 1479–1493.

Bridgwater, A.V., Czernik, S., Diebold, J., Meier, D., Oasmaa, A., Peacocke, C., Piskorz, J., Radlein, D., 1999b. Fast Pyrolysis of Biomass: A Handbook. CPL Press, Newbury, UK, ISBN: 1-872691-07-2, p. 188.

Bridgwater, A.V., Czernik, S., Piskorz, J., 2002a. The status of biomass fast pyrolysis. In: Fast Pyrolysis of Biomass: A Handbook. vol. 2. CPL Press, Newbury, UK, ISBN: 1-872691-07-2, p. 188. Reprint 2008, 1999.

Bridgwater, A.V., Czernik, K., Piskorz, J., 2002b. The status of biomass fast pyrolysis. In: Bridgwater (Ed.), Fast Pyrolysis of Biomass. Handbook 2. CPL Press, Newbury, UK, ISBN: 978-1-872691-47-3. Reprint 2008.

Brown, J.N., Brown, R.C., 2012. Process optimization of an auger pyrolyzer with heat carrier using response surface methodology. Bioresour. Technol. 103, 405–414.

Brown, R.C., Holmgren, J., 2006. Fast pyrolysis and biooil upgrading. In: National Program 207: Bioenergy and Energy Alternatives—Distributed Biomass to Diesel Workshop. Richland, WA, USA. 2006.

Bruins, M.E., Sanders, J.P.M., 2012. Small-scale processing of biomass for biorefinery. Biofuels Bioprod. Biorefin. 6 (2), 135–145. https://doi.org/10.1002/bbb.1319.

Buffi, M., Cappellettia, A., Rizzob, A.M., Francesco Martellia, F., Chiaramontia, D., 2018. Combustion of fast pyrolysis bio-oil and blends in a micro gas turbine. Biomass Bioenergy 115, 174–185. https://doi.org/10.1016/j.biombioe.2018.04.020.

Butler, E., Devlina, G., Meier, D., McDonnell, K., 2011. A review of recent laboratory research and commercial developments in fast pyrolysis and upgrading. Renew. Sust. Energ. Rev. 15 (2011), 4171–4186.

Canham, H.O., 2010. The Wood Chemical Industry in the Northeast: An Old Industry With New Possibilities. Northern Woodlands. http://www.faqs.org/photodict/phrase/10393/kilns.ht.

Carlson, T.R., Vispute, T.P., Huber, G.W., 2008. Green gasoline by catalytic fast pyrolysis of solid biomass derived compounds. ChemSusChem 1, 397–400.

Centeno, A., David, O., Vanbellinghen, C., Maggi, R., Delmon, B., 1997. In: Bridgwater, A.V., Boocock, D.G.B. (Eds.), Developments in Thermochemical Biomass Conversion. Blackie Academic & Professional, London, pp. 589–601.

Chang, C., Silvestri, A., 1977. The conversion of methanol and other O-compounds to hydrocarbons over zeolite catalysts. J. Catal. 47, 249.

Chong, K.J., Bridgwater, A.V., 2017. Fast pyrolysis oil fuel blend for marine vessels. Environ. Prog. Sustain. Energy 36, 677–684.

Chum, L.H., 2015. U.S. DOE-BETO-2.4.2.303 Brazil Bilateral: Petrobras-NREL CRADA 2015 Project Peer Review Excerpt, National Renewable Energy Laboratory, Andrea Pinho, Petrobras, March 25. http://www.energy.gov/eere/bioenergy/2015-project-peer-review). http://www.ensyn.com/uploads/6/9/7/8/69787119/_beto_presentation_summary_-_public_document.pdf.

Corma, A., Iborra, S., Velty, A., 2007. Chemical routes for the transformation of biomass into chemicals. Chem. Rev. 107 (6), 2411–2502.

Dahmen, N., Dinjus, E., Kolb, T., Arnold, U., Leibold, H. Ralph Stahl, R., 2011. State of the art of the bioliq® process for synthetic biofuels production. Environ. Prog. Sustain. Energy, 31 (2), 176–181 (Special Issue: TC-Biomass 2011).

Das, P., 2004. Studies on Pyrolysis of Sugarcane Bagasse and Cashew nut shell for Liquid Fuels. PhD thesis, Energy Systems Engineering, IIT Bombay, India.

Das, P., 2014. 'Multi-Feed Distributed Biomass Pyrolysis' in the Session—Carbon Sources, Supply Chains and Business Models Presented in 4th International Symposium on Biofuels and Bio-energy: 'Enablers for Sustainable and Scalable Solutions' by PetroFed& UOP, November 17–18, New Delhi.

Das, P., Bhatnagar, A., 2018. Different feedstocks and processes for production of methanol and DME as alternate transport fuels. In: Singh, A., Agarwal, R., Agarwal, A., Dhar, A., Shukla, M. (Eds.), Prospects of Alternative Transportation Fuels. Energy, Environment, and Sustainability. Springer, Singapore, pp. 131–165. 2017.

Das, P., Ganesh, A., 2003. Bio–oil from pyrolysis of cashew nut shell—a near fuel. Biomass Bioenergy 25 (1), 113–117.

Das, P., Kalita, P., 2021. Author's unpublished Results From Ongoing CHT (MoPNG) Bio-Oil Upgradation Project "Stabilization and Up Gradation of Biomass Derived Bio-Oils Over Tailored Multifunctional Catalysts in a Dual Stage Catalytic Process to Produce Liquid Hydrocarbon Fuels and its Application Studies". https://cht.gov.in/en/pages/ongoing-projects.

Das, P., Sreelatha, T., Ganesh, A., 2004a. Bio-oil from pyrolysis of cashew nut shell—characterization and related properties. Biomass Bioenergy 27 (3), 265–275.

Das, P., Ganesh, A., Wangikar, P., 2004b. Influence of pretreatment for deashing of sugarcane bagasse on pyrolysis products. Biomass Bioenergy 27 (5), 445–457.

Das, P., Mukherjee, A., Minu, K., 2013. Techno economics of pilot pyrolysis plant utilising jatropha and karanja residues, the wastes from Indian biodiesel industries. In: Paper Presented in 6th International Biomass Conference and Expo, Minneapolis, Minnesota, USA.

Das, P., Mukherjee, A., Minu, K., Indian Patent published, 2016. An Automated Programmable (Pneumatic) Logic Controller (PLC) Based Dual Heating Mode Equipped Pyro Reactor for Biofuel Generation and Method of Working for Same (Application No248/DEL/2014) http://ipindiaservices.gov.in/PublicSearch/PublicationSearch/ApplicationStatus.

Das, P., Dr Raman, P., Ram, N.K., Indian Patent granted, 2020. A Pyrolysis—Based Bioreactor and Method of Working for Same. (Appl.No. 4053/DEL/2012). Principal Inventor http://ipindiaservices.gov.in/PublicSearch/PublicationSearch/ApplicationStatus.

Dayton, C.D., 2017. U.S. DOE BETO-2017 Project Peer Review Report Catalytic Upgrading of Thermochemical Intermediates to Hydrocarbons, WBS 2.4.1.403. RTI International. https://projects.ncsu.edu/mckimmon/cpe/opd/tcs2016/pdf/oral/Session%203.2/1-Mante-Session%203.2.pdf.

Dayton, D.C., Carpenter, J.R., Kataria, A., Peters, J.E., Barbee, D., Mante, O.D., Gupta, R., 2015. Design and operation of a pilot-scale catalytic biomass pyrolysis unit. Green Chem 17, 4680–4689. https://doi.org/10.1039/c5gc01023c.

de Pinho, A.R., Almeida, B.B., Mendes, F.L., Casavechia, L.C., Talmadge, M.S., Kinchin, M.C., Chum, H.L., 2017. Fast pyrolysis oil from pinewood chips co-processing with vacuum gas oil in an FCC unit for second generation fuel production. Fuel 188, 462–473.

Diebold, J.P., 2002. A review of the chemical and physical mechanisms of the storage stability of fast pyrolysis bio-oils. In: Bridgwater, A. (Ed.), Fast Pyrolysis of Biomass: A Handbook. vol. 2. CPL Press, Newbury, U.K, pp. 243–292.

Diebold, J.P., Czernik, S., 1997. Additives to lower and stabilize the viscosity of pyrolysis oils during storage. Energy Fuel 11, 1081–1091.

Elliot, D.C., 1986. Analysis and Comparison of Biomass Pyrolysis/Gasification Condensates, Final Report, No. PNL-5943. Pacific Northwest Laboratory, Richland, WA.

Elliott, D.C., 2007. Historical developments in hydroprocessing bio-oils. Energy Fuel 1, 1792–1815.

Emrich, W., 1985. Handbook of Charcoal Making. The Traditional and Industrial Methods. D. Reidel Publishing Company.

Freel, B.A., Graham, R.G., Huffman, D.R., 1996. Commercial aspects of rapid thermal processing (RTM™). In: Bridgwater, A.V., Hogan, E. (Eds.), Bio-Oil Production and Utilization. CPL Press, Newbery, UK, pp. 86–95.

French, R.J., Milne, T.A., 1994. Vapor phase release of alkali species in the combustion of biomass pyrolysis oils. Biomass Bioenergy 7 (1–6), 315–325.

Garcia, P.M., Adams, T.T., Goodrum, J.W., Geller, D.P., Das, K.C., 2007. Production and fuel properties of pine chip bio-oil/biodiesel blends. Energy Fuel 21, 2363–2372.

Garcia, P.M., Shen, J., Wang, X.S., Li, C.Z., 2010. Production and fuel properties of fast pyrolysis oil/bio-diesel blends. Fuel Process. Technol. 91, 296–305.

Garcia-Nunez, J.A., Pelaez-Samaniego, M.R., García-Pérez, M.E., Fonts, I., Abrego, J., Westerhof, R.J.M., Garcia-Perez, M., 2017. Historical developments of pyrolysis reactors: a review. Energy Fuel 31 (6), 5751–5775. https://doi.org/10.1021/acs.energyfuels.7b00641.

Garg, S., Das, P., 2018. High-grade activated carbon from pyrolytic biochar of Jatropha and Karanja oil seed cakes—Indian biodiesel industry wastes. In: Biomass Conversion and Biorefinery. Springer, https://doi.org/10.1007/s13399-018-0308-8. Online published on 12 April https://link.springer.com/article/10.1007/s13399-018-0308-8.

Garg, S., Das, P., 2019. Microporous carbon from cashew-nut shell pyrolytic bio char and its potential application as CO2 adsorbent. Biomass Convers. Biorefin. https://doi.org/10.1007/s13399-019-00506-1.

Geraedts, S., 2017. Potential for Pyrolysis in the Marine Market. Goodfuels.

Greg Perkins, G., Bhaskar, T., Konarova, M., 2018. Process development status of fast pyrolysis technologies for the manufacture of renewable transport fuels from biomass. Renew. Sust. Energ. Rev. 90, 292–315. https://doi.org/10.1016/j.rser.2018.03.048.

Gust, S., 1997. Combustion experiences of flash pyrolysis fuel in intermediate size boilers. In: Bridgwater, A.V., Boocock, D.G.B. (Eds.), Developments in Thermochemical Biomass Conversion. Blackie Academic & Professional, London, pp. 481–488.

Hsieh, C.W.C., Felby, C., 2017. Biofuels for Marine Shipping Sector. IEA Bioenergy Report.

Anon., IEA Bioenergy Report. https://demoplants21.bioenergy2020.eu/projects/displaymap/twhWV.

Anon., 2015. IEA-Tas 34 Bioenergy Report. https://task34.ieabioenergy.com/pyrolysis-reactors.

Ikura, M., Slamak, M., Sawatzky, H., 1998. Pyrolysis liquid-in-diesel oil microemulsions. US Patent 5,820,640.

International Energy Agency, 2011. Technology Roadmap: Biofuels for Transport., p. 56. Available at: http://www.iea.org/publications/freepublications/publication/technology-roadmap-biofuels-for-transport.html.

Jackson, M.A., Compton, D.L., Boateng, A.A., 2009. Screening of heterogeneous catalysts for the pyrolysis of lignin. J. Anal. Appl. Pyrolysis 85, 226–230.

Karatzos, S., McMillan, J., Jack Saddler, J., 2014. The Potential and Challenges of "Drop-in" Biofuels., ISBN: 978-1-910154-09-0. International Energy Agency Bioenergy IEA Task 39 (liquid biofuels) Report.

Karumuri, S., 2002. Liquid Fuels From Thermochemical Conversion of Biomass. PhD thesis, Energy Systems Engineering, IIT Bombay, India.

Khodier, A., Kilgallon, P., Legrave, N., Simms, N., Oakey, J., Bridgwater, T., 2009. Pilot scale combustion of fast pyrolysis bio-oil: ash deposition and gaseous emissions. Environ. Prog. Sustain. Energy 28 (3), 397–403.

Kumar, A., Saini, K., Bhaskar, T., 2020. Hydrochar and biochar: production, physicochemical properties and techno-economic analysis. Bioresour. Technol. 310, 123442. https://doi.org/10.1016/j.biortech.2020.123442.

Leech, J., 1997. Running a dual fuel engine on pyrolysis oil. In: Kaltschmitt, M., Bridgwater, A.V. (Eds.), Biomass Gasification and Pyrolysis, State of Art and Future Prospects. CPL Press, Newbury, UK, pp. 495–497.

Lin, C.Y., 2013. Effects of biodiesel blend on marine fuel characteristics for marine vessels. Energies 6, 4945–4955.

Lupandin, V., Nikolayev, A., Thamburai, R., 2005. Test results of the OGT2500 gas turbine engine running on alternative fuels: bio-oil, ethanol, biodiesel and crude oil. In: Proceedings of GT2005 ASME Turbo Expo.

Marker, F.T.L., Petri, J.A., 2009. Gasoline and Diesel Production From Pyrolytic Lignin Produced From Pyrolysis of Cellulosic Waste. U S Patent 7,578,927.

Martin Beran, M., Axelsson, L.-U., 2013. Application of Pyrolysis Oil in the OP16 Gas Turbine-Feasibility Study. PyNe Newsletter No 33, pp. 12–13.

McMurry, J., 1998. Fundamentals of Organic Chemistry. Brooks/Cole Publishing Co., Pacific Grove, Albany, p. 566.

Meier, D., Schöll, S., Klaubert, H., Markgraf, J., 2007. Practical Results From Pytec's Biomass to-Oil (BTO) Process With Ablative Pyrolyser and Diesel CHP Plant, Success & Visions for Bioenergy. http://www.pytecsite.de/pytec_eng/publikationen.htm.

Meier, D., van de Beld, B., Bridgwater, A.V., Elliott, D., Oasmaa, A., Preto, F., 2013. State of the art of fast pyrolysis in IEA bioenergy member countries. Renew. Sust. Energ. Rev. 20, 619–641.

Mesa-Pérez, J.M., Cortez, L.A.B., Marín-Mesa, H.R., Rocha, J.D., Pelaez-Samaniego, M.R., Cascarosa, E., 2014. A statistical analysis of the auto thermal fast pyrolysis of elephant grass in fluidized bed reactor based on produced charcoal. Appl. Therm. Eng. 65 (1–2), 322–329.

Mohan, D., Pittman, C.U., Steele, P.H., 2006. Pyrolysis of wood/biomass for bio-oil: a critical review. Energy Fuel 20, 848–889.

Motasemi, F., Afzal, M.T., 2013. A review on the microwave-assisted pyrolysis technique. Renew. Sust. Energ. Rev. 28, 317–330.

Mukherjee, A., Das, P., Minu, K., 2014. Thermogravimetric analysis and kinetic modelling studies of selected agro-residues and biodiesel industry wastes for pyrolytic conversion to bio-oil. Biomass Convers. Biorefin. 4 (3), 259–268. https://doi.org/10.1007/s13399-013-0107-1 (Springer) https://link.springer.com/article/10.1007%2Fs13399-013-0107-1.

Mukherjee, A., Bruijnincx, P., Junginger, M., 2020. A perspective on biofuels use and CCS for GHG mitigation in the marine sector. iScience 23, 101758. http://creativecommons.org/licenses/by/4.0/.

Mura, E., Debono, O., Villot, A., Paviet, F., 2013. Pyrolysis of biomass in a semi-industrial scale reactor: study of the fuel-nitrogen oxidation during combustion of volatiles. Biomass Bioenergy 2013 (59), 187–194.

Mushtaq, F., Mat, R., Nasir-Ani, F., 2014. A review on microwave assisted pyrolysis of coal and biomass for fuel production. Renew. Sust. Energ. Rev. 39, 555–574.

Nguyen, D., Honnery, D., 2008. Combustion of bio-oil ethanol blends at elevated pressure. Fuel 87, 232–243.

Nguyen, N.Q., Choi, Y.S., Choi, S.K., Jeong, W.Y., Han, S.Y., 2021. Co-pyrolysis of coffee-grounds and waste polystyrene foam: synergistic effect and product characteristics analysis. Fuel Process. Technol. 175, 64–75. https://doi.org/10.1016/j.fuel.120375 (2018).

Oasmaa, A., 2003. Fuel Oil Quality Properties of Wood-Based Pyrolysis Liquids Academic Dissertation. Research Report Series, Report; 99., ISBN: 951-39-1572-7.

Oasmaa, A., Czernik, S., 1999. Fuel oil quality of biomass pyrolysis oils—state of the art for the end users. Energy Fuel 13 (4), 914–921.

Oasmaa, A., Kuoppala, E., 2003. Fast pyrolysis of forestry residue. 3. Storage stability of liquid fuel. Energy Fuel 17 (3), 1075–1084.

Oasmaa, A., Meier, D., 2002. Analysis, characterisation and test methods of fast pyrolysis liquids. In: Bridgwater, A. (Ed.), et al., Fast Pyrolysis of Biomass. Handbook 2. CPL Press, Newbury, UK (Reprint 2008).

Oasmaa, A., Peacocke, C., 2001. VTT Publication 450., ISBN: 951-38-6365-4. http://www.vtt.fi/inf/pdf.

Oasmaa, A., Peacocke, C., 2010. Properties and Fuel Use of biomass Derived Fast Pyrolysis Liquids-A Guide. VTT Publications, 731 VTT-PUBS-731, ISBN: 978-951-38-7384-4. (URL: ttp://www.vtt.fi/publications/index.jsp) ISSN 1455-0849 (URL: http://www.vtt.fi/publications/index.jsp), Copyright © VTT 2010.

Oasmaa, A., LeppaÈmaÈki, E., Koponen, P., Levander, J., Tapola, E., 1997. Physical Characterisation of Biomass-Based Pyrolysis Liquids. Application of Standard Fuel Oil Analyses. VTT Publications 306. VTT Energy, Espoo.

Oasmaa, A., Kyto, M., Sipila, K., 2001a. Pyrolysis liquid combustion tests in an industrial boiler. In: Bridgwater, A.V. (Ed.), Progress in Thermochemical Biomass Conversion. vol. 2. Blackwell Science, UK, pp. 1468–1481.

Oasmaa, A., Kyto, M., Sipila, K., 2001b. Pyrolysis oil combustion tests in an industrial boiler. In: Bridgwater, A.V. (Ed.), Progress in Thermo chemical Biomass Conversion. Blackwell Science, Oxford, UK, pp. 1468–1481.

Oasmaa, A., Kuoppala, E., Selin, J.F., Gust, S., Solantausta, Y., 2004a. Fast pyrolysis of forestry residue and pine. 4 Improvement of the product quality by solvent addition. Energy Fuel 18, 1578–1583.

Oasmaa, A., Kuoppala, E., Johan-Fredrik Selin, J.F., Steven Gust, S., Solantausta, Y., 2004b. Fast pyrolysis of forestry residue and pine. 4. Improvement of the product quality by solvent. 10.1021/ef040038n CCC, American Chemical Society, Energy Fuel 18, 1578–1583.

Oasmaa, A., van de Beld, B., Saari, P., Elliott, D.C., Solantausta, Y., 2015. Norms, standards, and legislation for fast pyrolysis bio-oils from lignocellulosic biomass. Energy Fuel 29 (4), 2471–2484.

Olah, G.A., 2013. Towards oil independence through renewable methanol chemistry. Angew. Chem. Int. Ed. 52 (1), 104–107. https://doi.org/10.1002/anie.201204995.

Olarte, M., Zacher, A., Padmaperuma, A., Burton, S., Job, H., Lemmon, T., Swita, M., 2016. Stabilization of softwood-derived pyrolysis oils for continuous bio-oil hydroprocessing. Top. Catal. 59, 55–64.

Ormod, D., Webster, A., 2000. Progress in Untilisation of Bio-Oil in Diesel Engines. PyNe Newsletter, 10, Aston University, Birmingham, UK, p. 15.

Pham, T.N., Shi, D., Resaseo, D.E., 2014. Kinetics and mechanism of ketonization of acetic acid on Ru/TiO$_2$ catalyst. Appl. Catal. B 2014 (145), 10–23.

PyNE 41, 2017. IEA BioEnergy Task 34.

Raveendran, K., 1995. Studies on influence of biomass composition on pyrolysis. PhD thesis, Energy Systems Engineering, IIT Bombay.

Resaseo, D.E., 2011. What should we demand from the catalysts responsible for upgrading biomass pyrolysis oil? J. Phys. Chem. Lett. 2, 2294–2295.

Scahill, J.W., Diebold, J.P., Feik, C.J., 1997. In: Bridgwater, A.V., Boocock, D.G.B. (Eds.), Developments in Thermochemical Bio mass Conversion. Blackie Academic & Professional, London, pp. 253–266.

Scott, D.S., Majerski, P., Piskorz, J., Radlein, D., 1999. A second look at fast pyrolysis of biomass—the RTI process. J. Anal. Appl. Pyrolysis 51, 23–37.

Seljak, T., Katrašnik, T., 2016. Designing the microturbine engine for waste-derived fuels. Waste Manag. 47. https://doi.org/10.1016/j.wasman.2015.06.004.

Shen, J., Wang, X.-S., Garcia-Perez, M., Mourant, D., Rhodes, M.J., Li, C.-Z., 2009. Effects of particle size on the fast pyrolysis of oil mallee woody biomass. Fuel 88, 1810–1817.

Shihadeh, A.L., 1998. Rural electrification from local resources: biomass pyrolysis oil combustion in a direct injection diesel engine. PhD thesis, Massachusetts Institute of Technology, USA.

Solantausta, Y., Nylund, N.O., Westerholm, M., Koljonen, T., Oasmaa, A., 1993. Wood pyrolysis oil as fuel in a diesel power plant. Bioresour. Technol. 46, 177–188.

Solantausta, Y., Nylund, N.O., Gust, S., 1994. Use of pyrolysis oil in a test diesel engine to study the feasibility of a diesel power plant concept. Biomass Bioenergy 7, 297–306.

Steele, P.H., Pittman, C.U., Ingram, L.L., Gajjela, S., 2011. Method to Upgrade Bio-Oils to Fuel and Bio Crude. US Patent 20110192072.

Sturzl, R., The Commercial Co-Firing of RTP™ Bio-Oil at the Manitowoc Public Utilities Power Generation Station. Available at http://www.ensyn.com.

Sundqvist, T., Oasmaa, A., Koskinen, A., 2015. Upgrading fast pyrolysis bio-oil quality by esterification and azeotropic water removal. Energy Fuel 29, 2527–2534.

SUPRABIO Newsletter, 2013. Sustainable Products From Economic Processing of Biomass in Highly Integrated Biorefineries. pp. 10–12. Available at: https://www.ifeu.de/landwirtschaft/pdf/SUPRABIO_Newsletter_3_final.pdf.

Suriapparao, V.D., Ojha, K.D., Ray, T., Vinu, R., R., 2014. Polypropylene. J. Therm. Anal. Calorim. 117 (3), 1441–1451.

Suriapparao, V.D., Pradeep, N., Vinu, R., 2015. Bio-oil production from *Prosopis juliflora* via microwave pyrolysis. Energy Fuel 29 (4), 2571–2581. https://doi.org/10.1021/acs.energyfuels.5b00357.

Suriapparao, V.D., Boruah, B., Raja, D., Vinu, R., 2018. Microwave assisted co-pyrolysis of biomasses with polypropylene and polystyrene for high quality bio-oil production. Fuel Process. Technol. 175, 64–75.

Thilakaratne, R., Brown, T., Hu, L.G., Brown, R., 2014. Mild catalytic pyrolysis of biomass for production of transportation fuels: a techno-economic analysis. Green Chem. 16, 627–636.

van de Beld, B., Holle, E., Florijn, J., 2011. An experimental study on the use of pyrolysis oil in diesel engines for CHP applications. In: 19th European Biomass Conference and Exhibition, Berlin-Germany, pp. 1181–1187.

van de Beld, B., Holle, E., Florijn, J., 2013. The use of pyrolysis oil and pyrolysis oil derived fuels in diesel engines for CHP applications. Appl. Energy 102 (C), 190–197.

van de Beld, B., Holle, E., Florijn, J., 2017. The use of fast pyrolysis oil in diesel engines for CHP application. In: 25th European Biomass Conference and Exhibition, Stockholm-Sweden, pp. 1932–1937.

Venderbosch, R.H., Prins, W., 2010. Fast Pyrolysis technology development. Biofuels Bioprod. Biorefin. 4, 178–208.

Venderbosch, R.H., Prins, W., 2011. Fast pyrolysis of biomass. In: Brown, R.C. (Ed.), Thermochemical Processing of Biomass. Wiley, pp. 124–156.

Vennestrøm, P.N.R., Osmundsen, C.M., Christensen, C.H., Taarning, E., 2011. Beyond petrochemicals: the renewable chemicals industry. Angew. Chem. Int. Ed. 50, 10502–10509. https://doi.org/10.1002/anie.201102117.

Vermeire, M.B., 2007. Everything You Need to Know About Marine Fuels. Chevron Global Marine Products. http://docplayer.net/14610783-Everything-you-need-to-know-about-marine-fuels.html.

Wang, H., Male, J., Wang, Y., 2013. Recent advances in hydrotreating of pyrolysis bio-oil and its oxygen-containing model compounds. ACS Catal. 3, 1047–1070.

Westerhof, R.J.M., Nygard, H.S., van Swaaij, W.P.M., Kersten, S.R.A., Brilman, D.W.F., 2012. Effect of particle geometry and microstructure on fast pyrolysis of beech wood. Energy Fuel 26 (4), 2274.

Williams, P.T., Nugranad, N., 2000. Comparison of products from the pyrolysis and catalytic pyrolysis of rice husks. Energy 25, 493–513.

Xhou, X., Broadbelt, L.J., Vinu, R., 2016. Mechanistic understanding of thermochemical conversion of polymers and lignocellulosic biomass. Adv. Chem. Eng. 49, 95–198.

Xiong, W.-M., Zhu, M.-Z., Deng, L., Fu, Y., Guo, Q.-X., 2009. Esterification of organic acid in bio-oil using acidic ionic liquid catalysts. Energy Fuel 23, 2278–2283.

Yang, Y., He, T., Liu, K., Wu, J., Fang, Y., 2012. From biomass to advanced bio-fuel by catalytic pyrolysis/hydroprocessing: hydrodeoxygenation of bio-oil derived from biomass catalytic pyrolysis. Bioresour. Technol. 108, 280–284.

Zacher, A., 2015. PNNL, DOE-BETO 2015 Project Peer Review Report. https://www.energy.gov/sites/prod/files/2015/04/f21/thermochemical_conversion_zacher_242402.pdf.

Zacher, H.A., Olarte, V.M., Santosa, N.D., Elliott, C.D., Jones, C., 2014a. A review and perspective of recent bio-oil hydrotreating research. Green Chem. 16, 491–515.

Zacher, H.A., Elliott, C.D., Olarte, V.M., Santosa, D.M., Preto, F., 2014b. Pyrolysis of woody residue feedstocks: upgrading of bio-oils from mountain-pine-beetle-killed trees and hog fuel. Energy Fuel 28, 7510–7516.

Zhang, Q., Chang, J., Wang, T.J., Xu, Y., 2006. Upgrading bio-oil over different solid catalysts. Energy Fuel 20, 2710–2717.

Zhang, H., Xaio, R., Huang, H., Xiao, G., 2009. Comparison of non-catalytic and catalytic fast pyrolysis of corn cob in a fluidized bed reactor. Bioresour. Technol. 100, 1428–1434.

Zhang, Z.J.Q., Wang, W., Yang, X.L., Chatterjee, S., Pittman Jr., C.U., 2010. Sulfonic acid resin-catalyzed addition of phenols, carboxylic acids, and water to olefins: model reactions for catalytic upgrading of bio-oil. Bioresour. Technol. 101, 3685–3695.

Zhang, A.H., Xiao, R., Jin, B., Xiao, G., Chen, R., 2013. Biomass catalytic pyrolysis to produce olefins and aromatics with a physically mixed catalyst. Bioresour. Technol. 140, 256–262.

12

Biomass gasification: Thermochemical route to energetic bio-chemicals

S. Dasappa[a] and Anand M. Shivapuji[b]

[a]Center for Sustainable Technologies, Interdisciplinary Centre for Energy Research, Indian Institute of Science, Bangalore, India [b]Center for Sustainable Technologies, Indian Institute of Science, Bangalore, India

1 Overview

Carbon and hydrogen are the two constituent elements of typical hydrocarbons based energetic chemicals. The synthetic approach to manufacturing such energetic chemicals involves Carbon Monoxide (CO) and molecular hydrogen (H_2) as the building blocks (Sheldon, 1983). At present, fossilized hydrocarbons like Natural Gas and Naphtha are used as the base raw material/chemical to generate a mixture of CO and H_2 (Rostrup-Nielsen, 2002). This mixture is used for downstream processing to synthesize desired energetic chemicals like alcohols, waxes, gasoline, diesel, and other higher hydrocarbons (Abbasi and Abbasi, 2010). However, the critical need of eliminating carbon from the energy/chemicals chain demands the use of alternative resources/compounds in place of fossilized resources like Natural Gas and Naphtha. With a CHO complex with carbon–hydrogen–oxygen as its constituent elements, Biomass is the only natural resource with the potential for displacing fossilized hydrocarbons (Srivastava et al., 2021). Analogous to fossilized hydrocarbon reforming, Biomass can also be reformed under appropriate conditions to generate the right mixture of CO and H_2 (Maschio et al., 1994), building blocks for developing energetic and industrial chemicals. Being a solid fuel and having different constituent structures as compared to fossilized hydrocarbons, Biomass reforming process varies from the fossilized hydrocarbons reforming process. Biomass is reformed through a process designated as gasification, an auto-thermal, sub-stoichiometric, thermochemical conversion process (Ahrenfeldt, 2012). The process results in the generation of a gaseous mixture known as syngas containing hydrogen, Carbon Monoxide, Methane, carbon Dioxide, and, depending on the process condition, Nitrogen (Reed, 1981). The H_2 to CO ratio can vary between 1.0 and

4.0 depending on the gasification process condition rendering the syngas suitable for generating a range of energetic chemicals, from Hydrogen to Alcohol and higher hydrocarbons. Prior to using generated syngas as feed for the downstream synthesis; the gas needs conditioning (treatment processes to remove unwanted components and species) to meet the specifications of the processes involved. Adjusting the syngas composition requires adopting multi-component separation techniques that can concurrently trap the designated contaminants.

This chapter reviews the state of the art corresponding to the syngas generation through the biomass thermochemical conversion route followed by a detailed discussion on biomass to bio-chemicals through the use of the fixed-bed gasification route. It is envisaged that this process will provide the appropriate gas composition and meet the specifications of the established processes in the industry. The fixed-bed gasification system coupled with the gas separation system, developed and tested at IISc as an example, throws open the possibility of surpassing carbon neutrality and progressing toward realizing a carbon-negative solution. The bio-chemical through biomass gasification route is potentially the only carbon-negative technology option.

2 Introduction and technology state of the art

Historically, biomass has been the first fuel used by humans and has been the primary contributor to the global fuel economy till about the mid-18th century, before fossilized fuels took over (Abbasi and Abbasi, 2010). While the energy supply/use pattern has shifted to coal and liquid hydrocarbons, biomass, primarily as fuelwood and agricultural residues, continues to contribute significantly to over 10%, of the primary energy supply (Fritsche et al., 2014). Beyond domestic applications of cooking and heating, the commonly adopted approach for biomass to energy use has been predominantly through combustion, over the complete regime from super stoichiometry, which involves direct combustion of biomass, typically in excess air (Rosendahl, 2013) to sub-stoichiometric thermochemical conversion (Reed, 1981). Using the combustion process, various technology packages to address heat, power, and combined heat and power needs have been developed and are catering to specific requirements (Koppejan and Van Loo, 2012; Van den Broek et al., 1996). Direct combustion of biomass under stoichiometric/excess air primarily caters to thermal applications like process heat and power generation (through the Rankine cycle route) (Van den Broek et al., 1996; Demirbas, 2005). The sub-stoichiometric combustion route through pyrolysis/gasification generate compounds like producer gas/syngas, bio-oil, and biochar with the producer gas/syngas being the most prominent of all the compounds (Kirkels and Verbong, 2011; Koppejan and Van Loo, 2012). The sub-stoichiometric, auto-thermal, thermochemical conversion process is designated as gasification (Reed, 1981). The gasification technology was developed basically for power generation, to substitute fossil fuel in internal combustion engines and gas turbines. The process results in the generation of a gaseous mixture known as syngas containing hydrogen, Carbon Monoxide, Methane, carbon Dioxide, and depending on the process condition, Nitrogen (Sandeep and Dasappa, 2014). With inherent biomass C:H in the ratio of 1:1.4, the gasification process can result in H_2 to CO ratio between 1.0 and 4.0

depending on the reactants and subsequent downstream processes. The syngas generated has similar applications like in the fossil fuels sector, toward developing a range of energetic chemicals, from Hydrogen to Alcohols and higher hydrocarbons apart from direct combustion applications.

Need-based technological advancement of the gasification process dates back to the Second World War with the use of charcoal as fuel for generating syngas and its subsequent use in internal combustion engines and for motive power applications (Reed and Das, 1988; Stassen, 1995). Producer gas, generated from air gasification of biomass, has found exhaustive use, primarily for power generation through the internal combustion engine route with spark-ignited engines (Reed and Das, 1988; Dasappa et al., 2004; Shivapuji and Dasappa, 2014). Regarding biomass to energy through the engine route, downdraft gasification has been the obvious choice due to relatively lower levels of tar than updraft and cross draft gasification systems (Asadullah, 2014). Even with the downdraft configuration, realizing engine quality gas has been a challenge, and treating the syngas for quality improvement has been an important field of research intervention.

2.1 Thermochemical conversion: India perspective

India has researched small gasification systems to address the 1970s oil crisis by developing biomass gasification systems. Both academic institutions and industries have played a critical role in this activity (Dasappa et al., 2004; Knoef, 2002). Major efforts across the technology platform were focused on improving the closed top design developed during the Second World War and adopting a range of gas treatment processes for post generation clean up toward meeting the engine quality gas. At the Indian Institute of Science, India (Dasappa et al., 2004), there have been attempts toward the reduction of higher molecular weight compounds, "tar" at the source. The open-top reburn downdraft gasifier developed at the Indian Institute of Science (Dasappa, 2011) is a unique design which by virtue of dual air entry streams (~70% from the top and ~30% from the side) ensures a rather extended high-temperature thermal bed resulting in in situ thermal cracking of the produced tars (Dasappa et al., 2004). In comparing nine different gasifier models, it has been reported that the open-top design has the lowest tar and particulate levels in the raw gas (Hasler, 1997). In the open-top design, tar levels under $50 \, mg/Nm^3$ have been reported in the raw gas (Prando et al., 2016). Coupled with downstream, predominantly physical cleaning elements, it has been possible to realize engine quality gas (Sridhar et al., 2001, 2005). While technically power generation from biomass-derived syngas is now a well-established process (Sridhar et al., 2001, 2005; Dasappa et al., 2012; Shivapuji and Dasappa, 2014), the electrical energy route through IC Engines is losing traction primarily due to a strong policy enabling support other renewables enjoyed over a period of time in India.

Concurrent to using bio-derived producer gas for engine applications, attempts have also been reported at generating liquid fuels through the biomass pyrolysis route. While biomass pyrolysis per se is a well-established process in terms of the fundamental process, the broad mechanisms involved (Sharma et al., 2015; Kan et al., 2016) and literature reports on the availability and performance of commercial pyrolysis plants (Cai and Liu, 2016; Chen et al., 2015), the challenge is in respect of the bio-oil produced from pyrolysis. The bio-oil obtained from pyrolysis is basically unprocessed and has undesired physicochemical properties like high viscosity, high water, and ash contents, low heating value, low pH, instability, high corrosiveness, poor ignition, and combustion properties, etc. (Prajitno et al., 2016; Gollakota et al., 2016) which render it unsuitable for direct application in engines and require further downstream processing. A range of processes like catalytic cracking, hydro-deoxygenation, esterification, supercritical extraction, steam reforming is being considered for upgrading bio-oil to engine quality (Xiu and Shahbazi, 2012; Gollakota et al., 2016). It is important to note that almost all the options being considered to remain handicapped by techno-economic challenges resulting in dwindling interest in the biomass to liquid route through pyrolysis (Zhang et al., 2013).

While the thermochemical processing of biomass from the perspective of power generation through the internal combustion route is a matured technology, it has taken a back seat, at least for now. Given the current status of power generation, interest is emerging in synthesizing chemicals like hydrogen (for proton exchange membrane fuel cell applications and other process industries), ethanol, methanol, di-methyl ether, etc., which find use in a range of industrial applications. Further, with the emphasis on ongoing "green," there is a renewed interest and potential for utilizing thermochemical conversion of biomass. The interest is largely toward the generation of syngas that can be used, after appropriate conditioning, in some of the well-established downstream processes for the production of the indicated chemicals. It is argued that such an approach potentially overcomes the limitations associated with the upgradation of bio-oil generated from biomass pyrolysis (Guruviah et al., 2019). Currently, the majority of hydrogen, as well as other chemicals like ethanol, methanol, di-methyl ether, etc., are produced using natural gas (Iulianelli et al., 2016); the route being syngas generation (Luyben, 2016) using the established steam-methane reforming process and subsequent, additional downstream processing like water-gas shift reaction (Pal et al., 2018) followed by separation, catalytic reforming (Ali et al., 2015), etc. It is also of interest to note that while biomass boiler power and biomass gasification-based power generation cycle have a conversion efficiency in the range of 30% and 40%, respectively, the efficiencies for biomass-based hydrogen, methanol, DME production are in excess of 50% (Huang and Zhang, 2011). With an initiative for going "green" and biomass-based chemical generation providing higher conversion efficiencies, there is a significant interest in adopting the gasification route for the generation of chemicals.

2.2 Thermochemical conversion: Europe and USA perspective

The developments in Europe and the USA are summarized in key technology, capacity, and process conditions.

Austria	
Facilities	2400 biomass boilers and over 140 CHP gasification
Capacities	44–880 kWth (18–550 kWel)
Key players	Lock Okoenergie, Hargassner, Froling, Syncraft, Gresco Power and URBAS

Austria

Case description 1	Gussing plant configured as a fluid bed reactor with steam as a reactant (Kirnbauer and Hofbauer, 2011). Initial attempts toward power generation as a part of CHP now diversified to use a split-stream for hydrogen as well as biofuel production
Case description 2	Comet working on waste to value, 1 MW dual fluidized bed gasification system combined with 300 kW Fischer–Tropsch synthesis plant (Hrbek, 2020)

Germany

Facilities	14 R&D centers, 8 small-scale and 3 industry large-scale plants dedicated for gasification
Case description 1	Karlsruhe Institute of Technology has developed a pilot plant (TRL-6) for the production of synthetic biofuels and base chemicals from biogenic residues. Has 2 MW fast pyrolysis system for bio-syncrude and 5 MW entrained flow gasifier (40/80 bar) with high-temperature gas cleaning fuel synthesis. A DME to gasoline at 100 L/h capacity was also installed (Kolb et al., 2013)
Case description 2	An entrained flow gasifier at Niederaußem uses sewage sludge, sludge ash and lignite (130 kg/h) for synthesis gas generation (Bondarenko et al., 2021)

Italy

Facilities	As of 2019, there are 267 gasification plants installed
Capacities	Total output of 56.7 MW$_{el}$; plan to include waste to methanol of 110 kTons/year capacity Electricity produced by Waste2grids plant using Biomass with LHV input 50,104 (kW), ranges from 21,862 to 26,986 kW and the methane produced ranges from 0.77 to 1.36 kg/s (power stored in methane: 38588–67,879 kW)
Key players	NextChem and ENI; Working on circular district model for converting non-recyclable waste into hydrogen and methanol through the gasification route
Case description 1	Waste2grids (Carbone et al., 2021) working on a single, dual-mode plant for power storage and chemical generation by converting waste. Power generation is through highly- integrated, efficient solid-oxide plants. Considered modes: a) Via biomass gasification and use of the syngas in SOFC and b) the produced syngas is converted to CH_4 by addition of H_2 from electrolyzer (SOEC)

The Netherlands

Facilities	12 Gasification plants
Key players	Synova renewable technologies, Gasunie, RWE
Case description 1	Synova renewable technology (Hrbek, 2020) uses MSW, agri, and forest waste in fluidized bed gasification followed by oil-based gas cleaning (OLGA). The rich gas is used to power turbines/gas engines for methane generation and liquid chemical production (hydrogen, BTX, diesel, etc.) Currently, there is a 30–60 kWth lab unit, 1 MWth pilot/demo, and 4 MWth demo/commercial being extensively tested
Case description 2	Gasunie (Smith, 1998) is working on torrgas process where heterogenous, low-quality waste streams are converted to homogenous high-quality biofuels. The liquid undergoes low-temperature followed by high-temperature gasification and reportedly does not produce tar and slag
Case description 3	RWE (Vreugdenhil, 2011) is working to generate 40 kton per year of hydrogen (from 1000 tpd of waste)

Continued

Sweden

Key players	Cortus Energy, MEVA, Bioshare
Case description 1	Cortus Energy (Mirzaei et al., 2021) has developed a 6 MW Wood Roll gasifier. Cortus has also initiated a biomass-to-hydrogen project in Bordeaux, France, and adopts a patented PSA configuration that results in about 15% more hydrogen output than other conventional systems
Case description 2	MEVA Energy (Chishty et al., 2021) has developed an entrained flow cyclone gasifier with an output of 1.2 MWel and 2.4 MWth using small fraction fuels (sawdust, wood fibers, and agricultural residues) as feedstock for applications in CHP, fossil process gas replacements and industrial drying processes

France

Key players	BioTfuel, Gaya, Xylowatt
Case description 1	BioTfuel is working on developing, demonstrating, and commercializing a full B-XtL chain: from R&D to market / from biomass to final products.
Case description 2	Xylowatt is working on NOTAR reactor (Milhé et al., 2013), a compact multi-stage downdraft gasifier that produces clean gas (99.95% tars destroyed). Biochar is obtained as a value-added product and power range of 0.1–2 MW.

Spain

Key players	Eqtec, WtEnergy, Neoelectra
Case description 1	Eqtec has developed a gasifier technology (5.9 MWe IBGPP) with 64% total efficiency, using olive mill pomace as feedstock and 4 ton/h throughput and 111,000+ operating hours
Case description 2	WtEnergy (González-Vázquez et al., 2017) is working on bubbling fluidized bed technology (0.3 bar-g/ 800 °C) with varying feedstocks and at various locations: • Zaragoza plant is rated 7.4 MWth (2 MWe) using Wood chips—Electricity production • Salamanca plant is rated 10 MWth using Meat and bone—12 T/h saturated steam production at 10 bar • Portugal plant is rated 16 MWth using Meat and bone—20 T/h saturated steam production at 10 bar
Case description 3	Neoelectra has developed a gasifier of 20 MWth capacity and driving a 25 t/h steam boiler for industrial processes using forest biomass, plastics, RDF, and waste biomass. Another wood gasifier (forest biomass) plant in Toledo generates 8 MWel of power through steam turbine (López et al., 2018)

United States of America

Key players	Fulcrum Bioenergy/Sierra Biofuels, Red Rock Bio
Case description 1	Fulcrum Bioenergy/Sierra Biofuels (Gershman and Hammond, 2012) is currently constructing a gasification/FT plant with a 175 kT/y of prepared MSW as input fluid bed steam reforming gasifier. Scaling to over 500 kT/y is planned in parallel
Case description 2	Red Rock Bio is constructing a 15 MGY gasification to Fischer–Tropsch plant using forest residue (136,000 tons per year). The end products are cellulosic renewable jet, diesel, and naphtha fuels (Nyström et al., 2019)

2.3 Biofuel production from the thermochemical route (excluding hydrogen)—Research and development

Thermochemical conversion of biomass is a promising option to produce energy-dense liquid biofuels. In this section, some of the thermochemical routes and their performance are investigated based on catalysts, type of biomass, pre-treatment of biomass, bioreactor design, and integrated systems. A comparative analysis was conducted to give an overview of the technologies for the thermochemical conversion of lignocellulosicBtL (Biomass to Liquid) fuels. Syngas obtained from gasification is subjected to Fischer–Tropsch synthesis that produces syncrude, a mixture of linear hydrocarbons such as n-alkanes and n-alkenes with similarities to crude oil. However, it also contains aromatics and oxygenates like 1-alkanols, aldehydes, ketones, carboxylic acids, and therefore, a different upgrade than the crude oil is needed in this process. In fast pyrolysis, liquid (bio-oil), is obtained in yields of up to 80 wt% on dry feed, together with the by-product char and gas. This bio-oil is a complex mixture of water (15–35 wt%), solid particles (0.01–3 wt%), and hundreds of organic compounds (acids, aldehydes, ketones, phenolics, alcohols, ethers, esters, anhydrous sugars, furans, nitrogen compounds as well as large molecular oligomers), which make bio-oil a low-grade liquid fuel. Hydrothermal liquefaction (HTL) was also examined as a promising route for the production of transportation fuels, but the HTL bio-oils are semi-liquid, viscous, dark-colored, and have a smoke-like smell; their typical viscosity is 10–10,000 times higher than that of diesel and biodiesel and therefore further upgrading is required. Compared to the other conversion methods mentioned, the supercritical fluid extraction technology was found to have advantages like fast kinetics, higher biomass conversion, ease of continuous operation, and elimination of the use of catalysts (Ibarra-Gonzalez and Ben-GuangRong., 2019). Song et al. (2020) in a review, outlined the use of agricultural and forestry wastes (AFW) and thermochemical liquefaction as a route for producing biofuels and/or chemicals. Direct liquefaction of AFW in the presence of various solvents noted that the oil yield followed the order of water/ethanol mixed solvent > pure water > pure ethanol regardless of the feedstocks. Another study presents an overview of the co-pyrolysis of biomass and waste plastics for high-grade biofuel. Results showed that using the catalyst, HZSM-5, is the most effective technique in the pyrolysis of biomass and plastics by transforming the waste products into aromatics. Volatile matter and ash contents are the important factors during pyrolysis: the increase in volatiles generation favors the production of a large amount of pyrolysis oil and results in high reactivity, but high ash content contributes to a decrease in oil yield, resulting in the production of more char. Blending ratio and temperature (400–600 °C) were found to be vital for the production of liquid fuel from co-pyrolysis (Uzoejinwa et al., 2018). Hassan et al. (2016) also summarized the research progress on co-pyrolysis and catalytic co-pyrolysis. It was found that co-pyrolysis produces a high calorific value of the liquid between 26.78 and 34.79 MJ/kg. The introduction of a catalyst into pyrolysis was found to decrease the catalytic temperature and alter product distribution with higher selectivity or enhance the yield of desirable hydrocarbon with commercial values. Also, ZSM-5 is generally used as a catalyst in co-pyrolysis of biomass and plastics because of its strong acidity for cracking and deoxygenating biomass-derived oxygenates as well as its unique pore structure for aromatization but it leads to coke formation. Another research was conducted on a different kind of pyrolysis for the generation of biofuel from microalgal biomass. The yield of bio-oil generation is maximum in fast pyrolysis of microalgae (HHV around 42 MJ/kg). Catalytic pyrolysis was also reported as an option to improve the quality of microalgae bio-oil. Microwave-assisted pyrolysis eliminates the need for agitation or fluidization of biomass particles during the process. However, it is concluded that a much higher yield of hydrocarbons with much better structural preservation can be obtained from hydropyrolysis compared to conventional pyrolysis. Also, higher volatiles and more liquid products are formed when the sample is heated to moderate temperature quickly, while lower heating gives more char for the same temperature (Azizi et al., 2018). To overcome the energy involved in the drying of biomass, studies are being carried out on using wet biomass in thermochemical conversion. A novel system concept for the thermochemical conversion of very wet biomasses to biofuel was proposed by Clausen (2017). *Thermodynamic modeling and comparison with the more "conventional" system were* conducted. It is observed from the simulated results that the novel system can handle 82%–84% more water than the conventional system and 50 wt% ash, while the conventional system can handle 10 wt% ash. High tolerable ash content is an advantage because very wet biomasses, such as sewage sludge and manure, have a high ash content. The analysis also shows that the system's total efficiency is 69%–70% depending on the biomass ash content (1–50 wt%). Because electricity is used to convert steam to hydrogen, almost all the carbon in the biomass ends up in the SNG product, resulting in biomass to SNG energy ratio of 165%. Similarly, another study investigated defatted biomasses of *Scenedesmus obliquus* BR003, with moisture contents of 2.5 and 32.5 wt%, subjected to a slow pyrolysis process at a temperature of 450 °C under inert atmosphere and ambient pressure. The bio-oil yield was 13.14 ± 0.87 and 16.75 ± 1.88 wt%, char yield was 45.79 ± 0.56 and 46.22 ± 0.42 wt%, the gas yield was 32.95 ± 0.72 and 31.33 ± 1.52 wt% and HHV was 34.8 and 36.05 MJ kg^{-1} for the treatments with lower and higher moisture content, respectively. Although the treatments showed a negative energy balance, the use of biomass with higher moisture content is an alternative in improving the production of microalgae biofuels since it significantly reduces the energy required for drying before the pyrolysis process (Rocha et al., 2020). Few other studies have worked on the effect of AAEMs (Alkali and Alkaline Earth Metals) and catalysts in various thermochemical conversion routes. Ong et al. (2019) present a comprehensive review on the effects of catalysts on the thermochemical conversion of biomass. The presence of the AAEMs salt during torrefaction increased the thermal reactivity and decreased the thermal decomposition temperature of biomass. AAEMs in slow pyrolysis promoted biochar generation and gas yield, thereby reducing liquid yield and increasing the generation of lighter hydrocarbons. In fast pyrolysis, AAEMs further bio-oil generation with more aromatics and phenols as the end products. Zeolites also increase the gas yield, thereby decreasing bio-oil yield with the upgraded quality of produced bio-oil. Homogeneous catalysts in hydrothermal liquefaction promote bio-oil production by decreasing char and tar formation, with K_2CO_3 being the most and NaOH being the least effective catalyst. A similar investigation on the effects of AAEMs was conducted by Isahak et al. (2012) and it was reported that higher content of cellulose in feedstock gave a higher amount of bio-oil and the presence of inorganic species like AAEMs in bio-oil greatly influences the subsequent use of bio-oil. For example, the alkali and alkaline earth metallic (AAEM) species are associated with the accelerated aging of bio-oil. In addition, a comparison of bioreactors was also carried out, and rotating cone pyrolyzer design demonstrated yields of 70% on a consistent basis, and in the case of vacuum pyrolyzer, the oil yield is up to 30–45 wt%, which is lower than that of the fluidized bed. With regards to bioreactors used and integrated systems to achieve energy efficiency, a study conducted by Butera et al. (2020)

investigated five systems, coupling solid-oxide cells (SOC) and thermochemical wood conversion via different gasification technologies (two-staged electro-gasifier, two-staged electrically heated gasifiers, bubbling fluidized bed, pressurized entrained flow gasifier (EFG) without fuel pre-treatment and pressurized entrained flow gasifier with pyrolysis) to produce methanol. The two-stage electro-gasifier and the EFG with pyrolysis outperform the other systems, exceeding 70% efficiency. But, the EFG with pyrolysis as fuel pre-treatment offers higher overall carbon conversion (97%), taking advantage of more carbon in the biomass. However, EFG with no fuel pre-treatment with slightly lower efficiency and overall carbon conversion (68% and 96%), represents an effective solution, especially at very low prices. Another work on integrated systems examined the potential for thermochemical conversion of biomass residues from an integrated sugar–ethanol process. Results showed that the cane straw (LHV—16.4 ± 0.1 MJ/kg, db), which is left in the field, contains one-third of the unutilized energy while the vinasse (final residue from sugar–ethanol process) contains 2% of the energy. However, the high levels of K and Cl make the straw a challenging fuel as K and Cl in biomass fuels cause corrosion and fouling problems in boilers and gasifiers. However, over 85% of these elements in the straw are water-soluble, indicating that water leaching would improve it for utilization in thermochemical conversion (Dirbeba et al., 2017).

2.4 Hydrogen production through thermochemical route—Research and development

Thermochemical conversion of biomass to hydrogen or H_2-enriched syngas has recently gained light with the shift to a low-carbon economy. This segment highlights the current advances in various thermochemical processes and the effect of operating conditions, catalysts, type of biomass, and integrated systems on the overall hydrogen production. Wang et al. (2015) used a two-stage gasifier (a separated pyrolyzer and throated downdraft gasifier) to produce clean syngas from pellets of pinewood sawdust and cotton stalk. Different oxygen concentrations as a gasifying agent were employed. As the oxygen concentration increased, the H_2 content in syngas obtained from the cotton stalk and pine sawdust increased from 17.17 to 35.89 vol% and 16.22 to 36.45 vol%, respectively. It was found that a gasification temperature of 900 °C and even above 1200 °C can be achieved without additional steam by manipulating oxygen level and gasification parameters for syngas with higher hydrogen content. Another study conducted energy and exergy analysis on the effect of air, steam, and air/steam as the gasifying agents for horse manure, pinewood, and sawdust as biomass materials for hydrogen production. Pinewood generated a more desired product gas composition with a much higher hydrogen exergy efficiency than other materials. It was observed that the higher STBM (steam to biomass molar ratio) and temperature with lower moisture content generate higher energy and exergy efficiencies of the product gas and lower exergy efficiencies of unreacted carbon. A hydrogen exergy efficiency of 44% was obtained when the product gas exergy efficiency reached the highest value (88.26%) and destruction and unreacted carbon efficiencies exhibited minimum values of 7.96% and 1.9% (Fereshteh et al., 2020). Supercritical water gasification of biomass for hydrogen production is another route that is being investigated. At near supercritical temperatures and high pressures, ionic mechanisms are dominant, resulting in the enhancement of the hydrolysis rate. Also, the presence of catalysts reduces the operating costs along with improved selectivity for H_2 by reducing the tar and char formation in SCWG. Another added advantage of SCWG of biomass is its high-pressure H_2 production which cuts down the compression energy costs during its storage (Reddy et al., 2014). Recently, extensive studies on integrated systems have been carried out for improved energy efficiency (Ishaq and Dincer, 2020) developed an integrated energy system comprising of an entrained flow gasifier, a Cryogenic Air Separation unit, a double-stage Rankine cycle, Water-Gas Shift Reactor, a combined gas-steam power cycle, and a Proton Exchange Membrane (PEM) electrolyzer, for the three useful outputs of electricity, heat, and hydrogen. When syngas produced from dry olive pits leaves the heat exchanger, the remaining available heat is integrated with a double-stage Rankine cycle where high-pressure turbine (HPT) provides a 1.37 MW of power output. In comparison, the low-pressure turbine (LPT) generates 1.4 MW of power. The steam is used as a working fluid in the Rankine cycle combined with the Brayton cycle. The electricity produced by the steam turbine is fed to the PEM electrolyzer to produce hydrogen. As a result, the electrical power output of the designed energy system is 1.4 MW for the biomass flow rate of 0.4 kg/s, and hydrogen production rate of the designed biomass-based integrated system is 10.74 mol/s. Another novel decoupled dual loop gasification system (DDLG), where both the fuel reactor and the reformer are separately interconnected with the combustor, forming two-bed material circulation loops, one for fuel pyrolysis/gasification and the other for tar cracking/reforming, was developed. With pine sawdust as feedstock and calcined olivine as solid heat carrier and in situ tar cracking/reforming catalyst, the steam gasification performance of the system has been investigated. Olivine was found to have a superior catalytic activity for tar cracking/reforming, ascribed to the olivine's FeOx species. As a result, a product gas with hydrogen concentration of 40.8 vol%, tar content as low as 14.1 g/Nm^3, and dry gas yield of 1.0 Nm^3/kg daf was obtained (Xiao et al., 2017a, b). Mahdi and Majid (2021) also investigated integrated pyrolysis and air gasification configuration for the production of hydrogen-rich syngas using different algal biomass (Algal waste, *Chlorella vulgaris*, *Rhizoclonium* sp., and *Spirogyra*). The optimum operating condition was found to be: gasifier temperature of 600 °C, gasifier pressure of 1 atm, and airflow rate of 0.01 m^3/h (for high hydrogen content syngas production). According to the achieved results, *C. vulgaris* had the highest H_2/CO ratio and H_2 mole% (48.9) among other feeds. A lot of studies on gasification and/or pyrolysis coupled with catalysts have shown improved results on the thermochemical conversion of various biomass with emphasis on CaO and char-based catalysts. The research focused on biomass gasification with CO_2 removal by CaO sorbent. It is known that apart from the capture of CO_2, CaO simultaneously plays a role as a catalyst in catalytic cracking of tars obtained from biomass gasification. The results demonstrated a decrease in tar yield from 15.07 to 6.68 g/Nm^3, increasing CaO/B from 0.0 to 2.0. An increase in CaO/B from 1.5 to 2 sharply increased the hydrogen yield approximately 1.55 times (Zhou et al., 2019). Anniwaer et al. (2021) worked on steam co-gasification of the banana peel with other biomass, i.e., Japanese cedar wood, rice husk, and their mixture for hydrogen-rich gas production in a fixed-bed reactor. It was found that the banana peel with a high content of alkali and alkaline earth metal (AAEM) species exhibited high gasification reactivity and had a significant catalytic effect on the co-gasification process at the low-temperature with cedarwood. Also, the addition of calcined seashell (CS) as CaO resource was found to offset the negative effect of Si species in the rice husk, resulting in an improved H_2-rich gas production yield and decreased CO yield. In this study conducted by Yao et al. (2016), hydrogen production from biomass (wheat straw) steam gasification with biochar or Ni-based biochar was investigated. Three bio-chars were obtained from fast pyrolysis of wheat straw (WC), rice husk (RC), and cotton stalk (CC), and commercial activated carbon was used as a comparison. Cotton char displayed the most efficient catalytic activity in terms of the reforming of pyrolysis volatiles, as the gas yield was 81.9 wt% which is much higher than the biomass gasification without biochar (34 wt%) and hydrogen production of 39.42 mg/g biomass. It was suggested that there were clear interactions between the pyrolysis volatiles and the cotton char. The addition of Ni to CC (Ni/CC catalyst) showed much higher hydrogen production (~90 mg/g biomass) due to higher content of AAEMs. Similarly, Duman and Yanik (2017) evaluated different char-based catalysts to increase hydrogen production from the steam pyrolysis of olive pomace in a two-stage fixed-bed reactor system. Biomass char, nickel-loaded biomass char, coal char, and nickel or iron-loaded coal chars were used as a catalyst. Results showed impregnation of Ni and/or Fe on char enhanced its catalytic activity. Even though the highest hydrogen (315.3 mL/g biomass) was yielded over nickel-based biomass char, the char itself was gasified. Hence, Ni-based brown coal char can be considered as a promising catalyst to produce a high amount of H_2 gas with relatively high thermal stability. Another study investigated the pyrolysis-catalytic steam reforming of six agricultural biomass waste samples (rice husk, coconut shell, sugarcane bagasse, palm kernel shell, cotton stalk, and wheat straw) as well as the three main components of biomass (cellulose, hemicellulose (xylan), and lignin) for hydrogen/syngas production. The catalyst used for steam reforming was a 10 wt% nickel-based alumina catalyst ($NiAl_2O_3$). Hydrogen yield from pyrolysis (550 °C) of the biomass types ranged from 5.81 mmol/g for the wheat straw to 6.77 mmol/g for cotton stalks and for pyrolysis-catalytic steam reforming (750 °C), the hydrogen yield ranged from 16.38 mmol/g for the wheat straw to 25.35 mmol/g for palm shell kernels. It was also found that pyrolysis of the biomass components, lignin produced the highest yield of hydrogen, whereas cellulose and hemicellulose favored CO and CO_2 production (Akubo et al., 2019); Jin et al. (2019) studied steam gasification of tableted biomass (sawdust) for H_2 production in molten salts under different conditions. Increasing the gasification temperature increased the concentration and yield of H_2 in the product gas from 28.26 vol% and 78.14 mL/g biomass to 55.47 vol% and 681.77 mL/g biomass, respectively. It was mainly because higher temperature enhanced the heat and mass transfer between molten salts and biomass, and promoted the penetration of Li+, Na+, and K+ of the molten salts into the pores of the biomass to break C—C, C—O, and C—O—C bonds and formed more stable small molecular substances, thereby increasing the concentration and yield of H_2. It was also found to simultaneously decrease the concentrations of CO and CH_4 in the product gas and decrease the yield of condensable tar. The highest H_2 yield reached up to 807.53 mL/g biomass. Thermochemical conversion to high-purity hydrogen of typical lignite (BYH) and seaweed biomass (BS) with carbon capture via alkaline gasification was explored by Yuan et al. (2021). NaOH was mixed with the coal/biomass blends (1 g) at various coal blending ratios (0%, 25%, 50%, 75%, 100%) to serve as the reactant. As a result, the alkali (i.e., NaOH) can decompose the molecular structures of coal and biomass, suppress CO formation, in situ capture CO_2, and produce a large amount of H_2 (80.1%–93.2% purity). The H_2 production (486.91–801.5 mL/g) rises with the increased coal blending ratio and the CH_4 production decreases by 51.7%. Hence, it has great carbon capture and storage potential.

2.5 Summarizing

The above review provides an insight into the range of technology packages developed and being used for a range of end-use applications. As evident, the majority of the installations cater to CHP requirements and concentrate on woody biomass, except for a few specifically addressing technology packages for other fuels like agricultural residues. In light of the potential of biomass to chemicals through the thermochemical conversion route as an alternative to the fossilized hydrocarbon processing route, the current chapter reports on a novel fixed-bed oxy-steam biomass gasification route coupled with swing adsorption separation technique as a potential thermochemical route to bio-refinery and bio-hydrogen. It is important to note that the emphasis is on the generation of appropriate quality and composition feed gas for further downstream processing to generate chemicals through established technology/process to meet the industrial needs.

3 Biomass to bio-refinery quality syngas and hydrogen: Experience and challenges

The production of bio-chemicals like ethanol, methanol, di-methyl ether, etc., using syngas derived from biomass gasification has been attempted by various groups across the globe. The challenge is principally from the generation of requisite quality syngas composition for the downstream catalytic synthesis for a range of products. The process quality requirements are substantially more stringent than the quality and composition delivered by conventional gasification systems for power and thermal applications. Similar challenges also exist in respect of using bio-hydrogen in a PEM fuel cell. The challenges arise from the limited tolerance of the membrane to CO and other contaminants like sulfur species, ammonia, methane and inert gases, and PM10 (International Organization for Standardization, 2019). Similar restrictions on the quality exist in using hydrogen as a feed for the generation of other chemicals. The bio-derived gas quality challenge must be evaluated because the approach using biomass to biofuels is a complete process from raw material to product, unlike the fossil fuel where intermediate product constitutes the starting feed material with significant quality considerations already addressed. Bio-derived fuels experience the additional complexity of achieving the quality specifications met by fossil fuel-derived products. Further, the use of a range of biomass poses additional challenges considering that the contaminant levels and composition can be biomass specific. As such, the raw syngas delivered by a conventional gasification system demands a rather complex and extended treatment process coupled with the demand for the consistency of gas composition and high gasification efficiency. Together, the elaborate gas conditioning system required downstream of the conventional gasifier renders the bio-syngas-based bio-chemical synthesis route technically challenging. It is also significant to note that downsizing the gasification technology package, principally the downstream processing system, to scales more in line with biomass availability remains a fundamental challenge. Thus, while syngas (derived from natural gas) to value-added chemicals is a well-established process, primarily at a very large-scale (over 5 million liters per day), further research, development, and demonstration (RD&D) are required for adaptation and application to the bio-based feedstock at scales more compatible with biomass availability. Some experiences and challenges regarding attempts at bio-refinery through the bio-syngas to value-added chemicals route, as reported in the literature, are described below.

One of the fundamental challenges concerning the bio-refinery concept is the low volumetric energetic density of typical biomass feedstock, which critically limits the collection area and the transportation distance (Dahmen et al., 2012). Owing to the unfavorable transport economics associated with biomass, decentralized gasification is being explored as one of the options. However, downscaling of gasification technology in sync with biomass availability is a major challenge (Maity, 2015). An approach being explored as an alternative to small-scale gasification (catering to local availability of biomass) is the decentralized fast pyrolysis wherein biomass is converted predominantly into bio-oil and biochar collectively known as bio-syncrude (Dahmen et al., 2012). The bio-syncrude is transported to a centralized large-scale processing facility and is further processed to liquid fuels through its gasification generating syngas for liquid fuel synthesis. The bio-syncrude generated from biomass with a density of nearly $1200 \text{kg}/\text{m}^3$ and a lower heating value of around 20 MJ/kg is suitable for storage and transportation. Experience along the described route in Germany suggests that the decentralized pyrolysis systems are typically about 100 MWth input capacity, catering to biomass supply in a 30 km radius, while the syngas to liquid plants is of several GW thermal input capacity (Trippe et al., 2010). At Karlsruhe Institute of Technology (KIT), Germany has developed what is described as the Bioliq plant involving biomass preparation, fast pyrolysis and bioslurry production at a decentralized level and subsequent gasification, cleaning-conditioning, and use of syngas in a centralized large-scale system (Dahmen et al., 2012). The fast pyrolysis plant is a 2 MW thermal, 500 kg/h pilot plant for gasification of lignocellulosic materials toward bio-syncrude preparation. The fast pyrolysis plant recovers about 90% of the inlet bioenergy in 20%–40% char and 50%–70% oil. The oil and char are mixed to form a dense slurry with a density of about $1200 \text{kg}/\text{m}^3$ and a higher heating value of about $18–25 \text{MJ}/\text{m}^3$. A 5 MW thermal, 1000 kg/h high-pressure gasification system (up to 80 bar) is used to gasify the bio-syncrude. The bio-syncrude is fed as a preheated slurry in the pressurized entrained flow gasifier in which near tar-free syngas is produced under temperatures excess of 1473 K and pressures of up to 80 bar. The syngas generated are cleaned (using water scrubbing, particle filtering, etc.) and conditioned (using catalytic water-gas shift reaction, pressure swing adsorption, etc.) to realize the desired H_2/CO ratio in a $700 \text{Nm}^3/\text{h}$ system. The conditioned syngas is used to generate gasoline from the DME route in a 2 MW thermal, 50 kg/h catalytic reactor (Dahmen et al., 2012). Along similar lines, under the project titled "Sustainable Products from Economic Processing of Biomass in Highly Integrated Bio Refineries" or SUPRABIO, ETC from Sweden and BTG from The Netherlands developed a two-stage unit for conversion of biomass to DME through bio-oil gasification in an entrained flow reactor and conversion of syngas directly to DME.

In Sweden, a consortium named BioDME comprising of 7 partners (CHEMREC, Delphi Diesel Systems, Energy Technology Centre, HaldorTopsøe, PREEM PETROLEUM AB, Total, and Volvo Group) has been established toward generating DME from the gasification of black liquor. Landälv et al. (2014) reported that the black liquor generated from the SmurfitKappaKraftliner pulp mill is gasified in a high-pressure gasifier operating at 30 bar and 1323.15 K to generate syngas. The syngas generated is conditioned for subsequent synthesis of methanol in a 130 bar reactor followed by synthesis of DME at 15 bar. Going beyond black liquor, it has been reported that glycerol obtained from biodiesel production could be utilized to generate syngas after reaction with superheated steam and oxygen. In China, Wang et al.

(2011) reported the techno-economic analysis of two biomasses to DME plants rated at 100 and 1000 tons per annum, respectively. The oxy-steam gasification system used dried corncob as the biomass feedstock, which was pyrolyzed at 673.15 K. The products of pyrolysis are subsequently gasified to produce raw bio-syngas at temperatures of about 1373.15 K. The typical composition of the cooled syngas has been reported to be H_2: 25%–38%; CO: 25%–38%; CO_2: 16%–25%; CH_4: under 2% along with the other trace gases and an overall H_2/CO ratio of 0.98–1.17. Boateng et al. (2015) have reported on the gasification of bio-oil (derived from fast pyrolysis of biomass) using entrained flow gasifiers for syngas production toward methanol and FT–diesel synthesis. They argue that bio-oil gasification is advantageous over direct biomass gasification, considering that liquids are easy to handle and are ash-free in addition to their inexpensive pre-treatment process.

Moving beyond gasification of derived liquids and into direct biomass gasification, it is well-established that conventional air gasification of biomass results in syngas with hydrogen volume fraction of about 20% with the presence of over 20% carbon dioxide (Sandeep and Dasappa, 2014), rendering it rather unsuitable for downstream methanol and DME synthesis. This challenge is overcome by the transition from air gasification to oxy-steam gasification. Due to nitrogen elimination from the feed and some in situ reactions. Regarding Oxy-steam gasification of biomass for generating high hydrogen-rich syngas, very little information is available in the literature. The available literature primarily reports on the lab-scale investigations (Sandeep and Dasappa, 2014), with minimal reporting on commercial-scale operations. The biomass combined heat and power plant at Gussing, Austria is a classical success story of generating high hydrogen syngas by using steam as a gasifying media in a fluidized bed gasifier. The gas generated from a typical oxygen/steam gasification of biomass typically has H_2, CO, CO_2, and CH_4 as the principal gaseous species. Toward using a similar approach as in the SMR process, the gas is further processed by subjecting it to water-gas shift reaction followed by swing adsorption separation to generate pure hydrogen or a controlled approach to achieve the desired H_2 to CO ratio.

It is evident from the above discussion that in respect of the thermochemical route to bio-refinery, the most exhaustively addressed route is the bio-liquid gasification route for syngas generation. There are no attempts to use fixed-bed gasification as the primary thermochemical route coupled with downstream separation systems for generating either bio-hydrogen or bio-syngas of a requisite quality suitable for the generation of chemicals like methanol and DME. Fixed-bed gasification can be of particular interest in the context of bio-refinery considering that the process itself is very simple (can operate near ambient pressure), ability to generate syngas with low tar levels, higher efficiency, and stable operations even as a distributed generating system. Before discussing the fixed-bed gasification system, the downstream processing approach conventionally adopted for hydrogen enhancement in syngas and control of contaminants is discussed.

4 Biomass gasification with downstream gas processing and highlights of work at Indian Institute of Science

Biomass gasification is a solid gas conversion process involving thermochemical reactions of pyrolysis, oxidation, and reduction. In the pyrolysis process, biomass, a C–H–O complex, loses almost all of the volatiles producing residual char, about 20% of biomass by weight. A portion of the volatiles is burned in the presence of oxygen, enabling auto-thermal reactions in the pyrolysis zone. In downdraft gasifiers, below the pyrolysis zone, the oxidizer is introduced (as secondary air in case of air gasification and oxygen in case of oxy-steam gasification) resulting in extended combustion of volatiles, breakdown of high molecular compounds, and some char reactions. The products of oxidation pass through the hotbed of char where a part of steam (from inherent moisture only) in air gasification and externally supplied steam in case of oxy-steam gasification) and CO_2 is reduced by reacting with char according to reactions 1 and 2. In the case of oxy-steam gasification, superheated steam supplied along with oxygen enhances the partial pressure of steam, favoring the water-gas shift reaction in the forward direction (Eq. 3), resulting in additional H_2 in the generated syngas.

$$C + H_2O \rightarrow CO + H_2 - 131\,kJ/mol \qquad (1)$$

$$C + CO_2 \rightarrow 2CO - 172\,kJ/mol \qquad (2)$$

$$CO + H_2O \leftrightarrow CO_2 + H_2 + 41.1\,kJ/mol \qquad (3)$$

The reactor's products are made to pass through a cleaning and cooling system to remove moisture, particulates, tars, etc. The resulting cold gas is typically a mixture of H_2, CO, and small amounts of CH_4 with CO_2 (and N_2 in air gasification). The typical arrangement of the fixed-bed open-top downdraft reactor configuration for air and oxy-steam gasification system is presented in Fig. 1 (Reactor A and Reactor B). It can be observed that the reactor is an open system drawing in air from the top and side nozzles in the case of the air gasification system, whereas the reactor is closed to the ambient in the case of the oxy-steam gasification system. The cold gas composition for the reactor configurations is presented in Table 1.

The gasification process is, in general, feed agnostic for a range of densities. In the current chapter, results corresponding to gasification using two types of feeds, Casuarina wood chips, and coconut shells, are briefly described. The important properties of the two biomass resources are consolidated in Table 2. Ultimate analysis has also been carried out on the biomass samples for quantifying the C–H–N–S–O composition and the results are consolidated in Table 3. It is important to note that the particle density of wood chips is nearly half that of coconut shells, while the ash content is almost double. However, the ultimate analysis indicates similar fractions for C–H–O and hence similar chemical formulae and molecular weights for the two biomass samples. The complete absence (of lower than detectable limits) of sulfur in coconut shells is interesting and can provide a distinct advantage, especially for PEM fuel cell operation where the total sulfur content has to be less than 4 ppbV. Biomass of varying particle density is chosen to evaluate and establish the ability of the gasifier to handle biomass of varying thermophysical properties.

The transition from air gasification to oxy-steam gasification and preliminary experimental investigations have been reported by Sandeep and Dasappa (2014), and the documented data/experience forms the baseline for the current set of investigations toward the generation of high hydrogen syngas. Reviewing the published article (Sandeep and Dasappa, 2014), initial efforts at enhancing the syngas hydrogen fraction basically involved using two approaches of (a) wet wood chips and (b) dry wood chips with steam toward introducing moisture into the reactor. Hydrogen generation quantity and efficiency are reported to be higher when using steam with dry biomass than using only wet biomass. It has been argued

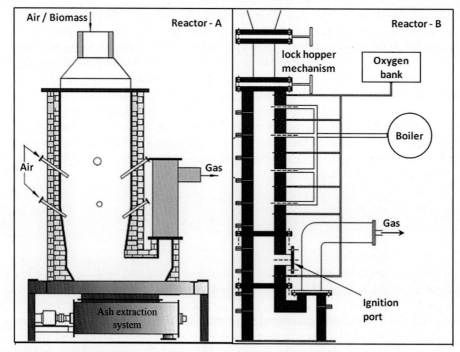

FIG. 1 Typical arrangement of air gasification (Reactor A) and oxy-steam gasification (Reactor B) (Sandeep and Dasappa, 2014).

TABLE 1 Cold gas composition for air gasification and oxy-steam gasification (Sandeep and Dasappa, 2014).

Component	Air gasification		Oxy-steam gasification (SBR 2, Φ 0.3)	
	Volume %	Mass %	Volume %	Mass %
H_2	~20.0	~1.7	~47.1	~4.4
CO	~20.0	~22.9	~11.4	~14.9
CH_4	~2.0	~1.3	~3.4	~2.5
CO_2	~12.0	~21.3	~38.1	~78.2
N_2	~46.0	~52.6	Not present	

to be due to the energetics involved. The Steam to Biomass ratio (SBR) has been varied from 0.75 to 2.7 (mole basis), and the equivalence ratio (ER) has been varied from 0.18 to 0.30. Steam has been superheated to 650 K prior to injection in the reactor. The volumetric gas composition variation as a function of SBR is indicated in Fig. 2.

It can be observed from Fig. 2 that, in the regime of SBR investigated, hydrogen concentration continues to increase. A peak hydrogen yield of 104 g per kg of biomass has been reported at SBR of 2.7, the peak SBR considered. The reported analysis suggests the potential

TABLE 2   Physical properties of the biomass used.

Fuel type	Wood chips	Coconut shells
	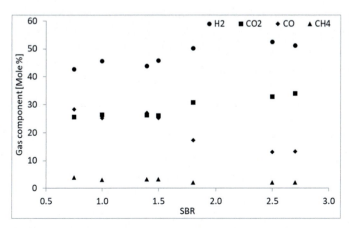	
Density (kg/m^3)	650±10	1100±100
Bulk density (kg/m^3)	400	400–450
Moisture content (%)	11	8
Ash content (%)	<3	<1.5

TABLE 3   Ultimate analysis of dry Casuarina wood chips and Coconut shells.

Item	Casuarina	Coconut shells
Carbon (mass fraction, %)	52.0	50.4
Nitrogen (mass fraction, %)	0.1	1.5
Sulfur (mass fraction, %)	0.4	0
Hydrogen (mass fraction, %)	6.6	6.7
Oxygen (mass fraction, %)	41.4	40.7
Chemical formula	$CH_{1.6}O_{0.6}$	$CH_{1.6}O_{0.6}$
Molecular weight (kg/k-mol)	23.2	23.6

FIG. 2   Variation of cold gas composition with SBR as reported by Sandeep and Dasappa (2014).

for further increase in the hydrogen yield by increasing the SBR beyond 2.7. However, detailed gas quality characterization was not attempted, and the impact of homogenous water-gas shift reaction has been sparsely explored by Sandeep and Dasappa (2014). Gas composition in a downdraft configuration closely approaches equilibrium composition, as reported by Reed and Das (1988). Toward optimizing the performance, equilibrium composition for an oxy-steam gasifier has been estimated for a range of SBR at an equivalence ratio (ER) of 0.3 using the NASA CEA (Chemical Equilibrium Application) program (Gordon and McBride, 1994). In deciding the temperature corresponding to the equilibrium analysis, it is important to note that, over a range of operating conditions, the bed temperature in the vicinity of the oxy-steam mixture injection and up to about 400 mm depth remains in the 973–1073 K range and consistently touching the 1073 K mark. This is evident from Fig. 3. A similar temperature (>1000 K) has also been reported by Sandeep and Dasappa (2014). Fig. 3 shows the temperatures as measured from the bottom of the reactor for the configuration presented in Fig. 1B.

Based on equilibrium analysis, the mass of hydrogen generated per kilogram of biomass as a function of SBR is presented in Fig. 4. It can be observed from Fig. 4 that the mass of

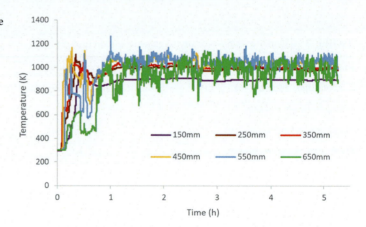

FIG. 3 Variation of bed temperature with time at different locations.

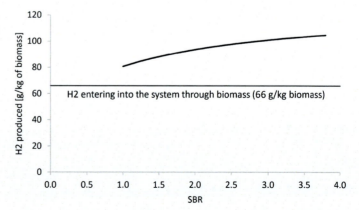

FIG. 4 Equilibrium analysis hydrogen generation per unit biomass input as a function of SBR.

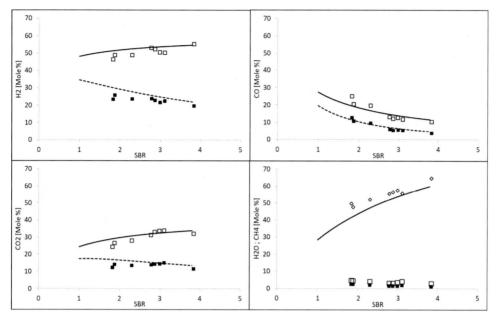

FIG. 5   Variation of individual syngas species as a function of SBR.

hydrogen produced per kg of biomass continues to increase with SBR. It should be noted that hydrogen present in dry biomass amounts to about 66 g per kg of biomass (refer to Table 5), also shown in Fig. 4. With a continuous increase in the SBR, due to the presence of excess steam, extended oxidation of char and reduction of H_2O results in the generation of the additional hydrogen in the system.

The results from the current experiments for a range of biomass (Casuarina wood chips and coconut shells) in terms of gas composition with SBR are presented in Fig. 5. Solid and dashed lines correspond to equilibrium numbers, cold/hot gas, respectively. Markers correspond to the experimental values, unfilled and filled for cold and hot gas, respectively. Along expected lines, the hydrogen fraction in the cold exit syngas increases with increasing SBR and at just under SBR of 4, the experimentally observed SBR almost abuts the equilibrium analysis prediction. The variation of CO and CO_2 are also consistent with equilibrium trends. Moisture content in the hot gas increases with SBR as indicated in Fig. 5. It is important to note that at higher SBR, the partial pressure of steam prevailing in the reactor bed is high which can drive the water-gas shift reaction in the forward direction as well as the heterogeneous carbon reactions with steam and carbon dioxide resulting in additional hydrogen in the product gas. Additionally, the effect of higher SBR is also evident in the hot gas where moisture content can be as high as 60% at SBR 3.8. Fig. 5 also presents the hot and cold gas composition with respect to each species with corresponding equilibrium composition for a range of SBR.

The performance of the gasifier as a function of SBR depicted in Fig. 5 is quantified in Table 4 where, apart from the volumetric composition of the cold gas, the mass of hydrogen

TABLE 4 Consolidated composition, efficiency, hydrogen yield as a function of ER and SBR.

SBR	H_2	CO	CO_2	CH_4	H_2	Eff.	ER
mol/mol	mole%				g/kg	%	
1.9	48.9	20.2	26.5	4.4	80.6	80.4	0.30
2.8	53.0	13.0	31.2	3.0	89.8	64.5	0.24
2.9	52.3	11.9	32.9	2.9	87.3	61.9	0.25
3.0	50.4	12.6	33.7	3.4	84.3	61.1	0.30
3.1	50.2	11.6	33.8	4.0	91.8	68.0	0.29
3.8	55.2	10.1	31.9	2.7	95.0	55.2	0.30

produced per unit biomass input and the gasification efficiency is also presented. Efficiency in the present case refers to the cold gas efficiency which is defined as presented in Eq. (4).

$$\text{Cold gas efficiency} = \frac{E_{\text{cold gas}}}{E_{\text{biomass}} + E_{\text{steam}} + E_{\text{oxygen}} + E_{\text{cooling-cleaning}}} \tag{4}$$

where E_{coldgas}, E_{biomass}, E_{steam}, E_{oxygen}, $E_{\text{cooling-cleaning}}$ refer to the energy associated with cold gas, biomass, steam generation, oxygen separation, and cooling and cleaning system.

It is of significance to note that with an increase in SBR, while the hydrogen mole % and the corresponding hydrogen yield increase, the cold gas efficiency decreases due to the penalty associated with steam generation. A gain of 15 g was realized at a conversion efficiency loss of 25%.

Based on the presented results, while higher SBR, in general, is able to provide a higher mass of hydrogen per unit biomass input, considering that the efficiency of gasification drops down substantially, it was decided to operate the oxy-steam gasification system at SBR of about 2.0 and ER of 0.3 with the oxygen-steam mixture inlet temperature at 948 ± 25 K.

After the generation of syngas, the same is subjected to swing adsorption-based separation toward generating the desired chemical or feed suitable for generating bio-chemical through the appropriate downstream process. The details are presented in the following section.

Various approaches have been attempted toward separating gaseous species generated from biomass gasification. Carbon Dioxide is typically removed using a sour gas treatment like using specific liquid media like rectisol, selexsol, etc., with subsequent regeneration of the base media. While these options have merited at large capacity ranges, there are challenges with small capacity systems. In the context of energetic bio-chemicals, the components of interest are typically H_2 and CO. One of the industrially established mixture separation processes is the pressure swing adsorption–desorption process with working pressures typically in the 10–30 bar range. Transiting into bio-derived gaseous fuels from the thermochemical conversion of biomass, considering that the base technologies generate these gases at near ambient pressures, using industrial separation processes would require external pressurization, entailing significant infrastructure creation (capital expenditure) and energy input (operational expenditure) with potential adverse techno-economic implications. To cater to the requirement of low-pressure gas separation, an innovative system has been designed

TABLE 5 Gas composition and thermophysical properties at different stages of downstream processing of syngas (Ahrenfeldt, 2012).

Items	Only PSA				CO₂ removal+PSA				WGS+CO₂+PSA						WGS+PSA			
	Input to PSA		O/P of PSA		Input to PSA		O/P of PSA		After WGS		After CO2 Sep		After PSA		After WGS		After PSA	
	vol %	mass %	vol %	mass %	vol %	mass %	vol %	mass %	vol %	mass %	vol %	mass %	vol %	mass %	vol %	mass %	vol %	mass %
H₂	40.20	4.55	17.00	1.22	60.53	10.55	23.70	2.30	55.60	5.71	92.67	56.87	71.06	20.35	55.65	5.71	20.05	1.20
CO₂	25.70	56.88	38.62	58.86	0.00	0.00	0.00	0.00	39.91	89.97	0.00	0.00	0.00	0.00	39.91	89.92	21.98	94.27
CO	24.90	35.07	37.41	34.29	33.51	31.33	65.29	35.83	1.01	1.44	1.63	14.39	0.45	26.41	1.01	1.44	1.82	1.51
CH₄	3.00	2.41	4.51	2.50	4.04	5.60	7.87	6.12	2.43	1.99	4.04	19.82	15.61	36.36	2.43	1.99	4.38	2.08
H₂O	1.20	1.09	1.80	1.12	1.62	2.52	3.15	2.75	0.97	0.89	1.62	8.92	6.24	16.36	0.97	0.89	1.75	0.54
Molecular weight	15.88		28.57		11.54		20.58		19.52		3.26		0.87		19.52		33.59	
Gas cal value (MJ/kg)	11.32		0.78		26.26		15.61		9.34		93.09		52.44		9.34		3.00	
Mass flow rate (Kh/h)	150.57		145.89		65.10		59.60		183.31		18.39		10.02		183.31		174.94	
Energy (MJ/s)	0.47		0.27		0.47		0.26		0.48		0.48		0.15		0.48		0.15	

at the Indian Institute of Science. The system, adopting the swing adsorption–desorption philosophy, generates desired pure gases/gaseous mixture while restricting both the peak pressure and working pressure range to sub 5 bar levels.

The low-pressure swing adsorption system design has been put into practice in the form of a prototype with the ability to handle 50 standard liters per minute of feed gas. The input gas with the composition of 13.9% CO; 35.5% CO_2, 3.5% CH_4, and 47.1% H_2 is subjected to swing adsorption at sub 5 bar absolute adsorption pressure and is subjected to two independent separation processes. One process is designed to generate a pure stream of H_2 by selective adsorption of all other species, while the other process is designed to selectively adsorb CO_2 to generate a mixture with H_2/CO ratio of 2.0 ± 0.1, the feed mixture for downstream methanol synthesis.

5 Downstream processing for hydrogen enhancement and contaminant removal

The Carbon Monoxide present in syngas offers potential for further enhancement of hydrogen through the H_2O reduction route. Theoretically, every mole of CO can reduce a mole of H_2O to generate 1 mole of hydrogen in what is conventionally known as the water-gas shift route. Considering that equilibrium prevents the complete conversion of CO to H_2, further downstream separation techniques are adopted for explicitly controlling the syngas composition. The downstream sub-processes adopted are described as below.

5.1 Water-gas shift unit

The high-temperature and the low-temperature WGS catalyst based on steam reforming enhances the hydrogen content in the gas by converting CO using the water-gas shift reaction along the stoichiometric reaction described below (Newsome, 1980). The water-gas shift reaction converts the CO into CO_2 and H_2 in the presence of steam. The reaction is mildly exothermic and reversible.

$$CO + H_2O \leftrightarrow H_2 + CO_2.$$

Water-gas shift reactions are equilibrium controlled at high temperatures and kinetically limited at lower temperatures; hence the process needs a catalyst for the reactions to be driven forward at lower temperatures. The water-gas shift processes usually involve two catalysts, one working at a high-temperature range (350–450°C) and the other at lower temperatures (250–300°C). Recently a third catalyst working at the mid-temperature range has also been developed (Pal et al., 2018). These catalysts have a CO conversion efficiency of about 50% and have been evaluated in the lab for their efficacy. Various parameters related to steam to CO/H_2 concentration, residence time, temperature dependence have been studied and assessed. With the syngas from the gasifier, when the H_2:CO_2:CO is in the 3:2:1 ratio, a conversion of 50% could be achieved with 9% H_2 enrichment in the shift unit. Further enrichment could be achieved when the CO_2 from the upstream of the WGS reactor is separated.

For efficient CO conversion, care must be taken such that the input gas has lower concentrations of carbon dioxide leading to increased CO conversion. Carbon dioxide removal can be achieved with a standard process available but adapted for small-scale operation. Ease of maintenance needs to be established at scales commensurate with biomass gasification systems.

5.2 Gas conditioning

Further gas conditioning is required depending upon the feed material concentration of inorganics in the biomass. This would be necessary if traces of H_2S, NH_3, and any other hydrocarbon that affect the gas separation process are present in the gas. The treatment process should address the removal of such unwanted compounds. Other requirements of the gas conditioning include removing particulates, if any, and higher molecular weight compounds. Standard processes involved are using activated carbon filters with specific chemical doping.

5.3 Gas separation

Various techniques are adopted toward adopting/establishing the gas separation techniques toward enhancing hydrogen gas composition and purity. CO_2 separation using amine solutions and other methods are adopted in the industry toward cost-effective solutions. Industrial-scale operations using specific commercially developed liquids like rectisol are being practiced and used in simulation studies. Membranes are also used for gas separation based on polymer from commercial outfits. Molecular sieves are also used for the separation of gases depending upon the concentration and the gaseous species.

5.4 Pressure swing adsorption (PSA)

From the gas separation system, the hydrogen-rich product gas is processed by a PSA for hydrogen purification. Product purity (>99.85%) and recovery depend on the operating conditions. Literature suggests that about 75% of hydrogen recovery is possible using a PSA (Yáñez et al., 2020). The standard processes as mentioned earlier will be the starting point to reach the desired techno-economic solution of the process of conditioning the gas for PEM fuel cell applications. Some of the adopted/planned approaches correspond to:

- CO conversion using water-gas shift reaction (WGS)—Several commercial catalysts are available and these would be tried initially. There has been some experience at IISc based on the commercial catalysts (low and high-temperature) from Alfa-Aesar. Depending upon the conversion process, the necessary rearrangement of the staging elements can be considered.
  - Current experience on the ratio of steam to CO and CO_2 would be used as a starting point. Steam and gas flow rates are measured to arrive at the superficial velocity for the reactor and compared with the available values from the literature. Depending upon the choice of catalyst, the gas hourly space velocity (GHSV) can vary between 400 to $1200H^{-1}$. The reactors are also operated at elevated pressures.

- Clean gas to be used for WGS and on further processing, some of the issues like coking and other aspects are expected to become less relevant.
- Possibility of using CaO/MnO_3 based catalyst is high in the case of the biomass-derived gas due to the presence of higher sulfur compounds.
- Based on the current experiment on oxy-steam reaction, in situ shift reactions inside the reactor are being explored for taking advantage of the reactor configuration and temperature.
- Unwanted species like H_2S, NH_3 and any other hydrocarbon (BTX—benzene, toluene, and xylenses) are to be removed by using activated carbon filters doped with chemicals—for example phosphoric acid to remove NH_3, iron filings/zinc oxide for H_2S.

- Hydrogen enrichment with CO_2 removal and separation of gases—While a few processes are available for separation depending upon the operating conditions, it is envisaged that these processes will be evaluated and integrated with the gas conditioning system after due diligence. While Polymer-based membrane is efficiently used for CO_2 and CH_4 separation, the possibility of using the membrane will be mainly for removing traces of unwanted species along with other processes.
- Hydrogen separation to generate pure hydrogen—PSA is one process that has been advocated for high-purity hydrogen generation. While very limited experience is available for producer gas synthesized hydrogen generation, some attempts have been made using activated charcoal. The important aspect that needs to be addressed is related to hydrogen purity and recovery. Some bench-scale results have indicated single layer of activated carbon was preferred over a mixed bed of zeolite for the desorption process. In the case of PSA, the recovery is fairly insensitive to feed pressure while 13.7–27.7 MPa is the optimum range. The tail gas pressure has the greatest effect on recovery, with low-pressure of 0.034 MP tail gas pressure having 15%–20% better recovery than 0.41 MPa tail gas. However, the cost to compress low-pressure tail gas to enter the 0.41 MPa fuel gas system can be significant and the operating pressure of a PSA system must be optimized. PSA systems are insensitive to changes in feed composition, giving constant product purity and recovery. PSA also has a good turndown ratio and is very reliable despite its complex valve system.

6 Summary

Summarizing, this chapter provides insight into the biomass to value-added products like liquid fuel and hydrogen through the thermochemical route, which can decarbonize the current fossilized hydrocarbon-based activities. In addressing the bioenergy options, details about the various process being developed toward achieving a few product portfolios in meeting the needs of the bio-refinery program have been highlighted. Experience in terms of sub-stoichiometric thermochemical conversion of biomass to syngas and further downstream processing toward generating bio-hydrogen and feed for bio-methanol synthesis has also been reported. The document, in principle, addresses the current status of various technological options and arrives at a broader conclusion that each of these needs demonstration and large-scale deployment to achieve the claimed techno-economic visibility.

Acknowledgments

The authors wish to thank the funding agencies: Ministry of New and Renewable Energy (MNRE), Department of Biotechnology (DBT), Department of Science and Technology DST), Indian Oil Corporation Limited (IOCL), and other agencies, who have supported a range of activities in the thermochemical conversion of Biomass at IISc.

References

Abbasi, T., Abbasi, S.A., 2010. Biomass energy and the environmental impacts associated with its production and utilization. Renew. Sust. Energ. Rev. 14 (3), 919–937.

Ahrenfeldt, J., 2012. In: Knoef, H. (Ed.), Handbook Biomass Gasification. BTG Biomass Technology Group, Enschede.

Akubo, K., Nahil, M.A., Williams, P.T., 2019. Pyrolysis-catalytic steam reforming of agricultural biomass wastes and biomass components for production of hydrogen/syngas. J. Energy Inst. 92 (6), 1987–1996.

Ali, K.A., Abdullah, A.Z., Mohamed, A.R., 2015. Recent development in catalytic technologies for methanol synthesis from renewable sources: a critical review. Renew. Sust. Energ. Rev. 44, 508–518.

Anniwaer, A., Chaihad, N., Zhang, M., Wang, C., Yu, T., Kasai, Y., Abudula, A., Guan, G., 2021. Hydrogen-rich gas production from steam co-gasification of banana peel with agricultural residues and woody biomass. Waste Manag. 125, 204–214.

Asadullah, M., 2014. Barriers of commercial power generation using biomass gasification gas: a review. Renew. Sust. Energ. Rev. 29, 201–215.

Azizi, K., Moraveji, M.K., Najafabadi, H.A., 2018. A review on bio-fuel production from microalgal biomass by using pyrolysis method. Renew. Sust. Energ. Rev. 82, 3046–3059.

Boateng, A.A., Garcia-Perez, M., Mašek, O., Brown, R., del Campo, B., 2015. Biochar production technology. In: Biochar for Environmental Management. Routledge, pp. 95–120.

Bondarenko, B.I., Rudyka, V.I., Soloviov, M.A., Malyna, V.P., Kurylko, S.Y., Abdullin, S.Y., 2021. Actual problems of gasification of fossil raw materials, biomass and waste with the receipt of fuels and chemical products. In: Bondarenko, B.I. (Ed.), Analytical Review of Conference Materials "Gasification India 2019", "from waste to energy 2019". Energy Technology and Resource Saving, pp. 27–39.

Butera, G., Fendt, S., Jensen, S.H., Ahrenfeldt, J., Clausen, L.R., 2020. Flexible methanol production units coupling solid oxide cells and thermochemical biomass conversion via different gasification technologies. Energy 208, 118432.

Cai, W., Liu, R., 2016. Performance of a commercial-scale biomass fast pyrolysis plant for bio-oil production. Fuel 182, 677–686.

Carbone, C., Gracceva, F., Pierro, N., Motola, V., Zong, Y., You, S., Agostini, A., 2021. Potential deployment of reversible solid-oxide cell systems to valorise organic waste, balance the power grid and produce renewable methane: a case study in the southern Italian peninsula. Front. Energy Res. 9, 15.

Chen, D., Yin, L., Wang, H., He, P., 2015. Reprint of: pyrolysis technologies for municipal solid waste: a review. Waste Manag. 37, 116–136.

Chishty, M.A., Umeki, K., Risberg, M., Wingren, A., Gebart, R., 2021. Numerical simulation of a biomass cyclone gasifier: effects of operating conditions on gasifier performance. Fuel Process. Technol. 218, 106861.

Clausen, L.R., 2017. Energy efficient thermochemical conversion of very wet biomass to biofuels by integration of steam drying, steam electrolysis and gasification. Energy 125, 327–336.

Dahmen, N., Henrich, E., Dinjus, E., Weirich, F., 2012. The bioliq® bioslurry gasification process for the production of biosynfuels, organic chemicals, and energy. Energy Sustain. Soc. 2 (1), 1–44.

Dasappa, S., 2011. Potential of biomass energy for electricity generation in sub-Saharan Africa. Energy Sustain. Dev. 15 (3), 203–213.

Dasappa, S., Paul, P.J., Mukunda, H.S., Rajan, N.K.S., Sridhar, G., Sridhar, H.V., 2004. Biomass gasification technology—a route to meet energy needs. Curr. Sci., 908–916.

Dasappa, S., Sridhar, G., Paul, P.J., 2012. Adaptation of small capacity natural gas engine for producer gas operation. Proc. Inst. Mech. Eng. C J. Mech. Eng. Sci. 226 (6), 1568–1578.

Demirbas, A., 2005. Potential applications of renewable energy sources, biomass combustion problems in boiler power systems and combustion related environmental issues. Prog. Energy Combust. Sci. 31 (2), 171–192.

Dirbeba, M.J., Brink, A., DeMartini, N., Zevenhoven, M., Hupa, M., 2017. Potential for thermochemical conversion of biomass residues from the integrated sugar-ethanol process—fate of ash and ash-forming elements. Bioresour. Technol. 234, 188–197.

Duman, G., Yanik, J., 2017. Two-step steam pyrolysis of biomass for hydrogen production. Int. J. Hydrog. Energy 42 (27), 17000–17008.

Fereshteh, S., Marzoughi, T., Rahimpour, M.R., 2020. Energy and exergy analysis and optimization of biomass gasification process for hydrogen production (based on air, steam and air/steam gasifying agents). Int. J. Hydrog. Energy 45, 33185–33197.

Fritsche, U.R., Gress, H.W., Iriarte, L., Coelho, S., Escobar, J., 2014. Possibilities of Sustainable Woody Energy Trade and Impacts on Developing Countries. IINAS, International Institute for Sustainability Analysis and Strategy, Darmstadt, Madrid.

Gershman, H.W., Hammond, M., 2012, April. The latest updates on waste-to-energy and conversion technologies; plus projects under development. In: North American Waste-to-Energy Conference. vol. 44830. American Society of Mechanical Engineers, pp. 57–62.

Gollakota, A.R., Reddy, M., Subramanyam, M.D., Kishore, N., 2016. A review on the upgradation techniques of pyrolysis oil. Renew. Sust. Energ. Rev. 58, 1543–1568.

González-Vázquez, M.P., García, R., Pevida, C., Rubiera, F., 2017. Optimization of a bubbling fluidized bed plant for low-temperature gasification of biomass. Energies 10 (3), 306.

Gordon, S., McBride, B.J., 1994. Computer Program for Calculation of Complex Chemical Equilibrium. 1311 NASA Reference Publication.

Guruviah, K.D., Sivasankaran, C., Bharathiraja, B., 2019. Thermochemical conversion: bio-oil and syngas production. In: Rastegari, A.A. (Ed.), Prospects of Renewable Bioprocessing in Future Energy Systems. Springer, p. 251.

Hasler, P., 1997, October. Producer gas quality from fixed bed gasifiers before and after gas cleaning. In: Proceedings of the IEA Thermal Gasification Seminar–IEA Bioenergy and Swiss Federal Office of Energy, Zurich.

Hassan, H., Lim, J.K., Hameed, B.H., 2016. Recent progress on biomass co-pyrolysis conversion into high-quality bio-oil. Bioresour. Technol. 221, 645–655.

Hrbek, J., 2020. Past, present and future of thermal gasification of biomass and waste. Acta Innov. 35, 5–20.

Huang, W.D., Zhang, Y.P., 2011. Energy efficiency analysis: biomass-to-wheel efficiency related with biofuels production, fuel distribution, and powertrain systems. PLoS One 6 (7), e22113.

Ibarra-Gonzalez, P., Ben-GuangRong, 2019. A review of the current state of biofuels production from lignocellulosic biomass using thermochemical conversion routes. Chin. J. Chem. Eng. 27, 1523–1535.

International Organization for Standardization, 2019. Hydrogen Fuel Quality—Product Specification (ISO 14687:2019). Retrieved from http://www.iso.org/iso/catalogue_detail?csnumber=63787.

Isahak, W.N.R.W., Hisham, M.W., Yarmo, M.A., Hin, T.Y.Y., 2012. A review on bio-oil production from biomass by using pyrolysis method. Renew. Sust. Energ. Rev. 16 (8), 5910–5923.

Ishaq, H., Dincer, I., 2020. A new energy system based on biomass gasification for hydrogen and power production. Energy Rep. 6, 771–781.

Iulianelli, A., Liguori, S., Wilcox, J., Basile, A., 2016. Advances on methane steam reforming to produce hydrogen through membrane reactors technology: a review. Catal. Rev. 58 (1), 1–35.

Jin, K., Ji, D., Xie, Q., Nie, Y., Yu, F., Ji, J., 2019. Hydrogen production from steam gasification of tableted biomass in molten eutectic carbonates. Int. J. Hydrog. Energy 44 (41), 22919–22925.

Kan, T., Strezov, V., Evans, T.J., 2016. Lignocellulosic biomass pyrolysis: a review of product properties and effects of pyrolysis parameters. Renew. Sust. Energ. Rev. 57, 1126–1140.

Kirkels, A.F., Verbong, G.P., 2011. Biomass gasification: still promising? A 30-year global overview. Renew. Sust. Energ. Rev. 15 (1), 471–481.

Kirnbauer, F., Hofbauer, H., 2011. Investigations on bed material changes in a dual fluidized bed steam gasification plant in Gussing, Austria. Energy Fuels 25 (8), 3793–3798.

Knoef, H., 2002. Review of small-scale biomass gasification. In: Pyrolysis and Gasification of Biomass and Waste Expert Meeting, Strasbourg, France.

Kolb, T., Eberhard, M., Leibold, H., Seifert, H., Zimmerlin, B., Dahmen, N., Neuberger, M., 2013. BtL. The Bioliq {Sup Registered} Process at KIT. Karlsruhe Institute of Technology.

Koppejan, J., Van Loo, S. (Eds.), 2012. The Handbook of Biomass Combustion and Co-Firing. Routledge.

Landälv, I., Gebart, R., Marke, B., Granberg, F., Furusjö, E., Löwnertz, P., Salomonsson, P., 2014. Two years experience of the BioDME project—a complete wood to wheel concept. Environ. Prog. Sustain. Energy 33 (3), 744–750.

López, R., Díaz, M.J., González-Pérez, J.A., 2018. Extra CO2 sequestration following reutilization of biomass ash. Sci. Total Environ. 625, 1013–1020.

Luyben, W., 2016. Control of parallel dry methane and steam methane reforming processes for Fischer–Tropsch syngas. J. Process Control 39, 77–87. https://doi.org/10.1016/j.jprocont.2015.11.007.

Mahdi, F., Majid, S., 2021. Hydrogen-rich syngas production via integrated configuration of pyrolysis and air gasification processes of various algal biomass: process simulation and evaluation using Aspen plus software. Int. J. Hydrog. Energy 46, 18844–18856.

Maity, S.K., 2015. Opportunities, recent trends and challenges of integrated biorefinery: part I. Renew. Sust. Energ. Rev. 43, 1427–1445.

Maschio, G., Lucchesi, A., Stoppato, G., 1994. Production of syngas from biomass. Bioresour. Technol. 48 (2), 119–126.

Milhé, M., Van de Steene, L., Haube, M., Commandré, J.M., Fassinou, W.F., Flamant, G., 2013. Autothermal and allothermal pyrolysis in a continuous fixed bed reactor. J. Anal. Appl. Pyrolysis 103, 102–111.

Mirzaei, N., Toffolo, A., Engvall, K., Kantarelis, E., 2021. Flexible production of liquid biofuels via thermochemical treatment of biomass and olefins oligomerization: a process study. Chem. Eng. Trans. 86, 187–192.

Newsome, D.S., 1980. The water-gas shift reaction. Catal. Rev. Sci. Eng. 21 (2), 275–318.

Nyström, I., Bokinge, P., Franck, P.-Å., 2019. Production of Liquid Advanced Biofuels-Global Status. CIT IndustriellEnergi AB.

Ong, H.C., Chen, W.-H., Farooq, A., Gana, Y.Y., Lee, K.T., Ashokkumar, V., 2019. Catalytic thermochemical conversion of biomass for biofuel production: a comprehensive review. Renew. Sustain. Energy Rev. 113, 109266.

Pal, D.B., Chand, R., Upadhyay, S.N., Mishra, P.K., 2018. Performance of water gas shift reaction catalysts: a review. Renew. Sust. Energ. Rev. 93, 549–565.

Prajitno, H., Insyani, R., Park, J., Ryu, C., Kim, J., 2016. Non-catalytic upgrading of fast pyrolysis bio-oil in supercritical ethanol and combustion behavior of the upgraded oil. Appl. Energy 172, 12–22.

Prando, D., Ail, S.S., Chiaramonti, D., Baratieri, M., Dasappa, S., 2016. Characterisation of the producer gas from an open top gasifier: assessment of different tar analysis approaches. Fuel 181, 566–572.

Reddy, S.N., Nanda, S., Dalai, A.K., Kozinski, J.A., 2014. Supercritical water gasification of biomass for hydrogen production. Int. J. Hydrog. Energy 39 (13), 6912–6926.

Reed, T.E., 1981. Biomass Gasification: Principles and Technology. Advances in Solar Energy. Springer, pp. 125–174.

Reed, T.B., Das, A., 1988. Handbook of Biomass Downdraft Gasifier Engine Systems. Biomass Energy Foundation.

Rocha, D.N., Barbosa, E.G., dos Santos Renato, N., Varejao, E.V.V., da Silva, U.P., de Araujo, M.E.V., Martins, M.A., 2020. Improving biofuel production by thermochemical conversion of defatted *Scenedesmus obliquus* biomass. J. Clean. Prod. 275, 124090.

Rosendahl, L. (Ed.), 2013. Biomass Combustion Science, Technology and Engineering. Elsevier.

Rostrup-Nielsen, J.R., 2002. Syngas in perspective. Catal. Today 71 (3–4), 243–247.

Sandeep, K., Dasappa, S., 2014. Oxy–steam gasification of biomass for hydrogen rich syngas production using downdraft reactor configuration. Int. J. Energy Res. 38 (2), 174–188.

Sharma, A., Pareek, V., Zhang, D., 2015. Biomass pyrolysis—a review of modelling, process parameters and catalytic studies. Renew. Sust. Energ. Rev. 50, 1081–1096.

Sheldon, R.A., 1983. Chemicals From Synthesis Gas: Catalytic Reactions of CO and H2. vol. 2 Springer Science & Business Media.

Shivapuji, A.M., Dasappa, S., 2014. Selection and thermodynamic analysis of a turbocharger for a producer gas-fuelled multi-cylinder engine. Proc. Inst. Mech. Eng. A 228 (3), 340–356.

Smith, P., 1998. Coordinated Combustion Research Between Netherlands Gasunie and the Groningen University, Netherlands. GasunieenGroningseuniversiteitwerkensamen in verbrandingsonderzoek. Boundary-Layer Meteorology.

Song, C., Cheng, Z., Zhang, S., Lin, H., Kim, Y., Ramakrishnan, M., Du, Y., Zhang, Y., Zheng, H., Barcelo, D., 2020. Thermochemical liquefaction of agricultural and forestry wastes into biofuels and chemicals from circular economy perspectives. Sci. Total Environ. 749, 141972.

Sridhar, G., Paul, P.J., Mukunda, H.S., 2001. Biomass derived producer gas as a reciprocating engine fuel—an experimental analysis. Biomass Bioenergy 21 (1), 61–72.

Sridhar, G., Sridhar, H.V., Dasappa, S., Paul, P.J., Rajan, N.K.S., Mukunda, H.S., 2005. Development of producer gas engines. Proc. Inst. Mech. Eng. D 219 (3), 423–438.

Srivastava, R.K., Shetti, N.P., Reddy, K.R., Kwon, E.E., Nadagouda, M.N., Aminabhavi, T.M., 2021. Biomass utilization and production of biofuels from carbon neutral materials. Environ. Pollut. 276, 116731.

Stassen, H.E., 1995. Small-Scale Biomass Gasifiers for Heat and Power: A Global Review. The World Bank.

Trippe, F., Fröhling, M., Schultmann, F., Stahl, R., Henrich, E., 2010. Techno-economic analysis of fast pyrolysis as a process step within biomass-to-liquid fuel production. Waste Biomass Valoriz. 1 (4), 415–430.

Uzoejinwa, B.B., He, X., Wang, S., Abomohra, A.E.-F., Hu, Y., Wang, Q., 2018. Co-pyrolysis of biomass and waste plastics as a thermochemical conversion technology for high-grade biofuel production: recent progress and future directions elsewhere worldwide. Energy Convers. Manag. 163, 468–492.

Van den Broek, R., Faaij, A., van Wijk, A., 1996. Biomass combustion for power generation. Biomass Bioenergy 11 (4), 271–281.

Vreugdenhil, B.J., 2011. Country Report the Netherlands. ECN, Petten.

Wang, T., Li, Y., Ma, L., Wu, C., 2011. Biomass to dimethyl ether by gasification/synthesis technology—an alternative biofuel production route. Front. Energy 5 (3), 330–339.

Wang, Z., He, T., Qin, J., Jingli, W., Li, J., Zi, Z., Liu, G., Jinhu, W., Sun, L., 2015. Gasification of biomass with oxygen-enriched air in a pilot scale two-stage gasifier. Fuel 150, 386–393.

Xiao, Y., Shaoping, X., Song, Y., Shan, Y., Wang, C., Wang, G., 2017a. Biomass steam gasification for hydrogen-rich gas production in a decoupled dual loop gasification system. Fuel Process. Technol. 165, 54–61.

Xiao, Y., Wang, X., Pinson, P., Wang, X., 2017b. A local energy market for electricity and hydrogen. IEEE Trans. Power Syst. 33 (4), 3898–3908.

Xiu, S., Shahbazi, A., 2012. Bio-oil production and upgrading research: a review. Renew. Sust. Energ. Rev. 16 (7), 4406–4414.

Yáñez, M., Relvas, F., Ortiz, A., Gorri, D., Mendes, A., Ortiz, I., 2020. PSA purification of waste hydrogen from ammonia plants to fuel cell grade. Sep. Purif. Technol. 240, 116334.

Yao, D., Qiang, H., Wang, D., Yang, H., Chunfei, W., Wang, X., Chen, H., 2016. Hydrogen production from biomass gasification using biochar as a catalyst/support. Bioresour. Technol. 216, 159–164.

Yuan, D., Zhang, K., Wang, L., Jin, L., Guo, X., Zhang, G., 2021. Gasification investigations of coal and biomass blends for high purity H_2 production with carbon capture potential. J. Ind. Eng. Chem. 103, 42–48. https://doi.org/10.1016/j.jiec.2021.05.014.

Zhang, L., Liu, R., Yin, R., Mei, Y., 2013. Upgrading of bio-oil from biomass fast pyrolysis in China: a review. Renew. Sust. Energ. Rev. 24, 66–72.

Zhou, L., Yang, Z., Tang, A., Huang, H., Wei, D., Erlei, Y., Wei, L., 2019. Steam-gasification of biomass with CaO as catalyst for hydrogen-rich syngas production. J. Energy Inst. 92, 1641–1646.

Progress and trends in renewable jet fuels

Saleem A. Farooqui, Anil K. Sinha, and Anjan Ray

CSIR-Indian Institute of Petroleum, Dehradun, India

Abbreviations

ASTM	American Society for Testing and Materials
ATF	aviation turbine fuel
ATJ	alcohol-to-jet
APU	auxiliary power unit
BIS	Bureau of Indian Standards
BPD	barrels per day
CH	catalytic hydrothermolysis
CORSIA	Carbon Offsetting and Reduction Scheme for International Aviation
DSHC	direct sugars to hydrocarbons
DARPA	Defense Advanced Research Projects Agency
FT	Fischer–Tropsch
FT-SKA	Fischer–Tropsch synthetic kerosene with aromatics
FAME	fatty acid methyl esters
FRL	fuel readiness level
GHG	green house gas
HEFA	hydroprocessed ester of fatty acids
HTL	hydro thermal liquefaction
HDO-SK	hydro-deoxygenated synthesized kerosene
HDO-SAK	hydro-deoxygenated synthesized aromatic kerosene
HDCJ	hydrotreated depolymerized cellulosic jet
HTFT	high-temperature Fischer–Tropsch
HVO	hydrogenated vegetable oil
HRJ	hydroprocessed renewable jet
HDO	hydrodeoxygenation
HDCN	hydrodecarbonylation
HDCX	hydrodecarboxylation
HC-HEFA	synthesized paraffinic kerosene from hydroprocessed hydrocarbons, esters, and fatty acids

ICAO	International Civil Aviation Organization
IRENA	International Renewable Energy Agency
LLDCs	landlocked developing countries
LTFT	low-temperature Fischer–Tropsch
MSW	municipal solid waste
MMTPA	million metric ton per annum
MFSP	minimum fuel selling price
OEM	original equipment manufacturers
PtL	power to liquid
RPK	revenue passenger kilometer
RWGS	reverse water gas shift reaction
SDGs	sustainable development goals
SPK	synthetic paraffinic kerosene
SAF	sustainable aviation fuel
SIP	synthesized iso-paraffins
SIDS	small island developing states
TRI	Thermochemical Recovery International, Inc.
UCO	used cooking oil
UPM	UPM Paper Mill Company

1 Introduction

1.1 The need for alternative aviation fuel

With depleting fossil fuel reserves, increasing energy needs due to industrialization and modernization, and growing awareness of global warming, an increasing thrust is provided toward securing alternatives to petroleum-based fuels (Rye et al., 2013; https://www.statista.com/statistics/655057/fuel-consumption-of-airlines-worldwide, n.d.; IPCC, 2014; World Energy Outlook, 2012; Girardet and Mendonça, 2009; Air Transport Action Group, 2014; Sustainable Aviation CO2 Road-MAP, 2012). The aviation industry consumes approximately 98 billion gallons/yr. in the pre-Covid era (https://www.statista.com/statistics/655057/fuel-consumption-of-airlines-worldwide, n.d.). The typical increase in global annual aviation fuel demand is ~2.9 billion gallons/yr. from 2009 until 2019. According to the Air Transport Action Group (ATAG), the global aviation industry is responsible for almost 12% of the worldwide transport sector's CO_2 emissions or ~2% of all human-induced CO_2 emissions (https://www.iatag.org/facts-figures.html, n.d.). About 80% of these emissions are from long-distance flights (>1500 km) due to the ever-increasing aviation load and no practical alternative (https://www.statista.com/statistics/655057/fuel-consumption-of-airlines-worldwide, n.d.; IPCC, 2014). Conventional Aviation Turbine Fuel (ATF) derived from crude-based sources must meet internationally accepted ASTM D1655 specifications, while liquid aviation turbine fuel obtained from non-crude-based sources needs to meet ASTM D7566 specifications (Standard Specification for Aviation Turbine Fuel Containing Synthesized Hydrocarbons, 2016a, b).

Albeit globally available today, fossil-derived jet fuel is subject to significant market volatility and contributes 30%–50% of an airline's operating cost. A massive change in current engine design and complex, lengthy regulatory and management approvals will be needed if one has to move away from this type of fuel to other energy sources for propulsion. Air transport depends on high energy density, liquid hydrocarbon fuels, and risk appetite is restricted mainly to the quest for alternative, affordable renewable fuels with similar properties to existing liquid hydrocarbon fuels (Girardet and Mendonça, 2009).

Stepping away from this established paradigm requires significant effort and investment. Hydrogen-powered airplanes need a much larger, heavier, and high-pressure tank for fuel storage, which would be too capital intensive and would arguably reduce efficiency due to increased tank volume and weight. Photovoltaics are at a nascent stage and a long way off from meeting range and performance criteria to be commercially viable (Sustainable Aviation CO2 Road-MAP, 2012). Hence efforts have been focused mainly on a transition from fossil-based fuels to more sustainable, carbon-neutral liquid fuels, such as second and third-generation biofuels. A less disruptive, smoother transition is likely to reduce CO_2 emissions more effectively over an extended period (Sustainable Aviation CO2 Road-MAP, 2012; Standard Specification for Aviation Turbine Fuel Containing Synthesized Hydrocarbons, 2016a, b; https://www.icao.int/environmental-protection/pages/a39_corsia_faq2.aspx, n.d.; Chuck, 2016).

Since, chemically, no two aviation fuels are entirely identical. Hence, various standards (such as ASTM D 1655 and Def Stan 91-091 are made to limit different properties in range to meet the required specifications. Aviation range hydrocarbons are typically in the boiling range of 110–280 °C, i.e., carbon range of C8 to C16 range hydrocarbons. There are four different types of aviation fuel: Kerosene (Jet A-1), Kerosene-gasoline blend (Jet B), Aviation gasoline (avgas), and Biokerosene.

Jet fuel (Jet A-1 type aviation fuel, also known as JP-1A) is used globally in turbine engines (jet engines, turboprops) in civil aviation. Jet A/A1 is a light, refined petroleum kerosene-type fuel. Jet A-1 has a flash point above 38 °C and a freezing point above −47 °C. Jet A is a similar type of kerosene fuel that is usually only available in the United States. Due to aircraft engines' very high jet fuel requirements, this fuel is subject to comprehensive, internationally standardized quality specifications.

Kerosene-gasoline mixture (Jet B) is used in military jets. It is a blend of approximately 65% gasoline and 35% kerosene, generally used in regions with notably lower temperatures. It is more flammable at a flashpoint of 20 °C, and its freezing point is as low as −72 °C (compared to −47 °C for Jet A-1). The engines must, however, be suitable for the use of these aviation fuels.

Aviation gasoline is also known as avgas. It is used in small private aircraft piston engines that require high-octane leaded fuel. Avgas with 100 octane meets the requirements. Gasoline engine aircraft can be operated with avgas, while turbine-powered or diesel-powered aircraft require kerosene as fuel.

1.2 The carbon offsetting and reduction scheme for international aviation (CORSIA)

The Carbon Offsetting and Reduction Scheme for International Aviation (CORSIA), introduced in October 2016, ushered in a new international cooperation model to regulate and mitigate the total CO_2 emissions of this sector above 2020 baseline levels. A staggered implementation model has been envisaged that addresses the specific circumstances and adoption capabilities of ICAO member states, particularly Least Developed Countries and island nations, with the intent of mitigating potential market distortions. As of January 1, 2021, 88 states had volunteered to engage in CORSIA (https://www.icao.int/environmental-protection/CORSIA/Documents/CORSIA_States_for_Chapter3_State_Pairs_Jul2020.pdf, n.d.), the pilot phase running from 2021 to 2023, followed by the first phase running from 2024 to 2026.

The second, effectively obligatory process would extend from 2027 to 2035 to those states whose share of foreign air traffic in Revenue Ton Kilometers (RTKs) is 0.5% or more of total global RTKs as of 2018. It is also applicable to the countries that fall within the list of those states whose average share from the highest to the lowest RTKs is within 90% of total RTKs. Least developed countries (LDCs), Small Island Developing States (SIDS), and Landlocked Developing Countries (LLDCs) are exempted unless they elect to participate in phase-2 voluntarily. All states that are not signed up have the option to voluntarily join the scheme from the beginning of a given year but need to notify ICAO of their decision to join more than 6 months in advance of the year in which their participation is to commence (https://www.icao.int/environmental-protection/pages/a39_corsia_faq2.aspx, n.d.).

Paragraph 11 of the Assembly Resolution (https://www.icao.int/environmental-protection/pages/a39_corsia_faq2.aspx, n.d.) lays out the formula for offsetting requirements in the given year, defined as equivalent to the cumulative additional GHG emissions above the average baseline emissions of 2019 and 2020; it also outlines how these offsetting requirements are distributed among aircraft operators participating in the scheme. This advocated for a "dynamic approach" to offset apportionment, steadily shifting from 100% sectoral (and 0% private) from 2021 to 2029 in terms of pollution growth factor, to increasingly more assignment to individual operators of 20% or more from 2030 to 2032. Further raising operator-level offsetting standards to 70% from 2033 to 2035, hoping that the market-based process will be matured by then and total industry responsibility devolved more and more to the operators themselves (https://www.icao.int/environmental-protection/pages/a39_corsia_faq2.aspx, n.d.).

1.3 Available pathways

Biomass-based feedstocks, such as non-edible vegetable oil, are among the most promising feed for converting to sustainable drop-in aviation fuel. For a standalone unit, the fuels derived from non-edible oils and fats (lipids) via hydroprocessing HEFA-SPK, FT-SPK, Pyrolysis, and upgrading and HTL and upgrading have higher capital costs compared to alcohol-to-jet (ATJ) and Advanced fermentation of sugars to hydrocarbons (Table 1). The minimum fuel selling price for HEFA-SPK is the lowest (Table 2).

Ever since the first biofuel-powered Virgin Atlantic flight took off from London's Heathrow in 2008, significant advancements have been made in the field of aviation biofuels. Liquid biofuels can be produced via different routes, such as hydroprocessing of lipids, processing of bio-oils (pyrolysis oils) (Basu, 2013), processing of syngas via Fisher–Tropsch synthesis (Kubickova and Kubicka, 2010; Gnanamani et al., 2013) coupled with hydroprocessing (Rye et al., 2013; Chuck, 2016; https://www.icao.int/environmental-protection/CORSIA/Documents/CORSIA_States_for_Chapter3_State_Pairs_Jul2020.pdf, n.d.; https://www.nedo.go.jp/content/100890890.pdf, n.d.; Tiwari et al., 2011; Veriansyah et al., 2012; Cheng et al., 2014; Simacek et al., 2011; Bezergianni et al., 2009a, 2010a, b, 2011;

TABLE 1   Summary of technologies, their status, and estimated capital cost (Biofuels for Aviation Technology Brief, 2017).

Conversion process	Status	Capital cost in crores INR
HEFA	Commercial	1855–5985
FT	Demonstration	3038–11,025
Pyrolysis and upgrading	Pilot/demo	1449–4480
HTL and upgrading	Pilot/demo	2534–4767
Alcohol-to-jet (ATJ)	Demo	476–504
Advanced fermentation of sugars to hydrocarbons	Small commercial	2716

TABLE 2   MFSP of bio-jet fuel based on technical and economic analysis (Biofuels for Aviation Technology Brief, 2017).

Conversion process	Feedstock	MFSP bio-jet produced in multiple plants per ton in INR
HEFA	UCO	10,620
FT	Forest residue/wheat straw	141,680–208,600
HEL	Forest residue/wheat straw	70,840–102,200
Pyrolysis	Forest residue/wheat straw	102,200–145,600
ATJ	Forest residue/wheat straw	189,000–275,450
DSHC	Forest residue/wheat straw	377,790–503,720

Huber et al., 2007; Fan et al., 2014; Šimácek et al., 2009; Walendziewski et al., 2009; Meller et al., 2014; Liu et al., 2015; Rozmyslowicz et al., 2010; Verma et al., 2011; Murata et al., 2010; Anand and Sinha, 2012; Sharma et al., 2012; Anand et al., 2014; Verma et al., 2015; Rivard et al., 2019; Pinto et al., 2014; Wyman, 1994; Huber et al., 2006), hydrothermal (Huber et al., 2006; Li et al., 2010; Kim et al., 2014; Lynd et al., 1991; Huber and Corma, 2007; *Review of Biojet Fuel Conversion Technologies*, 2016), transesterification (Chuck, 2016; https://www.icao.int/environmental-protection/CORSIA/Documents/CORSIA_States_for_Chapter3_State_Pairs_Jul2020.pdf, n.d.; https://www.nedo.go.jp/content/100890890.pdf, n.d.; Tiwari et al., 2011; Veriansyah et al., 2012; Cheng et al., 2014), or by alcohols to aviation fuels (ATJ) process using the dehydration-oligomerization route. All the methods have their advantages and disadvantages, and the most sustainable and economical path to produce sustainable aviation biofuels would eventually be expected to prevail (Rye et al., 2013; Sustainable Aviation CO2 Road-MAP, 2012). Since about 2005, the US Defense Advanced Research Projects Agency (DARPA) has funded millions of dollars in bringing green jet fuel to the US military. The project's objective was to develop a surrogate fuel for petroleum-based military jet fuel (Jet Propellant 8; JP-8) from oil-rich crops. The research and developments carried out by various organizations worldwide led to the successful synthesis of renewable bio-aviation fuel, having specifications similar to those for fossil-based jet fuels (Standard Specification for

Aviation Turbine Fuel Containing Synthesized Hydrocarbons, 2016a). As the American Society for Testing and Materials, ASTM International used data from the extensive engine and component testing from multiple sources and stakeholders for several years to finally approve fuel specifications in 2011 for synthesized hydrocarbons from renewable sources for blending in the Jet A1 pool. In June 2011, the ASTM committee gave technical approval for 50% biofuel/kerosene mix for use as Jet A1 fuel. Detailed specifications and test methods for these specifications were listed as a new ASTM standard, ASTM D7566, for aviation turbine fuel containing synthesized hydrocarbons (Standard Specification for Aviation Turbine Fuel Containing Synthesized Hydrocarbons, 2016a, b). Since 2011, there have been six additions (7 annexures) to the ASTM documents. A comparison of various specifications approved by ASTM under D7566 to date is listed in Table 3.

TABLE 3 Key difference of specification for the different approved processes as per ASTM D 7566 processes (Standard Specification for Aviation Turbine Fuel Containing Synthesized Hydrocarbons, 2016a).

Test properties	Jet A1 D 1655	FT-SPK	HEFA-SPK	SIP-SPK	SPK/A	ATJ-SPK	CHJ	HC-HEFA-SPK
Annex name		A1	A2	A3	A4	A5	A6	A7
Blending ratio approved-Max.%		50	50	10	50	50	50	10
Composition								
Acidity, total mg KOH/g (max)	0.1	0.015	0.015	0.015	0.015	0.015	0.015	0.015
Aromatics, vol% (max)	25	–	–	–	20	–	–	–
Sulfur, mercaptan, C mass % (max)	0.003	–	–	–	–	–	–	–
Sulfur, total mass % (max)	0.03	–	–	–	–	–	–	–
Volatility								
Distillation temperature (°C)								
10% recovered, °C (max)	205	205	205	250	205	205	205	205
50% recovered (°C)	Report	Report	Report	Report	Report	Report	Report	Report
90% recovered (°C)	Report	Report	Report	Report	Report	Report	Report	Report
Final boiling point, °C (max)	300	300	300	255	300	300	300	300
T50-T10, °C (min)	–	–	–	–	–	–	15	–
T90-T10, °C (min)	40	22	22	5	22	21	40	22
Distillation residue, % (max)	1.5	1.5	1.5	1.5	1.5	1.5	1.5	1.5
Distillation loss, % (max)	1.5	1.5	1.5	1.5	1.5	1.5	1.5	1.5

TABLE 3 Key difference of specification for the different approved processes as per ASTM D 7566 processes (Standard Specification for Aviation Turbine Fuel Containing Synthesized Hydrocarbons, 2016a)—cont'd

Test properties	Jet A1 D 1655	FT-SPK	HEFA-SPK	SIP-SPK	SPK/A	ATJ-SPK	CHJ	HC-HEFA-SPK
Flash point, °C (min)	38	38	38	100	38	38	38	38
Density at 15°C (kg/m^3)	**775–840**	**730–770**	**730–770**	**765–780**	**755–800**	**730–770**	**775–840**	**730–800**
Smoke point, mm (min)	25.0	–	–	–	–	–	–	25.0
Fluidity								
Freezing point, °C (max)	−47	−40	−40	−60	−40	−40	−40	−40
Viscosity −20°C, mm^2/s (max)	8							
Combustion								
Net heat of combustion, MJ/kg (min)	42.8	–	–	**43.5**	–	–	–	–
Naphthalenes, vol, % (max)	3	–	–	–	–	–	–	–
Smoke point, mm (min)	18	–	–	–	–	–	–	–
Corrosion								
Copper strip, 2h at 100°C (max)	No. 1	–	–	–	–	–	–	–
Thermal stability (2.5h at control temperature)								
Temperature, °C (min)	260	325	325	355	**355**	325	325	325
Filter pressure drop, mmHg (max)	25	25	25	25	25	25	25	25
Tube deposits less than (max)	3	3	3	3	3	3	3	3
Existent gum, mg/100mL (max)	7	–	7	7	**4**		7	7
MSEP, min					**90**			
FAME, ppm (max)	–	–	**<5**	–	–	–	**<5**	**<5**
Additives								
Anti-oxidants, mg/L (min–max)	–	17–24	17–24	17–24	17–24	17–24	17–24	17–24
Hydrocarbon composition								
Cycloparaffins, mass % (max)	–	15	15	15	15	15	**report**	50

Continued

TABLE 3 Key difference of specification for the different approved processes as per ASTM D 7566 processes (Standard Specification for Aviation Turbine Fuel Containing Synthesized Hydrocarbons, 2016a)—cont'd

Test properties	Jet A1 D 1655	FT-SPK	HEFA-SPK	SIP-SPK	SPK/A	ATJ-SPK	CHJ	HC-HEFA-SPK
Aromatics, mass % (max)	–	0.5	0.5	0.5	20	0.5	21.2	0.5
Aromatics, mass % (min)		–	–	–	–	–	8.4	–
Paraffins, mass %	–	Report	Report	Report	Report	Report	Report	Report
Carbon and hydrogen, mass% (min)	–	99.5	99.5	99.5	99.5	99.5	99.5	99.5
Non-hydrocarbon composition								
Saturated Hydrocarbons, mass percent	–	–	–	98	–	–	–	–
Farnesene, mass percent, max	–	–	–	97	–	–	–	–
Hexahydrofarnesol, mass percent	–	–	–	1.5	1.5	–	–	–
Olefins, mgBr$_2$/100 g	–	–	–	300	300	–	–	–
Nitrogen, mg/kg (max)	–	2	2	2	2	2	2	2
Water, mg/kg (max)	–	75	75	75	75	75	75	75
Sulfur, mg/kg (max)	–	15	15	2	2	15	15	15
Metals (Al, Ca, Co, Cr, Cu, Fe, K, Mg, Mn, Mo, Na, Ni, P, Pb, Pd, Pt, Sn, Sr, Ti, V, Zn), mg/kg (max)	–	0.1 per metal	0.1 per metal	0.1 per metal	0.1 per metal	0.1 per metal	0.1 per metal	0.1 per metal
Halogens, mg/kg (max)	–	1	1	1	1	1	1	1

FT-SPK, Fischer–Tropsch hydroprocessed synthesized paraffinic kerosene; HEFA-SPK, hydroprocessed esters and fatty acids synthesized paraffinic kerosene; SIP, synthesized iso-paraffins from hydroprocessed fermented sugars; ATJ, alcohol-to-jet synthetic paraffinic kerosene; SPK/A), synthesized paraffinic kerosene plus aromatics, CHJ, synthesized catalytic hydrothermolysis jet; HC-HEFA-SPK, synthesized paraffinic kerosene from hydroprocessed hydrocarbons, esters, and fatty acids.

The significant pathways certified by ASTM or in consideration for Bio—ATF production are (a) Fischer–Tropsch Synthetic Paraffinic Kerosene (FT-SPK, annex A1), (b) Hydroprocessed Ester of Fatty Acids—Synthetic Paraffinic Kerosene (HEFA- SPK, Annex A2), (c) Synthesized Iso-Paraffins (SIP) from Hydroprocessed Fermented Sugars (Annex A3), (d) Alcohols-to-Jet Synthetic Paraffinic Kerosene (ATJ-SPK, Annex A4), (e) Fischer–Tropsch Synthetic Kerosene with Aromatics (FT-SKA, Annex A5), (f) Catalytic Hydrothermolysis (CHJ, Annex A6), (g) Synthesized paraffinic kerosene from Hydroprocessed hydrocarbons, esters, and fatty acids (HC-HEFA-SPK, Annex A7) (h) Hydro-Deoxygenated Synthesized

Aromatic Kerosene (HDO-SAK), (I) Hydrotreated Depolymerized Cellulosic Jet (HDCJ), (j) Hydrothermal Liquefaction (ARA-CLG), (k) Hydro-Deoxygenated Synthesized Kerosene (HDO-SK).

Table 3 details the differences in the limits for different specifications of the approved pathways. SIP-SPK (A3) has a different limit for volatility, i.e., 10% boiling, and FBP, compared to other approved processes. The limit for the density for different approved processes is in the range of 730–770 kg/m^3, except SIP-SPK (765–780 kg/m^3), SPK/A (755–800 kg/m^3), CHJ (775–840 kg/m^3) and HC-HEFA-SPK (730–800 kg/m^3). The sulfur limit for all other approved processes is 15 PPM except SIP-SPK and SPK/A, for which the limit is 2 PPM. The maximum allowable existent gum for A4 is 4, while it is 7 mg/100 mL. There is no limit to Microseparometer (MSEP), except SPK/A, where the maximum allowable MSEP is 90. For all the approved processes, the limits for neat fuel for aromatics is set at 0.5% by weight); the limit for A4 is 20 and A 6 (minimum limit 8.4%, maximum 21.2%).

This chapter presents the progress and trends in renewable jet fuels, the technological advances in different routes being researched for the production of bio-aviation fuels, and their ASTM specifications. Fig. 1 shows the various conversion pathways for the production of renewable jet fuel based on biomass availability. In the biochemical route, genetic modifications to organisms for creating efficient and amplified pathways are directed toward selectively

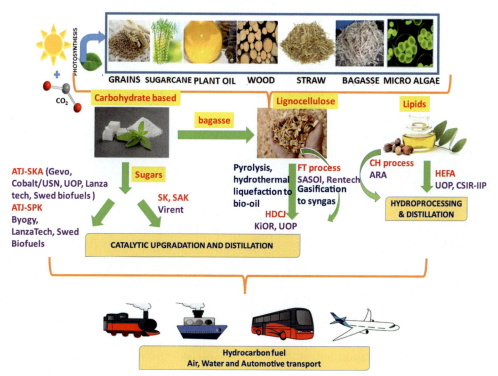

FIG. 1   Pathways for renewable bio-jet production (http://www.caafi.org/resources/pdf/CQ_Breakout_Session.pdf, n.d.).

producing plant sugars or amplified quantities of lipids as feedstock, and even to produce hydrocarbon molecules that can be converted to aviation fuel (https://www.nedo.go.jp/content/100890890.pdf, n.d.). Phototrophic pathways involve selecting and genetic modification using algae as feedstocks to produce a molecule of interest (https://www.nedo.go.jp/content/100890890.pdf, n.d.).

A list of various bio-jet fuel production processes and their status of approval is provided in Table 4. Since these fuels cannot be used directly in aviation engines, ASTM also provides a

TABLE 4 R&D technologies for Bio-aviation fuel and their status.

Technology pathway	Technology provider	STATUS	Feedstock	Aromatics	ASTM status
Fischer–Tropsch Synthetic Paraffinic Kerosene (FT-SPK)	Sasol, Shell, Syntroleum, TRI (FT-SPK)	Commercial/ demo	Mature technology Economics at a very large-scale	Low	Annex A1
Hydroprocessed Ester of Fatty Acids—Synthetic Paraffinic Kerosene (HEFA-SPK)	CSIR-IIP	Demonstration-scale-0.5 TPD processing capacity	Used cooking oil, Jatropha, Palm waste oil (Palm stearin and Palm fatty acid distillate), Pongamia, animal fat, and algae oil	Moderate 6%–10%	Annex A2 after aromatics removal Fuel in clearing house
	Honeywell UOP	Commercial	Vegetable oil, animal fat, recycled oil	Low	Annex A2
	Axens	In design		Low	
	Neste	Commercial		Low	
Synthesized Iso-Paraffins (SIP) from Hydroprocessed Fermented Sugars	Amyris, total	Demo	Sugar feed, flight, proven	Low	Annex A3
Fischer–Tropsch Synthetic Kerosene with Aromatics FT-SKA	Sasol		Coal, natural gas, biomass	High	Annex A4
Alcohols-to-Jet Synthetic Paraffinic Kerosene (ATJ-SPK)	Byogy/Swedish Biofuels/ LanzaTech	Pilot	Lignocellulosic feed, ATJ PROCESS, Flight-proven	Low	Annex A5
Catalytic Hydrothermolysis (CHJ)	Chevron Lummus Global, Applied Research Associates, Blue Sun Energy		Vegetable oil, animal fat, recycled oils	Low	Annex A6

TABLE 4 R&D technologies for Bio-aviation fuel and their status—cont'd

Technology pathway	Technology provider	STATUS	Feedstock	Aromatics	ASTM status
Hydro-Deoxygenated Synthesized Kerosene HC-HEFA-SPK	Virent		Starch, sugar, cellulosic biomass	Low	Annex A7
Hydrothermal Liquefaction (ARA-CLG)	Applied Research Associates (ARA) and Chevron Lummus Global (CLG)	Demo	Lignocellulosic feed, HTL process, flight-proven	Low	
Hydro-Deoxygenated Synthesized Aromatic Kerosene HDO-SAK	Virent		Starch, sugar, cellulosic biomass	High	
Hydrotreated Depolymerized Cellulosic Jet HDCJ	Honeywell UOP, Licella, KiOR		Cellulosic biomass	High	

maximum allowable blend percentage with a 50% conventional ATF cut for HEFA-SPK, FT-SPK, ATJ-SPK, FT-SKA, and CHJ. For SIP and HC-HEFA-SPK, it is as low as 10% of the overall mix. This is because bio-jet typically contains low aromatic content, often below 0.5%. In comparison, many aircraft designs require aviation fuels of aromatic content in the range of 8–22% to attain critical physical properties for ASTM specifications (Standard Specification for Aviation Turbine Fuel Containing Synthesized Hydrocarbons, 2016b). The ability to add 50% of HEFA-SPK, SPK/A, or FT-SPK blending components (SPK) to Jet A or Jet A-1 is limited by the fuel's physical properties with which it is blended. The practice has shown that the density, or aromatic content, or both, of the refined fuel, often limits the amount of SPK that can be added to the final blend to less than 50% (Standard Specification for Aviation Turbine Fuel Containing Synthesized Hydrocarbons, 2016b). The ability to add 10% of SIP blending components to Jet A or Jet A-1 may be limited by the physical properties of the fuel with which it is blended. In extreme cases, the refined fuel's viscosity may limit the amount of SIP that can be added to the final blend to less than 10% (Standard Specification for Aviation Turbine Fuel Containing Synthesized Hydrocarbons, 2016b).

In the commercial distribution system, aromatics blending adds to another entry in the supply and distribution system, making supply chain management even more complex. From Table 4, hydro-deoxygenated synthesized aromatic kerosene (HDO-SAK), hydro-treated depolymerized cellulosic jet (HDCJ), Single-Step-HEFA-SPK, and Fischer–Tropsch

Synthetic aromatic kerosene (FT-SKA) are some of the processes producing aromatics in the fuel. Since aromatics in the renewable jet limit the blending percentage with Jet-A/Jet A1, a process that produces aromatics along with other components may completely replace the fossil aviation fuel.

2 Fuel standards for synthetic aviation fuel

2.1 Aviation turbine fuel containing synthesized hydrocarbons (ASTM D7566)

ASTM D 1655 specifications specify the conventional fossil-based aviation turbine fuel. The ASTM D7566 (Standard Specification for Aviation Turbine Fuel Containing Synthesized Hydrocarbons, 2016b) covers the manufacturing of aviation turbine fuel that consists of traditional fossil-based fuels and synthetic blending components. Table 4 details the R&D technologies, technology provider, status, feedstock, aromatics in their product, and their status in ASTM approval. All the processes currently developed fall under three major categories. They are oleochemical, thermochemical, and biochemical processes. Oleochemical processes include the hydroprocessing of lipid feedstocks obtained from lipids sources such as palm oil, tallow oil, and jatropha oil. The thermochemical process converts biomass to fluid intermediates, which are then catalytically upgraded to hydrocarbon fuels. In biochemical processes, biological sources such as starch, sugar, and lignocellulosic feedstocks are converted by biological processes to hydrocarbons and longer chain alcohols.

ASTM D4054 provides the steps to be followed in the certification process. The qualification process by ASTM is rigorous and time-consuming, and hence it takes a much longer time for new processes developed to reach the industry. Currently, there are only seven pathways for bio-aviation fuel production which ASTM and separate specialized standards approved are provided for each pathway in the annexure of ASTM D566. But a significant advantage of the ASTM D7566 standards is that they can be directly interlinked with ASTM D1655, which provides specifications for Jet-A and Jet-A1 grades from petroleum feedstocks. Thus BIO–ATF meeting ASTM D7566 specifications also fulfill the ASTM D1655 specifications and can be directly used as a replacement for conventional ATF.

The test properties for the characterization of fuel containing synthesized hydrocarbons are broadly classified into nine classifications: composition, volatility, fluidity, combustion, corrosion, thermal stability, hydrocarbon composition, non-hydrocarbon composition, and additives. Among these hydrocarbon compositions, non-hydrocarbon composition and additives were added explicitly for synthesized hydrocarbons and were previously not included in ASTM D 1655 turbine fuel classification. The hydrocarbon composition class details the amounts of cycloparaffins, aromatics, paraffin, and combined carbon and hydrogen mass percentages (Tables 3 and 4). The non-hydrocarbon composition class details the specific quantities of heterogeneous atoms such as nitrogen, oxygen, sulfur, metals, and halogens present in these synthesized hydrocarbons (Table 3). As the origin of these hydrocarbons is from an oxygen-containing renewable compound and the slightest amounts of oxygen in the final processed products may lead to oxidation reactions and gumming. Hence anti-oxidants are a must in these synthesized hydrocarbons.

Additives in another class included explicitly limiting the number of anti-oxidants present in these synthesized hydrocarbons (Table 3). There is a relaxation in terms of fluidity for batches of both FT-SPK and HEFA-SPK with freezing point specifications of $-40\,°C$ compared to $-47\,°C$ in Jet-A1. The temperature for thermal stability evaluation has also been increased to $325\,°C$ (Annex A1–A7) from $260\,°C$ (Jet-A1) so as to provide a recurring, batch-by-batch verification of process stability and compositional consistency for this synthesized paraffinic kerosene (Table 3). Also, in addition, a limit to fatty acid methyl esters (FAME) content in the HEFA-SPK has been set to $<5\,ppm$ (Table 3). In totality, the final blended turbine fuel needs to meet the specifications of Jet-A1 (Table 3) as mentioned in ASTM D7566 and ASTM D1655.

2.2 ASTM approval process ASTM D4054

For every new candidate fuel to be added as a new annexure in ASTM D7566, the ASTM has defined a specific procedure. The detailed procedure is provided in the ASTM D4054 user guide. The document provides guidance to alternative jet fuel producers for testing and evaluation. It is an iterative process where the fuel developer tests the subject fuel samples for their properties, composition, and performance. The tests are classified as Tier 1 and Tier 2 tests, which include basic specification properties, expanded properties known as fit-for-purpose (FFP) properties, engine rig tests, and component tests. Full-scale engine testing is also carried out if necessary. This is a rigorous mandatory process required for the approval of alternative fuels.

ASTM D4054 was created as a reference by engine and airplane OEMs with ASTM International members' help. It informs the manufacturer of an alternative fuel about what is needed in terms of required testing and OEM participation. Fig. 2 depicts an outline of the ASTM D4054 assessment and acceptance process. The fuel manufacturer can work with the leading organizations in the international aviation jet fuel group who generally promote the assessment and acceptance of new fuels. ASTM International, the Coordinating Research Council (CRC), and Commercial Aviation Alternative Fuels Initiative (CAAFI) are leading organizations promoting alternative aviation fuel.

A task force is organized within ASTM's Emerging Fuel Subcommittee to solicit stakeholder comments on testing and preparing an ASTM study paper. OEMs will be included in the task force. The manufacturer submits an ASTM study report to the ASTM committee to secure final approval for the new fuel. Engine and airplane OEMs, airworthiness associations, and international fuel specification requirements organizations will review the study report.

The D4054 requires that the specification properties of the subject fuel being tested are not affected by the process variability during large-scale production. The D4054 data are then used to develop a proposed annex for inclusion in D7566 as drop-in synthetic jet fuel. This D4054 user's guide details the list of facilities to conduct the D4054 aviation fuel property testing. It also provides a list of rigs and test facilities of aircraft and engine OEMs. OEMs are actively interested in researching new alternative jet fuels, with many test facilities at their laboratories. For updates on their research facilities' availability, one can contact the aircraft and engine suppliers directly.

FIG. 2 Overview of ASTM D 4054 fuel and additive approval process.

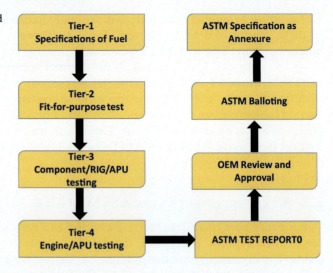

3 Fischer–Tropsch (FT) synthesis to renewable jet fuel (FT-SPK)

Fischer–Tropsch (FT) synthesis is an effective process for producing clean hydrocarbon fuels from syngas from different routes (Table 5). The carbonaceous material is converted into CO and H_2 by gasification by the thermochemical process. At an industrial scale, primarily autothermal gasification route is preferred. Biomass is reacted with a sub-stoichiometric quantity of oxygen at temperatures in the range 800–1200 °C. Many varieties of gasification technologies are available around the globe (Basu, 2013).

Long-chain paraffin is produced by Fischer–Tropsch (FT) using syngas with an H_2/CO ratio of ~2. The reaction occurs as follows: $CO + 2H_2 \rightarrow -[CH_2]- + H_2O$ ($\Delta Hrxn = 159\,MJ/kmol$). In FT synthesis, cobalt and iron-based catalysts are known to be effective (Kubickova and Kubicka, 2010; Gnanamani et al., 2013). Many consider Cobalt FT synthesis catalysts to have

TABLE 5 R&D technologies for Fisher–Tropsch based jet fuel.

Institution	Status
Shell	Commercial scale F.T. process
SASOL	Commercial scale F.T. process
Chevron	R&D & pilot
Conoco-Phillips	R&D & pilot
BP	R&D & pilot
Exxon	R&D & pilot
Solena in collaboration with Rentech	Plasma gasification coupled with FT
Syntroleum's bio-synfining technology	Air-force trial

advantages over iron-based catalysts, such as high per-pass conversion, long lifespan, and greater hydrocarbon selectivity. With only one product slate, FT synthesis is not a single technology. Many different combinations of catalysts, operating parameters, and reactor designs have been produced since the invention of FT synthesis to achieve demonstration-scale or commercial implementation. Despite the variety of technologies, only two specific forms of industrially generated FT synthesis products are viewed when FT synthesis is viewed from a product perspective. The first is a high-temperature Fischer–Tropsch (HTFT) synthesis product using Fe-based catalysts operating at a temperature of 320 °C and above. The second is the low-temperature Fischer–Tropsch (LTFT) synthesis product primarily using Fe- and Co-based catalysts. Operating temperatures for LTFT depend on the catalyst and reactor configuration and vary from about 170 to 270 °C in practice. It must be known that F-T synthesis can be manipulated to promote the refining of aviation turbine fuel supply. It is important to control the composition of the oil product from F-T synthesis (Kubickova and Kubicka, 2010; Gnanamani et al., 2013), unlike crude oil, where the natural resource determines the composition. To achieve commercial fuel specifications, the liquid mixture of hydrocarbons must be properly separated, and conversion steps are needed to transform molecules of high molecular weight into molecules of lower molecular weight of hydrocarbons or add other compounds. Fischer–Tropsch synthesis does not produce isomers and aromatic compounds.

4 Hydroprocessing of lipids (HEFA-SPK)

Hydroprocessed esters and fatty acids (HEFA), also referred to as hydrogenated vegetable oil (HVO) or hydroprocessed renewable jet (HRJ), covers aviation hydrocarbon fuel provided by the hydroprocessing of animal or vegetable oils (triglycerides). Lipids (triglycerides, diglycerides, monoglycerides), free fatty acids, and their derivatives hydroprocessed produce liquid hydrocarbons.

The ASTM D7566 specification is designed in its appendices to support different groups of alternative fuels, and HEFA has been licensed for use in blends with traditional kerosene in ASTM D7566 at up to 50% volume. In a two-step process, the product formed in the first step (hydrodeoxygenation step) is subsequently processed by cracked and isomerized to meet the required cold flow standards of aviation fuel. Although the triglyceride form is common to almost all oils and fats, the chain lengths and degree of unsaturation vary significantly. Also, it is required to remove all alkali metals and impurities before the process since the metals may cause catalyst deactivation and coke formation due to preferential adsorption on the catalyst surface (Sinha et al., 2013). Feed oils such as jatropha (Tiwari et al., 2011; Verma et al., 2011, 2015; Murata et al., 2010; Anand and Sinha, 2012; Sharma et al., 2012; Anand et al., 2014, 2016a, b; Gnanamani et al., 2013; Sinha et al., 2013; Kumar et al., 2010; Bezergianni and Kalogianni, 2009; Bezergianni et al., 2009b; Liu et al., 2009; Rana et al., 2013), soybean oil (Bezergianni et al., 2010b, 2011; Šimácek et al., 2009), sunflower oil (Simacek et al., 2011; Krár et al., 2010), palm oil (Cheng et al., 2014; Walendziewski et al., 2009; Vonortas et al., 2014; Sirifa et al., 2014), rapeseed oil (Šimácek et al., 2009), pomace oil, (Pinto et al., 2014), tall oil (Rozmyslowicz et al., 2010), algal oil (Kubicka et al., 2009; Zhang et al., 2010), castor oil, (Meller et al., 2014; Liu et al., 2015), and many more have been used to produce hydrocarbon fuel. These feedstocks have been hydroprocessed directly or co-processed with crude-based gas oils to produce hydrocarbons (Rana et al., 2013; Anand et al., 2016b; Batts and Fathoni, 1991; Krár et al., 2010; Vonortas et al., 2014; Sirifa et al., 2014; Singh et al., 2011).

Lipids generally contain oxygen and nitrogen. Hydroprocessing reactions remove oxygen over non-acidic support such as γ-Al_2O_3, activated carbon (Fan et al., 2014; Rozmyslowicz et al., 2010; Pinto et al., 2014; Kubicka et al., 2009) or acidic support such as zeolites, silica-aluminum, silica-aluminophosphates, titanosilicates, etc. (Tiwari et al., 2011; Verma et al., 2011, 2015; Murata et al., 2010; Anand and Sinha, 2012; Sharma et al., 2012; Anand et al., 2014, 2016a, b; Gnanamani et al., 2013; Sinha et al., 2013; Kumar et al., 2010; Bezergianni and Kalogianni, 2009; Bezergianni et al., 2009b; Liu et al., 2009; Rana et al., 2013). Acidic/non-acidic supports are provided with strong hydrogenation functionality provided by mono-metallic Pd, Pt, Ni, etc., or by bimetallic catalysts such as Pt-Re, NiW, Ni-Mo, Co-Mo catalysts. The hydrogenation feature is used to conduct hydrocracking, hydrogenation, and hydroisomerization reactions (Gnanamani et al., 2013; Sinha et al., 2013; Kumar et al., 2010; Bezergianni and Kalogianni, 2009; Bezergianni et al., 2009b; Liu et al., 2009; Anand et al., 2016a, b; Rana et al., 2013; Zhang et al., 2010). Catalyst material properties, such as the functionality of hydrogenation, acidity, porosity, surface area, hydrothermal stability, etc., can be tuned and surface morphology regulated to benefit a specific set of reactions and increase the performance of catalyst life. The reactions occurring during the process are (1) deoxygenation, (2) isomerization, (3) cracking, and (4) isomerization/aromatization in a single-step/multiple steps. The catalysts must handle the conditions inside the reactor caused by the formation of CO, H_2S, which inhibits the deoxygenation reaction. The product obtained in the process must contain normal/ isomerized paraffin, cyclic, and aromatics. The problem of high normal-paraffin content in the products leading to low cold flow properties also has to be addressed.

These hydroprocessing reactions require hydrogen as an input along with a lipid source; in fact, nearly 300–420 m^3 of H_2/m^3 of vegetable oil is needed to obtain desirable hydrocarbons (Cheng et al., 2014; Simacek et al., 2011; Bezergianni et al., 2009a, 2010a, b, 2011, 2009a, 2010a, b, 2011; Huber et al., 2007, 2006; Fan et al., 2014; Šimácek et al., 2009; Walendziewski et al., 2009; Meller et al., 2014; Liu et al., 2015, 2009; Rozmyslowicz et al., 2010; Verma et al., 2011, 2015; Murata et al., 2010; Anand and Sinha, 2012; Sharma et al., 2012; Anand et al., 2014, 2016a, b; Rivard et al., 2019; Pinto et al., 2014; Wyman, 1994; Li et al., 2010; Kim et al., 2014; Lynd et al., 1991; Huber and Corma, 2007; *Review of Biojet Fuel Conversion Technologies*, 2016; Basu, 2013; Kubickova and Kubicka, 2010; Gnanamani et al., 2013; Sinha et al., 2013; Kumar et al., 2010; Bezergianni and Kalogianni, 2009; Rana et al., 2013). Commercially Neste Oil's Next (Neste Oil Corporation, 2012) and UOP/Eni's EcofiningTM™ processes (www.eni.com, n.d.) are processing non-edible oils to produce biofuels in a two-step reaction comprising of hydro-deoxygenation and hydro-isomerization/selective cracking. CSIR-Indian Institute of Petroleum, a research laboratory in India, is producing bio-aviation fuels at the pilot-plant scale from non-edible oils such as jatropha, palm stearin, palm fatty acid distillates, used cooking oil, and Karanja, meeting the ASTM specifications by single-step catalytic process (Tiwari et al., 2011; Verma et al., 2011, 2015; Anand and Sinha, 2012; Anand et al., 2014, 2016a,b; Sinha et al., 2013; Kumar et al., 2010; Rana et al., 2013). Table 6 shows various catalyst hydroprocessing technologies available in the Indian and Global scenarios.

TABLE 6 R&D technologies for hydroprocessing.

Institution	Status
IIP, Dehradun	Pilot-scale testing (scale 50 kg/day feed processing)
IOCL, India/Haldor Topsoe, Denmark	Co-processing catalyst at refinery scale
UOP/Neste oil REG USA/Axens	Oils & fats hydroprocessing at refinery scale

TABLE 7 Heat of reactions for different reactions occurring during hydroprocessing of lipids.

Reaction $\Delta H_r^o = \sum \Delta H_{fProducts}^o - \sum \Delta H_{fReactants}^o$	ΔH_r^o (MJ/mole)
Depropanation $C_{57}H_{110}O_6(v) + 3H_2(g) \xrightarrow{\Delta Catalyst} C_3H_8(g) + 3C_{18}H_{36}O_2(v)$	−1.02
Decarboxylation $C_{18}H_{36}O_2(v) \xrightarrow{\Delta Catalyst} CO_2(g) + C_{17}H_{36}(v)$	+0.03
Decarboxylation $C_{18}H_{36}O_2(v) + H_2(g) \xrightarrow{\Delta Catalyst} CO(g) + H_2O(v) + C_{17}H_{36}(v)$	+0.07
Hydrodeoxygenation $C_{18}H_{36}O_2(v) + 3H_2(g) \xrightarrow{\Delta Catalyst} C_{18}H_{38}(v) + 2H_2O(v)$	−0.079
Hydrocracking $C_{18}H_{38} + H_2 \xrightarrow{\Delta} C_{12}H_{26} + C_6H_{14}$	−0.04
Hydrocracking $C_{17}H_{36} + H_2 \xrightarrow{\Delta} C_8H_{18} + C_9H_{20}$	−0.04

High hydrogen consumption and deactivation of catalysts due to coking are the main factors delaying lipids' commercial success hydroprocessing into bio-aviation fuels. Researchers have also investigated single-step processes for biofuel production with reduced hydrogen consumption (Anand and Sinha, 2012; Anand et al., 2014). The primary reactions involved during the lipids' hydroconversion are deprotonation ($-C_3H_8$) (Table 7). The lipid molecule glycerol linkage is hydrogenated, and propane is formed along with the corresponding acid/aldehyde molecule (Anand and Sinha, 2012; Anand et al., 2014; Verma et al., 2015). The oxygen removal in hydrocracking takes place by three methods—hydrodeoxygenation (HDO), Hydrodecarbonylation (HDCN), and hydro-decarboxylation (HDCX). Hydrogen consumption, product yield, catalyst inhibition, and heat balance are affected by the extent of these three mechanisms. While the HDO process consumes 16 mol of H_2, HDCX consumes only 7 mol of H_2. But the subsequent conversion of CO_2 to CO and to methanol may lead to the consumption of 35 mol of H_2 (Al-Sabawi and Chen, 2012). In HDCX, the HC yield is 94% of HDO. Hence conditions must be maintained such that HDO is favored. HDCN and HDCX mechanisms increase with increasing reaction temperature and decreasing pressure. After removing the oxygen molecule, the corresponding hydrocarbon is further cracked and isomerized into aviation range hydrocarbons (Table 7). Researchers focus on maximizing the decarboxylation pathway to minimize hydrogen consumption and reduce water formation (Anand and Sinha, 2012; Anand et al., 2014; Verma et al., 2015). Favoring the HDO pathway has some advantages, such as decreasing coking as it occurs in HDCX and HDCN mechanisms to a greater extent. Coking is a significant problem since it is the primary reason for reducing the catalyst life and activity in hydroprocessing units. Coking also causes plugging of the hydrocracking units' channels, causing wall effect and undesirable pressure drops in the system. Thus, bimetallic catalysts are more predominantly used as HDO is predominant in such catalysts.

5 Alcohol to aviation fuel (ATJ-SPK)

Alcohol-based jet (ATJ) fuel is produced from alcohol with the aid of a thermochemical route. Many companies (Table 8) use various alcohols and oxygenated intermediates to convert ethanol to jet fuel (Lynd et al., 1991; LanzaTech Bioethanol Platform, 2015). The pathways can be classified according to the chemistry involved: (1) production of aviation fuel by ethylene as an intermediate, (2) production by intermediate propylene, (3) production by intermediate higher alcohol, or (4) production by intermediate carbonyl. ATJ is developed to form jet fuel range hydrocarbons through alcohol dehydration/oligomerization, which involves linking short-chain alcohol molecules (e.g., methanol, ethanol, and others). The alcohol molecules lose water and oxygen, and hydrogen is added to the starting volume of alcohol is decreased to create a slightly more valuable hydrocarbon jet fuel (at current market prices). Methanol, ethanol, butanol, isopropanol, other alcohols, or a combination of them can be the alcohol intermediates. With the minimization of the operation steps of this process, a significant cost advantage can be achieved. In all cases, following the oligomerization process, the products are converted to primarily jet range paraffin components through a hydrotreating process (LanzaTech Bioethanol Platform, 2015).

6 Pyrolysis of biomass to aviation fuel

There are numerous technologies currently being investigated (Table 9) to convert renewable biomass feedstocks into liquid mixtures suitable for aviation; however, jet fuel's stringent

TABLE 8 R&D technologies for alcohols-to-jet fuels.

Institution	Status
INEOS	Two-step process; Oligomers from ethylene and alkylation/isomerization of oligomers to jet fuel; Demonstration phase
Chevron-Philips	Via methanol to jet fuel; R&D scale
LanzaTech	Ethanol to jet fuel
Gevo Inc.	In demonstration phase with Alaska airlines
NAWCWD and Cobalt Technologies	Dehydration of biobutane to 1-butene and oligomerization of butane to jet fuel

TABLE 9 R&D technologies for upgradation of fast pyrolysis oils.

Institution	Status
KiOR, USA	ASTM has not yet approved pyrolysis jet fuel
UOP LLC, PNNL, Ensyn, Tesoro	R&D demonstration stage of pyrolysis technology in integration with hydroconversion

international standards lead to a relatively narrow range of molecules. A significant research area that concentrates on producing liquid fuels from biomass is thermochemical conversion (Li et al., 2010; Kim et al., 2014; Basu, 2013; Kalnes et al., 2009). Bio-oil obtained by lignocellulosic biomass pyrolysis can be used as a fuel, such as gasoline, diesel, or jet fuel. Still, it needs to be upgraded by oxygen removal before widespread use can be found (Wei et al., 2017). A catalytic process is found to be the most powerful of the many oxygen removal techniques. Different kinds of catalysts, both mesoporous and microporous, have already been studied. It seems that a promising catalyst is a ZSM-5 catalyst. Bio-oil upgrades have also been studied using hydrodeoxygenation and catalytic cracking processes. The research challenge for deoxygenation is to develop novel catalysts with improved activity and selectivity and, in particular, better stability for deactivation. As co-processing with petroleum can be further upgraded, partially deoxygenated bio-oil is possibly the best option. The challenge is to design catalysts with less coke formation or use bio-oil with fewer phenolic components for the catalytic cracking process. Catalytic deoxygenation (HDO) processes include atmospheric catalytic rapid pyrolysis and high-pressure hydrodeoxygenation. The use of microporous zeolites (HZSM 5, HY, HBeta, etc.) and mesoporous materials (Al-SBA-15, Al-MCM-48, Al-MCM-41, etc.) requires catalytic fast pyrolysis (Radich, 2015).

7 Algae to jet fuel

Biofuels of algae may provide a viable alternative to fossil fuels, but this technology must overcome several obstacles before competing in the fuel market and be widely deployed (Singh et al., 2011; Stephens et al., 2010; Mascarelli, 2009). Further research and development are required before microalgae technologies can be used for large-scale, cost-effective, energy-efficient fuels, and chemicals (Mascarelli, 2009). All algae can generate oils that are rich in energy, and several species of microalgae have been found to accumulate high oil levels in total dry biomass naturally. Optimizing algae growth in open ponds is a key component of achieving economic sustainability and remains a major industry challenge. In several laboratories, finding species that grow well under these conditions is a focal point of the study. In a wide range of temperatures, algae can expand, with growth being restricted mainly by the availability of nutrients and light. Light provides the energy for carbon fixation and is transformed by photosynthesis into chemical energy. Providing the building blocks for biofuel production—pathways for producing bio-aviation fuel share many standard features regardless of the biomass feedstock being used. Oil-producing crops, such as soybean, jatropha, and camelina, are harvested, and the oil is separated for subsequent processing. It is the same pathway to produce biofuel using microalgae. Sulfide hydrocracking (such as Ni-W/SiO$_2$-Al$_2$O$_3$) and hydrotreatment (such as Ni-Mo/Al$_2$O$_3$, Co-Mo/Al$_2$O$_3$) catalysts are widely used for the deoxygenation of triglyceride and fatty acid molecules into pure hydrocarbons (Tiwari et al., 2011; Verma et al., 2011, 2015; Murata et al., 2010; Anand and Sinha, 2012; Sharma et al., 2012; Anand et al., 2014, 2016a,b; Gnanamani et al., 2013; Sinha et al., 2013; Kumar et al., 2010; Bezergianni and Kalogianni, 2009; Bezergianni et al., 2009b; Liu et al., 2009; Rana et al., 2013). Fuels with the necessary viscosity, low oxygen content, better cold flow properties, and lubricity would result in such a process. There are many benefits to the production of algal lipids by hydrotreating. Algae are rich in oil content than other crops, easier to crack, hence giving higher aviation range hydrocarbons during its conversion (Mawhood et al., 2016).

As the emphasis moves from the development of alcohols and esters and toward the production of 'drop-in' hydrocarbon fuels, thermochemical pathways for the conversion of biomass to fuels have garnered interest. A benefit of thermochemical technologies is that they are primarily agnostic feedstock and can tolerate any kind of biomass, including aquatic microalgal and macroalgal species biomass. The production of algal biomass for biofuels production is highly promising because algae produce higher energy yields and need far less space to grow than conventional feedstocks. The production of algal biomass does not require fertile or arable land. Green Fuel Technologies Company named algae the fastest growing plant in the world so that algae will not compete with food and could be developed with limited inputs using a variety of nutrient and carbon sources (Stephens et al., 2010). Fast growth rates, substantial growth densities, and high oil content were all cited as to why significant capital was invested in transforming algae into biofuels. However, there is a range of hurdles to overcome for algae to mature as an economically viable platform to offset petroleum and, ultimately, mitigate CO2 release, ranging from how and where to cultivate these algae to improve oil extraction fuel processing (Kalnes et al., 2009). Strain isolation, nutrient sourcing and utilization, production management, harvesting, co-product growth, fuel extraction, refining, and residual biomass utilization are significant challenges. Of course, liquid algae fuels are technically feasible but costly compared to petroleum fuels. The pretreatment (dewatering) cost for algae is energy-intensive and hence less economical (https://www.nedo.go.jp/content/100890890.pdf, n.d.). The vulnerability of petroleum prices to significant and volatile fluctuations is a significant impediment to investment in fuel-from-algae technologies (Stephens et al., 2010; Mascarelli, 2009).

8 Biomass conversion pathway to renewable jet through biotechnology platform

A sustainable approach creates alternative energy sources that emit fewer greenhouse gases and partner oil and gas to meet global energy demand. Biomass accounts for about 10% of the world's energy consumption and is primarily used for heating and cooking at the moment. For biomass conversion processes, resourcing feedstock is a significant challenge. Sustainability pathways will be produced through lignocellulose, a non-edible component of the plant, and non-edible plant-like jatropha. Biofuels include charcoal, corn-derived ethanol (maize), methane-rich biogas, non-edible plants such as jatropha, wood, and straw.

Biomass conversion through the biochemical route induces micro-organisms (yeast and bacteria) to convert biomass into fuel and chemicals (www.eni.com, n.d.; LanzaTech Bioethanol Platform, 2015; Holladay, 2012). While many other countries have vast biomass resources, they do not make the most of their resources. For example, millions of tons of straw are still burned in fields every year after harvesting in Australia and India; this 'waste' could generate electricity. Biomass conversion pathways have an industrial and technological challenge that must balance technical efficiency, environmental performance, social acceptability, and economic viability (https://www.statista.com/statistics/655057/fuel-consumption-of-airlines-worldwide, n.d.; Sustainable Aviation CO2 Road-MAP, 2012; Chuck, 2016). The aviation industry has identified biofuels' development as one of the significant ways to reduce its greenhouse gas emissions and has set a goal of halving its greenhouse gas emissions by 2050 (Chuck, 2016). Many commercial aviation industries such as Total and Amyris have developed breakthrough aviation fuel blend with up to 10% firesafe, leading to a meaningful reduction of greenhouse gas emissions. Farnesane is produced by yeast by fermenting sugars (SteveLicht, 2014; Wall and Rios-Solis, 2020; http://www.biodieselmagazine.com/articles/105324/total-amyris-biojet-fuel-ready-for-use-in-commercial-aviation, n.d.). Goal genes are selected to alter the yeast's metabolism, transforming the yeast into a hydrocarbon-producing organism from an ethanol-producing organism. The Amyris technology allows selected molecules to be generated at high purity levels (SteveLicht, 2014; Wall and Rios-Solis, 2020). Table 10 shows the production yields from various pathways with specifications in terms of hydrocarbon type. Through biochemical pathways using algal biomass as feedstock, depending on the processing conditions, jet fuel yield is 3 times more than that from sugar fermentation, approximately 1.5 times more than FT synthesis and alcohol to the jet route.

TABLE 10 Production yields from various pathways with specifications discussed in this chapter.

	Pathways	Biomass	Jet fuel yield gal/BDT	Hydrocarbon type analysis Aromatics vol%	paraffins mass%
1	Biochemical (Stephens et al., 2010; Mascarelli, 2009)	Microalgae biomass	8–122	Report	Report
2	Alcohol-to-jet (LanzaTech Bioethanol Platform, 2015)	Corn, wood straw	11–79	0.01	97.45
3	Sugar fermentation (SteveLicht, 2014; http://www.biodieselmagazine.com/articles/105324/total-amyris-biojet-fuel-ready-for-use-in-commercial-aviation, n.d.)	Sugarcane, wood, straw	24–43	2.6	40
4	FT (Kubickova and Kubicka, 2010; Gnanamani et al., 2013)	Biomass coal and wood	9–88	<1	<70
5	Pyrolysis (Basu, 2013; Gnanamani et al., 2013)	Corn Stover, wood, biomass	19	report	report
6	Hydroprocessing (Sinha et al., 2013)	Animal and plant lipids	30–50	>8	92

Thus, the biochemical pathway though still in infancy but is the most promising future route to produce aviation biofuel. Pyrolysis route provides the least amount of jet yield among the different discussed routes in Table 3. But based on the availability of feed, FT synthesis is the most promising pathway for the production of aviation jet fuel. The production yield of jet fuel via FT synthesis is approximate 2 times more than that of the sugar fermentation pathway and almost the same as the alcohol-to-jet route. But paraffinic composition is less through the FT route (<70%) compared to the alcohol-to-jet route (97%). Based on biomass, technology, food security, and production yield, FT Synthesis and hydroprocessing routes are attractive routes in producing renewable Jet fuel.

9 Power to liquid (PtL)

9.1 CO_2 conversion technologies

CO_2 is a thermodynamically stable and chemically inert molecule. The conversion of CO_2 to hydrocarbon via hydrogenation usually favors short-chain formation rather than desirable long-chain hydrocarbons. Hence the research in this area is focused on light hydrocarbon production rather than long-chain hydrocarbons. Most of the available literature deals with selective hydrogenation of CO_2 into methane, methanol, formic acid, light olefins, or other lighter oxygenates (XiaowaNi et al., 2019; Zhong et al., 2019; Marlin et al., 2018; Fegade and Jethave, 2020). Long-chain hydrocarbon from CO2 conversion is very limited (Wei et al., 2017). There are basically two pathways, to convert CO_2 to hydrocarbon molecule (a) an indirect route in which CO2 is converted to CO or methanol and subsequently into liquid hydrocarbons and (b) Direct CO_2 hydrogenation route, in which CO_2 is converted to CO via the reverse water gas shift (RWGS) reaction and then hydrogenation of CO to long-chain paraffin via Fischer–Tropsch synthesis (FTS) (Kubickova and Kubicka, 2010; Gnanamani et al., 2013). The direct route involves fewer steps and is expected to be more economically viable.

The reverse water gas shift reaction:

$$CO_2 + H_2 \rightarrow CO + H_2O \ (\Delta H°298 = +41\,Jk/mol)$$

The Fischer–Tropsch synthesis reaction:

$$CO_2 + 2H_2 \rightarrow -(CH_2) - + H_2O \ \ (\Delta H°298 = +166\,Jk/mol)$$

Recently a group of researchers from Oxford University has converted carbon dioxide (CO_2) to aviation fuel via the catalytic process. Targeting CO_2 as feedstock will for sure have a significant impact on global greenhouse gas reduction. This process could be a game-changer if implemented on a commercial scale. Yao et al. (2020) have used Fe-Mn-K (iron-manganese-potassium) catalyst by the organic combustion method (OCM). The reported conversion for CO_2 was 38.2%, with a yield of 17.2% for jet fuel and a selectivity of 47.8%. Selectivity toward carbon monoxide (5.6%) and methane selectivity (10.4%) was reported. Light olefins ethylene, propylene, and butenes, totaling a yield of 8.7%, are formed as a by-product, which is the critical feedstock to the petrochemical industry. The two significant challenges in making the process economical are (a) the process of atmospheric carbon capture and (b) the synthesis of hydrocarbons by CO_2 hydrogenation typically favors the formation of short-chain rather than the desirable long-chain needed for aviation fuel synthesis. This new method is a significant social advance that highlights the recycling of CO_2 and the conservation of resources as a vital, pivotal component of greenhouse gas management and sustainable growth. This catalytic process is supposed to be the path to the aviation industry's near-future net-zero carbon emissions, that is, before we as a society are entirely prepared to operate on eco-friendly electric aircraft. In this process, carbon dioxide extracted from the air is used for conversion and, when combusted in flight, is later re-emitted from jet fuels. Consequently, a carbon-neutral fuel is the ultimate result of this process.

9.2 Power to liquid (PtL)

The main constituents of PtL are electricity, water, and carbon dioxide (CO_2) (Schmidt et al., n.d.). PtL consists of hydrogen production by electrolysis using solar power, followed by CO_2 and H_2 combination to hydrocarbon fuel. There are two different pathways for the production of PtL: the Fischer–Tropsch (FT) pathway and the methanol pathway, as shown in Fig. 3. Currently, only a few small-scale PtL plants in operation produce liquid hydrocarbon products such as kerosene. Water electrolysis, hydrogen storage tanks, FT or methanol synthesis, and refining processes are the relevant process stages in PtL supply chains. Only the reverse water gas shift reaction (RWGS) reactor has yet to be demonstrated at the relevant scale. The cost of renewable energy is the primary driver of overall PtL production costs. The heat produced by the exothermic reactions in FT synthesis/methanol conversion can be used to supply heat to the temperature swing adsorption plant for the CO_2 capture cell and the heat demand of the solid oxide electrolysis plant in case of high-temperature electrolysis. Large-scale implementation's primary requirement is a continuous cost reduction of renewable hydrogen production from water electrolysis powered by solar and wind energy (Schmidt et al., n.d.).

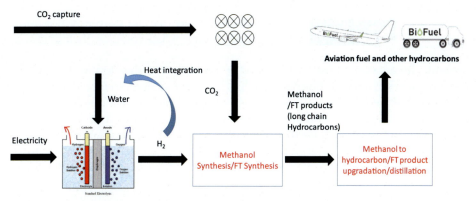

FIG. 3   Power to liquid (PtL) pathways (Schmidt et al., n.d.).

10 Techno-commercial analysis of various technologies available for bio-jet fuel production

The primary difficulties associated with the commercialization of any bio-jet technology are feedstock cost and availability, apart from the feed cost and availability, refinery infrastructure, logistics, supply chain development, higher operational costs, difficulties in the certification of technologies, lack of sufficient subsidies, and investments to meet global targets, and lack of international level playing field for investment by the aviation sector.

Hydroprocessing is a mandatory step in HEFA-based pathways (HEFA-SPK, FT-SPK). HEFA pathways require hydrogen to upgrade oxygen-rich lignin, lipids, and carbohydrates to hydrogen-rich compounds. There are also non-hydrogen processes available such as catalytic or thermal cracking. They remove carbon as coke and tar. But these processes consume feedstock and decrease the overall yield of the process. However, they can be favorable alternatives if the loss due to cracking processes is much lesser than the cost requirement for H_2 production.

Logistics and supply chain development is complex as current products use only bio-jet as blends with conventional ATF. Since they become indistinguishable, current infrastructure owners must agree with the new players in sharing the infrastructure. Otherwise, separating the components until the delivery points, i.e., the airports, will complicate the process and compromise product quality.

HEFA—diesel produced during bio-jet production has more market demand and does not require any additional blending or rigorous specifications. Hence most hydroprocessing plants currently in operation concentrate more on producing HEFA-diesel. This challenge for bio-jet to compete with HEFA-diesel reduces the net output of bio-jet from the existing plants.

The cost of producing bio-jet is not comparable with that of conventional jet fuels. Wang et al. (2016) reviewed several studies and normalized the values based on 2013 and an assumed output of 500 tons of fuel per day (Tables 1 and 2). The cost of bio-jet fuels is about two to seven times that of fossil-derived jet fuel as of 2015, according to IATA (https://www.iatag.org/facts-figures.html, n.d.). However, in the United States, it is possible to considerably reduce the cost through government subsidies that are currently not possible in other countries. Tables 1, 2, and 11 provides a summary of technologies, their status, and estimated capital cost.

At present, the aviation sector is expected to boom in the developing regions of Asia, Africa, South America, and the Middle East. These countries need to focus on the abundant cheap oleochemical feedstocks available in their regions, such as using waste coconut oil in the Philippines via the HEFA pathway to meet the demands in the short term.

Cost estimates for bio-jet fuels are difficult to achieve due to less availability of data. Hence many assumptions are being made in the cost calculations. Although HEFA is an industrially mature technology, its economic feasibility is still a question since; historically, vegetable oil rates were consistently higher than fossil fuels as well as jet fuels themselves. For example, in 2016, the conventional jet fuel cost was about 400 USD per ton. At the same time, crude palm oil was at 727 USD per ton. Similar conditions apply to most vegetable oils where non-edible

TABLE 11 Drivers for R&D collaboration for the production of bio-jet fuels.

Drivers for R&D	Status/Requirements
Hydroprocessed vegetable oil	
Abundant oils at low costs are required	Identification of cultivable arable land for cultivation on a crop rotation basis for optimized land utilization Creation of value chains for biomass supply and the establishment of collection networks
Extraction and pretreatment of oil	Technology and investments for establishing contaminants removal technology
Valorization of different types of feed	Standardization of feedstock
Hydrogen is expensive and not available everywhere	Cheaper/more abundant hydrogen source
Alcohol-to-jet fuel	
Technology for increased selectivity and efficiency for improving product yield	Improvement in technology required for selective catalysis, which would convert the alcohols more efficiently to jet fuels, more investment
Complexity and economic viability	Requires simpler technology with reduced operation steps and complications for improved economics
Fischer–Tropsch fuel	
Technological improvements for efficient catalysis and operational convenience	Better catalysts are needed, at present too expensive. Requires efficient high-pressure equipment/ Reduction in pressure
Abundant Biomass availability for reducing costs	Requires more concentrated/denser feedstock and optimization of costs
Bio-oil upgrading	
Technological improvements with cheaper catalysts	Expensive catalysts used/improvement in catalyst activity and life
Efficient hydrogen management	Molecular hydrogen uses and requires hydrogen transfer from cheap sources

vegetable oils such as jatropha oil will significantly cut down the cost as they do not demand other industries such as food and cosmetics. But at present, the commercial production of such oils is deficient, and hence their costs are high.

de Jong et al. (2015) in 2015, the minimum fuel selling price (MFSP) was calculated for bio-jet via different conversion routes (Table 7). But his estimates are based on modeling studies since no actual data was available. Lignocellulosic feedstocks were chosen for the modeling purpose except for HEFA waste cooking oil is used. The feedstock rates were fixed at USD 106 per ton for forest residues and USD 190 per ton for wheat straw. Since the MFSP was calculated based on European costs for feedstocks, the estimated values may vary depending on the type of feedstock and the region's local geography and economy. Table 1 shows the MFSP of bio-jet fuel based on technical and economic analysis.

The MFSP estimated is very high when compared to that of conventional bio-jet fuels. Hence more focus is needed on reducing the feedstock costs and capital investments as well as government subsidies to bring down the rates to economically feasible levels. In this regard, IEA bioenergy put forward a set of recommendations for steps to be taken to meet future targets.

Targets put forward for R&D by IEA bioenergy to increase the commercial viability of drop-in bio-jet are as follows:

- The advancement of commercially viable processes for the production of renewable hydrogen of industrial-grade.
- Establish more cost-effective and area-efficient systems for the development of oleochemical feedstocks based on land or water.
- Improve the cost, performance, lifetime, and recyclability of oxygenated biomass feedstock hydroprocessing catalysts.
- Creating commercially feasible processes of small-scale gasification and syngas cleanup.
- Produce intermediates for biofuels that are miscible and can be better co-processed with petroleum feed.
- Establish the possible synergies of bio-oil and fossil liquids co-processing in thermochemical processes and established oil refineries.

For a better economy of the process, the use of the side-products plays crucial importance. Aviation fuel range hydrocarbon, which is middle-range boiling range components as defined in Section 1.2. Since most of the available pathways, the final stage of the production chain involves distillation. The other component produced is mostly lighter hydrocarbons, i.e., LPG and gasoline, and heavier fractions, i.e., diesel. All the processes defined in Table 3, except hydro-deoxygenated synthesized aromatic kerosene (HDO-SAK), hydrotreated depolymerized cellulosic jet (HDCJ), Single-Step-HEFA-SPK, and Fischer–Tropsch Synthetic aromatic kerosene (FT-SKA), have low aromatics. Hence, they will produce the paraffinic component in gasoline/diesel components along with aviation range hydrocarbons. HEFA-SPK and FT-SPK are expected to produce a significant yield of diesel and FT waxes. The high demand for high cetane (low sulfur) diesel and low sulfur waxes makes these two processes more viable than other processes (Tables 1 and 2). Similarly, higher alcohols and platform chemicals produced from the ATJ process and green aromatics from HTL make these two processes attractive and potentially economical.

Among all the research pathways, FT-SPK, and HEFA-SPK are the only ones undergoing commercial production processes. Both the processes utilize fossil fuel-derived natural gas in their processes. In HEFA, natural gas is the potential and cheaper source for hydrogen production. Most FT plants utilize natural gas as their source for syngas. Although FT-SPK was certified earlier than HEFA, it is not commercialized more than HEFA. This may be attributed to difficulties in syngas cleanup, catalyst contamination, and economies of scale. The number of steps involved in the FT pathway is more than all other available alternative aviation fuel production processes. This subsequently increases the capital and operating costs required for FT-SPK production. An FT plant with a capacity of 200,000 tpa will require the same investment as a HEFA plant with four times the same capacity (Biofuels for Aviation Technology Brief, 2017). By 2016 total HEFA production capacity in the world was 4.3 billion liters/year (Biofuels for Aviation Technology Brief, 2017). There are two leading HEFA technologies, Neste's NEXBTL and UOP, and ENI's Ecofining™. CSIR-IIP has recently patented its single-step HEFA process for bio-aviation fuel production. Companies producing renewable jet fuel on a commercial scale are provided in Tables 4–6.

Fuel readiness level (FRL) indicates the maturity level of technology for being introduced into commercial production. Mawhood et al. (2016) categorized all the available pathways for renewable fuel production under these criteria. HEFA has the highest rating with FRL9, followed by FT with FRL between 7 and 8 and SIP with a range FRL5 to FRL7. All other pathways fall in the range of FRL4 to FRL7. As much as these methods show the maturity or readiness of these pathways, they do not indicate the processes' commercial viability. Vegetable oil-based HEFA bio-jet was concluded as the only economically viable option in the future in a recent report by France's Académie de I 'air et de l'espace.

11 An alternative to liquid aviation fuel

A recent article published in Forbes in September 2019 by B. Morgan (*How Far Are We from Flying Zero-Emission Airplanes?*, n.d.) discussed that electric planes are ideal for trips less than 1000 miles. In 2018, the two-seater electric plane (*E-Genius*) traveled at 20,000 ft. in the French Alps, and it reached a maximum speed of 140 miles per hour for 300 miles in a single battery charge. Engineers designed the electric plane at the University of Stuttgart. It has no emissions and costs only $3 in energy to fly, and it released just a fifth of the energy that a traditional two-seater fuel-powered plane would use to fly the same distance. Aviation giants like Boeing, Airbus, and Raytheon are also developing such airplanes. Boeing is in the development of hybrid planes, which use both electricity and fuel. Boeing is working with NASA to deliver electric planes by 2040. Airbus is building E-Fan X, a battery-powered plane expected to take its first flight in 2021.

Experts believe that electric planes could be a reality within the next 20 years. The only drawback with these planes is they will be small planes for approximately 100 passengers. Battery power is a significant challenge in their development. The electric aircraft industry is projected to reach more than $22 billion by 2035. Regulations and approval of these electric planes will take time and may have an administrative hurdle in their acceptance.

12 Conclusions

This chapter describes the progress and trends in renewable jet fuel production via various pathways. Feedstock availability remains a significant challenge for sustainable renewable jet fuel production. We have expressed different possible commercial routes for the production of renewable jet fuel, challenges related to catalytic requirements and technology implementation, as well as concerns from a sustainability point of view. The oleaginous biomass and lignocellulose biomass are processed via thermal or catalytic or combination of these two routes to produce fuel. In comparison, the sugars and starch biomass are converted via enzymatic or catalytic or various these two routes. The hydroprocessing route to producing alternative biofuels has become a well-established technology, though economically not yet cost-competitive due to animal and plant-derived lipids' cost. Inherently, hydroprocessing of lipids requires hydrogen (Radich, 2015; Biofuels for Aviation Technology Brief, 2017; *Review of Biojet Fuel Conversion Technologies*, 2016; de Jong et al., 2015).

Moreover, despite various catalysis developments and processes, specifically for converting lipids into fuels, widespread production and use of renewable aviation fuels are yet to be realized. The molecular approach (predominantly using genetically modified micro-organisms) would identify and maximize the yield of the molecules of interest and identify the molecular properties required to meet the main performance characteristics and specifications of renewable jet fuels. Renewable isoparaffinic fuel (hydroprocessing and FT route), renewable farnesane hydrocarbon, and alcohol-to-jet fuel are some of the options available to produce aviation fuel and are included in the approved ASTM standard.

13 Future outlook

Feedstock availability is a significant challenge for biomass conversion processes to produce aviation biofuels economically. Biotechnological interventions in increasing the yield of feeds such as lipids and other molecules of interest appear to be the most promising way to solve the feedstock availability issues. Biotechnological interventions have continued to improve the yields of non-edible plant oils such as jatropha, thus facilitating the pathway to sustainability. Similarly, biotechnological interventions can improve the yields and production of sugars and starch from biomass and their further conversion via the enzymatic route. Biotechnologically it may be possible to maximize the yield of the molecules of interest (predominantly using genetically modified micro-organisms) to produce chemicals that meet the specification of renewable jet fuels. The cultivation of algal biomass for biofuels production has excellent promise due to higher energy yield and low space requirement to grow than conventional feedstocks. It could be developed with minimal inputs using a variety of nutrient and carbon sources. However, for algae to mature as an economically viable platform, the significant challenges involved include strain isolation, nutrient sourcing and utilization, production management, harvesting, co-product development, fuel extraction, refining, and residual biomass utilization. Genetic engineering may play a key role in enhancing the algal lipid yield and hence competitive aviation biofuel production. CSIR-IIP has developed an indigenous oleaginous yeast (Rhodotorulamucilaginosa, IIPL32), a soil isolate, and an in-house strain for lipid production. Oleaginous micro-organisms are micro-organisms that can accumulate lipid to more than 20% of their dry weight. In this process, the xylose-rich sugar hydrolysate stream was recovered using mechanical-hydrothermal pretreatment and used as a feed-in fermentation and yeast lipid development. The process developed by CSIR-IIP has high cell biomass efficiency and lipid accumulation. The process is scalable for lipid production (Banerjee et al., 2020).

Sustainable aviation fuels (SAF) are considered a significant measure of the International Civil Aviation Organization (ICAO) to reduce aviation emissions, which also includes the Carbon Offsetting and Reduction Scheme for International Aviation (CORSIA). SAF needs to be developed and deployed in an economically feasible and socially, and environmentally acceptable manner. Approaches to assess the sustainability of all alternative fuels in aviation

TABLE 12 Domestic carbon source (India).

Commodity	Scope (MMT/yr)	%C, approx.	Potential C (MT/yr)
Agri-residue (surplus)	120	40%	48
Forest residue	150	42%	63
Biogas excl. landfill	800	45%	360
MSW	40	25%	10
Used cooking oil (UCO)	5	85%	4

should: achieve net GHG emissions reduction on a life cycle basis; while maintaining biodiversity conservation and benefits for people from ecosystems, following international and national regulations; and contribute to local social and economic development, and avoid competition with food and water. The sustainability of alternative fuels for aviation must be ensured, by monitoring, at a national level, the existing approaches or combination of approaches. Solid waste, agriculture waste, wood chips, forest residue, corn Stover, miscanthus, and switchgrass are some of the alternative feedstocks (Table 12) that do not compete with food production and provide effective biological waste disposal. The same feedstock would also be suitable for the syngas route to aviation biofuels.

References

Air Transport Action Group, 2014. Facts & Figures. www.atag.org/facts-and-figures.html.

Al-Sabawi, M., Chen, J., 2012. Hydroprocessing of biomass-derived oils and their blends with petroleum feedstocks: a review. Energy Fuels 26 (9), 5373–5399.

Anand, M., Sinha, A.K., 2012. Temperature-dependent reaction pathways for the anomalous hydrocracking of triglycerides in the presence of sulfided co-Mo-catalyst. Bioresour. Technol. 126, 148–155.

Anand, M., Sibi, M.G., Verma, D., Sinha, A.K., 2014. Anomalous hydrocracking of triglycerides over CoMo-catalyst–influence of reaction intermediates. J. Chem. Sci. 126, 473–480.

Anand, M., Farooqui, S.A., Kumar, R., Joshi, R., Kumar, R., Sibi, M.G., Singh, H., Sinha, A.K., 2016a. Kinetics, thermodynamics and mechanisms for hydroprocessing of renewable oils. Appl. Catal. A Gen. 516, 144–152.

Anand, M., Farooqui, S.A., Kumar, R., Joshi, A.R., Kumar, R., Sibi, M.G., Singh, H., Sinha, A.K., 2016b. Optimizing renewable oil hydrocracking conditions for aviation bio-kerosene production. Fuel Process. Technol. 151, 50–58.

Banerjee, A., Sharma, T., Nautiyal, A.K., Dasgupta, D., Hazra, S., Bhaskar, T., 2020. Debashish Ghosh; scale-up strategy for yeast single cell oil production for Rhodotorula mucilagenosa IIPL32 from corn cob derived pentosane. Bioresour. Technol. 309, 123329.

Basu, P., 2013. Biomass Gasification, Pyrolysis, and Torrefaction, second ed. Academic Press, London, United Kingdom.

Batts, B.D., Fathoni, A.Z., 1991. A literature review a fuel stability studies with particular biphasic on diesel oil. Energy Fuel 5, 2–21.

Bezergianni, S., Kalogianni, A., 2009. Catalytic hydro-cracking of bio-oil to bio-fuel. Bioresour. Technol. 100, 3927–3932.

Bezergianni, S., Kalogianni, A., Vasalos, I.A., 2009a. Hydrocracking of vacuum gas oil-vegetable oil mixtures for biofuels production original. Bioresour. Technol. 100, 3036–3042.

Bezergianni, S., Voutetakis, S., Kalogianni, A., 2009b. Catalytic hydrocracking of fresh and used cooking oil. Ind. Eng. Chem. Res. 48, 8402–8406.

Bezergianni, S., Dimitriadis, A., Kalogianni, A., Pilavachi, P.A., 2010a. Hydrotreating of waste cooking oil for biodiesel production. Part I: effect of temperature on product yields and heteroatom removal. Bioresour. Technol. 101, 6651–6656.

Bezergianni, S., Dimitriadis, A., Sfetsas, T., Kalogianni, A., 2010b. Hydrotreating of waste cooking oil for biodiesel production. Part II: effect of temperature on hydrocarbon composition. Bioresour. Technol. 101, 7658–7660.

Bezergianni, S., Dimitriadis, A., Kalogianni, A., Knudsen, K.G., 2011. Toward hydrotreating of waste cooking oil for biodiesel production. Effect of pressure, H_2/oil ratio, and liquid hourly space velocity. Ind. Eng. Chem. Res. 50, 3874–3879.

Biofuels for Aviation Technology Brief, January 2017. www.irena.org.

Cheng, J., Li, T., Huang, R., Zhou, J., Cen, K., 2014. Jet fuel production from palm oil by catalytic hydrocracking. Bioresour. Technol. 158, 378–382.

Chuck, C.J., 2016. Biofuels for Aviation. Elsevier book, UK.

de Jong, S., Hoefnagels, R., Slade, A.F.R., Mawhood, R., Junginger, M., 2015. The feasibility of short-term production strategies for renewable jet fuels—a comprehensive techno-economic comparison. Biofpr 9 (6), 778–800. https://doi.org/10.1002/bbb.1613.

Fan, K., Liu, J., Yang, X., Rong, L., 2014. Hydrocracking of Jatropha oil over Ni-H3PW12O40/nano-hydroxyapatite catalyst. Int. J. Hydrog. Energy 39, 3690–3697.

Fegade, U., Jethave, G., 2020. Conversion of carbon dioxide into formic acid. In: Inamuddin, A.A., Lichtfouse, E. (Eds.), Conversion of Carbon Dioxide Into Hydrocarbons Vol. 2 Technology. Environmental Chemistry for a Sustainable World. vol. 41. Springer, Cham, https://doi.org/10.1007/978-3-030-28638-5_4.

Girardet, H., Mendonça, M., 2009. A Renewable World: Energy, Ecology, Equality—A Report for the World Future Council. Green Books Ltd, UK.

Gnanamani, M.K., Jacobs, G., Shafer, W.D., Davis, B.H., 2013. Fischer–Tropsch synthesis: activity of metallic phases of cobalt supported on silica. Catal. Today 215, 13–17.

Holladay, J., 2012. Workshop on Advanced Bio-based Jet Fuel Cost of Production. US Department of Energy, Energy Efficiency and Renewable Energy. http://www1.eere.energy.gov/bioenergy/pdfs/holladay_caafi_workshop.pdf (Accessed August 2015).

How Far Are We from Flying Zero-Emission Airplanes? https://www.forbes.com/.

Anon. http://www.biodieselmagazine.com/articles/105324/total-amyris-biojet-fuel-ready-for-use-in-commercial-aviation.

http://www.caafi.org/resources/pdf/CQ_Breakout_Session.pdf.

https://www.iatag.org/facts-figures.html.

https://www.icao.int/environmentalprotection/CORSIA/Documents/CORSIA_States_for_Chapter3_State_Pairs_Jul2020.pdf.

https://www.icao.int/environmental-protection/pages/a39_corsia_faq2.aspx.

https://www.nedo.go.jp/content/100890890.pdf.

https://www.statista.com/statistics/655057/fuel-consumption-of-airlines-worldwide.

Huber, G.W., Corma, A., 2007. Synergies between bio- and oil refineries for the production of fuels from biomass. Angew. Chem. Int. Ed. 47, 7184–7201.

Huber, G.W., Iborra, S., Corma, A., 2006. Synthesis of transportation fuels from biomass: chemistry, catalysts, and engineering. Chem. Rev. 106, 4044.

Huber, G.W., Connor, P.O., Corma, A., 2007. Processing biomass in conventional oil refineries: production of high-quality diesel by hydrotreating vegetable oils in heavy vacuum oil mixtures. Appl. Catal. A Gen. 329, 120–129.

IPCC, 2014. The Intergovernmental Panel on Climate Change's (IPCC) Fifth Assessment Report (AR5).

Kalnes, T.N., Koers, K.P., Marker, T., Shonnard, D.R., 2009. Characterization of biochar from fast pyrolysis and gasification systems. Environ. Prog. Sustain. Energy 28, 111–120.

Kim, S.K., Han, J.Y., Lee, H., Yuma, T., Kim, Y., Kim, J., 2014. Effect of the hydrothermal pre-treatment for the reduction of NO emission from sewage sludge combustion. Appl. Energy 116, 199–205.

Krár, M., Kovács, S., Kalló, D., Hancsók, J., 2010. Fuel porpose of hydrotreating of sunflower oil on Co-Mo/Al$_2$O$_3$. Bioresour. Technol. 101, 9287–9293.

Kubicka, D., Simacek, P., Zilkova, N., 2009. Transformation of vegetable oils into hydrocarbons over mesoporous-alumina-supported CoMo catalysts. Top. Catal. 52, 161–168.

Kubickova, I., Kubicka, D., 2010. Utilization of triglycerides and related feedstocks for production of clean hydrocarbon fuels and petrochemicals: a review. Waste Biomass Valoriz. 1, 293–308.

Kumar, R., Rana, B.S., Tiwari, R., Verma, D., Kumar, R., Joshi, R.K., Garg, M.O., Sinha, A.K., 2010. Hydroprocessing of jatropha oil and its mixtures with gas oil. Green Chem. 12, 2232–2239.

LanzaTech Bioethanol Platform, 2015. http://www.lanzatech.com/innovation/markets/fuels/. accessed August 2015.

Li, L., Coppola, E., Rine, J., Miller, J.L., Walker, D., 2010. Catalytic hydrothermal conversion of triglycerides to non-ester biofuels. Energy Fuel 24, 1305–1315.

Liu, Y., Sotelo-Boyas, R., Murata, K., Minowa, T., Sakanishil, K., 2009. Optimizing renewable oil hydrocracking conditions for aviation. Chem. Lett. 38, 552–553.

Liu, S., Zhu, Q., Guan, Q., He, L., Li, W., 2015. Bio-aviation fuel production from hydroprocessing castor oil promoted by the nickel-based bifunctional catalysts. Bioresour. Technol. 183, 93–100.

Lynd, L.R., Cushman, J.H., Nichols, R.J., Wyman, C.E., 1991. Fuel ethanol from cellulosic biomass. Science 251, 1318.

Marlin, D.S., Sarron, E., Sigurbjörnsson, Ó., 2018. Process advantages of direct CO_2 to methanol synthesis. Front. Chem. https://doi.org/10.3389/fchem.2018.00446.

Mascarelli, A.L., 2009. Gold rush for algae. Nature 461, 460–461.

Mawhood, R., Gazis, E., de Jong, S., Hoefnagels, R., Slade, R., 2016. Production pathways for renewable jet fuel: a review of commercialization status and future prospects. Biofuels, Bioprod. Bioref. 10, 462–484. https://doi.org/10.1002/bbb.1644.

Meller, E., Green, U., Aizenshtat, Z., Sasson, Y., 2014. Catalytic deoxygenation of castor oil over Pd/C for the production of cost-effective biofuel. Fuel 133, 89–95.

Murata, K., Liu, Y., Inaba, M., Takahara, I., 2010. Production of synthetic diesel by hydrotreatment of Jatropha oils using Pt – re/H-ZSM-5 catalyst. Energy Fuel 24, 2404 2409.

Neste Oil Corporation, 2012. Neste Oil. Neste Oil Corporation, Finland.

Patrick Schmidt, Valentin Batteiger, Arne Roth, Werner Weindorf, and Tetyana Raksha; Power-to-liquids as renewable fuel option for aviation: a review; Chem. Ing. Tech.; 90, 1–2 p. 127–140; https://doi.org/10.1002/cite.201700129.

Pinto, F., Varela, F.T., Goncalves, M., Andre, R.N., Costa, P., Mendes, B., 2014. Production of bio-hydrocarbons by hydrotreating of pomace oil. Fuel 116, 84–93.

Radich, T., 2015. The Flight Paths for Biojet Fuel, Independent Statistics & Analysis. www.eia.gov.

Rana, B.S., Kumar, R., Tiwari, R., Kumar, R., Joshi, R.K., Garg, M.O., Sinha, A.K., 2013. Transportation fuels from co-processing of waste vegetable oil and gas oil mixtures. Biomass Bioenergy 56, 43–52.

Anon., 2016. Review of Biojet Fuel Conversion Technologies. National Renewable Energy Laboratory.

Rivard, E., Trudeau, M., Zaghib, K., 2019. Hydrogen storage for mobility: a review. Materials 12, 1973. https://doi.org/10.3390/ma12121973.

Rozmyslowicz, B., Maki-Arvela, P., Lestari, S., Simakova, O.A., Eranen, K., Simakova, I.L., Murzin, D.Y., Salmi, T.O., 2010. Catalytic deoxygenation of tall oil fatty acids over a palladium-mesoporous carbon catalyst: a new source of biofuels. Top. Catal. 53, 1274–1277.

Rye, L., Blakey, S., Wilson, C.W., 2013. Sustainability of supply or the planet: a review of potential drop-in alternative aviation fuels. Energy Environ. Sci. 3, 17–27.

Sharma, R.K., Anand, M., Rana, B.S., Kumar, R., Farooqui, S.A., Sibi, M.G., Sinha, A.K., 2012. Jatropha-oil conversion to liquid hydrocarbon fuels using mesoporous titanosilicate supported sulfide catalysts. Catal. Today 198, 314–320.

Šimácek, P., Kubicka, D., Šebor, G., Pospíšil, M., 2009. Hydroprocessed rapeseed oil as a source of hydrocarbon-based biodiesel. Fuel 88, 456 460.

Simacek, P., Kubicka, D., Kubickova, I., Homola, F., Pospisil, M., 2011. Premium quality renewable diesel fuel by hydroprocessing of sunflower oil. Fuel 90, 2473–2479.

Singh, A., Nigam, P.S., Murphy, J.D., 2011. Mechanism and challenges in commercialization of algal biofuels. Bioresour. Technol. 102, 26–34.

Sinha, A.K., Anand, M., Rana, B.S., Kumar, R., Farooqui, S.A., Sibi, M.G., Kumar, R., Joshi, R.K., 2013. Development of hydroprocessing route to transportation fuels from non-edible plant-oils. Catal. Surv. Asia 17, 1–13.

Sirifa, A., Faungnawakiji, K., Itthibenchapong, V., Viriya-empikul, N., Charinpanitkul, T., Assabumrungrat, S., 2014. Production of bio-hydrogenated diesel by catalytic hydrotreating of palm oil over NiMoS2/γ-Al2O3 catalyst. Bioresour. Technol. 158, 81–90.

Standard Specification for Aviation Turbine Fuel Containing Synthesized Hydrocarbons, 2016a. https://compass.astm.org/Standards/HISTORICAL/D7566htm.

Standard Specification for Aviation Turbine Fuel Containing Synthesized Hydrocarbons, 2016b. https://compass. astm.org/Standards/HISTORICAL/D1655.htm.

Stephens, E., Ross, I.L., King, Z., Mussgnug, J.H., Kruse, O., Posten, C., 2010. An economic and technical evaluation of microalgal biofuels. Nat. Biotechnol. 28, 8–126.

SteveLicht, 2014. Chapter 5—Fermentation for biofuels and bio-based chemicals. In: Fermentation and Biochemical Engineering Handbook, third ed. Elsevier.

Sustainable Aviation CO2 Road-MAP, 2012. http://www.sustainableaviation.co.uk/. March 2012 (accessed August 2015).

Tiwari, R., Rana, B.S., Kumar, R., Verma, D., Kumar, R., Joshi, R.K., Garg, M.O., Sinha, A.K., 2011. Catalytic conversion of Jatropha oil to alkanes under mild. Catal. Commun. 12, 559–562.

Veriansyah, B., Han, J.Y., Kim, S.K., Hong, S.A., Kim, Y.J., Lim, J.S., Shu, Y.W., Oh, S.G., Kim, J., 2012. Green energy materials & process laboratory. Fuel 94, 578–585.

Verma, D., Kumar, R., Rana, B.S., Sinha, A.K., 2011. Aviation fuel production from lipids by a single-step route using hierarchial mesoporous zeolites. Energy Environ. Sci. 4, 1667–1671.

Verma, D., Rana, B.S., Kumar, R., Sibi, M.G., Sinha, A.K., 2015. Diesel and aviation kerosene with desired aromatics from hydroprocessing of jatropha oil over hydrogenation catalysts supported on hierarchical mesoporous SAPO-11. Appl. Catal. A Gen. 490, 108–116.

Vonortas, A., Kubika, D., Papayannakos, N., 2014. Catalytic co-hydroprocessing of gasoil–palm oil/AVO mixtures over a NiMo/γ-Al2O3 catalyst. Fuel 116, 49–55.

Walendziewski, J., Stolarski, M., Łużny, R., Klimek, B., 2009. Hydroprocessing of crude palm oil at pilot plant scale fuel process. Catal. Today 90, 686–691.

Wall, L.E., Rios-Solis, L., 2020. Sustainable production of microbial isoprenoid derived advanced biojet fuels using different generation feedstocks: a review. Front. Bioeng. Biotechnol. 8, 599560.

Wang, W.-C., Tao, L., Markham, J., Zhang, Y., Tan, E., Lia, B., Warner, E., Biddy, M., July 2016. Review of Biojet Fuel Conversion Technologies, National Renewable Energy Laboratory, Technical Report, NREL/TP-5100-66291.

Wei, J., Ge, Q., Yao, R., Wen, Z., Fang, C., Guo, L., Hengyong, X., Sun, J., 2017. Directly converting CO2 into a gasoline fuel. Nat. Commun. 8, 15174.

www.eni.com.

World Energy Outlook, 2012. Global Energy Trends. ISBN: 978-92-64-18084-0.

Wyman, C.E., 1994. Alternative fuels from biomass and their impact on carbon dioxide accumulation. Appl. Biochem. Biotechnol. 897, 45–46.

XiaowaNi, W.L., Jiang, X., Guo, X., Song, C., 2019. Chapter two—Recent advances in catalytic CO2 hydrogenation to alcohols and hydrocarbons. Adv. Catal. 65, 121–233. https://doi.org/10.1016/bs.acat.2019.10.002.

Yao, B., Xiao, T., Makgae, O.A., et al., 2020. Transforming carbon dioxide into jet fuel using an organic combustion-synthesized Fe-Mn-K catalyst. Nat. Commun. 11, 6395. https://doi.org/10.1038/s41467-020-20214-z.

Zhang, S.Z., Chen, S.L., Dong, P., 2010. Synthesis, characterization and Hydroisomerization performance of SAPO-11 molecular sieves with caverns by polymer spheres. Catal. Lett. 136, 126–133.

Zhong, H., Yao, G., Cui, X., Yan, P., Wang, X., Jin, F., 2019. Selective conversion of carbon dioxide into methane with a 98% yield on an in situ formed Ni nanoparticle catalyst in water. Chem. Eng. J. 357, 421–427. https://doi.org/10.1016/j.cej.2018.09.155.

CHAPTER

14

Recent advances in lignin valorization

Ipsita Chakravarty, Dipali Gahane, and Sachin Mandavgane
Department of Chemical Engineering, VNIT Nagpur, Nagpur, India

1 Introduction

In the year 1874, two German scientists named Wilhelm Haarmann and Ferdinand Tiemann synthesized vanillin from coniferin. It is the glucoside of coniferyl alcohol, the monomer of lignin obtained from the cambial juice of pine trees. This novel process paved the way for the industrial utilization of lignin for vanillin production. This invention was found to be a boon for the cosmetic and food industries. Nowadays, lignosulfonic acid (a by-product of the wood pulping industry) is used for vanillin production. In the global scenario, 3 billion tons of industrial lignin waste is produced annually and only 2%–5% is utilized. Irrational burning of industrial and agricultural lignin waste causes enormous air pollution (Araújo, 2008; Bajwa et al., 2019). Sustainable valorization of lignin through superior technology can lead to its industrial applications. Several commercially significant and value-added chemicals like ethylbenzene, phydroxyl acetophenone, vanillin, and carbon-based materials can be synthesized from lignin. However, the complex, irregular polymeric structure of lignin-containing methoxylated phenylpropanoid constituents is challenging. Generally, organosolvolysis and enzymatic pre-treatment techniques are used to extract pure lignin using hydrogenation or dehydrogenation processes to obtain the required aromatics and bio-oil. The physiochemical and structural characteristics of lignin can change the degree of recalcitrance and thus, its valorization methodology (Wang et al., 2019). Recent researches have focused upon the techniques to maintain native lignin with slight structural modifications for improved production of valuable products. It is evident that hydrogenolysis of pure lignin can produce 50% aromatic monomers which are considerably more than that of the chemically separated modified version of lignin (Amiri et al., 2019). Chemical and thermal techniques like hydrogenolysis, hydrothermal processing are typically utilized for lignin value-addition. The catalytic conversion strategies for lignin processing are based upon high-performance catalysts and advanced mechanical and chemical depolymerization processes (Cao et al., 2019). However, the stability and stringent mechanisms of catalysts need to be explored and analyzed. The biochemical methods for lignin fractionation and conversions are reliable, less toxic, and eco-friendly (Gadhave et al., 2018). Lignin valorization through the biorefinery approach would require the development of strategies for the synthesis of valuable chemicals as well as biofuels from agro-industrial lignin waste. For setting up such biorefineries, technological advancement is required with minimum investment. However, the upstream and downstream processing of lignin are real challenges for biorefinery (Beckham et al., 2016). Literature suggests that more and more emphasis has to be given to the importance of native structure and various processing technologies for the synthesis of application-based products (Chen and Wan, 2017). Lignin valorization is thus, a promising source of the economy (Ponnusamy et al., 2019; Ragauskas et al., 2014). Thorough knowledge is required to develop an understanding of the role of native lignin structure, fractionation processes and mechanism of lignin depolymerization, and current technologies for transformation of lignin to useful aromatics and fuels with a biorefinery approach for sustainable development.

2 Lignin structure and composition

The source of lignin is a decisive factor for the structural properties of lignin. A thorough understanding of lignin composition is the way to its efficient valorization. Lignin is a complex aromatic biopolymer that is renewable in nature (Lu et al., 2017). It has a non-crystalline three-dimensional framework of phenolic polymers. Preliminary lignin structure and characteristics have been elucidated by NMR spectroscopic methods (Yoo et al., 2016; Ghaffar and Fan, 2013). But, there are lacunae in the efficient detailed analysis of the intermediate products formed during lignin processing. Advanced characterization techniques would help to get an insight into the molecular reaction dynamics of the lignin-based biorefineries involving multiple products. The lignin content varies with plant variety. It is different for softwood, hardwood, types of coniferous trees, and herbaceous plants. About 40% of lignocellulosic agricultural residues constitute lignin in agricultural residues (Cao et al., 2018a, b). Lignin is cross-linked with cellulose as well as hemicellulose units via hydrogen and covalent linkages (Terrett and Dupree, 2019). The biomass of lignin is composed of phenyl propane. It is evident that S/G-rich lignin yields maximum monomeric units in biorefining processes. But, its adverse effect is the development of undesirable carbon–carbon bonds leading to condensation during the depolymerization procedure (Anderson et al., 2019). The structure of lignin consists of pcoumaryl alcohol, sinapyl alcohol, and coniferyl alcohol. The phenylpropyl alkyl units joined by the covalent bonds lead to the three-dimensional network of lignin (Zhang et al., 2019a, b; Adler, 1957). The structural recalcitrance hampers lignin utilization (Kazzaz and Fatehi, 2020). Lignin fractionation and structural modifications leading to fractionation is an important aspect for the generation of value-added products. Lignin valorization can be efficiently designed by understanding and utilizing the properties and functionalities of the lignin structure (Fig. 1).

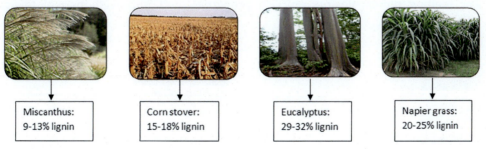

FIG. 1 Types of monomeric structures of lignin.

3 Present status of lignin valorization

Valorization of lignin can be done using physical, chemical, and biochemical techniques. Separation under high temperature and pressure such as steam explosion is the most common physical method of lignin extraction. In these techniques, the linkage between lignin and carbohydrate structures is broken to obtain a pure form of lignin. However, physical methods are difficult to apply at an industrial scale (Narron et al., 2016). Biological methods are non-stringent, non-toxic, and reliable in nature which helps to remove the chemical linkage between lignin and cellulose/hemicellulose by enzymatic approach. Low separation efficiency and cost-intensity of such processes need to be overcome for scale-up (Wong et al., 2020). The most popular method for large-scale application of lignin separation is the chemical method which has high separation efficiency (Xu et al., 2014). Lignin is a major derivative of papermaking and agricultural industries. This lignin is generally used in combustion. Several commercially viable products can be derived from lignin. Products like resin adhesive, rubber reinforcing agent, viscosity reducer for heavy oil, etc. can be synthesized using lignin (Cline and Smith, 2017) (Fig. 2).

Miscanthus:
9-13% lignin

Corn stover:
15-18% lignin

Eucalyptus:
29-32% lignin

Napier grass:
20-25% lignin

FIG. 2 Lignin composition in various feedstocks.

Lignin utilization technologies of the complex polymeric structure of lignin are still in their infancy. The perception of its structural heterogeneity is crucial to understand the basic properties of lignin for its valorization (Huang et al., 2019). The processes where fine chemicals synthesized from lignin can substitute petro-chemicals are yet to be explored. This would ensure reduced consumption of fossil fuels and a decline in environmental harm. In the present scenario, more emphasis has been given to the research of lignin monomers and their potential in lignin valorization. Recent research works reveal that lignin valorization processes have contributed immensely to industrial upliftment (Garlapati et al., 2020).

4 Fractionation/separation technologies

The pre-requisite of lignin valorization is the separation of the desired components. Lignin can either be hydrolyzed directly by dissolution keeping insoluble carbohydrate residues or by hydrolysis of lignocellulosic and leaving lignin in residual form (da Costa Lopes et al., 2013; Shen et al., 2019).The structural changes of lignin during separation processes often affect its characteristic properties. The organosolv lignin has an intact core structure but reduced molecular weight and partially cleaved β-aryl ether linkages (Li et al., 2019; El Hage et al., 2010). In the case of kraft lignin, ether linkages are cleaved during degradation reactions and new linkages are formed via condensation reactions (Chakar and Ragauskas, 2004; Lancefield et al., 2018). Biorefinery approach of lignin valorization involves pre-treatment of lignocellulosic waste or industrial waste stream containing lignin. To develop an economically feasible biorefinery, it is important to facilitate the production of lignin monomers, maximize delignification, and efficient fractionation of targeted components. Processes like organosolvolysis, enzymatic processing, and the like release structural constituents of lignin for further valorization (Morales et al., 2020) (Fig. 3).

Pure form of lignin can be extracted using very mild or neutral processing conditions. Alcohol and tetrahydrofuran are potential agents for lignin fractionation (Meng et al., 2019). Organosolv methods are less toxic and scalable. The self-condensation of lignin can be averted by using a high content of alcohol to enhance its yield (Lancefield et al., 2017). NMR HSQC analytical methods have shown that the original structure can be a source of several aromatic monomers. A pure form of lignin can be separated using *n*-butanol. Ionic liquids can improve lignin dissolution in an efficient and gentle manner. They have many benefits like low melting points, flexible physical properties, and improved thermal tolerance

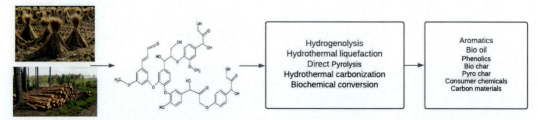

FIG. 3   Ligninvalorization from agro-wastes.

(Berthod et al., 2018). Bio-oil can be produced from native lignin produced without any re-condensation using triethylammonium hydrogen sulfate at 120°C (Yu et al., 2019). Milled-wood lignin avoids the use of any harsh chemicals. It is based on the recovery of lignin coupled with a large amount of carbohydrates. Such physical processes are eco-friendly but are time and energy consuming (Kumar et al., 2020). The milled lignin can be further treated with enzymes like cellulases to remove the carbohydrate residues to give native lignin (Rencoret et al., 2015). The native lignin has β-O-4 as well as β-β bonds which lead to C—O break down unlike the processed lignin. Modified lignin has its own boons and banes. The processing of modified lignin involves the use of harsh physical and chemical conditions. Lignin waste from paper and pulp industries often undergo either acid sulfite pulping process or base conditioned kraft process. Acid sulfite pulping process involves temperatures up to 160°C and an acidic environment that distorts the native form of kraft lignin (Hubbe et al., 2019). The improved water solubility of the modified lignin is an advantage. But, 4%–8% incorporated sulfonate causes harmful effects leading to catalytic transformations of lignin to synthesize aromatic compounds. Kraft lignin processing involves the application of hydrogen sulfide to break β-O-4 bonds at high temperatures along with the reformation of covalent bonds (Crestini et al., 2017). The outcome of such processes is lower molecular weight kraft lignin with improved water solubility due to low β-O-4 bonds and more hydroxyl groups. However, it is difficult to produce aromatic monomers from kraft lignin due to very stable C—C bonds (Gillet et al., 2017). In such challenging situations, technical advancement is needed for the transformation of lignin to produce value-added products.

5 Lignin depolymerization for aromatics and bio-oil

Lignin depolymerization is the second most important step toward lignin valorization after extraction or separation. Lignin depolymerization involves the cleavage of interlinking bonds with the help of suitable catalyzing agents. 50% of the bonds in native lignin are β-O-4. It has a low bond dissociation of 50–70 kcal/mol. The β-O-4 cleavage produced aromatic units and reactive intermediates for further subsequent repolymerization, leading to bio-oil and other useful products. Bio-oil is constituted of phenolic monomers like guaiacol, *p*-hydroxyacetophenon, hydroxyacetovanillon, and syringaldehyde. Depending on the method of depolymerization and monomers released, various aromatic products and bio-oil are synthesized. The oxidative depolymerization method entails oxidative cleavage of lignin resulting in the production of vanillin and syringaldehyde (Sun et al., 2018). The process is governed by oxygen present and pH optimized. Reductive depolymerization process generates aliphatic compounds like 4-propylcyclohexanol during the processing of lignin. This process was used to produce C8 and C9 cycloalkanes from birch wood sawdust (Stone et al., 2018). Reductive depolymerization maintains the selectivity of lignin monomers to a certain extent. But, elevated temperature and pressure required in such processes often release downgrade aromatic compounds. Hydrogenation of lignin catalyzed by metals distorts the fundamental nature. Compared to the reductive depolymerization process, oxidative depolymerization involves milder conditions to produce functionalized aromatic compounds for subsequent valorization (Picart et al., 2017). Aromatic monomer yield can be improved by developing strategic lignin conversion by preventing self-condensation and repolymerization. These are the key points for designing an efficient lignin valorization process. Lignin condensation can be prevented by the addition of formaldehyde which tends to inhibit benzylic positions (Schutyser et al., 2018). Methoxy groups generated during the depolymerization process can facilitate the functionalization of aromatic compounds into the desired products. Kraft lignin by-product contains 10%–12% methoxy groups which interact with carbon-di-oxide and water to generate acetic acid over rhodium chloride under a low-temperature range (Mei et al., 2017). This leads to about 88% acetic acid production. High-quality terephthalic acid can be produced from methoxy groups and carboxy groups obtained during the lignin depolymerization process (Zhang and Wang, 2020; Delidovich et al., 2016).

6 Catalytic depolymerization

The selection of the acid-catalyzed depolymerization method has a significant impact on the lignin structure (Zijlstra et al., 2019). The process of depolymerization can be done by treatment of dilute aqueous acids like sulfuric acid, hydrochloric acid, organic acid like formic acid, and zeolites under mild physical and chemical conditions (Forchheim et al., 2012; Chio et al., 2019). Phosphoric acid can lead to 40% of monomer yield. Dilute acids instead of crude acids can improve the extent of delignification maintaining the native structures of lignin (Shuai and Saha, 2017). Previous studies have shown that 60% of monomeric units are produced by using solid acid catalysts which have a profound influence on the deoxygenation process. FTIR and NMR studies have been used to identify the reaction sites as β-O-4 ether bond cleavage during depolymerization (Yang et al., 2016). Base-catalysis can be obtained through lignin dissolution via ester bond dissociations. The potential of basic sites has a key role in the improvement in the yield of aromatic compounds and bio-oil. Critical reassociation of macromolecular complexes is the main drawback in the base-catalyzed depolymerization process. Agents like sodium hydroxide, potassium hydroxide, calcium, and magnesium oxides are used (Toledano et al., 2014). These agents are helpful when Cα alcohol is oxidized to form a carbonyl group. Ester bonds are cleaved with the subsequent release of aromatic monomers (Lan et al., 2019). The process involves the use of a strong base at a raised temperature and prolonged incubation time. Though the amount of aromatic monomers increases due to the bond dissociations, it poses high chances of the generation of solid residues because of repolymerization. The yield and selectivity of lignin-based chemicals are governed by physico-chemical parameters, especially the concentration of catalyst and the incubation temperature (Wang et al., 2013) (Table 1).

7 Oxidative depolymerization

Propyl phenol units of lignin are bound by covalent bonds, predominated by α-O-4 as well as β-O-4 ether bonds. Oxidative dissociation of these linkages leads to the release of several functional groups such as aldehyde, ketone, and carboxyl groups (Ma et al., 2018). These functionalized compounds consist of several important aromatic compounds. The commonly used oxidative reagents are oxides and peroxides. Chemicals like sodium hydroxide and

TABLE 1 Catalytic depolymerization of lignin.

Lignin type	Catalyst	Value-added products	Reaction conditions	Yield wt%	References
Industrial lignin or kraft lignin	Sodium hydroxide	Pyrocatechol	270–315°C; 13 MPa	0.5–4.9	Beauchet et al. (2012)
Organosolv lignin	Cesium carbonate and water	Bio-oil	180°C for 8h	52–67	Dabral et al. (2018a, b)
Lignin from poplar sawdust	Pd/C-H₃PO₄, hydrogen and methanol	Monomeric phenolic compounds	200°C, 3h, 2 MPa	42	Renders et al. (2016)
Wheat straw derived lignin	Formic acid and ethanol	Methoxyphenols Catechols Phenols	380°C; 25 MPa	2, 1.7, 1.5	Forchheim et al. (2012)
Bamboo biomass	Methanol, hydrogen and Pt/C	Monomeric units	260°C, 4h	32.2	Zhang et al. (2019a, b)
Enzyme hydrolyzed lignin	Methanol, hydrogen and MoOx/CNT	Monomeric units	260°C, 4h	47	Xiao et al. (2017)
Paper industry waste	1,4-Dioxane and formic acid	Monomers Bio-oils	300°C, 2h	25.4; 55	Wu et al. (2019)
Isopropanol diphenyl ether	Ru/C, hydrogen	Mixed aromatic compounds	120°C, 10h	90–98	Wu et al. (2018a, b)

hydrogen peroxide help to improve the depolymerization process (Stärk et al., 2010). A cobalt-based catalyzing agent with nitrogen-modified carbon support led to 96% production of phenol with negligible chances of re-coupling (Mottweiler et al., 2015). Copper and vanadium-based catalysts were found to be useful for lignin oxidation without intermediate generation. Methyl ether intermediates are converted to benzoate using vanadium as a catalyzing agent. Similarly, benzaldehyde was synthesized using a copper catalyst (Sedai et al., 2011). Lignin waste from the paper and pulp industry contains interunit C—C bonds which have increased bond dissociation energy is required for C—C bonds is approximately 120 kcal/mol while that for C—O linkages is approximately 60 kcal/mol. Cobalt sulfide works at the site of covalent bonds with efficacy (Liu et al., 2020). Previous studies on novel strategic oxidative depolymerization of lignin substrate showed that carbon–carbon and carbon–oxygen linkages were cleaved by copper chloride polybenzoxazine composite catalysts with hydrogen peroxide as an oxidizing agent with 96% efficiency for aromatic monomeric units within a few hours (Ren et al., 2017). Oxidative depolymerization helped in bond energy minimization for smooth depolymerization (Guadix-Montero and Sankar, 2018).

8 Reductive depolymerization

The selective and efficient breakdown of lignin interlinkages is the benefit of using a reductive depolymerization strategy (Ren et al., 2017). Formic acid helps in the fragmentation of β-O-4 linkage as well as stabilization of reactive species generated within the process

(Rahimi et al., 2014). Nickel catalyzing agents are useful in selective delinkage of β-O-4 in the reductive depolymerization procedure. It has been revealed that H-BEA zeolite (Brønsted acid sites) aided nickel catalysts to obtain depolymerization of organosolv lignin (Kasakov et al., 2015). Nickel and iron alloy diminish the catalyst particle size and hence create available active reaction sites. The catalytic upstream bioprocessing technique based on early-stage catalytic transformation leads to two distinguishable fractions are formed, i.e., bio-oil and holocellulose. The holocellulose so generated is further harnessed via gasification. The bio-oil made by hydrodeoxygenation has aromatics or alkanes as per physico-chemical conditions optimized (Cao et al., 2018b). The pre-treatment of lignin is an economic and effective way for the utilization of lignocellulosic biomass. The catalytic upstream bioprocessing can be done using a noble metal catalyzing agent that leads to aldehyde groups being converted to alcohol. The accomplishment of reductive depolymerization is governed by the source of lignin, catalyst nature, and processing conditions (Beckham et al., 2016).

9 Mechanochemical depolymerization

Mechanochemical depolymerization creates high energy activation through localized pressure and frictional heating (Yao et al., 2018). This method is cost-effective, energy-efficient, and faster as compared to its conventional counterparts. Mechanochemical oxidation process forms a solvent-free condition for depolymerization of lignin combined with lignin catalysis under physical methods like the milling process (Dabral et al., 2018a, b). In a recent study conducted on depolymerization of lignin, researchers developed an efficient two-stage strategy for the breakdown of the lignin dimers and birchwood lignin. Stage one involved the conversion of β-O-4 bonds of lignin into aromatic acids and phenolic compounds via an oxidation process using DDQ/NaNO$_2$ catalyzing agent. The second stage was led by a sodium-hydroxide-based catalysis process along with ball-milling. This helped in the production of important aromatic products with a higher yield of syringate. Such methodology is a green solution for the effective generation of aromatic compounds under mild reaction conditions (Sun et al., 2020).

10 Microwave reaction

wMicrowave-assisted method is an energy-efficient and rapid process for lignin in depolymerization (Yunpu et al., 2016). It takes a few hours for reaction and yields a pure form of lignin, unlike the conventional methods. It involves direct heating leading to the production of several industrial products like bio-char, bio-oil, and syngas. Microwave-based depolymerization of lignin uses relatively mild reaction conditions. Higher yield of bio-oil can be achieved by microwave-assisted depolymerization technique as compared to the other conventional strategies (Bu et al., 2019). A recent study reported the use of microwave-assisted solvolysis for alkaline lignin using hydrotalcite-based copper-nickel aluminum mixed oxides. 60% of bio-oil production was attained at (<200 °C) in less than 2 h (Zhou et al., 2018). Microwave irradiation speeds up the oxidative degradation process in an effective manner. The increased vanillin production at 350 mg/L under the microwave aided depolymerization catalyzed by copper hydroxide and iron oxide is significant (Bjelić et al., 2018). The main drawback of this process is its difficulty to scale-up for lignin-based biorefinery purposes (Table 2).

TABLE 2 Comparison of various depolymerization methods.

Depolymerization technique	Advantages	Scalability challenges	References
Microwave-assisted	Fast process, mild conditions requirement	Process control difficulties and scale-up challenges	Yunpu et al. (2016)
Hydrothermal liquefaction	Mild reaction conditions, applicable to varied biomass	Ineffectual convection, heavy water requirements	Alhassan et al. (2020)
Pyrolysis/gasification	Applicable to varied biomass, fast rate, non-selective process	Harsh temperature, high yield, scale-up	Kawamoto (2017)
Photocatalytic method	Renewable and efficient process	More research is required to overcome technological barriers	Xiang et al. (2020)
Mechanochemical method	Rapid processing without chemical requirement, efficient, non-selective	Energy-consuming method	Agarwal et al. (2018)

11 Photocatalytic depolymerization

The production of lignin in higher plants takes place by the process of photosynthesis. Photocatalytic depolymerization is a reverse process and is, therefore, promising (Li et al., 2016). The oxidation hydrogenolysis sequential reaction can be performed to obtain ketone and phenol compounds via selective depolymerization using the dual light wavelength switching technique (Luo et al., 2016). Photocatalytic depolymerization combines visible light photo-redox and catalyzing agents. Solar energy could be used for photocatalysis of native lignin for the formation of valuable chemicals at room temperature. It is found that cadmium-based quantum dots improve lignin depolymerization or cleavage of interlinking bonds as well as help in the effective conversion of pure lignin into functionalized aromatic compounds (Wu et al., 2018a, b). The selection of suitable solvents and optimization of process parameters in photocatalytic lignin processing are important factors for lignin valorization (Xiang et al., 2020) (Fig. 4).

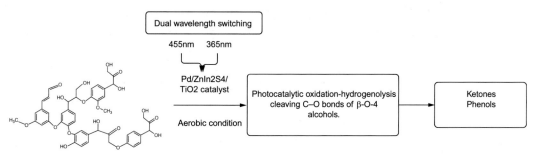

FIG. 4 Dual light wavelength switching strategy for lignin depolymerization (Luo et al., 2016).

12 Hydrothermal liquefaction

Hydrothermal liquefaction is sustainable technology for lignin valorization as it involves thermochemical transformation of biomass into a liquid using water (Kang et al., 2013). The method takes place under moderate temperature and increased pressure (Wang et al., 2013). Bio-oil is the key product of this method (Singh et al., 2014). Base-catalyzed hydrothermal liquefaction of kraft lignin using zirconium dioxide, potassium carbonate, and potassium hydroxide improve bio-oil productivity (Belkheiri et al., 2018). This process produces bio-oil with higher energy density as compared to pyrolysis oil. There are no pre-treatment or pre-drying requirements in the case of the hydrothermal liquefaction technique. This process is eco-friendly and has a high conversion rate at a faster rate (Parakh et al., 2020). The hot pressured water can act as a reactant as well as catalyzing agent in the transformation of lignin (Li et al., 2015). In the hydrothermal liquefaction technique, it is evident that during hydrothermal liquefaction, condensation, alkylation, and demethoxylation take place (Cao et al., 2017). The process is rapid under supercritical conditions obtaining a high yield of phenolic compounds but repolymerization with increased temperature is a major hindrance in the valorization process. The significance of the hydrothermal liquefaction technique has not been proven at an industrial scale due to the difficulty in depolymerization and downstream processing which need to be worked upon (Yang and Yang, 2019).

13 Hydrothermal carbonization for hydrochar

Hydrothermal carbonization is carried out by heating lignin in a high-pressure autoclaving system at subcritical water. Lignin carbonization involves hydrolytic processing, decarboxylation, aromatization followed by the process of re-condensation (Kumar et al., 2018). This method works at low carbonization temperature (180–300 °C) and does not require any pre-drying (Mäkelä et al., 2015). The hydrothermal carbonization process is governed by physicochemical properties, incubation time and type of substrate used which decide the fate and properties of hydrochar (Kalderis et al., 2014) (Fig. 5).

The hydrochar is an effective bio-fuel with calorific values up to 26 MJ/kg and its improved carbon content can be utilized for synthesizing activated carbon, bioadsorbent, catalyst, and supercapacitor (Chatterjee and Saito, 2015; Khiari et al., 2019). At temperatures above 300 °C, the structural changes and surface functionalities decomposed were modified (Ruan et al., 2018; Atta-Obeng et al., 2017). The activated carbon from lignin hydrochar has a large surface-to-volume ratio and is a potential raw material for various commercial commodities in cosmetic and pharmaceutical industries (Agrawal et al., 2014; Rodríguez Correa et al., 2017). It has also been used in the bioremediation of dyes (Supanchaiyamat et al., 2019). Fu et al. demonstrated the design of lignin-based carbon/zinc oxide composite. The process of electrostatic carbonization process leads to the formation of the crystalized ZnC_2O_4 particles. These particles behave as stimulators for carbonization processes. These unique composites have a distinctive three-dimensional structure with porosity containing zinc oxide nanoparticles covered by carbon derived from lignin-based nanosheets (Fu et al., 2019). The phenolic hydroxyl groups found on lignin hydrochar play a key role in the nitrogen-

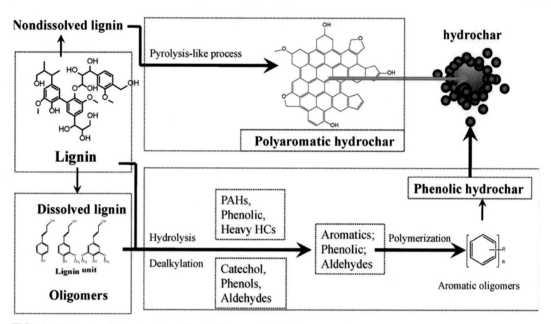

FIG. 5 Hydrochar formation from lignin (Wang et al., 2018).

doped carbon material production process through acetylation and aromatic nitration, which could further contribute to the redox reactions (Zhu and Xu, 2020).

14 Pyrolysis and gasification

The major thermochemical methods for lignin transformation into useful products are gasification and pyrolysis methods (De Wild et al., 2014; Chio et al., 2019). The gasification method helps in the generation of synthetic petroleum and syngas production by Fischer–Tropsch process gasification which involves thermophilic oxidation process (dos Santos and Alencar, 2020; Cao et al., 2018a, b). Hydrogen and carbon monoxide in gasification help in lignin depolymerization but the generation of coke and tar from aromatics is a serious drawback that can be overcome by suitable catalyzing agents (Sun et al., 2018). The use of oxidizing agent helps in the efficient formation of syngas from producer gas. Pyrolysis takes place at a temperature between 400 and 800 °C without oxygen supply (Suman and Gautam, 2017). Depending upon the temperature and time requirements, pyrolysis is of two types. Low-temperature-based pyrolysis taking prolonged residence time is called slow pyrolysis wherein, bio-char yield is high. Fast pyrolysis is just the opposite of slow pyrolysis and it gives rise to bio-oil production. Microwave aided pyrolysis helps in the ameliorated quality of bio-oil. The bio-char produced during the process is in low quantity (Mohamed et al., 2016). Some drawbacks of the pyrolysis process are low product selectivity and difficult separation techniques (Fig. 6).

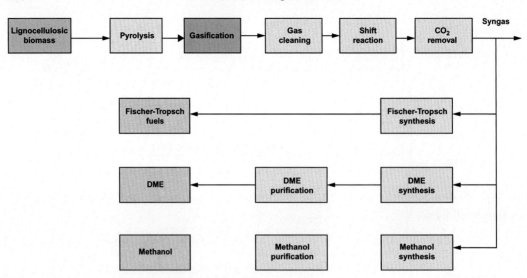

FIG. 6   Lignin valorization by pyrolysis and gasification.

14.1 Pyrolysis

Cellulose, hemicellulose, and lignin are the main components of biomass (Gholizadeh et al., 2019). These components are decomposed at different temperatures. Cellulose and hemicellulose are easily decomposed where lignin is difficult to decompose and biomass with high content of lignin produces bio-char in high amount (Jahirul et al., 2012). According to Kawamoto (2017) biomass with higher lignin content is a source of aromatic compounds. Lignin is a polymer of phenylpropane units that contains three different aromatic ring substitution patterns: *p*-hydroxyphenyl (H), syringyl (3,5-dimethoxy-4-hydroxyphenyl, S), and guaiacyl (4-hydroxy-3-methoxyphenyl, G) and it varies with different biomass (Kawamoto, 2017). Pyrolysis is the process of thermal degradation of biomass in absence of oxygen which produces various products. Bio-oil, bio-char, and syngas are the main products from the pyrolysis process and by applying some suitable upgrading process to these products value-added products, i.e., bio-chemicals and biofuels can be formed. Pyrolysis process can be divided into slow and fast depending upon operating conditions. Heating rate, particle size, gas residence time, and temperature are operating parameters in the pyrolysis process (Guedes et al., 2018). Other than these operating parameters drying and grinding of biomass are important steps before the pyrolysis process which affect the product yield. The pyrolysis reactor is the main part of the whole pyrolysis process. Choice of reactor affects product yield. Many reactors such as fixed bed reactor, fluidized bed reactor, rotary kiln, and auger reactor. Table 3 shows the relative distribution of various products which depend upon the type of pyrolysis process and pyrolysis operating parameters (Dhyani and Bhaskar, 2019). Lignin from palm kernel shell composed of *p*-hydroxyphenyl structural units which give phenol (13.49%) upon pyrolysis while wheat straw and pine sawdust lignin is composed of guaiacyl units which upon pyrolysis gives acetic acid as the main component from bio-oil (Chang et al., 2016).

TABLE 3   Different types of pyrolysis process and desired product.

Operating parameters	Slow pyrolysis	Fast pyrolysis
Temperature (°C)	400	500
Heating rate (°C/s)	0.1–1	10–200
Solid residence time (s)	Hours-days	1–2
Particle size (mm)	5–50	<1
Product yield (%)	Bio-oil: 25–30 Bio-char: 30–40 Syn gas: 25–35	Bio-oil: 60–75 Bio-char: 12–20 Syn gas:13–20

(1) *Slow pyrolysis*: The process of slow pyrolysis occurs at low temperature, low heat heating rate, and long residence time. As shown in Table 1 main products from slow pyrolysis of biomass are bio-oil, bio-char, and syngas. Among these products yield of bio-char is more than bio-oil and syngas. In slow pyrolysis residence time of vapors is large thus vapors get more time to react which results in the formation of a more solid product (bio-char) as compared to fast pyrolysis. Slow pyrolysis is recommended for the production of bio-char.

(2) *Fast pyrolysis*: Fast pyrolysis occurs at high temperatures as compared to slow pyrolysis. High temperature, high heating rate, and short residence time result in the high production of bio-oil. Temperature at about 500 °C and short residence time (<2 s) minimizes cracking of products which was done by removing pyrolysis products quickly from the reactor with rapid condensation of condensable gases for maximum liquid formation. Noncondensable gases are permanent gases which are CO, CO_2 and CH_4 called syngas used for heating purposes. Pyrolysis of lignin produces more H_2 and CH_4 because it has a methoxyl functional group and aromatic structure (Yang et al., 2007). Bio-oil is used for heating purposes, fuel blends, and also it contains various products such as phenol, aromatics, sugars, alcohols, acids, aldehydes, and ketones (Yorgun and Yıldız, 2015). Bio-oil has some deficiencies such as highly oxygenated, highly corrosive, low calorific value, highly viscous, and low stability as compared to conventional fossil fuel thus proper upgrading methods are used to improve the quality of bio-oil so that it can be used as a transportation fuel (Vienescu et al., 2018). In lignin pyrolysis, catalyst plays an important role to produce selective products and upgrade the yield of products. Some metallic compounds and zeolites as a catalyst are added with biomass during lignin pyrolysis (Banu et al., 2019).

15 Pyrolysis for bio-char

Process and mechanisms of pyrolysis: As shown in the table process of slow pyrolysis is usually done at a lower temperature (about 400 °C), lower heating rate (0.1–1 °C/s), and long residence time (hours-days) as compared to fast pyrolysis. Due to the longer residence time of gases products in reactor reactions such as repolymerization/recombination takes place to produce a higher amount of solid product called as bio-char (Fahmy et al., 2018). Longer residence time causes increasing pore size of bio-char and low heating rate causes no thermal cracking of biomass and increases yield of bio-char (Tripathi et al., 2016). Other than bio-char, slow pyrolysis also produces bio-oil and syngas. Yield of bio-char (12%–20%) from fast pyrolysis is low because of the short residence time of gases. Yield of bio-char decreases with increasing temperature (Wu et al., 2012). During pyrolysis, lignin was converted into monomer substituted phenol, and condensation of these phenolic compounds using phenypropane units of lignin gives a product called bio-char (Cao et al., 2018a, b).

Applications of bio-char (pyrochar): Bio-char has a high calorific value because of a high amount of carbon thus it was used for heating operations. Bio-char from pyrolysis contains some fractions of unconverted solids, carbonaceous residues, and inorganic materials and their content may be varied according to pyrolysis operating parameters (Dhyani and Bhaskar, 2019). By use of bio-char to the soil, fertility of soil increases as adsorption ability increases, loss of nutrients reduced, and carbon content increases and hence used as soil amendments (Wu et al., 2012). Bio-char also helps in increasing the water holding capacity of the soil. Bio-char is used as an adsorbent because of charged inorganic species and high surface area (Dhyani and Bhaskar, 2019). Bio-char was used for the removal of unwanted material (pollutants) from water, soil, and gas due to the presence of active functional groups such as –COOH, R-OH, and –OH and thus has the capacity to replace activated carbon (Oliveira et al., 2017). Bio-char is also used in aquaculture for productivity enhancement (Raul et al., 2019).

16 Biochemical processing of lignin

Bioprocessing of lignin is a promising green technology. Several studies have shown that the microorganisms and the enzymes secreted by them are able to depolymerize lignin naturally. The process is eco-friendly and sustainable. But, it requires huge initial investment and efforts to intensify scalability. The biochemical pathways inside microbial cells govern the fate of the lignin depolymerization and valorization. Lignin gets converted into oligomers or monomers which are further degraded into intermediates like protocatechuate. These intermediate products are converted to acetyl coenzyme A by β-ketoadipate biochemical mechanism to generate lipid. *Rhodococcus jostii* RHA1 is able to degrade lignin due to the presence of DyP-type peroxidase (DyP) (Ahmad et al., 2011). This enzyme cleaves C_β-O and produces vanillin (Rahmanpour and Bugg, 2015; Chen and Wan, 2017). Dyp-type peroxidases are expressed in *Pseudomonas fluorescens* to cause oxidation reaction for degradation of lignin (Bugg et al., 2020). Fermentative production of vanillin involves the enzymes, feruloyl-CoA synthetase as well as enoyl-CoA hydratase/aldolase produced by *Streptomyces* (Pour et al., 2019). Strategies like media optimization and elimination of product inhibition have been adopted for vanillin production from ferulic acid (Kaur and Chakraborty, 2013). TAG in *Streptomyces lividans* helps in lipid production from lignin as a carbon source. *Aspergillus fumigates* led to 26.2% lignin depolymerization in lignocellulosic biomass like wheat straw producing various fatty acids. Postia placenta Mad-698-R and other brown rot fungi can depolymerize lignin by demethylation (Lee et al., 2019). White-rot fungi are potential fungi capable of breaking down lignin and has been found in diverse applications like bioremediation of soil and water containing phenolic compounds delignification of biomass improving cellulose digestibility and improving paper quality (Zhi and Wang, 2014; Liu et al., 2014a, b; Freitas et al., 2009) (Table 4).

Enzymes are biological catalysts that are isolated from fungi and bacteria. They can help in the depolymerization and conversion of lignin into useful products. Enzymatic conversion is more specific than whole-cell utilization for lignin valorization (Beckham et al., 2016). Depolymerized lignin or lignin-derived compounds are channelized into intermediate compounds like catechol which undergo ring fission, followed by the Krebs cycle. Vanillin and 4-hydroxybenzoic acid are major products of these reactions. Laccase and laccase meditator-assisted processes play a key role in lignin degradation using oxygen as an electron acceptor (Roth and Spiess, 2015). Studies reveal that this enzyme is led by ABTS to reduce the chances of repolymerization of kraft lignin (Longe et al., 2018). Enzymes like laccase, dehydrogenases, and DyPs have significantly helped in the production of vanillin. DyP-type peroxidases (DyPs) catalyze phenolic as well as non-phenolic lignin and aromatic sulfide compound aldehydes and guaiacol (Chauhan, 2020). Dehydrogenases can cleave guaiacyl glycerol-guaiacyl ether and help in biotransformation of vanillin (Ayeronfe et al., 2018). Enzymatic transformations can be used in synergy with chemical methods for better results. The results for enzymatic transformations varied with the type of substrate and physicochemical conditions. In order to improve the productivity and scalability of biochemical valorization, strain selection and process optimization are the prerequisites. Metagenomic studies have shown that lignin-degrading bacterial genes (Wilhelm et al., 2019), and many other novel lignin decomposers are present in the digestate of lignocellulosic digestion (Kamimura et al., 2019). Metabolic engineering and genetic engineering of microorganisms can enhance the biochemical processing of lignin for valuable industrial applications.

TABLE 4 Biochemical valorization of lignin.

Microbial strain	Feedstock	Production details	References
Pleurotus ostreatus	Rice straw	41% lignin depolymerization	Taniguchi et al. (2005)
Rhodococcus jostii RHA1	Wheat straw	96 mg/L vanillin production	Sainsbury et al. (2013)
Phanerochaete chrysosporium	Maize stover	23% lignin depolymerization	Liu et al. (2014a, b)
Pseudomonas putida KT2440	P-coumerate and ferulic acid	0.85 g/g PHA	Linger et al. (2014)
Phanerochaete chrysosporium	Wheat straw	30% lignin depolymerization	Singh et al. (2011)
Rhodococcus opacus DSM 1069	Industrial lignin waste	Lipid 0.06 g/g	Wells Jr et al. (2015)
Pleutotus ostreatus PO45	Sugarcane bagasse	85% lignin depolymerization	Dong et al. (2013)

17　The biorefinery model

Biorefinery model refers to the setting up of integrated biorefineries with sequential processing facilities that use renewable biomass to produce value-added products. Biorefinery approach for lignin valorization would include the separation of lignin from its sources like agricultural residues and industrial waste. Further, pre-treatment would be done followed by lignin depolymerization, and transformation of lignin to useful commodities. The phenolic monomers of lignin would be channelized to develop valuable commodities at a large scale. Various products of commercial significance can be produced via lignin depolymerization process. These can be aromatic compounds, bio-oil, bio-char, and syngas. Further processing of syngas by Fischer–Tropsch process can lead to methanol, dimethyl ether, and the formation of other chemicals (Paone et al., 2020). Bio-char is a by-product of bioenergy production from lignin and can be used in a wide range of applications like electrochemical and catalytic reactions (Strassberger et al., 2014; De Wild et al., 2014). Aromatic monomers produced by lignin depolymerization like vanillin are currently produced at an industrial scale from lignin. This accounts for 3000 ton of vanillin per year (Fache et al., 2016). Apart from vanillin, many other renewable aromatic building blocks are produced via depolymerization that requires additional processing techniques to generate marketable products (Fig. 7).

The main challenge of developing lignin-based biorefinery is that the purification process of functional value-added products from complex streams. The downstream processing is challenging, cost-intensive as well as the yield of a single product is usually low. Advancement toward the selective functionalization process is required through improved separation techniques.

18　Conclusions

Lignin valorization has tremendous importance yet there are several technical challenges that need to be overcome to improve its scalability. Naturally, occurring lignin undergoes easy degradation and self-condensation reaction. The lignin found in the industrial waste streams is semi-processed and contains inconsistent aromatic structures which are difficult to harness. The quality of industrial lignin waste needs to be upgraded for further utilization. The heterogeneity of the lignin structure makes its depolymerization challenging. The structure of lignin must be deciphered properly to generate a wide spectrum of lignin-based commercial products. Lignin biorefinery is a holistic approach for the production of value-added products with the aid of various physical, chemical, and biochemical methods as discussed in this book chapter. The detailed knowledge of reaction mechanisms and degradation techniques like hydrogenolysis and organosolv methods need to be explored. The concept of lignin valorization has a promising future and technological advancement is required to make it scalable. Bio-char/hydrochar produced from lignin can be a potential catalyst. It can be used in a wide range of high-end applications through suitable functionalization. Comprehensive knowledge of lignin chemistry and reactivity can intensify technological advancements in lignin valorization for industrial applications. Lignin biorefinery is still in its infancy. Deep

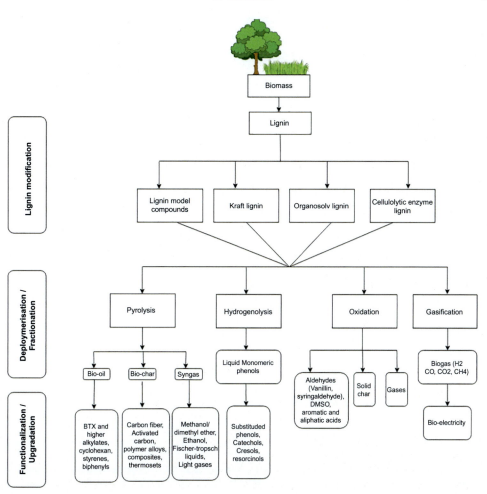

FIG. 7 Scheme for lignin biorefinery.

insights into the structure and reaction mechanism of lignin monomers are needed for efficient transformation and downstream processing. This would also help to establish an economically sustainable lignin biorefinery. The structure and functionality of the lignin monomers and the derived products are dependent upon the biomass source and the extraction techniques. The existing industrial processing methods often alter the lignin structure and affect productivity. More research is needed toward the development of new fractionation technologies to generate pure and potential products from lignin waste. This would also open up a new path for integration of lignin biorefinery with already existing cellulose and hemicellulose biorefinery schemes for tackling lignocellulosic agro-industrial waste. There is an urge to broaden the applicability of biochemical processing through metabolic engineering and genetic engineering to explore novel ways of lignin valorization. The utilization of lignin waste for industrial applications needs a better understanding of lignin chemistry to surmount the present technological challenges for wide-reaching applications. Biorefinery based on lignin is indeed an ecologically sustainable scheme with a low-carbon footprint that will be beneficial for economic growth.

References

Adler, E., 1957. Structural elements of lignin. Ind. Eng. Chem. 49 (9), 1377–1383.

Agarwal, A., Rana, M., Park, J.H., 2018. Advancement in technologies for the depolymerization of lignin. Fuel Process. Technol. 181, 115–132.

Agrawal, A., Kaushik, N., Biswas, S., 2014. Derivatives and applications of lignin–an insight. SciTech J. 1 (7), 30–36.

Ahmad, M., Roberts, J.N., Hardiman, E.M., Singh, R., Eltis, L.D., Bugg, T.D., 2011. Identification of DypB from Rhodococcus jostii RHA1 as a lignin peroxidase. Biochemistry 50 (23), 5096–5107.

Alhassan, Y., Hornung, U., Bugaje, I.M., 2020. Lignin hydrothermal liquefaction into bifunctional chemicals: a concise review. In: Biorefinery Concepts. IntechOpen.

Amiri, M.T., Dick, G.R., Questell-Santiago, Y.M., Luterbacher, J.S., 2019. Fractionation of lignocellulosic biomass to produce uncondensed aldehyde-stabilized lignin. Nat. Protoc. 14 (3), 921–954.

Anderson, E.M., Stone, M.L., Katahira, R., Reed, M., Muchero, W., Ramirez, K.J., Román-Leshkov, Y., 2019. Differences in S/G ratio in natural poplar variants do not predict catalytic depolymerization monomer yields. Nat. Commun. 10 (1), 1–10.

Araújo, J.D.P., 2008. Production of Vanillin From Lignin Present in the Kraft Black Liquor of the Pulp and Paper Industry. Universidade do Porto, Portugal.

Atta-Obeng, E., Dawson-Andoh, B., Seehra, M.S., Geddam, U., Poston, J., Leisen, J., 2017. Physico-chemical characterization of carbons produced from technical lignin by sub-critical hydrothermal carbonization. Biomass Bioenergy 107, 172–181.

Ayeronfe, F., Kassim, A., Ishak, N., Aripin, A., Hung, P., Abdulkareem, M., 2018. A review on microbial degradation of lignin. Adv. Sci. Lett. 24 (6), 4407–4413.

Bajwa, D.S., Pourhashem, G., Ullah, A.H., Bajwa, S.G., 2019. A concise review of current lignin production, applications, products and their environmental impact. Ind. Crop. Prod. 139, 111526.

Banu, J.R., Kavitha, S., Kannah, R.Y., Devi, T.P., Gunasekaran, M., Kim, S.H., Kumar, G., 2019. A review on biopolymer production via lignin valorization. Bioresour. Technol. 290, 121790.

Beauchet, R., Monteil-Rivera, F., Lavoie, J.M., 2012. Conversion of lignin to aromatic-based chemicals (L-chems) and biofuels (L-fuels). Bioresour. Technol. 121, 328–334.

Beckham, G.T., Johnson, C.W., Karp, E.M., Salvachúa, D., Vardon, D.R., 2016. Opportunities and challenges in biological lignin valorization. Curr. Opin. Biotechnol. 42, 40–53.

Belkheiri, T., Andersson, S.I., Mattsson, C., Olausson, L., Theliander, H., Vamling, L., 2018. Hydrothermal liquefaction of Kraft lignin in subcritical water: influence of phenol as capping agent. Energy Fuel 32 (5), 5923–5932.

Berthod, A., Ruiz-Ángel, M.J., Carda-Broch, S., 2018. Recent advances on ionic liquid uses in separation techniques. J. Chromatogr. A 1559, 2–16.

Bjelić, A., Grilc, M., Gyergyek, S., Kocjan, A., Makovec, D., Likozar, B., 2018. Catalytic hydrogenation, hydrodeoxygenation, and hydrocracking processes of a lignin monomer model compound eugenol over magnetic Ru/C–Fe2O3 and mechanistic reaction microkinetics. Catalysts 8 (10), 425.

Bu, Q., Chen, K., Xie, W., Liu, Y., Cao, M., Kong, X., Mao, H., 2019. Hydrocarbon rich bio-oil production, thermal behavior analysis and kinetic study of microwave-assisted co-pyrolysis of microwave-torrefied lignin with low density polyethylene. Bioresour. Technol. 291, 121860.

Bugg, T.D., Williamson, J.J., Rashid, G.M., 2020. Bacterial enzymes for lignin depolymerisation: new biocatalysts for generation of renewable chemicals from biomass. Curr. Opin. Chem. Biol. 55, 26–33.

Cao, L., Zhang, C., Chen, H., Tsang, D.C., Luo, G., Zhang, S., Chen, J., 2017. Hydrothermal liquefaction of agricultural and forestry wastes: state-of-the-art review and future prospects. Bioresour. Technol. 245, 1184–1193.

Cao, L., Iris, K.M., Liu, Y., Ruan, X., Tsang, D.C., Hunt, A.J., Zhang, S., 2018a. Lignin valorization for the production of renewable chemicals: state-of-the-art review and future prospects. Bioresour. Technol. 269, 465–475.

Cao, Z., Dierks, M., Clough, M.T., de Castro, I.B.D., Rinaldi, R., 2018b. A convergent approach for a deep converting lignin-first biorefinery rendering high-energy-density drop-in fuels. Joule 2 (6), 1118–1133.

Cao, Y., Chen, S.S., Zhang, S., Ok, Y.S., Matsagar, B.M., Wu, K.C.W., Tsang, D.C., 2019. Advances in lignin valorization towards bio-based chemicals and fuels: lignin biorefinery. Bioresour. Technol. 291, 121878.

Chakar, F.S., Ragauskas, A.J., 2004. Review of current and future softwood kraft lignin process chemistry. Ind. Crop. Prod. 20 (2), 131–141.

Chang, G., Huang, Y., Xie, J., Yang, H., Liu, H., Yin, X., Wu, C., 2016. The lignin pyrolysis composition and pyrolysis products of palm kernel shell, wheat straw, and pine sawdust. Energy Convers. Manag. 124, 587–597.

Chatterjee, S., Saito, T., 2015. Lignin-derived advanced carbon materials. ChemSusChem 8 (23), 3941–3958.

Chauhan, P.S., 2020. Role of various bacterial enzymes in complete depolymerization of lignin: a review. Biocatal. Agric. Biotechnol. 23, 101498.

Chen, Z., Wan, C., 2017. Biological valorization strategies for converting lignin into fuels and chemicals. Renew. Sust. Energ. Rev. 73, 610–621.

Chio, C., Sain, M., Qin, W., 2019. Lignin utilization: a review of lignin depolymerization from various aspects. Renew. Sust. Energ. Rev. 107, 232–249.

Cline, S.P., Smith, P.M., 2017. Opportunities for lignin valorization: an exploratory process. Energy Sustain. Soc. 7 (1), 26.

Crestini, C., Lange, H., Sette, M., Argyropoulos, D.S., 2017. On the structure of softwood kraft lignin. Green Chem. 19 (17), 4104–4121.

da Costa Lopes, A.M., João, K.G., Rubik, D.F., Bogel-Łukasik, E., Duarte, L.C., Andreaus, J., Bogel-Łukasik, R., 2013. Pre-treatment of lignocellulosic biomass using ionic liquids: wheat straw fractionation. Bioresour. Technol. 142, 198–208.

Dabral, S., Engel, J., Mottweiler, J., Spoehrle, S.S., Lahive, C.W., Bolm, C., 2018a. Mechanistic studies of base-catalysed lignin depolymerisation in dimethyl carbonate. Green Chem. 20 (1), 170–182.

Dabral, S., Wotruba, H., Hernandez, J.G., Bolm, C., 2018b. Mechanochemical oxidation and cleavage of lignin β-O-4 model compounds and lignin. ACS Sustain. Chem. Eng. 6 (3), 3242–3254.

De Wild, P.J., Huijgen, W.J., Gosselink, R.J., 2014. Lignin pyrolysis for profitable lignocellulosic biorefineries. Biofuels Bioprod. Biorefin. 8 (5), 645–657.

Delidovich, I., Hausoul, P.J., Deng, L., Pfutzenreuter, R., Rose, M., Palkovits, R., 2016. Alternative monomers based on lignocellulose and their use for polymer production. Chem. Rev. 116 (3), 1540–1599.

Dhyani, V., Bhaskar, T., 2019. Pyrolysis of biomass. In: Biofuels: Alternative Feedstocks and Conversion Processes for the Production of Liquid and Gaseous Biofuels. Academic Press, pp. 217–244.

Dong, X.Q., Yang, J.S., Zhu, N., Wang, E.T., Yuan, H.L., 2013. Sugarcane bagasse degradation and characterization of three white-rot fungi. Bioresour. Technol. 131, 443–451.

dos Santos, R.G., Alencar, A.C., 2020. Biomass-derived syngas production via gasification process and its catalytic conversion into fuels by Fischer Tropsch synthesis: a review. Int. J. Hydrog. Energy 45 (36), 18114–18132.

El Hage, R., Brosse, N., Sannigrahi, P., Ragauskas, A., 2010. Effects of process severity on the chemical structure of Miscanthus ethanol organosolv lignin. Polym. Degrad. Stab. 95 (6), 997–1003.

Fache, M., Boutevin, B., Caillol, S., 2016. Vanillin production from lignin and its use as a renewable chemical. ACS Sustain. Chem. Eng. 4 (1), 35–46.

Fahmy, T.Y., Fahmy, Y., Mobarak, F., El-Sakhawy, M., Abou-Zeid, R.E., 2018. Biomass pyrolysis: past, present, and future. Environ. Dev. Sustain., 1–16.

Forchheim, D., Gasson, J.R., Hornung, U., Kruse, A., Barth, T., 2012. Modeling the lignin degradation kinetics in a ethanol/formic acid solvolysis approach. Part 2. Validation and transfer to variable conditions. Ind. Eng. Chem. Res. 51 (46), 15053–15063.

Freitas, A.C., Ferreira, F., Costa, A.M., Pereira, R., Antunes, S.C., Gonçalves, F., et al., 2009. Biological treatment of the effluent from a bleached kraft pulp mill using basidiomycete and zygomycete fungi. Sci. Total Environ. 407, 3282–3289.

Fu, F., Yang, D., Wang, H., Qian, Y., Yuan, F., Zhong, J., Qiu, X., 2019. Three-dimensional porous framework lignin-derived carbon/ZnO composite fabricated by a facile electrostatic self-assembly showing good stability for high-performance supercapacitors. ACS Sustain. Chem. Eng. 7 (19), 16419–16427.

Gadhave, R.V., Mahanwar, P.A., Gadekar, P.T., 2018. Lignin-polyurethane based biodegradable foam. Open J. Polym. Chem. 8 (01), 1.

Garlapati, V.K., Chandel, A.K., Kumar, S.J., Sharma, S., Sevda, S., Ingle, A.P., Pant, D., 2020. Circular economy aspects of lignin: towards a lignocellulose biorefinery. Renew. Sust. Energy Rev. 130, 109977.

Ghaffar, S.H., Fan, M., 2013. Structural analysis for lignin characteristics in biomass straw. Biomass Bioenergy 57, 264–279.

Gholizadeh, M., Hu, X., Liu, Q., 2019. A mini review of the specialties of the bio-oils produced from pyrolysis of 20 different biomasses. Renew. Sust. Energ. Rev. 114, 109313.

Gillet, S., Aguedo, M., Petitjean, L., Morais, A.R.C., da Costa Lopes, A.M., Łukasik, R.M., Anastas, P.T., 2017. Lignin transformations for high value applications: towards targeted modifications using green chemistry. Green Chem. 19 (18), 4200–4233.

Guadix-Montero, S., Sankar, M., 2018. Review on catalytic cleavage of C–C inter-unit linkages in lignin model compounds: towards lignin depolymerisation. Top. Catal. 61 (3–4), 183–198.

Guedes, R.E., Luna, A.S., Torres, A.R., 2018. Operating parameters for bio-oil production in biomass pyrolysis: a review. J. Anal. Appl. Pyrolysis 129, 134–149.

Huang, J., Fu, S., Gan, L. (Eds.), 2019. Lignin Chemistry and Applications. Elsevier.

Hubbe, M.A., Alén, R., Paleologou, M., Kannangara, M., Kihlman, J., 2019. Lignin recovery from spent alkaline pulping liquors using acidification, membrane separation, and related processing steps: a review. Bioresources 14 (1), 2300–2351.

Jahirul, M.I., Rasul, M.G., Chowdhury, A.A., Ashwath, N., 2012. Biofuels production through biomass pyrolysis—a technological review. Energies 5 (12), 4952–5001.

Kalderis, D., Kotti, M.S., Méndez, A., Gascó, G., 2014. Characterization of hydrochars produced by hydrothermal carbonization of rice husk. Solid Earth 5 (1), 477.

Kamimura, N., Sakamoto, S., Mitsuda, N., Masai, E., Kajita, S., 2019. Advances in microbial lignin degradation and its applications. Curr. Opin. Biotechnol. 56, 179–186.

Kang, S., Li, X., Fan, J., Chang, J., 2013. Hydrothermal conversion of lignin: a review. Renew. Sust. Energ. Rev. 27, 546–558.

Kasakov, S., Shi, H., Camaioni, D.M., Zhao, C., Baráth, E., Jentys, A., Lercher, J.A., 2015. Reductive deconstruction of organosolv lignin catalyzed by zeolite supported nickel nanoparticles. Green Chem. 17 (11), 5079–5090.

Kaur, B., Chakraborty, D., 2013. Biotechnological and molecular approaches for vanillin production: a review. Appl. Biochem. Biotechnol. 169 (4), 1353–1372.

Kawamoto, H., 2017. Lignin pyrolysis reactions. J. Wood Sci. 63 (2), 117–132.

Kazzaz, A.E., Fatehi, P., 2020. Technical lignin and its potential modification routes: a mini-review. Ind. Crop. Prod. 154, 112732.

Khiari, B., Jeguirim, M., Limousy, L., Bennici, S., 2019. Biomass derived chars for energy applications. Renew. Sust. Energ. Rev. 108, 253–273.

Kumar, M., Oyedun, A.O., Kumar, A., 2018. A review on the current status of various hydrothermal technologies on biomass feedstock. Renew. Sust. Energ. Rev. 81, 1742–1770.

Kumar, A., Kumar, J., Bhaskar, T., 2020. Utilization of lignin: a sustainable and eco-friendly approach. J. Energy Inst. 93 (1), 235–271.

Lan, W., de Bueren, J.B., Luterbacher, J.S., 2019. Highly selective oxidation and depolymerization of α, γ diol protected lignin. Angew. Chem. Int. Ed. 58 (9), 2649–2654.

Lancefield, C.S., Panovic, I., Deuss, P.J., Barta, K., Westwood, N.J., 2017. Pre-treatment of lignocellulosic feedstocks using biorenewable alcohols: towards complete biomass valorisation. Green Chem. 19 (1), 202–214.

Lancefield, C.S., Wienk, H.L., Boelens, R., Weckhuysen, B.M., Bruijnincx, P.C., 2018. Identification of a diagnostic structural motif reveals a new reaction intermediate and condensation pathway in kraft lignin formation. Chem. Sci. 9 (30), 6348–6360.

Lee, S., Kang, M., Bae, J.H., Sohn, J.H., Sung, B.H., 2019. Bacterial valorization of lignin in biorefinery processes. Front. Bioeng. Biotechnol. 7, 209.

Li, C., Zhao, X., Wang, A., Huber, G.W., Zhang, T., 2015. Catalytic transformation of lignin for the production of chemicals and fuels. Chem. Rev. 115 (21), 11559–11624.

Li, S.H., Liu, S., Colmenares, J.C., Xu, Y.J., 2016. A sustainable approach for lignin valorization by heterogeneous photocatalysis. Green Chem. 18 (3), 594–607.

Li, J., Feng, P., Xiu, H., Li, J., Yang, X., Ma, F., Ji, Y., 2019. Morphological changes of lignin during separation of wheat straw components by the hydrothermal-ethanol method. Bioresour. Technol. 294, 122157.

Linger, J.G., Vardon, D.R., Guarnieri, M.T., Karp, E.M., Hunsinger, G.B., Franden, M.A., et al., 2014. Lignin valorization through integrated biological funneling and chemical catalysis. PNAS 111, 12013–12018.

Liu, S., Li, X., Wu, S., He, J., Pang, C., Deng, Y., et al., 2014a. Fungal pretreatment by Phanerochaete chrysosporium for enhancement of biogas production from corn stover silage. Appl. Biochem. Biotechnol. 174, 1907–1918.

Liu, S., Wu, S., Pang, C., Li, W., Dong, R., 2014b. Microbial pretreatment of corn stovers by solid-state cultivation of Phanerochaete chrysosporium for biogas production. Appl. Biochem. Biotechnol. 172, 1365–1376.

Liu, X., Bouxin, F.P., Fan, J., Budarin, V.L., Hu, C., Clark, J.H., 2020. Recent advances in the catalytic depolymerization of lignin towards phenolic chemicals: a review. ChemSusChem 13 (17), 4296.

Longe, L.F., Couvreur, J., Leriche Grandchamp, M., Garnier, G., Allais, F., Saito, K., 2018. Importance of mediators for lignin degradation by fungal laccase. ACS Sustain. Chem. Eng. 6 (8), 10097–10107.

Lu, Y., Lu, Y.C., Hu, H.Q., Xie, F.J., Wei, X.Y., Fan, X., 2017. Structural characterization of lignin and its degradation products with spectroscopic methods. J. Spectrosc. 2017.

Luo, N., Wang, M., Li, H., Zhang, J., Liu, H., Wang, F., 2016. Photocatalytic oxidation–hydrogenolysis of lignin β-O-4 models via a dual light wavelength switching strategy. ACS Catal. 6 (11), 7716–7721.

Ma, R., Guo, M., Zhang, X., 2018. Recent advances in oxidative valorization of lignin. Catal. Today 302, 50–60.

Mäkelä, M., Benavente, V., Fullana, A., 2015. Hydrothermal carbonization of lignocellulosic biomass: effect of process conditions on hydrochar properties. Appl. Energy 155, 576–584.

Mei, Q., Liu, H., Shen, X., Meng, Q., Liu, H., Xiang, J., Han, B., 2017. Selective utilization of the methoxy group in lignin to produce acetic acid. Angew. Chem. Int. Ed. 56 (47), 14868–14872.

Meng, X., Parikh, A., Seemala, B., Kumar, R., Pu, Y., Wyman, C.E., Ragauskas, A.J., 2019. Characterization of fractional cuts of co-solvent enhanced lignocellulosic fractionation lignin isolated by sequential precipitation. Bioresour. Technol. 272, 202–208.

Mohamed, B.A., Kim, C.S., Ellis, N., Bi, X., 2016. Microwave-assisted catalytic pyrolysis of switchgrass for improving bio-oil and biochar properties. Bioresour. Technol. 201, 121–132.

Morales, A., Hernández-Ramos, F., Sillero, L., Fernández-Marín, R., Dávila, I., Gullón, P., Labidi, J., 2020. Multiproduct biorefinery based on almond shells: impact of the delignification stage on the manufacture of valuable products. Bioresour. Technol. 315, 123896.

Mottweiler, J., Puche, M., Räuber, C., Schmidt, T., Concepción, P., Corma, A., Bolm, C., 2015. Copper-and vanadium-catalyzed oxidative cleavage of lignin using dioxygen. ChemSusChem 8 (12), 2106–2113.

Narron, R.H., Kim, H., Chang, H.M., Jameel, H., Park, S., 2016. Biomass pretreatments capable of enabling lignin valorization in a biorefinery process. Curr. Opin. Biotechnol. 38, 39–46.

Oliveira, F.R., Patel, A.K., Jaisi, D.P., Adhikari, S., Lu, H., Khanal, S.K., 2017. Environmental application of biochar: current status and perspectives. Bioresour. Technol. 246, 110–122.

Paone, E., Tabanelli, T., Mauriello, F., 2020. The rise of lignin biorefinery. Curr. Opin. Green Sustain. Chem. 24, 1–6.

Parakh, P.D., Nanda, S., Kozinski, J.A., 2020. Eco-friendly transformation of waste biomass to biofuels. Curr. Biochem. Eng. 6 (2), 120–134.

Picart, P., Liu, H., Grande, P.M., Anders, N., Zhu, L., Klankermayer, J., Schallmey, A., 2017. Multi-step biocatalytic depolymerization of lignin. Appl. Microbiol. Biotechnol. 101 (15), 6277–6287.

Ponnusamy, V.K., Nguyen, D.D., Dharmaraja, J., Shobana, S., Banu, J.R., Saratale, R.G., Kumar, G., 2019. A review on lignin structure, pretreatments, fermentation reactions and biorefinery potential. Bioresour. Technol. 271, 462–472.

Pour, R.R., Ehibhatiomhan, A., Huang, Y., Ashley, B., Rashid, G.M., Mendel-Williams, S., Bugg, T.D., 2019. Protein engineering of Pseudomonas fluorescens peroxidase Dyp1B for oxidation of phenolic and polymeric lignin substrates. Enzym. Microb. Technol. 123, 21–29.

Ragauskas, A.J., Beckham, G.T., Biddy, M.J., Chandra, R., Chen, F., Davis, M.F., Langan, P., 2014. Lignin valorization: improving lignin processing in the biorefinery. Science 344 (6185).

Rahimi, A., Ulbrich, A., Coon, J.J., Stahl, S.S., 2014. Formic-acid-induced depolymerization of oxidized lignin to aromatics. Nature 515 (7526), 249–252.

Rahmanpour, R., Bugg, T.D., 2015. Characterisation of Dyp-type peroxidases from Pseudomonas fluorescens Pf-5: oxidation of Mn (II) and polymeric lignin by Dyp1B. Arch. Biochem. Biophys. 574, 93–98.

Raul, C., Priyadarshi, S., Bharti, V.S., Prakash, S., 2019. Biochar: an emerging solution for sustainable aquaculture. World Aquacult. 50 (3), 64–65.

Ren, X., Wang, P., Han, X., Zhang, G., Gu, J., Ding, C., Cao, F., 2017. Depolymerization of lignin to aromatics by selectively oxidizing cleavage of C–C and C–O bonds using CuCl2/polybenzoxazine catalysts at room temperature. ACS Sustain. Chem. Eng. 5 (8), 6548–6556.

Rencoret, J., Prinsen, P., Gutiérrez, A., Martínez, Á.T., Del Río, J.C., 2015. Isolation and structural characterization of the milled wood lignin, dioxane lignin, and cellulolytic lignin preparations from brewer's spent grain. J. Agric. Food Chem. 63 (2), 603–613.

Renders, T., Van den Bosch, S., Vangeel, T., Ennaert, T., Koelewijn, S.F., Van den Bossche, G., Sels, B.F., 2016. Synergetic effects of alcohol/water mixing on the catalytic reductive fractionation of poplar wood. ACS Sustain. Chem. Eng. 4 (12), 6894–6904.

Rodríguez Correa, C., Stollovsky, M., Hehr, T., Rauscher, Y., Rolli, B., Kruse, A., 2017. Influence of the carbonization process on activated carbon properties from lignin and lignin-rich biomasses. ACS Sustain. Chem. Eng. 5 (9), 8222–8233.

Roth, S., Spiess, A.C., 2015. Laccases for biorefinery applications: a critical review on challenges and perspectives. Bioprocess Biosyst. Eng. 38 (12), 2285–2313.

Ruan, X., Liu, Y., Wang, G., Frost, R.L., Qian, G., Tsang, D.C., 2018. Transformation of functional groups and environmentally persistent free radicals in hydrothermal carbonisation of lignin. Bioresour. Technol. 270, 223–229.

Sainsbury, P.D., Hardiman, E.M., Ahmad, M., Otani, H., Seghezzi, N., Eltis, L.D., et al., 2013. Breaking down lignin to high-value chemicals: the conversion of lignocellulose to vanillin in a gene deletion mutant of Rhodococcus jostii RHA1. ACS Chem. Biol. 8, 2151–2156.

Schutyser, W., Renders, A.T., Van den Bosch, S., Koelewijn, S.F., Beckham, G.T., Sels, B.F., 2018. Chemicals from lignin: an interplay of lignocellulose fractionation, depolymerisation, and upgrading. Chem. Soc. Rev. 47 (3), 852–908.

Sedai, B., Diaz-Urrutia, C., Baker, R.T., Wu, R., Silks, L.P., Hanson, S.K., 2011. Comparison of copper and vanadium homogeneous catalysts for aerobic oxidation of lignin models. ACS Catal. 1 (7), 794–804.

Shen, X.J., Wen, J.L., Mei, Q.Q., Chen, X., Sun, D., Yuan, T.Q., Sun, R.C., 2019. Facile fractionation of lignocelluloses by biomass-derived deep eutectic solvent (DES) pretreatment for cellulose enzymatic hydrolysis and lignin valorization. Green Chem. 21 (2), 275–283.

Shuai, L., Saha, B., 2017. Towards high-yield lignin monomer production. Green Chem. 19 (16), 3752–3758.

Singh, D., Zeng, J., Laskar, D.D., Deobald, L., Hiscox, W.C., Chen, S., 2011. Investigation of wheat straw biodegradation by Phanerochaete chrysosporium. Biomass Bioenergy 35, 1030–1040.

Singh, R., Prakash, A., Dhiman, S.K., Balagurumurthy, B., Arora, A.K., Puri, S.K., Bhaskar, T., 2014. Hydrothermal conversion of lignin to substituted phenols and aromatic ethers. Bioresour. Technol. 165, 319–322.

Stärk, K., Taccardi, N., Bösmann, A., Wasserscheid, P., 2010. Oxidative depolymerization of lignin in ionic liquids. ChemSusChem 3 (6), 719–723.

Stone, M.L., Anderson, E.M., Meek, K.M., Reed, M., Katahira, R., Chen, F., Román-Leshkov, Y., 2018. Reductive catalytic fractionation of C-lignin. ACS Sustain. Chem. Eng. 6 (9), 11211–11218.

Strassberger, Z., Tanase, S., Rothenberg, G., 2014. The pros and cons of lignin valorisation in an integrated biorefinery. RSC Adv. 4 (48), 25310–25318.

Suman, S., Gautam, S., 2017. Effect of pyrolysis time and temperature on the characterization of biochars derived from biomass. Energy Sources, Part A 39 (9), 933–940.

Sun, Z., Fridrich, B., de Santi, A., Elangovan, S., Barta, K., 2018. Bright side of lignin depolymerization: toward new platform chemicals. Chem. Rev. 118 (2), 614–678.

Sun, C., Zheng, L., Xu, W., Dushkin, A.V., Su, W., 2020. Mechanochemical cleavage of lignin models and lignin via oxidation and a subsequent base-catalyzed strategy. Green Chem. 22 (11), 3489–3494.

Supanchaiyamat, N., Jetsrisuparb, K., Knijnenburg, J.T., Tsang, D.C., Hunt, A.J., 2019. Lignin materials for adsorption: current trend, perspectives and opportunities. Bioresour. Technol. 272, 570–581.

Taniguchi, M., Suzuki, H., Watanabe, D., Sakai, K., Hoshino, K., Tanaka, T., 2005. Evaluation of pretreatment with Pleurotus ostreatus for enzymatic hydrolysis of rice straw. J. Biosci. Bioeng. 100, 637–643.

Terrett, O.M., Dupree, P., 2019. Covalent interactions between lignin and hemicelluloses in plant secondary cell walls. Curr. Opin. Biotechnol. 56, 97–104.

Toledano, A., Serrano, L., Labidi, J., 2014. Improving base catalyzed lignin depolymerization by avoiding lignin repolymerization. Fuel 116, 617–624.

Tripathi, M., Sahu, J.N., Ganesan, P., 2016. Effect of process parameters on production of biochar from biomass waste through pyrolysis: a review. Renew. Sust. Energ. Rev. 55, 467–481.

Vienescu, D.N., Wang, J., Le Gresley, A., Nixon, J.D., 2018. A life cycle assessment of options for producing synthetic fuel via pyrolysis. Bioresour. Technol. 249, 626–634.

Wang, H., Tucker, M., Ji, Y., 2013. Recent development in chemical depolymerization of lignin: a review. J. Appl. Chem. 2013 (9).

Wang, T., Zhai, Y., Zhu, Y., Li, C., Zeng, G., 2018. A review of the hydrothermal carbonization of biomass waste for hydrochar formation: process conditions, fundamentals, and physicochemical properties. Renew. Sust. Energ. Rev. 90, 223–247.

Wang, H., Pu, Y., Ragauskas, A., Yang, B., 2019. From lignin to valuable products–strategies, challenges, and prospects. Bioresour. Technol. 271, 449–461.

Wells Jr., T., Wei, Z., Ragauskas, A., 2015. Bioconversion of lignocellulosic pretreatment effluent via oleaginous Rhodococcus opacus DSM 1069. Biomass Bioenergy 72, 200–205.

Wilhelm, R.C., Singh, R., Eltis, L.D., Mohn, W.W., 2019. Bacterial contributions to delignification and lignocellulose degradation in forest soils with metagenomic and quantitative stable isotope probing. ISME J. 13 (2), 413–429.

Wong, S.S., Shu, R., Zhang, J., Liu, H., Yan, N., 2020. Downstream processing of lignin derived feedstock into end products. Chem. Soc. Rev. 49 (15), 5510–5560.

Wu, W., Yang, M., Feng, Q., McGrouther, K., Wang, H., Lu, H., Chen, Y., 2012. Chemical characterization of rice straw-derived biochar for soil amendment. Biomass Bioenergy 47, 268–276.

Wu, H., Song, J., Xie, C., Wu, C., Chen, C., Han, B., 2018a. Efficient and mild transfer hydrogenolytic cleavage of aromatic ether bonds in lignin-derived compounds over Ru/C. ACS Sustain. Chem. Eng. 6 (3), 2872–2877.

Wu, X., Fan, X., Xie, S., Lin, J., Cheng, J., Zhang, Q., Wang, Y., 2018b. Solar energy-driven lignin-first approach to full utilization of lignocellulosic biomass under mild conditions. Nat. Catal. 1 (10), 772–780.

Wu, Z., Zhao, X., Zhang, J., Li, X., Zhang, Y., Wang, F., 2019. Ethanol/1,4-dioxane/formic acid as synergistic solvents for the conversion of lignin into high-value added phenolic monomers. Bioresour. Technol. 278, 187–194.

Xiang, Z., Han, W., Deng, J., Zhu, W., Zhang, Y., Wang, H., 2020. Photocatalytic conversion of lignin into chemicals and fuels. ChemSusChem 13 (17), 4199–4213.

Xiao, L.P., Wang, S., Li, H., Li, Z., Shi, Z.J., Xiao, L., Song, G., 2017. Catalytic hydrogenolysis of lignins into phenolic compounds over carbon nanotube supported molybdenum oxide. ACS Catal. 7 (11), 7535–7542.

Xu, C., Arancon, R.A.D., Labidi, J., Luque, R., 2014. Lignin depolymerisation strategies: towards valuable chemicals and fuels. Chem. Soc. Rev. 43 (22), 7485–7500.

Yang, J., Yang, L., 2019. A review on hydrothermal co-liquefaction of biomass. Appl. Energy 250, 926–945.

Yang, H., Yan, R., Chen, H., Lee, D.H., Zheng, C., 2007. Characteristics of hemicellulose, cellulose and lignin pyrolysis. Fuel 86 (12–13), 1781–1788.

Yang, X., Li, N., Lin, X., Pan, X., Zhou, Y., 2016. Selective cleavage of the aryl ether bonds in lignin for depolymerization by acidic lithium bromide molten salt hydrate under mild conditions. J. Agric. Food Chem. 64 (44), 8379–8387.

Yao, S.G., Mobley, J.K., Ralph, J., Crocker, M., Parkin, S., Selegue, J.P., Meier, M.S., 2018. Mechanochemical treatment facilitates two-step oxidative depolymerization of kraft lignin. ACS Sustain. Chem. Eng. 6 (5), 5990–5998.

Yoo, C.G., Pu, Y., Li, M., Ragauskas, A.J., 2016. Elucidating structural characteristics of biomass using solution-state 2 D NMR with a mixture of deuterated dimethylsulfoxide and hexamethylphosphoramide. ChemSusChem 9 (10), 1090–1095.

Yorgun, S., Yıldız, D., 2015. Slow pyrolysis of paulownia wood: effects of pyrolysis parameters on product yields and bio-oil characterization. J. Anal. Appl. Pyrolysis 114, 68–78.

Yu, X., Wei, Z., Lu, Z., Pei, H., Wang, H., 2019. Activation of lignin by selective oxidation: an emerging strategy for boosting lignin depolymerization to aromatics. Bioresour. Technol. 291, 121885.

Yunpu, W.A.N.G., Leilei, D.A.I., Liangliang, F.A.N., Shaoqi, S.H.A.N., Yuhuan, L.I.U., Roger, R.U.A.N., 2016. Review of microwave-assisted lignin conversion for renewable fuels and chemicals. J. Anal. Appl. Pyrolysis 119, 104–113.

Zhang, C., Wang, F., 2020. Catalytic lignin depolymerization to aromatic chemicals. Acc. Chem. Res. 53 (2), 470–484.

Zhang, K., Li, H., Xiao, L.P., Wang, B., Sun, R.C., Song, G., 2019a. Sequential utilization of bamboo biomass through reductive catalytic fractionation of lignin. Bioresour. Technol. 285, 121335.

Zhang, Y., He, H., Liu, Y., Wang, Y., Huo, F., Fan, M., Zhang, S., 2019b. Recent progress in theoretical and computational studies on the utilization of lignocellulosic materials. Green Chem. 21 (1), 9–35.

Zhi, Z., Wang, H., 2014. White-rot fungal pretreatment of wheat straw with Phanerochaete chrysosporium for biohydrogen production: simultaneous saccharification and fermentation. Bioprocess Biosyst. Eng. 37, 1447–1458.

Zhou, M., Xu, J., Jiang, J., Sharma, B.K., 2018. A review of microwave assisted liquefaction of lignin in hydrogen donor solvents: effect of solvents and catalysts. Energies 11 (11), 2877.

Zhu, Z., Xu, Z., 2020. The rational design of biomass-derived carbon materials towards next-generation energy storage: a review. Renew. Sust. Energ. Rev. 134, 110308.

Zijlstra, D.S., de Santi, A., Oldenburger, B., de Vries, J., Barta, K., Deuss, P.J., 2019. Extraction of lignin with high β-O-4 content by mild ethanol extraction and its effect on the depolymerization yield. J. Vis. Exp. 143, e58575.

C H A P T E R

15

Bio-waste to hydrogen production technologies

Triya Mukherjee[a,b] and S. Venkata Mohan[a,b]

[a]Bioengineering and Environmental Sciences Lab, Department of Energy and Environmental Engineering, CSIR-Indian Institute of Chemical Technology (CSIR-IICT), Hyderabad, India
[b]Academy of Scientific and Innovative Research (AcSIR), Ghaziabad, India

1 Introduction

Climate change is a challenge best measured in generations rather than years wherein energy plays a critical role. The demand for renewable energy is increasing due to the extinction of finite fossils and increasing environmental issues (Kalair et al., 2021; Antar et al., 2021). The need to replace the conventional fossil resources with alternative ones, which are renewable and green, is a major focus in the context of sustainability at present (Antar et al., 2021; Chandrasekhar et al., 2020). Therefore, a sustainable and clean energy source needs to thrive in the 21st century. Since H_2 gas does produce CO_2 on combustion, it carries significant quantities of energy per unit weight. H_2 can be utilized as a rocket propellant, vehicle fuel, and as a primary commodity in chemical processes like methanol processing, hydrodealkylation, and hydrocracking, among other processes (Mazloomi and Gomes, 2012).

With the fusion of H_2 into the fuel policies of all the nations, the concepts of blue H_2 (from reforming with CCUS), gray H_2 (from reforming with CO_2 emissions), turquoise H_2 (methane pyrolysis with solid carbon by-product), pink H_2 (electrolysis with nuclear power), yellow H_2 (electrolysis with solely solar energy), and green H_2 (water-splitting with renewable energy sources and biohydrogen) are coming into the limelight (Petrofac, International Renewable Energy Agency-IRENA). The electrolysis primarily requires fresh water and the mechanism of water-splitting is a highly endothermic reaction with a minimum of 237 kJ/mol energy (standard Gibbs free energy; ΔG_0 for water formation) in the form of electricity to split a mole of water and give rise to 1 mol of H_2 (Tsutsumi, 2009). The mechanism also requires additional

49 kJ/mol energy to subdue the entropy change (ΔH) in the reaction accounting for a total of 286 kJ/mol external energy supply for 1 mol of water (Eq. 1).

$$H_2O \rightarrow H_2 + 1/2 O_2 \tag{1}$$

$$\Delta G_0 = 237 kJ/mol$$

$$-\Delta H = -\Delta G - T\Delta S$$

The same amount of energy will be released in the combustion of H_2 as fuel, provided the efficiency of the engine is 100%. The overall sustainability of the H_2 production process depends upon the type of energy used for water dissociation and the efficiency of the respective procedure. Since the energy required for the water-splitting is provided with any renewable energy, the H_2 is considered as a low-carbon fuel. Production of H_2 can be done by thermochemical activities such as gasification, plasma technology, pyrolysis, hydrocarbon reforming, etc. (Sampath et al., 2020). The challenge in addressing both the energy crisis and pollution control issues is to establish a cost-effective and environmentally sustainable H_2 production process (Herzog and Tatsutani, 2005). Biological approaches such as indirect photobiolysis, dark or photo-fermentation, bacteriological electrolysis chambers, and other fermentative methods are green alternative routes (Fig. 1). As a result of their flexibility in using different raw materials and ability to derive energy from wastes, biological methods are being prioritized (Mohan, 2010; Ahorsu et al., 2018). H_2 produced from unwanted

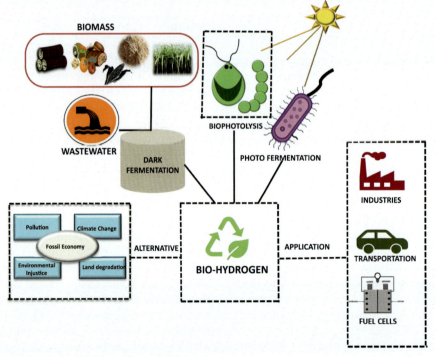

FIG. 1 Biohydrogen production and its application.

wastewater/biomass utilizing biotic processes as a sustainable source, on the other hand, is gaining popularity. In this manner, negatively priced organic wastes are converted to green energy thus, reducing pollution (Ghimire et al., 2015). The scope of this chapter is to discuss the potent waste feedstock for the production of bioH$_2$ through biological routes. The role of enzymes in the microbial metabolic pathways would be addressed along with the effect of integrating different biological and bio-electrochemical routes towards enhanced H$_2$ production. The process optimization, microbial metabolic routes, and sustainable and cost-effective biorefinery models are precisely discussed to acknowledge the importance of biological/green H$_2$.

2 Current process for biohydrogen production

Initial studies focused on the biological hydrogen (H$_2$) by Bio-photolysis (Green algae and cyanobacteria), Photo-fermentation (Photo-fermentative bacteria), Dark fermentation processes (fermentative bacteria), enzymatic and microbial electrolysis (Fig. 2). The low rate of H$_2$ production as well as the requirement of light as a source of energy makes the bio-photolysis and photo-fermentative process less effective compared to dark fermentation (Lee et al., 2010). The "dark fermentation" is a propitious approach toward bioH$_2$ production as it requires no source of light energy, can utilize waste as substrate at moderate conditions in turn delivering a good rate of BioH$_2$ generation (Mohan, 2010; Ahorsu et al., 2018; Ghimire et al., 2015).

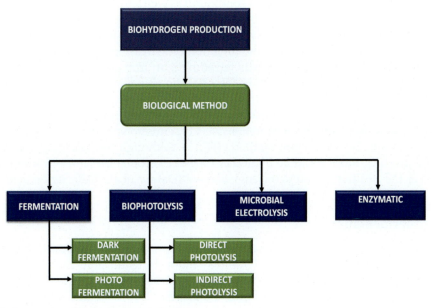

FIG. 2 Various routes of biohydrogen production.

2.1 Dark fermentation/acidogenesis

Due to its twin benefits of waste remediation and clean fuel generation, H_2 generation by dark fermentation using organic waste/wastewater is gaining popularity. Acidogenesis, hydrolysis, methanogenesis, and acetogenesis are all junctures of biochemical metabolic processes involved in dark fermentation (Dahiya et al., 2020, 2015). In hydrolysis, macromolecules are broken down into basic compounds, which then undergo fermentation to form organic acids in acidogenesis (Dahiya et al., 2020; Lee et al., 2010; Mohan, 2008; Mohan et al., 2008). The homoacetogens and acetogens produce hydrogen from acetic acid. Acetogens commonly cultivate in syntrophic partnership with hydrogenotrophic methanogens, maintaining partial burden to permit for thermodynamically constructive acidogenesis circumstances (Liu et al., 2020; Chandrasekhar et al., 2015). Methanogenic activity is therefore vanquished to boost bioH$_2$ yields to be a single metabolic substitute (Mohan, 2008; Chandrasekhar et al., 2015). During glycolysis, organic sugars are first metabolized to pyruvate, a crucial step for successive microbial fermentation (Müller, 2001; Arunasri et al., 2020). Pyruvate may also be used as a carbon source to produce variability of organic fatty acids and H_2 formed during the time of metabolism (Arunasri et al., 2020; Macfarlane and Macfarlane, 2003; Mukherjee and Mohan, 2021). Pyruvate is converted to acetyl-CoA through facultative anaerobes further forming formate by pyruvate lyase, and lastly forming hydrogen by formate hydrogen lyase (Arunasri et al., 2020; Mukherjee and Mohan, 2021). The pyruvate ferredoxin oxidoreductase assists the anaerobes to convert pyruvate to acetyl-CoA. The production of redox equivalents within the bacterium is increased by metabolites generated during anaerobic substrate metabolism (Sarkar and Mohan, 2020). In the occurrence of dehydrogenase enzyme (NADH), protons from redox mediators dissociate and are then reduced to H_2 using electrons generated by oxidized ferredoxin via hydrogenase enzyme (Buckel and Thauer, 2013; Lee et al., 2018). The membrane destined NADH dehydrogenase, cytochrome compound as well as other carrier proteins, network the electrons through the quinone pool (Brandt, 1997; Sarewicz and Osyczka, 2015). The transfer of electrons to cytochrome is aided by unbroken interconversions of quinone and protons named as b–c1 composite and then in the direction of cytochrome $aa3$. Lastly, electrons are transported from cytochrome $aa3$ towards ferredoxin, an iron-containing protein. In its reduced state, ferredoxin sends electrons to the catalytic site of hydrogenase, where protons are united to generate H_2 (Artz et al., 2019; Schuchmann et al., 2018).

The generation of bioH$_2$ through dark fermentation/acidogenesis can be considered as a low-carbon procedure where the comprehensive CO_2 emissions of the process are relatively lower than the other H_2 production methods (Fig. 3). Although the bioH$_2$ process results in CO_2 as a by-product, those could be treated as biogenic carbon emissions which may not be considered as the additional load to the environment (Dahiya et al., 2020; Sarkar et al., 2021a).

2.2 Photo-fermentation

Pretreatment procedure, raw material properties, photobioreactor configuration, light disposition and compactness, and light–heat–mass transmission materials all affect photo-fermentative bioH$_2$ processing, which is an intricate system connecting chemical, physical, and biological procedures (Zhang and Zhang, 2018). Volatile fatty acids such as

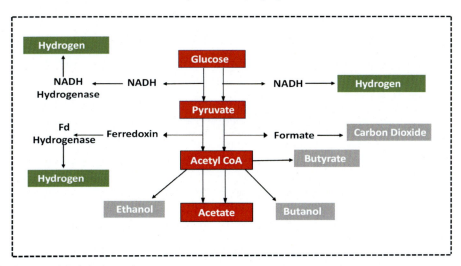

FIG. 3 Process of dark fermentation.

acetic or succinic acid, are used as e^- donors during photo-fermentation (Banu et al., 2021). From PSI, it is identified that electron counterparts are transported to nitrogenase enzyme which leads to the generation of H_2 (Veeravalli et al., 2019; Fang et al., 2020). Despite excellent yield of H_2, the foremost downsides of this biological technique are: (i) use of nitrogenase enzyme united with a high demand of energy, (ii) low-slung solar energy conversion efficiencies, further leading to less volumetric production rates, and (iii) requirement of anaerobic photobioreactors which covers huge area (Sagir and Alipour, 2021). The performance of photosynthetic bacteria is critical in the process of photo-fermentation, as it helps to control the substrate of explicit ranges and conversion competence (Dalena et al., 2017). As an end result, the photo-fermentation, process of H_2 generation necessitates the transmission of bacteria with a high dimension for H_2 synthesis (Liu et al., 2013). The decline of ferredoxins and formation of ATP in photo-fermentation is completed by the purple non-sulfur photosynthetic bacteria (PNS) in anaerobic circumstances using converse electron flow determined by captured solar energy. Instead of getting electrons through water-splitting processes, as in cyanobacteria or microalgae, the organic molecules, for example organic acid, function as a donor of the electron in the PNS bacterium under anaerobic circumstances (Basak and Das, 2007; Tiang et al., 2020). Some promising advantages, if H_2 is produced via this pathway, are basically: (a) Comprehensive substrate conversion to CO_2 and H_2; (b) Elimination of deleterious effect of O_2 which hinders the activity of [Fe-Fe] H_2ase, [NiFe] H_2ase, and nitrogenase enzyme donor; (c) effective consumption of sunlight in both contexts of visible (400–700 nm) and near-infrared (700–950 nm) areas of the solar light spectrum; (d) extensive availability of organic composites used as e^- donor; and (e) a relatively minor energy barrier to overpower, contrast to water-splitting in direct photolysis (Sun et al., 2019).

2.3 Bio-photolysis

2.3.1 *Direct photolysis*

An undeviating bio-photolysis of producing H_2 is a biological process, which exploits the use of solar energy and photosynthetic systems of algae towards generating chemical energy from water (Eq. 2) (Benemann, 1996, 2000).

$$2H_2O + Solar\ Energy \rightarrow 2H_2 + O_2 \qquad (2)$$

The components for photosynthesis mechanism are: (a) photosystem I (PSI) which harvests a reducing agent for CO_2 and (b) photosystem II (PSII) which breaches water to develop O_2 (Benemann, 1996, 2000). In presence of H_2ase, the two photons attained from H_2O splitting can either reduce CO_2 by PSI or yield H_2. Due to the lack of H_2ase enzyme, plants can only reduce CO_2 while green algae and blue-green algae (cyanobacteria) which contain H_2ase have the capability of producing H_2 (Kossalbayev et al., 2020). When PSII engrosses light energy in these organisms, the electrons are generated, which are subsequently transported to ferredoxin. With the existence of H_2ase, an irreversible hydrogenase takes all the electrons straight from the concentrated ferredoxin to produce H_2 (Benemann, 1996).

2.3.2 *Indirect photolysis*

The complexity of the H_2 evolving process to O_2 in indirect bio-photolysis can be sidestepped by extrication of O_2 and H_2 (Nagarajan et al., 2020). Within this process, CO_2 is periodically static and unconfined, serving as the electron transporter between O_2 generating (water-splitting) reactions and O_2 sensitive H_2ase reactions (Akhlaghi and Najafpour-Darzi, 2020). With such ideas, the algae go through a cycle of CO_2 fixation into storing carbohydrates such as starch and glycogen subsequently transforming into H_2 by the process of dark fermentation. Many species of green as well as blue-green algae, in addition to fixing CO_2 through photosynthesis, lead to fixing dinitrogen from the atmosphere and further create enzymes that catalyzes the second stage of the production of H_2 process (Sivaramakrishnan et al., 2021). Because these fixing of nitrogen enzymes, nitrogenase, are originated inside the heterocyst, they offer an oxygen-free environment for the evolution of H_2 events to take place (Sivaramakrishnan et al., 2021; Dincer, 2012).

2.4 Enzymatic action on biohydrogen production

A series of biological electrochemical reactions can be used to generate hydrogen via biological processes. A range of biocatalyst enzymes that are identified for their essential performance in $bioH_2$ synthesis helps to enable these processes (Sun et al., 2019; Srikanth and Mohan, 2012; Chandrasekhar et al., 2021; Chandrasekhar and Mohan, 2014). There are three chief $bioH_2$ consumption and production enzymes, which are accountable for the gross $bioH_2$ development (Baeyens et al., 2020). Reversible hydrogenase, membrane-bound uptake hydrogenase, and nitrogenase enzymes are the three enzymes in question. The two most important biocatalysts among them are nitrogenase and hydrogenase (Sun et al., 2019; Pandey et al., 2019). Hydrogenases are divided into three groups based on the metal composition of their active sites: (i) NiFe-hydrogenase, (ii) FeFe-hydrogenase, and (iii) Fe-hydrogenase

(Kim and Kim, 2011; Lubitz et al., 2014; Pandey et al., 2019). The NiFe-hydrogenases enzyme is majorly known for generating hydrogen or corrosion of catalyzing inactivity only in bacteria and in archaea, while FeFe-hydrogenases catalyze exclusively the production of hydrogen in eukaryotes and bacteria (Kim and Kim, 2011; Peters et al., 2015; Pandey et al., 2019). In hydrogenotrophic methanogenic archaea, on the other hand, Fe-hydrogenase is erratically distributed and utilizes hydrogen to produce reducing equivalents for transitional steps in CO_2 reduction to methane (Thauer et al., 2010; Constant and Hallenbeck, 2019; Pandey et al., 2019). The nitrogenase complex entails two subunits and they are dinitrogenase and nitrogenase reductase. Fe—S protein is originated in the reductase subunit, which is enclosable. A 65-kilodalton homodimer transports electrons from an external electron contributor to the dinitrogenasemultifacet. The dinitrogenase complex, a molecular weight of 230 kDa, and a heterotetramer with 22 units, which contains Mo-FeS protein (Mona et al., 2020). In a plodding process, it alters dinitrogen (N_2) into two different molecules of ammonia (Hallenbeck and Benemann, 2002). There is a synchronized proton drop into the molecular hydrogen. Nitrogenises are categorized as molybdenum, iron, or vanadium forms contingent on the metal cofactors in their catalytic spots (Hu et al., 2011). Mo–nitrogenase, V–nitrogenase, and Fe–nitrogenase are some of the distinct stoichiometry's of H_2 and ammonia synthesis by altered nitrogenases.

2.5 Microbial electrolysis cell (MEC)

Electro-hydrogenesis in microbial desalination cells (MECs) (Fig. 4) can synthesize H_2 with increased yield and energy competence than fermentation or water electrolysis (Call and Logan, 2008). Microbial electrolysis cells (MEC) are a unique and promising technique for

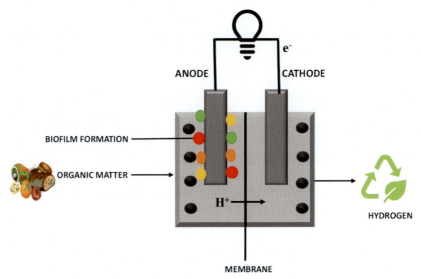

FIG. 4   Illustration of microbial electrolysis cell for biohydrogen production.

the production of H_2 from biogenic sources, such as sewerage and other renewable energy sources (Mohan, 2010; Kadier et al., 2016). The MEC is a modified version of the conventional microbial fuel cell (Sravan et al., 2021, 2019; Mohan et al., 2019a). Exclusive exoelectrogens transform degradable substrate and carry electrons to the anode in a microbial fuel cell (Sravan et al., 2021, 2019; Mohan et al., 2019a). Flowing transversely on an external load, the electrons amalgamated at the cathode, with protons and O_2, forms H_2O. Instead of generating voltage, an external voltage is applied to the MEC, which operates in anaerobic state (Sravan et al., 2019). The biocatalyst, membranes, electrode materials, applied potential, designing of the substrate, and loading rates are various optimization strategies of MEC systems, each of which shows a precarious role in recital competencies (Mohan et al., 2013a). For operating MEC, a variety of bacterial cultures, including pure strains, mixed strains, two strain combinations, etc., are incorporated. H_2-capture competencies from innumerable donor substrates (taking, e.g., of, glucose, cellulose, lactate, butyrate, ethanol, propionate, and acetate) which are ranged from 67% to 91%, indicating that MEC can provide high H_2 yields (Lee et al., 2010). Inspite of such exploration exertions, all endeavors to gage up the large-sized MECs have come across countless complications. Most of which MEC aviators with capacities of several liters exhibition at very low production rate in hydrogen, conforming cathode working at less than $1\,A.m^{-2}$ (Rousseau et al., 2020).

3 Waste feedstock for biohydrogen production

The production of $bioH_2$ primarily depends on the characteristics and composition of feedstock which determines the rate, yield, efficiency, and cost (Mohan, 2008; Mohan et al., 2008; Chozhavendhan et al., 2020; Ntaikou et al., 2010; Han et al., 2016). Plethora of options are available which qualify as feedstock materials for $BioH_2$ production ranging from wastewater (industrial effluent and sewage), agricultural crops (excessive lignocellulosic biomass), biogenic municipal solid waste, etc. In accumulation, to the other factors like carbon to nitrogen ratio (C/N), substrates volatile solid content (V/S), chemical oxygen demand (COD), the occurrence of some inhibitory composites and organic biodegradable fraction; determining the enactment of production that is $BioH_2$. The one technique to categorize feedstock is based on the composition of nutrients: (1) protein-rich; (2) carbohydrate-rich; (3) cellulose-rich and (4) fat-rich (Osman et al., 2020; Bartacek et al., 2007). Any kind of organic substance that encompasses any of these clusters could be considered for the production of biohydrogen. However, numerous studies have recommended that nitrogen-deficient and carbohydrate-rich feedstock harvest a superior extent of hydrogen (Liu et al., 2020). Biomass can be considered to be the major origin of farming (agricultural biomass); through the processing of biomass industry (emissions and agro-industrial residues); through the retail industry and consumer industry (unwanted food discarded from home-produced, restaurants and catering, etc.,) and larger percentage of the organic element of municipal dense waste (Osman et al., 2020; Bartacek et al., 2007).

The concept behind the cultivation of so-called energy crops is to exploit the crop (whole or a part) for the fabrication of energy (Paschalidou et al., 2016; Lopez-Bellido et al., 2014).

The energy production from such crops can be labeled as sustainable when they fulfill a set criterion such as (i) high biomass yield—for higher energy yield (ii) low cultivation cost (low H_2O & nutrient requirement)—as an economic factor is key for the success of any technology (iii) resistant to environmental stresses—which will reduce the losses which can be incurred due to environmental stress (Hawkes et al., 2002). The crops with great content of carbohydrates with low lignin contented are considered more suitable. The major drawback of the practice of energy crops that is for the production of $bioH_2$ is a debate for "Food vs Fuel" as the same land resources can be used to grow food crops instead of fuel crops and can be a source of alleviation of hunger and nutrition deprived population (Harlander, 2008). Crop or lignocellulosic residue originates through agricultural production and processing, for example, rice husk, wheat bran, sugarcane bagasse, etc. are a probable source for $bioH_2$ production as they are known to be rich in complex carbohydrate cellulose, hemicellulose, and lignin (Yadav et al., 2020; Cheng et al., 2011). The crop or lignocellulosic residues are one of the best contenders as a suitable feedstock because (i) they are present in abundance around the world (ii) low cost (iii) rich in carbohydrate (iv) sustainable over the long term (Ahorsu et al., 2018). The lignin content in these residues hinders the process of fermentation as they are closely bound and this conjugation makes the action of cellulolytic microorganisms difficult for the conversion (Cheng et al., 2011). The problem of $bioH_2$ production from these residues is degradation/removal of lignin content from the feedstock and for the removal of the lignin, a step of pretreatment for the feedstock is applied (Kumar and Sharma, 2017). This pretreatment of feedstock can be done using different mechanisms, i.e., mechanical, chemical, or biological (Kumar and Sharma, 2017; Kucharska et al., 2018; Mohan and Goud, 2012). This method of pretreatment facilitates the delignification of the residue which helps in the release of soluble carbohydrates (Panagiotopoulos et al., 2009). Generation of H_2 from fermentation process is possible with numerous forms of wastewater used with either pure or mixed cultures (Mohan, 2010, 2008, 2009; Mohan et al., 2008, 2007a,b, 2013; Hallenbeck and Benemann, 2002; Pasupuleti and Venkata Mohan, 2015). The use of various effluents, mainly wastewater containing cellulose, xylose glycerol, and pentose, filtrates from the production of biodiesel, overflow from cheese processing, processing of dairy wastewater by-products of wheat flour, black treacle, solid food wastes, effluent from the production of paper, and domestic sewage among others, were reported (Mohan, 2008; Mohan et al., 2007b, 2013; Han and Shin, 2004; Yang et al., 2007; Valdez-Vazquez et al., 2005; Pasupuleti and Venkata Mohan, 2015). Over the year a lot of efforts have been made on the solicitation of use of various wastewater from industrial and domestic sources as the budding feedstock for the manufacture of bio-H_2 over biological machinery, largely over light-independent & light-driven fermentation processes. This biotransformation of wastewater and wastes headed for H_2, can be considered alluring from the environmental (renewable energy and pollution control) and economical (low total cost waste supervision and resources recovery) point of view (Mohan et al., 2007b, 2013; Han and Shin, 2004). According to the criteria, a wastewater/waste must have a high concentration of degradable compounds, high percentage of voluntarily fermentable compounds, which are carbohydrates and sugars, and low concentration of compounds that are repressive towards microbial metabolism for being considered as effectual feedstock for H_2 production (Li et al., 2012).

4 Process integrations and hybrid systems for hydrogen production

The main restrictions associated with solid waste fermentation are low energy efficiency and energy conversion values (Keskin et al., 2019). It is hence preferable to improve current reactor configurations or integrate multiple systems rather than designing a novel apparatus setup. A two-phase system utilizing anaerobic digestion could improve overall performance and process dependability by reducing waste stabilization time (Mohanakrishna and Mohan, 2013; Nikhil et al., 2017). It has been already demonstrated by various researchers that integrating several processes (viz. acidogenic and photo-biological processes) can enhance the rate of H_2 production (Mohanakrishna and Mohan, 2013). The production of bioplastics from acidogenic fermentation from volatile fatty acids as a substrate and aerobic microbial consortia as a biocatalyst has been studied as well. This integration strategy can reduce the overall cost of the process and can enhance the treatment efficiency especially when wastewater is used as feedstock (Mohan et al., 2010). These technologies offer supplementary consistent feed rates in stage two by buffering the organic loading rate in stage one. Hybrid systems are most extensively used in recent years to increase bioH$_2$ yield from biogenic waste materials (Fig. 5).

4.1 Hythane (H-CNG)

Hythane ($H_2 + CH_4$) is gaining popularity due to its many benefits, including its use as a car fuel. Biohythane, which is made up of bioH$_2$ and bioCH$_4$ through a dual-stage fermentation process, is a possible high-value solution for the valorization of waste biomass resources and

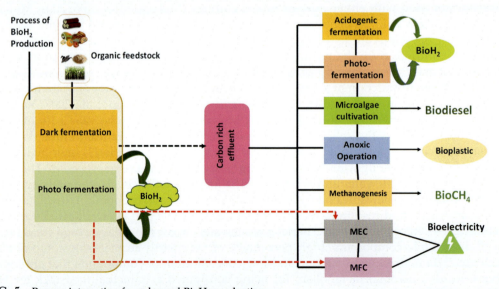

FIG. 5   Process integration for enhanced BioH$_2$ production.

is likely to be an alternative to the rising demand for compressed natural gas (CNG) as an engine fuel (Liu et al., 2013; Sarkar and Mohan, 2017; Sarkar and Venkata Mohan, 2016). Hythane generated organically from organic waste as a feedstock is more feasible and economical than hythane sourced from fossil-based feedstocks in terms of sustainability (Sarkar et al., 2021b). Biohythane was discovered in a single-stage bioreactor using spent-wash effluent in a recent study (Pasupuleti and Mohan, 2015). Another study described a novel technique for increasing acidogenesis and producing biohythane (Sarkar and Mohan, 2017; Sarkar and Venkata Mohan, 2016). An integrative approach to bio-hythane generation from waste has the capability to replicate the composition of H-CNG, resulting in fewer GHG emissions (Sarkar et al., 2021b).

4.2 Dark fermentation integrated with photo-fermentation

Dark fermentation generates volatile fatty acids and bioH$_2$ from biogenic feedstock comprising simple and complex carbohydrates. The resultant effluent is volatile fatty acids with a maximum theoretical production of 4 mol H$_2$/mol glucose, depending on the ultimate volatile fatty acid level (Dahiya et al., 2020; Keskin et al., 2019). Because photo-fermentative bacteria theoretically can yield 12 mol H$_2$/mol of glucose, which is a difficult feat to perform, the photo-fermentation system can be added into dark fermentation to boost bioH$_2$ production (Dahiya et al., 2020; Chandra and Mohan, 2014; Chandra et al., 2015). Due to light penetration issues, photo-fermentative bacteria cannot directly digest solid waste. As a result, combining this method with dark fermentation pre-degradation will improve the overall energy production (Chandra and Mohan, 2014; Chandra et al., 2015). Fatty acid inhibition can be overcome by combining a single-stage along with dual-stage dark and photo-fermentation while simultaneously overcoming limitations such as limited light, pH, and substrate inhibition (Chandra and Mohan, 2014; Chandra et al., 2015). These methods resulted in lower energy costs for bio-H$_2$ production, in turn fortifying the bioeconomy.

4.3 Dark fermentation integrated with microbial electrolysis cell

Because they can utilize the effluent from acidogenic reactors, bio-electrochemical systems can be integrated with dark fermentation systems. Liu et al. were the first to construct microbial electrolysis cells (MECs) (Logan et al., 2008). The MEC mechanism relies on microbial mediation to oxidize organic compounds in an anodic compartment, with a power supply provided by an external electrical circuit. This external connection allows electrons generated in the anode section to be transported to the cathode section (Wang et al., 2011). An ion-exchange membrane transports protons between the binary partitions. Protons produced by biological reactions in the cathode compartment are reduced by electrons. In conclusion, microbial electrolysis is a system that works similarly to chemical electrolysis (Nguyen et al., 2020). At the anode, the substrate can be oxidized, with protons reducing to bioH$_2$ at appropriate temperatures, and external electrical energy can be supplied. Because of its simplicity and benefits, the microbial electrolysis system is an excellent choice for stage two of a hybrid system.

4.4 Dark fermentation integrated with microbial fuel cells

The microbial fuel cell (MFC), which generates an electric current, is an alternative for a stage two treatment. The organic acids produced in the dark fermentation can be catalyzed by the microbes used in the MFC (Estrada-Arriaga et al., 2021). During the transfer of electrons from the anode to the cathode, the generated electricity may be observed. MFC systems have lately evolved as a viable renewable energy source that utilizes biogenic resources. MFC systems are carbon-neutral, which means they release only stable carbon into the oxidation-ending atmosphere of organic matter. This is one of the explanations why they are considered renewable energy sources (Bakonyi et al., 2018). Because MFCs combines microbial and electrochemical processes, they are also known as electrochemical hybrid systems (Wang et al., 2011). In a nutshell, MFCs can be thought of as chemical reactors capable of converting the chemical energy possessed by organic compounds into electrical energy. The potential of biogenic materials to serve as substrate for microbes in MFC will broaden its applications while also providing the benefit of overcoming the sludge purification problem in the future (Bundhoo, 2017).

4.5 Multistage bioreactor system

The multistage bioreactor system (Fig. 6) is made up of numerous stages that are blended together to produce biohydrogen (Sharma and Arya, 2017; Show et al., 2018). The first stage entails photolysis, which is the first step in photosynthesis and results in electron flow across the electron transport chain. The second stage is the photo-fermentation stage, in which light, as well as organic substrate, is required for the fermentation process, and the treated liquid is then transferred to the next reactor, where biomass is supplied as substrate in a dark

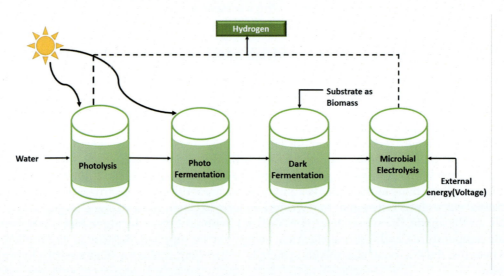

FIG. 6   A diversified multistage bioreactor system for biohydrogen production.

environment (Sharma and Arya, 2017). The bacteria in the third stage, known as the dark reactor, transform the substrate into H_2 and volatile fatty acids. The fourth stage reactor incorporates a MEC that uses external voltage to convert fatty acids into H_2 in the absence of light. There are complications that need to be addressed before multistage bioreactors may be used in the real world. Amalgamation of diverse biochemical reactions across multistage reactors is a crucial challenge. All aspects of system design, multireactor engineering, process control, and reactor operation and maintenance are affected (Show et al., 2018). Major obstacles related to the simultaneous synthesis of oxygen and hydrogen must be solved in photolytic hydrogen synthesis. These include the respiration/photosynthesis capacity ratio, coculture equilibrium, and microbial cell concentration and processing (Show et al., 2018).

5 Biorefinery approach

Significant research was reported in the past two decades towards understanding, optimization, and scale-up of the dark-fermentation process in terms of H_2 production. In this direction various integrated process were studied to enhance recovery efficiency and process feasibility (Mohanakrishna and Mohan, 2013). The integration of acidogenic with photobiological resulted in the higher substrate utilizing along with higher H_2 production (Srikanth et al., 2009). Further, to harness the lower substrate utilization, microbial fuel cell (MFC) was used with dark-fermentative process, which resulted in both H_2 as well as bioelectricity production (Mohankrishna et al., 2010). Acid rich effluent generated from the dark-fermentation process was used to produce poly-hydroxyalkanoate (PHA) [by employing aerobic consortia] (Mohan et al., 2010; Reddy and Mohan, 2013) and to cultivate microalgae (Mohan and Devi, 2012). Further, MEC was incorporated with dark fermentation where the acidogenic effluent was used as substrate which yielded additional H_2 with small inputs of current/poised potential (Babu et al., 2013). An integration of acidogenesis (dark-fermentation) and methanogenesis for green bio-hythane production from food waste in two stages (S–I and S-II) and phases (P–I and P-II) was studied in the framework of biogas upgradation in terms of H_2, CH_4, and H-CNG (Hythane) (Santhosh et al., 2021; Sarkar et al., 2021b). Electrofermentation was incorporated as an advanced process over conventional microbial fermentation, where multiple biobased products were yielded including volatile fatty acids, BioH$_2$, BioCH$_4$, and Hythane (Sravan et al., 2018). Another integrated approach was incorporated where the bio-H$_2$ and CO_2 generated during acidogenic fermentation was used in the production of succinic acid (Amulya and Mohan, 2022). Process integration was employed to dark fermentation with simultaneous bioethanol production in a biorefinery approach (Chatterjee and Mohan, 2021). In an integrated strategy, the bio-H$_2$ at pilot-scale ($10 \, m^3$) was demonstrated by CSIR-IICT with funding of MNRE and CSIR using food waste as the substrate (Sarkar et al., 2021a). The dark-fermentation process using waste-based feedstock produced 50,000 liters of H_2 per day using enriched mixed consortia as biocatalyst. Further, it was integrated with other bioprocesses such as methanogenesis, photosynthesis (algal cultivation), Ecological engineered system (electrochemical Phytoremediation) in a circular loop approach. The integrated system resulted in various biobased products such as methane, biomass, oxygen, and treated water in a biorefinery framework by enabling optimal resource recovery.

The Life cycle analysis (LCA) analysis represented that integrated biorefinery model aided maximal recovery with minimum environmental burdens (Sarkar et al., 2021a). Thus, the simultaneous production of green bioH$_2$ (major product) and other high valued products facilitated carbon neutrality and curtails the dependency on fossil based, non-renewable resources (Mohan et al., 2019b).

6 Conclusion

To enhance the "state of the art" in biohydrogen production, considerable research and development studies are needed. The slow rate and yields of H$_2$ production are major limitations of the selective feedstock. To scale-up of bio-H$_2$ and surpass the Thauer limit is need of the moment which can be achieved by the creation of more efficient processing systems (integrating multiple systems together), the optimization of environmental conditions, the enhancement of light consumption performance, and the development of more efficient photobioreactors with the selection and use of more crucial species or mixed cultures can enhance the quantity and rate of H$_2$ production. The role of the hydrogenase enzyme is crucial in the production of H$_2$, and efforts in the enhancement of this enzyme are much needed. Further, more emphasis should be given to the molecular details of H$_2$ producing metabolic pathways. The socioeconomic limitations can be hurdled by implementing circular models and thus hydrogen can be implemented as the "future fuel": the frontier towards development and sustainability.

Acknowledgment

The authors want to thank the Director of CSIR-IICT (IICT/Pubs./2021/334) for their immense support and also extend the gratitude towards DBT-TATA Innovative Fellowship "Self-Sustained Design of Photo-Biorefinery for the Closed Loop Production of Fuels and Chemicals," Department of Biotechnology (DBT) (BT/HRD/35/01/02/2018) for providing funds. TM wants to acknowledge Department of Science and Technology (DST) for providing INSPIRE fellowship for the pursual of Doctorate program.

References

Ahorsu, R., Medina, F., Constantí, M., 2018. Significance and challenges of biomass as a suitable feedstock for bioenergy and biochemical production: a review. Energies 11 (12), 3366.

Akhlaghi, N., Najafpour-Darzi, G., 2020. A comprehensive review on biological hydrogen production. Int. J. Hydrog. Energy 45 (43), 22492–22512.

Amulya, K., Mohan, S.V., 2022. Green hydrogen based succinic acid and biopolymer production in a biorefinery: adding value to CO$_2$ from acidogenic fermentation. Chem. Eng. J. 429, 132163.

Antar, M., Lyu, D., Nazari, M., Shah, A., Zhou, X., Smith, D.L., 2021. Biomass for a sustainable bioeconomy: an overview of world biomass production and utilization. Renew. Sust. Energ. Rev. 139, 110691.

Artz, J.H., Zadvornyy, O.A., Mulder, D.W., Keable, S.M., Cohen, A.E., Ratzloff, M.W., Peters, J.W., 2019. Tuning catalytic bias of hydrogen gas producing hydrogenases. J. Am. Chem. Soc. 142 (3), 1227–1235.

Arunasri, K., Yeruva, D.K., Krishna, K.V., Mohan, S.V., 2020. Monitoring metabolic pathway alterations in Escherichia coli due to applied potentials in microbial electrochemical system. Bioelectrochemistry 134, 107530.

Babu, M.L., Subhash, G.V., Sarma, P.N., Mohan, S.V., 2013. Bio-electrolytic conversion of acidogenic effluents to biohydrogen: an integration strategy for higher substrate conversion and product recovery. Bioresour. Technol. 133, 322–331.

Baeyens, J., Zhang, H., Nie, J., Appels, L., Dewil, R., Ansart, R., Deng, Y., 2020. Reviewing the potential of bio-hydrogen production by fermentation. Renew. Sust. Energ. Rev. 131, 110023.

Bakonyi, P., Kumar, G., Koók, L., Tóth, G., Rózsenberszki, T., Bélafi-Bakó, K., Nemestóthy, N., 2018. Microbial electrohydrogenesis linked to dark fermentation as integrated application for enhanced biohydrogen production: a review on process characteristics, experiences and lessons. Bioresour. Technol. 251, 381–389.

Banu, J.R., Ginni, G., Kavitha, S., Kannah, R.Y., Kumar, S.A., Bhatia, S.K., Kumar, G., 2021. Integrated biorefinery routes of biohydrogen: possible utilization of acidogenic fermentative effluent. Bioresour. Technol. 319, 124241.

Bartacek, J., Zabranska, J., Lens, P.N., 2007. Developments and constraints in fermentative hydrogen production. Biofuels Bioprod. Biorefin. 1 (3), 201–214.

Basak, N., Das, D., 2007. The prospect of purple non-sulfur (PNS) photosynthetic bacteria for hydrogen production: the present state of the art. World J. Microbiol. Biotechnol. 23 (1), 31–42.

Benemann, J., 1996. Hydrogen biotechnology: progress and prospects. Nat. Biotechnol. 14, 1101.

Benemann, J.R., 2000. Hydrogen production by microalgae. J. Appl. Phycol. 12 (3), 291–300.

Brandt, U., 1997. Proton-translocation by membrane-bound NADH: ubiquinone-oxidoreductase (complex I) through redox-gated ligand conduction. Biochim. Biophys. Acta Bioenerg. 1318 (1–2), 79–91.

Buckel, W., Thauer, R.K., 2013. Energy conservation via electron bifurcating ferredoxin reduction and proton/Na+ translocating ferredoxin oxidation. Biochim. Biophys. Acta Bioenerg. 1827 (2), 94–113.

Bundhoo, Z.M., 2017. Coupling dark fermentation with biochemical or bioelectrochemical systems for enhanced bio-energy production: a review. Int. J. Hydrog. Energy 42 (43), 26667–26686.

Call, D., Logan, B.E., 2008. Hydrogen production in a single chamber microbial electrolysis cell lacking a membrane. Environ. Sci. Technol. 42 (9), 3401–3406.

Chandra, R., Mohan, S.V., 2014. Enhanced bio-hydrogenesis by co-culturing photosynthetic bacteria with acidogenic process: augmented dark-photo fermentative hybrid system to regulate volatile fatty acid inhibition. Int. J. Hydrog. Energy 39 (14), 7604–7615.

Chandra, R., Nikhil, G.N., Mohan, S.V., 2015. Single-stage operation of hybrid dark-photo fermentation to enhance biohydrogen production through regulation of system redox condition: evaluation with real-field wastewater. Int. J. Mol. Sci. 16 (5), 9540–9556.

Chandrasekhar, K., Mohan, S.V., 2014. Induced catabolic bio-electrohydrolysis of complex food waste by regulating external resistance for enhancing acidogenic biohydrogen production. Bioresour. Technol. 165, 372–382.

Chandrasekhar, K., Lee, Y.J., Lee, D.W., 2015. Biohydrogen production: strategies to improve process efficiency through microbial routes. Int. J. Mol. Sci. 16 (4), 8266–8293.

Chandrasekhar, K., Kumar, S., Lee, B.D., Kim, S.H., 2020. Waste based hydrogen production for circular bioeconomy: current status and future directions. Bioresour. Technol. 302, 122920.

Chandrasekhar, K., Kumar, A.N., Kumar, G., Kim, D.H., Song, Y.C., Kim, S.H., 2021. Electro-fermentation for biofuels and biochemicals production: current status and future directions. Bioresour. Technol. 323, 124598.

Chatterjee, S., Mohan, S.V., 2021. Simultaneous production of green hydrogen and bioethanol from segregated sugarcane bagasse hydrolysate streams with circular biorefinery design. Chem. Eng. J. 425, 130386.

Cheng, C.L., Lo, Y.C., Lee, K.S., Lee, D.J., Lin, C.Y., Chang, J.S., 2011. Biohydrogen production from lignocellulosic feedstock. Bioresour. Technol. 102 (18), 8514–8523.

Chozhavendhan, S., Rajamehala, M., Karthigadevi, G., Praveenkumar, R., Bharathiraja, B., 2020. A review on feedstock, pretreatment methods, influencing factors, production and purification processes of bio-hydrogen production. Case Stud. Chem. Environ. Eng. 2, 100038.

Constant, P., Hallenbeck, P.C., 2019. Hydrogenase. In: Biohydrogen. Elsevier, pp. 49–78.

Dahiya, S., Sarkar, O., Swamy, Y.V., Mohan, S.V., 2015. Acidogenic fermentation of food waste for volatile fatty acid production with co-generation of biohydrogen. Bioresour. Technol. 182, 103–113.

Dahiya, S., Chatterjee, S., Sarkar, O., Mohan, S.V., 2020. Renewable hydrogen production by dark-fermentation: current status, challenges and perspectives. Bioresour. Technol. 124354.

Dalena, F., Senatore, A., Tursi, A., Basile, A., 2017. Bioenergy production from second-and third-generation feedstocks. In: Bioenergy Systems for the Future. Woodhead Publishing, pp. 559–599.

Dincer, I., 2012. Green methods for hydrogen production. Int. J. Hydrog. Energy 37 (2), 1954–1971.

Estrada-Arriaga, E.B., Hernández-Romano, J., Mijaylova-Nacheva, P., Gutiérrez-Macías, T., Morales-Morales, C., 2021. Assessment of a novel single-stage integrated dark fermentation-microbial fuel cell system coupled to proton-exchange membrane fuel cell to generate bio-hydrogen and recover electricity from wastewater. Biomass Bioenergy 147, 106016.

Fang, X., Kalathil, S., Reisner, E., 2020. Semi-biological approaches to solar-to-chemical conversion. Chem. Soc. Rev. 49 (14), 4926–4952.

Ghimire, A., Frunzo, L., Pirozzi, F., Trably, E., Escudie, R., Lens, P.N., Esposito, G., 2015. A review on dark fermentative biohydrogen production from organic biomass: process parameters and use of by-products. Appl. Energy 144, 73–95.

Hallenbeck, P.C., Benemann, J.R., 2002. Biological hydrogen production; fundamentals and limiting processes. Int. J. Hydrog. Energy 27 (11–12), 1185–1193.

Han, S.K., Shin, H.S., 2004. Biohydrogen production by anaerobic fermentation of food waste. Int. J. Hydrog. Energy 29 (6), 569–577.

Han, W., Yan, Y., Shi, Y., Gu, J., Tang, J., Zhao, H., 2016. Biohydrogen production from enzymatic hydrolysis of food waste in batch and continuous systems. Sci. Rep. 6 (1), 1–9.

Harlander, K., 2008. Food vs. fuel—a turning point for bioethanol? Acta Agronomica Hung. 56 (4), 429–433.

Hawkes, F.R., Dinsdale, R., Hawkes, D.L., Hussy, I., 2002. Sustainable fermentative hydrogen production: challenges for process optimisation. Int. J. Hydrog. Energy 27 (11–12), 1339–1347.

Herzog, A., Tatsutani, M., 2005. A Hydrogen Future? An Economic and Environmental Assessment of Hydrogen Production Pathways. Natural Resources Defense Council, p. 23.

Hu, Y., Lee, C.C., Ribbe, M.W., 2011. Extending the carbon chain: hydrocarbon formation catalyzed by vanadium/molybdenum nitrogenases. Science 333 (6043), 753–755.

Kadier, A., Simayi, Y., Abdeshahian, P., Azman, N.F., Chandrasekhar, K., Kalil, M.S., 2016. A comprehensive review of microbial electrolysis cells (MEC) reactor designs and configurations for sustainable hydrogen gas production. Alex. Eng. J. 55 (1), 427–443.

Kalair, A., Abas, N., Saleem, M.S., Kalair, A.R., Khan, N., 2021. Role of energy storage systems in energy transition from fossil fuels to renewables. Energy Storage 3 (1), e135.

Keskin, T., Abubackar, H.N., Arslan, K., Azbar, N., 2019. Biohydrogen production from solid wastes. In: Biohydrogen. Elsevier, pp. 321–346.

Kim, D.H., Kim, M.S., 2011. Hydrogenases for biological hydrogen production. Bioresour. Technol. 102 (18), 8423–8431.

Kossalbayev, B.D., Tomo, T., Zayadan, B.K., Sadvakasova, A.K., Bolatkhan, K., Alwasel, S., Allakhverdiev, S.I., 2020. Determination of the potential of cyanobacterial strains for hydrogen production. Int. J. Hydrog. Energy 45 (4), 2627–2639.

Kucharska, K., Rybarczyk, P., Hołowacz, I., Łukajtis, R., Glinka, M., Kamiński, M., 2018. Pretreatment of lignocellulosic materials as substrates for fermentation processes. Molecules 23 (11), 2937.

Kumar, A.K., Sharma, S., 2017. Recent updates on different methods of pretreatment of lignocellulosic feedstocks: a review. Bioresour. Bioprocess. 4 (1), 1–19.

Lee, H.S., Vermaas, W.F., Rittmann, B.E., 2010. Biological hydrogen production: prospects and challenges. Trends Biotechnol. 28 (5), 262–271.

Lee, S.H., Choi, D.S., Kuk, S.K., Park, C.B., 2018. Photobiocatalysis: activating redox enzymes by direct or indirect transfer of photoinduced electrons. Angew. Chem. Int. Ed. 57 (27), 7958–7985.

Li, Y.C., Liu, Y.F., Chu, C.Y., Chang, P.L., Hsu, C.W., Lin, P.J., Wu, S.Y., 2012. Techno-economic evaluation of biohydrogen production from wastewater and agricultural waste. Int. J. Hydrog. Energy 37 (20), 15704–15710.

Liu, Z., Zhang, C., Lu, Y., Wu, X., Wang, L., Wang, L., Xing, X.H., 2013. States and challenges for high-valuebiohythane production from waste biomass by dark fermentation technology. Bioresour. Technol. 135, 292–303.

Liu, C., Ren, L., Yan, B., Luo, L., Zhang, J., Awasthi, M.K., 2020. Electron transfer and mechanism of energy production among syntrophic bacteria during acidogenic fermentation: a review. Bioresour. Technol. 124637.

Logan, B.E., Call, D., Cheng, S., Hamelers, H.V., Sleutels, T.H., Jeremiasse, A.W., Rozendal, R.A., 2008. Microbial electrolysis cells for high yield hydrogen gas production from organic matter. Environ. Sci. Technol. 42 (23), 8630–8640.

Lopez-Bellido, L., Wery, J., Lopez-Bellido, R.J., 2014. Energy crops: prospects in the context of sustainable agriculture. Eur. J. Agron. 60, 1–12.

Lubitz, W., Ogata, H., Rudiger, O., Reijerse, E., 2014. Hydrogenases. Chem. Rev. 114 (8), 4081–4148.

Macfarlane, S., Macfarlane, G.T., 2003. Regulation of short-chain fatty acid production. Proc. Nutr. Soc. 62 (1), 67–72.

Mazloomi, K., Gomes, C., 2012. Hydrogen as an energy carrier: prospects and challenges. Renew. Sust. Energ. Rev. 16 (5), 3024–3033.

Mohan, S.V., 2008. Fermentative hydrogen production with simultaneous wastewater treatment: influence of pretreatment and system operating conditions. J. Sci. Ind. Res. 67, 950–961.

Mohan, S.V., 2009. Harnessing of biohydrogen from wastewater treatment using mixed fermentative consortia: process evaluation towards optimization. Int. J. Hydrog. Energy 34 (17), 7460–7474.

Mohan, S.V., 2010. Waste to renewable energy: a sustainable and green approach towards production of biohydrogen by acidogenic fermentation. In: Sustainable Biotechnology. Springer, Dordrecht, pp. 129–164.

Mohan, S.V., Devi, M.P., 2012. Fatty acid rich effluents from acidogenic biohydrogen reactor as substrates for lipid accumulation in heterotrophic microalgae with simultaneous treatment. Bioresour. Technol. 123, 471–479.

Mohan, S.V., Goud, R.K., 2012. Pretreatment of biocatalyst as viable option for sustained production of biohydrogen from wastewater treatment. In: Biogas Production: Pretreatment Methods in Anaerobic Digestion. Springer, pp. 291–311.

Mohan, S.V., Babu, V.L., Sarma, P.N., 2007a. Anaerobic biohydrogen production from dairy wastewater treatment in sequencing batch reactor (AnSBR): effect of organic loading rate. Enzym. Microb. Technol. 41 (4), 506–515.

Mohan, S.V., Bhaskar, Y.V., Sarma, P.N., 2007b. Biohydrogen production from chemical wastewater treatment in biofilm configured reactor operated in periodic discontinuous batch mode by selectively enriched anaerobic mixed consortia. Water Res. 41 (12), 2652–2664.

Mohan, S.V., Babu, V.L., Sarma, P.N., 2008. Effect of various pretreatment methods on anaerobic mixed microflora to enhance biohydrogen production utilizing dairy wastewater as substrate. Bioresour. Technol. 99 (1), 59–67.

Mohan, S.V., Reddy, M.V., Subhash, G.V., Sarma, P.N., 2010. Fermentative effluents from hydrogen producing bioreactor as substrate for poly (β-OH) butyrate production with simultaneous treatment: An integrated approach. Bioresour. Technol. 101 (23), 9382–9386.

Mohan, S.V., Chandrasekhar, K., Chiranjeevi, P., Babu, P.S., 2013. Biohydrogen production from wastewater. In: Biohydrogen. Elsevier, pp. 223–257.

Mohan, S.V., Srikanth, S., Velvizhi, G., Babu, M.L., 2013a. Microbial fuel cells for sustainable bioenergy generation: principles and perspective applications. In: Biofuel Technologies. Springer, Berlin, Heidelberg, pp. 335–368.

Mohan, S.V., Sravan, J.S., Butti, S.K., Krishna, K.V., Modestra, J.A., Velvizhi, G., Pandey, A., 2019a. Microbial electrochemical technology: emerging and sustainable platform. In: Microbial Electrochemical Technology. Elsevier, pp. 3–18.

Mohan, S.V., Dahiya, S., Amulya, K., Katakojwala, R., Vanitha, T.K., 2019b. Can circular bioeconomy be fueled by waste biorefineries—a closer look. Bioresour. Technol. Rep. 7, 100277.

Mohanakrishna, G., Mohan, S.V., 2013. Multiple process integrations for broad perspective analysis of fermentative H2 production from wastewater treatment: technical and environmental considerations. Appl. Energy 107, 244–254.

Mohankrishna, G., Mohan, S.V., Sarma, P.N., 2010. Utilizing acid-rich effluents of fermentative hydrogen production process as substrate for harnessing bioelectricity: an integrative approach. Int. J. Hydrog. Energy 35 (8), 3440–3449.

Mona, S., Kumar, S.S., Kumar, V., Parveen, K., Saini, N., Deepak, B., Pugazhendhi, A., 2020. Green technology for sustainable biohydrogen production (waste to energy): a review. Sci. Total Environ. 728, 13848.

Mukherjee, T., Mohan, S.V., 2021. Metabolic flux of *Bacillus subtilis* under poised potential in electrofermentation system: gene expression vs product formation. Bioresour. Technol. 342, 125854.

Müller, V., 2001. Bacterial fermentation. In: eLS. Nature Publishing Group.

Nagarajan, D., Dong, C.D., Chen, C.Y., Lee, D.J., Chang, J.S., 2020. Biohydrogen production from microalgae–major bottlenecks and future research perspectives. Biotechnol. J. 2000124.

Nguyen, P.K.T., Das, G., Kim, J., Yoon, H.H., 2020. Hydrogen production from macroalgae by simultaneous dark fermentation and microbial electrolysis cell. Bioresour. Technol. 315, 123795.

Nikhil, G.N., Sarkar, O., Mohan, S.V., 2017. Biohydrogen production: an outlook of fermentative processes and integration strategies. In: Optimization and Applicability of Bioprocesses. Springer, Singapore, pp. 249–265.

Ntaikou, I., Antonopoulou, G., Lyberatos, G., 2010. Biohydrogen production from biomass and wastes via dark fermentation: a review. Waste Biomass Valoriz. 1 (1), 21–39.

Osman, A.I., Deka, T.J., Baruah, D.C., Rooney, D.W., 2020. Critical challenges in biohydrogen production processes from the organic feedstocks. Biomass Convers. Biorefin., 1–19.

Panagiotopoulos, I.A., Bakker, R.R., Budde, M.A.W., de Vrije, T., Claassen, P.A.M., Koukios, E.G., 2009. Fermentative hydrogen production from pretreated biomass: a comparative study. Bioresour. Technol. 100, 6331–6338.

Pandey, A., Mohan, S.V., Chang, J.S., Hallenbeck, P.C., Larroche, C. (Eds.), 2019. Biomass, Biofuels, Biochemicals: Biohydrogen. Elsevier.

Paschalidou, A., Tsatiris, M., Kitikidou, K., 2016. Energy crops for biofuel production or for food?-SWOT analysis (case study: Greece). Renew. Energy 93, 636–647.

Pasupuleti, S.B., Mohan, S.V., 2015. Single-stage fermentation process for high-value biohythane production with the treatment of distillery spent-wash. Bioresour. Technol. 189, 177–185.

Pasupuleti, S.B., Venkata Mohan, S., 2015. Acidogenic hydrogen production from wastewater: process analysis with the function of influencing parameters. Int. J. Energy Res. 39 (8), 1131–1141.

Peters, J.W., Schut, G.J., Boyd, E.S., Mulder, D.W., Shepard, E.M., Broderick, J.B., Adams, M.W., 2015. [FeFe]-and [NiFe]-hydrogenase diversity, mechanism, and maturation. Biochim. Biophys. Acta Mol. Cell Res. 1853 (6), 1350–1369.

Reddy, M.V., Mohan, S.V., 2013. Influence of aerobic and anoxic microenvironments on polyhydroxyalkanoates (PHA) production from food waste and acidogenic effluents using aerobic consortia. Bioresour. Technol. 103 (1), 313–321.

Rousseau, R., Etcheverry, L., Roubaud, E., Basséguy, R., Délia, M.L., Bergel, A., 2020. Microbial electrolysis cell (MEC): strengths, weaknesses and research needs from electrochemical engineering standpoint. Appl. Energy 257, 113938.

Sagir, E., Alipour, S., 2021. Photofermentative hydrogen production by immobilized photosynthetic bacteria: current perspectives and challenges. Renew. Sust. Energ. Rev. 141, 110796.

Sampath, P., Reddy, K.R., Reddy, C.V., Shetti, N.P., Kulkarni, R.V., Raghu, A.V., 2020. Biohydrogen production from organic waste–a review. Chem. Eng. Technol. 43 (7), 1240–1248.

Santhosh, J., Sarkar, O., Mohan, S.V., 2021. Green hydrogen-compressed natural gas (bio-H-CNG) production from food waste: organic load influence on hydrogen and methane fusion. Bioresour. Technol. 340, 125643.

Sarewicz, M., Osyczka, A., 2015. Electronic connection between the quinone and cytochrome c redox pools and its role in regulation of mitochondrial electron transport and redox signaling. Physiol. Rev. 95 (1), 219–243.

Sarkar, O., Mohan, S.V., 2017. Pre-aeration of food waste to augment acidogenic process at higher organic load: valorizingbiohydrogen, volatile fatty acids and biohythane. Bioresour. Technol. 242, 68–76.

Sarkar, O., Mohan, S.V., 2020. Synergy of anoxic microenvironment and facultative anaerobes on acidogenic metabolism in a self-induced electrofermentation system. Bioresour. Technol. 313, 123604.

Sarkar, O., Santhosh, J., Dhar, A., Mohan, S.V., 2021b. Green hythane production from food waste: integration of dark-fermentation and methanogenic process towards biogas up-gradation. Int. J. Hydrog. Energy 46 (36), 18832–18843.

Sarkar, O., Venkata Mohan, S., 2016. Deciphering acidogenic process towards biohydrogen, biohythane, and short chain fatty acids production: multi-output optimization strategy. Biofuel Res. J. 3 (3), 458–469.

Sarkar, O., Katakojwala, R., Mohan, S.V., 2021a. Low carbon hydrogen production from a waste-based biorefinery system and environmental sustainability assessment. Green Chem. 23 (1), 561–574.

Schuchmann, K., Chowdhury, N.P., Müller, V., 2018. Complex multimeric [FeFe] hydrogenases: biochemistry, physiology and new opportunities for the hydrogen economy. Front. Microbiol. 9, 2911.

Sharma, A., Arya, S.K., 2017. Hydrogen from algal biomass: a review of production process. Biotechnol. Rep. 15, 63–69.

Show, K.Y., Yan, Y., Ling, M., Ye, G., Li, T., Lee, D.J., 2018. Hydrogen production from algal biomass–advances, challenges and prospects. Bioresour. Technol. 257, 290–300.

Sivaramakrishnan, R., Shanmugam, S., Sekar, M., Mathimani, T., Incharoensakdi, A., Kim, S.H., Pugazhendhi, A., 2021. Insights on biological hydrogen production routes and potential microorganisms for high hydrogen yield. Fuel 291, 120136.

Sravan, J.S., Butti, S.K., Sarkar, O., Mohan, S.V., 2018. Electrofermentation of food waste—regulating acidogenesis towards enhanced volatile fatty acids production. Chem. Eng. J. 334, 1709–1718.

Sravan, J.S., Butti, S.K., Sarkar, O., Mohan, S.V., 2019. Electrofermentation: chemicals and fuels. In: Microbial Electrochemical Technology. Elsevier, pp. 723–737.

Sravan, J.S., Tharak, A., Modestra, J.A., Chang, I.S., Mohan, S.V., 2021. Emerging trends in microbial fuel cell diversification-critical analysis. Bioresour. Technol. 124676.

Srikanth, S., Mohan, S.V., 2012. Regulatory function of divalent cations in controlling the acidogenic biohydrogen production process. RSC Adv. 2 (16), 6576–6589.

Srikanth, S., Mohan, S.V., Devi, M.P., Sarma, P.N., 2009. Effluents with soluble metabolites generated from acidogenic and methanogenic processes as substrate for additional hydrogen production through photo-biological process. Int. J. Hydrog. Energy 34 (4), 1771–1779.

Sun, Y., He, J., Yang, G., Sun, G., Sage, V., 2019. A review of the enhancement of bio-hydrogen generation by chemicals addition. Catalysts 9 (4), 353.

Thauer, R.K., Kaster, A.K., Goenrich, M., Schick, M., Hiromoto, T., Shima, S., 2010. Hydrogenases from methanogenic archaea, nickel, a novel cofactor, and H2 storage. Annu. Rev. Biochem. 79, 507–536.

Tiang, M.F., Hanipa, M.A.F., Abdul, P.M., Jahim, J.M., Mahmod, S.S., Takriff, M.S., Wu, S.Y., 2020. Recent advanced biotechnological strategies to enhance photo-fermentative biohydrogen production by purple non-sulphur bacteria: an overview. Int. J. Hydrog. Energy 45 (24), 13211–13230.

Tsutsumi, A., 2009. Thermodynamics of water splitting. In: Energy Carriers and Conversion Systems (EOLSS). vol. 1. UNESCO-EOLSS, pp. 136–145.

Valdez-Vazquez, I., Sparling, R., Risbey, D., Rinderknecht-Seijas, N., Poggi-Varaldo, H.M., 2005. Hydrogen generation via anaerobic fermentation of paper mill wastes. Bioresour. Technol. 96 (17), 1907–1913.

Veeravalli, S.S., Shanmugam, S.R., Ray, S., Lalman, J.A., Biswas, N., 2019. Biohydrogen production from renewable resources. In: Advanced Bioprocessing for Alternative Fuels, Biobased Chemicals, and Bioproducts. Woodhead Publishing, pp. 289–312.

Wang, A., Sun, D., Cao, G., Wang, H., Ren, N., Wu, W.M., Logan, B.E., 2011. Integrated hydrogen production process from cellulose by combining dark fermentation, microbial fuel cells, and a microbial electrolysis cell. Bioresour. Technol. 102 (5), 4137–4143.

Yadav, M., Paritosh, K., Vivekanand, V., 2020. Lignocellulose to bio-hydrogen: an overview on recent developments. Int. J. Hydrog. Energy 45 (36), 18195–18210.

Yang, P., Zhang, R., McGarvey, J.A., Benemann, J.R., 2007. Biohydrogen production from cheese processing wastewater by anaerobic fermentation using mixed microbial communities. Int. J. Hydrog. Energy 32 (18), 4761–4771.

Zhang, Q., Zhang, Z., 2018. Biological hydrogen production from renewable resources by photofermentation. In: Advances in Bioenergy. vol. 3. Elsevier, pp. 137–160.

Biorefinery approach for production of some high-value chemicals

Andrea Komesu[a], Johnatt Oliveira[b], Débora Kono Taketa Moreira[c], Ali Hassan Khalid[d], João Moreira Neto[e], and Luiza Helena da Silva Martins[f]

[a]Department of Marine Sciences, Federal University of São Paulo (UNIFESP), Santos, SP, Brazil
[b]Institute of Health Sciences, Faculty of Nutrition, Federal University of Pará (UFPA), Belém, PA, Brazil [c]Department of Food Technology, Federal Institute of Brasilia (IFB), Campus Gama, Brasília, DF, Brazil [d]National University of Science and Technology (NUST), Islamabad, Pakistan [e]Department of Engineering, Federal University of Lavras, Lavras, MG, Brazil [f]Institute of Animal Health and Production, Federal Rural University of the Amazon (UFRA), Belém, PA, Brazil

1 Introduction

The main fraction of worldwide energy carriers and material products, especially chemicals, is derived from fossil fuels, mainly oil and natural gas (Cherubini et al., 2009). The nonsustainable nature of fossil resources and increasing concerns about the environment have stimulated extensive research interest in producing chemicals and fuels from alternative starting materials (Deng et al., 2016). Biomass is one of the few resources that have the potential to meet the challenges of sustainable and green energy systems (Fernando et al., 2006). Renewable biomass is virtually inexhaustible and relies mostly on terrestrial plants with an estimated global production of 1.7×10^{11} tons per year (Gerardy et al., 2020).

The transformation of biomass into platform molecules, which can be used directly or serve as building blocks for synthesizing other useful materials, is a promising route for biomass utilization (Deng et al., 2016). A system similar to a petroleum refinery called a "biorefinery" has been proposed to produce useful chemicals and fuels from biomass (Fernando et al., 2006). According to National Renewable Energy Laboratory (NREL): "a

TABLE 1   Potential bio-based chemicals selected by the US Department of Energy in 2004 and 2010 (de Jong et al., 2012a; Chandel et al., 2018).

Potential bio-based chemicals selected by the US Department of Energy in 2004	Potential bio-based chemicals selected by the US Department of Energy in 2010
• 1, 4-Dicarboxylic acid (succinic, fumaric, and malic) • 2, 5-Furan dicarboxylic acid • 3-Hydroxy propionic acid • 3-Hydroxybutyrolactone • Aspartic acid • Glucaric acid • Glutamic acid • Glycerol • Itaconic acid • Levulinic acid • Sorbitol • Xylitol/arabinitol	• Ethanol • Furans (furfural, 5-hydroxymethylfurfural, 2, 5-FDCA) • Glycerol and its derivatives (propanediol, glycerol carbonate, epichlorohydrin) • Hydrocarbons • Lactic acid • Succinic acid • Aldehyde/3-hydroxy propionic acid • Levulinic acid • Sorbitol • Xylitol

biorefinery is a facility that integrates biomass conversion processes and equipment to produce fuels, power and (organic)chemicals from biomass" (Fernando et al., 2006). Another definition provided by the International Energy Agency (IEA): "biorefinery is the sustainable processing of biomass into a spectrum of marketable products (food, feed, materials, chemicals) and energy (fuels, power, heat)" (de Jong et al., 2012a). Therefore, biorefineries are emerging industrial systems that aim at sustainable and efficient utilization of biomass, valorize potentials lying in biomass resources, and deliver multiple useful bioenergies and bio-based products (Budzianowski, 2017).

Globally, bio-based chemicals (excluding biofuels) amount is estimated to be 50 billion kilos per year with a growth of 3%–4% per year (de Jong et al., 2012a; Chandel et al., 2018). In 2004 and 2010, the US Department of Energy issued a report which listed the chemical's building blocks which are considered as potential building blocks for the future (Table 1). Sorbitol, furfural, glucaric acid, hydroxymethylfurfural (HMF), and levulinic acid are among promising bio-based chemicals.

This chapter will summarize information about potential building blocks, such as sorbitol, furfural, glucaric acid, hydroxymethylfurfural (HMF), and levulinic acid in the biorefinery approach.

2 Biorefinery and sugar platform

A biorefinery is not totally a new concept because industries such as sugar, starch, pulp, and paper have used technologies in a similar way to a biorefinery approach. Nevertheless, economic and environmental factors have been the drivers guiding the development of the biorefinery concept (Aristizábal-Marulanda and Cardona Alzate, 2019). Biofuels and Bio-based products (chemicals, materials) can be produced in single product processes; however, the production in integrated biorefinery processes producing both bio-based products and secondary energy carriers (fuels, power, heat), in analogy with oil refineries, is probably a more efficient approach for the sustainable valorization of biomass resources in a future bio-based economy (De Jong et al., 2020).

A biorefinery is the integrated upstream, midstream, and downstream processing of biomass into a range of products. In the classification system, we have differentiated between mechanical pretreatments (extraction, fractionation, and separation), thermochemical conversions, chemical conversions, enzymatic conversions, and microbial fermentation (both aerobic, anaerobic) conversions (de Jong et al., 2009).

Biorefineries have been classified in different ways. According to the literature, the classifications are (i) systems or models, (ii) status of technological implementation, (iii) size, and (iv) feedstocks, platforms, processes, and products (Aristizábal-Marulanda and Cardona Alzate, 2019). A summary is provided in Table 2.

Based on IEA Bioenergy—Task 42, biorefineries can be classified using the four main features: feedstocks, platforms, processes, and products. A feedstock is a renewable raw material (biomass) that is converted into marketable products in a biorefinery (Cherubini et al., 2009). Major feedstocks for biorefineries include starch crops (e.g., wheat and maize), sugar crops (e.g., beet and cane), perennial grasses and legumes (e.g., ryegrass and alfalfa), lignocellulosic crops (e.g., managed forest, short rotation coppice, switchgrass), lignocellulosic residues (e.g., forest residues, stover, and straw), oil crops (e.g., palm and oilseed rape), aquatic biomass (e.g., algae and seaweeds), and organic residues (e.g., industrial, commercial and postconsumer waste) (De Jong et al., 2020). Lignocellulose, being the major nonfood component of biomass, proves an almost unlimited source of C5 and C6 sugars without interfering with food demand (Fiorentino et al., 2017).

Platforms are intermediates, which link feedstocks and final products. The platform concept is similar to that used in the petrochemical industry, where the crude oil is fractionated into a large number of intermediates that are further processed to final energy and chemical products (Cherubini et al., 2009). Among platforms, the sugar platform is one of the key platforms and considered by volume currently the largest platform for the production of chemicals based on biomass. Sugar is the basis for a large number of traditional biorefinery processes and well-established industries (De Jong et al., 2020). C6 sugars (e.g., glucose, fructose, galactose: $C_6H_{12}O_6$), are obtained from hydrolysis of sucrose, starch, cellulose, and hemicellulose. C5 sugars (e.g., xylose, arabinose: $C_5H_{10}O_5$), are obtained from hydrolysis of hemicellulose, food, and feed side streams.

Biorefinery technological processes can be divided into thermochemical, biochemical, mechanical/physical, and chemical processes. Thermochemical processes convert biomass mainly by gasification and pyrolysis. Biochemical process is focused on the fermentation of carbohydrates. Mechanical processes are used for size reduction and separation of the components of biomass with no change of the state or composition of the feedstock. Chemical processes, such as hydrolysis and transesterification, are performed by the use of acids, alkalis, or enzymes to depolymerize polysaccharides and proteins into their component monomers (e.g., glucose from cellulose) or derive chemicals (e.g., furfural from xylose) (Souza Filho and Taherzadeh, 2020).

TABLE 2 Classification of biorefineries.

Classification	Subgroups	Description
Systems or models	• Conventional biorefineries • Green biorefineries • Whole crop biorefineries • Ligno-cellulosic feedstock biorefineries • Marine biorefineries • Two platform concept biorefineries • Thermochemical biorefineries	According to Status Report Biorefinery 847, biorefineries can be classified as systems in seven different ways. The basic concept is the conventional biorefinery. Its production process concentrates on the main product and does not make great efforts to generate other value-added products (Aristizábal-Marulanda and Cardona Alzate, 2019).
Status of technological implementation	• First (simple or conventional) • Second (advanced) • Third (advanced)	In first-generation, the feedstocks used are agricultural and forestry biomass (sugar, starch, vegetable oil, or animal fats). In the second generation, the feedstocks used are lignocellulosic biomass (residues from agriculture, forestry, and industry, and dedicated lignocellulosic crops). In the third generation, the feedstocks used are agricultural, organic waste streams, and algae.
Size	• Small and medium-sized production facilities • Large production facilities • Very large production facilities	The optimal size of a biorefinery generally depends on factors such as the nature of the feedstock, the type of products that will be obtained, the feedstock flow that will be processed, the location, and the technologies employed (Aristizábal-Marulanda and Cardona Alzate, 2019).
Feedstocks, platforms, processes, and products	• Feedstocks (sugar crops, starch crops, lignocellulosic crops, oil-based crops, grasses, marine biomass, residues) • Platforms (biogas, syngas, hydrogen, C6 sugars, C5 sugars, lignin, pyrolysis liquid, oil, organic juice, electricity, and heat) • Processes (mechanical/physical, biochemical, chemical processes, thermochemical) • Products (fertilizers, biohydrogen, glycerin, chemicals, and building blocks, polymers, and resins, food, animal feed, biomaterials)	According to IEA Bioenergy—Task 42, biorefineries can be classified using the four main features: feedstocks, platforms, processes, and products.

Biorefineries produce both energetic and nonenergetic products. Bioproducts can enhance the business development of biorefineries. They are wholly or at least partly derived from biomass. They are biodegradable, nontoxic, and preferably durable. Biorefineries will benefit from the co-production of bioenergies and high-value bioproducts (Budzianowski, 2017).

Building blocks include numerous biochemicals with a proven place in industries. Examples are as follows: sorbitol, furfural, glucaric acid, hydroxymethylfurfural (HMF), and levulinic acid. These building blocks will be discussed below and mostly already have their place in industries and their greater use mainly depends on sufficiently low production costs. Most of them can be obtained via technologies coupled to bioenergies within biorefineries (Budzianowski, 2017).

3 Sorbitol

Polyols are defined as sugar alcohols, as they are types of carbohydrates that have a group of alcohol attached to each carbon. These products have a high sweetening power, and because they are absorbed incompletely and slower in the body, they end up providing less calories when compared to conventional sugar, having a positive impact on health, as their impact is minimal on the blood. Polyols also act as anticariogenic, as oral bacteria cannot metabolize these products. However, the excessive consumption of polyols ends up causing diarrhea and gases due to the slow absorption of sugar alcohol followed by its (Fang et al., 2020).

Sorbitol or glucitol is a well-known polyol is a six-carbon sugar alcohol. It was first extracted from the fruit of *Sorbus aucuparia* (family Rosaceae). Later, other fruits were reported for having sorbitol, such as apples, pears, and peaches. The vegetables through the leaves synthesize this product, which via phloem transports sugars to the fruits. Sorbitol represents the main photosynthetic product in Rosaceae, where its metabolism is involved with the quality and production of the fruit. Such a compound is not just a source of energy and a carbon base for the maturation and development of the plant, but it is also a molecule that is involved in signaling for the regulation of growth, development, and response to the environmental stress of the plant (Fang et al., 2020; Zhang et al., 2020).

Sorbitol is the epitome of mannitol at the C-2 position. Such products are similar but have different properties and applications. Sorbitol is physically characterized as a white crystalline substance, and in the solid-state, such product can present four crystalline forms that are α, β, γ, and δ, sorbitol presents a glass transition form (Ɛ). Thus, all these different forms of this polyol will have different properties, such as water absorption, stability, hardness, and compressibility, among others. Sorbitol also has a solubility in water, slight solubility in ethanol, acetic acid, methanol, and has insolubility in chloroform. It also presents slow hygroscopicity. It is also considered a sugar alcohol that has good stabilization and does not act on the Maillard reaction. Its sweetening power is 60% when compared to sucrose, sorbitol is smooth on the palate and has a pleasant taste, so it is widely used in the industry of food, provides the energy of 2.6 kcal/g to the body (1/3 less than the calories provided by sucrose) (Zhang et al., 2020).

According to the Global Market Insights cited by Kumar et al. (2020a, 2020b) the sorbitol market in 2015 was around 1.85 million tons and this estimate is expected to increase to 2.4 million tons by 2023 to a value of $ 4 billion. The companies Roquette Freres, Cargill, and SPI Polyols are the largest manufacturers of sorbitol, with 70% of the global market share (Sheldon, 2014).

Industrial sorbitol production uses batch process technology through reactions that use catalysts (Raney nickel) and generally use glucose as the main substrate, and for this production, the inputs used can be corn, cassava, or wheat, and it has as a problem the competition with the food market, in addition, this transformation reaction is conducted under very severe conditions at high temperatures (403 and 423 K) and pressures (40 and 120 bar) (Werpy and Petersen, 2004a, 2004b).

Moreover, the conventional chemical production of these polyols still has some bottlenecks that hinder their industrial and commercial production, as they present stages with overly complex reactions, as well as not very efficient chemical catalysts, in addition to the enormous amount of by-products of no economic interest that may arise during the occurrence of chemical reactions. For this reason, the biotechnological production of these polyols is very promising, as it is possible and much simpler and with clean energy when compared to more conventional methods (Chen et al., 2020).

Obtaining value-added chemical products from renewable sources is a very interesting approach for the environment and in the energy issue of the use of nonrenewable sources of energy (derived from petroleum) (Chin et al., 2018).

Cellulose is a homopolysaccharide (only glucose unit) very abundant on the planet, it is part of the constitution of the cell wall of vegetables and is responsible for 35 and 50% by weight of the lignocellulosic biomass (that constituted of cellulose, hemicellulose, and lignin), this polysaccharide is an input of great importance for the production of biofuels and chemical compounds with high added value, such as bio-based polyols that can be derived from cellulose, and such products have wide application in the area of food and pharmacy as shown in Fig. 1 (Kumar et al., 2020a, 2020b).

The catalytic production of C6 (cellulosic) and C5 (hemicellulosic) sugars has already been carried out using physicochemical pretreatments already consolidated in the literature, such treatments are able to promote the deconstruction of lignocellulose (Lange, 2007). However, the fragmentation of lignin, which is one of the bottlenecks for the polyol production block, has received more attention in recent years, when we remove lignin, we have more freely the fractions of hemicelluloses and cellulose that will give rise to pentoses (xylitol production) and hexoses (sorbitol production), respectively. We know that polyols have as their main characteristic the presence of at least two hydroxyl groups, being an important class of chemical compounds and intermediates that are sorbitol (polyols C6), xylitol (polyols C5), and glycerol (polyol C3) are considered the 12 molecules interesting features of the bio-derived platform (Xu et al., 2020).

Chin et al. (2019) produced sorbitol photosynthetically using the cyanobacterium *Synechocystis* sp. PCC 6803 by apple enzymes (S6P dehydrogenase, whose acronym is S6PDH) and sorbitol-6-phosphate (S6P). These authors concluded that there is a possibility of the production of sorbitol by using cyanobacterial cells, which could present itself as an effective metabolic engineering strategy in the near future for the production of value-added chemical products. The use of cyanobacteria to produce sorbitol has attracted a lot of attention lately because of its facilities to improve and modify the substrate to optimize sorbitol production.

The production of sorbitol can also be done using lactose as a substrate, and this approach has been widely used since lactose in addition to sorbitol can produce other products with high added value. In general, this method uses the hydrolysis of glucose and galactose,

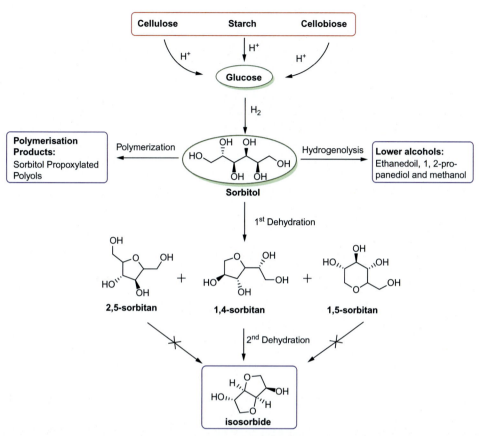

FIG. 1 Sorbitol and its derivatives products were obtained from cellulosic biomass. *This figure was extracted from the work of Kumar, S., Ali, H., Kansal, S.K., Pandey, A., Saravanamurugan, S., 2020a. Sustainable production of sorbitol—a potential hexitol. In Biomass, Biofuels, Biochemicals. Elsevier, pp. 259–281; Kumar, A., Shende, D.Z., Wasewar, K.L., 2020b. Extractive separation of levulinic acid using natural and chemical solvents. Chem. Data Collect. 28, 100417 with the permission of Elsevier.*

the hydrogenation of lactitol, the oxidation of lactobionic acid, and the isomerization of lactulose and epilactose, and the transglycosylation reactions of galactooligosaccharides and lactusucrose (Cheng and Martínez-Monteagudo, 2019; Zhang et al., 2020).

Currently, with the search for a better quality of life, healthy habits have become the focus of the public in modern society. Therefore, the use of polyols, which are low-calorie sweeteners and have peculiar physiological functions, comes into focus. Such products have been widely used in industries as flavor enhancers, wetting agents, refrigeration, and other benefits that have been widely used in the food industry. Sorbitol falls into the polyol class and can be used for the purposes mentioned above (Chen et al., 2020).

Not only these functions but also sorbitol can be used as a building block in derivatives of high added value (as shown in Fig. 1, in the production of isosorbide). Sorbitol can produce isosorbide, which is already commercially available and is used as a monomer for high-performance materials, for this reason, it has become a value-added product of great interest, due to the renewable issue of its production (Saxon et al., 2020). Sorbitol can also be used as a substrate to produce ascorbic acid and a starting substrate for numerous products such as ethylene glycol [EG], 1,2-propanediol, and glycerol Kumar et al. (2020a, 2020b).

Burt (2006) carried out a study of the inhibitory action of dental caries using polyols, here we will address the sorbitol that is the focus of this chapter, and this author carried out sweetening of chewing gums with sorbitol and tested these gums in study participants that were chosen randomly. As a result, this researcher observed that when compared to the gum sweetened with conventional sugar, the gum with sorbitol showed low cariogenicity, indicating that polyols may have applications to assist in dental treatments and prevention of cavities, which would be interesting as a preventive measure public health.

4 Furfural

Furfural is a chemical product resulting from the hydrolysis of pentoses or other polysaccharides rich in pentoses classified into four main groups: xylans, mannans, xyloglucans, and β-glucans (Ebringerová, 2005) with subsequent dehydration of the pentoses. In a commercial process, it is produced through acid hydrolysis of biomass, usually with sulfuric acid, and its production is not economically viable through fossil-based routes (Mariscal et al., 2016).

The economic and political pressure on the need to reduce the dependence on compounds derived from petroleum has expanded the creation and use of innumerable compounds of interest generated from bioderivated compounds. Chemical companies convert renewable biological resources into biofuels, into platform molecules for fine chemicals, agrochemicals, and special chemicals, such as biolubricants, natural fibers, and bio-based solvents (Schieb et al., 2015).

Furfural is a chemical compound with a large number of industrial applications and with growing market interest, this chemical compound usually comes from biomass used for example in the production of compounds such as ethanol and other processes involved in energy generation. Among the three main components in biomass, hemicellulose is a promising resource to produce furfural.

Thus, the use of biomass provides, in addition to the final products, such as fuels, a series of compounds of interest, considered valuable for the chemical industry and which can be obtained by biological transformations of carbohydrates (C5 and C6 sugars present in biomass) in the processes chemical and biological hydrolysis of biomass. This type of conversion has attracted a lot of interest due to the low cost of the raw material and the high availability of the material in nature.

Furfural is currently identified as a possible chemical substitute for some of the petroleum by-products, and it is possible to list around 80 furfuraldehyde chemicals used in numerous chemical industries. Among the compounds of interest derived from furfural, it is possible to find: furfulylic acid, 2-methylfuran, formic acid, levulinic acid, and acetic acid (Luo et al., 2018).

Conventionally, furfural is produced by reactions in aqueous systems using mineral acids through the hydrolysis of lignocellulosic material. However, due to cellulose crystallinity, hemicellulose hydrolysis is much more favorable than cellulose hydrolysis (Kumar et al., 2017).

There are several mechanisms proposed for the dehydration of xylose in furfural. These mechanisms can be divided into open-chain and closed-chain dehydration of xylose in furfural. Binder et al. (2010) proposed an open-chain mechanism for the dehydration of xylose in furfural catalyzed by CrCl3 or CrCl2 (Lewis acids) via xylulose as an intermediate product. In this procedure, it is proposed that Cr3 + promotes changes in the formal charge of the carbon atom C1, which, in turn, leads to a change of 1,2-hydride producing xylulose, with the xylulose subsequently dehydrated to produce furfural.

Furfural derivatives are in great demand for use in the plastics, herbicides, food, pharmaceutical, and agricultural industries (Ribeiro et al., 2012). According to the DOE (Department of energy), furfuraldehyde is among the bio-based products with research priority, which further reinforces the interest in this chemical compound (Werpy et al., 2004).

Also, they can serve as a production base for several chemicals of interest, including pharmaceuticals, polymers, resins, solvents, fungicides, and biofuels (Rosatella et al., 2011; van Putten et al., 2013; Galaverna and Pastre, 2017; Kong et al., 2018). It is possible to highlight the production of alkanes with high-molecular-weight (>C10) through aldolic condensation reactions and hydrodeoxygenation of 5-HMF derivatives, such as 2-methylfuran, furfural, and DMF, it appears as a promising and attractive methodology for the production of liquid fuels.

The development of efficient methods of pretreatment of hemicellulose is of great importance for the effective use of this constituent of biomass, in the production of furfuraldehyde. Some of the challenges in furfural production include the use of homogeneous catalysts, which can lead to serious operational, safety, and environmental problems during furfural production in several ways.

According to Delbecq et al. (2018), it is very difficult to separate mineral acids for recycling, which can lead to some contamination of the product. Besides, these acid catalysts are highly corrosive and can cause damage to the equipment and break the process, in addition to being toxic and can cause environmental problems during disposal (Delbecq et al., 2018).

The solvent systems used in the production of furfuraldehyde from lignocellulosic biomass have an important role in furfuraldehyde yield and the sustainability of the process. According to Hu et al. (2014) the solvent–catalyst, solvent–biomass, and biomass–catalyst interactions must be taken into account in the production of furfural, in which good biomass–catalyst interaction is essential, as well as the selection of a suitable solvent is important because the interaction biomass - catalyst is largely affected by the solvent.

The addition of catalysts promotes the cleavage of inter and intrabonds in the reaction process, which generates a process of selective dissolution of hemicellulose even more effective in biomass and increasing its conversion into furfural. According to Karinen et al. (2011) the addition of catalysts promoted the formation of furfural through the following effects: (1) increased cracking of the chemical bond of xylose and/or xylose oligomers to form furfural; (2) Increased rearrangement of different species to form furfural; (3) Increased selectivity for furfural. Thus, it is possible to realize the great importance of the presence of the catalyst for the conversion of hemicellulose into furfural.

Xing et al. (2011) reported that furfural yields > 90% can be achieved by conducting the catalyst acid in the dehydration of pentoses in a system containing sufficient tetrahydrofuran (THF) and sodium chloride so that the THF will form a second phase in which the furfural is extracted.

A slightly improved furfural yield (55%) was obtained in the continuous process developed by Quaker Oats. The company used a one-hour residency process. The process was successful, but the plant ended up closing due to the high cost of maintaining the continuous reactor system. Improving the yield of furfuraldehyde production beyond 55% has been the objective of many studies in the last 100 years (Hurd and Isenhour, 1932; Dunlop, 1948; Fulmer et al., 1936; Brownlee and Miner, 1948). The fact that furfural degrades quickly through resinification and condensation reactions are what makes this improvement difficult.

In the resinification reaction, furfural reacts with itself, while in condensation reactions, furfural reacts with xylose or one of the intermediates for converting xylose to furfural to form furfural pentose or difurfural pentose. This promotes a great loss of furfural by condensation, which is significantly greater than the loss by resinification (Zeitsch, 2000). Over the past few years, much research has been carried out to reduce this degradation and improve the yield of furfuraldehyde produced (Choudhary et al., 2011; Gürbüz et al., 2013).

In the biorefinery process, hydrolysates with a high percentage of sugar are desired, as well as offering hydrolysates with added value industrial products, in addition to industrial advantages, such as reduced reactor size, product recovery with reduced energy and water costs (Mittal et al., 2017).

In this sense, a rich in sugar hydrolyzed pentose was obtained from corn straw using 1 ton NREL per day of pretreatment with dilute reactor acid. This hydrolyzate relevant to the process contained xylose, arabinose, glucose, acetic acid, and sugar degradation products and was representative of a hydrolyzate sample that could be produced from a pretreatment process with diluted acid in a biorefinery.

As more recent studies on furfural production, it is possible to find new processes like the one proposed by Yemis and Mazza (2012) that reported a maximum furfural yield of 51.3%, with the use of wheat straw subjected to microwave irradiation at 155°C with hydrochloric acid, HCl, as a catalyst for 31 min.

According to Raman and Gnansounou (2015), microwave irradiation as an energy source is capable of generating higher furfural yields compared to those found in the literature, due to the ability of microwaves to interact with biomass components at the molecular level. Although generally higher temperatures are beneficial to furfural production, these authors reported a drastic decrease in furfural production was recorded at extreme temperatures and low pH values. It was observed that above 170°C, furfural performance decreased due to thermal instability of furfural at extreme temperatures.

5 Glucaric acid

D-Glucaric acid (GlucA) is an aldaric acid, its crystalline structure has a sickle conformation, is naturally occurring in vegetables (Denton et al., 2011), and can be obtained by the nitric

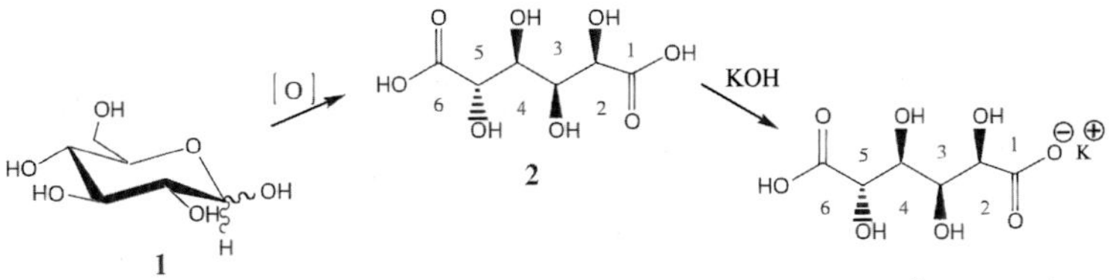

FIG. 2 Oxidative conversion of D-glucose (1) to D-glucaric acid (2), isolated as monopotassium D-glucarate (3). Fonte: Denton et al. (2011).

acid oxidation of D-glucose and isolated as its monopotassium salt (Sohst and Tollens, 1888; Mehltretter and Rist, 1953), as shown in Fig. 2. In addition, its metabolic pathway can potentially provide several advantages, including moderate processing conditions and high selectivity for the product of interest (Reizman et al., 2015).

Glucaric acid is an emerging platform-derived chemical derived from glucose with promising applications as a biodegradable and biocompatible product in the manufacture of plastics, detergents, and medicines. According to Petroll et al. (2020), glucaric acid has numerous applications and can be used as a biodegradable and biocompatible chelating agent for the degradation of organic contaminants, removal of heavy metals in the soil, calcium sequestration for a phosphate-free detergent component, as an adhesive, anticorrosive, anti -plasticizer and plastic reinforcer, starting material for biodegradable polymers and hydroxylated nylon, and also has biological properties such as anticarcinogenic. The wide application of glucaric acid and the growing consumer concern about detergents containing phosphates, which are harmful to the aquatic environment, led to the growth of the global glucaric acid market and according to the new research report by Global Market Insights, the estimate for 2024 is to exceed $440 M (Prnewswire, 2019).

The United States Department of Energy in 2004 identified glucaric acid, along with 1,4-diacids (succinic, fumaric, and malic), 2,5-furan dicarboxylic acid, 3-hydroxy propionic acid, aspartic acid, glutamic acid, itaconic acid, levulinic acid, 3-hydroxybutyrolactone, glycerol, sorbitol, and xylitol/arabinitol, like the 12 building block chemicals that can be produced from sugars via biological or chemical conversions and that can be subsequently converted to a number of high-value bio-based chemicals or materials (Werpy and Petersen, 2004a, 2004b).

Biorefineries invest in the transformation of glucose into glucaric acid because this is the starting point for the production of a wide range of products with applicability in high volume markets and for the development of efficient processes of production of glucaric acid that may also be applicable to efficient oxidation of other sugars, such as xylose or arabinose. Glucaric acid can be obtained by different techniques, such as the one-step oxidation of nitric acid from starch or by the catalytic oxidation of starch with bleach (basic), as well as by biotransformation. Glucaric acid derivatives can be obtained by means of primary transformation by dehydration, resulting in lactones (Solvents); and by amination and direct polymerization to obtain polyglucaric esters and amides (Nylons) (Werpy and Petersen, 2004a, 2004b).

GlucA is mainly produced through glucose oxidation using nitric acid or 2,2,6, 6-tetramethylpiperdinyl-1-oxyl. However, these processes have some technical barriers such as the generation of large amounts of by-products, low product yields, and the use of expensive catalysts (Mehltretter and Rist, 1953; Werpy and Petersen, 2004a, 2004b). In addition, growing environmental concerns and resource shortages have required the investment of "greener" synthetic routes for biochemicals and biocommodities, which has made GlucA microbial production an attractive target for the bio-based industry, since fermentation-based manufacturing is considered economical and environmentally friendly (Pellis et al., 2018).

According to Petroll et al. (2020), there has been a significant focus on microbial production of GlucA from renewable materials, such as glucose, sucrose, and myo-inositol. However, these production processes generally lack efficiency due to toxicity problems, competition of metabolites, and suboptimal enzyme reasons and proposed the use of synthetic biology and cell-free biocatalysis as a viable approach to overcome many of these limitations. The authors also reported that cell-free biocatalysis also has its own limitations for industrial applications, due to high enzyme costs and the consumption of cofactors, which can be mitigated when GlucA is produced from glucose-1-phosphate using a combination of thermostable and mesophilic enzymes, incorporation of a cofactor regeneration system and immobilization and recycling of pathway enzymes.

After the production of GlucA by microorganisms, purification has been a very important and delicate step in this process. Yuan et al. (2017) studied the adsorption of GlucA using seven ion exchange resins and three nonionic adsorbents. The ion exchange resins showed superior adsorption capacity (350 mg/mL a pH 3,9) and diffusivity (pore diffusion of $18,3 \times 10^{-10}$ m^2/s and the surface diffusion coefficient of $1,29 \times 10^{-11}$ m^2/s), being the best option for purifying GlucA.

An alternative for the chemical route is the use of glucose electrooxidation in glucaric acid in two consecutive steps using a pure gold electrode in a batch reactor, according to Moggia et al. (2021). The first stage, with low potential, enabled the effective formation of the intermediate gluconic acid, promoted by the oxidation of the aldehyde group in C1, under the optimum conditions of pH 11.3, temperature of 5°C and 0.04 M of initial glucose, obtaining maximum selectivity of 97.6%, with the conversion of 25%. At the highest potential, gluconic acid was converted to glucaric acid by oxidation of the hydroxymethyl group at C6, obtaining the maximum selectivity of 89.5%. In all experiments, regardless of the conditions and reaction time, a maximum concentration of approximately 1.2 mM of glucaric acid was achieved, and a drastic decrease in current density was observed in the first hours of electrolysis, as well as a process of poisoning caused by GlucA, which inactivates the catalyst, remains strongly bonded to the catalyst's active sites.

6 Hydroxymethylfurfural (HMF)

Once the conversion of lignocellulosic biomass to glucose is obtained, it can be converted into a series of other platform molecules for the production of value-added fuels and chemicals. Among the many products derivable, 5-hydroxymethylfurfural (5-HMF) plays a key role as a platform molecule. The chemical structure of 5-HMF, consisting of a furan ring with two functional groups: one aldehyde group and one alcohol group, makes 5-HMF an

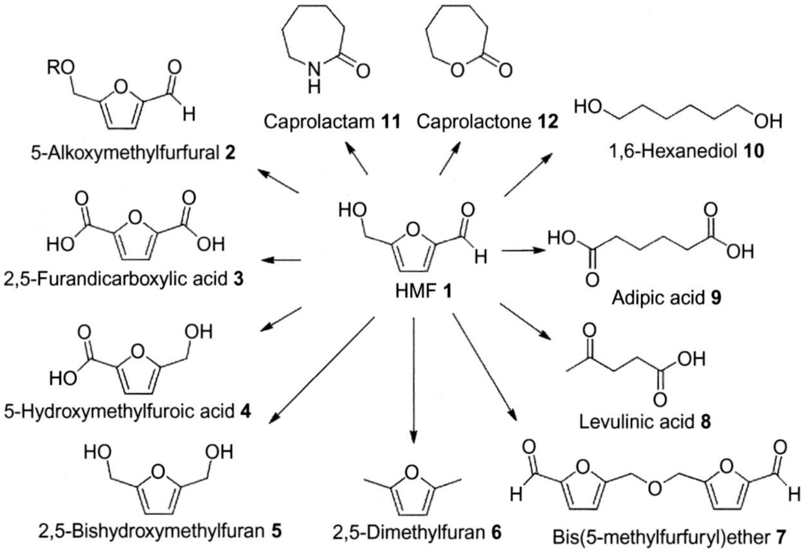

FIG. 3   Potential chemical compounds derived from 5-HMF (van Putten et al., 2013).

important bio-based commodity for the synthesis of various chemicals and polymers in the chemical and pharmaceutical industry (Trivedi et al., 2020). As shown in Fig. 3, important furan derivatives with a high potential in fuel or polymer applications can be formed through HMF: 5-alkoxymethylfurfural (2), 2,5-furandicarboxylic acid (3), 5-hydroxymethylfuroic acid (4), 2,5-bishydroxymethylfuran (5), 2,5-dimethylfuran (6) and bis(5-methylfurfuryl)ether. Some important nonfuranic compounds can also be produced from HMF, namely, levulinic acid (8), adipic acid (9), 1,6-hexanediol (10), caprolactam (11), and caprolactone (12) (van Putten et al., 2013).

5-HMF can be synthesized through the acid-catalyzed dehydration of C6 sugars by the elimination of three water molecules (Fig. 4). Glucose, which is the most abundant and the cheapest monosaccharide, has been considered as the preferred feedstock for the production of 5-HMF (Gomes et al., 2015). However, fructose is much more reactive and selective toward 5-HMF than glucose. According to Kuster (1990) glucose shows lower selectivity and yields for HMF formation because of its stable pyranoside ring structure, which hinders its ability to form the acyclic enediol intermediate. A previous isomerization of glucose to fructose is a strategy to overcome low yields from the direct conversion of lignocellulose-derived glucose to 5-HMF (Steinbach et al., 2018). The isomerization step can be done either using aqueous bases or by enzymatically or solid catalysts, such as hydrotalcites, zeolites, and different oxides (Gomes et al., 2015). The catalyst ZrO_2, under high-temperature conditions of hot compressed water (200°C), can promote glucose and fructose isomerization via a base-catalyzed route. Anatase TiO_2, which possesses both acid and base character, promotes glucose isomerization and dehydration into HMF (Wilson and Lee, 2014).

In order to achieve high productivity, most of the studies on 5-HMF production have been carried out in catalytic systems. A large number of inorganic acids (sulfuric acid, phosphoric acid, hydrochloric acid, etc.), organic acids (oxalic acid, maleic acid, levulinic, etc.), and

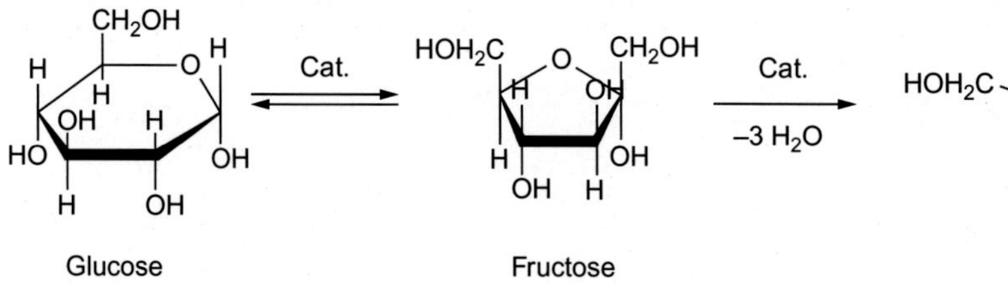

FIG. 4　Conversion of glucose into 5-HMF (Aresta and Dibenedetto, 2019).

heterogeneous catalysts have also been used to catalyze the production of 5-HMF (Ge et al., 2018).

Different strategies have been used in 5-HMF synthesis to decrease the formation of by-products. On the basis of the solvent system used, 5-HMF synthesis from carbohydrates can be divided into three types of processes: single-phase systems, biphasic systems, and ionic liquid-based systems.

Single-Phase Systems: A well-known and established synthetic strategy for 5-HMF production is the homogeneously catalyzed dehydration of hexoses in aqueous solutions with mineral acids as catalysts (mainly sulfuric and hydrochloric acid). Water is an obvious choice of solvent because it dissolves the majority of the sugars in high concentrations unlike most organic solvents; it is a cheap and environmentally friendly solvent and the process is easy to scale up (van Putten et al., 2013; Steinbach et al., 2018). However, 5-HMF rehydrates in a water-based medium under acidic conditions to form levulinic acid and formic acid. Furthermore, the self-condensation of HMF and condensation with other compounds lead to the formation of high-molecular-weight polymeric substances (humins) (Gomes et al., 2015; Steinbach et al., 2018). It is noticed that prolonged reaction time increases levulinic acid yield at the cost of the HMF yield (van Putten et al., 2013). Due to the relatively low 5-HMF yields in aqueous systems, the use of organic solvents such as dimethylsulfoxide (DMSO), 2,5-dimethylfuran (DMF) acetone, acetic acid, and methanol have been reported in the literature. Among these solvents, DMSO is the most widely used for 5-HMF synthesis. One of the reasons is due to preferential salvation of 5-HMF carbonyl groups by DMSO, thus preventing the generation of by-products, such as lactic acid and formic acid (Ge et al., 2018). However, the separation of HMF from these high-boiling point solvents is difficult, leading to thermal degradation of the HMF product (Gomes et al., 2015). The solubility of glucose in organic solvents is generally low, and therefore miscible aqueous–organic systems are used to solve this problem.

Biphasic Solvent Systems: In a biphasic system, the aqueous phase acts as a reactive phase where the dehydration of substrates to 5-HMF occurs, while an organic phase is used as an extracting phase to immediately remove the 5-HMF formed in the reactive phase, thus preventing the further rehydration of HMF (Ge et al., 2018). By continuously removing HMF, these undesired side reactions can be suppressed to a large extent. For the biphasic system, the partition coefficient is determined as the ratio of solubility of 5-HMF in the organic phase to that in the aqueous phase. A high partition coefficient leads to the high efficiency of 5-HMF extraction to the organic phase. The addition of inorganic salt may increase the partition coefficient for 5-HMF. In the work of Gomes et al. (2015), the NaCl addition to the aqueous phase promoted the extraction of 5-HMF to the organic phase, with an HMF yield of 80% in the case of 2:1 acetone/water medium.

Ionic liquid-based systems: Ionic liquids consist of a combination of organic and inorganic ions, which have low vapor pressure and are typically liquid below 100°C (van Putten et al., 2013). In general, the yields of 5-HMF in ionic liquids are higher than those in conventional solvents, because ionic liquids are able to stabilize 5-HMF, preventing the rehydration of 5-HMF to by-products. The ability of ionic liquids to dissolve cellulose makes it suitable for the synthesis of 5-HMF directly from lignocellulosic biomass (Ge et al., 2018). A 5-HMF yield of 89% was reported from cellulose by Zhang et al. (2010) by reacting 17 wt% cellulose in 1-ethyl-3-methylimidazolium chloride ([EMIm]Cl) in the presence of 10 mol% $CrCl_2$ at 120°C after 6h. Tao et al. (2011) reported a high HMF yield of 48% from cellulose in SO_3H-functionalized ionic liquids in combination with $MnCl_2$.

The ionic liquid also shows some drawbacks as the excess loading ionic liquid may increase the collision of reactive compounds and cause cross polymerization that leads to the formation of products such as furfural, 2-furancarboxyaldehyde, organic acids, and phenolics (Ab Rasid et al., 2020). The major disadvantage of the use of ionic liquids as solvents is the high cost that still limits their industrial application (Ge et al., 2018).

7 Levulinic acid

Levulinic acid (LA) is an organic compound and it has great versatility, as it has two functional groups; it fits within the 12 chemical platform products with high added value and can have many applications as in the pharmaceutical, chemical, plastic, and fuel industries. LA is obtained through the rehydration reaction of 5-hydroxymethylfurfural (5-HMF), which is a product of glucose degradation in an acidic medium (Ozsel, 2020; Kumar et al., 2020a, 2020b).

LA can be produced on an industrial scale using the synthesis process. GF Biochemical Ltd. in Italy is the world's largest producer of this compound. Some recent research has considered that levulinic acid has potential for production as it can be used for the manufacture of ethyl, butyl, and methyl levulinates, 5-aminolevulinic acid, 5-methyloxycyclopentan-2-ol, butanedioic acid, etc. LA can also be a starting material for the production of a series of valuable products such as succinic acid, MTHF, valeric acid, δ-aminolevulinic acid, pentanols, diphenolic acid, b-acetylacrylic acid, etc. (Kumar et al., 2020a, 2020b).

The interest in the economic area in the use of renewable energies and the biorefinery has grown a lot in recent years, but it has faced bottlenecks to achieve economic competitiveness, due to the fact that biorefineries have processes for converting biomass that is not yet efficient enough to leverage. In addition, there is also the generation of waste and lack of co-production of other products that have added value, in addition to bioethanol. Therefore, a way to overcome these bottlenecks would be to make this production of other products from biomass, which have numerous applications (Thakkar et al., 2020).

Levulinic acid is part of the group of gamma-keto acids, which are produced by the dehydration reaction that uses acid as a catalyst during the hydrolysis of C6 sugars (hexoses). The most common method for the production of LA is by acid, with a production yield of 64.5% by weight when it is obtained using lignocellulosic biomass, and when we have the formation of inhibitors, which are undesirable products during the process reaction, this yield is even lower (Thakkar et al., 2020).

LA also has another nomenclature, that of 4-oxopentanoic acid, from the group of carboxylic acids that have a ketone group in its structure. Physically, LA is present in yellow, pale white form with a melting point of 37°C; its solubility is good in the water, chloroform, alcohol, and ether, being insoluble in paraffin oil and aliphatic hydrocarbons (Kumar et al., 2020a, 2020b).

LA is a ketoacid that has high versatility and is produced from cellulose residues and/or agricultural by-products. Its thermodynamic properties present data of critical points of $T_C = 755\,K$, $\rho_C = 285.4\,kg/m^3$, and $P_C = 30.57\,bar$. Simulation studies can use such data as input to the state equations to assist in the industrial separation processes of biorefineries (Chakraborti et al., 2018).

Some applications of LA have already been discussed in this topic and some of them can be shown in Table 3. However, one of the most promising is the use of as an input for the production of diphenolic acid (DPA), a direct substitute for bisphenol A, which has numerous applications as shown in Table 3, adapted of the work of Bozell et al. (2000).

The main achievement of LA is through chemical synthesis, this production is the most conventional, and it can also be produced by bioprocesses, where lignocellulosic materials are used as a starting point for this production. In this use (with lignocelluloses), the process goes through a series of steps during its conversion and goes through the use of a series of catalysts (such as sulfuric acid, hydrochloric acid, etc.). The consolidated method for producing LA is through precipitation with sulfuric acid, which is a strong acid, in this process, the use of a high concentration of this acid is necessary and this brings the inconvenience of producing inhibitors, material residues, and greenhouse gases. The production of LA by fermentation requires the development of environmentally friendly technologies that have a biological basis. Fermentation is a low-cost process when compared to the chemical synthesis

TABLE 3 Some add value was obtained by LA.

Product	Use
MTHF	Fuel extender
DALA	Biodegradable herbicide
Diphenolic acid	*Solvent* *Thermoplastics* *Polysulfones* *Polyphenylene* *Ethers* *Hyperbranched and dendrimeric* *Polyesters* *Thermally reversible isocyanates* *Phenolic and polyester resins*, etc.
BDO	Monomer

Source: Bozell, J.J., Moens, L., Elliott, D.C., Wang, Y., Neuenscwander, G.G., Fitzpatrick, S.W., …, Jarnefeld, J.L., 2000. Production of levulinic acid and use as a platform chemical for derived products. Resour. Conserv. Recycl. 28(3–4), 227–239 with the permission of Elsevier.

process, but it is still somewhat limited and needs to be better consolidated, as there is difficulty in recovering the final products (Kumar et al., 2020a, 2020b).

The production of high added value products such as LA from lignocellulosic biomass can be a promising idea since the products of interest in this route are very limited to ethanol and furfural, which does not make the process profitable, but when we amplify the quantity of other products, we can resolve that issue (Singh et al., 2019).

Singh et al. (2019) produced LA from pretreated lignocellulosic (palmerous) biomass (with p-cymene-2-sulfonic acid (p-CSA) synthesized from D-limonene from renewable sources) from reflux temperatures of 140–200°C, all over The process used renewable energy sources, being a green and environmentally friendly process. This study demonstrated that the tested process had advantages such as not being corrosive, using p-CSA instead of mineral acids, is considered a green reagent, in addition to using water as a solvent instead of toxic or flammable organic solvents.

Ozsel (2020) carried out a study of the use of textile waste to produce LA from the cellulose contained in this residue, using different catalysts (Pt/AC and BT500S) producing about 655.3 mg/mL of LA at a temperature of 200°C and the catalyst BT500S. This author showed promising results and an alternative to minimizing solid waste from this industry through the production of a product with high added value.

Thakkar et al. (2020) carry out a study to improve this issue of inefficient conversion of biomass, thus, these authors proposed conversion of corn straw (lignocellulosic biomass) in LA, pretreating this biomass with the hydrothermal process under the conditions of 0.45% K_2CO_3, which was able to remove 76% lignin and 85% xylans, and preserved 83% by weight of glucans. After this treatment, the material was used to produce levulinic acid using 2% (w/w) H_2SO_4 in a discontinued reactor at 190°C for 5 min, thus, 30%–35.8% (w/w) of glucans in pretreated corn straw were converted into levulinic acid. Thus, these authors were able to approach the concept of integrated biorefinery proposing the production of several value-added products for greater financial and environmental sustainability.

In addition to being produced from lignocellulosic bioassay, levulinic acid can also be produced using microalgae as an input, such study was carried out by Jeong and Kim (2021) who managed to produce LA using microalgae (*Chlorella vulgaris*) through hydrothermal pretreatment catalytic analysis of the residue extracted by lipid from these microalgae that was used as a renewable input. These authors optimized the process through a statistical design using a Box–Behnken project, reaching yields of 39.27% LA in a solids concentration of 5% biomass treated using 0.95 M HCl at 170°C for 30 min. In conclusion, the microalgae residue extracted by lipid acid catalysts was promising for the production of LA. In this way, microalgae represented a sustainable raw material for the production of LA.

8 Conclusions

Biorefineries can provide a significant contribution to sustainable development. As discussed, various platform chemicals can be produced in biorefineries and are utilizing renewable biomass as a resource. The sugar platform is the largest platform for the production of chemicals based on biomass, such as sorbitol, furfural, glucaric acid, hydroxymethyl-furfural (HMF), and levulinic acid. These chemicals are important in several industrial sectors, such as the chemical industry, food industry, pharmaceutical industry, and others. Therefore, the chemicals discussed in this chapter can be used as a building block for various chemicals, by different methods and ways of obtaining, where the tendency is to use biomass and green processes to obtain a cheaper and more environmentally friendly product. This requires optimal biomass conversion efficiency, thus minimizing feedstock requirements while at the same time strengthening the economic viability of market sectors.

References

Ab Rasid, N.S., Zainol, M.M., Amin, N.A.S., 2020. Pretreatment of agroindustry waste by ozonolysis for synthesis of biorefinery products. In: Praveen Kumar, R., Nansounou, E., Raman, J.K., Baskar, G. (Eds.), Refining Biomass Residues for Sustainable Energy and Bioproducts. Academic Press, pp. 303–336.

Aresta, M., Dibenedetto, A., 2019. Beyond fractionation in the utilization of microalgal components. In: Magalhães Pires, J.-C., Cunha-Gonçalves, A.L.D. (Eds.), Bioenergy with Carbon Capture and Storage. Academic Press, pp. 173–193.

Aristizábal-Marulanda, V., Cardona Alzate, C.A., 2019. Methods for designing and assessing biorefineries. Biofuels Bioprod. Biorefin. 13 (3), 789–808.

Binder, J.B., Blank, J.J., Cefali, A.V., Raines, R.T., 2010. Synthesis of furfural from xylose and xylan. ChemSusChem 2010 (3), 1268–1272.

Bozell, J.J., Moens, L., Elliott, D.C., Wang, Y., Neuenscwander, G.G., Fitzpatrick, S.W., Jarnefeld, J.L., 2000. Production of levulinic acid and use as a platform chemical for derived products. Resour. Conserv. Recycl. 28 (3–4), 227–239.

Brownlee, H.J., Miner, C.S., 1948. Industrial development of furfural. Ind. Eng. Chem. 40 (2), 201–204.

Budzianowski, W.M., 2017. High-value low-volume bioproducts coupled to bioenergies with potential to enhance business development of sustainable biorefineries. Renew. Sust. Energ. Rev. 70, 793–804.

Burt, B.A., 2006. The use of sorbitol-and xylitol-sweetened chewing gum in caries control. J. Am. Dent. Assoc. 137 (2), 190–196.

Chakraborti, T., Desouza, A., Adhikari, J., 2018. Prediction of thermodynamic properties of levulinic acid via molecular simulation techniques. ACS Omega 3 (12), 18877–18884.

Chandel, A.K., Garlapati, V.K., Singh, A.K., Antunes, F.A.F., da Silva, S.S., 2018. The path forward for lignocellulose biorefineries: bottlenecks, solutions, and perspective on commercialization. Bioresour. Technol. 264, 370–381.

Chen, M., Zhang, W., Wu, H., Guang, C., Mu, W., 2020. Mannitol: physiological functionalities, determination methods, biotechnological production, and applications. Appl. Microbiol. Biotechnol. 104, 1–11.

Cheng, S., Martínez-Monteagudo, S.I., 2019. Hydrogenation of lactose for the production of lactitol. Asia Pac. J. Chem. Eng. 14 (1), e2275.

Cherubini, F., Jungmeier, G., Wellisch, M., Willke, T., Skiadas, I., Van Ree, R., de Jong, E., 2009. Toward a common classification approach for biorefinery systems. Biofuels Bioprod. Biorefin. 3 (5), 534–546. https://doi.org/10.1002/bbb.172.

Chin, T., Okuda, Y., Ikeuchi, M., 2018. Sorbitol production and optimization of photosynthetic supply in the cyanobacterium Synechocystis PCC 6803. J. Biotechnol. 276, 25–33.

Chin, T., Okuda, Y., Ikeuchi, M., 2019. Improved sorbitol production and growth in cyanobacteria using promiscuous haloacid dehalogenase-like hydrolase. J. Biotechnol. 306 (Suppl), 100002.

Choudhary, V., Pinar, A.B., Sandler, S.I., Vlachos, D.G., Lobo, R.F., 2011. Xylose isomerization to xylulose and its dehydration to furfural in aqueous media. ACS Catal. 1 (12), 1724–1728.

De Jong, E., van Ree, R., Kwant, I.K., 2009. Biorefineries: Adding Value to the Sustainable Utilisation of Biomass. 1 IEA Bioenergy, pp. 1–16.

de Jong, E., Higson, A., Walsh, P., Wellisch, M., 2012a. Bio-based Chemicals Value Added Products from Biorefineries. IEA Bioenergy, Task42 Biorefinery, p. 34.

De Jong, E., Stichnothe, H., Bell, G., Jørgensen, H., 2020. Bio-Based Chemicals: A 2020 Update. IEA Bioenergy, Paris.

Delbecq, F., Wang, Y., Muralidhara, A., El Ouardi, K., Marlair, G., Len, C., 2018. Hydrolysis of hemicellulose and derivatives—a review of recent advances in the production of furfural. Front. Chem. 6, 146. https://doi.org/10.3389/fchem.2018.00146.

Deng, W., Wang, Y., Yan, N., 2016. Production of organic acids from biomass resources. Curr. Opin. Green Sustain. Chem. 2, 54–58.

Denton, T.T., Hardcastle, K.I., Dowd, M.K., Kiely, D.E., 2011. Characterization of d-glucaric acid using NMR, X-ray crystal structure, and mm3 molecular modeling analyses. Carbohydr. Res. 346 (16), 2551–2557. 29 November 2011 ISSN 0008-6215. Disponível em: http://www.sciencedirect.com/science/article/pii/S0008621511004113.

Dunlop, A., 1948. Furfural formation and behavior. Ind. Eng. Chem. 40 (2), 204–209.

Ebringerová, A., 2005. Structural diversity and application potential of hemicelluloses. Macromol. Symp. 232, 1–12. https://doi.org/10.1002/masy.200551401.

Fang, T., Cai, Y., Yang, Q., Ogutu, C.O., Liao, L., Han, Y., 2020. Analysis of sorbitol content variation in wild and cultivated apples. J. Sci. Food Agric. 100 (1), 139–144.

Fernando, S., Adhikari, S., Chandrapal, C., Murali, N., 2006. Biorefineries: current status, challenges, and future direction. Energy Fuel 20 (4), 1727–1737.

Fiorentino, G., Ripa, M., Ulgiati, S., 2017. Chemicals from biomass: technological versus environmental feasibility. A review. Biofuels Bioprod. Biorefin. 11 (1), 195–214.

Fulmer, E.I., Christensen, L., Hixon, R., Foster, R., 1936. The production of furfural from xylose solutions by means of hydrochloric acid–sodium chloride systems. J. Phys. Chem. 40 (1), 133–141.

Galaverna, R., Pastre, J.C., 2017. Produção de 5-(Hidroximetil) furfural a partir de biomassa: desafiossintéticos e aplicaçõescomobloco de construçãonaprodução de polímeros e combustíveislíquidos. Rev. Virtual Quim. 9 (1), 248–273. https://doi.org/10.21577/1984-6835.20170017.

Ge, X., Chang, C., Zhang, L., Cui, S., Luo, X., Hu, S., Qin, Y., Li, Y., 2018. Conversion of lignocellulosic biomass into platform chemicals for biobased polyurethane application. In: Li, Y., Ge, X. (Eds.), Advances in Bioenergy. vol. 3. Elsevier, pp. 161–213.

Gerardy, R., Debecker, D.P., Estager, J., Luis, P., Monbaliu, J.C.M., 2020. Continuous flow upgrading of selected C2–C6 platform chemicals derived from biomass. Chem. Rev. 120 (15), 7219–7347.

Gomes, F.N.D.C., Pereira, L.R., Ribeiro, N.F.P., Souza, M.M.V.M., 2015. Production of 5-hydroxymethylfurfural (HMF) via fructose dehydration: effect of solvent and salting-out. Braz. J. Chem. Eng. 32 (1), 119–126.

Gürbüz, E.I., Gallo, J.M.R., Alonso, D.M., Wettstein, S.G., Lim, W.Y., Dumesic, J.A., 2013. Conversion of hemicellulose into furfural using solid acid catalysts in γ-valerolactone. Angew. Chem. Int. Ed. 52 (4), 1270–1274. https://doi.org/10.1002/anie.201207334.

Hu, X., et al., 2014. Acid-catalyzed conversion of xylose in 20 solvents: insight into interactions of the solvents with xylose, furfural, and the acid catalyst. ACS Sustain. Chem. Eng. 2 (11), 2562–2575.

Hurd, C.D., Isenhour, L.L., 1932. Pentose reactions. I. Furfural formation. J. Am. Chem. Soc. 54 (1), 317–330.

Jeong, G.T., Kim, S.K., 2021. Statistical optimization of levulinic acid and formic acid production from lipid-extracted residue of *Chlorella vulgaris*. J. Environ. Chem. Eng. 9, 105142.

Karinen, R., Vilonen, K., Niemela, M., 2011. Biorefining: heterogeneously catalyzed reactions of carbohydrates for the production of furfural and hydroxymethylfurfural. ChemSusChem 4, 1002–1016. https://doi.org/10.1002/cssc.201000375.

Kong, X., Zhu, Y., Fang, Z., Kozinski, J.A., Butler, I.S., Xu, L., Song, H., Wei, X., 2018. Catalytic conversion of 5-hydroxymethylfurfural to some value-added derivatives. Green Chem. 20 (16), 3657–3682. https://doi.org/10.1039/C8GC00234G.

Kumar, D., Singh, B., Korstad, J., 2017. Utilization of lignocellulosic biomass by oleaginous yeast and bacteria for production of biodiesel and renewable diesel. Renew. Sust. Energ. Rev. 73, 654–671.

Kumar, S., Ali, H., Kansal, S.K., Pandey, A., Saravanamurugan, S., 2020a. Sustainable production of sorbitol—a potential hexitol. In: Biomass, Biofuels, Biochemicals. Elsevier, pp. 259–281.

Kumar, A., Shende, D.Z., Wasewar, K.L., 2020b. Extractive separation of levulinic acid using natural and chemical solvents. Chem. Data Collect. 28, 100417.

Kuster, B.F.M., 1990. 5-Hydroxymethylfurfural (HMF). A review focusing on its manufacture. Starch/Staerke 42 (8), 314–321.

Lange, J.P., 2007. Lignocellulose conversion: an introduction to chemistry, process and economics. Biofuels Bioprod. Biorefin. 1 (1), 39–48.

Luo, Y., Li, Z., Li, X., Liu, X., Fan, J., Clark, J.H., Hu, C., 2018. The production of furfural directly from hemicellulose in lignocellulosic biomass: a review. Catal. Today. https://doi.org/10.1016/j.cattod.2018.06.042.

Mariscal, R., Maireles-Torres, P., Ojeda, M., Sadaba, I., Lopez Granados, M., 2016. Furfural: a renewable and versatile platform molecule for the synthesis of chemicals and fuels. Energy Environ. Sci. 9, 1144–1189.

Mehltretter, C.L., Rist, C.E., 1953. Sugar oxidation, saccharic and oxalic acids by the nitric acid oxidation of dextrose. J. Agric. Food Chem. 1 (12), 779–783. https://doi.org/10.1021/jf60012a005. 1 September 1953. ISSN 0021-8561. Disponívelem:.

Mittal, A., Black, S.K., Vinzant, T.B., O'Brien, M., Tucker, M.P., Johnson, D.K., 2017. Production of furfural from process-relevant biomass-derived pentoses in a biphasic reaction system. ACS Sustain. Chem. Eng. https://doi.org/10.1021/acssuschemeng.7b00215.

Moggia, G., Schalck, J., Daems, N., Breugelmans, T., 2021. Two-steps synthesis of D-glucaric acid via D-gluconic acid by electrocatalytic oxidation of D-glucose on gold electrode: influence of operational parameters. Electrochim. Acta, 137852. 28 January 2021. ISSN 0013-4686. Disponível em: http://www.sciencedirect.com/science/article/pii/S0013468621001420.

Ozsel, B.K., 2020. Valorization of textile waste hydrolysate for hydrogen gas and levulinic acid production. Int. J. Hydrog. Energy 46, 4992–4997.

Pellis, A., Cantone, S., Ebert, C., Gardossi, L., 2018. Evolving biocatalysis to meet bioeconomy challenges and opportunities. New Biotechnol. 40 (Part A), 154–169. https://doi.org/10.1016/j.nbt.2017.07.005. https://www.sciencedirect.com/science/article/pii/S1871678416325882?via%3Dihub.

Petroll, K., Care, A., Bergquist, P.L., Sunna, A., 2020. A novel framework for the cell-free enzymatic production of glucaric acid. Metab. Eng. 57, 162–173. 1 January 2020. ISSN 1096-7176. Disponívelem: http://www.sciencedirect.com/science/article/pii/S1096717619300345.

Prnewswire, 2019. Glucaric acid market to exceed $440M by 2024. Focus Surfactants 2019 (1), 3–4. 1 January 2019. ISSN 1351-4210. Disponível em: http://www.sciencedirect.com/science/article/pii/S1351421019300149.

Raman, J.K., Gnansounou, E., 2015. Furfural production from empty fruit bunch—a biorefinery approach. Ind. Crop. Prod. 69, 371–377.

Reizman, I.M.B., Stenger, A.R., Reisch, C.R., Gupta, A., Connors, N.C., Prather, K.L.J., 2015. Improvement of glucaric acid production in E. coli via dynamic control of metabolic fluxes. Metab. Eng. Commun. 2, 109–116. 1 December 2015. ISSN 2214-0301. Disponívelem: http://www.sciencedirect.com/science/article/pii/S2214030115300080.

Ribeiro, P.R., Carvalho, J.R.M., Geris, R., Queiroz, V., Fascio, M., 2012. Furfural—da biomassaaolaboratório de químicaorgânica. Quim Nova 35 (5), 1046–1051. https://doi.org/10.1590/S0100-40422012000500033.

Rosatella, A.A., Simeonov, S.P., Frade, R.F.M., Afonso, C.A.M., 2011. 5-Hydroxymethylfurfural (HMF) as a building block platform: biological properties, synthesis and synthetic applications. Green Chem. 13 (4), 754–793. https://doi.org/10.1039/c0gc00401d.

Saxon, D.J., Luke, A.M., Sajjad, H., Tolman, W.B., Reineke, T.M., 2020. Next-generation polymers: isosorbide as a renewable alternative. Prog. Polym. Sci. 101, 101196.

Schieb, P.A., Lescieux-Katir, H., Thénot, M., Clément-Larosiere, B., 2015. Biorefinery 2030: Future Prospects for the Bioeconomy. Springer-Verlag, Berlin, Heidelberg.

Sheldon, R.A., 2014. Green and sustainable manufacture of chemicals from biomass: state of the art. Green Chem. 16 (3), 950–963.

Singh, M., Pandey, N., Dwivedi, P., Kumar, V., Mishra, B.B., 2019. Production of xylose, levulinic acid, and lignin from spent aromatic biomass with a recyclable Brønsted acid synthesized from d-limonene as renewable feedstock from citrus waste. Bioresour. Technol. 293, 122105.

Sohst, O., Tollens, B., 1888. Über krystallisirte Zuckersäure (Zuckerlactonsäure). Justus Liebigs Ann. Chem. 245 (1–2), 1–27. ISSN 0075-4617. Disponível em: https://chemistry-europe.onlinelibrary.wiley.com/doi/abs/10.1002/jlac.18882450102.

Souza Filho, P.F., Taherzadeh, M.J., 2020. Integrated biorefineries for the production of bioethanol, biodiesel, and other commodity chemicals. In: Green Energy to Sustainability: Strategies for Global Industries. Wiley, pp. 465–488.

Steinbach, D., Kruse, A., Sauer, J., Vetter, P., 2018. Sucrose is a promising feedstock for the synthesis of the platform chemical hydroxymethylfurfural. Energies 11, 645–660.

Tao, F., Song, H., Chou, L., 2011. Hydrolysis of cellulose in SO_3H-functionalized ionic liquids. Bioresour. Technol. 102 (19), 9000–9006.

Thakkar, A., Shell, K.M., Bertosin, M., Rodene, D.D., Amar, V., Bertucco, A., Kumar, S., 2020. Production of levulinic acid and biocarbon electrode material from corn Stover through an integrated biorefinery process. Fuel Process. Technol. 213, 106644.

Trivedi, J., Bhonsle, A.K., Atray, N., 2020. Processing food waste for the production of platform chemicals. In: Praveen Kumar, R., Gnansounou, E., Raman, J.K., Baskar, G. (Eds.), Refining Biomass Residues for Sustainable Energy and Bioproducts. Academic Press, pp. 427–448.

van Putten, R.-J., van der Waal, J.C., de Jong, E., Rasrendra, C.B., Heeres, H.J., de Vries, J.G., 2013. Hydroxymethyl-furfural, a versatile platform chemical made from renewable resources. Chem. Rev. 113 (3), 1499–1597. https://doi.org/10.1021/cr300182k. PMid: 23394139.

Werpy, T., Petersen, G., 2004a. Top value added chemicals from biomass: volume I—results of screening for potential candidates from sugars and synthesis gas no. DOE/GO-102004-1992. National Renewable Energy Laboratory, Golden, CO (US).

Werpy, T., Petersen, G., 2004b. Top Value Added Chemicals from Biomass Volume I—Results of Screening for Potential Candidates from Sugars and Synthesis Gas. Department of Energy, United States, p. 67. Disponívelem: https://www.osti.gov/biblio/15008859.

Werpy, T., Petersen, G., Aden, A., Bozell, J., Holladay, J., White, J., Manheim, A., 2004. Top Value Added Chemicals from Biomass. Vol. 1—Results of Screening for Potential Candidates from Sugars and Synthesis Gas. Department of Energy, Washington, DC.

Wilson, K., Lee, A.F., 2014. Bio-based chemicals from biorefining: carbohydrate conversion and utilisation. In: Waldron, K. (Ed.), Advances in Biorefineries—Biomass and Waste Supply Chain Exploitation. Woodhead Publishing, pp. 624–658.

Xing, R., Qi, W., Huber, G.W., 2011. Production of furfural and carboxylic acids from waste aqueous hemicellulose solutions from the pulp and paper and cellulosic ethanol industries. Energy Environ. Sci. 4 (6), 2193–2205.

Xu, C., Paone, E., Rodríguez-Padrón, D., Luque, R., Mauriello, F., 2020. Reductive catalytic routes towards sustainable production of hydrogen, fuels and chemicals from biomass derived polyols. Renew. Sust. Energ. Rev. 127, 109852.

Yemis, O., Mazza, G., 2012. Optimization of furfural and 5-hydroxymethylfurfural production from wheat straw by a microwave-assisted process. Bioresour. Technol. 109, 215–223.

Yuan, W., Ding, R.-H., Ge, H., Zhu, P.-L., Ma, S.-S., Zhang, B., Song, X.-M., 2017. Solid-phase extraction of d-glucaric acid from aqueous solution. Sep. Purif. Technol. 175, 352–357. 24 March 2017. ISSN 1383-5866. Disponívelem: http://www.sciencedirect.com/science/article/pii/S1383586616319116.

Zeitsch, K.J., 2000. The Chemistry and Technology of Furfural and its Many by-Products. Elsevier Science.

Zhang, Y., Du, H., Qian, X., Chen, E.Y.X., 2010. Ionic liquid – water mixtures: enhanced kw for efficient cellulosic biomass conversion. Energy Fuel 24 (4), 2410–2417.

Zhang, W., Chen, J., Chen, Q., Wu, H., Mu, W., 2020. Sugar alcohols derived from lactose: lactitol, galactitol, and sorbitol. Appl. Microbiol. Biotechnol. 104, 1–9.

Biofuels and bioproducts from seaweeds

Karuna Nagula[a,b], *Himanshu Sati*[a], *Nitin Trivedi*[a], *and C.R.K. Reddy*[a,c]

[a]DBT—ICT Centre for Energy Biosciences, Institute of Chemical Technology, Mumbai, Maharashtra, India [b]Parul Institute of Pharmacy & Research, Parul University, Vadodara, Gujarat, India [c]Indian Centre for Climate and Societal Impacts Research, Shri Vivekanand Research and Training Institute (VRTI), Kutch, Gujarat, India

1 Introduction

Seaweeds represent a diverse group of photosynthetic marine organisms that are exceptionally unique in their form, function, structure, and biochemical composition, offering unique opportunities for their commercial exploration. They are also termed as macroalgae. They have higher growth rates, less dependency on land for cultivation and increased CO_2 sequestration potential, which makes them a preferred choice for energy applications in all those countries where their utilization for food and food supplements is nonexistent. Seaweeds are considered a potential feedstock for biofuel production and categorized as third-generation biofuels along with microalgae (Tabassum et al., 2017) and non-lignocellulosic feedstock. The main advantage of macroalgae over microalgae is that they offer more biomass productivity. As compared to the lignocellulosic biomass they allow milder processing and shorter processing time due to the absence of recalcitrant lignin (Tabassum et al., 2017). As per the National Bio-Fuel Policy of India 2018 (Press Information Bureau, Government of India, Ministry of Petroleum and Natural Gas, 2018), there is a need to increase the biofuel production capacities to meet the 20% blending of biofuels with fossil fuels by 2030. It has been estimated that 460 million tons of fresh seaweed biomass needs to be harvested annually over a 10 million hectare area of seaweed farming to generate 6.66 billion liters of ethanol and meet the targeted petrol blending (Baghel et al., 2016). According to the Indian biofuel policy, the following components are considered as biofuels like bioethanol, biodiesel, advanced biofuels like second-generation (2G) ethanol, drop-in fuels, algae-based 3G biofuels, BioCNG, bio-methanol, dimethyl ether (DME) derived from bio-methanol, biohydrogen, and BioCNG unless they meet the specification as per global and Indian fuel standards. Thus, government policies have considered seaweeds as potential alternatives for fossil fuels. To achieve these targets, industrial production of biofuel from seaweeds needs to be given due consideration; the vast global sea coast can be utilized for growing macroalgae without encroaching on the farm and forest lands. The macroalgal research for biofuel is in the developmental stage with a growth history of just a decade (Michalak, 2018).

Due to high carbohydrate content of seaweeds, biogas and bioethanol are considered the most suitable biofuels for extraction. However, exploration of seaweeds solely for biofuel production seems unprofitable and hence, integrated biorefinery is considered a more profitable technique. From the biofuel perspective, Ulva spp. scores more from the point of view of cosmopolitan distribution, climate resilience, higher productivity, short turnover period, etc. and negligible waste stream. Integration of seaweed production with aquaculture systems for bioremediation will also help in pollution abatement. With the growing awareness on seaweeds, there is a huge surge in the research on biofuels from seaweeds. The present chapter briefly reviews chemical characterization of feedstock, potential species of biofuel interest, sources of seaweeds, experimental conditions, pros and cons of seaweeds as a potential bioresource for energy production.

2 Potential seaweed feedstocks and their characteristics

Globally, as many as 221 seaweed species have been explored either for food or for the extraction of products of commercial interest (Pereira, 2011); out of which only a few species like *Gelidium pusillum, G. elegans, Ulva fasciata, U. lactuca, Sargassum latifolium* have largely been explored for biofuel production, especially bioethanol production along with the recovery of other value-added products like carbohydrates and proteins (Baghel et al., 2016; Soliman et al., 2018; Hessami et al., 2019). Macroalgal species selection is very crucial for enhanced bioproducts and biofuel production successfully. According to FAO 2020 report, the species that are farmed on a large scale are confined to less than a dozen and includes the species such as *Ulvalactuca, Porphyra vietnamensis, Gracilaria pusillum, Gracilaria dura, Kappaphycus alvarezii, Sargassum decurrens, Laminaria janonica,* and *Undaria pinnatifida*. Table 1 shows various seaweed species with their proximate biomass composition and biofuels; that have been reported in literature.

TABLE 1 Data on seaweeds production, compositional analysis, bioproducts, and biofuels.

Macroalgal biomass Green seaweeds	Value added bioproduct	Compositonal analysis (% w/w) dry weight basis			Biofuel reported	Biomass production[a] (tons per year)	References
		Carbohydrate	Lipid	Protein			
Ulva lactuca	Ulvan	58.1	11.2	13.6	Bioethanol, biogas, HTL	3000 tons per year (Japan)	Ohno and Largo (2006), Mhatre et al. (2019), Mahtre et al. (2018), Rasyid (2017)
Ulva faciata	Ulvan	20.02	6.28	9.3	Bioethanol, biogas, HTL	NA	Baghel et al. (2016), Suthar et al. (2019)
Monostroma oxyspermum	Sulfated polysaccharide (SP)	92	0	0	Not reported	1500	Seedevi et al. (2015), Ohno and Largo (2006)
Enteropmorpha intestinalis	SP	35.8 (cellulose)	11.6	31.6	Bio-oil	NA	Kim et al. (2014)
Enteropmorpha prolifera	SP	35.8 (cellulose)	11.6	13	Bio-oil	NA	Yang et al. (2014)
Cladophora spp	Cellulose	33.6	0.78	14.45	Bio-oil	NA	Parsa et al. (2018), Michalak and Messyasz (2021)
Red Seaweeds							
Porphyra tenera, *Porphyra vietnamensis*	Cellulose	31.598–31.662	0.45–0.53	42.939–43.041	Food, pyrolysis	855.0 lakhs tons (2018)	FAO (2020), Kim et al. (2018)
Gracilaria pusillum, *Gracilaria dura*, *Gracilaria gracilis*	Agar	63.13	0.18	10.86	Bio-oil	3454.8 lakhs tons (2018)	FAO (2020), Parsa et al. (2018), Rasyid et al. (2019)
Kappaphycus alvarezii	Lignin, carrageenan, agar	53%–51%	0.2	3.8–2.3	Food / bioproducts	1597.3 lakhs tons (2018)	Das et al. (2017), Masarin et al. (2016), FAO (2020)

Continued

TABLE 1 Data on seaweeds production, compositional analysis, bioproducts, and biofuels—cont'd

Macroalgal biomass	Value added bioproduct	Compositonal analysis (% w/w) dry weight basis			Biofuel reported	Biomass production[a] (tons per year)	References
		Carbohydrate	Lipid	Protein			
Green seaweeds							
Porphyra purpurea, Porphyra umbilicalis	Vitamins, pigments, phycobilliprotein source	21.7	1.0	33.2	Food	2017.8 lakhs tons (2018)	FAO (2020), Taboada et al. (2012)
Brown seaweeds							
Undaria pinnatifida	Alginate, fucoidan	21.4	2.7	16.8	Bioethanol	2320.4 lakhs tons (2018)	FAO (2020), Kanda et al. (2014), Taboada et al. (2012)
Sargassum fusiforme, Sargassum decurrens	Antioxidants, carotenoids, and phenols (pharmaceutical relevant)	Cellulose 22.2	0.3		Bioethanol	268.7 lakhs tons (2018)	FAO (2020), Borines et al. (2013), Milledge et al. (2016)
Laminaria janonica Laminaria hyperborea Laminaria digita	Laminarin, mannitol, alginate	34–55		5–12	Bioethanol, biogas	11,448.3 lakhs tons (2018)	FAO (2020), Tabassum et al. (2017)
Phaeophyceae	Alginate, cellulose	64–54	12–9	5–10	Food	891.5 lakhs tons (2018)	FAO (2020), Vizetto-Duarte et al. (2016)

[a] *Biomass production for the period 2018–19 has been reported.*
NA data not reported on a commercial scale.

3 Cultivation, harvesting and pretreatment of seaweed biomass

In 2018, the global production of seaweeds was estimated to be 32.4 million tons with a market value of US$ 13.3 billion. The farmed seaweeds represented 97% by volume. Recently, seaweed farming is gaining increasing attention to be promoted and monitored for climate and environment-friendly bioeconomy development. Of the 32.4 million tons of farmed seaweeds cultivated in 2018, the major share represented the genera *Undaria, Porphyra/Pyropia,* and *Cladophora* produced in East and Southeast Asia (Table 1) are produced primarily as human food. According to NCDC report 2020 and SOFIA 2020, the seaweed production in India has been estimated as 25,000 t (wet) in 2020 and a bulk of this production was represented by merely a few species such as *Gelidiella acerosa, Gracilaria edulis, G. dura, G. debilis,* and *Ulva.*

Seaweeds can be sourced from natural beds or augmented through artificial cultivation, including near and offshore farming. The integrated multitrophic aquaculture (IMTA) system has gained recent advancements wherein nitrogenous compounds from fish are being sequestered by cultivated seaweeds. Another efficient source of seaweed biomass can be offshore or near-shore wastewater treatment plants where seaweeds utilize the nutrients from the human and animal source, mineral fertilizers; thus, these are not suitable for food purposes. The seaweeds have been reported to utilize high amounts of nitrogen and phosphorous from water reservoirs like lakes, ponds, seas, and even oceans. *U. lactuca* has been widely used for this purpose to clean the shrimp wastewater and ponds in which it is being cultivated, and successful IMTA technology has been developed and utilized efficiently (Elizondo-González et al., 2018; Laramore et al., 2018). Also, as compared to the other seaweeds, Ulva has no or less toxic metals adsorption (Fig. 1).

4 Thermochemical conversion

The thermochemical process of biofuel production involves subjecting the biomass to increasing temperatures leading to decomposition of organic matter of the algal biomass to produce biofuels, including liquid, gaseous, and solid fuels (Halder and Azad, 2019), as depicted in Fig. 2. Thermochemical conversion is an efficient and one-step method to convert biomass

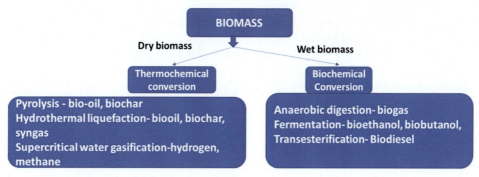

FIG. 1 Various technologies converting seaweeds to biofuels.

FIG. 2 Overview of thermochemical conversion of macroalgae.

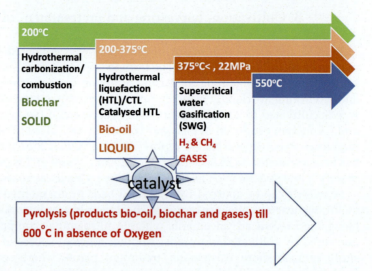

into biofuels. It includes two different categories: dry (nonaqueous) techniques like combustion and pyrolysis and wet biomass processing hydrothermal techniques such as hydrothermal liquefaction (bio-oil), hydrothermal carbonization (to produce biochar), hydrothermal gasification (syngas), and supercritical water gasification (to produce syngas). The various steps of thermochemical conversion of dry seaweed biomass are mainly destruction of biomass structure, degrading to condensable vapors which can form a liquid. These vapors decompose to gaseous molecules by increasing the temperature. The production of syngas, predominantly methane, has gained importance even though with less energy density than ethanol since it can be directly used to operate turbines or used as fuel (Behera et al., 2015). It is a one-step process with a shorter processing time, taking few minutes to generate methane than biogas generation.

It is important to have a criteria in place for the selection of biomass for thermochemical conversion to biofuel. The requirements for lignocellulosic biomass are already well established. Choosing high energy density feedstock is essential in all cases. Along with this, the other parameters that are important include the availability of biomass feedstock, its characteristics, geographical conditions, demand, feedstock costs and oil prices, the energy content of biomass feedstock, CO_2 emissions of biomass end products, effects on soil, water, and biodiversity, job creation, and local community support, business, and economic growth, conversion technologies and sustainability, job creation and local community support. Biomass suitable for thermochemical conversion based on HHV values and elemental composition is important in supercritical water gasification, HTL.

4.1 Pyrolysis

"Pyrolysis is defined as the incomplete combustion of biomass producing a wide range of products which are carbon-rich, and generally O-H-S-N-poor compounds." "Pyrolysis is the dry chemical process of decomposing or converting biomass into one or more recoverable compounds by heating it to very higher temperatures in the absence of oxygen to obtain products like bio-oil, syngas, biochar" (Basu, 2018). Several modifications of pyrolysis such as flash pyrolysis, fast pyrolysis, and microwave-assisted pyrolysis were reported in the literature (Lee et al., 2020; Yingbin et al., 2012). Macroalgal biomass shows excellent potential as pyrolysis feedstock in generating energy-dense and valuable pyrolytic products. The qualities of the pyrolytic products depend on the carbohydrates, lipids, and protein contents of macroalgae. From the established conventional pyrolysis, the products produced from macroalgae show moderate energy contents (<34 MJ/kg). To enhance the production of biofuels from algal biomass, advanced or nonconventional pyrolysis techniques have been employed. Catalytic pyrolysis on algal biomass could reduce the nitrogenates and oxygenates in the biofuels.

In hydropyrolysis of algal biomass, the generated biofuels can produce up to 48 MJ/kg with a high product yield of bio-oil up to 50 wt%, comparable to conventional fuels. Therefore, the formation of biofuel fractions suitable for energy generation highly depends on the selected pyrolysis technologies. Guda et al. (2015) has reviewed fast pyrolysis techniques for algal biomass, which can be an appropriate process for biofuel production from seaweeds to obtain an energy density liquid product (~17 MJ/kg). Over more than two decades, numerous research and development efforts have led to the improvement of various pyrolysis reactor design systems; this has resulted in an enhanced bio-oil yield and quality. A commercial-scale catalytic pyrolysis facility to produce hydrocarbons at a scale of 500 ton/day production capacity is currently in the test stage, which can be utilized for processing seaweeds and the commercialization of other pyrolysis systems to produce transportation fuels (Guda et al., 2015).

Poryphyra tenera with a higher heating value (HVV) of 16.46 MJ/kg among seaweeds suggests a potential fuel candidate reported by Kim et al. (2018). The carbohydrates content decomposed to levoglucosan and deoxygenated sugars in the temperature range of 200–300°C producing glycerol, cholesterol, pyroles, and indoles (Kim et al., 2018). Inorganic metals in the pyrolysis of algae are widely studied by Bae et al. (2011) where the author suggested that the metals can enhance the catalytic effect on the pyrolysis reaction at lab-scale batch reactor since their concentration is low. But they may inhibit the reaction on a pilot plant operation leading to more ash deposition. Bae et al. (2011) obtained the highest bio-oil yield from pyrolysis of various seaweeds like *Undaria pinnatifida*, *Laminaria japonica*, and *Poryphyra* at 500°C and found that a large number of inorganic metals contained in algae can act as a catalyst in pyrolysis. These metals enhance the catalytic decompose of the tar vapor in *Porphyra* species which makes it a potential feedstock for pyrolysis. But the major disadvantage of *Poryphyra* is the high protein content which produces nitrogenous compounds, thus making the direct use of bio-oil infeasible. Therefore, such bio-oil obtained by pyrolysis needs to be catalytic upgraded. Verma et al. (2021) reported catalytic pyrolysis of *U. lactuca* macroalgae studied over a series of ZrO_2 supported metals such as Co, Ni, and Co—Ni metal catalysts at a temperature range of 300–500°C. The highest bio-oil yield (47.8 wt%) was found with Co-Ni/ZrO_2 (10 wt%) catalyst, while noncatalytic conversion yielded 42.5 wt% bio-oil. Moreover, with an increase in the metal amount to 15 wt%, the bio-oil yield further increased to 49.2 wt% with improved quality. Introducing Co—Ni into the ZrO_2 must have resulted in higher surface area and increased active sites. Catalytic bio-oils consisted of mainly long-chain hydrocarbon in the range of C6–C16 and showed a higher "high heating value"

(HHV) 38.1 MJ/kg as compared to noncatalytic bio-oils (29.4 MJ/kg). Catalysts have been shown excellent recyclability on bio-oil yield and compounds selectivity. Acid-catalyzed pyrolysis is more suitable for seaweeds with high carbohydrate contents since there is an increased solubilization of sugars due to acid. TGA, GCMS, FTIR, NMR, and elemental analysis techniques are fast methods to predict the rate of degradation, mechanism, and product analysis for chemical compounds and functional groups. The pyrolytic conditions and type of catalyst change according to the seaweed employed.

As reported by Ma et al. (2020), pyrolysis of *Ulva prolifera* biomass was carried out in a fixed-bed reactor in the presence of three zeolite-based catalysts (ZSM-5, Y-Zeolite, and Mordenite) with the different catalyst-to-biomass ratio. A comparison between the noncatalytic and catalytic behavior of ZSM-5, Y-Zeolite, and Mordenite catalysts in the conversion of *Ulva* biomass showed that it is affected by the properties of zeolites. The bio-oil yield was increased in the presence of Y-Zeolite while decreased in the presence of ZSM-5 and Mordenite catalysts. Catalysts' properties significantly affect product distribution and product properties. All the catalysts showed higher gas yield because catalytic pyrolysis caused the cracking of pyrolytic volatiles and promoted the conversion of larger to small molecules. The phenolic content obtained in catalytic pyrolysis of bio-oil was observed to be reduced by 16.1% since the catalyst enhanced the reaction of oxygen with carbon and produced (CO and CO_2) and water.

The co-pyrolysis with suitable feedstock shows excellent enhancement in the bio-oil yield. Co-pyrolysis of acid treated *Ulva prolifera* and Rice husk (Hao et al., 2021), cellulose and *Enteromorpha clathrata* (using ZSM-5 and MCM-41 (Hu et al., 2021) catalysts) removed a significant amount of ash and yielded bio-oil of 46.6% wt at 500°C. This study implied that the synthesized Ga/NiFe-LDO/AC could be considered as a promising catalyst for bio-oil upgrading, which reduced coke formation (Hao et al., 2021). Microwave-assisted pyrolysis of algal biomass has significantly shortened the processing time through advanced heating with low bio-oil yields and high syngas yields of 84 wt% depending on the feedstock used (Kostas et al., 2019). The bio-oil by microwave-assisted pyrolysis of seaweeds *Ulva prolifera* (Zhuang et al., 2012) had HHV 7.1 MJ/kg mainly composed of methyl ester which proves that potential exploration is needed. The seaweed biomass is a suitable feedstock for pyrolysis due to the negligible amount of lignin and high content of soluble carbohydrates; thus, temperatures required for pyrolysis are lower as compared to lignocellulosic biomass. The presence of alkali metals and salts in seaweeds behaves like catalyst and favors pyrolysis with less processing time (Bae et al., 2011).

4.2 Hydrothermal liquefaction (HTL) (catalyzed and uncatalyzed)

Hydrothermal liquefaction (HTL) is a thermochemical conversion process in which the macromolecules of biomass are hydrolyzed or degraded by means of water, usually carried out in temperature ranges between 280°C and 370°C and pressures in the range from 10 to 25 MPa. Hydrothermal liquefaction (HTL) can also be referred to as "hydrous pyrolysis." It is a thermochemical depolymerization process in an enclosed reactor to convert wet biomass into crude oil and chemicals at moderate temperature (typically 200–400°C) and high pressure (typically 10–25 MPa) (Zhang and Chen, 2018). Micro- and macroalgae having high moisture content make HTL an appropriate process. In HTL processes, water acts as an important reactant and serves as a catalyst. As the reaction condition approaches the critical point of water, several properties of water drastically change, and it is able to bring about fast, homogeneous, and efficient reactions. HTL is relatively independent of particle size of biomass or heating rates because the subcritical water has a major role as a heat transfer and extracting medium. This fact makes seaweeds or macroalgal suitable as feedstock due to their larger size. The product yield and physicochemical properties of HTL products are basically affected by the types of biomass, processing conditions (primarily reaction temperature and time), and the existence of catalyst.

Review on red seaweeds for biofuel production by pyrolysis and HTL is well documented by Kumar et al. (2020). The biomass of *U. lactuca* has been recently studied as a feedstock for producing crude oil using hydrothermal liquefaction (HTL), catalyst-based hydrothermal liquefaction (CTL), and pyrolysis. The catalyzed-based hydrothermal liquefaction is very well reported for MSW and lignocellulosic biomass, which can be further utilized for seaweeds also (Sreenivasan et al., 2020). The partially processed biomass (residual biomass remained after recovery of sap and protein) has shown enhanced conversion efficiency to oil using the CTL process. Thermo-liquefaction at the temperature of 250°C without a catalyst for 60 min of reaction time results in a greasy material. Pyrolysis turns the major part of the biomass into a liquid "bio-oil." This oil could be used as a bio-lubricant, or it can be further refined to liquid fuel, which typically recovers 40%–80% of the biomass energy content (Sjöhag, 2020). The pyrolysis carried out at environment-friendly conditions and pressures <100 bar produces bio-oil, which contains 78% of the energy content of the biomass. The sap and biochar, as leftovers of this process, are rich sources of minerals and plant growth stimulants that could be used in the agro-industry. A techno-economic assessment of three options, bio-oil, bioethanol, and biogas, has been developed by various researchers with the purpose of identifying which biofuel is the best for which type of biomass (Sjöhag, 2020).

Seaweed selection for the HTL process needs to have a substantial amount of carbohydrates since these carbohydrates convert to bio-oil. Protein-rich biomass or seaweeds are not suitable feedstock since they produce nitrogen on degradation, so they can be removed from the seaweed before processing to HTL. These proteins have high commercial value, which will add to the benefit of the circular economy. Solid catalysts, especially alkaline catalysts (e.g., Na_2CO_3, $CaOH_2$, $BaOH_2$, and LaC3), are also applied to accelerate the reactions in HTL and maximize the bio-oil production. Calcium ions present in seaweeds are very reactive (Bae et al., 2011). Hydrothermal processing includes hydrothermal liquefaction (wet process, bio-oil) and hydrothermal gasification (methane or hydrogen gases). This processing can be conducted using a catalyst or without a catalyst. The mechanism of hydrothermal liquefaction follows three basic steps: depolymerization of biomass, decomposition of biomass monomers, and steam reforming of gases to form bio-oil.

Co-liquefaction of *E. clathrata* (EN) and lignocellulosic biomass rice husk (RH) studied by Yuan et al. (2019) gave enhanced bio-oil yield by 71.7% over HTL of *E. clathrate* (26.0%) and rice husk (45.6%), respectively. The enhanced conversion rate and bio-oil characteristics open a new exploration of the optimized conditions of hydrothermal co-liquefaction (co-HTL) of the green seaweeds and the lignocellulosic agricultural waste. Nevertheless, the conversion ratio of co-HTL showed a 10.6% significant increase over that of RH. GC–MS results showed that the main compounds of EN and RH bio-oil lump into the C15–C20 and C5–C12 regions, mainly representing the carbon range of diesel and gasoline, respectively. Short-chain (C5–C12) and long-chain (C14–C20) compounds in the bio-oil obtained by co-HTL represented 72% and 28%, respectively. In addition, the ratio of aromatic compounds in the bio-oil of RH was reduced by 9.3% as a result of co-HTL. In conclusion, results suggested 50% ethanol as a co-solvent, 300°C, and 45 min as optimum conditions for co-HTL of EN:RH (1:1 w/w). This study demonstrated an efficient route for co-HTL of third-generation feed-stocks with second-generation feedstocks which will have a significant impact on large-scale applications (He et al., 2016).

Among all the seaweed species evaluated for bio-oil production using pyrolysis and HTL, *Cladophora socialis, Ulva prolifera, Enteromorpha* species (Yang et al., 2014), *Laminaria* (Anastasakis and Ross, 2011), and *Fucus* have shown high yield. The macroalgal species can be classified, selected, and predicted for performance by HTL or pyrolysis using a van Krevelen diagram which is based on ultimate proximate analysis, inorganic content, and cal-orific value analysis. The production of bio-oil from macroalgal biomass through catalytic or noncatalytic pyrolysis is a one-step, cost-effective process. The role of catalyst is to deoxygen-ate and enhance hydrogenation of the bio-oil, which is mostly composed of C3 in HTL.

Pyrolysis technology is best suited for the conversion of dry feedstocks (<5% moisture), whereas HTL is ideal for processing high-moisture (i.e., >20% wet) biomass. The observed fact is that the bio-oil obtained from HTL has less oxygen than pyrolysis in the case of seaweeds with high oxygen content in initial biomass. HTL is still at its early development stage compared to pyrolysis even a thorough understanding of its reaction chemistry is still lacking. The energy input for liquefaction is lower than pyrolysis and gasification since it does not require dried biomass. The challenge relating to HTL is the requirement for high-pressure working conditions, which leads to severe and high-cost technical barriers. The aqueous phase of the liquid yield of the HTL can also be valorized by anaerobic digestion or catalytic hydrothermal gasification or can be upgraded by supercritical water gasification (Anastasakis and Ross, 2011). Advantages of HTL include no need of spending energy to dry the biomass, the presence of water enhances maximum solubilization of carbohydrates, sugars, and metal salts, and water acts like a catalyst at higher temperatures, thus accelerating the process. HTL provides a single-step conversion of biomass to biofuel, the process is environment-friendly since the aqueous phase has fewer toxins formed and can be used as fertilizer for soil amendment. The scale-up versions of HTL reactors are well developed, and thus technology transfer will be faster. The catalyzed HTL allows to generate selective bio-oil, which can be upgraded for biofuel purposes. The disadvantage can be ashing, fouling of catalysts due to sulfur and chlorine elements present in seaweeds (Bruhn et al., 2011) at pilot-scale operations.

4.3 Supercritical water gasification

Supercritical water gasification (SCWG) is an oxidation reaction carried out at 374.1°C and pressure of 22.1 MPa (220 bar) in the presence of water as gasifying medium which yields high hydrogen (H_2) and carbon dioxide (CO_2), low char and tar formation. But the separation of hydrogen is a major difficulty (Aziz, 2017). Till now, the operation is in few mL scales. Ionic inorganic salts are soluble in subcritical water as in the case of HTL and insoluble in supercritical water, so they form precipitate fouling plugging and corrosion of the reactors. There is a plethora of literature available for microalgae, but there is a lack of reports on seaweeds using SCWG. The influences of process variables containing feedstock concentration, temperature, and holding time have to be studied. Supercritical water gasification is reported for the hydrothermal processing of macroalgae to produce gaseous fuel—mainly hydrogen and methane. SCWG has been reported for upgrading the water obtained from hydrothermal liquefaction of microalgae for hydrogen production for biocrude hydrotreating.

The potential of supercritical water gasification (SCWG) of macroalgae for hydrogen and methane production has been investigated by Cherad et al. (2014). Cherad et al. (2014) widely studied the four macroalgae species *Saccharina latissima*, *Laminaria digitata*, *Laminaria hyperborea*, and *Alaria esculenta*. They produced a gas that mainly consisted of hydrogen, methane, and carbon dioxide with noncatalytic SCWG hydrogen yields of 3.3–4.2 mol/kg macroalgae and methane yields of 1.6–3.3 mol/kg macroalgae for all the species. The catalytic SCWG using ruthenium catalyst resulted in hydrogen yields of 7.8–10.2 mol/kg macroalgae and methane yields of 4.7–6.4 mol/kg macroalgae. The use of sodium hydroxide as catalyst (16.3 mol H_2/kg macroalgae) resulted in the yield of hydrogen which was three times higher compared to noncatalyzed SCWG for *L. hyperborea* (5.18 mol H_2 kg^{-1} macroalgae) (Onwudili and Williams, 2013). The energy recovery was 83% when sodium hydroxide was used as a catalyst, compared to 52% for the noncatalytic SCWG of *L. hyperborea*. The yield of methane was approximately 2.5 times higher (9.0 mol CH_4 kg^{-1} macroalgae) when using ruthenium catalyst compared to the noncatalyzed reaction (3.36 mol CH_4/kg macroalgae), and the energy recovery increased from 22% to 74% for all the species studied (Cherad et al., 2016). The study showed that the selectivity of methane or hydrogen production during the SCWG of macroalgae could be controlled using ruthenium or sodium hydroxide, respectively. Longer hold times and increased reaction temperature favored methane production when using ruthenium. Higher hydrogen yields were obtained by using higher concentrations of sodium hydroxide, lower algal feed concentration, and shorter hold times (30 min) for all species. The increase in reaction times (>30 min) with a base catalyst (sodium hydroxide) decreased the hydrogen yield. Thus, the overall energy recovery was highest at the lowest feed concentrations; 90.5% using ruthenium and 111% using sodium hydroxide. The bio-oil formed in HTL has more calorific value under high pressure. Duan et al. (2018) noted that gasification results varied significantly for the microalgae and macroalgae HTL-AP as feedstock. In particular, this study noted that under constant reaction conditions, HTL-AP derived from macroalgae consistently produced a greater amount of methane than hydrogen, whereas, for microalgae, the trend was not always consistent. This was attributed to the greater amount of organic matter in the HTL-AP of microalgae (~23.9 9 g/L) than macroalgae (9.8 g/L). This fact indicates that macroalgal biomass is a potential feedstock for supercritical gasification.

The SCWG step can be easily integrated with the biorefinery of seaweeds or even with the HTL process.(Aziz et al., 2014) The compositions of syngas from the catalytic SCWG of *L. hyperborea* under varying parameters, including catalyst loading, feed concentration, hold time, and temperature, have been investigated. Their effects on gas yields, gasification efficiency, and energy recovery are presented. Results show that the carbon gasification efficiencies increased with reaction temperature, reaction hold time, and catalyst loading but decreased with increasing feed concentrations. This indicates that the scale-up of supercritical gasification is difficult. In addition, the selectivity toward hydrogen and/or methane production from the SCWG tests can be controlled by the combination of catalysts and varying reaction conditions. For instance, Ru/Al_2O_3 gave the highest carbon conversion, and highest methane yield of up to 11 mol/kg, while NaOH produced the highest hydrogen yield of nearly 30 mol/kg under certain gasification conditions. Ash chemistry restricts the use of macroalgae for direct combustion and gasification.

5 Biochemical conversion

Biochemical conversion technologies include anaerobic digestion (AD) and fermentation of seaweeds. Before going for biochemical conversions, pretreatment steps play a vital role in the efficiency of biochemical processes. The pretreatment of seaweeds includes biomass hydrolysis either using thermochemical methods like steam and acid hydrolysis or enzymatic hydrolysis. Biomass hydrolysis is crucial for the fermentation process to get sugars from complex polysaccharides. The pretreatment techniques should generate minimum toxins and inhibitors. Anaerobic digestion (AD) for biomethane, ethanol fermentation (EF), and dark fermentation (DF) (Radha and Murugesan, 2017), which is for hydrogen generation, are processes used for the production of biofuel from seaweeds.

5.1 Anaerobic digestion (AD)

Anaerobic digestion of seaweeds is a more direct mode of bioenergy generation than bioethanol (Milledge et al., 2014). Some authors suggest that AD is technically more viable, and it is an excellent biofuel that can produce heat and electricity (Bruhn et al., 2011). Steps of AD include pretreatment of biomass via physical (mechanical, thermal), chemical (alkali, acids), and biological (microbes, enzymatic digestion) methods. The pretreatment of biomass plays a crucial role in anaerobic digestion since the biological processes are inhibited by high ash contents (mainly chlorine and sulfur), nitrogen, alkali earth metals, dietary fibers, and polyphenols (Jard et al., 2013). The second step is anaerobic digestion, which depends on seaweed biomass, freshwater liquid substrate, and the conventional microbial seed (co-digestion with organic waste products). Important parameters like C/N ratio, temperature and pH, and digestion time have to be optimized.

The reactors used for the biochemical processes also play a critical role (Mhatre et al., 2019). The various anaerobic reactors used are continuous stirred tank reactors (CSTR), upflow anaerobic sludge blanket reactors (UASB), and anaerobic sequential batch reactors (ASBR). The parameters that affect the anaerobic digestion processes are organic loading rate (OLR), mineral-rich liquid extract (MRLE), total solids (TS), volatile solids (VS), chemical oxygen demand (COD), and hydraulic retention time (HRT). These parameters influence biofuel generation from marine macroalgae and consider the conversion techniques used in biomethane generation. Akila et al. (2019) investigated the production of biogas and biofertilizer from green seaweed *Ulva* sp. using the anaerobic digestion method. The *Ulva* sp. was mixed with organic matter (cow dung). The solid residue was used as an organic fertilizer for the growth of mung bean. Their research provided an example of maximizing the utilization of green seaweeds in an eco-friendly and sustainable way.

Mhatre et al. (2019) explored strategies for improving biogas production from the green seaweed *U. lactuca*. The separate and sequential extraction studies were conducted to investigate the influence of the extraction of sap, which contains valuable minerals, ulvan, and protein, on methane yields. It was found that the biomethane production enhanced after the extraction, and the highest methane yield (408 mL/g) was obtained in the sap and ulvan-removed residue. This was due to the fact that high protein and sulfate content act as major inhibitors in the anaerobic digestion of *U. lactuca*. The extraction of sap, protein, and ulvan prior to anaerobic digestion not only improved methane productivity but also provided high-value products. This strategy makes the biomethane production process more efficient and sustainable. Vanegas and Bartlett (2013) reported that macroalgal species *Saccharina latissimi* gave a biomethane yield of 335 mL/g among all seaweeds studied. The methane yield for *S. polyschides*, *L. digita*, and *Ulva* was 255 mL/g VS, 246 mL/g VS and 191 mK/g VS respectively, after 109 days. The studies were scaled from batch of 120 mL–1 L using Bovine slurry.

Seaweeds have shown great potential for biomethane generation as per the report where biomethane yield is 140 mL for green seaweeds and 280 mL per gram of volatile solids for brown seaweeds (Allen et al., 2013). Some studies also report high methane potential range of 260–500 mL per g of volatile solids from macroalgae like *Gracilaria* sp., *Macrocystis* sp., and *Laminaria* sp. (Camus et al., 2016). A detailed study on biomethane potential of different segments of thalli from various species of brown macroalgae like *Ascophyllum nodosum*, *L. digitata*, *L. hyperborea*, *Saccharina latissima*, and *S. polyschides* has also been conducted (Tabassum et al., 2018). Pechsiri et al. (2016) carried out system analysis for biogas and fertilizer production using Kelp and concluded that GHG (greenhouse gases) and energy inputs are better if both are produced in the same system.

The beating pretreatment speeds up the start of digestion. After 3 days of digestion, beating pretreatment enhances the methane yield by 37%. Microwave and ball milling pretreatments impact the methane production negatively. At 25 days of digestion, the overall methane is not enhanced by beating pretreatment (Montingelli et al., 2016). The effect of beating (BT), ball milling (BM), and microwave pretreatment (MW) on the conversion of the macroalgae *Laminaria* spp. to biogas by anaerobic digestion (AD) has been evaluated by Montingelli et al. (2016). After 3 days of digestion, the BT pretreated samples yielded the best result by achieving a methane increase of up to 37% with respect to the raw seaweed. At 25 days, both BM and MW pretreatment lowered the methane yield with respect to the raw seaweed. Since BT produced higher methane yields with respect to the untreated sample, it was considered for energy balance analysis. After 3 days of digestion, the beating treatment resulted in an energy gain of 28%, while at the end of digestion, the break-even point was reached.

5.2 Bioethanol /biobutanol production

The main steps involved in bioethanol production from seaweeds involve pretreatment of biomass which includes (a) washing, cleaning, milling, and chopping, (b) hydrolysis/saccharification (acid/enzymatic hydrolysis) of macroalgal polysaccharides to fermentable sugars, and (c) fermentation using selected microbial strain. Then, there are few more processing steps of distillation, dehydration to get final product in the form of pure ethanol which can be used as clean biofuel. The fermentation process allows to produce a mixture of ethanol, butanol, and organic acids. The fermentation step in bioethanol production can be categorized as separate hydrolysis and fermentation (SHF), where hydrolysis and fermentation are carried out in two different reactors with different reaction conditions. The second is simultaneous saccharification and fermentation (SSF). In SSF, both hydrolysis and fermentation are carried out in a single vessel with the same reaction conditions. The third approach is consolidated biomass processing (CBP) production of cellulase enzyme, hydrolysis and fermentation takes place in a single reactor. The SHF, SSF, and CBP have been reported for *Gelidium amansi* where autoclaving of biomass gave ethanol yields of separate hydrolysis and fermentation processing (SHF) achieved a maximum ethanol concentration of 3.33 mg/mL, with a conversion yield of 74.7% after 6 h (2% substrate loading, w/v). In contrast, simultaneous saccharification and fermentation (SSF) produced an ethanol concentration of 3.78 mg/mL, with an ethanol conversion yield of 84.9% after 12 h. It also has recorded an ethanol concentration of 25.7 mg/mL from SSF processing of 15% (w/v) dry matter from autoclaved *Gelidium amansi* after 24 h. These results indicate that autoclaving can improve the glucose and ethanol conversion yield of *Gelidium amansi* and that SSF is superior to SHF for ethanol production (Kim et al., 2015). There are several reports of ethanol production from red and brown seaweeds but very few papers are on green seaweeds like *Enteromorhpa prolifera* (Zhou et al., 2010) and *U. lactuca*.

In case of red seaweeds, the acid hydrolysate of three most prominent agar-containing red seaweeds (agarophytes) containing agar 20%–51% (g $^{-1}$ dry weight) like *Gelidium amansii*, *Gracilaria tenuistipitata*, and *Gracilariopsis chorda*, the optimized conditions for acid hydrolysis were found to be 100 g of seaweed hydrolyzed at 130°C for 15 min with 0.2 M sulfuric acid. The 120 mL of a 1:2 mixture of the acid-treated hydrolysate broth and basal medium was fermented in a 200 mL bottle at 30°C for 96 h. *G. amansii* had the best ethanol yield, about 45% of the theoretical yield. This yield increased to 60% after detoxification of the hydrolysate with activated carbon was carried out (MDN Meinita et al., 2013). *Gelidium elegan* has also been studied where the concentration of 13.27 ± 0.47 g/L ethanol was obtained using 2.5% (w/v) H_2SO_4 at 120°C for 40 min was selected for hydrolysis of the seaweed biomass, followed by purification and fermentation to yield analytical grade ethanol. The best analytical GC method for ethanol from seaweeds was developed Hessami et al. (2019). Ethanol yields from the fermentation of seaweed are typically between 0.08 and 0.12 kg·kg^{-1} dry seaweed varying with the alga and method of pretreatment and saccharification.

Various parameters which affect the pretreatment process (acid hydrolysis) involve acid concentration, temperature, time, etc. The enzymatic hydrolysis is influenced by enzymes cocktails, temperature, biomass loading, and pH. For the fermentation process, the influencing factors include inoculum concentration and inoculum types (conventional or bioengineered yeast or bacteria strains employed), pH of the medium, fermentation time or residence time, solid loading, and temperature. There is vast literature available on bioethanol production from seaweeds as compared to other biofuels. In the case of green seaweeds, species, such as *Ulva faciata*, *Ulva rigida*, *U. lactuca*, and *Ulva reticulata*, have been reported (Trivedi et al., 2013).

Van der Wal et al. (2013) used *U. lactuca* for acetone, butanol, and ethanol (ABE) production by anaerobic fermentation. In this study, *U. lactuca* was subjected to enzymatic pretreatment without any chemical catalysts to get a hydrolysate containing 75%–93% of the sugars. Subsequently, the hydrolysate was fermented for the production of ABE by using *Clostridium acetobutylicum* and *Clostridium beijerinckii*. A high yield of 0.35 g ABE/g sugar was achieved in this process. Interestingly, *C. beijerinckii* produced 1,2-propanediol from rhamnose, whereas *C. acetobutylicum* produced mostly organic acids. These results demonstrate the great potential of *U. lactuca* as a feedstock for ABE fermentation. Potts et al. (2012) also reported butanol production from *U. lactuca* using ABE fermentation. The *U. lactuca* seaweeds were first manually and mechanically harvested, dried, and ground. Subsequently, the acid hydrolysis of ground algae was conducted to extract carbohydrates for the algal sugar solution preparation. Finally, butanol was produced by the fermentation of the pretreated algal sugar solution removing excess solids before fermentation using *C. beijerinckii* and *Clostridium saccharoperbutylacetonicum* resulted in butanol concentration of 4 g/L. Fermentation resulted in a 75% increase in productivity. Both the examples demonstrate the possibility and potential of the utilization of green seaweeds like *Ulva* as feedstocks for the production of biobutanol through ABE fermentation. ABE was proposed for the subsequent production of 1,2-propanediol (propylene glycol) in a seaweed biorefinery, replacing fossil fuel-derived product rather than as a source of butanol as fuel. In a study using the naturally occurring macroalgae (*Ulva*) from Jamaica Bay, New York City, it was found that butanol could be made on a pilot scale from algal sugars. Butanol production from biomass could also be more energy-efficient than ethanol as some bacteria used in butanol production digest not only starch and sugars but also cellulose. Due to these studies, the utilization of green seaweeds as feedstocks for the production of biobutanol has gained significant interest (Bikker et al., 2016).

5.3 Biodiesel production

Biodiesel is an alternative and efficient fuel to replace the use of fossil fuels. It is derived as monoalkyl esters of long-chain fatty acids and is usually produced from algal oils, waste cooking oil, and edible and nonedible oils (Demirbas, 2007). Typically, seaweed biomass has lower lipid content, 0.3%–6% and hence, a lower biodiesel potential is expected. The biodiesel yields for *Chaetomorpha linum* (Aresta et al., 2005), *U. lactuca* (Khan et al., 2016), and *Enteromorpha (Ulva) compressa* (Suganya et al., 2013) have been reported to be as low as 11% of the total dry macroalgal biomass (Khan et al., 2016; Suganya et al., 2013). Therefore, macroalgae would not appear to be a suitable feedstock for the production of biodiesel via transesterification.

The steps involved in biodiesel production from seaweeds involve oil/lipid extraction using presses or expellers, solvent-based extraction techniques, oil transesterification using processes like methanolysis or ethanolysis, acid/base catalysts, or enzymatic catalyst like lipase or heterogeneous metal oxide catalyst. The various solvent-based extraction techniques have been reported by different researchers for seaweeds processing like microwave, ultrasound for *Padina tetrastromatica* (Veeramuthu et al., 2017), enzyme-assisted extraction of algal oil using lipases of *Enteromorpha compressa* (Suutari et al., 2015), pressurized liquid extraction (Aresta et al., 2005), ultrasound-assisted extraction, supercritical fluid extraction, acid-based transesterification, etc.

In biodiesel production process, for seaweeds like *Ulva* species, the oil extraction methods and the types of catalysts used are crucial because both factors can influence the efficiency of the transesterification process. Kalavathy and Baskar (2019) used *U. lactuca* as feedstock to produce biodiesel using silica doped with zinc oxide as a heterogeneous nanocatalyst for the transesterification process. The lipid content of algal biomass was extracted by an autoclave followed by the ultrasonication method. The extraction of oil was carried out in the Soxhlet apparatus using a solvent mixture of n-hexane and methyl tertbutyl ether. It was found that the maximum oil was extracted at optimal conditions of 5% moisture content of algal biomass, 0.15 mm size of biomass, and solvent/solid ratio of 6:1 at 55°C in 140 min. The maximum biodiesel yield of 97.43% was obtained at optimized conditions of 800°C calcination temperature, 8% catalyst concentration, 9:1 methanol-to-oil ratio, 55°C reaction temperature, and 50 min reaction time (Kalavathy and Baskar 2019). Khan et al. (2016) reported *U. fasciata* as feedstock for the production of biodiesel. The oil was extracted with n-hexane. The transesterification was carried out by fast stirring using a 9:1 M ratio of methanol/oil in the presence of industrial waste catalysts for 6 h at 80–100°C. It was found that the maximum yield of biodiesel equal to 88% was achieved by using the waste brown dust from the steel converter as a catalyst in the transesterification process.

Biodiesel from seaweeds can be consumed by engines, and its combustion reduces emissions. The residual biomass after lipid extraction can be used for bioethanol production. High contents of metals, sulfur, and nitrogen inhibit the transesterification process, as reported by Suutari et al. (2015). The lipid extraction and purification from seaweeds involve cost and high energy demand, which makes them unfit for biodiesel production.

5.4 Biohydrogen production

Biohydrogen production from seaweeds is defined as "the fermentative production of hydrogen." Dark fermentation is the most suitable method for seaweeds which can be extended to biomethane generation(Kannah et al., 2019). The hydrogen obtained after catalytic upgradation of HTL aqueous phase and supercritical water gasification is also termed biohydrogen since it is obtained from seaweed biomass. Macroalgae is an efficient source of biomass for biohydrogen production. Biohydrogen (H_2) is believed as a sustainable and clean energy carrier with a high energy yield. The first report on biohydrogen production by using dark fermentation from green seaweed Ulva sp. is reported by Margareta et al. (2020). The Ulva biomass was subjected to the acid–thermal combined pretreatment to release fermentable sugars. The *Clostridium butyricum* CGS5 achieved the highest cumulative hydrogen production, equal to 2340 mL/L; the maximum hydrogen productivity, equal to (208.3 mL/L per hour); and hydrogen yield equal to 1.53 mol H_2 per mol of reducing sugar. The hydrogen productivity reported by Margareta et al. (2020) was better than that reported by (Jung et al., 2011), in which *Laminaria japonica* was used as feedstock. Besides the production of biobutanol and biomethane from green seaweeds, the production of biohydrogen from green seaweeds is also reported in the literature. Recent research has studied that both the red algae *G. amansii* (Park et al., 2011) and the brown algae *Laminaria japonica* (Yin and Wang, 2018) are potential biomass sources for biohydrogen production through anaerobic fermentation. Park et al. (2009) screened different macroalgal species *U. lactuca*, Poryphyra tenera,

Undaria pinnatifida, and *Laminari Japonica* with optimum fermentation temperature, substrate concentration, initial pH, and pretreatment condition were determined to be 35°C, 5%, 7.5, and BT120 (Ball mill and thermal treatments at 120°C for 30 min), heat-treated at 65°C for 20 min 4164 mL of hydrogen was produced from 50 g/L of dry algae 28 mL/g dry algae. Park et al. (2009) suggested that 5-hydroxymethyl furfural was the main inhibitor of biohydrogen fermentation and pretreatment method has significant effect on biohydrogen yields.

Macroalgal hydrogen production is still in the early stages of development. The pretreatment is essential to enhance the hydrolytic process during dark fermentation. During pretreatment, some inhibitory substances are formed and are controlled by detoxification techniques (Kumar et al., 2021). The biohydrogen from seaweeds is very recent. There are excellent reviews on this topic (Shobana et al., 2017). Mannitol which is a typical carbohydrate component of seaweeds has been reported to be used as a substrate for hydrogen fermentation.

The summary of research work on the biochemical conversion of seaweeds to biofuel demonstrates that the seaweeds are potential feedstocks for the production of biobutanol, biohydrogen, biogas, and biodiesel. However, in all the cases, the pretreatment using hydrolysis processes, such as acid and enzymatic hydrolysis, is crucial to the extraction of sugar solution from seaweeds, which consequently affects the productivity of biofuel production. It can also be concluded that the types of bacteria strains are also very important for the productivity of biobutanol and biohydrogen. In the biogas production process, the extraction of value-added products like sap and ulvan from macroalgae enhanced methane productivity due to the removal of inhibitors.

6 Integrated approach for biorefinery and biofuel production from waste streams of seaweed processing

Biorefinery is a holistic way to generate multiple products from a single macroalgal source and utilize all streams with a short processing time. There are two approaches for biorefinery, first, sequential extraction of bioproducts according to the bulk of the bioproducts. Second, a cascading approach where the bioproduct with the highest commercial value is extracted first. Cascading biorefinery approach involves designing of seaweed biorefineries so that optimized use of all valuable components can be done. Cascading of high-value but low-volume products, such as pharmaceuticals, bioactives, nutraceuticals, health-promoting food and feed supplements, and functional food ingredients, to lower value bulk volume products, such as plant stimulants, soil improvers, and finally bioenergy makes the entire biorefinery process viable. Thus, there is a need to focus on the recent advancement in the macroalgal biomass scaling-up, which could help in the growth of the macroalgal biorefinery industry in the near future, ultimately making biofuel generation more feasible.

By deep understanding, critically the processes involved in biofuel, value-added products, and chemical production utilizing macroalgal biomass as a feedstock along with zero waste formation are need of the hour. Apart from this, reviews summarizing the major issues linked to the microalgae-based biofuels and bioproducts generation processes and their possible

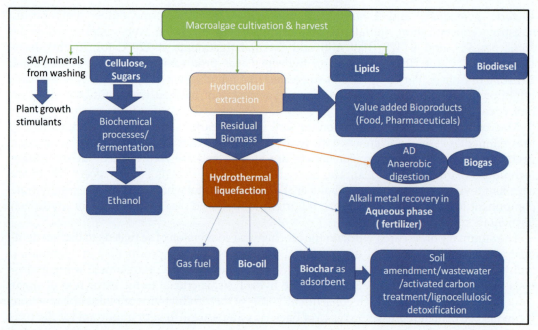

FIG. 3   Possible routes for biofuels and bioproducts from seaweeds.

corrective measures need to be given attention. The sequential biorefinery of *U. lactuca* is reported by Baghel et al. (2016). The cellulose obtained was converted to bioethanol by fermentation is one of the few reports of integration of biorefinery with biofuel production. There are few reports on the integration of HTL with value-added products, as shown in Figs. 2 and 3 (Deng et al., 2020). There is one report on the integration of pyrolysis and anaerobic digestion (Deng et al., 2020). The *Gracilaria dura* is one such report where the residual biomass after extraction of bioproducts was subjected to HTL, which yielded bio-oil of fuel quality and in a substantially larger amount (Francavilla et al., 2015). The integration of environmentally benign processes and zero waste biorefinery along with biofuel from the residues or waste streams is the future of sustainable development of the seaweeds industry. Rajak et al. (2020) has reviewed the importance of the development of the zero waste biorefinery concept to get complete utilization of macroalgal biomass. The author suggested bioethanol as the potential biofuel which can be scaled up to pilot scale (Ra et al., 2014). There are biorefinery approaches well established at laboratory scale for seaweed species like *U. lactuca* and *Gracilaria* (Gajaria et al., 2017; Baghel et al., 2016). The starting biomass and yield at each step and economic analysis of each step are very crucial in deciding the design of the sequential or cascading biorefinery. For example, the bioproduct yield reported by Baghel et al. (2016) for *U. lactuca* was calculated as the percentage of the initial starting weight of biomass. It was reported as 26% of its starting mass as sap, ~3% as lipid, ~25% as ulvan, and ~11% as cellulose. This cellulose was used to produce ethanol at a concentration of 0.45 g/g, reducing sugar. But a more detailed research is needed to explore the full potential of seaweeds.

The thermochemical and biochemical processes used for biofuel production from seaweeds itself can be integrated, as shown in Fig. 3 reported by Deng et al. (2020), where all the streams of biofuel generation have been utilized for energy production. Advanced biofuels include fuels derived from biomass sources that are grown without land, such as seaweed grown on the seashore. Seaweed biomethane may contribute significantly to carbon-neutral transport fuel for the future. The major hurdle lies in the limited biodegradability of seaweed via anaerobic digestion (AD). To solve this issue, the authors proposed a cascading circular bioenergy system incorporating pyrolysis for the production of biochar, bio-oil, and syngas. The primary use of biochar in AD was to enhance biomethane production through directing interspecies electron transfer. The feasibility of integrating a seaweed-based AD and a residue-based Py system was studied. The AD process showed that biochar achieved comparable to high-cost graphene in terms of enhancing biomethane production from seaweed. The species *L. digitata* and Saccharina latissima demonstrated that digestate could fulfill all the heat demand for the integrated AD-Py system and increased biomethane yield by 17% and bio-oil yield by 10%. Furthermore, a 26% decrease in digested biomass mass flow could be achieved, thereby reducing the demand for agricultural land for digestate application.

The cost of seaweed farming and harvesting is currently too expensive, so integration of this with biorefinery is important to make the process of bioenergy/biofuel production from seaweed sustainable. By utilizing the biorefinery approach, we can get high-value-added products from the biomass, then direct the streams to various processes for energy production. One of the proposed integration is shown in Fig. 3. The aqueous phase obtained in HTL has been utilized as fertilizer or media for growing microbes or even microalgae. It is more economical to use the waste streams of biorefinery of seaweeds for biofuel production. This will serve the dual purpose of wastewater treatment and as well as waste stream utilization. Integrated biorefinery with biofuel production is beneficial so that all the hurdles of technical and economical, and environmental feasibility can be removed.

7 Challenges of biofuel production from seaweed biomass

The major challenge for macroalgal biofuel commercialization is round the year availability of the biomass in a bulk quantity and high cost of production during scale-up from laboratory to pilot scale (Nikolaisen et al., 2008; Ganesan et al., 2019). However, there is a huge potential for commercialization if a holistic zero waste biorefinery approach is used to produce multiple value-added products with biofuel production. To achieve this, the research should aim at utilizing the seaweed waste streams along with focusing on targeted biofuels such as, ethanol, butanol, furan derivatives (as a fuel additive), and biogas. A breakthrough in biofuel production from macroalgae is needed by increasing the biomass supply by developing a rotating crop scheme for the cultivation of seaweed, using native, highly productive brown, red, and green seaweeds. The improvement of the pretreatment methods, storage techniques for seaweeds, increasing the biofuel production to economically viable concentrations with the help of enzyme cocktails that can degrade a wide variety of polysaccharides to fermentable sugars. Enhanced bio-butanol yield by developing novel fermenting organisms that metabolize all sugars in seaweeds; Increasing the biogas yield to convert available carbon in the residues by adapting the organisms to seaweed; Development of strains that can withstand salinity levels; developing the thermochemical conversion of sugars to fuel additives will be a value-added product of high carbohydrate content of seaweeds using catalytic conversions. Performing an integral techno-economic, sustainability, and risk assessment of the entire seaweed to biofuel chain is also very important in developing models to predict biofuel potential and bioproducts by exergy flow studies of macroalgal species for scaling up (Ingle et al., 2018).

8 Conclusions

The survey of research work in biofuels from seaweeds reveals that futuristic macroalgal biorefinery integrated with cascading product approach along with waste stream utilization will minimize the cost related issues of the whole process. Inclusion of co-utilization of mixed biomass feedstocks such as food waste, sewage, lignocellulosic, and agricultural waste can help in the continuous operation of the plant. Although all the biofuels have certain pros and cons, certain biofuels from seaweeds show promising scaled up potential, e.g., bioethanol, biogas, and bio-oil. Novel methods of cultivation, harvesting, pretreatment, processing and recovery of end products need to be developed that ensure more economic feasibility and scale up possibility.

Acknowledgment

Present work is an outcome of the literature survey conducted while working on a Research Project at DBT-ICT Centre for Energy Biosciences, funded by Department of Biotechnology DBT, Government of India (Grant Sanction number BT/PB/01/02/2005) dated 8.11.2017. The authors acknowledge the same.

References

Akila, V., Manikandan, A., Sahaya Sukeetha, D., Balakrishnan, S., Ayyasamy, P.M., Rajakumar, S., 2019. Biogas and biofertilizer production of marine macroalgae: an effective anaerobic digestion of Ulva sp. Biocatal. Agric. Biotechnol. 18, 101035.

Allen, E., Browne, J., Hynes, S., Murphy, J.D., 2013. The potential of algae blooms to produce renewable gaseous fue. Waste Manag. 33, 2425–2433. https://doi.org/10.1016/j.wasman.2013.06.017.

Anastasakis, K., Ross, A.B., 2011. Hydrothermal liquefaction of the brown macro-alga *Laminaria Saccharina*: effect of reaction conditions on product distribution and composition. Bioresour. Technol. 102 (7), 4876–4883. https://doi.org/10.1016/j.biortech.2011.01.031.

Aresta, M., Dibenedetto, A., Carone, M., Colonna, T., Fragale, C., 2005. Production of biodiesel from macroalgae by supercritical CO_2 extraction and thermochemical liquefaction. Environ. Chem. Lett. 3 (3), 136–139. https://doi.org/10.1007/s10311-005-0020-3.

Aziz, M., 2017. Combined supercritical water gasification of algae and hydrogenation for hydrogen production and storage. Energy Procedia 119, 530–535.

Aziz, M., Oda, T., Kashiwagi, T., 2014. Advanced energy harvesting from macroalgae—innovative integration of drying, gasification and combined cycle. Energies 7, 8217–8235. https://doi.org/10.3390/en7128217.

Bae, Y.J., Ryu, C., Jeon, J.K., Park, J., Suth, D.J., Suh, Y.W., Chang, D., Park, Y.Y., 2011. The characteristics of bio-oil produced from the pyrolysis of three marine macroalgae. Bioresour. Technol. 102 (3), 47–56.

Baghel, R.S., Trivedi, N., Reddy, C.R.K., 2016. A simple process for recovery of a stream of products from marine macroalgal biomass. Bioresour. Technol. 206, 160–165. https://doi.org/10.1016/j.biortech.2015.12.051.

Basu, P., 2018. Chapter 5—pyrolysis. In: Basu, P. (Ed.), Biomass Gasification, Pyrolysis and Torrefaction, third ed. Academic Press, pp. 55–187, https://doi.org/10.1016/B978-0-12-812992-0.00005-4.

Behera, S., Singh, R., Arora, R., Sharma, N.K., Shukla, M., Kumar, S., 2015. Scope of algae as third generation biofuels. Front. Bioeng. Biotechnol. 2, 90. https://doi.org/10.3389/fbioe.2014.00090.

Bikker, P., van Krimpen, M.M., van Wikselaar, P., et al., 2016. Biorefinery of the green seaweed Ulva lactuca to produce animal feed, chemicals and biofuels. J. Appl. Phycol. 28, 3511–3525. https://doi.org/10.1007/s10811-016-0842-3.

Borines, M.G., de Leon, R.L., Cuello, J.L., 2013. Bioethanol production from the macroalgae Sargassum spp. Bioresour. Technol. 138, 22–29.

Bruhn, A., Dahl, J., Nielsen, H.B., Nikolaisen, L., Rasmussen, M.B., Markager, S., Olesen, B., Arias, C., Jensen, P.D., 2011. Bioenergy potential of Ulva lactuca: biomass yield, methane production and combustion. Bioresour. Technol. 102, 2595–2604.

Camus, C., Ballerino, P., Delgado, R., Olivera-Nappa, Á., Leyton, C., Buschmann, A.H., 2016. Scaling up bioethanol production from the farmed brown macroalga Macrocystis pyrifera in Chile. Biofuels Bioprod. Biorefin. 10, 673–685.

Cherad, R., Onwudili, J.A., Williams, P.T., Ross, A.B., 2014. A parametric study on supercritical water gasification of *Laminaria hyperborea*: a carbohydrate-rich macroalga. Bioresour. Technol. 169, 573–580.

Cherad, R., Onwudili, J.A., Biller, P., Williams, P.T., Ross, A.B., 2016. Hydrogen production from the catalytic supercritical water gasification of process water generated from hydrothermal liquefaction of microalgae. Fuel 166, 24–28. https://doi.org/10.1016/j.fuel.2015.10.088.

Das, P., Mondal, D., Maiti, S., 2017. Thermochemical conversion pathways of Kappaphycus alvarezii granules through study of kinetic models. Bioresour. Technol. 234, 233–242. ISSN 0960-8524 https://doi.org/10.1016/j.biortech.2017.03.007.

Demirbas, A., 2007. Importance of biodiesel as transportation fuel. Energy Policy 35 (9), 4661–4670. https://doi.org/10.1016/j.enpol.2007.04.003.

Deng, C., Lin, R., Kang, X., Wu, B., O'Shea, R., Murphy, J.D., 2020. Improving gaseous biofuel yield from seaweed through a cascading circular bioenergy system integrating anaerobic digestion and pyrolysis. Renew. Sustain. Energy Rev. 128, 109895. ISSN 1364-0321 https://doi.org/10.1016/j.rser.2020.109895. https://www.sciencedirect.com/science/article/pii/S1364032120301878.

Duan, P., Yang, S., Xu, Y., Wang, F., Zhao, D., Weng, Y., et al., 2018. Integration of hydrothermal liquefaction and supercritical water gasification for improvement of energy recovery from algal biomass. Energy 155, 734–745.

Elizondo-González, et al., 2018. Use of seaweed Ulva lactuca for water bioremediation and as feed additive for white shrimp Litopenaeus vannamei. PeerJ 6, e4459. https://doi.org/10.7717/peerj.4459. 1-16.

FAO, 2020. The State of World Fisheries and Aquaculture 2020. FAO, Rome.

Francavilla, M., Manara, P., Kamaterou, P., Monteleone, M., Zabaniotou, A., 2015. Cascade approach of red macroalgae Gracilaria gracilis sustainable valorization by extraction of phycobiliproteins and pyrolysis of residue. Bioresour. Technol. 184, 305–313. ISSN 0960-8524 https://doi.org/10.1016/j.biortech.2014.10.147.

Gajaria, T., Suthar, P., Baghel, R., Balar, N., Sharnagat, P., Mantri, V., Reddy, C.R.K., 2017. Integration of protein extraction with a stream of byproducts from marine macroalgae: a model forms the basis for marine bioeconomy. Bioresour. Technol. 243, 867–873. https://doi.org/10.1016/j.biortech.2017.06.149.

Ganesan, M., Trivedi, N., Gupta, V., Madhav, S.V., Reddy, C.R.K., Levine, I.A., 2019. Seaweed resources in India—current status of diversity and cultivation: prospects and challenges. Bot. Mar. 62 (5), 463–482. https://doi.org/10.1515/bot-2018-0056.

Guda, V.K., Steele, P.H., Penmetsa, V.K., Li, Q., 2015. Chapter 7—fast pyrolysis of biomass: recent advances in fast pyrolysis technology. In: Pandey, A., Bhaskar, T., Stöcker, M., Sukumaran, R.K. (Eds.), Recent Advances in Thermo-Chemical Conversion of Biomass. Elsevier, pp. 177–211. ISBN 9780444632890 https://doi.org/10.1016/B978-0-444-63289-0.00007-7.

Halder, P., Azad, A.K., 2019. Recent trends and challenges of algal biofuel conversion technologies. Advanced Biofuels. Woodhead Publishing, United Kingdom, pp. 167–179.

Hao, J., Qi, B., Dong, L., Zeng, F., 2021. Catalytic co-pyrolysis of rice straw and *Ulva prolifera* macroalgae: effects of process parameter on bio-oil up-gradation. Renew. Energy 164, 460–471. ISSN 0960-1481 https://doi.org/10.1016/j.renene.2020.09.056.

He, Y., Liang, X., Jazrawi, C., Montoya, A., Yuen, A., Cole, A.J., Neveux, N., Paul, N.A., de Nys, R., Maschmeyer, T., Haynes, B.S., 2016. Continuous hydrothermal liquefaction of macroalgae in the presence of organic co-solvents. Algal Res. 17, 185–195. ISSN 2211-9264 https://doi.org/10.1016/j.algal.2016.05.010.

Hessami, M.J., Cheng, S.F., Ambati, R.R., et al., 2019. Bioethanol production from agarophyte red seaweed, Gelidium elegans, using a novel sample preparation method for analysing bioethanol content by gas chromatography. 3 Biotech 9, 25. https://doi.org/10.1007/s13205-018-1549-8.

Hu, Y., Wang, H., Lakshmikandan, M., 2021. Catalytic co-pyrolysis of seaweeds and cellulose using mixed ZSM-5 and MCM-41 for enhanced crude bio-oil production. J. Therm. Anal. Calorim. 143, 827–842. https://doi.org/10.1007/s10973-020-09291-w.

Ingle, K., Vitkin, E., Robin, A., Yakhini, Z., Mishori, D., Golberg, A., 2018. Macroalgae biorefinery from *Kappaphycus alvarezii*: conversion modeling and performance prediction for India and Philippines as examples. Bioenergy Res. 11, 22–32.

Jard, G., Dumas, C., Delgenes, J.P., Marfaing, H., Sialve, J.P., Steyer, J.P., Carrère, H., 2013. Effect of thermochemical pretreatment on the solubilization and anaerobic biodegradability of the red macroalga *Palmaria palmata*. Biochem. Eng. J. 79, 253–258. https://doi.org/10.1016/j.bej.2013.08.011.

Jung, K.-W., Kim, D.-H., Shin, H.-S., 2011. Fermentative hydrogen production from *Laminaria japonica* and optimization of thermal pretreatment conditions. Bioresour. Technol. 102, 2745–2750.

Kalavathy, G., Baskar, G., 2019. Synergism of clay with zinc oxide as nanocatalyst for production of biodiesel from marine *Ulva lactuca*. Bioresour. Technol. 281, 234–238.

Kanda, H., Kamo, Y., Machmudah, S., Diono, W., Goto, M., 2014. Extraction of fucoxanthin from raw macroalgae excluding drying and cell wall disruption by liquefied dimethyl ether. Mar. Drugs 12, 2383–2396. https://doi.org/10.3390/md12052383.

Kannah, R.Y., Kavitha, S., Sivashanmugham, P., Kumar, G., Nguyen, D.D., Chang, S.W., 2019. Biohydrogen production from rice straw: effect of combinative pretreatment, modelling assessment and energy balance consideration. Int. J. Hydrogen Energy 44, 2203–2215. https://doi.org/10.1016/j.ijhydene.2018.07.201.

Khan, A.M., Fatima, N., Hussain, M.S., Yasmeen, K., 2016. Biodiesel production from green seaweed Ulva fasciata catalyzed by novel waste catalysts from Pakistan Steel Industry. Chin. J. Chem. Eng. 24, 1080–1086.

Kim, D.H., Lee, S.B., Jeong, G.T., 2014. Production of reducing sugar from Enteromorpha intestinalis by hydrothermal and enzymatic hydrolysis. Bioresour. Technol. 161, 348–353. https://doi.org/10.1016/j.biortech.2014.03.078. Epub 2014 Mar 25 24727694.

Kim, H., Wi, S.G., Jung, S., Song, Y., Bae, H.-J., 2015. Efficient approach for bioethanol production from red seaweed Gelidium amansii. Bioresour. Technol., 175. https://doi.org/10.1016/j.biortech.2014.10.050.

Kim, Y.-M., Han, T.U., Lee, B., Watanabe, A., Teramae, N., Kim, J.-H., Park, Y.-K., Park, H., Kim, S., 2018. Analytical pyrolysis reaction characteristics of *Poryphyra tenera*. Algal Res. 32, 60–69. https://doi.org/10.1016/j.algal.2018.03.003.

Kostas, E.T., Williams, O.S.A., Duran-Jimenez, G., Tapper, A.J., Cooper, M., Meehan, R., Robinson, J.P., 2019. Microwave pyrolysis of *Laminaria digitata* to produce unique seaweed-derived bio-oils. Biomass Bioenergy. https://doi.org/10.1016/j.biombioe.2019.04.006.

Kumar, S., Roat, P., Hada, S., Chechani, B., Kumari, N., Ghodke, P., Rawat, D.S., 2020. Catalytic approach for production of hydrocarbon rich bio-oil from red seaweed species. In: Kumar, N. (Ed.), Biotechnology for Biofuels: A Sustainable Green Energy Solution. Springer Nature Singapore Ltd, pp. 109–128, https://doi.org/10.1007/978-981-15-3761-5_5.

Kumar, M.D., Kavitha, S., Tyagi, V.K., et al., 2021. Macroalgae-derived biohydrogen production: biorefinery and circular bioeconomy. Biomass Convers. Biorefin. https://doi.org/10.1007/s13399-020-01187-x.

Laramore, S., Baptiste, R., Wills, P., et al., 2018. Utilization of IMTA-produced Ulva lactuca to supplement or partially replace pelleted diets in shrimp (Litopenaeus vannamei) reared in a clear water production system. J. Appl. Phycol. 30, 3603–3610. https://doi.org/10.1007/s10811-018-1485-3.

Lee, X.J., Ong, H.C., Gan, Y.Y., Chen, W.-H., Mahlia, T.M.I., 2020. State of art review on conventional and advanced pyrolysis of macroalgae and microalgae for biochar, bio-oil and bio-syngas production. Energ. Conver. Manage. 210, 112707. ISSN 0196-8904 https://doi.org/10.1016/j.enconman.2020.112707.

Ma, C., Geng, J., Dong, Z., Ning, X., 2020. Noncatalytic and catalytic pyrolysis of Ulva prolifera macroalgae for production of quality bio-oil. J. Energy Inst. 93 (1), 303–311. ISSN 1743-9671 https://doi.org/10.1016/j.joei.2019.03.001.

Margareta, W., Nagarajan, D., Chang, J.-S., Lee, D.-J., 2020. Dark fermentative hydrogen production using macroalgae (Ulva sp.) as the renewable feedstock. Appl. Energy 262, 114574. https://doi.org/10.1016/j.apenergy.2020.114574.

Masarin, F., Cedeno, F.R., Chavez, E.G., de Oliveira, L.E., Gelli, V.C., Monti, R., 2016. Chemical analysis and biorefinery of red algae Kappaphycus alvarezii for efficient production of glucose from residue of carrageenan extraction process. Biotechnol. Biofuels 9, 122. https://doi.org/10.1186/s13068-016-0535-9.

Meinita, M.D.N., Marhaeni, B., Winanto, T., et al., 2013. Comparison of agarophytes (Gelidium, Gracilaria, and Gracilariopsis) as potential resources for bioethanol production. J. Appl. Phycol. 25, 1957–1961. https://doi.org/10.1007/s10811-013-0041-4.

Mhatre, A., Navel, M., Trivedi, N., Pandit, R., Lali, A., 2018. Pilot scale flat panel photobioreactor system for mass production of *Ulva lactuca* (Chlorophyta). Bioresour. Technol. 249, 582–591.

Mhatre, A., Gore, S., Mhatre, A., Trivedi, N., Sharma, M., Pandit, R., Anil, A., Lali, A., 2019. Effect of multiple product extractions on bio-methane potential of marine macrophytic green alga *Ulva lactuca*. Renew. Energy 132, 742–751. ISSN 0960-1481 https://doi.org/10.1016/j.renene.2018.08.012.

Michalak, I., 2018. Experimental processing of seaweeds for biofuels. WIREs Energy Environ., 1–25. https://doi.org/10.1002/wene.288.

Michalak, I., Messyasz, B., 2021. Concise review of *Cladophora* spp.: macroalgae of commercial interest. J. Appl. Phycol. 33, 133–166. https://doi.org/10.1007/s10811-020-02211-3.

Milledge, J.J., Smith, B., Dyer, P.W., Harvey, P., 2014. Macroalgae-derived biofuel: a review of methods of energy extraction from seaweed biomass. Energies 7, 7194–7222. https://doi.org/10.3390/en7117194.

Milledge, J.J., Nielsen, B.V., Bailey, D., 2016. High-value products from macroalgae: the potential uses of the invasive brown seaweed, *Sargassum muticum*. Rev. Environ. Sci. Biotechnol. 15, 67–88. https://doi.org/10.1007/s11157-015-9381-7.

Montingelli, M.E., Benyounis, K.Y., Stokes, J., Olabi, A.G., 2016. Pretreatment of macroalgal biomass for biogas production. Energ. Conver. Manage. 108, 202–209. ISSN 0196-8904 https://doi.org/10.1016/j.enconman.2015.11.008.

Nikolaisen, L., Jensen, P.D., Bech, K.S., 2008. Energy production from marine biomass (*Ulva lactuca*). In: Technical Report, pp. 1–72.

Ohno, M., Largo, D.B., 2006. The seaweed resources of Japan. World Seaweed Resources (DVD). Springer, Amsterdam, The Netherlands, p. 26.

Onwudili, J.A., Williams, P.T., 2013. Hydrogen and methane selectivity during alkaline supercritical water gasification of biomass with ruthenium-alumina catalyst. Appl. Catal. Environ. 132–133, 70–79.

Park, J.-I., Lee, J., Sim, S.J., Lee, J.-H., 2009. Production of hydrogen from marine macro-algae biomass using anaerobic sewage sludge microflora. Biotechnol. Bioprocess Eng. 14, 307–315. https://doi.org/10.1007/s12257-008-0241-y.

Park, J.-H., Yoon, J.-J., Park, H.-D., Kim, Y.J., Lim, D.J., Kim, S.-H., 2011. Feasibility of biohydrogen production from *Gelidium amansii*. Int. J. Hydrogen Energy 36 (21), 13997–14003. https://doi.org/10.1016/j.ijhydene.2011.04.003.

Parsa, M., Jalilzadeh, H., Pazoki, M., Ghasemzadeh, R., Abduli, M.A., 2018. Hydrothermal liquefaction of *Gracilaria gracilis* and *Cladophora glomerata* macro-algae for biocrude production. Bioresour. Technol. 250, 26–34. ISSN 0960-8524 https://doi.org/10.1016/j.biortech.2017.10.059.

Pechsiri, J., Jean-Baptiste E, T., Emma, R., Mauricio, R., Malmström, M., Nylund, G., Jansson, A., Welander, U., Pavia, H., Gröndahl, F., 2016. Energy performance and greenhouse gas emissions of kelp cultivation for biogas and fertilizer recovery in Sweden. Sci. Total Environ. 573, 347–355. https://doi.org/10.1016/j.scitotenv.2016.07.220.

Pereira, L., 2011. A review of the nutrient composition of selected edible seaweeds. Seaweed: Ecology, Nutrient Composition and Medicinal Uses, 1st. Nova Science Publishers, United States, pp. 15–47.

Potts, T., Du, J., Paul, M., May, P., Beitle, R., Jamie, H., 2012. The production of butanol from Jamaica bay macro algae. Environ. Prog. Sustain. Energy 31 (1), 29–36. https://doi.org/10.1002/ep.10606.

Press Information Bureau, Government of India, Ministry of Petroleum & Natural Gas, 2018. 16 MAY 2018 3:27PM by PIB Delhi, Cabinet Approves National Policy on Biofuels.

Ra, C.H., Kang, C.H., Jeong, G.T., Kim, S.K., 2014. Bioethanol production from the waste product of salted Undaria pinnatifida using laboratory and pilot development unit (PDU) scale fermenters. Biotechnol. Bioprocess Eng. 19, 984–988.

Radha, M., Murugesan, A.G., 2017. Enhanced dark fermentative biohydrogen production from marine macroalgae *Padina tetrastromatica* by different pretreatment processes. Biofuel Res. J. 13, 551–558. https://doi.org/10.18331/BRJ2017.4.1.5.

Rajak, R.C., Jacob, S., Kim, B.S., 2020. A holistic zero waste biorefinery approach for macroalgal biomass utilization: a review. Sci. Total Environ. 716, 137067. ISSN 0048-9697 https://doi.org/10.1016/j.scitotenv.2020.137067.

Rasyid, A., 2017. Evaluation of nutritional composition of the dried seaweed Ulva lactuca from Pameungpeuk Indonesia. Trop. Life Sci. Res. 28 (2), 119–125. https://doi.org/10.21315/tlsr2017.28.2.9.

Rasyid, A., Ardiansyah, A., Pangestut, R., 2019. Nutrient composition of dried seaweed Gracilaria gracilis. Indones. J. Mar. Sci. 24 (1), 1–6.

Seedevi, P., Moovendhan, M., Sudharsan, S., Vasanthkumar, S., Srinivasan, A., Vairamani, S., Shanmugam, A., 2015. Structural characterization and bioactivities of sulfated polysaccharide from *Monostroma oxyspermum*. Int. J. Biol. Macromol. 72, 1459–1465. ISSN 0141-8130 https://doi.org/10.1016/j.ijbiomac.2014.09.062.

Shobana, S., Kumar, G., Bakonyi, P., Saratale, G.D., Al-Muhtaseb, A.'a.H., Nemestóthy, N., Bélafi-Bakó, K., Xia, A., Chang, J.-S., 2017. A review on the biomass pretreatment and inhibitor removal methods as key-steps towards efficient macroalgae-based biohydrogen production. Bioresour. Technol. 244 (Part 2), 1341–1348. https://doi.org/10.1016/j.biortech.2017.05.172.

Sjöhag, E., June 2020. Algae Hydrocarbons Designed for Bio-Based Lubricants. Thesis https://www.divaportal.org/smash/get/diva2:1459455/FULLTEXT01.pdf.

Soliman, R.M., Younis, S.A., El-Gendy, N.Sh., Mostafa, S.S.M., El-Temtamy, S.A., Hashim, A.I., 2018. Batch bioethanol production via the biological and chemical saccharification of some Egyptian marine macroalgae. J. Appl. Microbiol. 125, 422–440. https://doi.org/10.1111/jam.13886.

Sreenivasan, S., Ukarde, T.M., Pandey, P.H., Pawar, H.S., 2020. BAILs mediated Catalytic Thermo Liquefaction (CTL) process to convert municipal solid waste into carbon densified liquid (CTL-Oil). Waste Manag. 113, 294–303. https://doi.org/10.1016/j.wasman.2020.06.001.

Suganya, T., Gandhi, M.N., Renganathan, S., 2013. Production of algal biodiesel from marine macroalgae *Enteromorpha compressa* by two step process: optimization and kinetic study. Bioresour. Technol. 128, 392–400.

Suthar, P., Gajaria, T.K., Reddy, C.R.K., 2019. Production of quality seaweed biomass through nutrient optimization for the sustainable land based cultivation. Algal Res. 42, 101583.

Suutari, M., Leskinen, E., Fagerstedt, K., Kuparinen, J., Kuuppo, P., Blomster, J., 2015. Macroalgae in biofuel production. Phycol. Res. 63, 1–18. https://doi.org/10.1111/pre.12078.

Tabassum, M.R., Xia, A., Murphy, J.D., 2017. Potential of seaweed as a feedstock for renewable gaseous fuel production in Ireland. Renew. Sustain. Energy Rev. 68 (Part 1), 136–146. ISSN 1364-0321 https://doi.org/10.1016/j.rser.2016.09.111.

Tabassum, M.R., Xia, A., Murphy, J.D., 2018. Biomethane production from various segments of brown seaweed. Energ. Conver. Manage. 174, 855–862.

Taboada, M.C, Millán, R., Miguez, M.I., 2012. Nutritional value of the marine algae wakame (*Undaria pinnatifida*) and nori (*Porphyra purpurea*) as food supplements. J. Appl. Phycol. 25, 1271–1276. https://doi.org/10.1007/s10811-012-9951-9.

Trivedi, N., Gupta, V., Reddy, C.R.K., Jha, B., 2013. Enzymatic hydrolysis and production of bioethanol from common macrophytic green alga *Ulva fasciata* Delile. Bioresour. Technol. 150, 106–112.

van der Wal, H., Sperber, B.L.H.M., Houweling-Tan, B., Bakker, R.R.C., Brandenburg, W., López-Contreras, A.M., 2013. Production of acetone, butanol, and ethanol from biomass of the green seaweed *Ulva lactuca*. Bioresour. Technol. 128, 431–437.

Vanegas, C.H., Bartlett, J., 2013. Green energy from marine algae: biogas production and composition from the anaerobic digestion of Irish seaweed species. Environ. Technol. 34 (15), 2277–2283. https://doi.org/10.1080/09593330.2013.765922.

Veeramuthu, A., Mohd, S., Zainal, S., Chong, C.T., Elumalai, S., Veeraperumal, S., Ani, F., 2017. Production of liquid biofuels (biodiesel and bioethanol) from brown marine macroalgae *Padina tetrastromatica*. Energy Convers. Manag. 135, 131–161.

Verma, R., Verma, S., Verma, V., Verma, S., Vaishnav, Y., Jena, V., Kumar, A., 2021. Catalytic pyrolysis of ulva lactuca macroalgae: effects of mono and bimetallic catalysts and reaction parameters on bio-oil up-gradation. Bioresour. Technol. 324, 124594. https://doi.org/10.1016/j.biortech.2020.124594.

Vizetto-Duarte, C., Custódio, L., Barreira, L., Moreira da Silva, M., Rauter, A.P., Albericio, F., Varela, J., 2016. Proximate biochemical composition and mineral content of edible species from the genus Cystoseira in Portugal. Bot. Mar. 59 (4), 251–257. https://doi.org/10.1515/bot-2016-0014.

Yang, W., Li, X., Liu, S., Feng, L., 2014. Direct hydrothermal liquefaction of undried macroalgae *Enteromorpha prolifera* using acid catalysts. Energ. Conver. Manage. 87, 938–945.

Yin, Y., Wang, J., 2018. Pretreatment of macroalgal *Laminaria japonica* by combined microwave-acid method for biohydrogen production. Bioresour. Technol. 268, 52–59. https://doi.org/10.1016/j.biortech.2018.07.126.

Yingbin, Z., Guo, J., Chen, L., Li, D., Liu, J., Ye, N., 2012. Microwave-assisted direct liquefaction of Ulva prolifera for bio-oil production by acid catalysis. Bioresour. Technol. 116, 133–139. https://doi.org/10.1016/j.biortech.2012.04.036.

Yuan, C., Wang, S., Cao, B., Hu, Y., Abomohra, A.E.-F., Wang, Q., Qian, L., Liu, L., Liu, X., He, Z., Sun, C., Feng, Y., Zhang, B., 2019. Optimization of hydrothermal co-liquefaction of seaweeds with lignocellulosic biomass: merging 2nd and 3rd generation feedstocks for enhanced bio-oil production. Energy 173, 413–422. ISSN 0360-5442 https://doi.org/10.1016/j.energy.2019.02.091.

Zhang, Y., Chen, W.-T., 2018. Book chapter hydrothermal liquefaction of protein-containing feedstocks. In: Direct Thermochemical Liquefaction for Energy Applications. Woodhead Publishing, pp. 127–168.

Zhou, D., Zhang, L., Zhang, S., Fu, H., Chen, J., 2010. Hydrothermal liquefaction of macroalgae enteromorpha prolifera to bio-oil. Energy Fuel 24 (7), 4054–4061. https://doi.org/10.1021/ef100151h.

Zhuang, Y., Guo, J., Chen, L., Demao, L., Junhai, L., Naihao, Y., 2012. Microwave-assisted direct liquefaction of *Ulva prolifera* for bio-oil production by acid catalysis. Bioresour. Technol. 116, 133–139. https://doi.org/10.1016/j.biortech.2012.04.036.

Anaerobic gas fermentation: A carbon-refining process for the production of sustainable fuels, chemicals, and food

Melvin Moore, Vicki Z. Liu, Chih-Kai Yang, Zachary Cowden, and Sean D. Simpson

LanzaTech Inc., Skokie, IL, United States

1 Introduction

The deepening climate crisis presents humanity with three distinct challenges. First, we must rapidly transition away from activities and processes that result in the release of greenhouse carbon dioxide (CO_2) into the atmosphere. Second, we must actively work to reduce the concentration of CO_2 in our planet's atmosphere. Finally, we must achieve these goals while continuing to supply the globe with the energy, materials, and food required to thrive economically. In summary, we need to urgently adopt new industrial processes and technical solutions that leverage sustainable resources to avoid the release of CO_2 at an unprecedented scale.

In nature, the carbon cycle is a continuum facilitated by biological organisms, in which CO_2 is consumed, fixed into more complex molecules, and decomposed, resulting in CO_2 release for re-consumption. Therefore, it is only logical that biological processes applied at an industrial scale can play a crucial role in allowing society to close our own carbon cycle and implement a climate-safe circular economy. Biological processes have evolved to be specialists in the conversion of chemically simple, but compositionally chaotic inputs into defined, ordered, and more complex outputs. Consider a deep-sea thermal vent—this system continuously spews varying streams of carbon-rich gases into the ocean. Microorganisms feed off this emission, precisely converting the released gases into the complex but ordered combination of molecules required for life. Indeed, it is in the context of such systems that life on earth was thought to have first emerged.

Today, this ability is being harnessed by highly scalable gas fermentation processes. Gas fermentation leverages specialized classes of microorganisms that consume mixtures of gases including carbon dioxide (CO_2), hydrogen (H_2), carbon monoxide (CO), and methane (CH_4) as their sole source of carbon and energy for product synthesis. Using gaseous inputs, these technologies can leverage an unprecedented array of high-volume, low-cost feedstocks that are available globally. Specifically, waste gas streams from hard-to-abate industrial processes such as steelmaking, syngas produced from agricultural or societal residues, and even concentrated streams of CO_2 mixed with green hydrogen can all be inputs to gas fermentation processes to produce a spectrum of sustainable fuel, chemical, and nutritional products.

The potential for gas fermentation to offer a solution to large-scale sustainable fuels, materials, and protein production has come into greater focus in recent years. The first commercial-scale gas fermentation facility was recently commissioned and operates in China. This pioneering facility is operated by the Beijing Shougang LanzaTech New Energy Technology Co. Ltd. and uses an anaerobic gas fermentation process developed by the US-based cleantech start-up company, LanzaTech Inc. The plant is located on the site of a primary steel mill converting the carbon-rich gas residues from steelmaking into sustainable ethanol and protein.

Waste resources from industry, society, and agriculture are available at a vast scale globally and can displace fossil resources as the basis for producing the materials, chemicals, and fuels that societies rely upon. These waste streams are excellent sources of renewable carbon for sustainable manufacturing technologies. In the use of these compositionally varied resources, biological systems are inherently advantaged over the traditional thermochemical processes used within the petrochemical industry to make fuel and chemical molecules. This advantage stems from both their innate ability to convert chaotic inputs into defined outputs, and their elevated tolerance to the typical "contaminating compounds" found in these waste streams. These two traits allow commercial production of sustainable molecules in so-called carbon refineries at much smaller industrial scales than is possible using traditional oil refining processes. This is commercially important because while waste streams are plentiful globally, they are dispersed, and only available at delivered volumes of 100s or low 1000s of tons per day in any given location. Fossil feedstocks, however, are delivered to refineries at rates of 100s of thousands of tons per day. With a reduced need to either harmonize the composition of waste streams or remove contaminating molecules to achieve commercial production rates and yields, the capital and operating costs of gas fermentation processes are economically viable at these smaller scales.

The ability to replace fossil oil and coal as the primary inputs for the manufacture of fuels, materials, and proteins offers an opportunity to nations that are resource-poor in oil and coal to establish a domestic infrastructure for making sustainable fuels, materials, and protein. For example, in India, the combination of municipal solid waste (MSW), agricultural waste, and gaseous waste from heavy polluting industries such as steelmaking and oil refining presents a resource to produce over 35 billion liters of sustainable ethanol (*Central Pollution Control Board, Ministry of Urban Development, and Planning Commission. India*, n.d.; *Ministry of New and renewable energy, Government of India*, n.d.; *Ministry of Petroleum and Natural Gas;*

government of India, n.d.; Ministry of Steel, Government of India, n.d.). Leveraging green H_2 in combination with concentrated streams of CO_2 could elevate this potential even further.

The products from such carbon refineries can offer a path to replace not just the volume but also the functional diversity of products produced from today's petrochemical industry. The primary use for ethanol today is as fuel for displacing gasoline. However, ethanol can be the basis to produce an array of more complex products from jet fuel to plastics, and antifreeze to fibers. This is because ethanol can be readily converted to ethylene, a highly versatile molecule, using a well-established and decades-old process technology. Ethylene is a globally traded chemical commodity with over 150 million tons of annual production from oil (Lewandowski, 2016). Over the years to come, there will be a rapid shift away from gasoline to electric-powered vehicles (Rezvani et al., 2015). It is anticipated that simultaneous with this shift to electric vehicles will be a growing interest in the use of sustainable materials and products. In the first instance, sustainable ethylene production from ethanol could form the basis for a carbon-refining industry. Additionally, through accelerated development of synthetic biology tools and capabilities seen in the past decade, gas fermenting organisms are now "programmable," allowing an even greater diversity of products to become targets for direct production by this approach. The production of molecules such as acetone, used in the manufacture of acrylic glass, or propanol which can be readily dehydrated to the propylene monomer used in polypropylene plastics, has already been reported in the literature.

This chapter will focus on the biochemical basis, industrial state of the art, and future potential of the most advanced field of gas fermentation, i.e., anaerobic gas fermentation. Through the ability to accept various mixtures of CO_2, H_2, and CO, anaerobic gas fermentation offers access to the greatest volume of sustainable feedstocks for large-scale carbon refining for the manufacture of sustainable fuels, materials, and protein.

2 Biochemistry of gas fermentation

There is a broad range of autotrophic bacteria that possess the native ability to convert inorganic gaseous compounds into liquid products such as ethanol. In anaerobic gas fermentation, the bacteria employed are part of a group of prokaryotic single-cell organisms called "acetogens." Acetogenic bacteria are defined by their use of the Wood–Ljungdahl pathway (WLP), or the reductive acetyl-CoA pathway, as their main mechanism for synthesizing acetyl-CoA from CO_2 for the conservation of energy and incorporating carbon from CO_2 into biomass (Daniell et al., 2012). From a biotechnological perspective, anaerobic acetogens are of considerable interest as they employ the most energetically efficient of all known carbon fixation pathways (Bar-Even et al., 2012; Fuchs, 2011; Gaddy, 1997; Tracy et al., 2012).

A diverse set of prokaryotes use the WLP to fix CO_2, making the pathway an integral part of the global carbon cycle. Over 100 acetogenic species spanning 22 genera have been isolated to date (Daniel et al., 2008) from a highly diverse array of ecosystems including the gastrointestinal tracts of animals and termites (Breznak, 1994), surface soils (Peters and Conrad, 1995), deep subsurface sediments (Liu and Suflita, 1993), and hydrothermal vents (Bar-Even et al., 2012). It is theorized that life emerged in hydrothermal vents long before the emergence of oxygen in Earth's atmosphere using the WLP—a pathway theorized to be the first autotrophic process on earth directing the early evolution of the cell (Russel and Martin, 2004). As earth's early atmosphere lacked O_2, anaerobic acetogens instead used the reduced gases CO and H_2 as energy-rich substrates for the WLP (Daniell et al., 2012). These prokaryotes are now known to play a pivotal role in carbon fixation in each of their respective ecosystems.

The WLP, first characterized by and subsequently named after Dr. Harland Wood and Dr. Lars Ljungdahl, has been studied for over 70 years, resulting in several comprehensive reviews that describe its biochemistry. A detailed description of the key stages in the elucidation of the WLP from a historical perspective is given in an extensive review by Daniel et al. (2008); additionally, Ragsdale and Pierce (2008) provide a detailed account of the enzymology of the WLP. In this stepwise pathway depicted in Fig. 1, simple inorganic substrates of CO, H_2, and CO_2 are converted first to acetyl-CoA and then to organic products—most commonly acetic acid, butyric acid, 2,3-butanediol, lactate, and ethanol. Acetyl-CoA, the central intermediate in this pathway, can be utilized catabolically to yield products, or can also be redirected anabolically to build biomass and cell components such as proteins and lipids (Phillips et al., 2017).

The WLP consists of two branches, the methyl (also referred to as "eastern") and carbonyl (or "western") branch, and merges at the synthesis of acetyl-CoA. CO can enter the pathway through both branches—one can enter directly through the carbonyl branch or be oxidized to CO_2 through a biological water gas shift reaction in the methyl branch (Daniell et al., 2012). In bacteria grown on CO alone, such that CO is the sole source of both carbon and energy, this reaction produces the CO_2 required for the methyl branch catalyzed by carbon monoxide dehydrogenase (Drake et al., 1980). In the methyl branch, one molecule of CO_2 undergoes a series of enzymatic reactions to yield the methyl group of acetyl-CoA. In brief, CO_2 is reduced to formate, where formate is condensed with tetrahydrofolate (THF) to form formyl-THF, thereby consuming one molecule of ATP (Liew et al., 2016). Formyl-THF is then reduced to methyl-THF where in the final step, the methyl group is transferred to a corrinoid iron–sulfur-containing protein, then fused to a molecule of CO from the carbonyl branch to form acetyl-CoA (Liew et al., 2016). The carbonyl group of acetyl-CoA is supplied by the CO bound to carbon monoxide dehydrogenase and is transferred within the enzyme to acetyl-CoA synthase's active site (De Tissera et al., 2019). There it is condensed with the methyl group and coenzyme A to form acetyl-CoA.

Acetyl-CoA can either be incorporated into cell components or converted to acetic acid in the cell via acetyl-phosphate (Phillips et al., 2017). In some acetogens, such as *Clostridium autoethanogenum*, acetic acid is then further reduced to ethanol in two steps, first via the action of aldehyde dehydrogenase followed by the reduction by alcohol dehydrogenase. These enzymes are also functional for the reduction of other acids to their respective alcohols such as the reduction of butyric acid to butanol (Phillips et al., 2015).

While the enzymatic steps of the WLP have been thoroughly elucidated, the energetics of gas fermentation is still the subject of ongoing research. As an outcome of the pathway, one molecule of ATP is created by substrate-level phosphorylation during the formation of acetic acid while one ATP is consumed for formyl-THF formation. The net balance of ATP thus far is zero, so an additional energy conservation mechanism is needed to support cell growth and the formation of products (De Tissera et al., 2019). Energy flows via the transfer of electrons which can be supplied by the oxidation of CO or H_2 and carried to their reaction sites by ferredoxin and NADH. Additional energy is theorized to come from electron transport

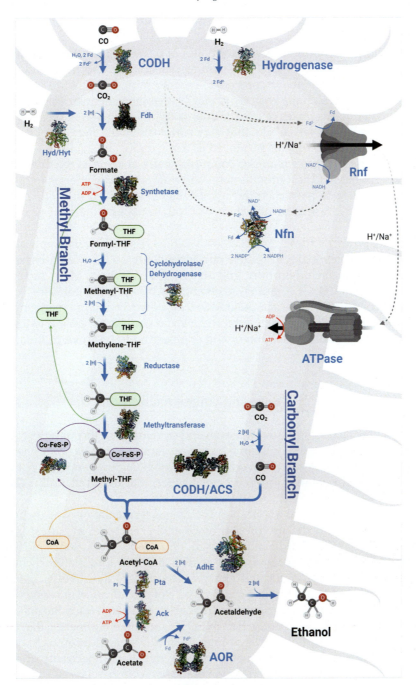

FIG. 1 Schematic of the WLP depicting both the methyl (eastern) and carbonyl (western) branches and the enzymes involved (taken from Fackler et al., 2021). Figure generated using Biorender.com.

phosphorylation, chemiosmosis, and the generation of ion gradients across the cell membrane (Phillips et al., 2017).

The key to acetogen's commercial versatility lies in the acetogen's ability to utilize the biological water–gas shift reaction. This unassuming reaction has enormous industrial significance as it allows a gas fermentation process to operate using both a wide range of inlet gas compositions, as well gas streams with dynamically changing chemistries. Being able to use dynamically fluctuating gas mixes opens the door to applying this technology to a myriad of gas streams produced from feedstocks that vary compositionally over time, such as syngas produced from MSW to process gas produced as a by-product of steel production or petroleum refining. Specifically, gas streams as compositionally extreme as the CO-rich emissions from steelmaking, where CO becomes the sole source of carbon and energy, to hydrogen-rich waste gas streams found in petrochemical plants where H_2 and CO are sources of reducing power and CO and CO_2 are sources of carbon, can both be used by the same gas fermentation process.

Harnessing the power of these remarkable acetogens has significant implications for developing innovative technologies to reduce our dependence on petroleum-based chemicals, while simultaneously reducing harmful greenhouse gas emissions. Many factors affect the microbial conversion of CO, CO_2, and H_2 to ethanol such as pH, temperature, concentration of nutrients, and conditions inside and outside the cell; however, commercializing syngas fermentation has several challenges such as bioreactor and equipment design to improve mass transfer and microbial strain development and engineering to improve production rates.

3 Scale-up and commercialization

Developing a new technology from research and development through to scaling up and commercialization is challenging. Research on syngas fermentation began in the 1980s but it was not until the 1990s that research leveraging anaerobic microorganisms to produce valuable chemicals showed potential for commercialization (Fackler et al., 2021; Gaddy, 1997, 2000; Gaddy and Chen, 1998; Gaddy and Claussen, 1992).

The first potential commercial application for syngas fermentation was demonstrated in 1989 by James Gaddy with the production of ethanol (Fackler et al., 2021; Gaddy, 1997, 2000; Gaddy and Chen, 1998; Gaddy and Claussen, 1992; Gaddy et al., 2007). In 2008, the technology developed by Gaddy was acquired by INEOS, a British multinational chemical company, that established it as a dedicated entity called INEOS Bio. INEOS Bio focused on developing the technology further using gas derived from gasification of waste streams from construction, MSW, forestry, and agricultural industries. In 2009, INEOS Bio and New Planet Energy Florida formed a joint venture, called INEOS New Planet BioEnergy, that used INEOS Bio's syngas fermentation and New Planet Energy Florida's gasification technology to convert vegetative and wood waste into ethanol and renewable electricity (Niederschulte et al., n.d.). In 2013, INEOS Bio began operation of the first commercial-scale biorefinery with a nameplate capacity of 8 million gallons of ethanol per year (30 million liters per year) and a renewable power generation capacity of up to 6 MW at the Indian River Bioenergy Center near Vero Beach, Florida (Lane, 2014b; Niederschulte et al., n.d.). However, in 2016, operations ceased. A report by the Department of Energy (DOE) stated that the facility shut down due to challenges related to elevated concentrations of contaminating compounds present in the syngas such as hydrogen cyanide, excessive moisture in the feedstocks, and various equipment and power failures (Daprile, 2017; Lane, 2014b; Niederschulte et al., n.d.). In 2017, a Chinese company named JupengBio acquired ownership of the INEOS Bio's technology and the INEOS Bio R&D center in Fayetteville, Arkansas (Jupeng Bio, 2017). In 2018, INEOS Bio sold the commercial plant in Florida to Frankens Energy, but the plant was back on the market in 2019 (Sapp, 2019; Voegele, 2019).

Another major commercial initiative in the field of gas fermentation was Coskata, a US-based start-up company. Formed in 2006, Coskata initially focused on the production of ethanol from syngas derived from a variety of biomass feedstocks such as woodchips, bagasse, corn stover, and switchgrass (Lane, 2012, 2014a; Sobolik, 2008). In 2009, Coskata began operation of its Lighthouse facility, located in Pennsylvania, for ethanol production from syngas derived from woodchips. Sorted MSW and natural gas were also tested as potential feedstocks (Blanco, 2009). This facility had a nameplate capacity of 50,000 gal (189,270 L) of ethanol per year. In 2012, Coskata shifted its focus from biomass to natural gas conversion to produce ethanol stating greater market opportunities due to the abundance of natural gas and its lower cost compared to biomass processing (Lane, 2012, 2014a). To further expand on market opportunities, in 2013 Coskata was awarded funds from the U.S. DOE's ARPA-E (Advanced Research Projects Agency-Energy) to engineer anaerobic microorganisms for the conversion of methanol into chemicals and fuels (Coskata, Inc, 2013). However, in 2015, Coskata ceased operations and subsequently, their technology platform resurfaced as Synata Bio (Lane, 2016). In 2016, Synata Bio purchased a cellulosic ethanol plant from Abengoa Bioenergy Biomass in Hugoton, Kansas (Leon, 2017).

LanzaTech has become the foremost successful company within the field of gas fermentation. Subsequently, their success has been met with numerous opportunities and partnerships with major entities within the oil, gas, and renewable energy sectors. In 2005, LanzaTech was founded by Sean Simpson and Richard Forster in New Zealand. After relocating from New Zealand in 2014, LanzaTech is currently headquartered in Illinois, USA. LanzaTech's technology focuses on the anaerobic bioconversion of industrial waste gases from steel mills, refineries, chemical manufactures, and syngas derived from the gasification of biomass or MSW feedstocks to produce valuable chemicals and fuels. Since 2018, the joint venture Beijing Shougang LanzaTech New Energy Science & Technology Company achieved commercial-scale operation and continuous production of ethanol at their facility in Caofeidian, China. This first-of-its-kind biorefinery has a nameplate capacity of 16 million gallons (60 million liters) of ethanol per year (Burton, 2021). To date, LanzaTech has reported the successful operation of this biorefinery, having demonstrated continuous ethanol production for 100 days (Fackler et al., 2021). This pioneering outcome solidifies gas fermentation as a scalable and commercially viable technology for carbon capture and recycling. A second commercial-scale biorefinery from the joint venture is scheduled to be mechanically complete in the second half of 2021, also located in China (Fig. 2).

As described previously, the biological water–gas shift reaction allows the microbe to convert gas mixtures with different ratios of $H_2:CO:CO_2$ into ethanol. This flexibility and the robustness of gas fermentation as a bioconversion process enable the diversity of feedstocks represented within LanzaTech's commercial pipeline. In 2021, two additional biorefineries

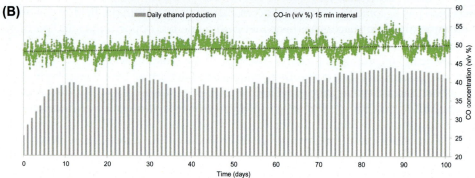

FIG. 2   (A) Image of the Beijing Shougang LanzaTech New Energy Science & Technology Company biorefinery in Caofeidian, China. This pioneering biorefinery as a nameplate capacity of 16 million gallons (60 million liters) of ethanol per year. (B) Plot of the daily ethanol production levels and inflow CO gas concentration over 100 days of continuous operation from the commercial Beijing Shougang LanzaTech New Energy Science & Technology Company biorefinery in Caofeidian, China in 2020 (Fackler et al., 2021).

are due for mechanical completion: one at an ArcelorMittal steel mill in Ghent, Belgium (capacity: 21.1 million gallons or 80 million liters of ethanol per year) and the other at an Indian Oil Company (IOC) refinery in Panipat, India (capacity: 10.6 million gallons or 40 million liters of ethanol per year) (Burton, 2020a; India Oil Corporation, 2021; McWard, 2017b). The composition of the gases that will be used in each of these commercial-scale biorefineries is quite different, highlighting the versatility of syngas fermentation. With H_2:CO:CO_2 ratios >2:1, CO_2 is consumed as a carbon source by the gas fermenting acetogens, further increasing the ratio of H_2:CO can ensure that CO_2 accounts for 50% of the carbon consumed by the process. Put simply, under these circumstances 50% of the carbon in the ethanol product can come from CO_2. This has tremendous significance for the future of the sustainable fuel and chemical industry as it opens the door to opportunities to use CO_2 directly as a resource in commercial fuel and chemical production.

As observed by INEOS Bio, syngas streams produced from biomass and to a greater extent, MSW, are known to contain compounds such as hydrogen cyanide, acetylene, and aromatic compounds such as benzene that are inhibitory or toxic to gas fermenting acetogens. The types of inhibitory compounds and their concentrations depend on many factors such as the source of the gas stream, intrinsic characteristics of the feedstock used to generate the gas (organic, inorganic, moisture content, etc.), gasifier type or technology, and operating conditions during gasification (Liew et al., 2016; Monir et al., 2020; Piatek et al., 2020; Stoll et al., 2019; Xu et al., 2011). For scale-up and commercialization of syngas bioconversion to valuable liquid products, the syngas must be well characterized and understood. This enables measures to be put in place that sufficient syngas cleanup operations for removal of inhibitory compounds can be designed to reduce plant downtime, and capital and operating costs.

LanzaTech has so far overcome the challenge of syngas cleanup to enable the use of gas streams produced from biomass as well as MSW. In 2017, Sekisui Chemical (Japan) and LanzaTech announced a joint project to produce ethanol from syngas derived from MSW at a pilot plant facility in Japan (McWard, 2017a; Sekisui Chemical Co., Ltd, 2017). In April 2020, Sekisui announced the formation of a joint venture: SEKISUI Bio REFINERY CO., LTD to build a larger scale verification plant with a capacity of approximately one-tenth that of a full commercial plant (nameplate capacity: 15,877 gal or 60,102 L of ethanol per year) (Sekisui Chemical Co., Ltd, 2020). Within the same year, LanzaTech announced a partnership with one of India's largest refiners, Mangalore Refinery and Petrochemical Limited (MRPL), and Ankur Scientific to convert syngas derived from agricultural waste into ethanol. This ethanol production facility will use the gasification technology developed by Ankur Scientific and the facility will have a production capacity of 5.3 million gallons (20 million liters) of ethanol per year (Burton, 2020a).

4 Current/future potential of synthetic biology and new process concepts

As commercial-scale gas fermentation finds success across multiple platforms, there is an increasing number of groups engaging in acetogen strain engineering efforts to enable the production of an array of alternative chemical products that will replace ethanol as the outcome of gas fermentation processes (India Oil Corporation, 2021; Lane, 2020; Sekisui Chemical Co., Ltd, 2020; Simpson et al., 2019; Yasin et al., 2019). While acetogenic metabolism inherently allows flexibility in feedstocks for gas fermentation, developing a system with robust product flexibility will require targeted modification of the chassis strain. Where implemented successfully, this strategy will yield a portfolio of "drop-in" strains capable of thriving under process conditions at existing gas fermentation plants, but each producing a specific product. Traditional methods of genetic engineering often fall short when attempting to rewire complex, interconnected metabolic networks, and many of the standard genetic tools that have been developed in model organisms simply do not function well in acetogens (Karim et al., 2020a). In order to achieve the level of biological control required for designing uniquely tailored carbon refining microbes, strain engineers look to synthetic biology as an enabling technology.

Strain engineering of LanzaTech's proprietary microbe guided by computer modeling and synthetic biology approaches offers various advantages to the carbon refinery process, such as limiting unwanted byproducts through deletion of competing pathways, increasing yields for products of interest by selecting for pathway enzymes with higher product selectivity, and expanding the portfolio of carbon refinery products by introducing nonnative biosynthetic pathways. To date, LanzaTech has demonstrated *C. autoethanogenum's* capability to produce more than 20 unique chemicals via native and nonnative biosynthetic pathways through successful integration of synthetic biology tools and continues to explore the potential of synthetic biology in improving LanzaTech's carbon refinery organism (Fackler et al., 2021; Köpke et al., 2014a,b, 2011; Rosenfeld et al., 2019). A current overview and future potential of synthetic biology tools and applications developed for metabolic engineering of *C. autoethanogenum* will be discussed with examples. Notably, LanzaTech's project funded by the US Department of Energy focused on developing strains for acetone production using synthetic metabolic pathways, highlights the successful application and future potential of synthetic biology— including screening of combinatorial gene libraries to identify gene combinations for higher acetone selectivity, genetic knockouts to reduce carbon loss to competing pathways, and genomic integration of pathway genes to improve pathway stability (Simpson et al., 2019).

Adapting common synthetic biology tools to function in a non-model, industrial organism such as *Clostridium autoethanogenum* is not a trivial task. Nevertheless, a variety of genetic tools have been made functional in this acetogen, including CRISPR-based systems for gene deletion, insertion, and activation; reliable plasmid transformation, and cell-free metabolic engineering (CFME) (Krüger et al., 2020; Nagaraju et al., 2016; Woods et al., 2019) These tools have recently been developed to a scale robust enough to support working with large combinatorial libraries, greatly amplifying the design space of genetic engineering efforts, and reducing the time and resources required to investigate complex metabolic networks. Examples include CRISPRi libraries, combinatorial pathway assemblies of varying promoter strength, and random mutagenesis libraries (Fackler et al., 2021; Karim et al., 2020b). As CFME technology continues to develop, it unlocks the possibility to conduct first-pass screening of engineered biological systems without the burden of cell transformation or keeping cells alive. This is especially useful in the context of anaerobic systems, as it removes the requirement to handle cells in oxygen-free environments, allowing experiments to be carried out in standard laboratory setups where large-scale automation and other research tools can be interfaced with more easily. These capabilities drastically reduce the time required to engineer new strains and product pathways. The synergistic pairing of these experimental capabilities with high-throughput advancements such as anaerobic laboratory automation and microfluidics handling is currently another exciting area of growth. As these tools at both the macro- and microscale continue to develop alongside each other, it is expected that the pace of discovery and engineering at the cellular level will drastically improve.

Key applications of synthetic biology in acetogens today include strain characterization, improving strain fitness, and engineering synthetic product pathways. Recent examples include defining gene essentiality through saturation transposon mutagenesis, introduction of nonnative genes in *Clostridium autoethanogenum* to enable vitamin prototrophy, and engineering industrial-scale production of nonnative metabolites, such as acetone (Annan et al., 2019; Cartman and Minton, 2010; Simpson et al., 2019). Newer genetic tools being investigated include extra-long sgRNA arrays (ELSA's) composed of numerous, nonrepetitive targeting sequences that allow for multiplexed regulation of genetic elements using CRISPR-based systems (Reis et al., 2019). Other recent work involves the validation of anaerobic fluorescent reporters and protein tags, including tags that can target proteins to specific loci within the cell (Charubin et al., 2020). These tools are useful both in probing basic cellular functions as well as engineering novel synthetic circuits. Combining these capabilities with the enhanced throughput and lower cost of next-gen omics workflows widens the scope of experimental potential even further. Experiments can be designed to probe or perturb cellular systems in a precise manner while capturing large swaths of omics data, which can then be algorithmically mined to yield novel insights. With improvements like these rapidly increasing the pace of discovery in synthetic biology research, and in turn propelling the development of even more new tools in the synthetic biology arsenal, the promise of harnessing biological systems as purpose-engineered carbon refineries continues to grow.

The successful introduction of the acetone biosynthesis into an acetogen provides an excellent case study of the potential of synthetic biology to engineer a nontraditional, yet industrially relevant microbial strain. For this effort, pathway combinations extrapolated from promoter and gene variant libraries are constructed onto plasmids, heterologously expressed, and screened for acetone production. For the native acetone biosynthesis pathway dependent on thiolase (*thlA*), CoA transferase subunit A (*ctfA*), CoA transferase subunit B (*ctfB*), and acetoacetate decarboxylase (*adc*), 236 unique strains were constructed based on combinations derived from 3 unique, nonrepetitive promoters, and 5–13 gene variants for each of the former steps (Simpson et al., 2019). Of those strains, the highest acetone-producing strain showed a tenfold improvement in acetone titer compared to the control strain with genes from *C. acetobutylicum* type strain ATCC824 (Fig. 3). This volume of strain testing is made possible by the recent advancements in DNA synthesis technology and the development of robust plasmid transformation in *C. autoethanogenum* (Woods et al., 2019). Furthermore, the 236 strains tested merely account for a subset of the 4,680 possible permutations from the *thlA-ctfA,B-adc* pathway, in which future screening of remaining combinations may have the potential to further increase acetone titers and selectivity (Simpson et al., 2019).

In addition to screening pathway combinations to identify robust pathway enzymes, multiple rounds of genetic knockouts identified by Genome Scale Modeling improved combined acetone and ethanol productivity from 26.3% to 95.3% and increased acetone selectivity by 400% by limiting byproduct formation (Fig. 3) (Simpson et al., 2019). Genetic knockouts can improve acetone selectivity by eliminating competing pathways that act as carbon sinks. For instance, knocking out a *secAdh* gene which encodes for a primary: secondary alcohol dehydrogenase that specifically converts acetone to isopropanol, was demonstrated to minimize acetone loss to isopropanol (Simpson et al., 2019). Similarly, subsequent knockout efforts have been associated with limiting carbon loss to endogenous 2,3-butanediol, lactate, and 3-hydroxybutyrate biosynthesis pathways. In conjunction with the reported use of CRISPR-based genome modification in *C. autoethanogenum* (Nagaraju et al., 2016), a variety of genome modification methods have been developed to provide options that may be more

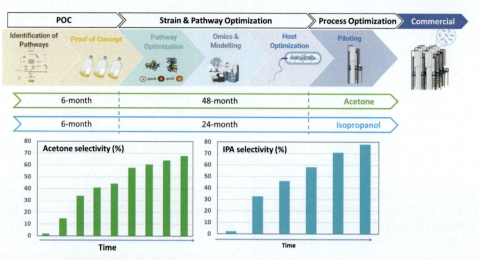

FIG. 3 Strain engineering improves acetone and isopropanol selectivity over time. Process and timeline of strain engineering are depicted for the development of acetone and isopropanol-producing *C. autoethanogenum* strains. Following proof-of-concept for acetone and isopropanol production, strain and pathway are optimized to improve titer and selectivity for chemicals of interest. Strain and pathway optimization consists of screening enzyme variants to improve pathway performance, modeling, analysis of omics data to identify potential bottlenecks, and gene editing to limit byproduct formation and improve pathway stability. Acetone and isopropanol selectivity were drastically improved by applying the strain and pathway optimization process. Figure generated using Biorender.com.

suitable for genome modification at different loci. Not only can genomic modification tools assist in the generation of loss-of-function strains, but they can also be used to increase pathway stability by integrating pathway genes into the chromosome. Based on transcriptomic and proteomic analysis from the acetone project, pathway stability of strains that heterologously express pathway genes were observed to decrease over time (Simpson et al., 2019). As acetone production increases, transcript and protein levels of *thlA, ctfA, ctfB, and adc* decrease significantly (Simpson et al., 2019). By integrating the acetone pathway under constitutive promoters in the genome, improvement of acetone pathway stability was observed (Simpson et al., 2019). Furthermore, genomic integration of pathway genes eliminates the use of plasmids, and thereby limits the use of antibiotics required for plasmid maintenance.

Ongoing efforts utilizing the growing repertoire of synthetic biology are underway to both improve base strain performance and enable the production of an increasingly diverse spectrum of chemicals from sustainable waste streams using *C. autoethanogenum* in dedicated carbon refineries. For example, LanzaTech has publicized internal programs exploring the production of long-chain alcohols and mono ethylene glycol. In combination with new and enhanced tools such as CRISPRi multiplexing, automated high-throughput workflows, and improved Genome Scale Modeling, synthetic biology has continued to unlock the biological potential behind *C. autoethanogenum* to revolutionize LanzaTech's carbon capture technology.

New process concepts also offer an opportunity to increase the breadth of sustainable products that can be produced at scale by gas fermentation. One such disruptive process under development is a collaboration between DBT-IOC Centre for Advanced Bio-Energy Research (an entity co-funded by India's Department of Biotechnology and IOC) and LanzaTech, offering the potential to enable a universal CO_2-based platform to produce fuels, chemicals, and food. This novel concept uses acetate as the energy carrier to link renewable energy with the synthesis of sustainable products. The project partners have demonstrated that acetate can be produced at industrially relevant rates by feeding acetogens gaseous mixtures of CO_2 and H_2. This acetate is then efficiently converted to high-value products in a second bioreactor by a proprietary industrial strain that was isolated and developed by DBT-IOC researchers. This can be a net-zero process in which CO_2 is the sole carbon source for product synthesis and renewable electricity can be used to generate a green hydrogen feed stream. Early product targets include diesel-range lipids, high-value omega-3 fatty acids, and nutrient protein for animal feed. In a recent public announcement, DBT-IOC and LanzaTech showcased the successful operation of a dedicated industrial pilot plant (Fig. 4) within the IOC Research campus in Faridabad, India, with a capacity to recycle 10 kg of CO_2 per day (Burton, 2020b; McWard, 2017b).

FIG. 4 A pilot facility converting CO_2 to fuels, food, and high value lipids at the DBT-IOC Centre for Advanced Bio-Energy Research in Faridabad, India.

5 Conclusion

The famous Saudi Arabian Oil Minister of the 1970s and 80s Sheikh Yamani is attributed to the quote "the stone age did not end because we ran out of stone, and the oil age will not end because we run out of oil." Advances in technology and a global momentum to avert a deepening climate crisis have brought us to the cusp of a new industrial era in which large volumes of above-ground carbon sources such as waste streams from industry, society, and agriculture, and even concentrated streams of CO_2 when combined with clean electricity can all be used to produce the climate-safe materials and fuels that are today, made from oil.

The dramatic fall in the cost of clean electricity (International Renewable Energy Agency, n.d.) is a key enabler in the transition away from oil refining. Electricity will not only power everyday appliances but ultimately many forms of transport, as well as chemical and materials manufacture. Since the 19th century, electrical power has been used to produce hydrogen gas from water. As highlighted in this review gas fermentation processes can now harness the energy in this gas to convert CO_2 into ethanol.

Using gas fermentation, distributed low-cost manufacture of sustainable products from resources available in virtually every nation is possible. Rural communities are likely to emerge as the major beneficiaries of a transition away from oil. Solar energy is already being harvested for sustainable production of power, food, and fuels in these communities. Large-scale wind and solar farms can ensure the availability of low-cost sustainable power. These zero-emission power generation facilities are being deployed increasingly in these regions. Add to this the fact that in countries like India some of the most highly concentrated sources of CO_2 are produced by the rural bioethanol industry, and we have the basis for a new "carbon refining" infrastructure to displace the oil refining infrastructure that we depend upon today.

Carbon refining is a combination of process technologies that ultimately enable greenhouse CO_2 to be used as a resource for making sustainable fuels and materials therefore replacing virgin oil production. Today, gas fermentation is one such technology that is used for sustainable ethanol production, but with advances in synthetic biology will very soon be capable of offering a direct path to the manufacture of a broad array of chemical intermediates in the production of fuels, polymers, fibers, and even food.

For example, an existing bioethanol plant producing 150,000 tons of ethanol per year from sugar also produces around 143,656 tons of concentrated CO_2. Using gas fermentation and 1500 MW of clean electricity, this CO_2 can be turned into an additional 75,000 tons of ethanol. One perceived barrier to the potential for bioprocesses to meaningfully displace products made via oil refining is the almost inconceivable scale of production needed. To match this scale, Carbon Refinery plants need to look far beyond increasing the carbon efficiency of bioethanol plants. There are clear logical phases to ramp up production capacity centered around delivering available sources of renewable carbon to refining locations.

Biomass residues from farms and forests, as well as the nonrecyclable fraction of MSW from cities and towns, are large-volume resources that offer the most immediate path to large-scale production. Established gasification technologies can already convert these resources to gas streams suitable for fermentation. Blending these gas streams with additional green H_2 would ensure that all of the renewable carbon in these resources is recycled into sustainable products. Similarly, carbon-rich gases from other hard-to-abate industrial processes like cement and steel production could provide high volumes of additional carbon molecules for a carbon refining infrastructure.

By far the largest reservoir of waste carbon is in our atmosphere. The challenge of selectively pulling CO_2 out of the air and delivering it as a concentrated stream is the focus of Direct Air Capture (DAC) technologies being developed by several start-up companies. The CO_2 from these processes is prohibitively expensive today. However, with improvements in energy efficiency and mass-scale production, these processes may well become a competitive method for delivering carbon directly to gas fermentation processes in carbon refineries of the future.

Gas fermentation is poised to become a pivotal technology, enabling a transition to climate-friendly infrastructure for fuel and materials production. The past decade has witnessed dramatic advances in our ability to apply gas fermentation technology at scale as well as our understanding of how to maximize the array of products we can make by this approach. In parallel there has been an acceleration in the deployment of zero-emission power generation capacity, and a growing demand for green hydrogen produced from this renewable power. In combination, this emerging capability and infrastructure may form the basis of a carbon refining industry that brings the age of oil to a close.

References

Annan, F.J., Al-Sinawi, B., Humphreys, C.M., Norman, R., Winzer, K., Köpke, M., Simpson, S.D., Minton, N.P., Henstra, A., M., 2019. Engineering of vitamin prototrophy in *Clostridium ljungdahlii* and *Clostridium autoethanogenum*. Appl. Microbiol. Biotechnol. 103, 4633–4648.

Bar-Even, A., Noor, E., Milo, R., 2012. A survey of carbon fixation pathways through a quantitative lens. J. Exp. Bot. 63, 2325–2342.

Blanco, S., 2009, October 16. Coskata's new lighthouse cellulosic ethanol plant. In: Depth. Autoblog. https://www.autoblog.com/2009/10/16/coskatas-new-lighthouse-cellulosic-ethanol-plant-in-depth/.

Breznak, J.A., 1994. Acetogenesis from carbon dioxide in termite guts. In: Drake, H.L. (Ed.), Acetogenesis. Chapman & Hall Microbiology Series (Physiology/Ecology/Molecular Biology/Biotechnology). Springer, Boston, MA.

Burton, F., 2020, September 15. Advanced Biofuel Facility Gets Green Light in India. LanzaTech. https://www.lanzatech.com/2020/09/15/advanced-biofuel-facility-gets-green-light-in-india/.

Burton, F., 2020, December 14. Universal Carbon Dioxide Platform for Production of Sustainable Food, Fuels and Chemicals. LanzaTech. https://www.lanzatech.com/2020/12/14/universal-carbon-dioxide-platform-for-production-of-sustainable-food-fuels-and-chemicals/.

Burton, F., 2021, January 8. Commercial CCU Plant Using LanzaTech Technology Receives RSB Advanced Production Certification. LanzaTech. https://www.lanzatech.com/2021/01/28/commercial-ccu-plant-using-lanzatech-technology-receives-rsb-advanced-products-certification/.

Cartman, S.T., Minton, N.P., 2010. A mariner-based transposon system for in vivo random mutagenesis of Clostridium difficile. Appl. Environ. Microbiol. 76, 1103–1109.

Central Pollution Control Board, Ministry of Urban Development, and Planning Commission. India. n.d.: https://cpcb.nic.in/.

Charubin, K., Streett, H., Papoutsakis, E.T., 2020. Development of strong anaerobic fluorescent reporters for *Clostridium acetobutylicum* and *Clostridium ljungdahlii* using HaloTag and SNAP-tag proteins. Appl. Environ. Microbiol. 86, 1–19.

Coskata, Inc, 2013, September 20. Awarded Funds from ARPA-E's Transformational Energy Technology Report. Advanced Biofuels Association. http://advancedbiofuelsassociation.com/blog/coskata-inc-awarded-funds/.

Daniel, S.L., Drake, H.L., Gössner, A.S., 2008. Old acetogens, new light. Ann. N. Y. Acad. Sci. 1125, 100–128.

Daniell, J., Köpke, M., Simpson, S.D., 2012. Commercial biomass syngas fermentation. Energies 5, 5372–5417.

Daprile, L., 2017, January 17. Investigation: INEOS Failed Despite $129 Million in Taxpayer Subsidies. TC Palm. https://www.tcpalm.com/story/news/2017/01/17/ineos-closes-vero-beach-biofuel-plant/96412616/.

De Tissera, S., Köpke, M., Simpson, S.D., Humphreys, C., Minton, N.P., Dürre, P., 2019. Syngas biorefinery and syngas utilization. Adv. Biochem. Eng. Biotechnol. 166, 247–280.

Drake, H.L., Hu, S.I., Wood, H.G., 1980. Purification of carbon monoxide dehydrogenase, a nickel enzyme from *Clostridium thermocaceticum*. J. Boil. Chem. 255, 7174–7180.

Fackler, N., Heijstra, B.D., Rasor, B.J., Brown, H., Martin, J., Ni, Z., Shebek, K.M., Rosin, R.R., Simpson, S.D., Tyo, K.E., Giannone, R.J., Hettich, R.L., Tschaplinski, T., Leang, C., Brown, S.D., Jewett, M.C., Köpke, M., 2021. Stepping on the gas to a circular economy: accelerating development of carbon-negative chemical production from gas fermentation. Annu. Rev. Chem. Biomol. Eng. 12, 439–470.

Fuchs, G., 2011. Alternative pathways of carbon dioxide fixation: insights into the early evolution of life? Annu. Rev. Microbiol. 65, 631–658.

Gaddy, G.L., 1997. Clostridium Strain Which Produces Acetic Acid From Waste Gases. US5593886, U.S. Patent and Trademark Office, Washington, DC.

Gaddy, J.L., 2000. Clostridium Strains Which Produce Ethanol From Substrate-Containing Gases. WO2000068407, U.S. Patent and Trademark Office, Washington, DC.

Gaddy, J.L., Chen, G., 1998. Bioconversion of Waste Biomass to Useful Products. US5821111, U.S. Patent and Trademark Office, Washington, DC.

Gaddy, J.L., Claussen, E.C., 1992. Clostridiumm ljungdahlii, an Anaerobic Ethanol and Acetate Producing Microorganism. US5173429, U.S. Patent and Trademark Office, Washington, DC.

Gaddy, J.L., Dinesh, K.A., Ko, C.W., Phillips, J.R., Basu, R., Wikstrom, C.V., et al., 2007. Methods for Increasing the Production of Ethanol From Microbial Fermentation. US7285402, U.S. Patent and Trademark Office, Washington, DC.

India Oil Corporation, 2021, March 22. Our Major Projects. India Oil Corporation Ltd. Retrieved March 2021, from https://iocl.com/AboutUs/Majorprojects.aspx.

International Renewable Energy Agency. n.d.. https://www.irena.org/newsroom/articles/2020/Jun/How-Falling-Costs-Make-Renewables-a-Cost-effective-Investment.

Jupeng Bio, 2017, June 7. Jupeng Bio has Acquired INEOS Bio for the Manufacture of Biofuels From Renewable Carbon Sources. Jupeng. http://www.jupengbio.com/blog/jupeng-bio-has-acquired-ineos-bio-for-the-manufacture-of-biofuels-from.

Karim, A.S., Dudley, Q.M., Juminaga, A., Yuan, Y., Crowe, S.A., et al., 2020a. In vitro prototyping and rapid optimization of biosynthetic enzymes for cellular design. Nat. Chem. Biol. 16, 912–919.

Karim, A.S., Liew, F.M., Garg, S., Vögeli, B., Rasor, B., Gonnot, A., Pavan, M., Juminaga, A., Simpson, S.D., Köpke, M., Jewett, M.C., 2020b. Modular cell-free expression plasmids to accelerate biological design in cells. Synth. Biol. 5. https://doi.org/10.1093/synbio/ysaa019.

Köpke, M., Mihalcea, C., Liew, F.M., Tizard, J.H., Ali, M.S., Conolly, J.J., Al-Sinawi, B., Simpson, S.D., 2011. 2,3-Butanediol production by acetogenic bacteria, an alternative route to chemical synthesis using industrial waste gas. Appl. Environ. Microbiol. 77, 5467–5475.

Köpke, M., Gerth, M.L., Maddock, D.J., Mueller, A.P., Liew, F.M., Patrick, W.M., Simpson, S.D., 2014a. Reconstruction of the acetogenic 2,3-butanediol pathway involving a novel NADPH-dependent primary:secondary alcohol dehydrogenase. Appl. Environ. Microbiol. 80, 3394–3403.

Köpke, M., Hill, R.E., Jensen, R.O., Dürre, P., 2014b. Production of biobutanol, from ABE to syngas fermentation. In: Lu, X. (Ed.), Biofuels, From Microbes to Molecules. Horizon Press, Poole, UK, ISBN: 978-1-908230-45-4, pp. 136–162.

Krüger, A., Mueller, A.P., Rybnicky, G.A., Engle, N.L., Yang, Z.K., Tschaplinski, T.J., Simpson, S.D., Köpke, M., Jewett, M.C., 2020. Development of a clostridia-based cell-free system for prototyping genetic parts and metabolic pathways. Metab. Eng. 62, 95–105.

Lane, J., 2012, July 20. Coskata Switches Focus From Biomass to Natural Gas; To Raise $100M in Natgas-Oriented Private Placement. Biofuels Digest. https://www.biofuelsdigest.com/bdigest/2012/07/20/coskata-switches-from-biomass-to-natural-gas-to-raise-100m-in-natgas-oriented-private-placement/.

Lane, J., 2014, March 25. Coskata: Biofuels Digest's 2014 5-Minute Guide. Biofuels Digest. https://www.biofuelsdigest.com/bdigest/2014/03/25/coskata-biofuels-digests-2014-5-minute-guide/.

Lane, J., 2014, September 5. On the Mend: Why INEOS Bio Isn't Producing Ethanol in Florida. Biofuels Digest. https://www.biofuelsdigest.com/bdigest/2014/09/05/on-the-mend-why-ineos-bio-isnt-reporting-much-ethanol-production/.

Lane, J., 2016, January 24. Coskata's Technology Re-emerges as Synata Bio. Biofuels Digest. https://www.biofuelsdigest.com/bdigest/2016/01/24/coskatas-technology-re-emerges-as-synata-bio/.

Lane, J., 2020, May 18. New Lords of Circular Carbon: ArcelorMittal, EU Complete Financing of Steelanol Project. Biofuels Digest. https://www.biofuelsdigest.com/bdigest/2020/05/18/new-lords-of-circular-carbon-arcelormittal-eu-complete-financing-of-steelanol-project/.

Leon, G., 2017. Integrated Biorefinery for Conversion of Biomass to Ethanol, Synthesis Gas, and Heat. OSTI, US OSTI Report, United States. https://doi.org/10.2172/1364372.

Lewandowski, S., 2016. Big changes ahead for ethylene implications for Asia. In: IHS Markit, Asia Chemical Conference presentation. https://cdn.ihs.com/www/pdf/Steve-Lewandowski-Big-Changes-Ahead-for-Ethylene-Implications-for-Asia.pdf.

Liew, F.M., Martin, M.E., Tappel, R.C., Heijstra, B.D., Mihalcea, C., Kopke, M., 2016. Gas fermentation—a flexible platform for commercial scale production of low-carbon-fuels and chemicals from waste and renewable feedstocks. Front. Microbiol. 7, 694.

Liu, S., Suflita, J.M., 1993. H_2-CO_2-dependent anaerobic O-demethylation activity in subsurface sediments and by an isolated bacterium. Appl. Environ. Microbiol. 59, 1325–1331.

McWard, S., 2017, December 7. From Trash to Tank. Upcycling of Landfill to Fuel Demonstrated in Japan. LanzaTech. https://www.lanzatech.com/2017/12/07/trash-tank-upcycling-landfill-fuel-demonstrated-japan/.

McWard, S., 2017, July 10. Indian Oil and LanzaTech Sign a Statement of Intent to Construct World's First Refinery Off Gas-to-Bioethanol Production Facility in India. LanzaTech. https://www.lanzatech.com/2017/07/10/indianoil-lanzatech-sign-statement-intent-construct-worlds-first-refinery-gas-bioethanol-production-facility-india/.

Ministry of New and Renewable Energy, Government of India. n.d.. https://mnre.gov.in/.

Ministry of Petroleum and Natural Gas; government of India. n.d.. https://mopng.gov.in/en.

Ministry of Steel, Government of India. n.d.. https://steel.gov.in/.

Monir, M.U., Yousuf, A., Aziz, A.A., 2020. Chapter 6—Syngas fermentation to bioethanol. In: Lignocellulosic Biomass to Liquid Biofuels. Academic Press, pp. 195–216.

Nagaraju, S., Davies, N.K., Walker, D.F.J., Köpke, M., Simpson, S.D., 2016. Genome editing of *Clostridium autoethanogenum* using CRISPR/Cas9. Biofuels Biotechnol. 9, 219.

Niederschulte, M., et al., DOE-INES New Planet Bioenergy Technical Report Final Public Version 7-22-16. United States. https://www.osti.gov/biblio/1268295.

Peters, V., Conrad, R., 1995. Methanogenic and other strictly anaerobic bacteria in desert soil and other oxic soils. Appl. Environ. Microbiol. 61, 1673–1676.

Phillips, J.R., Atiyeh, H.K., Tanner, R.S., Torres, J.R., Saxena, J., Wilkins, M.R., Huhnke, R.L., 2015. Butanol and hexanol production in *Clostridium carboxidivorans* syngas fermentation: medium development and culture techniques. Bioresour. Technol. 190, 114–121.

Phillips, J.R., Huhnke, R.L., Atiyeh, H.K., 2017. Syngas fermentation: a microbial conversion process of gaseous substrates to various products. Fermentation 3, 28.

Piatek, P., Olsson, L., Nygard, Y., 2020. Adaptation during propagation improves *Clostridium autoethanogenum* tolerance towards benzene, toluene, and xylenes during gas fermentation. Bioresour. Technol. Rep. 12, 1–5.

Ragsdale, S.W., Pierce, E., 2008. Acetogenesis and the Wood-Ljungdahl pathway of CO_2 fixation. Biochim. Biophys. Acta 1784, 1873–1898.

Reis, A.C., Halper, S.M., Vezeau, G.E., Cetnar, D.P., Hossain, A., Clauer, P.R., Salis, H.M., 2019. Simultaneous repression of multiple bacterial genes using nonrepetitive extra-long sgRNA arrays. Nat. Biotechnol. 37, 1294–1301.

Rezvani, Z., Jansson, J., Bodin, J., 2015. Advances in consumer electric vehicle adoption research: A review and research agenda. Transp. Res. Part D: Transp. Environ. 34, 122–136.

Rosenfeld, D.C., Liew, F.M., Köpke, M., Gao, A., Harris, A., Mueller, A., Nagaraju, S., Smith, A., Tran, L., O'Brien, J., Tyo, K., Martin, J., Strutz, J., Metz, J., 2019. Bio-Syngas to Fatty Alcohols (C6-14) as a Pathway to Fuels. US OSTI Report, United States, OSTI, https://doi.org/10.2172/1604947.

Russel, M.J., Martin, W., 2004. The rocky roots of the acetyl-CoA pathway. Trends Biochem. Sci. 29, 358–363.

Sapp, M., 2019, January 3. INEOS Bio Ethanol and Biomass Plants Up for Immediate Sale. Biofuels Digest. https://www.biofuelsdigest.com/bdigest/2019/01/03/ineos-bio-ethanol-and-biomass-plants-up-for-immediate-sale/.

Sekisui Chemical Co., Ltd, 2017, December 6. Turning "Garbage" into Ethanol Establishing a First-in-the-World Innovative Production Technology. Sekisui Chemical. https://www.sekisuichemical.com/whatsnew/2017/1325318_29675.html.

Sekisui Chemical Co., Ltd, 2020, April 16. Establishment of Joint Venture to Commercialize "Waste to Ethanol" Technology. Sekisui Chemical. https://www.sekisuichemical.com/whatsnew/2020/1349043_36556.html.

Simpson, S.D., Abdalla, T., Brown, S.D., Canter, C., Conrado, R., Daniell, J., Dassanayake, A., Gao, A., Jensen, R.O., Köpke, M., Leang, C., Liew, F.M., Nagaraju, S., Nogle, R., Tappel, R.C., Tran, L., Charania, P., Engle, N., Giannone, R., Hettich, R., Klingeman, D., Poudel, S., Tschaplinski, T., Yang, Z., 2019. Development of a Sustainable Green Chemistry Platform for Production of Acetone and Downstream Drop-in Fuel and Commodity Products directly from Biomass Syngas via a Novel Energy Conserving Route in Engineered Acetogenic Bacteria. US OSTI Report, United States, OSTI, https://doi.org/10.2172/1543199.

Sobolik, J., 2008, June 2. Anaerobic Organisms Key to Coskata's Rapid Rise. Ethanol Producer Magazine. http://www.ethanolproducer.com/articles/4268/anaerobic-organisms-key-to-coskata's-rapid-rise/.

Stoll, I.K., Boukis, N., Sauer, J., 2019. Syngas fermentation to alcohols: reactor technology and application perspective. Chem. Ing. Tech. 92, 125–136.

Tracy, B.P., Jones, S.W., Fast, A.G., Indurthi, D.C., Papoutsakis, E.T., 2012. Clostridia, the importance of their exceptional substrate and metabolite diversity for biofuel and biorefinery applications. Curr. Opin. Biotechnol. 23, 364–381.

Voegele, E., 2019, January 11. Assets of Former INEOS Bio Plant to be Sold Piecemeal Via Auction. Biomass Magazine. http://biomassmagazine.com/articles/15871/assets-of-former-ineos-bio-plant-to-be-sold-piecemeal-via-auction.

Woods, C., Humphreys, C.M., Rodrigues, R.M., Ingle, P., Rowe, P., Henstra, A.M., Köpke, M., Simpson, S.D., Winzer, K., Minton, N.P., 2019. A novel conjugal donor strain for improved DNA transfer into Clostridium spp. Anaerobe 59, 184–191.

Xu, D., Tree, D.R., Lewis, R.S., 2011. The effects of syngas impurities on syngas fermentation to liquid fuels. Biomass Bioenergy 35, 2690–2696.

Yasin, M., Cha, M., Chang, I.S., Atiyeh, H.K., Munasinghe, P., Khanal, S.K., 2019. Chapter 13—Syngas fermentation into biofuels and biochemicals. In: Biomass, Biofuels, Biochemicals, Biofuels: Alternative Feedstocks and Conversion Processes for the Production of Liquid and Gaseous Biofuels (Second Edition). Academic Press, pp. 301–327.

Cyanobacteria as a renewable resource for biofuel production

Deepti Sahasrabuddhe[a,b,c], Annesha Sengupta[a], Shinjinee Sengupta[b], Vivek Mishra[a], and Pramod P. Wangikar[a,b,c]

[a]Department of Chemical Engineering, Indian Institute of Technology Bombay, Mumbai, India
[b]DBT-Pan IIT Center for Bioenergy, Indian Institute of Technology Bombay, Mumbai, India
[c]Wadhwani Research Center for Bioengineering, Indian Institute of Technology Bombay, Mumbai, India

1 Introduction

Increased global fuel demand and rapidly depleting resources have resulted in increased interest in alternative fuels such as biofuels (Johnson et al., 2016). However, the industrial biofuel sector is majorly dominated by processes that utilize biomass to produce biofuels. Unless the process is based on waste biomass, it may compete with agricultural cropland (Kaygusuz, 2009; Zahra et al., 2020). Photosynthetic organisms such as cyanobacteria can utilize inorganic CO_2 and convert it to several important organic molecules (Nozzi et al., 2013). In light of this, these organisms have emerged as attractive candidates to efficiently capture renewable solar energy and fix atmospheric CO_2 for the production of sustainable fuels (Nozzi et al., 2013; Quintana et al., 2011).

Cyanobacteria evolved around 2.6–3.5 billion years ago and are considered to be responsible for the oxygenic environment on Earth (Hedges et al., 2001; Lau et al., 2015). They are found in varied ecological habitats and exhibit metabolic as well as morphological diversity to accommodate extreme environmental changes (Garcia-Pichel, 2009; Haselkorn, 2009). Besides carbon fixation, some cyanobacterial species are capable of fixing atmospheric nitrogen. These diazotrophic cyanobacteria have evolved a mechanism to separate the incompatible processes of carbon and nitrogen fixation by either developing specialized cells called heterocysts or by temporal separation in the single-celled nitrogen fixers (Issa et al., 2014). Unlike heterotrophs, these photosynthetic prokaryotes exhibit circadian rhythm enabling them to modulate their metabolism with respect to the phase of the day (Jaiswal and Wangikar, 2020). Additionally, cyanobacteria also have comparatively higher photosynthetic efficiency and biomass conversion rates compared to higher plants. Therefore, these can be effectively engineered to redirect the carbon flux towards the production of desired molecules by engineering a specific pathway (Lau et al., 2015). Moreover, the prokaryotic cellular structure of cyanobacteria facilitates genetic engineering. Further, advanced genome editing techniques like CRISPR (clustered regularly interspaced short palindromic repeats) have been developed for cyanobacteria (Sengupta et al., 2018). Moreover, cyanobacteria can grow in large, naturally available water bodies making their cultivation economically feasible.

Commercial production of biofuels from cyanobacteria essentially is comprised of four major steps, characterization of cyanobacterial strains and genetic elements, designing a producer strain, large-scale cultivation, and commercialization (Fig. 1). The first two steps are research-intensive and are the focus of the current chapter. We have consolidated information about the potential cyanobacterial hosts and briefly comment on their desirable properties. Specific, tunable regulatory elements to create novel genetic circuits, modular assembly of the gene constructs and advanced genome editing techniques for redesigning the cell have also been discussed in detail. We further elaborate on empirical genetic engineering strategies

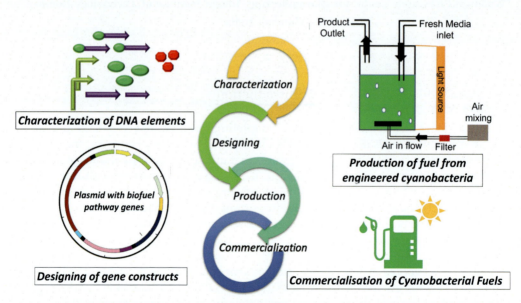

FIG. 1   Converting cyanobacterial model strain into a biofuel production factory: the steps involved in developing a cyanobacterial-based biofuel production factory include characterization of genetic elements followed by designing strategies to introduce new pathways or modify existing pathways. These are followed by optimizing production in a large-scale reactor and later commercialization of the biofuel. The annotations used for synthetic genetic parts are similar to BioBricks annotations. *The designed plasmid has been adapted from the previous report by Sengupta, S., Jaiswal, D., Sengupta, A., Shah, S., Gadagkar, S., Wangikar, P.P., 2020a. Metabolic engineering of a fast-growing cyanobacterium Synechococcus elongatus PCC 11801 for photoautotrophic production of succinic acid. Biotechnol. Biofuels 13 (1), 1–18. https://doi.org/10.1186/s13068-020-01727-7.*

for some representative fuel molecules and platform chemicals. However, despite multiple efforts, the biofuel titers from engineered strains remain lower compared to the industrial benchmark achieved by their heterotrophic counterparts, such as *E. coli* (Noreña-Caro and Benton, 2018). Advanced techniques like Metabolic Flux Analysis (MFA) can help to identify a potential bottleneck and guide systemic genetic rewiring (Babele and Young, 2020). This chapter also discusses flux analysis guided metabolic engineering examples in cyanobacterial biofuels domain. Easy accessibility of synthetic biology parts, protocols, and host strains along with translating the foundational efforts to commercial bioprocesses is required for utilizing the complete potential of cyanobacteria as a renewable resource for biofuel production.

2 Cyanobacterial host systems

For developing commercially viable bioprocesses, significant improvements are required in the current expanse of model host strains. Isolating cyanobacteria with unique characteristics and deploying it as host strain is critical for industrial-scale production (Hitchcock et al., 2020; Mukherjee et al., 2020).

2.1 Essential parameters for cyanobacterial hosts

- A cyanobacterial host needs to be robust and stress-tolerant to be able to survive in open pond conditions for a commercially viable production process (Kitchener and Grunden, 2018).
- The host strains should have faster growth rates which would imply faster metabolic turnover and may enable greater flux towards the product of interest (Kim et al., 2016). Such competitive host strains can either be isolated from natural habitats or can be engineered in the laboratory (Jaiswal et al., 2018, 2020; Yu et al., 2015).
- The cyanobacterial hosts should be extremely agile for multiple metabolic pathway alterations. Currently, cyanobacterial pathway engineering relies on the regulatory elements derived from a heterologous host (Hitchcock et al., 2020). However, inadequate knowledge of their portability can limit their efficiency in cyanobacteria (Till et al., 2020). Moreover, more advanced genome editing techniques and tightly regulated expression systems need to be optimized for these polyploid photoautotrophs (Ramey et al., 2015). This has been discussed in detail in subsequent sections in the chapter.
- The critical bottleneck in the process of genetic engineering of the host cell is the toxicity imparted by the fuel molecules during production (Atsumi et al., 2010). The strategy of improving tolerance using Adaptive Laboratory Evolution (ALE) by evolving over multiple generations in the presence of higher concentrations of fuel molecule has been tried in the cyanobacterial domain (Matsusako et al., 2017; Song et al., 2014; Srivastava et al., 2021). The adaptability of the host to improve the tolerance towards fuel molecules would be important.

2.2 Examples of cyanobacterial host systems

The search for a suitable cyanobacterial host strain has been an ongoing process. *Synechococcus elongatus* PCC 7942 and *Synechocystis* sp. PCC 6803 (henceforth *S. elongatus*

7942 and *Synechocystis* 6803, respectively) have been extensively used as model strains for genetic engineering (Kaneko et al., 1996; Golden, 2019). However, slower growth rates and lack of tolerance to bright light may limit the application of these cyanobacteria at industrial scale and under outdoor settings (Sengupta et al., 2018). Additionally, *Synechococcus* sp. PCC 7002 (henceforth *Synechococcus* 7002), a robust, marine euryhaline cyanobacterium, has emerged as a faster growing host strain with a doubling time of ∼3.5h (Berla et al., 2013; Xu et al., 2011). The strain has been engineered for the production of free fatty acids as well as ethanol (Kopka et al., 2017; Ruffing, 2014). More recently, *S. elongatus* UTEX 2973 (henceforth *S. elongatus* 2973) was isolated as a fast-growing, high light, and temperature-tolerant strain having a doubling time of around 1.9h and is being developed as a new bio-production chassis (Lin et al., 2020; Yu et al., 2015). Recently, two cyanobacterial strains, *S. elongatus* PCC 11801 and *S. elongatus* PCC 11802 (henceforth *S. elongatus* 11,801 and 11,802, respectively) were isolated from a local water body in Mumbai, India. These strains are fast-growing, genetically amenable, and show greater tolerance to various abiotic stresses (Jaiswal et al., 2018, 2020). These strains can prove to be attractive candidates as commercial host strains and have been demonstrated for photoautotrophic production of platform chemical, succinate (Sengupta et al., 2020a). Another strain isolated from Singaporean estuary, *Synechococcus* sp. PCC 11901 proved to be a faster growing strain producing higher biomass and larger fatty acid pool (Włodarczyk et al., 2020).

Additionally, the available model strains can be engineered to further improve the photosynthetic and carbon capture efficiency, pool of nodal precursors, and growth rates (Song et al., 2021; Ungerer et al., 2018). Such engineered strains need to be made easily available and then can be deployed for pathway engineering efforts (Abernathy et al., 2017; Hirokawa et al., 2020; Luan et al., 2020). Sophisticated synthetic biology tools and techniques are prerequisites for developing and exploring the potential of cyanobacterial hosts.

3 Advances in synthetic biology of cyanobacteria

Developing a cyanobacterial production host necessitates genetic modifications which can be introduced into the cell quickly and predictably by using a synthetic biology approach. Synthetic biology incorporates the traditional design, build, and test principles used by engineers. Modularity, abstraction, and standardization are the basic requirements for developing predictable and standardized biological devices (Canton et al., 2008; Endy, 2005). The synthetic biology toolbox comprises promoters, ribosomal binding sites (RBSs), terminators, genes, assembly methods, and genome editing tools (Fig. 2). Though efforts are being made to develop synthetic biology tools for cyanobacteria, the repertoire is still sparse and needs more augmentation (Sengupta et al., 2018). In this section, advancements in synthetic biology of cyanobacteria have been discussed. The toolbox can be broadly categorized into two: the characterized biological parts and genome editing techniques.

3.1 Characterized biological parts

3.1.1 cis-*Acting regulatory elements: Promoters, RBSs, and terminators*

Promoters are one of the key players in gene expression and are the most highly studied biological parts. The BioBricks Registry of standard biological parts currently holds a few

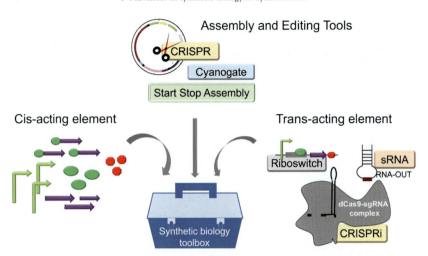

FIG. 2 Developing effective synthetic biology toolbox for cyanobacteria: a schematic showing a synthetic biology toolbox of cyanobacteria. A toolbox includes species-specific *cis* and *trans* regulatory elements for modulating transcription of specific genetic element along with optimized genome engineering tools and modular gene assembly strategies including CRISPR and Cyanogate, respectively. The annotations used for genetic parts are similar to BioBricks annotations. *Adapted from Endy, D., 2005. Foundations for engineering biology. Nature 438 (7067), 449–453. https://doi.org/10.1038/nature04342.*

cyanobacterial promoters (Sengupta et al., 2018). A greater percentage of cyanobacterial promoters are responsive to circadian and diurnal rhythm (Dong and Golden, 2009; Vijayan and O'Shea, 2013). Besides light, some of the promoters tune their gene expression with respect to the CO_2 levels or metal concentrations (Berla et al., 2013). Therefore, these promoters have huge potential to be used under outdoor conditions for regulated bioproduction. Some of the highly characterized promoters of cyanobacteria are psbA, cpcB, rbcL, and some native metal inducible promoters (Deng and Coleman, 1999; Liu and Pakrasi, 2018; Ruffing, 2014; Sengupta et al., 2019; Zhou et al., 2015). (Englund et al., 2016; Liu and Pakrasi, 2018; Ng et al., 2015; Sengupta et al., 2018; Wang et al., 2018; Berla et al., 2013). Recently, a large dynamic mutant promoter library of rbcL and cpcB was developed and characterized under different light and CO_2 conditions across three *S. elongatus* strains (Sengupta et al., 2020b). While there has been advancement in increasing the inventory of cyanobacterial promoters, the heterotrophic chemically induced promoters (P_{tet}, P_{trc}, P_{lac}) or their variants are still favored for bioproduction due to their vast knowledge and predictable nature (Berla et al., 2013; Huang et al., 2010; Huang and Lindblad, 2013; Lin et al., 2017; Oliver et al., 2013). However, there are several disadvantages of using chemical inducers on large scale under outdoor conditions (Liu and Pakrasi, 2018). Thus, to develop these photosynthetic prokaryotes as potential cell factories, a detailed characterization of novel and mutant promoters would be required to have a dynamic range of tightly regulated promoters.

Ribosomal binding sites (RBSs) in prokaryotes initiate the translation process and the efficiency of the process is dependent on the core Shine-Dalgarno sequence (Heidorn et al., 2011). Modified RBS sequences have been effective in expressing heterologous genes leading to a higher yield of products of interest (Xiong et al., 2015; Englund et al., 2018). A few native RBSs from cyanobacteria have been listed in the BioBrick Registry such as psbA2, cpcB, and rbcL. Recently, the repertoire of RBSs has been increased effectively by characterizing 20 native RBSs from *Synechocystis* 6803 which are responsible for expressing photosynthetic genes (Liu and Pakrasi, 2018). Though several predictive tools are available for predicting synthetic RBS sequences having dynamic range of translation initiation rates, there is a discrepancy between the predicted and experimentally obtained rates in cyanobacteria (Markley et al., 2015). Thus, a hybrid strategy was incorporated that complemented the RBS calculator predicted sequences with modifications derived from experimental knowledge in *Synechocystis* 6803 (Wang et al., 2018). This approach showed significant improvement in the production of ethylene (Wang et al., 2018; Xiong et al., 2015).

Similar to promoters, RBSs, the terminator sequences also play a vital role in controlling the expression of genes, such that its transcript is not carried forward to the adjacent gene construct. A large array of terminator sequences has been extensively characterized in *E. coli* and cyanobacterial strains have successfully borrowed some of the Rho-dependent terminators (Chen et al., 2013; Heidorn et al., 2011; Wang et al., 2012). Some of the native terminators have also been characterized and more awaits (Huang et al., 2010; Wang et al., 2012).

3.1.2 trans-*Acting regulatory elements*

The *trans*-acting elements are a category of parts that finds applications in regulating gene expression. Some of the trans-acting elements optimized for cyanobacteria are CRISPR interference (CRISPRi), synthetic small RNA (sRNA), and riboswitches. While CRISPRi regulates the transcription, sRNA and riboswitches are posttranscriptional regulators (Copeland et al., 2014; Nakahira et al., 2013). The CRISPRi involves a dCas protein (lacks the nuclease activity) and sgRNA (single guide RNA) to facilitate multiple gene repression in a cell simultaneously (Hentschel et al., 2013; Knoot et al., 2020; Yao et al., 2016). This technique has successfully repressed both heterologous and native photosynthetic genes of cyanobacteria (Gordon et al., 2016; Liu et al., 2020; Yao et al., 2016). Further, an advanced titratable CRISPRi repression system was developed by varying inducer concentration which enabled more controlled repression of gene expression (Gordon et al., 2016; Knoot et al., 2020). Besides the transcriptional regulation, sRNA and riboswitches have been reported as an effective alternative. The basic working principle is to make RBS inaccessible for ribosome binding. While the sRNA system showed maximum repression of 70% in *Synechococcus* sp. PCC 7002, around 98% repression was observed for theophylline-induced riboswitches (Taton et al., 2017; Zess et al., 2016). Riboswitches have been reported as a better candidate for regulation than the existing trans-acting systems known (Taton et al., 2017).

3.2 Genome editing and assembly tools used in cyanobacteria

Cyanobacteria are polyploid with varying genome copy numbers depending on the growth phase (Ramey et al., 2015). Thus, genome editing requires several rounds of chromosomal segregation and hence necessitates efficient genome editing tools (Ng et al., 2015). To that end, the CRISPR-Cas system was optimized for cyanobacteria. It is a simple technology that facilitates genome editing by introducing a double-strand DNA break (Ma et al., 2016). This CRISPR/Cas technology has been demonstrated for targeted deletion of single gene as well as large regions up to 118kb (Li et al., 2016; Niu et al., 2019; Wendt et al., 2016). Further, the addition of a knock-in template would facilitate the insertion of desired genes as observed for enhancing succinate production in *S. elongatus* 7942 and ethylene-succinate co-production in *S. elongatus* 11,801 (Li et al., 2016; Sengupta et al., 2020a, 2020c). The ongoing research focuses on identifying different categories of Cas proteins to address issues such as nuclease toxicity, easy large deletion, and high-throughput single mutagenesis (Cai et al., 2013; Li et al., 2016; Ungerer and Pakrasi, 2016; Wendt et al., 2016). Since CRISPR/Cas editing is markerless, this process eradicates the issue of antibiotic resistance when grown in an outdoor environment.

The DNA assembly tools and the vectors for delivering the DNA of interest into the host cell are equally important steps in synthetic biology. A widely used tool for gene assembly is Gibson Assembly, which is available as a commercial kit (Gibson et al., 2009; Ungerer and Pakrasi, 2016). Some modular cloning approach has also been developed for cyanobacteria such as CyanoGate, Start Stop Assembly which facilitates easy swapping of biological parts (Taylor et al., 2019; Vasudevan et al., 2019). Cyanobacteria have witnessed the development of modular vectors, SyneBrick vectors which will expedite the metabolic engineering of cyanobacteria as these can regulate multiple gene expression (Kim et al., 2017) (Fig. 2).

3.3 Challenges and future perspective for improving the synthetic biology approach for cyanobacteria

Developing cyanobacteria as a production host similar to *E. coli* or yeast would require significant advancement in the synthetic biology approach. Detailed understanding of cyanobacterial parts under different environmental conditions and maintaining a standardized datasheet would serve as a global database of parts (Canton et al., 2008). However, the major challenge in developing such datasheets is the context-dependent response of these biological parts. Therefore, universal standardized protocols and reporting systems are necessary for the characterization. The next challenge is the modularity of parts, which ensures easy integration of bio parts to create a high-level device with predictable outcomes (Sengupta et al., 2018). Though some modular vectors are available, domestication of these parts might be challenging (Taylor et al., 2019; Vasudevan et al., 2019). Heterotrophic parts have been borrowed for use in cyanobacteria, limited reports are available showing the degree of portability (Lin et al., 2020; Oliver et al., 2014; Sengupta et al., 2020b). Among the different parts, the portability of promoters has shown a very weak correlation between photoautotrophs and heterotrophs (Englund et al., 2016; Markley et al., 2015). The markerless genome editing technique CRISPR/Cas has improved the efficiency of the process of genetic modification in cyanobacteria and facilitated the assembly of complex pathways (Behler et al., 2018; Sengupta et al., 2020c; Ungerer and Pakrasi, 2016). However, a more sophisticated technique is necessary for high-throughput modifications, similar to the Multiplex Automated Genome Engineering (MAGE) and trackable multiplex recombineering, optimized for *E. coli* (Ramey et al., 2015; Sengupta et al., 2020a; Wang et al., 2009). The augmentation of these high-throughput techniques with CRISPR and optimizing it for carbon-neutral prokaryotes would expand the horizon for efficient bioproduction (Ronda et al., 2016).

Along with empirical techniques, it is also essential to develop in silico model-based tools for comprehensive understanding of cyanobacterial metabolism. Most of these tools and software are optimized for well-studied hosts such as *E. coli*, yeast, plants, and humans, and hence predictions are not correlated with the experimental data for cyanobacteria

(Sengupta et al., 2018). Therefore, it will be important to improve our understanding further with respect to regulatory pathways by integrating metabolomics studies with synthetic biology. Metabolic flux analysis can provide valuable insights into the potential bottlenecks and guide pathway engineering.

4 Metabolic flux analysis of cyanobacteria

Metabolic control analysis of lactate-producing, engineered *Synechocystis* 6803 showed that despite the use of promoters with differential strengths, production is controlled by the inherent limitations of the pathway and decided mainly by the carbon flux. Therefore, a detailed understanding of metabolic flux map will provide useful insights in obtaining controlled gene expression. This in turn will directly help design efficient genetic constructs aiming at higher productivity (Angermayr and Hellingwerf, 2013). Thus, in this section, we discuss the advancement in the field of understanding the metabolic flux distribution of cyanobacteria for biofuel production.

Direct biosynthesis of biofuels from CO_2 brings a great prospect from an environmental point of view (Hays and Ducat, 2015). [13]C Metabolic flux analysis (MFA) is a prime technique used to estimate intracellular fluxes and is preferred over the constrained-based analysis such as Flux Balance Analysis (FBA) because of its precise measurement of in vivo metabolic fluxes (Buescher et al., 2015). ^{13}C-MFA exploits stable ^{13}C labeled tracers to estimate the fluxes since the labeling pattern of metabolites is based on the metabolic fluxes and its pool size (Zamboni et al., 2009). The isotopic-steady state ^{13}C-MFA approach uses selectively labeled carbon substrates such as glucose, glycerol, etc. and is broadly used in the heterotrophic as well as mixotrophic metabolism (He et al., 2014; Jahn et al., 2013; Wasylenko and Stephanopoulos, 2015). However, the approach of isotopically nonstationary-^{13}C MFA is applied for elucidating the flux values in photoautotrophic cyanobacteria as all the carbon atoms are coming from the same source (atmospheric CO_2). In INST-MFA, dynamic labeling patterns emerged from the introduction of ^{13}C-labeled substrates (Shastri and Morgan, 2007; Young et al., 2011). Labeling experiments are performed where a stable ^{13}C isotope tracer is fed to cells growing in the steady state (mid-to-late exponential phase) followed by measurement of labeling pattern analysis using analytical techniques such as LCMS or GCMS (Dange et al., 2020; Prasannan et al., 2019). In general, amino acids, organic acids, fatty acids, and sugars can be analyzed using GC–MS after the chemical derivatization while sugar phosphates and acyl-CoA molecules using LC–MS to avoid degradation of thermally labile metabolites at higher temperature in GC–MS (Jazmin and Young, 2013). Flux is then estimated by comparing the observed labeling pattern from the experiment with a simulated labeling pattern (based on isotopomer model) and recursively fit (Fig. 3).

[13]C-MFA can be potentially applied for identifying bottleneck reaction/enzymes and wasteful by-product pathways. Several studies have been reported that showed improvement in titer in heterotrophic organisms using ^{13}C-MFA guided metabolic engineering. In a recent report, the acetol production was improved using ^{13}C-MFA in *E. coli* by identifying bottlenecks in the conversion of acetol from glycerol (Yao et al., 2019). In addition to the bottleneck pathway, ^{13}C-MFA also revealed rigidity of upper glycolysis, TCA cycle, and Pentose

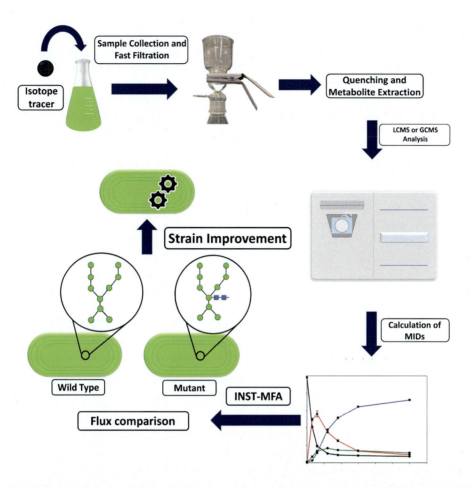

FIG. 3 A stepwise strategy for metabolic flux analysis of wild-type and mutant cyanobacterial cells: The method initiates with collecting a sample and metabolite extraction ultimately generating flux maps using MIDs calculated by LC–MS. The comparative flux map between the wild-type and engineered strain eventually paves a path for rational metabolic engineering of the cyanobacteria.

Phosphate pathway (Yao et al., 2019). ^{13}C-MFA was also used to improve biofuel chemicals in *Saccharomyces cerevisiae* like butanol after identifying the bottleneck step in the acetyl-CoA supply (Guo et al., 2016).

Few studies have been reported in cyanobacteria utilizing ^{13}C-MFA guided metabolic engineering. Typical flux maps of cyanobacteria grown under photoautotrophic conditions reported the low flux values in the TCA cycle and related reactions in comparison to heterotrophic organisms (Hendry et al., 2017; Xiong et al., 2015). In an early report, ^{13}C-MFA-guided strategy was used to guide genetic engineering of *Synechocystis* 6803 to produce ethylene (Xiong et al., 2015). Later *S. elongatus* 7942 was engineered for isoprene production from

CO_2. The strain design was guided by dynamic flux analysis and metabolite profiling of the cell that helped in directing almost 40% of the fixed carbon towards the production of isoprene (Gao et al., 2016). Likewise, INST-MFA has been used to identify pyruvate kinase as a bottleneck in isobutyraldehyde (IBA) production in engineered strain of *S. elongatus* 7942 (Jazmin et al., 2017). Overexpression of bottleneck enzymes increased the IBA-specific productivity significantly. Moreover, in a subsequent report, the potential use of INST-MFA for alleviating the pathway bottlenecks in the aldehyde production showed increased productivity. Flux analysis identified a negative correlation between aldehyde production and genes such as pyruvate dehydrogenase (PDH) and phosphoenolpyruvate carboxylase (PPC) and downregulation of both provided significant improvement (Cheah et al., 2020). Therefore, metabolic engineering guided by ^{13}C metabolic flux analysis can systematically identify and remove bottleneck pathways in cyanobacteria host cells. Similarly, ^{13}C-MFA can be potentially implemented to estimate the rates of Calvin–Benson–Bassham (CBB), pentose phosphate pathway, and TCA pathway reactions, intermediates of which can be used for the heterologous production of biofuels to achieve the improved titer/specific productivity of biofuels.

5 Engineering cyanobacteria for the production of biofuels

While bioethanol remains an attractive fuel candidate from the cyanobacteria, longer chain alcohols like butanol, isobutanol, and isopropanol carry some additional advantages. Similarly, acetone, alkanes, isoprene, and ethylene serve as energy substitutes whereas 2,3-butanediol, 1,3-propanediol, and succinate are industrially important platform chemicals and included in the top 10 chemicals produced from biological origin as per DOE report (Bozell and Petersen, 2010). These are traditionally produced by a complex chemical synthesis or biologically produced via various heterotrophs which require carbohydrate feedstocks. Cyanobacteria possess all the essential precursors required for the synthesis of these chemicals and can be engineered for production. Here we focus on platform chemicals and fuel substitutes that can be produced by engineered pathways in cyanobacteria and otherwise demand very heavy energy input for their chemical synthesis.

In the current chapter, we segregate the engineering strategies for representative biofuels and platform chemicals, based on the cyanobacterial pathway from where their precursors originate. The details of the fuel product/platform chemicals, engineered enzymes, promoters used, and maximum titer achieved have been summarized in Tables 1 and 2.

5.1 Glycolytic pathway derivatives

Pyruvate produced from phosphoenolpyruvate (PEP) towards the end of the glycolytic pathway is a vital node for engineering biofuel pathways along with some other key compounds like glyceraldehyde 3-phosphate.

5.1.1 Ethanol

First report of cyanobacteria-based ethanol production was in 1999 by Deng et al., where pyruvate decarboxylase (pdc) and alcohol dehydrogenase (adh) were heterologously

TABLE 1 List of other representative biofuels engineered from cyanobacterial strains along with modified genes, promoters used, and final titers.

Biofuel molecule	Cyanobacterial strain	Genes over expressed/ deleted	Promoters used	Cumulative titer (g/L)	Cultivation time (in days)	References
Ethanol	*Synechococcus elongatus* sp. PCC 7942	*adh, pdc, ΔglgC*	P_{sc}	3.8	20	Velmurugan and Incharoensakdi (2020)
	Synechocystis sp. PCC 6803	*pdc, adh, ΔfbaA, ΔtktA*	P_{nrsB}, P_{sbaII}	1.2	20	Roussou et al. (2021)
Isobutanol	*S. elongatus* sp. PCC 7942	*alsS, kivD, yqhD, ilvCD, ΔglgC*	PL_{lac}	0.550	8	Li et al. (2014)
	Synechocystis sp. PCC 6803	*alsS, livCD, kivD^{S286T}, adh (slr1192)*	$P_{trcCore}$	0.911	40	Miao et al. (2018)
2,3-Butanediol	*S. elongatus* sp. PCC 7942	*alsS, alsD, adh*	PL_{lacO1}	0.496	3	Oliver et al. (2014)
	Synechococcus sp. PCC 7002	*alsS, alsD, adh*	$P_{clac143}$ ($lacI^{W220F}$)	1.6	16	Nozzi et al. (2017)
1,3-Propanediol	*S. elongatus* sp. PCC 7942	*gpd1, hor2, gdrA-B, dhaB1–3, yqhD Δndh-1*	PL_{lacO1}	0.338	20	Hirokawa et al. (2017b)
1-Butanol	*S. elongatus* sp. PCC 7942	*nphT7, phaB, phaJ, ter, pduP, yqhD*	PL_{lacO1}, P_{trc}	0.404	12	Lan et al. (2013)
	Synechocystis sp. PCC 6803	*pduP, slr1192, ccr, phaJ, pK, ΔphaCE, Δach*	P_{psbA2}, P_{trc}	4.8	30	Liu et al. (2019)
Isopropanol	*S. elongatus* sp. PCC 7942	*thl, atoA, atoD, adc, sadh, pta*	PL_{lacO1}	0.033	14	Hirokawa et al. (2017a)

expressed in cyanobacterium *S. elongatus* 7942 for converting pyruvate into ethanol (Deng and Coleman, 1999) (Fig. 4). Subsequently, increased ethanol productivity was reported in another model cyanobacterium, *Synechocystis* 6803 by introducing *pdc* and *adh* genes (Dexter and Fu, 2009). Efficient, homologous, ethanol-producing enzymes were screened from other organisms and were introduced in the cyanobacteria resulting in a cumulative ethanol titer of 5.5 g/L in a *Synechocystis* 6803 mutant (Gao et al., 2012). In few recent reports, a systemic genetic rewiring of CBB cycle enzymes in *Synechocystis* 6803 showed improved productivity of 1.2 g/L (Liang et al., 2018; Roussou et al., 2021; Velmurugan and Incharoensakdi, 2020). An approach of cofactor supplementation and knocking out glycogen sink pathway has been used in *S. elongatus* 7942 and *Synechococcus* 7002 showing improved productivity

TABLE 2   List of other representative platform chemicals engineered from cyanobacteria along with modified genes, promoters used, and final titers.

Platform chemical	Cyanobacterial strain	Genes over expressed/ deleted	Promoters used	Cumulative titer (g/L)	Cultivation time (in days)	References
Acetone	*Synechococcus elongatus* sp. PCC 7942	*atoB, atoDA, adc, adc, ald6, acs (L641P), pepc, mms*	P_{trc}	0.41	8	Lee et al. (2020)
Isoprene	*Synechocystis* sp. PCC 6803	*fni, IpsS, mk, pmk, pmd, hmS,hmgR, atoB, ΔglgX/ ΔglgA*	P_{psbA2}	250 µg/DCW/D	–	Bentley et al. (2014)
	S. elongatus sp. PCC 7942	*ipsS, dxsS, ispG, idi-E*	P_{trc}, P_{psbA2}, P_{cpcB}	1.26	–	Gao et al. (2016)
Ethylene	*Synechocystis* sp. PCC 6803	*efe, ppsa, pepc*	P_{nrsB}, P_{psbA2}	18.3±3.3 µg/mL/OD	1	Durall et al. (2020)
	S. elongatus sp. PCC 11801	*efe*	$P_{cpcB300}$	338.26 µmole/g dry cell weight/h	12	Sengupta et al. (2020c)
Isobutaraldehyde	*S. elongatus* sp. PCC 7942	*Kivd, alsS, alvC, alvD, pck ΔαPdhB*	P_{trc}, P_{smtA}, P_{trc}	∼0.13 mmol/gDCW/h	2	Cheah et al. (2020)
Succinate	*S. elongatus* sp. PCC 11801	*ogdA, ssaD, ppc, gltA, SBPase, Yjjp, YjjB, Δsdh, ΔglgA*	P_{psbA1}, P_{psbA3}, P_{rbcL}, $P_{cpcB300}$	0.93	5	Sengupta et al. (2020a)
	S. elongatus sp. PCC 7942	*Kgd, gabD, ppc, gltA*	P_{trc}	0.430	8	Lan and Wei (2016)

(Velmurugan and Incharoensakdi, 2020; Wang et al., 2020). Deploying these pathway engineering efforts on a commercial-scale production would still require translational research.

5.1.2 Isobutanol

Production of isobutanol also requires pyruvate as a precursor (Fig. 4). An enzyme ketoisovalerate decarboxylase (kivd) has been studied to be a bottleneck in the pathway and has been subsequently engineered to improve the substrate-binding pocket for higher isobutanol production. The resulting *Synechocystis* 6803 strain expressing heterologous pathway produced around 0.911 g/L isobutanol in 40 days (Miao et al., 2018). The glycogen deficient mutant of another model organism *S. elongatus* 7942 expressing the heterologous

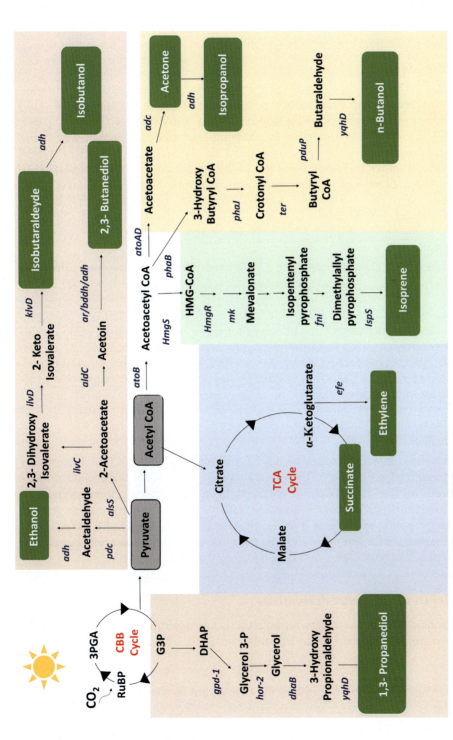

FIG. 4 Systemic overview of various biofuels produced by cyanobacteria: some representative examples of engineered heterologous production pathways are shown along with the enzymes catalyzing the particular reaction (name of the gene encoding the enzyme in *blue color*). *Gray boxes highlight key nodal points in the cyanobacterial metabolic network and green boxes represent the potential biofuels produced. These biofuels are segregated based on the pathways from which their precursors originate in metabolic network and are therefore depicted as differentially colored boxes. Orange: fatty acid synthesis pathway, Green: MVA pathway, Blue: TCA cycle pathway, Pink: glycolysis derived pathway. RuBP, ribulose 1,5-bisphosphate; 3PGA, 3-phosphoglyceric acid; G3P, glyceraldehyde 3-phosphate; DHAP, dihydroxyacetone phosphate; CoA, coenzyme A; HMG, 3-hydroxy-3-methylglutaryl. Adapted from Arora, N., Jaiswal, D., Sengupta, S., Wangikar, P.P., 2020. (Chapter 8) Metabolic engineering of cyanobacteria for production of platform chemicals: a synthetic biology approach. In: Handbook of Algal Science, Technology and Medicine; pp. 127–145). Academic Press. https://doi.org/10.1016/B978-0-12-818305-2.00008-5.*

pathway produced 0.550 g/L isobutanol (Li et al., 2014). However, isobutanol titer needs further pathway modifications for enhanced production.

5.1.3 Isobutaraldehyde

Isobutaraldehyde (IBA) is a precursor for an attractive fuel substitute Isobutanol. *S. elongatus* 7942 has been engineered earlier using ketoisovalerate decarboxylase (kdc) dependent IBA pathway and overexpressing isobutaraldehyde (Atsumi et al., 2009). Later, pyruvate kinase (PK) reaction was identified as potential bottleneck of the pathway using INST-MFA and engineering the PK-bypass improved the specific productivity (Jazmin et al., 2017). Recently, another INST-MFA-guided study predicted the downregulation of pyruvate dehydrogenase (PDH) and phosphoenolpyruvate carboxylase (PPC) as an efficient strategy to relieve the PK bottleneck which improved productivity to around 0.13 mmol/gDCW/h (Cheah et al., 2020).

5.1.4 2,3-Butanediol

Cyanobacteria can be engineered to produce 2,3-butanediol (2,3-BDO) by overexpression of three genes including *alsS* (acetolactate synthase), *aldc* (acetolactate decarboxylase), and *ar* (acetoin reductase) or *bddh* (butanediol dehydrogenase) (Fig. 4). 2,3-BDO titer in *S. elongatus* 7942 was enhanced by optimizing the RBS in the construct (Oliver et al., 2014). *S. elongatus* 7942 have been further engineered to utilize alternative carbon source along with overexpression of heterologous 2,3-BDO pathway, produced 0.761 g/L of 2,3-BDO in 48 h (Kanno et al., 2017). Another model host *Synechococcus* 7002 has also been engineered to produce 2,3-BDO which titers up to 1.6 g/L in 16 days by the screening of various operonic arrangements, light strength, copy number of genes, initial cell density, and nutrient concentration (Nozzi et al., 2017).

5.1.5 1,3-Propanediol

There has been limited research in producing 1,3-propanediol (1,3-PDO) in *S. elongatus* 7942 and *Synechococcus* 7002 (Chin et al., 2014; Hirokawa et al., 2016, 2017b). The heterologous pathway introduced in the cyanobacteria initiates from dihydroxyacetone phosphate (DHAP), an intermediate from Pentose Phosphate pathway and involves four enzymatic steps including the formation of glycerol (Fig. 4). Highest titer of 1,3-PDO of 0.338 g/L was obtained with the genes under the regulation of stronger inducible promoter *S. elongatus* 7942 (Hirokawa et al., 2017b).

5.2 Fatty acid synthesis pathway derivatives

Acetyl-CoA is a main precursor of fatty acids, and also an essential precursor for longer chain alcohol production and acetate-based biofuels.

5.2.1 1-Butanol

Bio-production of butanol has been traditionally carried out by *Clostridium acetobutylicum* via a CoA-dependent pathway (Jones and Woods, 1986). The key enzyme butyryl CoA dehydrogenase is difficult to express in oxygenic, photoautotrophic environment of cyanobacteria. Therefore, production of 1-butanol in cyanobacteria was achieved by engineering a modified CoA-dependent pathway into *S. elongatus* 7942 (Lan and Liao, 2011). The genes of this pathway were assembled from different organisms such as beta-hydroxybutyryl-CoA dehydrogenase (*hbd*), 3-hydroxybutyryl-CoA dehydratase (Crotonase: *crt*), and aldehyde-alcohol dehydrogenase 2 (*adhE2*) from *C. acetobutylicum*, *trans*-2-enoy CoA reductase (*ter*) from *Treponema denticola*, and Thiolase (*atoB*) from *E. coli*. These were integrated into *S. elongatus* 7942 genome for successful proof-of-concept butanol production initiating from acetyl-CoA as a precursor molecule (Fig. 4). The enzymatic activity of these heterologous genes revealed that Ter enzyme could potentially be a rate-limiting step in the butanol production process (Shen et al., 2011). Flux analysis of butanol producing cyanobacterial strain revealed another rate-limiting step of aldehyde dehydrogenase (*PduP*), a penultimate step in the pathway (Lan et al., 2013; Noguchi et al., 2016). There have been a series of genetic engineering trials to overcome these limitations. Screening homologous enzymes from other organisms, engineering enzymes, modifying feeder pathways and nutrient modulations have proved to be attractive strategies for improved productivity in *Synechocystis* 6803 (Anfelt et al., 2015; Liu et al., 2019). With systematic and modular genetic engineering strategy, the highest titers of butanol were reported at around 4.8 g/L after 28 days (Liu et al., 2019).

5.2.2 Isopropanol

Isopropanol is an acetate-derived product produced from acetyl-CoA via four reaction pathways (Fig. 4). Since the production of isopropanol requires acetate as one of the starting metabolites, in many trials the photosynthetically grown cells are transferred in dark conditions for final production. Recently, this challenge was overcome by introducing *pta* gene encoding phosphate acetyltransferase from *E. coli* (Hirokawa et al., 2017a). The cumulative titer of isopropanol achieved in *S. elongatus* 7942 was 33.1 mg/L with intermediate production of another fuel substitute acetone at 12.2 mg/L (Hirokawa et al., 2017a).

5.2.3 Acetone

Acetone can be synthesized using either ATP-driven malonyl-CoA synthesis pathway or by heterologous phosphoketolase (PHK)-phosphotransacetylase (Pta) pathway (Fig. 4). *Synechocystis* 6803 has been engineered for the production of acetone using rational bioengineering to produce 36 mg/L acetone (Zhou et al., 2012). Recently, a group of scientists designed an acetate-acetyl CoA bypass to improve acetyl-CoA pool in *S. elongatus* 7942 and thereby increase the production of acetone to 0.41 g/L which is the highest production reported so far (Lee et al., 2020).

5.3 TCA cycle derivatives

Intermediates of TCA cycle become precursors for important building blocks in the cell including amino acids and are well regulated naturally. However, lower carbon flux towards TCA would eventually limit the titer of the intermediates. Significant background engineering would be required to enhance their pool (Xiong et al., 2017; Zhang and Bryant, 2011).

5.3.1 Succinate

S. elongatus 7942 was engineered for the production of succinate by introducing enzymes from different organisms which yielded 430 mg/L succinate (Lan and Wei, 2016). Recently, an improvement in succinate titers was proposed by engineering a novel fast-growing strain *S. elongatus* 11,801 (Sengupta et al., 2020a). The seven genes including α-ketoglutarate decarboxylase (*OgdA*), succinate semialdehyde dehydrogenase (*SsaD*), PPC, SBPase, citrate synthase (*gltA*), and succinate transporters were overexpressed along with the deletion of two genes to enhance the succinate titer. The final mutant strain produced 0.93 g/L succinate in 5 days, which is the highest photoautotrophic titer (Sengupta et al., 2020a).

5.3.2 Ethylene

Cyanobacteria can produce ethylene from 2-oxoglutarate by overexpression of the ethylene-forming enzyme (efe) (Fig. 4). The ethylene productivity was reported as 2463 μL/L/h/OD_{730} with overexpression of four copies of *efe* gene with deletion of nitrogen control A transcription factor (ntcA) in *Synechocystis* 6803 (Mo et al., 2017). Optimization of regulatory elements like promoters and RBS along with the deletion of glycogen sink pathway was observed to increase ethylene productivity in *Synechocystis* 6803 (Veetil et al., 2017). However, the anaplerotic reactions were observed to be the rate-limiting step for ethylene production and can be partially overcome by the overexpression of phosphoenolpyruvate carboxylase. Enhanced carbon assimilation leads to an increase in productivity to ~18 μg/mL/OD/day in *Synechocystis* 6803 (Durall et al., 2020). Recently, a novel fast-growing strain *S. elongatus* 11,801 was engineered to overexpress *efe* gene using CRISPR-Cpf1 markerless genome editing technique to co-produce ethylene (338.26 μmole/g dry cell weight/h) and succinate (1044.18 μmole/g dry cell weight/h) (Sengupta et al., 2020c).

5.4 MEP and MVA pathway derivatives

MEP pathway refers to an alternate route of synthesis of terpenoids that uses pyruvate and glyceraldehyde-3-phosphate as substrates to form isopentenyl diphosphate (IPP) and dimethylallyl diphosphate (DMAPP), precursors of terpenoids via the intermediate, methylerythritol-4-phosphate (Pattanaik and Lindberg, 2015). Whereas MVA pathway requires acetyl-CoA as a feeder molecule and IPP and DMAPP are produced via a series of six enzymatic steps (Fig. 4) (Chaves and Melis, 2018).

5.4.1 Isoprene

The production of isoprene was reported in *Synechocystis* 6803 by the heterologous expression of isoprene synthase (IspS) from *Pueraria montana* (Lindberg et al., 2010). The recombinant *Synechocystis* 6803 strain was able to produce ~50 μg isoprene/g dry cell weight/day. In another report, the expression of *IspS* gene along with the mevalonic acid pathway genes was observed to enhance the isoprene yield by 2.5-fold when compared to recombinant strains with only *IspS* gene (Bentley et al., 2014). Metabolomics guided rational pathway engineering of *S. elongatus* 7942 improved the titer to 1.26 g/L and involved overexpression of IspS along with other background engineering to relieve the metabolic bottleneck in MEP pathway (Gao et al., 2016). Very recently, the energy saved from the impaired photorespiration pathway was used to increase the production of isoprene. The strain with deletion of glycolate dehydroge-nase showed a twofold increase in isoprene productivity (Zhou et al., 2021).

While cyanobacteria exhibit the tremendous potential of producing a wide range of fuel substitutes and platform chemicals, research in the interdisciplinary field of synthetic biology, host engineering, and metabolic flux analysis will be required for improving titers of the chemicals produced.

6 Indian perspective on bioenergy

Indian scientists have been actively working towards the identification of novel cyanobacterial hosts, developing synthetic biology techniques as well as metabolomics of these hosts for converting these organisms into a cellular biofactory. Realizing the unparalleled advantage of searching novel cyanobacterial strains, DBT, India had sponsored the Algal Network Program. Approximately 2000 strains of cyanobacteria and microalgae were collected from diverse regions of India by a team of scientists from several National institutes and Universities.

There have been multiple efforts in engineering model cyanobacterial strains for producing other essential compounds like carotenoid zeaxanthine and heparosan (Sarnaik et al., 2018, 2019). Another research group identified Type I-D CRISPR-Cas system in a photosynthetic bacteria *Anabaena* PCC 7120 for the first time, which can be explored further for efficient genome editing (Kalwani et al., 2020). A group of scientists engineered *Synechococcus* 7002 to produce higher glycogen at ambient CO_2 conditions, which can be useful for improved sugar and biofuel production (Gupta et al., 2020). The robust, fast-growing cyanobacterial strains have been isolated from Powai lake in Mumbai, India and engineered for production of a platform chemical like succinic acid and ethylene (Jaiswal et al., 2018, 2020; Sengupta et al., 2020a, 2020c). The group performed metabolite profiling of these hosts for rational pathway engineering and identification of bottleneck reaction (Sengupta et al., 2020a). These cyanobacterial research groups in India are working in collaboration with each other, industries, and other eminent scientists across the globe for green energy solutions.

7 Challenges and future outlook

The cyanobacterial potential of biofuel production needs further exploration. There have been some inherent challenges in the light-harvesting potential and CO_2 fixation efficiency of cyanobacteria (Knoot et al., 2018; Seth and Wangikar, 2015). Challenges for enhanced production include overcoming metabolic and carbon flux distribution constraints. Photosynthetic microorganisms show a limited pool of acetyl-CoA and need extensive rewiring of central carbon pathway with state-of-art genome editing technologies (Abernathy et al., 2017; Xiong et al., 2015). Multiple genome copies, difficulty in natural transformability when dealing with large size vectors, and resistance to foreign DNA also pose challenges in efficient genetic engineering of these photoautotrophs (Stucken et al., 2013). Therefore, significant efforts are required in developing synthetic biology tools, protocols, and techniques specific to cyanobacteria and making them easily accessible. Another bottleneck for increased biofuel production is the toxicity of the end product. Adaptive Laboratory Evolution has been used as a strategy for improving product tolerance in some cyanobacteria (Srivastava et al., 2021).

Moreover, only a limited number of cyanobacterial species have been adapted in open pond successfully on a larger scale. Therefore, more research needs to be focused on developing genetically engineered strains compatible with industrial bioreactors or open ponds (Zahra et al., 2020). The other aspect which needs attention is the downstream processing for harvesting the product of interest such that the cost is minimized considerably. The introduction of transporters facilitates increased out-flux of products formed in the cell (Lin et al., 2020; Sengupta et al., 2020a). This strategy might not only minimize the product separation cost but will also reduce the product toxicity. Engineering the cells for utilizing wastewater will develop these photoautotrophs as industrially relevant cell factories and in turn, would also help in bioremediation of water bodies.

There have been systematic efforts from scientists and governing bodies across the globe to develop cyanobacterial bio-factories for the production of fuel substitutes. In parallel, there needs to be a research directed towards scaling up and techno-economic analysis of cyanobacterial biofuel production process for translating the lab-scale research to commercial scale. Overall, photoautotrophic renewable production of the essential fuel molecules at a minimalistic cost and in open water bodies by cyanobacteria can be economically and environmentally advantageous.

Acknowledgments

The authors acknowledge grants from the Department of Biotechnology (DBT), Government of India, awarded to P.P.W. towards DBT-PAN IIT Centre for Bioenergy (Grant No: BT/EB/PAN IIT/2012), Mission Innovation Grant (BT/PR31330/PBD/26/721/2019), and the Infrastructure Facility for Advanced Research and Education in Diagnostics (BT/INF/22/SP23026/2017). The authors acknowledge the donation grant from Praj Industries Limited, Pune, India.

Conflict of interest

The authors declare they have no conflicting interests.

References

Abernathy, M.H., Yu, J., Ma, F., Liberton, M., Ungerer, J., Hollinshead, W.D., Gopalakrishnan, S., He, L., Maranas, C.D., Pakrasi, H.B., Allen, D.K., Tang, Y.J., 2017. Deciphering cyanobacterial phenotypes for fast photoautotrophic growth via isotopically nonstationary metabolic flux analysis. Biotechnol. Biofuels 10 (1), 1–13. https://doi.org/10.1186/s13068-017-0958-y.

Anfelt, J., Kaczmarzyk, D., Shabestary, K., Renberg, B., Rockberg, J., Nielsen, J., Uhlén, M., Hudson, E.P., 2015. Genetic and nutrient modulation of acetyl-CoA levels in synechocystis for n-butanol production. Microb. Cell Factories 14 (1), 1–12. https://doi.org/10.1186/s12934-015-0355-9.

Angermayr, S.A., Hellingwerf, K.J., 2013. On the use of metabolic control analysis in the optimization of cyanobacterial biosolar cell factories. J. Phys. Chem. B 117 (38), 11169–11175. https://doi.org/10.1021/jp4013152.

Atsumi, S., Higashide, W., Liao, J.C., 2009. Direct photosynthetic recycling of carbon dioxide to isobutyraldehyde. Nat. Biotechnol. 27 (12). https://doi.org/10.1038/nbt.1586.

Atsumi, S., Wu, T.Y., MacHado, I.M.P., Huang, W.C., Chen, P.Y., Pellegrini, M., Liao, J.C., 2010. Evolution, genomic analysis, and reconstruction of isobutanol tolerance in Escherichia coli. Mol. Syst. Biol. 6 (449), 1–11. https://doi.org/10.1038/msb.2010.98.

Babele, P.K., Young, J.D., 2020. Applications of stable isotope-based metabolomics and fluxomics toward synthetic biology of cyanobacteria. Wiley Interdiscip. Rev. Syst. Biol. Med. 12 (3), 1–19. https://doi.org/10.1002/wsbm.1472.

Behler, J., Vijay, D., Hess, W.R., Akhtar, M.K., 2018. CRISPR-based technologies for metabolic engineering in cyanobacteria. Trends Biotechnol. 36 (10), 996–1010. https://doi.org/10.1016/j.tibtech.2018.05.011.

Bentley, F.K., Zurbriggen, A., Melis, A., 2014. Heterologous expression of the mevalonic acid pathway in cyanobacteria enhances endogenous carbon partitioning to isoprene. Mol. Plant 7 (1), 71–86. https://doi.org/10.1093/mp/sst134.

Berla, B.M., Saha, R., Immethun, C.M., Maranas, C.D., Moon, T.S., Pakrasi, H.B., 2013. Synthetic biology of cyanobacteria: unique challenges and opportunities. Front. Microbiol. 4 (August), 1–14. https://doi.org/10.3389/fmicb.2013.00246.

Bozell, J.J., Petersen, G.R., 2010. Cutting-edge research for a greener sustainable future technology development for the production of biobased products from biorefinery carbohydrates—the US Department of Energy's "top 10" revisited. Green Chem. 12 (4). https://doi.org/10.1039/b922014c.

Buescher, J.M., Antoniewicz, M.R., Boros, L.G., Burgess, S.C., Brunengraber, H., Clish, C.B., DeBerardinis, R.J., Feron, O., Frezza, C., Ghesquiere, B., Gottlieb, E., Hiller, K., Jones, R.G., Kamphorst, J.J., Kibbey, R.G., Kimmelman, A.C., Locasale, J.W., Lunt, S.Y., Maddocks, O.D.K., Fendt, S.M., 2015. A roadmap for interpreting ^{13}C metabolite labeling patterns from cells. Curr. Opin. Biotechnol. 34, 189–201. https://doi.org/10.1016/j.copbio.2015.02.003.

Cai, F., Axen, S.D., Kerfeld, C.A., 2013. Evidence for the widespread distribution of CRISPR-cas system in the phylum cyanobacteria. RNA Biol. 10 (5), 687–693. https://doi.org/10.4161/rna.24571.

Canton, B., Labno, A., Endy, D., 2008. Refinement and standardization of synthetic biological parts and devices. Nat. Biotechnol. 26 (7), 787–793. https://doi.org/10.1038/nbt1413.

Chaves, J.E., Melis, A., 2018. Engineering isoprene synthesis in cyanobacteria. FEBS Lett. 592 (12), 2059–2069. https://doi.org/10.1002/1873-3468.13052.

Cheah, Y.E., Xu, Y., Sacco, S.A., Babele, P.K., Zheng, A.O., Johnson, C.H., Young, J.D., 2020. Systematic identification and elimination of flux bottlenecks in the aldehyde production pathway of *Synechococcus elongatus* PCC 7942. Metab. Eng. 60 (February), 56–65. https://doi.org/10.1016/j.ymben.2020.03.007.

Chen, Y.-J., Liu, P., Nielsen, A.A.K., Brophy, J.A.N., Clancy, K., Peterson, T., Voigt, C.A., 2013. Characterization of 582 natural and synthetic terminators and quantification of their design constraints. Nat. Methods 10 (7), 659–664. https://doi.org/10.1038/nmeth.2515.

Chin, J.W., Anderson, M.A., Cui, J., Spieker, M., 2014. Production of 1,3-propanediol in cyanobacteria (Patent No. WO 2014/062997 Al).

Copeland, M.F., Politz, M.C., Pfleger, B.F., 2014. Application of TALEs, CRISPR/Cas and sRNAs as trans-acting regulators in prokaryotes. Curr. Opin. Biotechnol. 29, 46–54. https://doi.org/10.1016/j.copbio.2014.02.010.

Dange, M.C., Mishra, V., Mukherjee, B., Jaiswal, D., Merchant, M.S., Prasannan, C.B., Wangikar, P.P., 2020. Evaluation of freely available software tools for untargeted quantification of ^{13}C isotopic enrichment in cellular metabolome from HR-LC/MS data. Metab. Eng. Commun. 10 (December 2019). https://doi.org/10.1016/j.mec.2019.e00120, e00120.

Deng, M.D., Coleman, J.R., 1999. Ethanol synthesis by genetic engineering in cyanobacteria. Appl. Environ. Microbiol. 65 (2), 523–528. https://doi.org/10.1128/aem.65.2.523-528.1999.

Dexter, J., Fu, P., 2009. Metabolic engineering of cyanobacteria for ethanol production. Energy Environ. Sci. 2 (8), 857–864. https://doi.org/10.1039/b811937f.

Dong, G., Golden, S.S., 2009. How cyanobacteria tells time. Curr. Opin. Microbiol. 11 (6), 541–546. https://doi.org/10.1016/j.mib.2008.10.003.How.

Durall, C., Lindberg, P., Yu, J., Lindblad, P., 2020. Increased ethylene production by overexpressing phosphoenolpyruvate carboxylase in the cyanobacterium *Synechocystis* PCC 6803. Biotechnol. Biofuels 13 (1), 1–13. https://doi.org/10.1186/s13068-020-1653-y.

Endy, D., 2005. Foundations for engineering biology. Nature 438 (7067), 449–453. https://doi.org/10.1038/nature04342.

Englund, E., Liang, F., Lindberg, P., 2016. Evaluation of promoters and ribosome binding sites for biotechnological applications in the unicellular cyanobacterium *Synechocystis* sp. PCC 6803. Sci. Rep. 6 (1), 36640. https://doi.org/10.1038/srep36640.

Englund, E., Shabestary, K., Hudson, E.P., Lindberg, P., 2018. Systematic overexpression study to find target enzymes enhancing production of terpenes in *Synechocystis* PCC 6803, using isoprene as a model compound. Metab. Eng. 49 (April), 164–177. https://doi.org/10.1016/j.ymben.2018.07.004.

Gao, Z., Zhao, H., Li, Z., Tan, X., Lu, X., 2012. Photosynthetic production of ethanol from carbon dioxide in genetically engineered cyanobacteria. Energy Environ. Sci. 5 (12), 9857–9865. https://doi.org/10.1039/c2ee22675h.

Gao, X., Gao, F., Liu, D., Zhang, H., Nie, X., Yang, C., 2016. Engineering the methylerythritol phosphate pathway in cyanobacteria for photosynthetic isoprene production from CO_2. Energy Environ. Sci. 9 (4), 1400–1411. https://doi.org/10.1039/c5ee03102h.

Garcia-Pichel, F., 2009. Cyanobacteria. In: Schaechter, M. (Ed.), Encyclopedia of Microbiology, third ed. Academic Press, pp. 107–124, https://doi.org/10.1016/B978-012373944-5.00250-9.

Gibson, D.G., Young, L., Chuang, R.-Y., Venter, J.C., Hutchison III, C.A., Smith, H.O., 2009. Enzymatic assembly of DNA molecules up to several hundred kilobases. Nat. Methods 6 (5), 343–345. https://doi.org/10.1038/nmeth.1318.

Golden, S.S., 2019. The international journeys and aliases of *Synechococcus elongatus*. N. Z. J. Bot. 57 (2), 70–75. https://doi.org/10.1080/0028825X.2018.1551805.

Gordon, G.C., Korosh, T.C., Cameron, J.C., Markley, A.L., Begemann, M.B., Pfleger, B.F., 2016. CRISPR interference as a titratable, trans -acting regulatory tool for metabolic engineering in the cyanobacterium *Synechococcus* sp. strain PCC 7002. Metab. Eng. 38, 170–179. https://doi.org/10.1016/j.ymben.2016.07.007.

Guo, W., Sheng, J., Feng, X., 2016. ^{13}C-metabolic flux analysis: an accurate approach to demystify microbial metabolism for biochemical production. Bioengineering 3 (1). https://doi.org/10.3390/bioengineering3010003.

Gupta, J.K., Rai, P., Jain, K.K., Srivastava, S., 2020. Overexpression of bicarbonate transporters in the marine cyanobacterium *Synechococcus* sp. PCC 7002 increases growth rate and glycogen accumulation. Biotechnol. Biofuels 13 (1), 1–12. https://doi.org/10.1186/s13068-020-1656-8.

Haselkorn, R., 2009. Cyanobacteria. Curr. Biol. 19 (7), 277–278. https://doi.org/10.1016/j.cub.2009.01.016.

Hays, S.G., Ducat, D.C., 2015. Engineering cyanobacteria as photosynthetic feedstock factories. Photosynth. Res. 123 (3), 285–295. https://doi.org/10.1007/s11120-014-9980-0.

He, L., Xiao, Y., Gebreselassie, N., Zhang, F., Antoniewicz, M.R., Tang, Y.J., Peng, L., 2014. Central metabolic responses to the overproduction of fatty acids in Escherichia coli based on ^{13}C-metabolic flux analysis. Biotechnol. Bioeng. 111 (3), 575–585. https://doi.org/10.1002/bit.25124.

Hedges, S.B., Chen, H., Kumar, S., Wang, D.Y.C., Thompson, A.S., Watanabe, H., 2001. A genomic timescale for the origin of eukaryotes. BMC Evol. Biol. 1. https://doi.org/10.1186/1471-2148-1-4.

Heidorn, T., Camsund, D., Huang, H., Lindberg, P., Lindblad, P., Oliveira, P., Stensjo, K., 2011. Synthetic biology in cyanobacteria: engineering and analyzing novel functions. In: Methods in Enzymology. vol. 497. Elsevier, https://doi.org/10.1016/B978-0-12-385075-1.00024-X.

Hendry, J.I., Prasannan, C., Ma, F., Möllers, K.B., Jaiswal, D., Digmurti, M., Allen, D.K., Frigaard, N.U., Dasgupta, S., Wangikar, P.P., 2017. Rerouting of carbon flux in a glycogen mutant of cyanobacteria assessed via isotopically non-stationary ^{13}C metabolic flux analysis. Biotechnol. Bioeng. 114 (10), 2298–2308. https://doi.org/10.1002/bit.26350.

Hentschel, E., Will, C., Mustafi, N., Burkovski, A., Rehm, N., Frunzke, J., 2013. Destabilized eYFP variants for dynamic gene expression studies in *Corynebacterium glutamicum*. Microb. Biotechnol. 6 (2), 196–201. https://doi.org/10.1111/j.1751-7915.2012.00360.x.

Hirokawa, Y., Maki, Y., Tatsuke, T., Hanai, T., 2016. Cyanobacterial production of 1,3-propanediol directly from carbon dioxide using a synthetic metabolic pathway. Metab. Eng. 34, 97–103. https://doi.org/10.1016/j.ymben.2015.12.008.

Hirokawa, Y., Dempo, Y., Fukusaki, E., Hanai, T., 2017a. Metabolic engineering for isopropanol production by an engineered cyanobacterium, *Synechococcus elongatus* PCC 7942, under photosynthetic conditions. J. Biosci. Bioeng. 123 (1), 39–45. https://doi.org/10.1016/j.jbiosc.2016.07.005.

Hirokawa, Y., Matsuo, S., Hamada, H., Matsuda, F., Hanai, T., 2017b. Metabolic engineering of *Synechococcus elongatus* PCC 7942 for improvement of 1,3-propanediol and glycerol production based on in silico simulation of metabolic flux distribution. Microb. Cell Factories 16 (1), 1–12. https://doi.org/10.1186/s12934-017-0824-4.

Hirokawa, Y., Kubo, T., Soma, Y., Saruta, F., Hanai, T., 2020. Enhancement of acetyl-CoA flux for photosynthetic chemical production by pyruvate dehydrogenase complex overexpression in *Synechococcus elongatus* PCC 7942. Metab. Eng. 57 (July 2019), 23–30. https://doi.org/10.1016/j.ymben.2019.07.012.

Hitchcock, A., Hunter, C.N., Canniffe, D.P., 2020. Progress and challenges in engineering cyanobacteria as chassis for light-driven biotechnology. Microb. Biotechnol. 13 (2), 363–367. https://doi.org/10.1111/1751-7915.13526.

Huang, H.-H., Lindblad, P., 2013. Wide-dynamic-range promoters engineered for cyanobacteria. J. Biol. Eng. 7 (1), 10. https://doi.org/10.1186/1754-1611-7-10.

Huang, H.-H., Camsund, D., Lindblad, P., Heidorn, T., 2010. Design and characterization of molecular tools for a synthetic biology approach towards developing cyanobacterial biotechnology. Nucleic Acids Res. 38 (8), 2577–2593. https://doi.org/10.1093/nar/gkq164.

Issa, A., Hemida, M., Ohyam, T., 2014. Nitrogen fixing cyanobacteria: future prospect. In: Advances in Biology and Ecology of Nitrogen Fixation. IntechOpen, https://doi.org/10.5772/56995.

Jahn, S., Haverkorn van Rijsewijk, B.R., Sauer, U., Bettenbrock, K., 2013. A role for EIIANtr in controlling fluxes in the central metabolism of *E. coli* K12. Biochim. Biophys. Acta, Mol. Cell Res. 1833 (12), 2879–2889. https://doi.org/10.1016/j.bbamcr.2013.07.011.

Jaiswal, D., Wangikar, P.P., 2020. Dynamic inventory of intermediate metabolites of cyanobacteria in a diurnal cycle. iScience 23 (11). https://doi.org/10.1016/j.isci.2020.101704, 101704.

Jaiswal, D., Sengupta, A., Sohoni, S., Sengupta, S., Phadnavis, A.G., Pakrasi, H.B., Wangikar, P.P., 2018. Genome features and biochemical characteristics of a robust, fast growing and naturally transformable cyanobacterium *Synechococcus elongatus* PCC 11801 isolated from India. Sci. Rep. 8 (1), 1–13. https://doi.org/10.1038/s41598-018-34872-z.

Jaiswal, D., Sengupta, A., Sengupta, S., Madhu, S., Pakrasi, H.B., Wangikar, P.P., 2020. A novel cyanobacterium *Synechococcus elongatus* PCC 11802 has distinct genomic and metabolomic characteristics compared to its neighbor PCC 11801. Sci. Rep. 10 (1), 1–15. https://doi.org/10.1038/s41598-019-57051-0.

Jazmin, L.J., Young, J.D., 2013. Isotopically non-stationary 13C MEtabolic flux. Analysis 985 (4), 113–121. https://doi.org/10.1007/978-1-62703-299-5.

Jazmin, L.J., Xu, Y., Cheah, Y.E., Adebiyi, A.O., Johnson, C.H., Young, J.D., 2017. Isotopically nonstationary[13]C flux analysis of cyanobacterial isobutyraldehyde production. Metab. Eng. 42 (October 2016), 9–18. https://doi.org/10.1016/j.ymben.2017.05.001.

Johnson, T.J., Zhou, R., Johnson, J.G., 2016. Outlook on the potential of cyanobacteria to photosynthetically produce high-value chemicals and biofuels at an industrial scale. Bioenergetics: Open Access 5 (2). https://doi.org/10.4172/2167-7662.1000142.

Jones, D.T., Woods, D.R., 1986. Acetone-butanol fermentation revisited. Microbiol. Rev. 50 (4), 484–524. https://doi.org/10.1128/mr.50.4.484-524.1986.

Kalwani, P., Rath, D., Ballal, A., 2020. Novel molecular aspects of the CRISPR backbone protein "Cas7" from cyanobacteria. Biochem. J. 477 (5), 971–983. https://doi.org/10.1042/BCJ20200026.

Kaneko, T., Sato, S., Kotani, H., Tanaka, A., Asamizu, E., Nakamura, Y., Miyajima, N., Hirosawa, M., Sugiura, M., Sasamoto, S., Kimura, T., Hosouchi, T., Matsuno, A., Muraki, A., Nakazaki, N., Naruo, K., Okumura, S., Shimpo, S., Takeuchi, C., Tabata, S., 1996. Sequence analysis of the genome of the unicellular cyanobacterium synechocystis sp. strain PCC6803. II. Sequence determination of the entire genome and assignment of potential protein-coding regions (supplement). DNA Res. 3 (3), 185–209. https://doi.org/10.1093/dnares/3.3.185.

Kanno, M., Carroll, A.L., Atsumi, S., 2017. Global metabolic rewiring for improved CO_2 fixation and chemical production in cyanobacteria. Nat. Commun. 8, 1–11. https://doi.org/10.1038/ncomms14724.

Kaygusuz, K., 2009. Bioenergy as a clean and sustainable fuel. Energy Sources Pt A 31 (12), 1069–1080. https://doi.org/10.1080/15567030801909839.

Kim, J., Salvador, M., Saunders, E., González, J., Avignone-Rossa, C., Jiménez, J.I., 2016. Properties of alternative microbial hosts used in synthetic biology: towards the design of a modular chassis. Essays Biochem. 60 (4), 303–313. https://doi.org/10.1042/EBC20160015.

Kim, W.J., Lee, S.-M., Um, Y., Sim, S.J., Woo, H.M., 2017. Development of SyneBrick vectors as a synthetic biology platform for gene expression in *Synechococcus elongatus* PCC 7942. Front. Plant Sci. 8 (March), 1–9. https://doi.org/10.3389/fpls.2017.00293.

Kitchener, R.L., Grunden, A.M., 2018. Methods for enhancing cyanobacterial stress tolerance to enable improved production of biofuels and industrially relevant chemicals. Appl. Microbiol. Biotechnol. 102 (4), 1617–1628. https://doi.org/10.1007/s00253-018-8755-5.

Knoot, C.J., Ungerer, J., Wangikar, P.P., Pakrasi, H.B., 2018. Cyanobacteria: promising biocatalysts for sustainable chemical production. J. Biol. Chem. 293 (14), 5044–5052. https://doi.org/10.1074/jbc.R117.815886.

Knoot, C.J., Biswas, S., Pakrasi, H.B., 2020. Tunable repression of key photosynthetic processes using Cas12a CRISPR interference in the fast-growing cyanobacterium *Synechococcus* sp. UTEX 2973. ACS Synth. Biol. 9 (1), 132–143. https://doi.org/10.1021/acssynbio.9b00417.

Kopka, J., Schmidt, S., Dethloff, F., Pade, N., Berendt, S., Schottkowski, M., Martin, N., Dühring, U., Kuchmina, E., Enke, H., Kramer, D., Wilde, A., Hagemann, M., Friedrich, A., 2017. Systems analysis of ethanol production in the genetically engineered cyanobacterium *Synechococcus* sp. PCC 7002. Biotechnol. Biofuels 10 (1), 1–21. https://doi.org/10.1186/s13068-017-0741-0.

Lan, E.I., Liao, J.C., 2011. Metabolic engineering of cyanobacteria for 1-butanol production from carbon dioxide. Metab. Eng. 13 (4), 353–363. https://doi.org/10.1016/j.ymben.2011.04.004.

Lan, E.I., Wei, C.T., 2016. Metabolic engineering of cyanobacteria for the photosynthetic production of succinate. Metab. Eng. 38 (October), 483–493. https://doi.org/10.1016/j.ymben.2016.10.014.

Lan, E.I., Ro, S.Y., Liao, J.C., 2013. Oxygen-tolerant coenzyme A-acylating aldehyde dehydrogenase facilitates efficient photosynthetic n-butanol biosynthesis in cyanobacteria. Energy Environ. Sci. 6 (9), 2672–2681. https://doi.org/10.1039/c3ee41405a.

Lau, N.S., Matsui, M., Abdullah, A.A.A., 2015. Cyanobacteria: photoautotrophic microbial factories for the sustainable synthesis of industrial products. Biomed. Res. Int. 2015. https://doi.org/10.1155/2015/754934.

Lee, H.J., Son, J., Sim, S.J., Woo, H.M., 2020. Metabolic rewiring of synthetic pyruvate dehydrogenase bypasses for acetone production in cyanobacteria. Plant Biotechnol. J. 18 (9), 1860–1868. https://doi.org/10.1111/pbi.13342.

Li, X., Shen, C.R., Liao, J.C., 2014. Isobutanol production as an alternative metabolic sink to rescue the growth deficiency of the glycogen mutant of *Synechococcus elongatus* PCC 7942. Photosynth. Res. 120 (3), 301–310. https://doi.org/10.1007/s11120-014-9987-6.

Li, H., Shen, C.R., Huang, C., Sung, L., Wu, M., Hu, Y., 2016. CRISPR-Cas9 for the genome engineering of cyanobacteria and succinate production. Metab. Eng. 38 (August), 293–302. https://doi.org/10.1016/j.ymben.2016.09.006.

Liang, F., Englund, E., Lindberg, P., Lindblad, P., 2018. Engineered cyanobacteria with enhanced growth show increased ethanol production and higher biofuel to biomass ratio. Metab. Eng. 46 (November 2017), 51–59. https://doi.org/10.1016/j.ymben.2018.02.006.

Lin, P.C., Saha, R., Zhang, F., Pakrasi, H.B., 2017. Metabolic engineering of the pentose phosphate pathway for enhanced limonene production in the cyanobacterium *Synechocystis* sp. PCC. Sci. Rep. 7 (1), 1–10. https://doi.org/10.1038/s41598-017-17831-y.

Lin, P.C., Zhang, F., Pakrasi, H.B., 2020. Enhanced production of sucrose in the fast-growing cyanobacterium *Synechococcus elongatus* UTEX 2973. Sci. Rep. 10 (1), 1–8. https://doi.org/10.1038/s41598-019-57319-5.

Lindberg, P., Park, S., Melis, A., 2010. Engineering a platform for photosynthetic isoprene production in cyanobacteria, using synechocystis as the model organism. Metab. Eng. 12 (1), 70–79. https://doi.org/10.1016/j.ymben.2009.10.001.

Liu, D., Pakrasi, H.B., 2018. Exploring native genetic elements as plug-in tools for synthetic biology in the cyanobacterium *Synechocystis* sp. PCC 6803. Microb. Cell Factories 17 (1), 48. https://doi.org/10.1186/s12934-018-0897-8.

Liu, X., Miao, R., Lindberg, P., Lindblad, P., 2019. Modular engineering for efficient photosynthetic biosynthesis of 1-butanol from CO_2 in cyanobacteria. Energy Environ. Sci. 12 (9), 2765–2777. https://doi.org/10.1039/c9ee01214a.

Liu, D., Johnson, V.M., Pakrasi, H.B., 2020. A reversibly induced CRISPRi system targeting photosystem II in the cyanobacterium *Synechocystis* sp. PCC 6803. BioRxiv, 1–16. https://doi.org/10.1101/2020.03.24.005744.

Luan, G., Zhang, S., Lu, X., 2020. Engineering cyanobacteria chassis cells toward more efficient photosynthesis. Curr. Opin. Biotechnol. 62, 1–6. https://doi.org/10.1016/j.copbio.2019.07.004.

Ma, X., Zhu, Q., Chen, Y., Liu, Y.-G., Y-g, L., 2016. CRISPR/Cas9 platforms for genome editing in plants: developments and applications. Mol. Plant 9 (April), 1–32. https://doi.org/10.1016/j.molp.2016.04.009.

Markley, A.L., Begemann, M.B., Clarke, R.E., Gordon, G.C., Pfleger, B.F., 2015. Synthetic biology toolbox for controlling gene expression in the cyanobacterium *Synechococcus* sp. strain PCC 7002. ACS Synth. Biol. 4 (5), 595–603. https://doi.org/10.1021/sb500260k.

Matsusako, T., Toya, Y., Yoshikawa, K., Shimizu, H., 2017. Identification of alcohol stress tolerance genes of *Synechocystis* sp. PCC 6803 using adaptive laboratory evolution. Biotechnol. Biofuels 10 (1), 1–9. https://doi.org/10.1186/s13068-017-0996-5.

Miao, R., Xie, H., Lindblad, P., 2018. Enhancement of photosynthetic isobutanol production in engineered cells of *Synechocystis* PCC 6803. Biotechnol. Biofuels 11 (1), 1–9. https://doi.org/10.1186/s13068-018-1268-8.

Mo, H., Xie, X., Zhu, T., Lu, X., 2017. Effects of global transcription factor NtcA on photosynthetic production of ethylene in recombinant *Synechocystis* sp. PCC 6803. Biotechnol. Biofuels 10 (1), 1–13. https://doi.org/10.1186/s13068-017-0832-y.

Mukherjee, B., Madhu, S., Wangikar, P.P., 2020. The role of systems biology in developing non-model cyanobacteria as hosts for chemical production. Curr. Opin. Biotechnol. 64, 62–69. https://doi.org/10.1016/j.copbio.2019.10.003.

Nakahira, Y., Ogawa, A., Asano, H., Oyama, T., Tozawa, Y., 2013. Theophylline-dependent riboswitch as a novel genetic tool for strict regulation of protein expression in cyanobacterium *Synechococcus elongatus* PCC 7942. Plant Cell Physiol. 54 (10), 1724–1735. https://doi.org/10.1093/pcp/pct115.

Ng, A.H., Berla, B.M., Pakrasi, H.B., 2015. Fine-tuning of photoautotrophic protein production by combining promoters and neutral sites in the cyanobacterium *Synechocystis* sp. strain PCC 6803. Appl. Environ. Microbiol. 81 (19), 6857–6863. https://doi.org/10.1128/AEM.01349-15.

Niu, T.C., Lin, G.M., Xie, L.R., Wang, Z.Q., Xing, W.Y., Zhang, J.Y., Zhang, C.C., 2019. Expanding the potential of CRISPR-Cpf1-based genome editing technology in the cyanobacterium *Anabaena* PCC 7120. ACS Synth. Biol. 8 (1), 170–180. https://doi.org/10.1021/acssynbio.8b00437.

Noguchi, S., Putri, S.P., Lan, E.I., Laviña, W.A., Dempo, Y., Bamba, T., Liao, J.C., Fukusaki, E., 2016. Quantitative target analysis and kinetic profiling of acyl-CoAs reveal the rate-limiting step in cyanobacterial 1-butanol production. Metabolomics 12 (2), 1–10. https://doi.org/10.1007/s11306-015-0940-2.

Noreña-Caro, D., Benton, M.G., 2018. Cyanobacteria as photoautotrophic biofactories of high-value chemicals. J. CO_2 Util. 28 (September), 335–366. https://doi.org/10.1016/j.jcou.2018.10.008.

Nozzi, N.E., Oliver, J.W.K., Atsumi, S., 2013. Cyanobacteria as a platform for biofuel production. Front. Bioeng. Biotechnol. 1 (August). https://doi.org/10.3389/fbioe.2013.00007.

Nozzi, N.E., Case, A.E., Carroll, A.L., Atsumi, S., 2017. Systematic approaches to efficiently produce 2,3-butanediol in a marine cyanobacterium. ACS Synth. Biol. 6 (11), 2136–2144. https://doi.org/10.1021/acssynbio.7b00157.

Oliver, J.W.K., Machado, I.M.P., Yoneda, H., Atsumi, S., 2013. Cyanobacterial conversion of carbon dioxide to 2,3-butanediol. Proc. Natl. Acad. Sci. 110 (4), 1249–1254. https://doi.org/10.1073/pnas.1213024110.

Oliver, J.W.K., Machado, I.M.P., Yoneda, H., Atsumi, S., 2014. Combinatorial optimization of cyanobacterial 2,3-butanediol production. Metab. Eng. 22, 76–82. https://doi.org/10.1016/j.ymben.2014.01.001.

Pattanaik, B., Lindberg, P., 2015. Terpenoids and their biosynthesis in cyanobacteria. Life 5 (1), 269–293. https://doi.org/10.3390/life5010269.

Prasannan, C.B., Mishra, V., Jaiswal, D., Wangikar, P.P., 2019. Mass isotopologue distribution of dimer ion adducts of intracellular metabolites for potential applications in ^{13}C metabolic flux analysis. PLoS One 14 (8), 1–16. https://doi.org/10.1371/journal.pone.0220412.

Quintana, N., Van Der Kooy, F., Van De Rhee, M.D., Voshol, G.P., Verpoorte, R., 2011. Renewable energy from cyanobacteria: energy production optimization by metabolic pathway engineering. Appl. Microbiol. Biotechnol. 91 (3), 471–490. https://doi.org/10.1007/s00253-011-3394-0.

Ramey, C.J., Barón-Sola, Á., Aucoin, H.R., Boyle, N.R., 2015. Genome engineering in cyanobacteria: where we are and where we need to go. ACS Synth. Biol. 4 (11), 1186–1196. https://doi.org/10.1021/acssynbio.5b00043.

Ronda, C., Pedersen, L.E., Sommer, M.O.A., Nielsen, A.T., 2016. CRMAGE: CRISPR optimized MAGE recombineering. Sci. Rep. 6 (1), 1–11. https://doi.org/10.1038/srep19452.

Roussou, S., Albergati, A., Liang, F., Lindblad, P., 2021. Engineered cyanobacteria with additional overexpression of selected Calvin-Benson-Bassham enzymes show further increased ethanol production. Metab. Eng. Commun. 12 (December 2020). https://doi.org/10.1016/j.mec.2021.e00161, e00161.

Ruffing, A.M., 2014. Improved free fatty acid production in cyanobacteria with *Synechococcus* sp. PCC 7002 as host. Front. Bioeng. Biotechnol. 2 (May), 1–10. https://doi.org/10.3389/fbioe.2014.00017.

Sarnaik, A., Nambissan, V., Pandit, R., Lali, A., 2018. Recombinant *Synechococcus elongatus* PCC 7942 for improved zeaxanthin production under natural light conditions. Algal Res. 36 (April), 139–151. https://doi.org/10.1016/j.algal.2018.10.021.

Sarnaik, A., Abernathy, M.H., Han, X., Ouyang, Y., Xia, K., Chen, Y., Cress, B., Zhang, F., Lali, A., Pandit, R., Linhardt, R.J., Tang, Y.J., Koffas, M.A.G., 2019. Metabolic engineering of cyanobacteria for photoautotrophic production of heparosan, a pharmaceutical precursor of heparin. Algal Res. 37 (November 2018), 57–63. https://doi.org/10.1016/j.algal.2018.11.010.

Sengupta, A., Pakrasi, H.B., Wangikar, P.P., 2018. Recent advances in synthetic biology of cyanobacteria. Appl. Microbiol. Biotechnol. 102 (13), 5457–5471. https://doi.org/10.1007/s00253-018-9046-x.

Sengupta, A., Sunder, A.V., Sohoni, S.V., Wangikar, P.P., 2019. Fine-tuning native promoters of *Synechococcus elongatus* PCC 7942 to develop a synthetic toolbox for heterologous protein expression [brief-report]. ACS Synth. Biol. 8 (5), 1219–1223. https://doi.org/10.1021/acssynbio.9b00066.

Sengupta, S., Jaiswal, D., Sengupta, A., Shah, S., Gadagkar, S., Wangikar, P.P., 2020a. Metabolic engineering of a fast-growing cyanobacterium *Synechococcus elongatus* PCC 11801 for photoautotrophic production of succinic acid. Biotechnol. Biofuels 13 (1), 1–18. https://doi.org/10.1186/s13068-020-01727-7.

Sengupta, A., Madhu, S., Wangikar, P.P., 2020b. A library of tunable, portable, and inducer-free promoters derived from cyanobacteria. ACS Synth. Biol. 9 (7), 1790–1801. https://doi.org/10.1021/acssynbio.0c00152.

Sengupta, A., Pritam, P., Jaiswal, D., Bandyopadhyay, A., Pakrasi, H.B., Wangikar, P.P., 2020c. Photosynthetic co-production of succinate and ethylene in a fast-growing cyanobacterium, *Synechococcus elongatus* PCC 11801. Meta 10 (6), 1–22. https://doi.org/10.3390/metabo10060250.

Seth, J.R., Wangikar, P.P., 2015. Challenges and opportunities for microalgae-mediated CO_2 capture and biorefinery. Biotechnol. Bioeng. 112 (7), 1281–1296. https://doi.org/10.1002/bit.25619.

Shastri, A.A., Morgan, J.A., 2007. A transient isotopic labeling methodology for 13C metabolic flux analysis of photoautotrophic microorganisms. Phytochemistry 68 (16–18), 2302–2312. https://doi.org/10.1016/j.phytochem.2007.03.042.

Shen, C.R., Lan, E.I., Dekishima, Y., Baez, A., Cho, K.M., Liao, J.C., 2011. Driving forces enable high-titer anaerobic 1-butanol synthesis in *Escherichia coli*. Appl. Environ. Microbiol. 77 (9), 2905–2915. https://doi.org/10.1128/AEM.03034-10.

Song, Z., Chen, L., Wang, J., Lu, Y., Jiang, W., Zhang, W., 2014. A transcriptional regulator Sll0794 regulates tolerance to biofuel ethanol in photosynthetic *Synechocystis* sp. PCC 6803. Mol. Cell. Proteomics 13 (12), 3519–3532. https://doi.org/10.1074/mcp.M113.035675.

Song, X., Diao, J., Yao, J., Cui, J., Sun, T., Chen, L., Zhang, W., 2021. Engineering a central carbon metabolism pathway to increase the intracellular acetyl-CoA Pool in *Synechocystis* sp. PCC 6803 grown under photomixotrophic conditions. ACS Synth. Biol. 10 (4), 836–846. https://doi.org/10.1021/acssynbio.0c00629.

Srivastava, V., Amanna, R., Rowden, S.J.L., Sengupta, S., Madhu, S., Howe, C.J., Wangikar, P.P., 2021. Adaptive laboratory evolution of the fast-growing cyanobacterium *Synechococcus elongatus* PCC 11801 for improved solvent tolerance. J. Biosci. Bioeng. 131 (5), 491–500. https://doi.org/10.1016/j.jbiosc.2020.11.012.

Stucken, K., Koch, R., Dagan, T., 2013. Cyanobacterial defense mechanisms against foreign DNA transfer and their impact on genetic engineering. Biol. Res. 46 (4), 373–382. https://doi.org/10.4067/S0716-97602013000400009.

Taton, A., Ma, A.T., Ota, M., Golden, S.S., Golden, J.W., 2017. NOT gate genetic circuits to control gene expression in cyanobacteria. ACS Synth. Biol. https://doi.org/10.1021/acssynbio.7b00203.

Taylor, G.M., Mordaka, P.M., Heap, J.T., 2019. Start-stop assembly: a functionally scarless DNA assembly system optimized for metabolic engineering. Nucleic Acids Res. 47 (3). https://doi.org/10.1093/nar/gky1182.

Till, P., Toepel, J., Bühler, B., Mach, R.L., Mach-Aigner, A.R., 2020. Regulatory systems for gene expression control in cyanobacteria. Appl. Microbiol. Biotechnol. 104 (5), 1977–1991. https://doi.org/10.1007/s00253-019-10344-w.

Ungerer, J., Pakrasi, H.B., 2016. Cpf1 is a versatile tool for CRISPR genome editing across diverse species of cyanobacteria. Sci. Rep. 6 (October), 39681. https://doi.org/10.1038/srep39681.

Ungerer, J., Wendt, K.E., Hendry, J.I., Maranas, C.D., Pakrasi, H.B., 2018. Comparative genomics reveals the molecular determinants of rapid growth of the cyanobacterium *Synechococcus elongatus* UTEX 2973. Proc. Natl. Acad. Sci. 115 (50), E11761–E11770. https://doi.org/10.1073/pnas.1814912115.

Vasudevan, R., Gale, G.A.R., Schiavon, A.A., Puzorjov, A., Malin, J., Gillespie, M.D., Vavitsas, K., Zulkower, V., Wang, B., Howe, C.J., Lea-Smith, D.J., McCormick, A.J., 2019. CyanoGate: a modular cloning suite for engineering cyanobacteria based on the plant MoClo syntax. Plant Physiol. 180 (1), 39–55. https://doi.org/10.1104/pp.18.01401.

Veetil, V.P., Angermayr, S.A., Hellingwerf, K.J., 2017. Ethylene production with engineered *Synechocystis* sp PCC 6803 strains. Microb. Cell Factories 16 (1), 1–11. https://doi.org/10.1186/s12934-017-0645-5.

Velmurugan, R., Incharoensakdi, A., 2020. Heterologous expression of ethanol synthesis pathway in glycogen deficient *Synechococcus elongatus* PCC 7942 resulted in enhanced production of ethanol and exopolysaccharides. Front. Plant Sci. 11 (February), 1–12. https://doi.org/10.3389/fpls.2020.00074.

Vijayan, V., O'Shea, E.K., 2013. Sequence determinants of circadian gene expression phase in cyanobacteria. J. Bacteriol. 195 (4), 665–671. https://doi.org/10.1128/JB.02012-12.

Wang, H.H., Isaacs, F.J., Carr, P.A., Sun, Z.Z., Xu, G., Forest, C.R., Church, G.M., 2009. Programming cells by multiplex genome engineering and accelerated evolution. Nature 460 (7257), 894–898. https://doi.org/10.1038/nature08187.

Wang, B., Wang, J., Meldrum, D.R., 2012. Application of synthetic biology in cyanobacteria and algae. Front. Microbiol. 3 (September), 1–15. https://doi.org/10.3389/fmicb.2012.00344.

Wang, B., Eckert, C., Maness, P.-C., Yu, J., 2018. A genetic toolbox for modulating the expression of heterologous genes in the cyanobacterium *Synechocystis* sp. PCC 6803. ACS Synth. Biol. 7 (1), 276–286. https://doi.org/10.1021/acssynbio.7b00297.

Wang, M., Luan, G., Lu, X., 2020. Engineering ethanol production in a marine cyanobacterium *Synechococcus* sp. PCC7002 through simultaneously removing glycogen synthesis genes and introducing ethanolgenic cassettes. J. Biotechnol. 317 (March), 1–4. https://doi.org/10.1016/j.jbiotec.2020.04.002.

Wasylenko, T.M., Stephanopoulos, G., 2015. Metabolomic and ^{13}C-metabolic flux analysis of a xylose-consuming *Saccharomyces cerevisiae* strain expressing xylose isomerase. Biotechnol. Bioeng. 112 (3), 470–483. https://doi.org/10.1002/bit.25447.

Wendt, K.E., Ungerer, J., Cobb, R.E., Zhao, H., Pakrasi, H.B., 2016. CRISPR/Cas9 mediated targeted mutagenesis of the fast growing cyanobacterium *Synechococcus elongatus* UTEX 2973. Microb. Cell Factories 15 (1), 115. https://doi.org/10.1186/s12934-016-0514-7.

Włodarczyk, A., Selão, T.T., Norling, B., Nixon, P.J., 2020. Newly discovered *Synechococcus* sp. PCC 11901 is a robust cyanobacterial strain for high biomass production. Commun. Biol. 3 (1). https://doi.org/10.1038/s42003-020-0910-8.

Xiong, W., Morgan, J.A., Ungerer, J., Wang, B., Maness, P.C., Yu, J., 2015. The plasticity of cyanobacterial metabolism supports direct CO_2 conversion to ethylene. Nat. Plants 1 (5), 1–36. https://doi.org/10.1038/NPLANTS.2015.53.

Xiong, W., Cano, M., Wang, B., Douchi, D., Yu, J., 2017. The plasticity of cyanobacterial carbon metabolism. Curr. Opin. Chem. Biol. 41, 12–19. https://doi.org/10.1016/j.cbpa.2017.09.004.

Xu, Y., Alvey, R.M., Byrne, P.O., Graham, J.E., Shen, G., Bryant, D.A., 2011. Expression of genes in cyanobacteria: adaptation of endogenous plasmids as platforms for high-level gene expression in *Synechococcus* sp. PCC 7002. Methods Mol. Biol. 684, 273–293. https://doi.org/10.1007/978-1-60761-925-3_21.

Yao, L., Cengic, I., Anfelt, J., Hudson, E.P., 2016. Multiple gene repression in cyanobacteria using CRISPRi. ACS Synth. Biol. 5 (3), 207–212. https://doi.org/10.1021/acssynbio.5b00264.

Yao, R., Li, J., Feng, L., Zhang, X., Hu, H., 2019. [13]C metabolic flux analysis-guided metabolic engineering of *Escherichia coli* for improved acetol production from glycerol. Biotechnol. Biofuels 12 (1), 1–13. https://doi.org/10.1186/s13068-019-1372-4.

Young, J.D., Shastri, A.A., Stephanopoulos, G., Morgan, J.A., 2011. Mapping photoautotrophic metabolism with isotopically nonstationary ^{13}C flux analysis. Metab. Eng. 13 (6), 656–665. https://doi.org/10.1016/j.ymben.2011.08.002.

Yu, J., Liberton, M., Cliften, P.F., Head, R.D., Jacobs, J.M., Smith, R.D., Koppenaal, D.W., Brand, J.J., Pakrasi, H.B., 2015. *Synechococcus elongatus* UTEX 2973, a fast growing cyanobacterial chassis for biosynthesis using light and CO_2. Sci. Rep. 5, 8132. https://doi.org/10.1038/srep08132.

Zahra, Z., Choo, D.H., Lee, H., Parveen, A., 2020. Cyanobacteria: review of current potentials and applications. Environments 7 (2). https://doi.org/10.3390/environments7020013.

Zamboni, N., Fendt, S.M., Rühl, M., Sauer, U., 2009. ^{13}C-based metabolic flux analysis. Nat. Protoc. 4 (6), 878–892. https://doi.org/10.1038/nprot.2009.58.

Zess, E.K., Begemann, M.B., Pfleger, B.F., 2016. Construction of new synthetic biology tools for the control of gene expression in the cyanobacterium *Synechococcus* sp. strain PCC 7002. Biotechnol. Bioeng. 113 (2), 424–432. https://doi.org/10.1002/bit.25713.

Zhang, S., Bryant, D.A., 2011. The tricarboxylic acid cycle in cyanobacteria. Science 334 (6062), 1551–1553. https://doi.org/10.1126/science.1210858.

Zhou, J., Zhang, H., Zhang, Y., Li, Y., Ma, Y., 2012. Designing and creating a modularized synthetic pathway in cyanobacterium *Synechocystis* enables production of acetone from carbon dioxide. Metab. Eng. 14 (4), 394–400. https://doi.org/10.1016/j.ymben.2012.03.005.

Zhou, J., Zhang, H., Meng, H., Zhu, Y., Bao, G., Zhang, Y., Li, Y., Ma, Y., 2015. Discovery of a super-strong promoter enables efficient production of heterologous proteins in cyanobacteria. Sci. Rep. 4 (1), 4500. https://doi.org/10.1038/srep04500.

Zhou, J., Yang, F., Zhang, F., Meng, H., Zhang, Y., Li, Y., 2021. Impairing photorespiration increases photosynthetic conversion of -CO_2 to isoprene in engineered cyanobacteria. Bioresour. Bioprocess. https://doi.org/10.1186/s40643-021-00398-y.

Current technical advancement in biogas production and Indian status

S.D. Sawale and A.A. Kulkarni

Praj Matrix—R & D Center (Division of Praj Industries Limited), Pune, Maharashtra, India

1 Introduction

Globally, many governments are trying to adapt the anaerobic digestion (AD) systems to overcome land and water pollution and greenhouse gases (GHG) emission, due to landfills using sustainable renewable sources. Energy obtained from AD system is used for number of applications such as compressed heat and power (CHP), cooking, electricity, transportation fuel, etc. (Weiland, 2010). In this regard, government bodies are laying down various policies, regulations, and incentives to foster and enhance the biogas project viability at varying scales of operation (Pfay et al., 2017; Scarlat et al., 2015; EC, 2009; Sawale et al., 2020).

Government of India (GOI) is committed to increase the use of large-scale sustainable power projects and promotion of green energy. In 2017, India attained 63% of overall energy self-sufficiency (IEA, 2018). It is aiming to achieve 227 GW of renewable energy capacity by 2022 and 500 GW by 2030, more than its target of 175 GW as per the Paris Agreement. Ministry of New and Renewable Energy (MNRE) targets to produce 227 GW energy capacity by 2022, out of which 114 GW is planned for solar, 67 GW for wind, and remaining 46 GW for hydro and biofuels (IEA, 2020). Thus, the renewable energy sector of India is poised to attract an investment of US$ 80 billion in 4 years. By 2040, nearly 49% of total power requirement needs to be generated using renewable energy. India will save 54,000 crore (US$ 8.43 billion) rupees annually using renewable energy. In 2030, 55% of the total installed power capacity will be contributed by renewable energy sector. As per the strategic plan, India has planned about 5000 compressed biogas (CBG) plants to be set up by 2023 (IBEF, 2021). In renewable energy sector in India, CBG is in focus due to technology advances. Global biogas production technologies market is expected to grow at a CAGR of 10.6% up to 2022, worth $10.1 billion. Wastewater/sludge and industrial application will lead the market with a CAGR of 11.4%, and landfill gas, agriculture, municipal waste, and food will capture a market of $5.8 billion by 2022 (BCC Research Report, 2018).

India, a fast-growing economy, has seen increased energy demand due to continuous urbanization and development of manufacturing sectors. Developed and developing countries are concerned about future energy security and better use of renewable natural energy sources due to increasing population. Currently, the required energy demand is being fulfilled by traditional energy sources such as coal, oil, electricity, natural gas, etc. The aim of reduction of fossil fuel due to its nonrenewable nature and tremendous utilization creates the need to find alternative renewable, abundant, and cost-effective energy sources (Kumar and Majid, 2020). Additionally, fossil fuel contributes to global warming due to GHG emission. India is focusing on developing alternative sources of energy such as solar, nuclear, wind energy, and biofuels as primary energy consumption of India grew by 2.3% by 2019, being third after China and USA with 5.8% global energy share (*BP Statistical Review of World Energy*, 2020).

The "National Biofuel Policy 2018" paved way to "Sustainable Alternative Towards Affordable Transportation" (SATAT) initiative launched by Ministry of Petroleum and Natural Gas (MoPNG) to promote CBG production plants in the market for use in automotive fuels. SATAT initiative has increased the scope to CBG plants based on biodegradable waste/feedstock such as cow dung, pressmud, spent wash, vegetable waste, and agricultural biomass. The projects are gaining popularity due to availability of technology to give sustainable solution for waste management and provide clean energy. The sustained quality of feedstock or raw materials is a key requirement for the success of commercially viable operation of CBG project (MoPNG, 2018).

Most energy-efficient and environment-friendly AD technology is one of the promising options of renewable energy for biogas production. AD is a complex process where the organic substrates are converted to biogas in the presence of various acidogenic, acetoclastic, and methanogenic bacteria. Biogas mainly contains methane as rich component followed by carbon dioxide and other trace gases such as ammonia, oxygen, hydrogen, hydrogen sulfite, and nitrogen. (Sun et al., 2015). The biogas obtained can be used as gaseous fuel in the form of CBG or compressed natural gas (CNG) instead of fossil fuel. CBG can replace natural gas as a feedstock for producing chemicals and materials (Saravanan et al., 2018). The digestate or biomethane slurry of AD process can be a superior alternative for mineral fertilizer (Herrmann et al., 2017; Zheng et al., 2017; Sogn et al., 2018).

The biogas can be used to meet futuristic energy demands. The interest in biogas as a renewable energy is increasing due to fluctuating cost and negative effect on environment by the use of traditional fossil fuels (Ighravwe and Babatunde, 2018). The success and sustainability of any biogas plant is dependent on the type of feedstock and process parameters such as temperature, pH, organic loading rate, hydraulic retention time (HRT), and digester design (Sarket et al., 2019; Nsair et al., 2020). Biogas composition and yield vary with different substrates depending on their overall compositional analysis such as total carbohydrates, fats, volatile solids, lignin, and other recalcitrants content (Xu et al., 2019). Thus, it is imperative to select the best available organic residues as such or in combination. The National Biogas and Manure Management Programme (NBMMP) promotes construction of domestic biogas plants for producing fuel for cooking and the use of slurry/digested material as organic biomanure for farming purpose. Biomanure is rich in various nutrients such as nitrogen, phosphorus, and potassium. 4.75 million farm size biogas plants are in existence from 2014 producing 10 billion M^3 biogas/year. The installed electricity capacity of biogas plants reached 179 MW in 2015 and 157 MW in 2016 on the basis of these plants (IREA, 2017; MNRE, 2016).

While the other feedstock's have been researched upon extensively and are routine sources of biogas production, biomass or lignocellulosic feedstock has attracted the attention of the world currently due to its large-scale availability. The application of this feedstock is currently in the animal industry as animal feed. However, not all the feedstock is consumed. The one that is not consumed is mostly burnt and causes a health hazard as has been the case in North India. The feedstock also needs to be vacated from fields very quickly and hence is disposed in a hurried manner without fetching any value. If this feedstock (in addition to the earlier one's) is captured in the biogas ambit, then the entire rural economy is set to change for the better. In addition to improvement of rural lifestyle through job and revenue creation, the biogas industry can contribute to better environment through value creation.

The current chapter delves into biogas production capacity of India, various pretreatment methods of lignocellulosic methods, types of digesters, latest in digester design for handling solid feedstock's as different from the earlier slurry or liquid feedstocks, biogas upgradation techniques, city gas distribution network, biogas policies of India and few Indian biogas players.

2 Biogas production capacity of India

The estimated biogas potential of India is about 40.734 MM^3/year (Rao et al., 2010). India can produce 17,000 MW power using biogas equivalent to 10% of its total energy requirement (EAI, 2009). Traditionally, household plants use animal waste such as cow dung in anaerobic digester for biogas generation limited to kitchen use. Advancement in waste collection technologies such as collection, segregation, and distribution in addition to advanced biogas production technologies can help to boost the biogas development program of India. New feedstock such as kitchen waste, municipal solid waste, agricultural waste, and processing industries waste can be used successfully for renewable energy generation in biogas form.

Organic waste obtained from livestock is considered as the superb feedstock for biogas generation. Total livestock population of India is about 535.8 million. Livestock includes bovine animals like cattle and buffalo (302.3 million) and other livestock such as sheep, goat, poultry, etc. (1075 million) (NDDB, 2019). The average dung yield of cattle and buffalo is about 4.5 kg/day and 10.2 kg/day, respectively. Thus, the total annual dung production is estimated to be about 1064 MT. In case, if the total dung is used for biogas generation, it can produce 22,343 MM^3 of energy for various applications.

Municipal solid waste (MSW) management is a global issue due to increased urbanization and population growth. MSW contributes directly to environmental pollution and ill-health. India generates about 127,486 tons/day MSW, which includes organic and nonorganic waste (CPCB, 2012). The organic waste after proper segregation can be used as a potential substrate for biogas production. MSW contains 42.19% organic matter and C:N ratio is 21:30, thus MSW acts as the best available substrate for biogas generation (Rao et al., 2010). Biogas potential of 9.29 MM^3/day is estimated by the use of MSW (Sharholy et al., 2008).

Sugar mill of 1000 ton crushing capacity per day produces nearly 40 ton/day pressmud. BioCNG of 1.6–1.7 per ton of press mud per day can be generated after purification. Additionally, 14–17 MT/day generated solids and liquid by-products can be sold at high selling cost as an organic fertilizer (Patil et al., 2020).

3 Pretreatment of lignocellulosic feedstock

Agricultural residue/waste is one of the major sources of lignocellulosic biomass for biogas generation. India annually produces about 686 MT of agricultural residue of which 234 MT is surplus (Hiloidhari et al., 2014). However, various residues like rice, straw, and maize can act as best suitable feedstock after chemical or biological pretreatment.

Lignocellulosic feedstock includes agro-industrial residues, forest industrial residues, energy crops, municipal solid waste, vegetable residues, etc. However, due to complex and recalcitrant nature lignocellulosic feedstocks are difficult to breakdown using enzymes or microbes (Hendriks and Zeeman, 2009). Lignocellulosic feedstock is a complex structure of cellulose (20%–50%), hemicellulose (15%–35%), and lignin (5%–30%) (Lynd et al., 2002, Kumar and Wymen, 2014). Hence, it is imperative to use chemical or biological or a combination of pretreatment methods including physical, chemical, and biological strategies for the disruption of complex biomass to C6 or C5 chain before AD process for biogas production. AD process also depends on process temperature, reaction pH, organic loading rate, and substrate to inoculum ratio. The substrate characteristics define the biogas yield. Maximal theoretical methane yields (Nm^3/t_{ts}) for substrate constituents such as carbohydrate, raw protein, raw fat, and lignin are 395–400, 497, 816–850, and 0, respectively (Weiland, 2010).

3.1 Chemical pretreatment

Acid pretreatment under superheated steam involves the use of sulfuric acid, hydrochloric acid, phosphoric acid, and sulfur dioxide where hemicellulose is converted to monosaccharides. Hydrothermal processing solubilizes hemicellulose without complete hydrolysis. The process does not require any catalysts and significant corrosion problems are not observed. Water penetrates into feedstock under high pressure and hydrates cellulose and removes hemicellulose and lignin to minor extent. Mild alkali treatment can be used at lower temperature compared to above methods using sodium hydroxide, ammonia or calcium hydroxide removes lignin and a minor amount of hemicellulose. Calcium hydroxide and ammonia can be used in lime pretreatment, ammonia recycled percolation (ARP), and ammonia fiber expansion (AFEX) (Yang and Wyman, 2008; Hu and Ragauskas, 2012) to separate constituents.

In case of oxidative pretreatment, partial removal of lignin and hemicellulose occurs by the use of oxidants like hydrogen peroxide and ozonolysis. Wet oxidation is carried out by treating feedstock with water and air or oxygen at high temperature and short time (HTST). Lignin is oxidized to aliphatic carboxylic acids and phenolic compounds (Klinke et al., 2002; Carlos et al., 2007). Ionic liquid has also been used as alternative solvent for biomass pretreatment where dissolution of lignocellulosic matter occurs. These ionic liquids break the non-covalent bonds of lignocellulose resulting in its disruption. (Karatzos et al., 2012). Chemical pulping processes such as Kraft pulping and sulfite pulping are used for degradation of lignocellulosic matter of softwoods and hardwoods. Sodium hydroxide and sodium sulfite convert lignin and hemicellulosic content to black liquor. The black liquor is used for further energy production (Pereira et al., 2013). Recently developed BALI™ and SPORL (sulfite pretreatment to overcome recalcitrance of lignocellulose) techniques followed by biomass size reduction convert softwood or hardwood to easily convertible cellulose (Rodsrud et al., 2012). Soda pulping is used mostly for non-wood plants containing higher inorganic matter. In case of soda pulping, both lignin and hemicellulose linkages are broken down to some extent.

3.2 Biological pretreatment

Biological pretreatment of biomass is preferred due to their specific selective nature in the disruption of crystalline nature of lignocellulosic, additionally lesser inhibitors are generated during process. The process operational conditions such as pH and temperature are milder and enhance selective hydrolysis and production of monomers for further anaerobic digestion for biogas. Net energy output is higher and relatively inexpensive in comparison to chemical and mechanical pretreatment processes (Wagner et al., 2018; Barik et al., 2000). Biological pretreatment does not produce any inhibitory molecules. However, improved biogas productivity and process conditions such as pH, rate of hydrolysis, and maximum substrate conversion rate are observed (Kalyani et al., 2013).

Microbes screened and selected need potential cellulose and hemicellulose degradation activities. The microbes isolated from natural environment help in hydrolysis of biomass during aerobic pretreatment. The microbial consortia isolated from thermophilic straw decomposing landfill showed 96% higher biomethanation at 55 °C in AD of distillery wastewater fortified with cassava refuse compared to untreated sample (Zieminski et al., 2012). Wood-decomposing fungus *Auricularia auricularjudae* increased yield by 15% compared to untreated biomass of chestnut leaves and hay at 1:2 ratio when pretreated for 2–3 weeks at 37 °C (Mackulak et al., 2012). White rot fungus enhances biomethanation process as they produce enzymes such as hemicellulases, cellulases, and xylanases other than essential lignin-degrading enzymes (Take et al., 2006; Amirta et al., 2006; Mackulak et al., 2012). However, enzymatic pretreatment of biomass is limited and not much popular due to substrate selectivity/specificity and the cost factor (Zheng et al., 2014). The efficiency of any kind of treatment is dependent on feedstock particle size, nature of biomass (softwood or hardwood), wettability, flowability, and homogenization.

Biological treatments are those wherein explorations have been closest to natural phenomenon. Lignocellulosic feedstock's are processed in nature in the soil wherein a number of fungi and bacteria act upon them to open up the fibers thus contributing to their degradation. These soil fungi have therefore been the focus point of research since many decades. Notable are the white, brown, and soft rot fungi along with basidiomycetes. These have been known to act on agricultural residues, chestnut leaves, and sisal leaves with the pretreatment occurring at ambient temperatures of 28–37°C for 12 days to 8 weeks. These pretreatments thought slow have shown a 15% increase in methanation of these difficult feedstock's (Amirta et al., 2006;

Mackulak et al., 2012; Take et al., 2006). Thereafter a number of complex microbial systems were reported to process the agricultural feedstocks in similar manner. Notable are mixtures of yeast and cellulolytic bacteria, sludge, Clostridium species as well as compost forming microbes (Bruni et al., 2010; Zhang et al., 2011; Zhong et al., 2011). These microbes are also slow acting but show a broad range of higher methanation (25%–95%) than without any pretreatment. A lot of research from the above microbes also led to usage of their enzymes in pretreatment wherein laccases and cellulases led the way. The enzymes shortened the time of pretreatment to a few hours with about 35% improvement in methanation (Bruni et al., 2010; Yunqin et al., 2010; Zieminski et al., 2012). Yet even till date they face challenges in scale up and commercialization due to multiple hurdles in fermentation and purification economics.

Chemical, physical, and physicochemical pretreatment processes required high energy demands for bioenergy output. These processes also produce various harmful by-products resulting in decreased biomethane production and also impacts negatively on environment (Paudel et al., 2017). To overcome such problems, available industrial, lignocellulosic-degrading hydrolytic enzymes can be used for biomethanation (Lopez et al., 2007). However, due to enzymes' high price, they are not desirable. Hence, use of microbial consortia for biomass pretreatment is the best option for enhanced biogas production.

4 Factors affecting biogas production

Feedstock particle size is an important factor that must be taken into consideration before proceeding with pretreatment. Larger particle size will restrict water diffusion, microbial surface colonization, and overall mass transfer of the substrate. However, too small particle size will lead to clogging and reduced biogas yields due to reduced mass transfer of substrates. Thus, appropriate particle size is of utmost importance to achieve optimized process conversion efficiencies for higher yields and improve the economy and sustainability of biogas plant (van Kuijk et al., 2016).

Apart from particle size, the other factor to be considered is homogenization of the feedstock slurry to be fed to the digester. A non-homogenized slurry leads to uneven treatment of the feedstock resulting in erratic gas production. Thus, a well-mixed or homogenized slurry is very critical especially at large commercial scale of operation.

Operational parameters such as process temperature, pH, and mixing of digester directly affect the microbial action on feedstock. Thus, a properly optimized process needs to be developed at small scale before its commercialization. Sensor and parametric monitoring are crucial. At scale up, it is all the more critical that mass transfers happen effectively since the biogas plants involve more resilient solids that also add buffering into the process. Techno-economic feasibility of biogas project must be taken into consideration before biogas plant development. Project capacity, planning and permission procedure, continuous and uniform feedstock logistics, feedstock supply chain management, plant location, funding, subsidies, plant operation and maintenance, investment cost, technology provider, and stakeholders involved in projects must be taken into consideration for successful running of a biogas plant.

4.1 Digester designs

Biogas production is a complex process that involves aerobic and anaerobic microbial biotransformation of substrates into methane. The steps involved are hydrolysis, acidogenesis, acetogenesis, and methanogenesis. Hydrolysis involves the conversion of polymeric organic matter into sugar monomers, alcohol, and other degradable matter. The formed organic matter undergoes acidogenesis and acetogenesis forming volatile fatty acids, acetic acid, carbon dioxide, and hydrogen. The formed acetic acid is converted into methane by methanogenesis process. The schematic AD process is shown in Fig. 1 (Paritosh et al., 2017).

To monitor the AD performance process, volatile fatty acids, alkalinity, pH, biogas volume, biogas concentration in terms of methane and carbon dioxide, and chemical oxygen demand are estimated. To ensure all the AD processes to be carried out successfully and efficiently, the type of digester used plays an important role. Different types of digester used are mentioned in detail as below.

4.2 Types of anaerobic digester

Based on feedstock developments in processing and treatment technologies, various digester designs are developed over decades. Some of them are discussed here due to their current and future relevance. Globally, different unit designs of digesters like continuous stirred tank reactor (CSTR), plug flow reactor (PFR), and upflow anaerobic sludge blanket (UASB) designs are evolved over the years of biogas history. Selection and construction of biogas plant also takes into consideration factors such as space availability, existing structure, cost, and substrate availability (Amaratunga, 1986). Commercial or household biogas plants are designed considering feedstock type, process parameters such as organic loading rate (OLR), hydraulic retention time (HRT), temperature, pH as well as climatic and socioeconomic conditions.

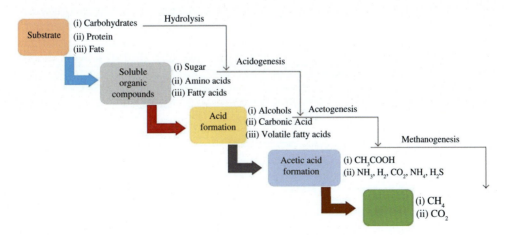

FIG. 1   Anaerobic digestion process.

4.2.1 Batch type biogas digester

In batch type biogas digester, all the contents including substrate, inoculum, water, and other required components are charged at once in air-tight reactor. In some cases, base is added to maintain the neutral pH over the process time. The digester is then sealed and anaerobic digestion is carried out for 30–180 days depending on climatic temperature. Throughout the process, the daily gas production reaches a maximum level and then starts to decline. The process can use solid content ranging from 6% to 10% or maximum of 20% depending on the nature of the substrate. The batch process which runs at >20% solid concentration is known as dry fermentation (Chiumenti et al., 2017). Fig. 2 shows various parts of a batch digester (Source: FAO, 1992, Reproduced with permission).

The dry fermentation process parameters need further improvement as it is seeming to be a viable technology due to water availability challenges and the gas production rates are comparable with semicontinuously fed reactors.

4.2.2 Fixed dome

The reactor consists of a gas-tight chamber constructed with bricks, stones, or concrete. Top and bottom are hemispherical in shape and are joined together by straight side walls. Inside surface is sealed to avoid gas leakage through dome. The digester is fed semicontinuously, and the inlet pipe is straight and ends at mid-level in the digester. A manhole is created at top to facilitate entrance for cleaning and gas outlet pipe exits from the manhole cover. Produced gas is stored under hemispherical dome and displaces some of the digester content into the effluent chamber thus maintaining the gas pressure between 1 and 1.5 m of water. Hence, to maintain the gas pressure, the reactor has hemispherical top and bottom.

Fixed dome digester is further improved for holding higher gas pressure, increased rate of gas production, gas storage, digester size, geometric forms to sustain loads, and forces created by biogas on four horizontal walls. At ambient pressure, the inside water level is at 95% of the total digester volume. Digester's diameter-to-height ratio is maintained as 2:1. The HRT for cow dung and pig manure that are common substrates is about 35–40 days at 5%–8% and 4%–7% of total solid concentration, respectively. The average gas production ranges between 0.15 and 0.6 m/day depending on ambient temperature. These digesters are widely used in China as they are improved over the period of time with respect to construction material quality, cost, suitable feedstock, and better gas production rates (Kaur et al., 2017). Fig. 3 shows various parts of a fixed dome digester (Source: FAO, 1992, Reproduced with permission).

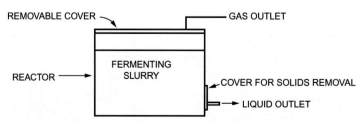

FIG. 2   Batch type biogas digester.

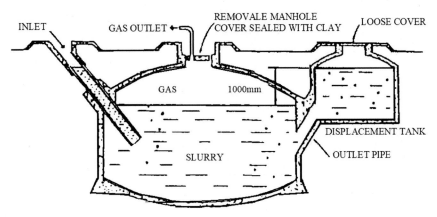

FIG. 3 Fixed dome digesters.

4.2.3 Floating dome (KVIC design)

In 1962, Khadi and Village Industries Commission (KVIC) started using floating dome anaerobic digesters originally designed in 1950 by Patel. The digester contains floating gas holder leading to increased interest in biogas in India. The digesters constructed have capacity of 6–8 M^3 gas production capacity and designed for 30, 42, and 55 days of retention time according to the seasonal and climatic temperature variation. Construction cost also varies with respect to the climatic conditions. However, Indian government has provided partial subsidies to cover the cost of construction. The major substrate is a mixture of cattle dung and night soil. At larger scale, digestate material also includes water hyacinth. The gas holders were initially made with mild steel, until fiberglass reinforced plastic (FRP) was implanted successfully to prevent metal corrosion. Currently, all KVIC type digesters are equipped with FRP gas holders. The biogas plant contains a digester for anaerobic fermentation and a floating drum for gas collection as shown in Fig. 4. The depth and diameter of digester are 3.5–6.5 m and 1.2–1.6 m, respectively. The partition wall divides digester vertically after complete slurry filling. Dung mixed with water (4:5 ratio) and filled into digester through inlet pipe serves as the starter culture that passes through outlet pipe. The outlet is mostly connected to a compost pit. Gas holder is cylindrical in shape with concave top. The gas holder sinks into slurry due to its weight and rests on a ring constructed for this purpose. To improve volumetric efficiency and to make KVIC type biodigesters economically feasible and strong structure, efforts are continued. Heating, mixing, and insulation have been introduced as well as geometric configurations and locations of inlet and outlet are modified (Sooch and Gautam, 2013). Fig. 4 shows design of floating dome Indian KVIC anaerobic digester.

4.2.4 Janta model

These digesters are 20%–30% cheaper compared to KVIC models. Janta type digesters are mostly designed for 60 days of retention time. Biogas production is obtained in a wide range

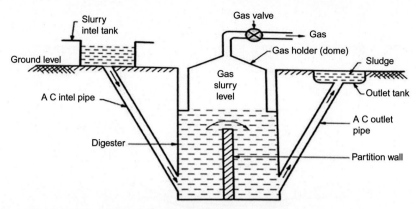

FIG. 4 Floating dome or Indian KVIC anaerobic digester.

from 2 to $30\,M^3$. Most common sizes are 2, 3, 4, and $6\,M^3$ gas/day. Janta models do not contain any steel during construction and do not have moving part maintenance resulting in lesser cost compared to KVIC model. The plant must be constructed using good quality construction materials to avoid any structural damage and gas leakage in the long run.

Substrates such as municipal waste and plant residue along with cattle dung can be used. Two rectangular openings act as inlet and outlet, while the dome-shaped roof remains below ground level. The gas outlet pipe is fitted on top of the dome-shaped roof. The biogas is collected in restricted space of fixed dome leading to high pressure of gas. Fig. 5 shows schematic view of Janta model.

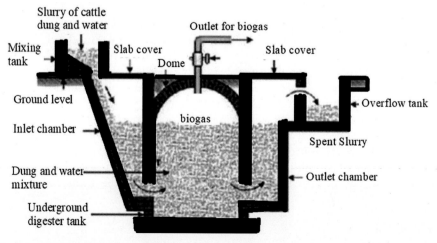

FIG. 5 Janta type anaerobic digester.

4.2.5 Deenbandhu biogas plant

The model was developed by Action for Food Production (AFPRO) in 1984 (Fig. 6). The cost of plant was half of KVIC model which brought biogas technology within reach for the lower economical population. The cost was reduced by minimizing the surface area by joining the segments of two spheres of different diameters at their base. It consists of a hemispherical fixed dome gas holder made of prefabricated cement and concrete. The slurry is passed to the digester via inlet pipe connected to digester from a mixing tank. The biogas is collected under the dome space. The gas can be taken out via a pipe connected to top of the dome. The sludge, a co-product, comes out through a side opening of the digester. In India, about 90% of household biogas plants are of Deenbandhu type.

4.3 Upflow anaerobic sludge blanket (UASB)

Globally, UASB is mostly used for wastewater treatment. The reactor contains a cylindrical tank of H/D=2 ratio. The wastewater passes upward through an anaerobic sludge blanket containing about half the volume of reactor. Solid–liquid separation is carried out by the inverted cone settler placed at the top of digester. The biological solids settle slowly during initial start. However, these biological solids are converted to granular sludge over the period of time and it settles at reactor bottom very well. Thus, retaining the active biomass in the reactor. Loading rate up to 40 kg/m/day COD is possible, with retention time as low as 3.5 h. Total solid content in the feed is up to 3% (Strezov and Evans, 2015). UASB process depends on the natural immobilization of anaerobic bacteria (Chan et al., 2009). In countries, where landfilling is commonly utilized for waste management, use of batch reactors is a good option to treat such biowaste due to its simplicity and portability. Fig. 7 shows schematic diagram of UASB reactor (Goncalves et al., 2005).

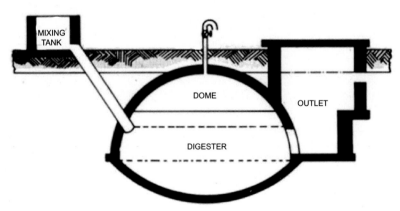

FIG. 6   Deenbandhu biogas plant.

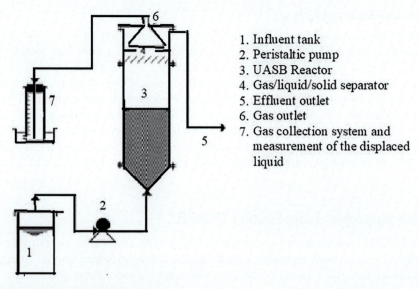

1. Influent tank
2. Peristaltic pump
3. UASB Reactor
4. Gas/liquid/solid separator
5. Effluent outlet
6. Gas outlet
7. Gas collection system and measurement of the displaced liquid

FIG. 7   Upflow anaerobic sludge blanket reactor.

4.4 Plug flow reactor (PFR)

The PFR produces biogas at a variable pressure with a constant working volume. These digesters consist of a narrow and long tank with an average length-to-width ratio of 5:1. Both inlet and outlet are kept above the ground and are in opposite directions, and the complete biodigester is built underground in inclined position. Due to inclination, the feed passes into the inlet and digestate passes toward outlet. The inclined position separates acidogenesis and methanogenesis phases longitudinally producing a two-phase system (Rajendran et al., 2012). The HRT under thermophilic conditions ranges between 15 and 20 days. The solid concentration of the feed is between 11% and 24% (Abbasi et al., 2012). PFR can be operated in mesophilic conditions even though they maintain optimal thermophilic conditions (Strezov and Evans, 2015). Fig. 8 shows the schematic view of PFR (Source: https://www. epa.gov/agstar/anaerobic-system-design-and-technology).

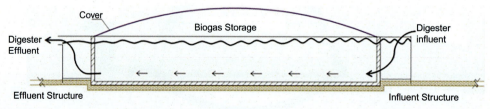

FIG. 8   Plug flow reactor. *Source: https://epa.gov, US Agstar.*

4.5 Continuous stirred tank reactor (CSTR)

CSTR is the most common digester used at commercial biogas plants for treating high-moisture organic waste in past decades (Mao et al., 2015). CSTR is easy to operate and most commonly used for wastewater treatment with high chemical oxygen demand (COD) and high solid content. CSTR stabilizes the sludge by conversion of biodegradable fractions into biogas (Massoud et al., 2007). To increase process rates, CSTR is operated at high temperatures. Mixing is performed mechanically or by flow recycle or from produced biogas. Perfect mixing is difficult in big CSTR volumes. Hence, mixing efficiency is an important factor to be considered during modeling of CSTR in regard to solid transport in reactor and evaluation of solid retention time. Materials with high COD loading rates (30 kg/M^3/day) can be treated using CSTR with an adequate treatment at lower HRT. In case of CSTR, a removal efficiency of 85%–95% of COD of inlet material and a produced methane content of 80%–85% have been reported (Chan et al., 2009; Wang et al., 2005). The digestate content is continuously mixed at homogenous state using agitator and impellers. CSTRs have been commonly used in middle to large-scale projects. The substrate is added continuously and after mixing and contact with anaerobic bacteria the organic matter is converted into biogas. The biogas is discharged from the top and purified further.

CSTR can be used in series to improve the biogas yield from manure. In case of serial CSTR, 11% higher biogas yield was achieved compared to single CSTR for 15 days HRT. Single-stage CSTR and a hybrid anaerobic digester were compared to investigate the methane production potential using dairy manure and co-digestion with maize silage. Six different organic loading rates were tested from 1.1 to 5.4 gVS/L/day. The specific biogas production of hybrid system was 440–320 mL/gVS with 81%–65% volatile solids destruction. The hybrid system provided 116% increase in specific biogas production and VS destruction improved by more than 14%. Hybrid system generated methane-enriched biogas (>75% methane) by enabling phase separation. Acidogenic conditions were observed in first two compartments followed by two segments as methanogenic conditions. The pH of acidogenic part varies between 4.7 and 5.5 and methanogenic part between 6.8 and 7.2 (Boe and Angelidaki, 2008).

4.6 Unique mixed plug flow digester

In such anaerobic digesters variable pressure is provided intermittently for AD mixing. PFR contains a narrow and long reactor with average length-to-diameter ratio of 5:1. The PFR is built completely underground and only inlet and outlet are kept above the ground at opposite end of the reactor. Thus, due to inclination, the substrate passes from inlet to outlet. Additionally, due to inclination, acidogenesis and methanogenesis phases are formed. HRT ranges between 15 and 20 days under thermophilic conditions. PFR can handle a feed concentration from 11% to 24% (Rajendran et al., 2012; Abbasi et al., 2012). The advantages of plug flow reactor over CSTR are mentioned in Table 1.

TABLE 1   Advantages of mixed duel PFR over CSTR.

Biomethanation	CSTR	PFR
Reaction mechanism	Inefficient—prone to bypassing substrates	Highly efficient reaction—plug flow ensures no bypassing
Reaction	Single phase—no separation of acidogenic and methanogenic stages	Allows multiple stages
Robustness	Sensitive to feed and parametric variations	Steady and robust operation, internal recycle of adapted bacteria
Conversion efficiency	Low (<60%)	High (65%–75%)
Gas yields	Low due to low degradation	30% higher than CSTR
Retention time	Very high—makes plant bigger and expensive	Low—small and compact plant
Mixing efficiency	Low due to agitators	High due to gas circulation which gives highly efficient and rigorous mixing
Flexibility and expansion	Low due to multiple reactors, difficult to synchronize, high cost expansion	High due to single plug flow design, easy, and low cost to expand capacity
Electricity consumption	High due to large multiple mechanical agitators	Low because one blower sparges biogas through multiple internal nozzles
Start-up and restart time	High due to slow reaction	Quick start-up and less restart time
Maintenance	Frequent and high cost—due to mechanical agitators and membrane domes	Rare and low cost—because there are no moving/plastic parts
Maintenance shutdowns	Complete shutdown every alternate year and loss of production	No shut downs—continuous full-scale operation over year to year
Operation	Complex to automate and high manpower	Easy to automate and very low manpower

5 Biogas purification system

Biogas composition in production differs depending on the feedstock used during anaerobic digestion (Gigot et al., 2012). Methane (55%–70%) is the main product of biogas anaerobic digestion including CO_2 (25%–45%), H_2S (0.002%–2%), and other gases such as nitrogen (<2%), water vapor (2%–7%), hydrogen and ammonia (<1%) (Hashimoto and Varriel, 1978; Weiland, 2010). The calorific value of biogas is directly proportional to its methane percentage. Biogas as such can be used to provide energy in the form of combined heat and power (CHP). Biogas purification is important prior to use as it contains H_2S, which may lead to formation of sulfuric acids resulting in failure and damage of engines. Dust and siloxanes are present in biogas obtained from landfills and wastewater treatment plants. Siloxanes are formed from the reaction of silicon, oxygen, and methyl groups. Siloxanes damage heat exchanges and pumping equipment resulting in reduction in heat transfer (Huertas et al., 2011).

Methane concentration of more than 90% is recommended for internal combustion of engines. Higher carbon dioxide content reduces the engine power output. Siloxane compounds present in biogas can cause abrasion problem due to deposition of silica on engine metallic surface. Thus, biogas must be purified/upgraded to remove other components and to enrich methane content more than 95% before its use as CHP or supply through national gas grid system to ensure further safety. The purified/upgraded biogas, i.e., biomethane can be sold to various industries in the form of CBG or CNG, liquefied natural gas (LNG) to be used for transport and cooking applications.

Biogas purification process includes absorption, adsorption, cryogenic separation, and membrane separation. Membrane purification technologies for biogas purification are in advanced stage due to their cost factor. The membrane processes are also combined with other processes to have further purification cost reduction. Currently, the biogas upgrading market and technologies are rapidly evolving to keep up with other bioenergy sources.

5.1 Absorption

Physical and chemical absorption techniques are employed for biogas upgradation. Physical absorption is carried out by high pressure water scrubbing (HPWS) and organic physical scrubbing (OPS). Whereas amine scrubbing (AS) and inorganic solvent scrubbing (ISS) are chemical adsorption processes.

In aqueous medium, polyvalent metal ions, such as chelates of iron is used for scrubbing hydrogen sulfide from biogas. The sulfur present in hydrogen sulfide is precipitated as elemental sulfur. The elemental sulfur can be sold as co-product. Further hydro scrubbing process is continued under high pressure. HPWS process is based on the difference in solubility of methane, hydrogen sulfide, and carbon dioxide in water. The process is intensified by pressurizing the absorption system using chilled water ensuring removal of carbon dioxide and traces of hydrogen sulfide. The moisture content of upgraded biogas is removed in methane gas dryer. The enriched biogas is compressed under pressure to fill up in cylinder as compressed biogas or BioCNG. The process is very simple with high methane recovery (>97%). Rotunno et al. (2017) simulated a pressurized water scrubbing process considering a biogas mixture of methane (60%) and carbon dioxide (40%) and reported 98.1% methane purity at 10 bar and 25°C, suitable for gas grid injection and BioCNG production. Energy efficiency of upgrading plant producing biomethane with quality of gas grid injection was 89.8%. Chandra et al. (2012) used improved automated water scrubbing system to enrich 97% methane at an operating column pressure of 1.0 MPa with 2.5 m³/h biogas inflow rate and 2.0 m³/h water inflow rate in the scrubbing column unit. However, high operating (OPEX) and capital (CAPEX) expenditure is required. Bacterial growth during process may lead to clogging, foaming, and lower flexibility toward variation of raw biogas feed, which are the disadvantages of HPWS.

Water is replaced by polyethylene glycol (PEG) in case of organic physical adsorption. Carbon dioxide has higher solubility in PEG compared to water, leading to lower scrubbing liquid circulation and less equipment for same biogas capacity (Harasimowicz et al., 2007). In case of amine scrubbing, different alkylamines such as diethanolamine (DEA), monoethanolamine (MEA), and methyldiethanolamine (MDEA) are used to remove hydrogen sulfide and carbon dioxide. In amine scrubbing system, the amine solution absorbs H_2S and CO_2. The formed amine solution is further loaded into amine regeneration column. The regenerated amine is recirculated to the amine scrubbing absorption column. Amine scrubbing system results in methane concentration $> 99\%$ with low operational cost. Disadvantages are high CAPEX, heat is required for regeneration, corrosion, poisoning of the amine by oxygen, salt precipitation, and foaming. Amine scrubbing can be carried out using sodium carbonate, potassium, and aqueous ammonia solution.

Capra et al. (2018) evaluated the energy and economic performance of a biogas upgrading process involving chemical absorption with two different aqueous solvent formulations: 50% w MDEA and a 20%w/20%w MDEA/MEA blend. Optimal process conditions are determined with a multi-objective optimization approach targeting the maximum efficiency and minimum specific equipment cost. To make the system energy self-sufficient, an internal combustion engine burns a fraction of the raw biogas stream to co-generate both the electric power for upgrading process and the thermal power required for solvent regeneration. Aqueous MDEA turns out to be more efficient and less expensive than the MDEA/MEA blend, as a result of the lower regeneration energy (0.94 vs 1.43 kJ/Nm³BM/h) which yields to lower energy consumptions and smaller engine sizes. Even though the cost estimates are subject to a higher degree of uncertainty, the MDEA option features an expected specific total equipment cost as low as 1550€/(Nm³BM/h)) vs 1850€/(Nm3BM/h)) for the MDEA/MEA blend.

5.2 Adsorption

Pressure swing adsorption, vacuum swing adsorption, temperature swing adsorption, and electrical swing adsorption are biogas techniques that use adsorption phenomenon.

5.2.1 Pressure swing adsorption (PSA) and vacuum swing adsorption (VSA)

The PSA method involves the adsorption of biogas under high pressure by adsorbents like zeolite, activated carbon, and silica. Under pressure, H_2S and CO_2 have more affinity toward these adsorbents thus increasing methane concentration up to 95%–99%. The atmospheric air is compressed and the PSA system will be pressurized for a predetermined period and depressurized to atmospheric pressure. The low sorbing gas will slowly leave the adsorption column first followed by other gases. To avoid downtime, PSA process involves the use of two or more columns during pressurized and depressurized processes. Thus, the upgraded biogas can be directly delivered at high pressure. If the same adsorption process occurs under vacuum, it will be called as vacuum swing adsorption (VSA). The disadvantages are high CAPEX and OPEX and extensive process control is needed (Kohlheb et al., 2020). EcoClean™ is a robust and economical commercial gas cleaning system supplied by Praj Industries Limited, India. EcoClean™ efficiently removes 0.02%–3% of H_2S and up to 40% CO_2 and other impurities. The system uses pressurized chilled water for CO_2 removal. The purified methane is then subjected to molecular sieve drying and compressed in cylinder. Praj supplies customized EcoClean™ designs varying from 100 to 200 M³/h of biogas flow (Fig. 9). The schematic diagram of pressure swing adsorption is shown in Fig. 10 (Zhao et al., 2010).

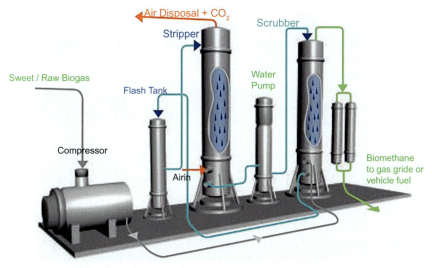

FIG. 9 High pressure water scrubbing system for biogas purification. *Source: https://www.praj.net/wp-content/up loads/2017/11/Ecoclean-BioCNG-Energy-from-Waste.pdf.*

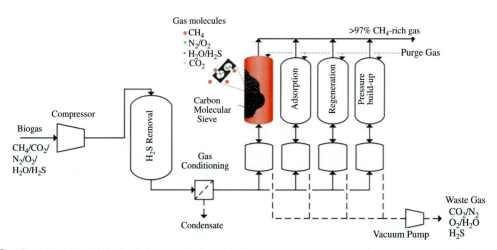

FIG. 10 Schematic diagram of pressure swing adsorption.

5.2.2 Temperature swing adsorption (TSA)

It consists of interconnected multistage fluidized bed column and utilizes solid amine sorbents for selective adsorption of carbon dioxide. TSA processes are not commonly used due to low thermal conductivity of adsorbent bed, which possesses challenges for desorption of impurities and regeneration of adsorbent. However, high heat and mass transfer coefficients possible with microchannel offer the potential for using the TSA process for gas purification. TSA disadvantages are low energy efficiency and thermal aging of the adsorbent (Vogtenhuber et al., 2018).

5.2.3 *Electrical swing adsorption (ESA)*

In ESA, voltage is applied to heat adsorbent and the adsorbed gas is released. Grande et al. (2009) used ESA technique to capture CO_2 from flue gases of power plant. Carbon dioxide rich streams with 89.7% purity and 70% recovery and energy consumption of 1.6GJ/ton of CO_2 were captured. ESA technique is not very common at industrial scale.

5.3 Cryogenic separation

Cryogenic process is a distillation process performed under low temperature ($-170\,°C$) and high pressure ($\sim 80\,bar$) resulting in liquefaction of crude biogas content under different temperature and pressure conditions. The process involves sequential condensation and distillation. Main impurity such as carbon dioxide is collected from the raw biogas by condensation as methane has a lower boiling point ($-165.1\,°C$) than carbon dioxide ($-78.2\,°C$) at atmospheric pressure. The methane purity of 90%–99% can be achieved in this process. However, it requires high OPEX and high CAPEX (Wellinger and Lindberg, 2000; Yousuf et al., 2016). Amine scrubbing based absorption has been considered as attractive option for biogas upgrading at industrial scale. Temperature increase is associated with amine regeneration in compatible with cooling requirement for subsequent liquefaction process. Thus, cryogenic biogas upgradation integrated with liquefaction can be an interesting alternative (Hashemia et al., 2019).

5.4 Membrane separation technology

Membranes are dense and nonporous or made up of hollow fibers. Biogas dissolves and diffuses into these polymeric membranes. The rate of diffusion depends on the molecule size. The driving force is a pressure gradient subjected to the membrane by compression of inflow. Carbon dioxide molecules are larger than methane. Membranes with certain pore size having high separation efficiency will concentrate methane content up to 98% in purified biogas and pure carbon dioxide is also obtained. Membrane technology is accepted globally due to good selectivity, low energy consumption, low level of mechanical wear, low maintenance and absence of moving parts, compact design, lower CAPEX and OPEX. However, multiple steps are required to achieve higher purity. Due to high pressure gradient, membrane disruption may occur. Clogging of membrane, wear and tear can happen after continuous exposure to certain solvents. These limitations are important as membranes are expensive. The membranes are made up of various polymeric materials such as polysulfone, polydimethylsiloxane, polyamide, poly vinyl alcohol (PVA), and poly ethylene glycol (PEG).

Biogas content varies depending upon the process conditions such as pH, temperature, feedstock, carbon-to-nitrogen ratio, and anaerobic digestion kinetics. Thus, biogas purification is of utmost importance. Biogas upgradation process involves the concentration of methane by removal of other gases such as carbon dioxide, nitrogen, hydrogen sulfide, etc. Various biogas upgradation techniques such as absorption, adsorption, cryogenic distillation, and membrane purification can be used considering the initial biogas composition, operating conditions, and operating cost. However, due to simplicity, membrane technologies are being used frequently due to their high separation efficiency. The compact design of membranes and easy operations have high potential to replace conventional biogas purification methods. However, as per all processes, membrane separation also needs to be optimized to reduce cost and achieve high purification. Thus, hybrid systems are being used to reduce both capital and operational costs compared to single systems. Additionally, the combustion of refined biogas produces less GHG compared to biodiesel and bioethanol (Numiuncharoen et al., 2015).

The upgraded biogas can be injected directly into natural gas network of interest as these have similar properties as of natural gas for different applications. Additionally, the pipeline networks also need to be of good quality to meet the standards of upgraded biogas to prevent corrosion. Currently, biogas upgradation techniques are rapidly evolving to cope up with other bioenergy sources such as bioethanol, biodiesel, biomethanol, etc. In Denmark, 25% of upgraded biogas containing 90% methane is injected into natural gas network and <60% is used after mixing with natural gas (Deublein and Steinhauser, 2011).

Membrane biogas upgradation system is simple in design, easy for installation, scale up and operation, low capital and maintenance cost, and uses low energy. The membrane system is classified into high and low pressure processes. Primary group is gas–gas based processes operated at 8–20 bars and the permeation depends on membrane properties, gas composition, and operating parameters. Other is an absorbent gas based process operated at atmospheric pressure and permeation depends on solvent affinity for the acid gases as well (Patterson et al., 2011). Membranes are classified as organic and inorganic membranes. Thus, as per the process conditions, membranes can be selected depending on their physical and chemical properties. Polysulfone, polyamide, or polydimethylsiloxane are commonly used for biogas upgradation. However, polyamide, cellulose acetate, perfluro polymers, silicon rubbers, and polysulfone can also be selected to remove acid gases from natural gas. The membranes' selectivity and permeability can be enhanced using mixed matrix membranes. Polymer membranes have low cost and good operational stability at high pressure and temperature with flat sheets and hollow fibers (Adewole et al., 2013).

Membrane biogas separation processes require larger surface area as per the higher gas capacity. Three types of membrane such as hollow fiber, spiral wound, and envelop are used for gas separation. Hollow fiber and spiral wound modules provide larger surface area compared to envelop module (Scholz et al., 2013). However, hollow fiber module is more economical due to its higher effective surface area per unit volume of membrane module compared with spiral wound module.

The number of modules and their configuration as well as transmembrane pressure highly affect separation efficiency. Hence, depending on the biogas composition and its application, single module for upgradation may be possible. However, single module is economically not attractive as methane loss contributes to 10%–15% and operates at low feed flow rate due to low capital and operation cost (Baker and Lokhandwala, 2008). Using permeate recycle, the methane recovery can be increased up to 95% and by reducing methane loss. Two or more membrane modules can highly improve methane purity and recovery compared to single-stage membrane module.

5.5 Hybrid systems for biogas upgradation

In case of hybrid systems, conventional methods (absorption, adsorption, and cryogenic distillation) and membrane technology methods are combined to increase methane concentration and reduce the overall operational cost. Conventional methods consume high energy, requires high capital and maintenance cost and space. On contrary, membrane requires lesser space, low maintenance and operational cost. Thus, a hybrid system probably provides a better biogas upgradation system to benefit from the advantage of each process to overcome the individual disadvantages. In a hybrid membrane/adsorption process, membrane unit removes the bulk of acid gases from crude biogas before injecting it into the absorber system. Thus, less quantity of absorbent is required. Similarly, integration with cryogenic process can also result in reduction of capital and operation cost by decreasing equipment and plant size.

Bhide et al. (1998) used first hybrid system including cellulose acetate membrane and amine scrubbing technology for biogas upgradation. The cellulose acetate membrane consisting of three membrane units with CO_2/CH_4 and H_2S/CH_4 selectivity was used simultaneously to remove acid gases and to minimize methane loss and purification cost. The retentate stream was passed to absorber unit using diethanolamine (DEA) for further purification. The permeate stream, enriched with CO_2 and H_2S, was in other application. The techno-economic analysis showed that the total purification cost was affected by the H_2S concentration in crude biogas. The hybrid system and membrane systems had the lowest total purification cost for the feed with and without H_2S, respectively. Hybrid system purified biogas with lower cost compared to gas adsorption process alone. Carbon dioxide ($\sim78\%$) was captured using membrane unit resulting in lower absorbent circulation rate and plant size. Total separation cost of single membrane unit changed with the feed gas composition, as higher acid gas concentration required higher membrane area and power to obtain higher methane purity and recovery. Table 2 shows advantages and disadvantages of biogas separation technologies (Chandra et al., 2012).

5.6 City gas distribution (CGD) in India

Governments should also provide more attention to lay down policies and invest in the upgradation of biogas production plants than just using crude biogas for domestic use. Free parking, lower tax on biogas vehicles, toll exemption, and financial support for investment in biogas vehicles are some of the initiatives made by Swedish government (Holmgren, 2013).

The GOI is planning to set 10,000 CNG stations to develop gas-based economy and to double city gas networks to 400 districts, quadruple pipe cooking gas connection to households to 2 crores. About 32 lakh gas-based vehicles are being served by 1500 CNG stations (Press Report, 2018a). Produced BioCNG can be transported to fuel station network of oil market companies as a green transport fuel for efficient distribution alternative for traditional vehicle fuel. Additionally, the BioCNG plants can be integrated with CGD network.

TABLE 2 Comparison of different methods of methane enrichment in biogas.

Method	Advantages	Disadvantages
Absorption in water	One of the easiest and cheapest methods for CO_2 removal. Recommended for rural application	Water pumping load is high and some loss of methane with washing water
Absorption by chemicals	More efficient in low pressure and can remove CO_2 at partial pressure in treated gas	Regeneration of the solvent requires a relatively high energy input. Disposal of by-product formed due to chemical reaction is a problem
Pressure swing adsorption	By proper choice of adsorbent, CO_2, H_2S, moisture, and other impurities can be removed	Adsorbent is accomplished at high temperature and pressure. Regeneration is carried out by vacuum. Costly process
Membrane separation	Modular in nature and separates CO_2 and CH_4 efficiently	High pressure required. Pressure cost is also high
Cryogenic separation	Allows recovery of pure component in the form of liquid which can be transported conveniently	High cost involved makes it impractical for biogas applications
Chemical conversion	Extremely high purity in the product gas	Process is extremely expensive and is not warranted in most cases of biogas applications

As per the GOI, 5000 CBG plants are expected to produce 15 million tonnes CBG/annum which is equivalent to 40% of current CNG consumption of 44 million tonnes/annum of country. Thus, an investment of around Rs. 1.7 lakh crore is expected to generate direct employment for 75 thousand people and produce 50 million ton of biomanure for crops. The potential for BioCNG production from various sources stands about 62 million ton/annum (Press Report, 2018b, c).

By 2030, the biogas share in energy sector will be increased from 6.2% to 15% helping to meet India's COP-21 commitment of reducing emission intensity by 33%–35% (Press Report, 2018a). The entrepreneurs will be allowed to market co-products of biogas plant to enhance returns on investment (EOI, 2018). As per the reports, CGD market of CNG of India is estimated to be 25,570 Million Metric Standard Cubic Meter (MMSCM) in 2030 at a CAGR of 10% compared to 2020. The increase in demand is due to demand from automotive, industrial, commercial, and residential end users. GOI is laying down various favorable policies to authorize the licenses for CGD due to which number of public and private corporates are interested in various geographies of India. Thus expanding CGD network to 228 geographical areas covering overall 70.86% of cumulative population and 52.80% of the countries area (Report Linker, 2020).

The first BioCNG plant with $85M^3$ capacity was installed in Rajasthan with collaboration from IIT, Delhi. Plant used cow dung as feedstock and production cost was about US\$ 0.23–0.24/kg which is much cheaper as compared to CNG (Press Report, 2016). However, the CBG supplied shall meet IS16087:2016 specification as per Bureau of Indian Standards (BIS). As per specification the compressed biogas should contain at least 90% methane, 4% carbon dioxide, 0.5% oxygen, H_2S not more than 20 mg/m^3.

Additionally, the CBG shall be free from liquids over the entire range of temperature and pressure encountered in storage and dispensing systems. It shall be free from particulate matter. Purified biogas is compressed at 250 bar and supplied through cascades at nearest CGD network plant (BIS Standards, 2016).

6 Biogas programme of India

The Ministry of New and Renewable Energy (MNRE) is the nodal ministry of the GOI for all matters relating to new and renewable energy. The broad aim of the ministry is to develop and deploy new and renewable energy to supplement the energy requirements of the country. MNRE focuses to develop new and renewable energy technologies, processes, materials, components, subsystems, products, and services at par with international specifications, standards, and performance parameters in order to make the country a net foreign exchange earner in the sector and deploy such indigenously developed and/or manufactured products and services in furtherance of the national goal of energy security.

6.1 Waste to energy programme

The objectives of waste to energy programme are to promote setting up of projects for recovery of energy in the form of Biogas/BioCNG/Biopower from urban, industrial, and agricultural waste and captive power and thermal use through gasification in industries. Setting up of projects for recovery of energy from municipal solid waste (MSW) for feeding power into the grid and for meeting captive power, thermal and vehicular fuel requirements. Biomass gasifier is used for feeding power into the grid or meeting captive power and thermal needs of rice mills/other industries and villages.

Waste to energy projects are mainly installed to treat effluent/waste generated for the biogas plant capacity in the size range of >2500 M^3/day and power generation capacity in the range of >250 kW. Mostly the projects are set up in industrial sectors namely distillery, paper and pulp, solvent extraction, rice mills, textiles, pharmaceutical industries, etc. India has set up 186 waste to energy projects with a cumulative capacity of 317.03 MW. Estimated potential of energy recovery from urban and industrial organic waste is up to 5690 MWeq (https://mnre. gov.in/waste-to-energy/current-status). Out of 186 projects, 5 projects are based on municipality solid waste generating a total energy capacity of 66.5 MW , 94 projects are using waste from urban, industrial and MSW for power generation, 78 projects are under biogas (off-grid) category and 16 are under BioCNG section (Press Report, 2019).

The programme also provides back-ended capital subsidy for setting up waste to energy plants. Waste to energy programme provides nearly 20%–30% of financial assistance for biogas plant installation and generation. Waste to energy program also offers up to 5% tax relaxation including concessional customs duty for initial setting up of grid-connected projects for power generation (MNRE, 2018).

Department of Biotechnology has been promoting R&D for Biofuel Technology development recognizing the need for clean and renewable energy. Various waste to energy projects were funded to develop/demonstrate novel and viable technologies for sustainable utilization of municipal solid waste (MSW) for cleaner and pollution-free environment as well as generation of the energy. The newly started DBT-BIRAC Clean Tech Demo Park will demonstrate innovative Waste-to-Value technologies with support from DBT and BIRAC. DBT supported two demonstration plants successfully commissioned and operated for converting municipal solid waste (bioorganic fraction) into energy by an improved biomethanation process at the scale of 3–7 tons/day. The trial is ongoing for operation at full capacity of plants located at Hyderabad and Goa (*DBT Annual Report* (2020-2021), 2021).

6.2 New national biogas and organic manure programme (NNBOMP)

NNBOMP promotes the use of biogas produced from cattle manure and other organic wastes available in rural areas. It has helped to establish multiple small-scale biogas plants with installed capacity ranging from 1 to 25 M^3/day. The biogas obtained is used mostly for cooking purpose and the digestate is used as organic fertilizer reducing the use of chemical fertilizers. Biogas Power Generation (off-grid) and Thermal energy application programme (BPGTP) also supports community-scale biogas plants and thermal energy with installed capacity ranging from 30 to 2500 M^3/day and power generation capacity of 3–250 kW.

Financial subsidies up to 30%–35% of total biogas plant installation cost are provided by MNRE. Till date, over 5 million small-scale biogas plants are installed in India under NNBMOP scheme producing about 7.2 MW of power. Additionally, under BPGTP, the community scale plants contribute power generation capacity of 8753 MW and biogas generation capacity of 86,595 M^3/day (MNRE, 2020).

6.3 National biofuel policy

In last decade, biofuels have caught the attention and to keep up with new developments in renewable sectors the national biofuel policy was relaunched in 2018. The policy sets new agenda consistent with the redefined role of emerging developments in the renewable sector in the context of international perspective and national scenario.

The policy targets the use of agricultural waste and surplus food grain. Policy emphasized on promotion of advanced biofuels such as ethanol (from 2G sources), algae-based (3G) biofuels, BioCNG, biomethanol, dimethyl ether (DME), biohydrogen, etc. As of now, only 5% ethanol and 1% biodiesel are blended with petrol and diesel. However, the relaunched national biofuel policy targets blending of ethanol up to 20% in petrol and biodiesel up to 5% in diesel by 2030. Ministry of Road Transport and Highways (MoRTH) also included the provision for the use of advanced biofuels and BioCNG in motor vehicle produced from waste in Central Motor Vehicle Rules and directed various oil manufacturing companies (OMC) accordingly (MoRTH, 2015).

Policy also directed Central Financial Assistance (CFA) to provide financial support in the form of capital subsidy and Grant-in-Aid for biogas produced from industrial waste, wastewater, municipal solid waste, and agricultural waste by biomethanation process. Policy will also assist in various advanced biofuel promotional activities including research and development, resource assessment, technology upgradation, and performance evaluation.

6.4 Sustainable alternative toward affordable transportation (SATAT)

SATAT initiative promotes CBG or BioCNG as an alternative green transport fuel. It also supports the form of marketing of BioCNG. As per the National Biofuel Policy 2018, the SATAT initiative aims to guarantee production offtake where public sector OMC buy CBG at fixed rate of Rs. 48.30/Kg in cascade form, set up CBG plants mainly by independent entrepreneurs, provide an additional revenue source to farmers, and reduce import of natural gas. SATAT aims to sell CBG initially at OMC fuel stations and later it can be integrated with gas grid. The potential for CBG production from various sources in India is estimated to be about 62 million ton per annum. Indian government expects to produce 15 million ton of CBG per annum by 2023 from 5000 CBG plants, which is about 40% of current CNG consumption of 44 million ton per annum (MoPNG, 2020).

6.5 Galvanizing organic bio-agro resources Dhan (GOBAR-Dhan)

GOBAR-Dhan was launched in 2018 as a part of Clean India Mission (Rural) to efficiently manage rural biowaste leading to better environment and public health. The programme will also lead to generate employment opportunities and household savings. Rural India generates enormous quantities of biowaste including kitchen waste, crop residues, animal waste, fecal sludge, market waste, etc. Biogas plants are to be set up by gram panchayat, self-help groups, bulk waste generators, and entrepreneurs. At least one project per district will be selected by every state to achieve effective biowaste management at rural level. The total financial support will be provided on the basis of total number of households in each gram panchayat (MoDDWS, 2018).

6.6 Fertilizer control order (FCO)

FCO plays an important role to maintain rules and regulations for statutory quality control requirements of biofertilizers and organic fertilizers including revision of standards and testing protocols considering advancement in research and technology and bringing remaining organic inputs under quality control regime. FCO helps for soil health assessment, organic input resource management and market development for fertilizers. FCO also helps in promotion or organic farming in India including technology development, technology transfer, promotion and production of quality organic and biological inputs, awareness creation, and publicity. The Ministry of Agriculture and Farmers' Welfare (MoAFW) issued fertilizer (inorganic, organic, or mixed) control second amendment order in 2020 to further amend Fertilizer Control Order, 1985 (MoAFW, 2020).

7 Indian biogas players

Small biogas systems of $1-10\,M^3$ capacity per day mostly exist in rural areas. Feedstock such as animal manure and agriculture waste are used for biogas production only for localized consumption. However, large- and industrial-scale biogas plants with a production capacity above $5000\,M^3$ per day use municipal or industrial organic waste. The large-scale biogas suppliers provide end-to-end solutions and few Indian domestic players are mentioned here.

7.1 Praj Industries Limited

Praj Industries Limited has always been at the forefront of introducing innovative technologies in the bioenergy space. In addition to multifeedstock to multiproduct Enfinity™—2G lignocellulosic to ethanol technology, Bioethanol (1G), Ecodiezel™, marine biofuels (3G), and sustainable aviation fuel (SAF) technology, Praj introduced RenGas™ technology—a renewable fuel technology to the circular bioenergy sector. These technologies complement each other and play a key role in India's pursuit of energy self-reliance and decarbonization through circular bioeconomy. RenGas™ technology will contribute to the MoPNG, GOI's SATAT initiative with an objective to promote CBG as an alternative, green transport fuel.

Praj Industries Limited, India offers RenGas™ technology to process agri-residues for the production of CBG from different feedstock such as biomass, pressmud, variety of agricultural waste, paper mill pith, etc., developed at Praj-Matrix, R&D Center. RenGas™ technology uses patented PMStab™ preservation technique for pressmud, BMSolve microbes for biomass pretreatment, and RC (rumen consortia) for AD to achieve higher biogas yields. RenGas™ technology has incorporated unique dual plug flow digester design, in collaboration with DVO Inc. of USA (www.dvoinc.com). This patented design technology has several advantages including higher efficiency, lower energy consumption, and near zero maintenance. Praj also provides customized gas cleaning system EcoClean™.

Praj also offers complete turnkey/EPC plant and technology supply with O&M options, commercially sound CBG projects with high IRR and low payback periods. The project will generate significant employment opportunities in surrounding farming and rural community. Agricultural waste in the form of rice straw procured as feedstock for the CBG project will provide additional income revenue stream to farmers, facilitating GOI's flagship program of doubling farmers' income.

7.2 ^{SLPP} Re-new LLP

^{SLPP} Re-New is a fast-growing bioenergy EPC company providing one stop solution for waste to energy advanced gas production projects for power generation. Company also provides biogas upgradation system to produce BioCNG. ^{SLPP} Re-New delivers cutting edge technology and holistic solutions for natural gas/biogas/sewage gas/landfill gas/coal bed methane based power plants and BioCNG systems. ^{SLPP} Re-New provides different modules depending on the feedstock quantity ranging from 0.5 TPD to more than 30 TPD. The company is specialist in biogas power and BioCNG systems (http://www.slppre-new.in/commercial_equipment1.html).

7.3 Primove Engineering Pvt. Ltd. (PEPL)

PEPL is a nationally recognized EPC company in the domain of gaseous fuels and energy started in 2008. A pilot plant of patented AgroGas® technology produces 100 kg of CNG per day. First commercial AgroGas® plant in India was established in Nagpur which has the capacity of 5 ton CNG production per day. AgroGas technology uses agricultural biomass and converts it to BioCNG by AD process. The BioCNG is used as transport fuel and for domestic and commercial cooking (http://www.primove.in/pdf/Brochure_AgroGas.pdf).

7.4 Thermax limited

Thermax is an engineering company founded in 1966. It is involved in energy and the environment sector. Thermax developed BioEnergen technology for processing organic biodegradable solid waste in an environment-friendly manner to generate biogas. It also generates nutrient-rich fertilizer. The technology offers an excellent alternative for decentralized processing of solid biodegradable waste and avoids the contamination of landfill sites. The units are standardized and are compact in size. The size of BioEnergen plant processes kitchen waste ranging from 250 to 1000 kg/day. The produced biogas is used for cooking purpose, process heating, and power generation through gas gensets (www.thermaxglobal.com; https://www.nikhiltechnochem.com/pdf/Bioenergen.pdf).

7.5 NobleExchange environment solutions Pvt. ltd.

NobleExchange (NEX) is a leader in anaerobic digestion technology of organic food waste processing. The company was established in 2013 and has been producing high-quality biogas and organic fertilizer. The biogas is compressed and commercialized to replace fossil fuel. They have developed a proprietary one-box biogas technology. The Pune-based plant processes 350 ton/day (TPD) of organic food waste and produces BioCNG of 17,500 kg/day. NEX also deals with the Bruhat Bangalore Municipal Corporation (BBMP) with a plant processing capacity of 250 TPD of organic food waste and produces about 12,500 kg/day BioCNG. NEX plants conform to European standards of safety and environment. Several technical agencies have analyzed the proposed NEX technology and have found it to be an environmentally efficient way of disposing the wet and organic/food waste. By-product generated from the process will be nitrogen-rich, organic fertilizer that replaces conventional chemical fertilizers for farming and gardening purposes. CBG produced at the processing plant can be marketed to institutional clients for replacing conventional energy sources such as LPG, diesel, kerosene, furnace oil, etc. (https://nobleexchangesolutions.com/about-us/about-noble-exchange-environment-solutions-nex/).

7.6 Sampurn Agri Ventures Pvt. Ltd. (SAVPL)

SAVPL has developed the novel retreatment process of rice crop stubble (paddy straw) to produce biogas and fermented manure. The technology for the fermentation of paddy straw on a commercial scale was indigenously derived from the basic anaerobic digestion process along with its pretreatment. SAVPL's plant at Fazilka is the first of its kind in the world to be able to process paddy straw on a commercial scale for gainful management of paddy straw. The biogas produced after its purification can be sold to a prospective buyer. Additionally, the remnant compost is processed further to produce bio-enriched manure which has a good potential for sustainable agriculture. The bio-enriched manure can be customized as per the soil condition such as acidic, alkaline, or neutral. The project is also helpful to create much required rural employment for the population.

7.7 Spectrum renewable energy Pvt. Ltd. (SREL)

SREL offers organic waste producers a comprehensive spectrum of services including project management, plant design, construction as well as maintenance and technological support. SREL implements AD technology to produce CBG from pressmud (100 TPD) while increasing agricultural profitability and sustainability through environmentally responsible practices at Warnanagar, Maharashtra. With an onsite combined heat and power unit, electrical power is also generated. The portion of gas is used for auxiliary power consumption to run the plant. The waste heat from the CHP unit is being used to maintain the digester temperature. The solids and liquid obtained after biomethanation process are used as organic fertilizer (http://srel.in/).

8 Future considerations

Biogas in India is of strategic importance with the ongoing initiatives of the Government of India such as Make in India and Swachh Bharat Abhiyan and offers great opportunity to integrate with the ambitious targets of doubling farmers' income, import reduction, employment generation, and waste to wealth creation. Simultaneously, the existing biodiversity of the country can be put to optimum use by utilizing dry lands for generating wealth for the local populous and in turn contribute to the sustainable development. With the technological advances that have exponentially gone up in the past decade, the biogas and BioCNG industry is well poised to grow and contribute to a sustainable planet.

India with its "Atmanirbhar Mission" is well poised to lead this change in bringing global technology to local value generation. For the betterment of the planet and conditions of living, most of the household biogas plants based on animal waste as a substrate started this change. With advancement in AD technology, various organic waste residues like agricultural residues, food, sugar industry, municipal solid waste, animal farms waste, etc. can be now utilized for biogas. The AD technique has various social advantages such as waste management, better health benefits and increased employment at rural level. The renewable sector will also create jobs for people in transport of feedstock and products. Improved health condition of biogas users indirectly helps to increase income by preventing expenses due to illness. With comprehensive policies, the transition may greatly boost overall employment in the renewable sector. All these positive events mark the beginning of the circular economy wherein this gas economy plays a crucial role. With India poised to take a lead in circular bioeconomy through supportive policy framework, technological advantages customized to the country's needs, and a progressive community to support, gas economy will show the path toward sustainable future to the world.

Praj Industries Limited: Praj, India's most accomplished industrial biotechnology company, is driven by innovation, integration, and delivery capabilities. Praj Industries Ltd. is

India's leading company in industrial biotechnology, globally known for its Technology, Engineering, Manufacturing, Project management, Operations (TEMPO) capabilities. Praj is ranked second in the list of top 50 hottest companies in Advanced Bioeconomy in 2020 released by the industry's leading publication Biofuels Digest, USA. Bio-Mobility™ and Bio-Prism™ are the mainstays of Praj's contribution to the global bioeconomy. The Bio-Mobility™ portfolio offers technology solutions globally to produce renewable transportation fuel, thus ensuring sustainable decarbonization through circular bioeconomy. The company's Bio-Prism™ portfolio comprises technologies for the production of renewable chemicals and materials solutions, promises sustainability, while reimagining nature.

For more information, visit www.praj.net.

References

Abbasi, T., Tauseef, S.M., Abbasi, S.A., 2012. Low-Rate and High-Rate Anaerobic Reactors/Digesters/Fermenters. Biogas Energy. Springer, New York.

Adewole, J., Ahmad, A., Ismail, S., Leo, C., 2013. Current challenges in membrane separation of CO_2 from natural gas: a review. Int. J. Greenhouse Gas Control 17, 46–65. https://doi.org/10.1016/j.ijggc.2013.04.012.

Amaratunga, M., 1986. Structural behaviour and stress conditions of fixed dome types of biogas units. In: El-Halwagi, M.M. (Ed.), Biogas Technology, Transfer and Diffusion. Springer, London, pp. 295–301.

Amirta, R., Tanabe, T., Watanabe, T., Honda, Y., Kuwahara, M., Watanabe, T., 2006. Methane fermentation of Japanese cedar wood pretreated with a white rot fungus, *Ceriporiopsis subvermispora*. J. Biotechnol. 123 (1), 71–77. https://doi.org/10.1016/j.jbiotec.2005.10.004.

Baker, R.W., Lokhandwala, K., 2008. Natural gas processing with membranes: an overview. Ind. Eng. Chem. Res. 47 (7), 2109–2121. https://doi.org/10.1021/ie071083w.

Barik, S.K., Mishra, S., Ayyappan, S., 2000. Decomposition patterns of unprocessed and processed lignocellulosics in freshwater fish pond ecosystem. Aquat. Ecol. 34 (2), 185–204. https://doi.org/10.1023/A:1009981319515.

BCC Research Report, 2018. Waste Derived Biogas: Global Market for Anaerobic Digestion Equipment. https://www.bccresearch.com/market-research/energy-and-resources/waste-derived-biogas-global-markets-for-anaerobic-digestion-equipment-report.html.

Bhide, B., Voskericyan, A., Stern, S., 1998. Hybrid processes for the removal of acid gases from natural gas. J. Membr. Sci. 140 (1), 27–49. https://doi.org/10.1016/S0376-7388(97)00257-3.

BIS Standards, 2016. http://services.bis.gov.in:8071/php/BIS/bisconnect/pow/is_details?IDS=MjA3.

Boe, K., Angelidaki, I., 2008. Serial digester configuration for improving biogas production from manure. Water Res. 43 (1), 166–172. https://doi.org/10.1016/j.watres.2008.09.041.

Anon., 2020. BP Statistical Review of World Energy. https://www.bp.com/content/dam/bp/business-sites/en/global/corporate/pdfs/energy-economics/statistical-review/bp-stats-review-2020-full-report.pdf.

Bruni, E., Jensen, A.P., Angelidaki, I., 2010. Comparative study of mechanical, hydro thermal, chemical and enzymatic treatments of digested biofibers to improve biogas production. Bioresour. Technol. 101 (22), 8713–8717. https://doi.org/10.1016/j.biort ech.2010.06.108.

Capra, F., Fettarappa, F., Magli, F., Gatti, M., Martelli, E., 2018. Biogas upgrading by amine scrubbing solvent comparison between MEDA and MDEA/MEA blend. Energy Procedia 148, 970–977. https://doi.org/10.1016/j.egypro.2018.08.065.

Carlos, M., Klinke, H.B., Marcet, M., Gracia, L., Hernandez, E., Thomsen, A.B., 2007. Study of the phenolic compounds formed during pretreatment of sugargane bagasse by wet oxidation and steam explosion. Holzforschung 61 (5), 523–530. https://doi.org/10.1515/HF.2007.106.

Chan, Y.J., Chong, M.F., Law, C.L., Hassell, D.G., 2009. A review on anaerobic–aerobic treatment of industrial and municipal wastewater. Chem. Eng. J. 155, 1–18. https://doi.org/10.1016/j.cej.2009.06.041.

Chandra, R., Vijay, V.K., Subbarao, P.M.V., 2012. Vehicular quality biomethane production from biogas by using an automated water scrubbing system. ISRN Renew. Energy 904167, 1–6. https://doi.org/10.5402/2012/904167.

Chiumenti, A., da Borso, F., Limina, S., 2017. Dry anaerobic digestion of cow manure and agricultural products in a full-scale plant: efficiency and comparison with wet fermentation. Waste Manag. 2017 (71), 704–710. https://doi.org/10.1016/j.wasman.2017.03.046 (Epub 2017 Apr 25).

CPCB, 2012. Central Pollution Control Board, Government of India.

Anon., 2021. DBT Annual Report (2020-2021). http://dbtindia.gov.in/sites/default/files/Final%20English%20Annual%20Report%202021.pdf.

Deublein, D., Steinhauser, A., 2011. Biogas From Waste and Renewable Resources: An Introduction. 2011 John Wiley & Sons.

EAI, 2009. Biomethanation in India—Biogas Potential, Trends and Prospects.

EC, 2009. DIRECTIVE 2009/28/EC of the European Parliament and of the Council on the Promotion of the Use of Energy from Renewable Sources and Amending and Subsequently Repealing Directives 2001/77/EC and 2003/30/EC, 2009.

EOI, 2018. EOI for Production and Supply of Compressed Bio Gas (CBG). https://www.bharatpetroleum.com/pdf/EOICBGFINAL-46978-4fd490.pdf.

Gigot, C., Ongena, M., Fauconnier, M.L., Muhovski, Y., Wathelet, J.P., du Jardin, P., Thonart, P., 2012. Optimization and scaling up of a biotechnological synthesis of natural green leaf volatiles using *Beta vulgaris* hydroperoxide lyase. Process Biochem. 47 (12), 2547–2551. https://doi.org/10.1016/j.procbio.2012.07.018.

Goncalves, M.M.M, Leite, S.G.F, SantaAnna Jr, G.L, 2005. The bioactivation procedure for increasing the sulphate-reducing bacteria in a UASB reactor. Braz. J. Chem. Eng. 22 (4), 565–571. https://www.scielo.br/scielo.php?script=sci_arttext&pid=S0104-66322005000400009.

Grande, C.A., Ribeiro, R.P.L., Oliveira, E.L.G., Rodrigues, A.E., 2009. Electric swing adsorption as emerging CO_2 capture technique. Energy Procedia 1, 1219–1225. https://doi.org/10.1016/j.egypro.2009.01.160.

Harasimowicz, M., Zakrzewska-Trznadel, G., Chmielewski, A.G., 2007. Application of polyimide membranes for biogas purification and enrichment. Hazard. Mater. 144, 698–702. https://doi.org/10.1016/j.jhazmat.2007.01.098.

Hashemia, S.E., Sarkera, S., Liena, K.M., Schnellb, S.K., Austbo, B., 2019. Cryogenic vs. absorption biogas upgrading in liquefied biomethane production—an energy efficiency analysis. Fuel 245, 294–304. https://doi.org/10.1016/j.fuel.2019.01.172.

Hashimoto, G., Varriel, H., 1978. Factors affecting methane yield and production rate. Am. Soc. Agric. Eng. 6 (2), 68–76.

Hendriks, A.T.W.M., Zeeman, G., 2009. Pretreatments to enhance the digestibility of lignocellulosic biomass. Bioresour. Technol. 100, 10–18. https://doi.org/10.1016/j.biortech.2008.05.027.

Herrmann, A., Kage, H., Taube, F., Sieling, K., 2017. Effect of biogas digestate, animal manure and mineral fertilizer application on nitrogen flows in biogas feedstock production. Eur. J. Agron. 91, 63–73. https://doi.org/10.1016/j.eja.2017.09.011.

Hiloidhari, M., Das, D., Baruah, D.C., 2014. Bioenergy potential from crop residue biomass in India. Renew. Sust. Energ. Rev. 32, 504–512. https://doi.org/10.1016/j.rser.2014.01.025.

Holmgren, K., 2013. Policies Promoting Biofuels in Sweden. https://www.osti.gov/etdeweb/servlets/purl/22138977.

Hu, F., Ragauskas, A., 2012. Pretreatment and lignocellulosic chemistry. Bioenergy Res. 5, 1043–1066. https://doi.org/10.1007/s12155-012-9208-0.

Huertas, J.L., Giraldo, N., Izquierdo, S., 2011. In: Markos, J. (Ed.), Mass Transfer in Chemical Engineering Processes. IntechOpen, pp. 133–135.

IBEF, 2021. https://www.ibef.org/industry/renewable-energy.aspx.

IEA, 2018. India Energy Balance. https://www.iea.org/sankey/#?c=India&s=Balance.

IEA, 2020. India 2020 Review Policy. https://niti.gov.in/sites/default/files/2020-01/IEA-India%202020-In-depth-EnergyPolicy_0.pdf.

Ighravwe, D.E., Babatunde, M.O., 2018. Determination of a suitable renewable energy source for mini-grid business: a risk-based multicriteria approach. J. Renew. Energy, 1–20. https://doi.org/10.1155/2018/2163262.

IREA, 2017. Renewable Capacity Statistics 2016. International Renewable Energy Agency.

Kalyani, D., Lee, K.M., Kim, T.S., Li, J., Dhiman, S.S., Kang, Y.C., Lee, J.K., 2013. Microbial consortia for saccharification of woody biomass and ethanol fermentation. Fuel 107, 815–822. https://doi.org/10.1016/j.fuel.2013.01.037.

Karatzos, S.K., Edye, L.A., Doherty, W.O.S., 2012. Sugarcane bagasse pretreatment using three imidazolium based ionic liquids: mass balances and enzyme kinetics. Biotechnol. Biofuels 5, 62. https://doi.org/10.1186/1754-6834-5-62.

Kaur, H., Sohpal, V., Kumar, S., 2017. Designing of small scale fixed dome biogas digester for paddy straw. Int. J. Renew. Energy Res. 7 (1), 422–431.

Klinke, H.B., Ahring, B.K., Schmidt, A.S., Thomsen, A.B., 2002. Characterization of degradation products from alkaline wet oxidation of wheat straw. Bioresour. Technol. 82 (1), 15–26. https://doi.org/10.1016/s0960-8524(01)00152-3.

Kohlheb, N., Wluka, M., Bezama, A., Thran, D., Aurich, A., Muller, R.A., 2020. Environmental-economic assessment of the pressure swing adsorption biogas upgrading technology. Bioenergy Res. https://doi.org/10.1007/s12155-020-10205-9.

Kumar, J.C.R., Majid, M.A., 2020. Renewable energy for sustainable development in India: current status, future prospects, challenges, employment, and investment opportunities. Energy Sustain. Soc. 10, 2. https://doi.org/10.1186/s13705-019-0232-1.

Kumar, R., Wymen, C.E., 2014. Strong cellulose inhibition by mannan polysaccharides in cellulose conversion to sugars. Biotechnol. Bioeng. 111 (7), 1341–1353. https://doi.org/10.1002/bit.25218.

Lopez, M.J., del Carmen Vargas-Garcia, M., Suarez-Estrella, F., Nichols, N.N., Dien, B.S., Moreno, J., 2007. Lignocellulose degrading enzymes produced by the ascomycete *Coniochaeta ligniaria* and related species: application for a lignocellulosic substrate treatment. Enzym. Microb. Technol. 40 (4), 794–800. https://doi.org/10.1016/j.enzmictec.2006.06.012.

Lynd, L.R., Weimer, P.J., van Zyl, W.H., Pretoris, I.S., 2002. Microbial cellulose utilization: fundamentals and biotechnology. Microbiol. Mol. Biol. Rev. 66 (3), 506–577. https://doi.org/10.1128/mmbr.66.3.506-577.2002.

Mackulak, T., Prousek, J., Svorc, L., Drtil, M., 2012. Increase of biogas production from pretreated hay and leaves using wood-rotting fungi. Chem. Pap. 66 (7), 649–653. https://doi.org/10.2478/s11696-012-0171-1.

Mao, C., Feng, Y., Wang, X., Ren, G., 2015. Review on research achievements of biogas from anaerobic digestion. Renew. Sust. Energ. Rev. 45, 540–555. https://doi.org/10.1016/j.rser.2015.02.032.

Massoud, K., George, T., Robert, C.B., 2007. Biomass Conversion Processes for Energy Recovery. Energy Conversion. CRC Press.

MNRE, 2016. Ministry of New and Renewable Energy for Rural Applications, Annual Report 2013–2014. Ministry of Renewable Energy, Government of India.

MNRE, 2018. Revised Guidelines of Waste to Energy Programme. https://mnre.gov.in/img/documents/uploads/file_f-1592674328017.pdf.

MNRE, 2020. Guidelines for Implementation of the Central Sector Scheme, New National Biogas and Organic Manure (NNBOMP) During the Period 2017–18 to 2019–20. https://mnre.gov.in/img/documents/uploads/file_s-1592215264726.pdf.

MoAFW, 2020. https://legalitysimplified.com/wp-content/uploads/2020/10/Minstry-of-Agriculture.pdf.

MoDDWS, 2018. https://drsjk.jk.gov.in/pdf/GOBAR%20DHAN%20guidelines.pdf.

MoPNG, 2018. http://petroleum.nic.in/sites/default/files/biofuelpolicy2018_1.pdf.

MoPNG, 2020. https://pib.gov.in/PressReleaseIframePage.aspx?PRID=1674428.

MoRTH, 2015. http://egazette.nic.in/WriteReadData/2015/164547.pdf.

NDDB, 2019. Livestock Population in India by Species. National Dairy Development Board, India. https://www.nddb.coop/information/stats/pop.

Nsair, A., Cinar, S.O., Alassali, A., Qdais, H.A., Kuchta, K., 2020. Operational parameters of biogas plants: a review and evaluation study. Energies 13, 3761. https://doi.org/10.3390/en13153761.

Numiuncharoen, T., Papong, S., Malakul, P., Mungcharoen, T., 2015. Life cycle GHG emissions of cassava based bioethanol production. Energy Procedia 79, 265–271. https://doi.org/10.1016/j.egypro.2015.11.477.

Paritosh, K., Kushwaha, S.K., Yadav, M., Pareek, N., Chawade, A., Vivekanand, V., 2017. Food waste to energy: an overview of sustainable approaches for food waste management and nutrient recycling. Biomed. Res. Int., 1–19. https://doi.org/10.1155/2017/2370927.

Patil, D., Laldas, S., Natu, A., Akade, J., Sawale, S., Kulkarni, A., 2020. Renewable natural gas technology (RenGas): sustainable solution. Chem. Ind. Digest, 40–43. https://www.researchgate.net/publication/340376454.

Patterson, T., Esteves, S., Dinsdale, R., Guwy, A., 2011. An evaluation of the policy and techno-economic factors affecting the potential for biogas upgrading for transport fuel use in the UK. Energy Policy 39 (3), 1806–1816. https://doi.org/10.1016/j.enpol.2011.01.017.

Paudel, S.R., Banjara, S.P., Choi, O.K., Park, K.Y., Kim, Y.M., Lee, J.W., 2017. Pretreatment of agricultural biomass for anaerobic digestion: current status and challenges. Bioresour. Technol. 245, 1194–1205. https://doi.org/10.1016/j.biortech.2017.08.182.

Pereira, S.R., Portugal-Numes, D.J., Evtuguin, D.V., Serafim, L.S., Xavier, A.M.R.B., 2013. Advances in ethanol production from hardwood spent sulphite liquors. Process Biochem. 48 (2), 272–282. https://doi.org/10.1016/j.procbio.2012.12.004.

Pfay, S.F., Hagens, J.E., Dankbaar, B., 2017. Biogas between renewable energy and bio-economy policies-opportunities and constraints resulting from a dual role. Energy Sustain. Soc. 7, 17. https://doi.org/10.1186/s13705-017-0120-5.

Press Report, 2016. New Source for Bio-CNG Production. The Hindu. http://www.thehindu.com/todays-paper/tp-national/tp-otherstates/new-source-for-biocng-production/article1283967.ece.

Press Report. 2018a. https://www.economictimes.indiatimes.com/articleshow/66752011.cms?from¼mdr&utm_source¼contentofinterest&utm_medium¼text&utm_campaign¼cppst.

Press Report, 2018b. https://energy.economictimes.indiatimes.com/news/oil-and-gas/indian-oil-companies-to-invite-interest-for-setting-up-bio-cng-projects/65995194.

Press Report, 2018c. http://www.sustainabilityoutlook.in/news/city-gas-distribution-see-big-investments-763567.

Press Report, 2019. https://mercomindia.com/india-has-set-up-waste-to-energy-projects-so-far.

Rajendran, K., Aslanzadeh, S., Taherzadeh, M., 2012. Household biogas digesters—a review. Energies 5 (8), 2911–2942. https://doi.org/10.3390/en5082911.

Rao, P.V., Baral, S.S., Dey, R., Mutnuri, S., 2010. Biogas generation potential by anaerobic digestion for sustainable energy development in India. Renew. Sust. Energ. Rev. 14 (7), 2086–2094. https://doi.org/10.1016/j.rser.2010.03.031.

Report Linker, 2020. India City Gas Distribution Market, by Type, by Source of Supply, by End Use, Competition Forecast & Opportunities, 2015–2030. https://www.reportlinker.com/p05879457/?utm_source=GNW.

Rodsrud, G., Lersch, M., Sjode, A., 2012. History and future of world's most advanced biorefinery in operation. Biomass Bioenergy 46, 46–59. https://doi.org/10.1016/j.biombioe.2012.03.028.

Rotunno, P., Lanzini, A., Leone, P., 2017. Energy and economic analysis of a water scrubbing based biogas upgrading process for biomethane injection into the gas grid or use as transportation fuel. Renew. Energy 102 (Part B), 417–432. https://doi.org/10.1016/j.renene.2016.10.062.

Saravanan, A.P., Mathimani, T., Deviram, G., Rajendran, K., Pugazhendhi, A., 2018. Biofuel policy in India: a review of policy barriers in sustainable marketing of biofuel. J. Clean. Prod. 193, 734–747. https://doi.org/10.1016/j.jclepro.2018.05.033.

Sarket, S., Lamb, J.J., Hielme, D.R., Lien, K.M., 2019. A review of the role of critical parameters in the design and operation of biogas production plants. Appl. Sci. 9, 1915. https://doi.org/10.3390/app9091915.

Sawale, S., Patil, D., Joshi, C., Rachappanavar, B., Mishra, D., Kulkarni, A., 2020. Biogas commercialization: commercial players, key business drivers, potential market, and fostering investment. In: Balagurusamy, N., Chandel, A.K. (Eds.), Biogas Production. Springer, Cham, https://doi.org/10.1007/978-3-030-58827-4_16.

Scarlat, N., Dallemand, J.F., Monforti-Ferrario, F., Banja, M., Motola, V., 2015. Renewable energy policy framework and bioenergy contribution in the European Union an overview from national renewable energy action plans and progress reports. Renew. Sust. Energ. Rev. 51, 969–985. https://doi.org/10.1016/j.rser.2015.06.062.

Scholz, M., Melin, T., Wessling, M., 2013. Transforming biogas into biomethane using membrane technology. Renew. Sust. Energ. Rev. 17, 199–212. https://doi.org/10.1016/j.rser.2012.08.009.

Sharholy, M., Ahmad, K., Mahmood, G., Trivedi, R.C., 2008. Municipal solid waste management in Indian cities—a review. Waste Manag. 28 (2), 459–467. https://doi.org/10.1016/j.wasman.2007.02.008.

Sogn, T.A., Dragicevic, I., Linjordet, R., Krogstad, T., Eijsink, V.G.H., Eich-Greatorex, S., 2018. Recycling of biogas digestates in plant production: NPK fertilizer value and risk of leaching. Int. J. Recycl. Org. Waste Agric. 7 (1), 49–58. https://doi.org/10.1007/s40093-017-0188-0.

Sooch, S.S., Gautam, A., 2013. PAU kutcha—Pucca model of biogas plants in Punjab—a case study. Agric. Eng. Today 37 (3), 6–11.

Strezov, V., Evans, T.J., 2015. Biomass Processing Technologies. CRC Press.

Sun, Q., Li, H., Yan, J., Liu, L., Yu, Z., Yu, X., 2015. Selection of appropriate biogas upgrading technology-a review of biogas cleaning, upgrading and utilisation. Renew. Sustain. Energy Rev. 51, 521–532. https://doi.org/10.1016/j.rser.2015.06.029.

Take, H., Andou, Y., Nakamura, Y., Kobayashi, F., Kurimoto, Y., Kuwahara, M., 2006. Production of methane gas from Japanese cedar chips pretreated by various delignification methods. Biochem. Eng. J. 28 (1), 30–35. https://doi.org/10.1016/j.bej.2005.08.036.

van Kuijk, S.J.A., Sonnenberg, A.S.M., Baars, J.J.P., Hendriks, W.H., Cone, J.W., 2016. The effect of particle size and amount of inoculum on fungal treatment of wheat straw and wood chips. J. Anim. Sci. Biotechnol. 7, 39. https://doi.org/10.1186/s40104-016-0098-4.

Vogtenhuber, H., Hofmann, R., Helminger, F., Schony, G., 2018. Process simulation of an efficient temperature swing adsorption concept for biogas upgrading. Energy 162, 200–209. https://doi.org/10.1016/j.energy.2018.07.193.

Wagner, A.O., Lackner, N., Mutschlechner, M., Prem, E.M., Markt, R., Illmer, P., 2018. Biological pretreatment strategies for second-generation lignocellulosic resources to enhance biogas production. Energies 11 (7), 1–14. https://doi.org/10.3390/en11071797.

Wang, L.K., Hung, Y.T., Lo, H.H., Yapijakis, C., 2005. Waste Treatment in the Food Processing Industry. CRC Press.

Weiland, P., 2010. Biogas production: current state and perspectives. Appl. Microbiol. Biotechnol. 85, 849–860. https://doi.org/10.1007/s00253-009-2246-7.

Wellinger, A., Lindberg, A., 2000. Biogas upgrading and utilization. In: IEA Bioenergy, Task 24: Energy from Biological Conversion of Organic Waste. IEA Bioenergy.

Xu, N., Liu, S., Xin, F., Zhou, J., Jia, H., Xu, J., Jiang, M., Dong, W., 2019. Biomethane production from lignocellulose: biomass recalcitrance and its impacts on anaerobic digestion. Front. Bioeng. Biotechnol. https://doi.org/10.3389/fbioe.2019.00191.

Yang, B., Wyman, C.E., 2008. Pretreatment: the key to unlocking low-cost cellulosic ethanol. Biofuels Bioprod. Biorefin. 2 (1), 26–40. https://doi.org/10.1002/bbb.49.

Yousuf, A.M.I., Eldrainy, Y.A., El-Maghlany, W.M., Attia, A., 2016. Upgrading biogas by a low-temperature CO_2 removal technique. Alex. Eng. J. 55 (2), 1143–1150. https://doi.org/10.1016/j.aej.2016.03.026.

Yunqin, L., Dehan, W., Lishang, W., 2010. Biological pretreatment enhances biogas production in the anaerobic digestion of pulp and paper sludge. Waste Manag. Res. 28 (9), 800–810. https://doi.org/10.1177/0734242X09358734.

Zhang, Q., He, J., Tian, M., Mao, Z., Tang, L., Zhang, J., Zhang, H., 2011. Enhancement of methane production from cassava residues by biological pretreatment using a constructed microbial consortium. Bioresour. Technol. 102 (19), 8899–8906. https://doi.org/10.1016/j.biortech.2011.06.061.

Zhao, Q., Leonhardt, E., MacConnell, C., Frear, C., Chen, S., 2010. Purification technologies for biogas generated by anaerobic digestion. In: Improving the Carbon Footprint of Agriculture in the Pacific Northwest. Climate Friendly Farming – Final Report. CSANR Research Report, pp. 1–24. http://csanr.wsu.edu/wp-content/uploads/sites/32/2013/02/CSANR2010-001.Ch09.pdf.

Zheng, Y., Zhao, J., Xu, F., Li, Y., 2014. Pretreatment of lignocellulosic biomass for enhanced biogas production. Prog. Energy Combust. Sci. 42, 35–53. https://doi.org/10.1016/j.pecs.2014.01.001.

Zheng, X., Fan, J., Xu, L., Zhou, J., 2017. Effects of combined application of biogas slurry and chemical fertilizer on soil aggregation and C/N distribution in an Ultisol. PLoS One 12 (1). https://doi.org/10.1371/journal.pone.0170491, e0170491.

Zhong, W., Zhang, Z., Luo, Y., Sun, S., Qiao, W., Xiao, M., 2011. Effect of biological pretreatments in enhancing corn straw biogas production. Bioresour. Technol. 102 (24), 11177–11182. https://doi.org/10.1016/j.biortech.2011.09.077.

Zieminski, K., Romanowska, I., Kowalska, M., 2012. Enzymatic pretreatment of lignocellulosic wastes to improve biogas production. Waste Manag. 32 (6), 1131–1137. https://doi.org/10.1016/j.wasman.2012.01.016.

FAO, 1992. Wastewater Treatment and Reuse in Agriculture. Food and Agriculture Organization, Rome (Irrigation and drainage paper).

21

Regulations and specifications for biofuels

T. Nagamani[a] and Sangita Kasture[b]

[a]Petroleum, Coal and Related Products Department (PCD), Bureau of Indian Standards, New Delhi, India [b]Department of Biotechnology, Ministry of Science and Technology, New Delhi, India

1 Introduction

Every nation accords a major part of its GDP to its fuel resources and their imports. The fact remained intangible over the last few decades. The need for fuel resources and latest technologies for conversions of various resources into fuel have also been increasing, as the days are passing by. Petroleum-based fuels are most widely used energy sources in the world and have anchored the growth of nations. However, with excessive use of these natural resources of crude oil, these ores started depleting. Crude oil distillation produces aviation gasoline, kerosene, gas oil/fuel oil, gasoline, diesel oil, petroleum coke, etc., used for heating purposes in different engines. With increased usage of petroleum fuels, internationally, there is a fast depletion of these natural resources of conventional fuels, and thus the necessity for alternate resources is increasing.

India, for example, has imported 63.579 MMT in 1998–99, compared to 151.360 MMT in 2008–09, 259.846 MMT in 2018–19, and 270.742 MMT in 2019–20 (Source: Data of PPAC, MoPNG). This clearly indicates that use of crude oil produce has increased multifold. It is also envisaged that crude oil reserves are drastically decreasing and would diminish after 2045. This data can be correlated to many other similar economies all over the world. Thus, it has become imperative for all the countries to look into alternative sources which can replace the crude oil-based fuel consumption.

Alternate sources for crude oil fuels include the hydrocarbon chains of biological origin or synthetic origin. Research on these alternative resources has gained momentum in the last few decades. Many new technologies have evolved for the production of similar biofuels or synthesized fuels. Ethanol, methanol, dimethyl ether, ter-butane, biodiesel, and biogas have become sources for blending purposes or use as alternative fuels, as such.

Biofuels are the alternate fuels derived from renewable biomass resources and wastes such as plastic, Municipal Solid Waste (MSW), waste gases, etc. They provide a higher degree of national energy security in an environment-friendly and sustainable manner by supplementing conventional energy resources, reducing dependence on imported fossil fuels and meeting the energy needs of India's urban and vast rural population (National Biofuel Policy of India, 2018).

2 Quality requirements

There may be few variations in the quality parameters of these fuels produced through different technologies. It is pertinent to say that certain testing is required to qualify all such fuels for a specific usage and so specifications play an important role in benchmarking a fuel for determining its efficacy. Specifications, if already exist, also help new players in the field while developing their technologies. Based on the technologies available in a country, different country national standards bodies have published their national standards.

According to Worldwide Fuel Charter (WWFC-2019), most of the countries, viz., European Union, USA, China, and Japan have adopted Euro-VI B emission norms and equivalent fuel quality. India has also skipped Bharat Stage V (BS V) and adopted BS VI emission norm in Delhi and NCR region from April 1, 2019 and the whole country from April 1, 2020, inspite of the existing pandemic. The fuels meeting these latest emission norms generally require blending of biofuels, so as to comply with the emission limits. For example, in India, with this changed emission norms to BS VI, blending of biodiesel in Automotive Diesel fuel was increased from 5% to 7%.

3 Indian scenario

3.1 Quality standards

Quality standards are the documents that provide specifications or a set of requirements defining the quality of a product/process so that they can be used consistently to ensure that the products/processes are fit for their purposes. Generally, these quality standards can be of industry level, association level, national level, and international level. For a country, they are defined by the national standards body of that country. In India, Bureau of Indian Standards (BIS) is the National Standards Body (NSB) that publishes relevant national standards (Indian Standards) for a variety of products. BIS has published a few Indian Standards for various biofuels. Standards are dynamic in nature and are modified with the changing technologies. Thus, the requirements mentioned in this document are referred to the standards existing in that period.

3.2 Regulations

Regulations are official documents defining the controls of something by using a set of well-defined rules. In most of the developing countries, the national standard (number) or some of the requirements of these standards are referred in the relevant regulation/s by the line ministries. For example, many of the Indian Standards on petrol (motor gasoline) and diesel (automotive diesel fuel) fuels are referred in the *Motor Spirit and High Speed Diesel (Regulation of Supply and Distribution and Prevention of Malpractices) Order, 1998 under the Essential Commodities Act 1955*. In some cases, based on the requirements specified in regulations, the national standards are framed. Para E of rule 115B, Annexures of *The Central Motor Vehicle Rules, 1989 as amended by the Central Motor Vehicles (Seventh Amendment)* (Rules, 2020) *and its subsequent amendments* include the quality requirements of petrol (motor gasoline), diesel (automotive diesel fuel) fuels including blended fuels of ethanol, biodiesel, methanol, Bio-CNG, DME, etc.

The quality of the biofuels and their requirements may vary from one country to other and so the specifications. Many of the countries have published national standards and regulations for fuels and biofuels, suitable to their country's environmental needs, population and effect on their health, availability of natural resources, availability of technologies, etc.

3.3 Indian standards for biofuels

National Biofuel Policy of India 2018 describes advanced biofuels as fuels which are (1) produced from lignocellulosic feedstocks (i.e., agricultural and forestry residues, e.g., rice and wheat straw/corn cobs and stover/bagasse, woody biomass), non-food crops (i.e., grasses, algae), or industrial waste and residue streams and (2) having low CO_2 emission or high GHG reduction and do not compete with food crops for land use.

Fuels such as second-generation (2G) ethanol, drop-in fuels, algae-based 3G biofuels, bio-CNG, bio-methanol, dimethyl ether (DME) derived from bio-methanol, biohydrogen drop in fuels with MSW as the source/feedstock material will qualify as "Advanced Biofuels."

The Indian Standards published for these biofuels are given as under:

(a) *Ethanol-based biofuels*: A number of blends of bio-ethanol are under consideration in different countries and India is not far away from them. Indian oil industry itself is gearing up with a plethora of 2G plants, which would be completing their commissioning in the next year. Ethanol-blended motor gasolines vary in terms of the requirements of reid vapor pressure (RVP), vapour lock index (VLI), oxygen content, oxygenates along with the changing ethanol content.

    (i) *IS 15464* is the Indian Standard for anhydrous ethanol, which is used in admixture with petrol and diesel as a fuel for automobile engines. This base standard of anhydrous ethanol for blending purposes in automotive fuels, is currently under revision, to include the requirements of 2G ethanol. The ethanol produced from the 1G ethanol has a purity of more than 99.5%, while that from 2G ethanol is around 97.5% only. Also there is a possibility of methanol or higher alcohols, which need to be addressed for 2G ethanol.

    (ii) *IS 2796 Motor Gasoline*—Specification covers the requirements of blending of 5% and 10% ethanol in motor gasoline and refers IS 15464 for the quality of ethanol. Oxygen content and RON increase with the increased blend percent of ethanol, from 0% to 5% and 10%, covered in this specification.

(iii) *IS 16629 Hydrous ethanol for use in ED 95 automotive fuel*—Specification prescribes the requirements, methods of sampling, and test for hydrous ethanol which is the main component to formulate ED95 automotive fuel for modified compression ignition engines. Total content of alcohols including ethanol is specified up to 92.4% minimum and water content as 4.5–7.4 mg/kg. This fuel is currently in use only by specific engine manufacturers and in very specific areas.

(iv) *IS 16634 E85 fuel blend of anhydrous ethanol and gasoline*—Specification prescribes requirements, methods of sampling, and test for E85 fuel which is used as a blend of anhydrous ethanol and gasoline for use in flex fuel vehicles equipped with spark ignition engines, specially designed for such fuel. Fuel produced to this standard contains 70–85 vol% of ethanol. This fuel is used in many countries in flexi-fuel vehicles (FFVs). FFVs are expected to be launched in India in the next few years with increase in the ethanol blend percentage.

(v) *IS 17021* prescribes requirements, methods of sampling, and test methods for two octane grades of E20 fuel, an admixture of anhydrous ethanol meeting IS 15464 at 20% by volume and ethanol free Motor Gasoline meeting IS 2796 at 80% volume, under each of BS IV and BS VI categories suitable for use as a fuel in the automobile spark-ignition internal combustion engines of vehicles complying with BS IV and BS VI emission norms, respectively. It has to be ensured that vehicles running on E20 have no adverse effects on fuel system materials, emission, on-board diagnostics (OBD) systems, or drivability. Original Equipment Manufacturers (OEMs) need to identify such vehicles with proper identifications to indicate E20 capability. The engine control unit adjusts engine fueling relating to the oxygen content and reduced energy content of E20 fuel in order to maintain the proper air/fuel ratio under the various engine operating loads and conditions. The vaporization characteristics of E20 tend to be different from normal Motor Gasoline and therefore the engine fueling strategies under engine cold-start and warm-up conditions need to be addressed for this fuel. If E20 is used in a conventional vehicle, fuel system materials and components may be affected over a time and lead to leaks. Drivability, performance, and emissions may also be different. However, since it is to be used in spark ignition engine vehicles with relevant modification, the fuel properties of E20 should be very similar to Motor Gasoline.

(vi) *IS 17586* prescribes requirements, methods of sampling, and test methods for E12 fuel, an admixture of anhydrous ethanol meeting IS 15464 at 12% by volume and ethanol free motor gasoline meeting IS 2796 at 88% volume and E15 fuel, an admixture of anhydrous ethanol meeting IS 15464 at 15% by volume and ethanol free motor gasoline meeting IS 2796 at 85% volume, for use as a fuel in the automobile positive-ignition internal combustion engines of vehicles complying with BS VI emission norms. As the country is planning to move toward gasoline fuel with increased proportion of ethanol blending, it was proposed to formulate specifications for E12 and E15 fuels, to help in handling excess ethanol available in some pockets of the country, though such specifications do not exist in other countries.

A comparative list of the requirements of the Indian Standards of ethanol blended fuels is given under:

Characteristic	Requirements		
	IS 2796 (includes 5% and 10% ethanol blending)	IS 17021 (20% ethanol blending)	IS 16634 (E 85)
Color	a) MG 91-Orange b) MG 95-Red	a) MG 95-Red b) MG 98-Green	Colorless
Density at 15°C (kg/m^3)	720–775	720–775	To report
Distillation	a) percent evaporated at 70°C (E70°C), percent v/v 　a) Motor gasoline—10–45 　b) E10 　c) —10–55 (Summer) 　d) 10–58 (Other months) b) percent evaporated at 100°C (E100°C), percent v/v—40–70 c) percent evaporated at 150°C (E150°C), percent v/v, Min—75 d) Final boiling point, °C, Max—210 E) Residue, percent by volume, Max—2.0	a) percent evaporated at 70°C (E70°C), percent v/v —10–55 (Summer) 10–58 (other months) b) percent evaporated at 100°C (E100°C), percent v/v—45–78 c) percent evaporated at 150°C (E150°C), percent v/v, min—85 d) Final Boiling Point, °C, Max—210 e) Residue, percent by volume, Max—2.0	—
Research octane number (RON), Min	MG 91–91 MG 95–95	MG 95–95 MG 98–98	–
Motor octane number (MON), Min	MG 91—81 MG 95—85	MG 95—95 MG 98—98	—
Gum content (solvent washed), g/m^3, Max	40	40	5mg/100mL
Total sulfur, mg/kg, Max	10	10	50
Reid vapor pressure (RVP) at 38°C, kPa, Max	MG (without ethanol)—60 Ethanol blended MG—70	70	Vapor pressure, kPa = 35–60
Vapor lock index (VLI), Max	Summer/(Other months) MG (without ethanol)—750/(950) Ethanol blended MG—900/(1050) E10—1050/(1100)	Summer—1050 Other Months—1100	—

Continued

Characteristic	Requirements		
	IS 2796 (includes 5% and 10% ethanol blending)	IS 17021 (20% ethanol blending)	IS 16634 (E 85)
Water tolerance of motor gasoline-alcohol blends, temperature for phase separation:	Winter, °C, Max—0 Other months, °C, Max—10		—
Olefin content, percent by volume, Max	MG 91—21 MG 95—18	18	—
Oxidation stability, minutes, Min	360	360	360
Aromatics content, percent by volume, Max	35	35	—
Oxygen content, percent by mass, Max	3.7	7.4	–
Oxygenates percent by volume, Max	Ethers containing 5 or more "C" atoms per molecule such as MTBE, ETBE, or TAME—15 Any other oxygenates—Not permitted	Ethers containing 5 or more "C" atoms per molecule such as MTBE, ETBE, or TAME—15 Any other oxygenates—Not permitted	—
Motor Gasoline Content, percent v/v		(100—Ethanol content and stabilizing agents/additives)	(100—Ethanol content)
Acidity (as acetic acid), percent by mass, Max			0.005
Copper content, mg/kg, Max			0.10
Electrical conductivity, 1 μS/cm, Max			1.50
Higher alcohols (C3–C8) content percent, v/v, Max			0.5
Inorganic chloride, mg/kg, Max			1.2
Methanol content percent, v/v, Max			0.5
pH			6.5–9.0
Phosphorus content, mg/L, Max			0.15
Water content, percent by mass, Max			1.0

3.3.1 *Biodiesel blended fuels*

Biodiesel basically was derived from plants like jatropha, karanja, etc. With the changing technologies, more number of plants are considered for the production of biodiesel. Biodiesel is currently made from virgin or used vegetable oils (both edible and nonedible), waste vegetable oils, by-products of edible oil manufacture like, fatty acids, stearin, etc., and animal fats through transesterification or esterification. Biodiesel, fatty acid methyl ester (FAME) is now derived even from used cooking oil. The raw material used also may change the specification of biodiesel produced to a little extent. For example, in USA, sunflower oil and soya been oil are used for biodiesel production, whereas Germany uses rape seed oil and used cooking oil. While Argentina is also using soya been seeds, many of the Asian countries are dependent on palm oil derivatives. In India, it is envisaged that *Jatropha Curcas* and *Pongamia Pinnata* ("Honge" or "Karanja") and such plants may be cultivated profitably. The seeds of these plants bear rich in oil which is expected to be used for production of biodiesel through transesterification in combination with methanol. Used cooking oil has become the latest alternate source for biodiesel production, worldwide, including India. Biodiesel is highly nonvolatile and hence its flash point is very high and is therefore considered as safe in handling. Further, as biodiesel contains polyunsaturated fatty acid chains, its degradation causes gum formation that may block the fuel filter system. The acid number of biodiesel is maintained low to ensure that no residual-free fatty acids or processing acids are present in the fuel. The presence of excess acids can lead to corrosion and deposits in fuel systems.

(i) *IS 15607* is the Indian Standard for biodiesel (B100)—fatty acid methyl esters (FAME) for use in compression ignition engines designed for using as standalone fuel and as a blend stock for diesel fuel. B100 standalone can also be used for heating applications and industrial engines. It is used for blending purposes in IS 1460 Automotive Diesel fuel up to 7%. Requirement of glycerides and phosphorus were included in the specification, for a check on the feedstock. The requirements of sodium, potassium, calcium, magnesium, and water content are included, as these cannot be separated from the oils derived from plants.

(ii) *IS 16531* covers fuel blend grades of 6%–20% (v/v) biodiesel (see IS 15607) with remainder being an automotive diesel fuel (see IS 1460), collectively designated as B6 to B20. This standard is currently under revision for updating it to 7%–20% as IS 1460 includes blending up to 7%. The properties of commercial B6 to B20 blends depend on the refining practices employed and the nature of the distillate fuel oils and biodiesel from which they are produced. Distillate fuel oils, for example, may be produced within the boiling range of 150–400 °C having many possible combinations of various properties, such as volatility, ignition quality, viscosity, and other characteristics. Biodiesel, for example, can be produced from a variety of animal fats or vegetable oils that produce similar volatility characteristics and combustion emissions with varying cold flow properties.

A comparative list of the requirements of the Indian Standards of biodiesel blended fuels is given under:

Characteristic	Requirements		
	IS 1460	IS 16531	IS 15607
Acidity, inorganic, mg of KOH/g	NIL		
Acid No, mg of KOH/g Max	0.20	0.2	0.50
Ash, percent by mass, Max	0.01	0.01	
Carbon residue (Ramsbottom or micro) on 10% residue by mass, Max	0.3	0.3	0.05
Cetane number, Min	51	51	51
Cetane index, Min	46	46	46
Pour point, Max	a) Winter—3°C b) Summer—15°C	a) Winter—3°C b) Summer—15°C	
Copper strip corrosion for 3h at 50°C	Not worse than No. 1	1	1
Distillation, 95% v/v, recovery, °C, Max	360	360	
Flash point, Abel, °C, Min	35	35	
Flash point (PMCC) °C, Min			101
Kinematic viscosity, cSt, at 40°C	2.0–4.5	2.0–4.62	3.5–5.0
Total contamination, mg/kg, Max	24	24	24
Density at 15°C, kg/m^3	810–845	820–860	860–900
Total sulfur, mg/kg, Max	50	50	10.0
Water content, mg/kg, Max	200		500
Water, ppm, Max		260	
Cold Filter Plugging Point (CFPP), Max	a) Winter—6°C b) Summer—18°C	a) Winter—6°C b) Summer—18°C	a) Winter—6°C b) Summer—8°C
Oxidation stability, g/m^3, Max	25		
Oxidation stability by Rancidity meter, hours, Min	20		
Oxidation stability, at 110°C, h, Min		20	8
Polycyclic Aromatic Hydrocarbon (PAH), percent by mass, Max	8		
PAH, m/m Max	11	11	
Lubricity corrected wear scar diameter (wsd 1.4) at 60°C, microns, Max	460	460	
FAME content, % v/v, Max	7.0		

Characteristic	Requirements		
	IS 1460	IS 16531	IS 15607
Biodiesel content, percent, v/v		6–20	
Sulfated ash, percent by mass, Max			0.02
Methanol, percent by mass, Max			0.20
Ester content, percent by mass, Min			96.5
Monoglycerides content, percent by mass, Max			0.7
Diglyceride content, percent by mass, Max			0.2
Triglyceride content, percent by mass, Max			0.2
Free glycerol, percent by mass, Max			0.02
Total glycerol, percent by mass, Max			0.25
Phosphorous, mg/kg, Max			4.0
Sodium + Potassium, mg/kg, Max			5
Calcium + Magnesium, mg/kg, Max			5
Iodine value, g iodine/100 g, Max			120
Linolenic acid methyl ester, percent m/m, Max			12
Polyunsaturated ($\geq$4 double bonds) methyl ester percent m/m, Max			1

3.4 A comparison with ASTM standards for biodiesel blended fuels

The first national biodiesel specification in the United States was the ASTM standard D6751, "*Standard Specification for Biodiesel Fuel (B100) Blend Stock for Distillate Fuels*," adopted in 2002.

In 2008, ASTM adopted two separate specifications: (1) for biodiesel blends of B5 or less, D 975 were to meet diesel fuel standard and (2) for biodiesel blends from B6 to B20 were by ASTMD 7467.

All biodiesel to be added to diesel must meet B100 specifications D 6751.

ASTMD 6751 Specifications for 100% Biodiesel.

Property	Reference test methods*	Grade no.1-B S15	Grade no. 1-B S500	Grade no.2-B S15	Grade no. 2-B S500
Sulfur, % mass (ppm), max	D5453	0.0015 (15)	0.05 (500)	0.0015 (15)	0.050 (500)
Cold soak filterability, s, max	D7501	200	200	360	360
Monoglycerides, % mass, max	D6584	0.40	0.40	–	–

Continued

ASTMD 6751 Specifications for 100% Biodiesel.—cont'd

Property	Reference test methods*	Grade no.1-B S15	Grade no. 1-B S500	Grade no.2-B S15	Grade no. 2-B S500
Requirements for all grades					
Calcium and magnesium, combined, ppm, Max	EN14538	5	5	5	5
Flash point (closed cup), °C, min	D93	93	93	93	93
Alcohol control: One of the following must be met					
(1) Methanol content, mass%, max	EN14110	0.2	0.2	0.2	0.2
(2) Flash point, °C, min	D93	130	130	130	130
Water and sediment, % volume, max	D2709	0.050	0.050	0.050	0.050
Kinematic viscosity, mm^2/s, 40°C	D445	1.9–6.0	1.9–6.0	1.9–6.0	1.9–6.0
Sulfated ash, % mass, max	D874	0.020	0.020	0.020	0.020
Copper strip corrosion	D130	No. 3	No. 3	No. 3	No. 3
Cetane number, min	D613	47	47	47	47
Cloud point, °C	D2500	Report	Report	Report	Report
Carbon residue[a], % mass, max	D4530	0.050	0.050	0.050	0.050
Acid number, mg KOH/g, max	D664	0.50	0.50	0.50	0.50
Free glycerin, % mass, max	D6584	0.020	0.020	0.020	0.020
Total glycerin, % mass, max	D6584	0.240	0.240	0.240	0.240
Phosphorus content, % mass, max	D4951	0.001	0.001	0.001	0.001
Distillation temperature, 90% recovered (T90)[b], °C, max	D1160	360	360	360	360
Sodium and potassium, combined, ppm, max	EN14538	5	5	5	5
Oxidation stability, hrs, min	EN15751	3	3	3	3

ASTM D 7467 Specifications for Biodiesel B6 to B20.

Property	Reference Test Methods*	Grade B6 to B20 S15	B6 to B20 S500	B6 to B20 S5000
Acid number, mg KOH/g, max	D664	0.3	0.3	0.3
Viscosity, mm^2/s at 40°C	D445	1.9–4.1[a]	1.9–4.1[a]	1.9–4.1[a]
Flash point, °C, min	D93	52[b]	52[b]	52[b]
Cloud point, °C, max or LTFT/CFPP, °C, max	D2500, D4539, D6371	c	c	c

ASTM D 7467 Specifications for Biodiesel B6 to B20.—cont'd

Property	Reference Test Methods*	Grade B6 to B20 S15	B6 to B20 S500	B6 to B20 S5000
Sulfur Content, (μg/g or ppm)	D5453	15	—	—
mass %, max	D2622	—	0.05	—
mass %, max	D129	—	—	0.50
Distillation temperature, °C, 90% evaporated, max	D86	343	343	343
Ramsbottom carbon residue on 10% bottoms, mass %, max	D524	0.35	0.35	0.35
Cetane number, min	D613	40	40	40
One of the following must be met:				
(1) Cetane index, min	D976–80	40	40	40
(2) Aromaticity, vol%, max	D1319–03	35	35	–
Ash Content, mass %, max	D482	0.01	0.01	0.01
Water and Sediment, vol%, max	D2709	0.05	0.05	0.05
Copper Corrosion, 3h @ 50°C, max	D130	No. 3	No. 3	No. 3
Biodiesel content, % (v/v)	D7371	6–20	6–20	6–20
Oxidation stability, hours, min	EN15751	6	6	6
Lubricity, HFRR @ 60°C, (micron μm), max	D6079	520	520	520
Conductivity (pS/m) or conductivity units (CU), min	D2624/D4308	25	25	25

As can be seen from these tables, Indian specifications for 100% biodiesel (IS 15607) are very similar to US specifications ASTM D 6751. Similarly Indian Specifications (IS 16531) for Biodiesel-Diesel blends of 6%–20% are as stringent as the corresponding US specifications, ASTM D 7467.

3.4.1 Significance of biodiesel specifications

Biodiesel is mostly blended in diesel for fuelling the engines. Therefore, it is ensured through biodiesel specifications that the basic specifications of the diesel are not affected. Biodiesel production process can have few carry over impurities like methanol, glycerine, and metals and these are specified so as to minimize their effect. Some of the properties of biodiesel and FAME are because of its chemical structure, e.g., lower oxidation and gum formation. Significance of some of the properties is given as under.

Flash point and alcohol control

The volatility of biodiesel is extremely low, and properly produced biodiesel presents minimal risk for handling and safety. To ensure safe handling of biodiesel, a minimum flash point of 130 °C is set to limit residual methanol. If there is residual methanol present then flash point will be low and danger of fire will be there.

Flow properties

Another issue is cold weather properties of biodiesel. These properties include cloud point, pour point, and cold soak filtration. Biodiesel can form cloud points at a much higher temperature than petro-diesel, close to the freezing point. The cloud point is the temperature that crystals begin to form; it can cause the biodiesel to gel and flow slower than it should. Once the pour point is reached (basically completely frozen), the fuel cannot move. It depends on the normal temperature of the climate as to whether the fuel can be used or blended with petrodiesel. What can complicate it more is the saturated or unsaturated fatty acid content. High saturated fatty acid content can lead to higher fuel stability but higher pour points.

Acid number

The acid number of biodiesel should be low to ensure no residual-free fatty acids or processing acids are present in the fuel. The presence of excess acids can lead to corrosion and deposits in fuel systems.

Oxidation stability

Because biodiesel contains polyunsaturated fatty acid chains, it will degrade by the well-known peroxidation mechanism and would result in gums which can block the fuel filter system.

Group I and II metals

Most biodiesel production is by alkali catalysts like sodium or potassium hydroxide. Metals in B100 are typically contaminants from the production and cleanup process and are limited to very low levels in specifications. Higher metals will lead to deposits in the combustion chamber.

Phosphorus

Higher levels of P will deactivate the exhaust catalytic converters which use noble metals.

(c) **Biogas** *(Biomethane/BioCNG)* is primarily methane gas which is generated through an anaerobic digestion of organic wastes by microorganisms. It is a relatively simple and economical method to produce a fuel from waste. The waste could comprise agricultural and crop waste, human waste, and animal waste (cow dung for instance). Biogas (biomethane) is an environment-friendly, clean fuel. CBG is widely used in European countries like United Kingdom, Sweden, Switzerland, and Germany for transportation, including in trains also. In India, the fuel is mostly used in rural areas for cooking purposes and for stationary engines. It is now-a-days gaining its usage as CBG, for transportation purposes. *IS 16807* prescribes the requirements and the methods of sampling and test for biogas (biomethane) use in stationary engines, automotive (bio-CNG/bio compressed gas [CBG]), and thermal and industrial applications as supplied in cylinders and through piped networks.

(d) **Dimethyl ether** is also used as heating fuel, industrial fuel, and to replace diesel fuel or gas oil. DME can also be a substitute for diesel in slow RPM diesel engines and hence promotion of industrial production of methanol is pertinent for widespread usage, industrial application and acceptance of DME as potential fuel. IS 16704:2018/ISO 16861 petroleum products—Fuels (class F)—specifications of dimethyl ether (DME) specify the characteristics of DME used as fuel of which the main component is the dimethyl ether synthesized from any organic raw materials. The fuel can also be used in specifically designed compressed ignition engines, in place of diesel. Due to no emissions of NO_X, SO_X, and carbon monoxide, it is promoted as an advanced biofuel.

(e) *Methanol* is also considered as an alternative fuel source that offers environmental friendly characteristics associated with its use. Though methanol is toxic, its use as an advance fuel is promoted as it is less cost and a more sustainable alternative to bioethanol fuels. ASTM D 1152:2016 Standard specification for methanol (methyl alcohol) and IMPCA specifications, 2015 (International Methanol Producers and Consumers Association) provide specification of methyl alcohol for wider application and not limited to automotive application. Methanol is considered as an alternative fuel source that offers environmental friendly characteristics associated with its use. Hence, in India, *IS 17075* specification for anhydrous methanol used as a blending component in methanol-based fuels, used in engines designed for running on such fuels for captive power generation, industrial, marine, locomotive, automotive, and off-road applications is published. This standard is applicable only for fuel applications and not for other industrial applications.

Another Indian Standard for 15% blending of this anhydrous methanol in motor gasoline *IS 17076* has also been published. This fuel is an admixture of anhydrous methanol meeting IS 17075 at 15% by volume and methanol-free motor gasoline meeting IS 2796 at 85% volume, for use in vehicles equipped with spark ignition engines, stationary and industrial engines specially designed for using such fuel. The engine control unit adjusts engine fueling relating to the oxygen content and reduced energy content of M15 fuel in order to maintain the proper air/fuel ratio under the various engine operating loads and conditions. The vaporization characteristics of M15 fuel tend to be different from normal petrol and therefore the engine fueling strategies under engine cold-start and warm-up conditions need to be addressed for this fuel. If M15 fuel has to be used in a conventional vehicle, fuel system materials and components may be affected over a time and may lead to concerns like material compatibility, drivability, performance, and emissions. Since this fuel has to be used in spark ignition engine vehicles with relevant modification, the fuel properties of M15 should be very similar to motor gasoline. The properties like Reid Vapor Pressure (RVP), Vapor Lock Index (VLI), research octane number, and motor octane number are important and shall be similar to motor gasoline, as this fuel is intended to be used in spark ignition engines.

(f) *Hydrogen fuel* is also one of the least polluting alternate advanced fuels. IS 16061 [P:1]:2013/ISO 14687-1:1999 Hydrogen fuel—Product specification—Part 1: All applications except proton exchange membrane (PEM) fuel cells for road vehicles and IS 16061 (Part 2): 2016/ISO 14687-2: 2012 Hydrogen Fuel—Product Specification—Part 2 Proton Exchange Membrane (PEM) Fuel Cell Applications for Road Vehicles which are the Indian Standards that are adoptions of ISO standards. These specifications include very low impurities in hydrogen. Ministry of New & Renewable Energy (MNRE), in its draft National Hydrogen Emission Mission document, mentioned that hydrogen has the potential in enabling greater share of renewables in the energy mix, especially in hydrogen-based transport- and industrial use by 2024 and 2030, respectively.

(g) In *aviation sector* also, *IS 17081* Aviation turbine fuel, kerosene type, Jet A-1 containing synthesized hydrocarbons—Specification (generally termed as biojet fuel) is published by BIS. This standard includes the test for synthetic blending components as well as aviation turbine fuel, kerosene type, Jet A-1 containing synthesized hydrocarbons for use in aircraft gas turbine engines designed to operate on such fuel. *Biojet fuel* prepared indigenously by CSIR-Indian Institute of Petroleum (CSIR-IIP) has been included in this specification. The synthetic components meeting this IS 17081 can be blended in ATF, kerosene type, Jet A-1 fuel, IS 1571. Aviation fuels specifications followed in India are generally adopted from relevant Def Stan specifications, as international refilling and coordination are highly required in aviation sector. Aviation Turbine Fuel (Regulation of Marketing) Order 2001; and subsequent orders are the prime regulation for aviation fuels in India.

(h) In *marine sector* also, India has adopted ISO standards as international refilling is required by naval coordination and Directorate General of shipping initiated actions to prepare for IMO 2020 orders to follow MARPOL conventions laid down by International Maritime Organization. IS 16731/ISO 8217 Petroleum products—Fuels (Class F)—Specifications of marine fuels and IS/ISO/PAS 22263: 2019 petroleum products—Fuels (class F)—Considerations for fuel suppliers and users regarding marine fuel quality in view of the implementation of maximum 0.50% sulfur in 2020 have been adopted.

Reference

The Central Motor Vehicle Rules, 1989 as Amended by the Central Motor Vehicles (Seventh Amendment) Rules, 2020.

22

Life cycle and techno-economic assessment of microalgal biofuels

Soumyajit Sen Gupta[a] and Yogendra Shastri[b]

[a]Department of Chemical Engineering, Indian Institute of Technology (Indian School of Mines) Dhanbad, Dhanbad, India [b]Department of Chemical Engineering, Indian Institute of Technology Bombay, Mumbai, India

1 Introduction

Transition to sustainable energy alternatives is one of the essential requirements for a sustainable future (Chen et al., 2019). Since most of the energy we currently use is derived from fossil fuels, rapidly rising energy consumption has resulted in problems such as climate change and regional air pollution. However, energy consumption is also strongly related to societal and human development (UNDP, 2013). Consequently, reduction in energy consumption is not really an option to address the sustainability issues. This is particularly true for a developing nation such as India. The per capita energy consumption in India is still much lower than that globally, and it is expected to increase significantly in the coming decades to achieve development targets. Consequently, the need for better access to more sustainable energy sources is imperative.

Transport sector is an important sector responsible for energy consumption. Since fossil-based fuels such as gasoline (petrol) and diesel are primarily used, this sector is also a key contributor to climate change. Globally, 16% of the total greenhouse gas (GHG) emissions are attributed to the transport sector. In India, 75%–80% of the GHG emissions from the transport sector are due to road transport (Singh et al., 2019). With increasing vehicle ownership and driving, the contribution will increase further, and quite rapidly.

Thus, if we want to transition toward a sustainable future while also meeting the expected energy demands, developing advanced fuels from biomass-based resources is imperative (Singh et al., 2019; Gomez et al., 2008; Shastri, 2017). Biomass-based fuels can potentially be carbon neutral. Moreover, biomass, unlike petroleum, is a local resource, and therefore, contributes toward energy security. Finally, biomass is produced through agriculture or aquaculture. Thus, impetus to biomass-based energy sector can provide opportunities for local entrepreneurship and job creation. It can thus be seen that biomass-based energy has a lot of advantages that can make it a truly sustainable option (Chung, 2013).

Microalgal biofuels are one potentially attractive advanced biofuel option (Wijffels et al., 2010; Wijffels and Barbosa, 2010; Lam and Lee, 2012). Microalgal biofuels are considered as the third-generation biofuels, in which the feedstocks (microalgae) are grown specifically for biofuel production. Microalgae are unicellular or simple multicellular species. The size of the species can range from a few microns to millimeters. Microalgae have been cultivated for a long time for nonenergy applications such as *Spirulina* for food supplements. The microalgae are made up of lipids, proteins, and carbohydrates, and the composition is specific to the species/strain being considered (Brennan and Owende, 2010). Microalgae also contain relatively large fraction of ash. From an energy perspective, neutral lipids found in microalgae can be converted to biodiesel through the transesterification route, and this route has been most commonly explored in the literature (Chisti, 2007). However, other options are also being explored (Delrue et al., 2013). The carbohydrates present in microalgal cells can be converted to ethanol using biochemical routes. Thermal routes have also been explored to convert complete microalgal cells into gaseous fuels such as syngas, methane, and hydrogen (Gholkar et al., 2019; Brown et al., 2010). Many studies have shown promising results at laboratory scale, but the scale-up and commercialization of these processes have been limited. Efforts are being made to improve the process performance and thereby, achieve commercial feasibility (Amer et al., 2011; Singh and Gu, 2010; Gholkar et al., 2021).

However, we need to systematically and quantitatively ascertain the sustainability of these advanced biofuels. Several processes, that looked promising at the laboratory scale, have failed to scale-up. Moreover, although biomass-based fuels are expected to be environmentally sustainable, it is not necessary (Shastri, 2017). The production and conversion steps may have significant impacts such as high energy consumption or water consumption, and they may completely negate the benefits obtained by using a biomass-based feedstock. Water consumption or footprint is a particularly key aspect since microalgae cultures are very dilute. Systematic assessment of the overall sustainability is thus essential (Subhadra, 2010; Zaimes and Khanna, 2013; Chamkalani et al., 2020).

In this work, we focus on two dimensions of sustainability, namely, techno-economic feasibility and environmental impacts, of the advanced biofuels derived from microalgae. Specifically, we focus on life cycle assessment (LCA) as a tool to quantify environmental impacts. The key novelty of this work is that it determines the optimal biorefinery configuration from an economic standpoint and then performs detailed LCA for the selected configurations. This ensures that the results are directly comparable and the trade-offs can be identified. Most studies in literature have performed either economic evaluation or LCA separately. This work has addressed those limitations.

The rest of the chapter is organized as follows. In the next section, we briefly summarize the methodology of techno-economic and life cycle assessments and explain the scope of our study. Section 3 discusses the results of techno-economic assessment and Section 4 discusses the results for life cycle assessment. The chapter ends with summary and important conclusions in Section 5.

2 Scope and methodology

From the point of view of microalgal biofuels, there are several decisions that can profoundly affect the sustainability of a commercial scale system. Microalgae cultivation is recognized as the most critical step, and there are broadly two methods of cultivation, namely, open raceway ponds and closed photobioreactors (PBR). Cultivation in raceway ponds requires less capital investment and the operating costs are low. However, the land cost can be prohibitive due to very low yields. The open pond is also affected by the vagaries of nature such as precipitation and invasion by predators. In contrast, PBRs offer controlled environment, lower land footprint, and higher yields. However, the cost of building and operating PBRs can be quite high. Similar trade-off can be observed for harvesting and drying. Natural methods, such as settling and solar drying, are cheap but take a lot of time and space. Alternative methods such as centrifugation and industrial drying are expensive. Lipid extraction can also be performed using multiple routes. There is an added layer of complexity introduced by the microalgal strain being cultivated. Different strains have different growth cycles, maturation periods, maximum yields, and compositions. Therefore, the desirable system design must take into consideration the strain as well. These factors highlight the need to perform systematic and holistic assessment of a microalgal biofuel system. The objective of this work is to perform such an analysis. Considering multiple potential options for each step, the objective is to identify the most desirable route and determine the life cycle impacts of that route.

We next explain the specific methodology followed to perform such analysis in this work.

2.1 Techno-economic assessment

Model based techno-economic assessment (TEA) is commonly performed in chemical process industries (Ribeiro and Silva, 2013). This approach is cost- and time-effective and provides very good estimates of the final economic performance. Even model-based approaches can be classified into two types. Simulation-based approaches assume a certain process configuration and operating strategy and determine the cost for the same. Such an approach has been frequently used in literature (Richardson et al., 2012; Davis et al., 2011; Tredici et al., 2016; Heo et al., 2019). The other approach is to formulate an optimization problem that determines the best possible solution to achieve a particular economic objective, such as production cost or profit. The optimization can be performed considering design and/or operating strategy as decision variables. The optimization problem focused on identifying the optimal processing route is popularly known as the process synthesis problem. Here, a network of different processing routes, known as the superstructure, is developed and optimization problem determines the optimal solutions (Gebreslassie et al., 2013; Slegers et al., 2014; González-Delgado et al., 2015; Rizwan et al., 2013). This work uses the superstructure-based optimization approach.

We have formulated a superstructure-based Mixed Integer Linear Programming (MILP) for an optimal design and operation scheduling of microalgae biorefinery. The details of the model, its scope, and assumptions are presented in our prior work (Sen Gupta et al., 2016, 2017a). The model minimizes the annualized life cycle cost (ALCC) for meeting a certain daily demand of biodiesel for any specific scenarios, while various alternative decisions are treated in the form of a superstructure. Since many of the process steps are batch in nature, the rate or throughput per batch determines the net production rate. The model variables that are optimized are of integer, continuous or binary in nature. Instances such as number of equipment are integer variables whereas flow variables are continuous and selection of any specific decision among multiple alternatives is binary. The constraint equations that define the feasible region of the optimization problem are based on mass balance across the process stages, connectivity among different stages of the process, selection of one among multiple alternative decisions, and demand for equipment and product which also relate to the process economics. The model has been used to study various scenarios that are explained in detail in Sen Gupta et al. (2016, 2017a,b). From a broad perspective, the salient features of our techno-economic analysis are incorporation of the growth dynamics' impact on the downstream processes, study on optimal equipment sizing, and different strategies of process scheduling.

Fig. 1 presents the different alternate options across all the processing steps considered in the model for biodiesel production. For instance, in case of growth, we consider two different systems—open pond and closed reactor/photobioreactor (of specified capacities)—different nutrient mediums, different durations of growth, and two types of microalgae strains. All these different decisions would have an impact on the downstream process performance, i.e., both the yield as well as infrastructure required. In other words, through our model, we capture the trade-off existing among different sets of combination of options for a certain objective.

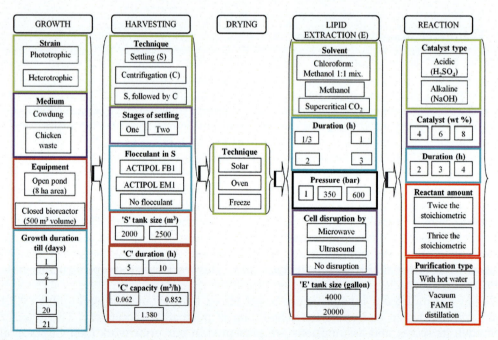

FIG. 1   Superstructure decisions for the steps mandatory for biodiesel production from microalgae.

2.2 Life cycle assessment

Life cycle assessment must be performed using the methodology reported in the literature. As per the methodology, the LCA exercise is divided into three stages, namely, goal and scope definition, life cycle inventory calculations, and impact assessment. Among these, goal and scope definition stage is conceptually important and decides the scope of the study. Fig. 2 shows the typical system boundary and product system for microalgal biodiesel production. For other routes for converting microalgae to energy products, similar product system can be developed. The stage of life cycle inventory calculation is the most challenging. In this stage, the data for all the steps in biodiesel production are required. Moreover, emission inventories for the inputs (e.g., chemicals, electricity) required in all the steps are also needed. Often, these inventories are obtained from commercial databases such as Ecoinvent© or GaBi©. The final stage converts the life cycle inventory calculations outputs into limited impact categories so that the results can be used for analysis and comparison.

LCA studies of microalgal biofuels have also been reported in the literature (Grierson et al., 2013; Adesanya et al., 2014; Collet et al., 2011, 2014; Soratana et al., 2014; Quinn et al., 2014). Among the two scopes shown in Fig. 2, cradle-to-gate scope is more commonly used in the reported literature. This is because the emissions due to the burning of biodiesel are relatively well understood. However, the activities and inputs for the production of biodiesel can vary significantly, thereby affecting the overall impacts. Some of these studies have also considered the benefit of producing co-products along with the primary energy product. Studies are also reported in the literature in which the LCA calculations are combined with economic calculations (Resurreccion et al., 2012; Dutta et al., 2016). Finally, Gong et al. (Gong and You,

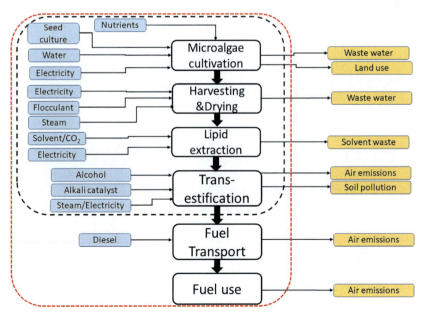

FIG. 2 Typical product system and system boundary for life cycle assessment of microalgal biofuels. Black line indicates cradle-to-gate system boundary while red line indicates cradle-to-grave system boundary.

2014) have used LCA within the optimization framework and the life cycle impacts are considered as the objective function.

In this work, we have performed LCA as a follow-up of the economic optimization work. The model-based optimization identified an optimal processing route that resulted in minimum cost of production. We subsequently performed detailed LCA of the route to determine the corresponding impacts. Additionally, we also performed detailed LCA of several other process configurations. This provided inputs regarding trade-offs between economic performance and life cycle impacts.

The model has been developed in OpenLCA 1.6 and Ecoinvent© 3.3 database has been used for getting the inventory. ILCD midpoint method (version August 2016) has been used for impact assessment. Wherever possible, data inventory pertaining to India has been used (e.g., electricity). However, for many chemicals, India specific inventory data are not available. In such cases, data either for RoW (Rest of the World) or GLO (Global) category have been used. It is to be noted that for the steps of lipid extraction, reaction, and hydrolysis, the necessary chemicals such as methanol, sodium hydroxide, and sulfuric acid appear at both input as well as output. Their inputs refer to the requirement in the specific processes whereas the outputs relate to the recycle loss. All these chemicals have been assumed to have a finite recycle efficiency by Sen Gupta et al. (2017a,b). The organic matter output refers to the excess biomass which has been cultivated on account of finite yield of any process step and presence of different constituents, not necessary for the target end product, in the microalgae cells as well as the co-production of glycerol for the case of reaction. Additional details regarding the LCA calculations are provided in Section 4.

3 Techno-economic assessment: Results

This section summarizes the techno-economic assessment performed by us for the integrated microalgal system. The objective here is to determine the optimal route for minimizing the cost of biodiesel production of specified demand among the various processing routes available.

3.1 Biodiesel as the only product

The base case considers biodiesel as the only product from microalgae, and does not include the steps of protein, carbohydrate, and polar lipid extraction. Moreover, only a subset of the options shown in Fig. 1 is considered. For instance, growth in closed reactor is not included. We study this base case of producing biodiesel since the potential for microalgae as a source of biodiesel is very high. The original optimization problem is first solved to determine the best solution. Subsequently, three additional scenarios are developed by enforcing some options that are not optimal. This quantifies the level of suboptimality if the optimal option is not selected. The three scenarios are cultivation of heterotrophic strain, lipid extraction with carbon dioxide, and lipid extraction with external aid of microwave or ultrasound energy as means for cell disruption. These results—the design and operational protocol for each scenario—are discussed in detail in Sen Gupta et al. (2016). Fig. 3 presents stepwise contribution of different steps toward the net ALCC for all these scenarios.

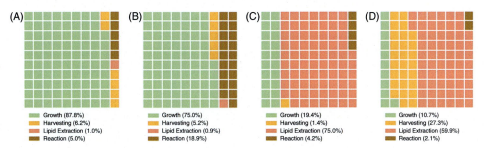

FIG. 3 Contribution of different steps of the process toward net ALCC for biodiesel as the only product: (A): base case with phototrophic strain, (B): case with heterotrophic strain, (C): case with supercritical carbon dioxide as lipid extraction solvent, (D): case with cell disruption-assisted lipid extraction.

For the base case, the minimized ALCC is US$ 13.286/L for producing 30 Mg (metric ton) of biodiesel per day. Phototrophic strain of *C. vulgaris* is cultivated for 20 days in open pond and the cells are then subjected to settling with flocculant ACTIPOL FB1 for concentrating the biomass. This is followed by lipid extraction which is conducted with conventional extraction solvent (chloroform/methanol 1:1 mixture) without any external energy. Growth contributed to 87.8% of the total cost as many of the downstream steps were inefficient in concentrating the solution (settling, as against centrifugation) and extracting the lipid (conventional solvent, as against supercritical). However, for the cases concerning enforced decisions on lipid extraction (solvent or cell disruption strategy), decision changes for that specific step would effectively increase the net cost (Sen Gupta et al., 2016). For supercritical fluid extraction of lipids, even though the growth and settling steps would have become cheaper owing to more efficient lipid extraction, the lipid extraction step would have become comparatively more expensive—75% (Fig. 3C). For lipid extraction facilitated with microwave-assisted cell disruption, the step of harvesting also has a significant share (27.3%) of the net ALCC (Fig. 3D) as efficient harvesting steps were necessary to reduce amount of residual water to a level in order to effectively reduce the cost of highly expensive lipid extraction step at the downstream. Moreover, cultivation of heterotrophic strain ensures a lipid-rich biomass for downstream steps, thereby facilitating a reduction in the expenses for all the steps, since final demand of biodiesel remains same. While all the decisions for the four steps remain same and their relative contribution to process economics remain almost similar to the base case (Fig. 3B), growth of heterotrophic strain would also require a substantially high amount of glucose as substrate which would make the overall process economically inferior. Moreover, using glucose also introduces undesirable competition with food resources. For all these scenarios, the decisions for transesterification reaction ensure highest reaction yield with lowest possible expenses for reactor, reactant, and catalyst.

Sensitivity analysis for model parameters shows that parameters such as biomass yield, lipid yield, extraction efficiency, and reaction yield have a greater impact on the economics since these have a direct impact on the availability of lipid in the reactant stream for the transesterification reaction step. A change of 10% in these parameter values results in around 10.05%–10.25% change in ALCC. Other parameters such as concentration factor for settling, lipid extraction solvent dose or necessary reactant, and catalyst amount do not have any appreciable impact on ALCC.

3.2 Co-production of multiple products

Since microalgae cells contain carbohydrate and protein alongside lipid, the prospect of co-generating all these products is studied. Moreover, whereas neutral lipid yields biodiesel on being reacted with alcohol, polar lipid of the microalgae cells can also be extracted as a separate product. In this section, we summarize some of the key studies and analyze those results. The detailed study can be found in Sen Gupta et al. (2017a, b). As the products other than biodiesel are of low-volume high-value as against high-volume low-value biodiesel, the cases are studied based on a minimum daily demand of 30 Mg for biodiesel, and the maximum annual demand of protein (11,650 Mg), reduced sugar (6657 Mg), and polar lipids (832 kg). These values are based on the ratio of global demand of diesel and these co-products. The selling price for all these co-products as used in the model is US$ 610.5/Mg for protein, US$ 416.25/Mg for reduced sugar, and US$ 3330/Mg for polar lipids.

Fig. 4 presents the optimized protocol of operation, yielding biodiesel and reduced sugar as co-products while no other co-products (protein and polar lipid) are produced. The net ALCC for biodiesel production for this base case study with phototrophic strain (with no user-enforced decisions and all parameters at their base values) is US$ 8.53/L, as a result of co-product reduced sugar aiding in economics by reducing the effective ALCC by 2.4%. Primarily, two rounds of lipid extraction to allow for possibility of polar lipid extraction in second round facilitate this sharp reduction in price from the previous case of having biodiesel as the only product. Due to higher amount of lipid extraction, the necessary volume per

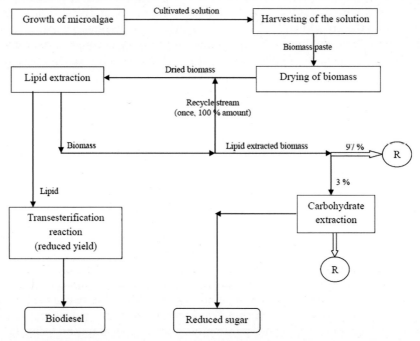

FIG. 4 Optimal design of the process steps in integrated microalgae biorefinery for the base case. The split fraction of streams at relevant nodes is also shown. "R" represents residual biomass.

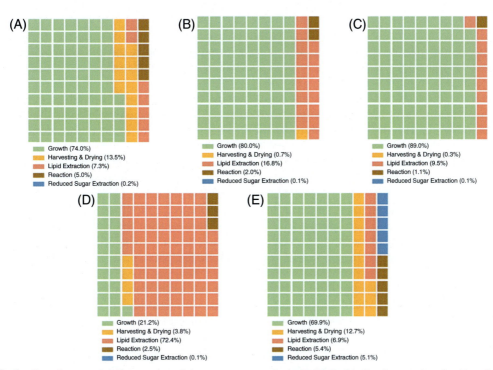

FIG. 5 Contribution of different steps of the process toward net ALCC for biodiesel co-produced with reduced sugar: (A): base case with phototrophic strain, (B): case with heterotrophic strain, (C): case with growth in closed reactor, (D): case with supercritical carbon dioxide as lipid extraction solvent, (E): case with co-cultivation of phototrophic and heterotrophic strains, with recycled residual biomass for growth of the latter.

batch drops, thereby reducing the cost at all the steps. The process decisions are almost similar to the previous case, as well. Fig. 5 presents the distribution of the total cost across the different operating steps for various scenarios (base case and with enforced decisions for growth and lipid extraction steps) for co-production of biodiesel and reduced sugar.

For the cultivation of heterotrophic strain, use of high amount of glucose (4.84×10^5 Mg/year) increased the cost to US\$ 18.85/L even though lipid-rich heterotrophic strain requires less volume of cultivation and accordingly, lesser cost at each of the downstream steps. If we choose to perform the cultivation in a closed system, heterotrophic strain and supercritical fluid extraction of lipid would be selected as the optimal decision. This would ensure the highest possible economic benefit of using a highly efficient growth system with higher productivity, albeit at a higher expense. The total cost for such a system would be US\$ 33.37/L, with reduced sugar being a co-product alongside biodiesel. For supercritical extraction of lipid being enforced upon the model as a design decision, the net ALCC increased to US\$ 16.32/L owing to the use of expensive solvent—supercritical carbon dioxide even though the efficiency of extraction would be enhanced. The co-cultivation of phototrophic and heterotrophic strains improved the process economics (as compared to phototrophic strain only) by around 10.2%. This ensures simultaneous benefit being harnessed from higher productivity of heterotrophic strain at relatively lower requirement of glucose since phototrophic strain, high in carbohydrate, would provide the same on being recycled after one round of lipid extraction. As against 3% of lipid extracted biomass that would be used in carbohydrate extraction in the case of only phototrophic cultivation (Fig. 4), this co-culture would utilize the entire lipid extracted biomass for this purpose. Accordingly, we would also notice a higher share of reduced sugar extraction step in process economics (Fig. 5E).

Similarly, if the entire lipid extracted biomass would be utilized for carbohydrate extraction in case of the phototrophic strain cultivation, the net ALCC for the biodiesel production would be US$ -27.23/L, with the negative sign indicating a net profit. Many of the design decisions of the biorefinery would also change (cultivation till 10 days rather than 20 days) to facilitate relatively higher carbohydrate content in the cell. Further, assuming a target biodiesel selling price of US$ 0.48/L (Bart et al., 2010), this profit would effectively reduce selling price of reduced sugar by US$ 51.1/kg, presenting a strong case for such co-cultivation. On the other hand, mandatory production of protein and polar lipid from the phototrophic strain would increase the net ALCC by 77% and 55%, respectively. The design decisions for the system would remain same for all the pre-extraction steps, for both these cases. An increase in selling price of protein by 11.5 times and a huge increase by orders of magnitude for that in polar lipid would facilitate their extraction in any scenario. Moreover, we analyzed the economic viability for direct use of residual biomass, obtained after lipid and co-products extraction from microalgae as fertilizers with a specified demand (4626.72 Mg/year); no downstream treatment of the leftover biomass was assumed. While for the specific case of phototrophic strain (base case), such an add-on strategy would reduce ALCC only by 4%, allowing for unlimited production of fertilizer would change many of the design decisions (for example, growth till only 7 days to ensure protein-rich biomass), resulting in a net profit of US$ 3.83/L.

All the scenarios mentioned thus far do not consider any storage between process steps. Provision of intermediate storage allows staggered operation for the faster steps and thus, reduces the number of equipment necessary at faster steps for a fixed demand of product. For the phototrophic strain cultivation (base case) to co-produce biodiesel and reduced sugar, such a strategy would reduce the net ALCC by 11.25%. Debottlenecking of slower steps would increase the number of batches over any period, thereby reducing the throughput per batch for a fixed demand of product. This would again essentially help in reducing the capital cost of the equipment for the faster steps. For our base case, this would reduce the cost by 20.5%. The design decisions would mostly remain same except for use of microwave as assistance for cell disruption prior to lipid extraction as against only in the second round of extraction, as in the base case. This additional expense, ensuring more lipid extraction, helps in reducing the throughput and, thus, capital expenditure even further. Both these strategies of scheduling together would be able to reduce the ALCC by 25% to a base value of US$ 6.4/L. While infinite storage is practically infeasible, we have obtained an upper bound on the reduction in ALCC through this analysis.

4 Life cycle assessment: Results

This section describes the life cycle assessment work performed on the same system. As per the economically optimized design of base case, reduced sugar was co-produced as a value-

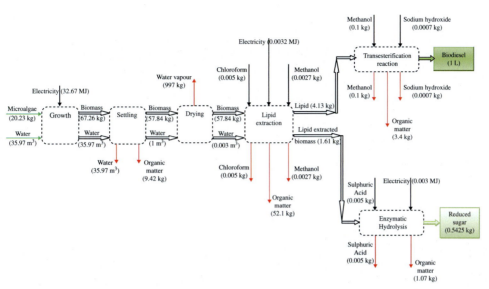

FIG. 6 Product system for the base case co-producing biodiesel and reducing sugar from microalgae. Numbers in parenthesis are references flows corresponding to functional unit of 1 L of biodiesel.

added product. Fig. 6 depicts the individual process steps with the input and output streams for which the impacts are to be computed in the base case. The numerical values in Fig. 6 correspond to the production of 1 L of biodiesel, the functional unit for the present study. It is to be noted that the flocculant added in the step of settling is not considered since its exact composition is not available. Moreover, animal waste used in microalgae cultivation medium is not considered as its cost is not considered in the economic calculations. In Fig. 6, the block arrows represent the product flow within the system boundary. The streams in green are the resource inputs as elementary flows, whereas the ones in black stand for the product flows. The waste flows as output streams from the system boundary are marked in red. Biodiesel, the main product, and reduced sugar, the co-product, are highlighted in green.

As can be noticed in Fig. 6, the process stream at the output of lipid extraction step gets bifurcated into two substreams. The lipid encounters the step of transesterification reaction to produce biodiesel, the target end product; while, lipid extracted biomass is sent for enzymatic hydrolysis to co-produce reduced sugar. Thus, it being a case of multiple end products, the LCA study is carried out by considering allocation strategies; two separate allocation strategies, physical and economic, with respect to these endproducts are studied. For the former, the allocation factors for the two output streams from the lipid extraction step are proportional to the mass of the final products obtained from these individual streams. For the case of economic allocation, these factors are proportional to the net economic value of the end products. The impacts for the steps upstream to inclusive of lipid extraction get distributed between the two products as per the allocation factor and the downstream steps of reaction and hydrolysis are mutually exclusive on the basis of the final product.

The allocation factor for the lipid stream was 0.619 for the case of physical allocation on the basis of mass of final products—biodiesel (0.88 kg) and reduced sugar (0.5425 kg). For the case of economic allocation between the two output streams from the lipid extraction step, the net economic values of the final products are considered, and accordingly, the economic allocation factor for the lipid stream is computed as 0.68 (biodiesel target selling price US$ 0.48/L, reduced sugar selling price US$ 416.25/Mg) (Sen Gupta et al., 2017a,b).

4.1 Base case scenario

Table 1 reports the life cycle impacts for mid-point indicators using physical as well as economic allocation. Since the economic allocation factor for biodiesel (0.68) is more than its corresponding factor for physical allocation (0.619), the impact values are higher for the case of economic allocation. Analysis of the distribution of the total impact (Fig. 7) shows that the major contributing factor for most of the categories of environmental impacts is the step of microalgae growth. Drying does not have any impact as completely natural (solar) drying is assumed. Lipid extraction is prominent only in terms of ozone depletion on account of use of chloroform in the solvent mix. In the step of microalgal growth, the impact is high mainly due to the use of electricity to drive paddlewheel. Due to low biomass and lipid

TABLE 1   Impacts of 1 L biodiesel production in the base case for different allocations.

Impact category and unit	Physical allocation	Economic allocation
Acidification (Mole H$^+$ eq.)	0.061	0.067
Climate change (kg CO_2 eq.)	6.990	7.675
Freshwater ecotoxicity (CTUe)	41.655	45.696
Freshwater eutrophication (kg P eq.)	0.004	0.005
Human toxicity—carcinogenics (CTUh)	4.6×10^{-7}	5.1×10^{-7}
Human toxicity—noncarcinogenics (CTUh)	1.6×10^{-6}	1.8×10^{-6}
Ionizing radiation—ecosystems (CTUe)	1.6×10^{-6}	1.8×10^{-6}
Ionizing radiation—human health (kg U235 eq.)	0.270	0.296
Land use (kg SOC)	4.026	4.408
Marine eutrophication (kg N eq.)	0.008	0.009
Ozone depletion (kg CFC-11 eq.)	2.4×10^{-6}	2.6×10^{-6}
Particulate matter/Respiratory inorganics (kg PM2.5 eq.)	0.021	0.023
Photochemical ozone formation (kg C_2H_4 eq.)	0.023	0.025
Resource depletion—minerals, fossils, and renewable (kg Sb eq.)	1.1×10^{-5}	1.2×10^{-5}
Resource depletion—water (m^3)	3.520	3.867
Terrestrial eutrophication (Mole N eq.)	0.082	0.090

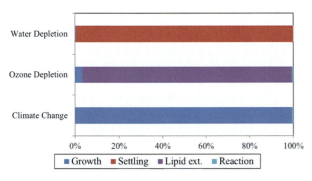

FIG. 7 Contribution of individual stages of the overall process toward the net impact for biodiesel production considering economic allocation for selected impact categories.

productivity, large volume of solution is required to be cultivated necessitating high electricity usage (Sen Gupta et al., 2017a). This underlines the necessity of developing strain with higher biomass and lipid productivity.

The water depletion potential is observed to be relatively low. That is because the source of water for growth phase is assumed to be natural water and hence an elementary flow. As per the LCA methodology, elementary flow inputs do not contribute to life cycle impacts. However, this may not be always true, and the consideration of water as elementary flow may perhaps be giving a misleading result. The water depletion is a function of exact characterization of water emission from the settling step. Thus, the proper characterization of individual inputs and outputs for each of the steps is very critical. Moreover, recycling water from the settling outlet back to the growth reduces water footprint for 1 L biodiesel production to $0.014\,\text{m}^3$ for both the allocations.

We notice a large variability in terms of results reported in the literature since each of the studies is on different design for the system with variations in process performance parameters, while, the methods for the individual life cycle studies also differ. Adesanya et al. (2014) reported the total climate change impact to be $2.137\,\text{kg CO}_2$ eq./kg of biodiesel. However, they also considered anaerobic digestion of algal residue to produce methane, subsequently converted to electricity and heat. Significant negative emissions in the form of avoided impact from this route resulted in the final number 2.137. In our work, we have not considered avoided impacts. Adesanya et al. (2014) further reported that growth phase was the major contributor to the overall impact and electricity consumption was the main factor. They also reported that the water requirement would be $16.5\,\text{L/kg}$ with recycling and it went up to $1700\,\text{L/kg}$ without recycling. Our results are higher since we have not considered recycling and have also completely relied on open pond cultivation as compared to hybrid system by Adesanya et al. (2014). Collet et al. (2014) reported the climate change impact to be about $2\,\text{kg}$ CO_2 eq./kg of biodiesel. They also reported that the majority of the impact was due to electricity and fertilizers used in the growth phase. They reported that increasing the algal productivity and switching to cleaner sources of electricity can reduce the total impacts. Water depletion was not calculated in that study. Soratana et al. (2014) showed that production of co-products such as ethanol and electricity provide significant reduction in the life cycle impacts. However, their flowsheet was quite complex and developing an industrial process with such a complex flowsheet will be quite challenging. Gnansounou and Kenthorai Raman (2016) employed Recipe midpoint method and reported the climate change impact to be

0.26 kg CO_2 eq. for biodiesel and protein co-production and 0.42 kg CO_2 eq. for co-production of biodiesel, protein, and succinic acid. Although these impacts were low, they assumed the lipid extraction efficiency to be 91%, almost three times the values considered for the present design. Thus, we notice a wide range of variability in terms of environmental impacts for the studies reported in literature, and the results obtained here are within such ranges.

Literature has shown that co-product formation can help in reducing the overall costs as well as climate change impacts. Here, the impacts for the co-production of reducing sugar are also calculated. The results show that the climate change impact for 1 kg of sugar is 4.286 and 3.605 kg CO_2 eq. for physical and economic allocation, respectively. Similar to the case of biodiesel production, growth is the significant contributor to any impact category except ozone depletion and water usage, as discussed earlier. At the same time, the step of lipid extraction has more impact with respect to all the indicators for the economic allocation since the corresponding factor (0.68) is greater than physical allocation factor (0.619) for the target end-product biodiesel.

4.2 Scenario studies

The system (Fig. 6) is modified as per the specific cases, whose optimized techno-economic analysis has been discussed in the previous section. These scenarios differed as per the choice of superstructure decisions as well as product demand. The LCA results for some of the scenarios are discussed here in terms of climate change, land use, and water depletion.

- *Cultivation in closed reactor*: Such a scenario involves higher productivity, thereby requiring lesser water while necessitating higher electricity. The LCA results (Table 2) show that water depletion impact reduces significantly by 73%, as expected. However, the downside is the significant increase in the climate change impact. This is due to the very high electricity requirement in the closed photobioreactor as compared to open raceway ponds. This option is not promising from an economic standpoint as well with ALCC for biodiesel production being 2.9 times the base case values.

TABLE 2 Impacts for 1 L biodiesel production when the microalgae cultivation is done in closed photobioreactors.

Impact category	Biodiesel		Reduced sugar	
	Physical allocation	Economic allocation	Physical allocation	Economic allocation
Acidification (Mole H$^+$ eq.)	1.515	1.664	0.932	0.783
Climate change (kg CO_2 eq.)	115.821	127.229	71.417	60.086
Land use (kg SOC)	210.359	231.075	129.454	108.768
Ozone depletion (kg CFC-11 eq.)	2.4×10^{-6}	2.7×10^{-6}	1.5×10^{-6}	1.3×10^{-6}
Resource depletion—water (m^3)	0.942	1.035	0.580	0.487

The results are reported for both products (biodiesel and reduced sugar) and for different allocation approaches.

- *Lipid extraction with carbon dioxide*: Supercritical carbon dioxide as the lipid extraction solvent has a higher extraction efficiency than conventional solvents used in base case, thereby requiring lesser biomass to be cultivated. Accordingly, lesser amount of water is required at the growth step. This scenario involves carbon dioxide as the lipid extraction solvent at the first stage; while, chloroform-methanol mixture is used at the second stage. Liquid carbon dioxide production for "Rest of the World" region is considered for this study. The results (Table 3) show that the climate change, land use, and water depletion impacts are lower than those for the base case for both types of allocations for almost all the cases, and growth is comparatively less dominant. This is due to higher impact on climate change (1.79 kg CO_2 eq.) by the lipid extraction step with carbon dioxide as against base case (0.013 kg CO_2 eq.). While reduction in water usage with respect to base case is almost similar (45%–46%) for both the allocation strategies for both the products, reduction in land use is more (27%) for biodiesel as product for both the allocation. With respect to reduced sugar, land usage was reduced by 17.5% for physical and 8.9% for economic allocation. Almost analogously, climate change impacts were reduced by 20% for both the allocation strategies with respect to biodiesel. However, for reduced sugar co-product, physical allocation facilitated a reduction in climate change by almost 4%, while economic allocation enhanced the same by 8.9%. It is further to be noted that similar to the base case, impact on ozone depletion for the reduced sugar co-production is more with economic allocation in comparison to physical allocation due to predominance of impact of lipid extraction over growth on ozone depletion. Ozone depletion increased by 188% and 180% with respect to biodiesel and reduced sugar, respectively, for both the allocation strategies. Such increase in climate change and ozone depletion is due to change in solvent for lipid extraction and use of two different solvents at two stages. Similar nonintuitive result is noticed for the case of depletion of minerals, fossils, and renewable, as well. It is facilitated by use of carbon dioxide as its production involves 10^4 times more impact than that due to generation of electricity for growth.

TABLE 3 Impacts for 1 L biodiesel production when the lipid extraction is done using supercritical carbon dioxide.

	Biodiesel		Reduced sugar	
Impact category	Physical allocation	Economic allocation	Physical allocation	Economic allocation
Acidification (Mole H⁺ eq.)	0.037	0.040	0.024	0.021
Climate change (kg CO_2 eq.)	5.589	6.135	4.108	3.916
Land use (kg SOC)	2.917	3.189	1.976	1.837
Ozone depletion (kg CFC-11 eq.)	6.8×10^{-7}	7.5×10^{-7}	6.5×10^{-7}	7.1×10^{-7}
Resource depletion—water (m³)	1.905	2.092	1.173	0.986

The results are reported for both products (biodiesel and reduced sugar) and for different allocation approaches.

In terms of biodiesel production, the present scenario involving carbon dioxide as lipid extraction solvent for the first stage fares better than the base case for all the impact indicators except depletion of minerals, fossils, and renewable. However, as per ALCC, this scenario is 91% dearer to the base case. Thus, it points to necessity of analyzing the system from environmental standpoint in addition to process economics.

These scenario studies have highlighted the necessity of the comprehensive life cycle assessment in terms of all the critical impact indicators over and above the process economics for deciding upon the long-term aspects of an integrated microalgae biorefinery.

5 Conclusion

The species of microalgae has presented itself as a potential source for bio-based economy with a potential of yielding multiple end products such as food and fuel as well as providing means of bio-mitigating flue gas and wastewater. In this context, a comprehensive analysis from economic, environmental as well as social perspective should be carried out to adjudge its long-term viability as a sustainable resource. The work on economic optimization highlighted the contribution of systematic optimization in designing the integrated biorefinery. The results showed that production of biodiesel is economically not feasible at this stage. While this is known even otherwise, the optimization work identified the key bottlenecks and focus areas for process improvements. The work has also identified the role of process engineering through optimal scheduling and debottlenecking to contribute toward economic feasibility.

For the life cycle assessment work of the economically optimized integrated biorefinery, both physical and economic allocation factors were considered on the basis of the two end products—biodiesel and reduced sugar. The step of growth, owing to the excessive usage of electricity was found to be the most significant contributor to most of the impact categories. Growth was also the most predominant contributor to the net cost. It, thus, highlights the importance of research, focused on growth for addressing this challenge through better strain development or improved equipment performance. The different scenarios studied here highlighted the necessity of an LCA study for such a system. For example, the scenario with carbon dioxide-assisted lipid extraction at the first stage was 91% more expensive than the base case; however, it involved lesser impact for almost all the categories except depletion of minerals, fossils, and renewable. The scenario with heterotrophic strain, though being expensive, was less impactful than the base case in terms of some of the categories, thereby highlighting the challenges in decision-making. Moreover, like any other computational study, the crucial importance of the exact characterization of flows and processes for individual steps cannot be overstated. The findings from LCA study help us to determine the suitability of a certain design over the others through mutual comparisons on the basis of a variety of environment-specific indicators, thereby aiding in a well-informed decision-making prior to large-scale implementation.

Continuous improvements are being made in the area of microalgae cultivation. This includes developing high yielding strains, strains with high lipid content, or strains with shorter growth period. The proposed optimization framework can be used to consider the

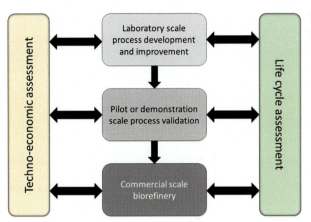

FIG. 8   Techno-economic and life cycle assessment: Vision.

recent data and reoptimize the system and subsequently study its environmental impacts. That is the benefit of developing a generic model-based framework.

Finally, Fig. 8 schematically shows the vision for performing techno-economic and life cycle assessment studies. Ideally, the techno-economic and life cycle assessment must be done in parallel, and must have a role throughout the process development and scale-up stages. Typically, the development starts at the laboratory scale based on small-scale experiments under highly idealized conditions. Promising results at laboratory scale lead to pilot or demonstration scale studies that include related aspects of reactor design and operating strategies. Finally, success at pilot or demonstration scale studies lead to commercial scale implementation. For the commercial scale implementation, aspects of heat integration, utility design, and so on are considered. As one moves from the laboratory scale to commercial scale, the uncertainty regarding process data reduces. However, the decisions made in the initial phase of the project (laboratory or pilot) majorly affect the future performance. In particular, it limits the scope of optimizing the performance of the process. Although traditionally this aspect has been highlighted from the point of view of process economics, it is also very relevant from the point of view of environmental performance. Therefore, environmental assessment using methods such as life cycle assessment (LCA) must be undertaken from the initial stages itself. The data are often preliminary and scarce at this stage and several assumptions are needed. However, it is still highly recommended. An early incorporation of LCA or similar other approach can help in eliminating options that are obviously unsustainable, and this can often save costs and time in future.

References

Adesanya, V.O., Cadena, E., Scott, S.A., Smith, A.G., 2014. Life cycle assessment on microalgal biodiesel production using a hybrid cultivation system. Bioresour. Technol. 163, 343–355.

Amer, L., Adhikari, B., Pellegrino, J., 2011. Technoeconomic analysis of five microalgae-to-biofuels processes of varying complexity. Bioresour. Technol. 102, 9350–9359.

Bart, J.C.J., Palmeri, N., Cavallaro, S., 2010. Processes for biodiesel production from unrefined oils and fats. Biodiesel Science and Technology: From Soil to Oil. Woodhead Publishing Series in Energy: Number 7. CRC Press, pp. 386–433.

Brennan, L., Owende, P., 2010. Biofuels from microalgae—a review of technologies for production, processing, and extractions of biofuels and co-products. Renew. Sust. Energ. Rev. 14 (2), 557–577.

Brown, T.M., Duan, P., Savage, P.E., 2010. Hydrothermal liquefaction and gasification of Nannochloropsis sp. Energy Fuel 24 (6), 3639–3646.

Chamkalani, A., Zendehboudi, S., Rezaei, N., Hawboldt, K., 2020. A critical review on life cycle analysis of algae biodiesel: current challenges and future prospects. Renew. Sust. Energ. Rev. 134, 110143.

Chen, B., Xiong, R., Li, H., Sun, Q., Yang, J., 2019. Pathways for sustainable energy transition. J. Clean. Prod. 228, 1564–1571.

Chisti, Y., 2007. Biodiesel from microalgae. Biotechnol. Adv. 25, 294–306.

Chung, J., 2013. Grand challenges in bioenergy and biofuel research: engineering and technology development, environmental impact, and sustainability. Front. Energy Res. 1, 4.

Collet, P., Hélias, A., Lardon, L., Ras, M., Goy, R., Steyer, J., 2011. Life-cycle assessment of microalgae culture coupled to biogas production. Bioresour. Technol. 102 (1), 207–214.

Collet, P., Lardon, L., Hélias, A., Bricout, S., Lombaert-Valot, I., Perrier, B., et al., 2014. Biodiesel from microalgae—life cycle assessment and recommendations for potential improvements. Renew. Energy 71, 525–533.

Davis, R., Aden, A., Pienkos, P.T., 2011. Techno-economic analysis of autotrophic microalgae for fuel production. Appl. Energy 88, 3524–3531.

Delrue, F., Li-Beisson, Y., Setier, P., Sahut, C., Roubaud, A., Froment, A., et al., 2013. Comparison of various microalgae liquid biofuel production pathways based on energetic, economic and environmental criteria. Bioresour. Technol. 136, 205–212.

Dutta, S., Neto, F., Coelho, M.C., 2016. Microalgae biofuels: a comparative study on techno-economic analysis & life-cycle assessment. Algal Res. 20, 44–52.

Gebreslassie, B.H., Waymire, R., You, F., 2013. Sustainable design and synthesis of algae-based biorefinery for simultaneous hydrocarbon biofuel production and carbon sequestration. AICHE J. 59 (5), 1599–1621.

Gholkar, P., Shastri, Y., Tanksale, A., 2019. Catalytic reactive flash volatilisation of microalgae to produce hydrogen or methane-rich syngas. Appl. Catal. B Environ. 251, 326–334.

Gholkar, P., Shastri, Y., Tanksale, A., 2021. Renewable hydrogen and methane production from microalgae: a techno-economic and life cycle assessment study. J. Clean. Prod. 279, 123726.

Gnansounou, E., Kenthorai Raman, J., 2016. Life cycle assessment of algae biodiesel and its co-products. Appl. Energy 161, 300–308.

Gomez, L.D., Steele-King, C.G., McQueen-Mason, S.J., 2008. Sustainable liquid biofuels from biomass: the writing's on the walls. New Phytol. 178 (3), 473–485.

Gong, J., You, F., 2014. Global optimization for sustainable design and synthesis of algae processing network for CO2 mitigation and biofuel production using life cycle optimization. AICHE J. 60 (9), 3195–3210.

González-Delgado, Á., Kafarov, V., El-Halwagi, M., 2015. Development of a topology of microalgae-based biorefinery: process synthesis and optimization using a combined forward–backward screening and superstructure approach. Clean Techn. Environ. Policy 17 (8), 2213–2228.

Grierson, S., Strezov, V., Bengtsson, J., 2013. Life cycle assessment of a microalgae biomass cultivation, bio-oil extraction and pyrolysis processing regime. Algal Res. 2 (3), 299–311.

Heo, H.Y., Heo, S., Lee, J.H., 2019. Comparative techno-economic analysis of transesterification technologies for microalgal biodiesel production. Ind. Eng. Chem. Res. 58 (40), 18772–18779.

Lam, M.K., Lee, K.T., 2012. Microalgae biofuels: a critical review of issues, problems and the way forward. Biotechnol. Adv. 30, 673–690.

Quinn, J.C., Smith, T.G., Downes, C.M., Quinn, C., 2014. Microalgae to biofuels lifecycle assessment—multiple pathway evaluation. Algal Res. 4, 116–122.

Resurreccion, E.P., Colosi, L.M., White, M.A., Clarens, A.F., 2012. Comparison of algae cultivation methods for bioenergy production using a combined life cycle assessment and life cycle costing approach. Bioresour. Technol. 126, 298–306.

Ribeiro, L.A., Silva, P.P.D., 2013. Surveying techno-economic indicators of microalgae biofuel technologies. Renew. Sust. Energ. Rev. 25, 89–96.

Richardson, J.W., Johnson, M.D., Outlaw, J.L., 2012. Economic comparison of open pond raceways to photo bioreactors for profitable production of algae for transportation fuels in the southwest. Algal Res. 1 (1), 93–100.

Rizwan, M., Lee, J.H., Gani, R., 2013. Optimal processing pathway for the production of biodiesel from microalgal biomass: a superstructure based approach. Comput. Chem. Eng. 58, 305–314.

Sen Gupta, S., Shastri, Y., Bhartiya, S., 2016. Model-based optimisation of biodiesel production from microalgae. Comput. Chem. Eng. 89, 222–249.

Sen Gupta, S., Shastri, Y., Bhartiya, S., 2017a. Optimization of integrated microalgal biorefinery producing fuel and value-added products. Biofuels Bioprod. Biorefin. 11 (6), 1030–1050.

Sen Gupta, S., Shastri, Y., Bhartiya, S., 2017b. Integrated microalgae biorefinery: impact of product demand profile and prospect of carbon capture. Biofuels Bioprod. Biorefin. 11 (6), 1065–1076.

Shastri, Y., 2017. Renewable energy, bioenergy. Curr. Opin. Chem. Eng. 17, 42–47.

Singh, J., Gu, S., 2010. Commercialization potential of microalgae for biofuels production. Renew. Sust. Energ. Rev. 14 (9), 2596–2610.

Singh, N., Mishra, T., Banerjee, R., 2019. Greenhouse gas emissions in India's road transport sector. In: Venkataraman, C., Mishra, T., Ghosh, S., Karmakar, S. (Eds.), Climate Change Signals and Response: A Strategic Knowledge Compendium for India Singapore. Springer Singapore, pp. 197–209.

Slegers, P.M., Koetzier, B.J., Fasaei, F., Wijffels, R.H., van Straten, G., van Boxtel, A.J.B., 2014. A model-based combinatorial optimisation approach for energy-efficient processing of microalgae. Algal Res. 5, 140–157.

Soratana, K., Barr, W.J., Landis, A.E., 2014. Effects of co-products on the life-cycle impacts of microalgal biodiesel. Bioresour. Technol. 159, 157–166.

Subhadra, B.G., 2010. Sustainability of algal biofuel production using integrated renewable energy park (IREP) and algal biorefinery approach. Energy Policy 38 (10), 5892–5901.

Tredici, M.R., Rodolfi, L., Biondi, N., Bassi, N., Sampietro, G., 2016. Techno-economic analysis of microalgal biomass production in a 1-ha Green Wall panel (GWP) plant. Algal Res. 19, 253–263.

UNDP, 2013. Human Development Report 2013. The Rise of the South: Human Progress in a Diverse World. United Nations Development Program, New York.

Wijffels, R.H., Barbosa, M.J., 2010. An outlook on microalgal biofuels. Science 329 (5993), 796.

Wijffels, R.H., Barbosa, M.J., Eppink, M.H.M., 2010. Microalgae for the production of bulk chemicals and biofuels. Biofuels Bioprod. Biorefin. 4 (3), 287–295.

Zaimes, G.G., Khanna, V., 2013. Environmental sustainability of emerging algal biofuels: a comparative life cycle evaluation of algal biodiesel and renewable diesel. Environ. Prog. Sustain. Energy 32 (4), 926–936.

Index

Note: Page numbers followed by *t* indicate tables.